SIXTH EDITION

MATHEMATICS
Applied to Electronics

James H. Harter
Mesa Community College

Wallace D. Beitzel
Northrop Grumman Space Technology

D0169189

PEARSON

Prentice
Hall

Upper Saddle River, New Jersey
Columbus, Ohio

Library of Congress Cataloging-in-Publication Data

Harter, James H.
    Mathematics applied to electronics/James H. Harter, Wallace D. Beitzel.—6th ed.
        p. cm.
    Includes index.
    ISBN 0-13-047600-5
    1. Electronics—Mathematics. I. Beitzel, Wallace D. II. Title.

TK7864.H37 2004
512'.1'0246213—dc21   2002037075

Editor in Chief: Stephen Helba
Executive Editor: Gary Bauer
Editorial Assistant: Natasha Holden
Production Editor: Louise N. Sette
Design Coordinator: Diane Ernsberger
Cover Designer: Thomas Borah
Production Manager: Brian Fox
Marketing Manager: Leigh Ann Sims
Illustrator: Marolyn Young

This book was set in Times Roman by Carlisle Communications, Ltd. It was printed and bound by
R. R. Donnelley & Sons Company. The cover was printed by Phoenix Color Corp.

Pearson Education Ltd.
Pearson Education Singapore Pte. Ltd.
Pearson Education Canada, Ltd.
Pearson Education—Japan
Pearson Education Australia Pty. Limited
Pearson Education North Asia Ltd.
Pearson Educación de Mexico, S.A. de C.V.
Pearson Education Malaysia Pte. Ltd.

10 9 8 7 6 5 4 3 2 1
ISBN: 0-13-047600-5

$$\text{relative error} = \frac{\text{absolute error}}{\text{nominal value}} \times 100\% \quad \textbf{(12–4)}$$

$$\eta = \frac{W_{out}}{W_{in}} \quad \textbf{(12–5)}$$

$$\eta = \frac{P_{out}}{P_{in}} \quad \textbf{(12–6)}$$

$$\eta_\% = \frac{P_{out}}{P_{in}} \times 100\% \quad \textbf{(12–7)}$$

$$1 \text{ hp} = 746 \text{ W} \quad \textbf{(12–8)}$$

$$\eta_{ov} = \eta_1 \times \eta_2 \times \eta_3 \times 100\% \quad \textbf{(12–9)}$$

$$\frac{R_1}{R_2} = \frac{l_1}{l_2} \quad \textbf{(12–10)}$$

$$\frac{R_1}{R_2} = \frac{A_2}{A_1} \quad \textbf{(12–11)}$$

$$\frac{R_1}{R_2} = \frac{d_2^2}{d_1^2} \quad \textbf{(12–12)}$$

$$\rho = \frac{RA}{l} \quad \text{(ohm-meters)} \quad \textbf{(12–13)}$$

$$\frac{R_2}{R_1} = \frac{234.5 + T_2}{234.5 + T_1} \quad \textbf{(12–14)}$$

$$y = mx + b \quad \textbf{(14–1)}$$

$$\text{Slope} = m = \frac{y_1 - y_2}{x_1 - x_2} \quad \textbf{(14–2)}$$

$$h_{fe} = \frac{\Delta i_c}{\Delta i_b} \bigg|_{v_{ce}} \quad \textbf{(15–1)}$$

$$h_{oe} = \frac{\Delta i_c}{\Delta v_{ce}} \bigg|_{i_b} \quad \text{(siemens)} \quad \textbf{(15–2)}$$

$$y_{fs} = \frac{\Delta i_d}{\Delta v_{gs}} \bigg|_{v_{ds}} \quad \text{(siemens)} \quad \textbf{(15–3)}$$

$$r_{ac} = \frac{\Delta v_f}{\Delta i_f} \quad \text{(ohms)} \quad \textbf{(15–4)}$$

$$E = V_F + V_{RL} \quad \textbf{(15–5)}$$

$$E = V_F + IR_L \quad \textbf{(15–6)}$$

$$I = \frac{E - V_F}{R_L} \quad \textbf{(15–7)}$$

$$I = \frac{-1}{R_L} V_F + \frac{E}{R_L} \quad \textbf{(15–8)}$$

$$\begin{vmatrix} a_1 & b_1 \\ a_2 & b_2 \end{vmatrix} = a_1 b_2 - a_2 b_1 \quad \textbf{(16–1)}$$

$$\begin{cases} a_1 x + b_1 y = k_1 \\ a_2 x + b_2 y = k_2 \end{cases} \quad \textbf{(16–2)}$$

$$\Delta = \begin{vmatrix} a_1 & b_1 \\ a_2 & b_2 \end{vmatrix} = a_1 b_2 - a_2 b_1 \quad \textbf{(16–3)}$$

$$x = \frac{\begin{vmatrix} k_1 & b_1 \\ k_2 & b_2 \end{vmatrix}}{\Delta} = \frac{k_1 b_2 - k_2 b_1}{\Delta} \quad \textbf{(16–4)}$$

$$y = \frac{\begin{vmatrix} a_1 & k_1 \\ a_2 & k_2 \end{vmatrix}}{\Delta} = \frac{a_1 k_2 - a_2 k_1}{\Delta} \quad \textbf{(16–5)}$$

$$\begin{vmatrix} a_1 & b_1 & c_1 \\ a_2 & b_2 & c_2 \\ a_3 & b_3 & c_3 \end{vmatrix} = \begin{aligned} &a_1 b_2 c_3 + b_1 c_2 a_3 + c_1 a_2 b_3 \\ &- a_3 b_2 c_1 - b_3 c_2 a_1 - c_3 a_2 b_1 \end{aligned}$$
$$\textbf{(16–6)}$$

$$\begin{cases} a_1 x + b_1 y + c_1 z = k_1 \\ a_2 x + b_2 y + c_2 z = k_2 \\ a_3 x + b_3 y + c_3 z = k_3 \end{cases} \quad \textbf{(16–7)}$$

$$\Delta = \begin{vmatrix} a_1 & b_1 & c_1 \\ a_2 & b_2 & c_2 \\ a_3 & b_3 & c_3 \end{vmatrix} \quad \textbf{(16–8)}$$

$$x = \frac{\begin{vmatrix} k_1 & b_1 & c_1 \\ k_2 & b_2 & c_2 \\ k_3 & b_3 & c_3 \end{vmatrix}}{\Delta} \quad \textbf{(16–9)}$$

ABOUT THE CHAPTER ART

The marvels of electricity and electronics are featured in the series of photographs used on the opening of each of the 32 chapters in this edition of *Mathematics Applied to Electronics*.

These photographs remind us of the range of devices, systems, structures, objects, and machines we have come to rely on to make ourselves comfortable, healthy, safe, and informed. Being able to communicate around the world in an instant, know what the weather will bring, monitor our health, explore the universe, transport ourselves, provide for our defense, develop new electrical energy sources, adapt technology for the good of all, harness machines to unburden us from repetitive physical labor, research and comprehend the unknown, understand the forces of nature, and simply toast bread—these and more are all made possible through the application of electricity and electronics to our universe.

PREFACE

The purpose of this book is to provide an understanding of mathematics as it is applied to electronics. The text may be used in a formal classroom setting or in a self-paced or self-study program. *Mathematics Applied to Electronics* is for those who are studying technology related to electronics, computers, electromechanics, or automation.

Modern curriculums, based on electronics, need the support of a large and diverse amount of mathematics, so the content of this text is a trade-off between a formal proof orientation and the need for expediency in developing a broad, general mathematics ability. The sequence of chapters and topics within each chapter have been planned to be compatible with the electric circuits books currently in use. The scientific calculator is an integral part of the text, and its introduction early in the book enhances the learning process.

NEW TO THE SIXTH EDITION

- Chapter 10, "Applying Fractions to Electrical Circuits," is rewritten with an emphasis on the product over the sum form of the two branch equivalent resistance equation.
- Chapter 24, "Vectors and Phasors," is rewritten with an emphasis on the use of the calculator's rectangular to polar and polar to rectangular keys to carry out mathematical operations with complex numbers.
- New Chapter Performance Objectives provide the learner with key outcomes for each chapter.
- New Section Challenges unify the reader's comprehension of key mathematical concepts spanning several chapters.
- New companion website with multiple choice and true/false review quizzes for each chapter.
- New Appendix A entry, the Development of the International System of Units.

TEXT ORGANIZATION

The book begins with selected topics in prealgebra, number notation, and units of measurement, which are followed by several chapters dealing with the fundamentals of algebra, including the evaluation of formulas. This series of chapters culminates with a chapter devoted to the solution of linear equations, which is followed by a chapter that applies mathematics to electronic circuits.

The text is structured so that each section of theoretical chapters is followed by one or more application chapters. The application chapters reinforce materials previously presented and provide the learner with an opportunity to transfer mathematical skills to electronics concepts. Interspersed throughout the book are chapters and topics dealing with graphing and graphical analysis. These chapters are essential because so much valuable information is presented in graphical form in handbooks and data sheets.

Following chapters dealing with quadratic equations and exponents and radicals are chapters covering logarithmic, exponential, and trigonometric functions. These topics are followed by a series of chapters covering the mathematics of alternating current. The text concludes with chapters dealing with math analysis, computer number systems, and computer logic.

FEATURES

This book has been designed to guide the reader through the learning process by providing a means of coordinating the instruction in the classroom with outside assignments. The reader is helped by hundreds of detailed examples, figures, graphs, and problems. The utilization of the SI system of measurement throughout the text enables the user to make an easy transition to any technology book in use today.

A companion website (www.prenhall.com/harter) is available for this text. It contains true/false and multiple choice questions for each chapter. This website also contains *Syllabus Manager,* which instructors can use to easily create and revise syllabi. *Syllabus Manager* includes direct links into the companion website and other online content.

ACKNOWLEDGMENTS

In closing, we wish to thank you, the adopters of this and the previous editions of *Mathematics Applied to Electronics,* for your helpful comments and suggestions and for your diligence in bringing oversights and omissions to our attention. We acknowledge all of you who have communicated with us, including Roger Harlow, Archie Gillespie, and Anna Spear. We also thank the reviewers: Michael Bezusko, Pima Community College; Nasser Hedayat, Valencia Community College; Barbara Miller, University of Alaska; Saeed A. Shaikh, Miami-Dade Community College; and Stephen Trudeau, Denver Technical College.

The authors take this opportunity to acknowledge and thank the following individuals at Pearson/Prentice Hall for their diligence in making the publication of this text possible: Gary Bauer, Executive Editor; Louise Sette, Production Editor; and Carol Mohr, Copy Editor.

Proposals for improvement, questions about problems, or comments on the content may be made by writing to us in care of the publisher.

SUPPLEMENTS

It is our sincere intention to provide high-quality materials for your use. To that end, adopters of the text may request (from the publisher) a complimentary package of materials, including a detailed Instructor's Solutions Manual, Transparency Masters, an extensive Text Item File, and a Windows-based TestGen (an electronic version of the Test Item File).

Again, thank you for your continued support.

James Harter
Surprise, Arizona
2003

Wallace Beitzel
Redondo Beach, California
2003

BRIEF CONTENTS

Introduction **xix**

1   Selected Prealgebra Topics   **1**

2   Number Notation and Operation   **25**

3   Quantities and Units of Measurement   **61**
Section Challenge for Chapters 1, 2, and 3   86

4   Algebra Fundamentals I   **89**

5   Algebra Fundamentals II   **102**

6   Solving Equations   **119**

7   Applying Mathematics to Electrical Circuits   **143**
Section Challenge for Chapters 4, 5, 6, and 7   172

8   Fractions   **174**

9   Equations Containing Fractions   **201**

10 Applying Fractions to Electrical Circuits   **215**
Section Challenge for Chapters 8, 9, and 10   239

11 Special Products, Factoring, and Equations   **240**

12 Applying Mathematics to Electrical Concepts   **256**
Section Challenge for Chapters 11 and 12   290

13 Relations and Functions   **292**

14 Graphs and Graphing Techniques   **307**

15 Applying Graphs to Electronic Concepts   **330**
Section Challenge for Chapters 13, 14, and 15   351

16 Solving Systems of Linear Equations   **353**

17   Applying Systems of Linear Equations to Electronic Concepts   **373**
Section Challenge for Chapters 16 and 17   396

18   Solving Quadratic Equations   **398**

19   Exponents, Radicals, and Equations   **423**
Section Challenge for Chapters 18 and 19   437

20   Logarithmic and Exponential Functions   **438**

21   Applications of Logarithmic and Exponential Equations to Electronic Concepts   **467**
Section Challenge for Chapters 20 and 21   490

22   Angles and Triangles   **492**

23   Circular Functions   **517**

24   Vectors and Phasors   **536**

25   The Mathematics of Phasors   **552**
Section Challenge for Chapters 22, 23, 24, and 25   567

26   Fundamentals of Alternating Current   **569**

27   Alternating-Current Circuits   **596**

28   Sinusoidal Alternating Current   **624**
Section Challenge for Chapters 26, 27, and 28   641

29   Additional Trigonometric and Exponential Functions   **642**

30   Mathematical Analysis   **656**
Section Challenge for Chapters 29 and 30   667

31   Computer Number Systems   **669**

32   Mathematics of Computer Logic   **704**
Section Challenge for Chapters 31 and 32   739

CONTENTS

Introduction   xix

1

Selected Prealgebra Topics   1

1–1	Natural Numbers and Number Systems	2
1–2	Signed Numbers	3
1–3	Numerical Expressions and Equations	5
1–4	Order of Operations	6
1–5	Symbols of Grouping	9
1–6	Double Meaning of $+$ and $-$	10
1–7	Absolute Value of a Signed Number	12
1–8	Combining Signed Numbers	13
1–9	Relational Operators	17
1–10	Multiplying with Signed Numbers	19
1–11	Dividing with Signed Numbers	21

2

Number Notation and Operation   25

2–1	Introduction to Exponents	26
2–2	Number Notation	28
2–3	Numeric Operations and Rounding	35
2–4	Operations with Approximate Numbers	40
2–5	Square Roots, Radicals, and Reciprocals	44
2–6	Combined Operations	50
2–7	Powers of Ten and Approximations	52

3

Quantities and Units of Measurement   61

3–1	International System of Units	62
3–2	Selected Physical Quantities	65

3–3 Forming Decimal Multiples and Submultiples of the SI Units 69
3–4 Unit Analysis and Conversion Between Systems 73
3–5 Applying Unit Analysis to Energy Cost 78
3–6 Units and Exponents 82
Section Challenge for Chapters 1, 2, and 3 86

4 Algebra Fundamentals I 89

4–1 Variables, Subscripts, and Primes 90
4–2 Indicating Multiplication 92
4–3 General Numbers 93
4–4 Algebraic Expressions 93
4–5 Products, Factors, and Coefficients 95
4–6 Combining Like Terms 96
4–7 Polynomials 97
4–8 Adding Polynomials 98

5 Algebra Fundamentals II 102

5–1 Multiplying Monomials 103
5–2 Multiplying a Monomial and a Binomial 105
5–3 Multiplying a Monomial and a Polynomial 107
5–4 Subtracting Polynomials 108
5–5 Additional Work with Polynomials 110
5–6 Division of Monomials 111
5–7 Dividing a Polynomial by a Monomial 113
5–8 Factoring Polynomials with a Common Monomial Factor 115
5–9 Evaluating Algebraic Expressions 116

6 Solving Equations 119

6–1 Equations 120
6–2 Finding the Root of an Equation 121
6–3 Using Addition to Transform Equations 121
6–4 Using Multiplication to Transform Equations 124
6–5 Additional Techniques 126
6–6 Equations Containing Parentheses 129
6–7 Solving Formulas 131
6–8 Evaluating Formulas 135
6–9 Forming Equations 138
6–10 Solving Word Problems 139

7   Applying Mathematics to Electrical Circuits   143

7–1   Current, Voltage, and Resistance   144
7–2   Ohm's Law   150
7–3   Resistance in a Series Circuit   154
7–4   Applying Ohm's Law   157
7–5   Summary of the Series Circuit   161
7–6   Power   165
     Section Challenge for Chapters 4, 5, 6, and 7   172

8   Fractions   174

8–1   Introductory Concepts   175
8–2   Forming Equivalent Fractions   177
8–3   Simplifying Fractions   178
8–4   Multiplying Fractions   180
8–5   Dividing Fractions   182
8–6   Complex Fractions   184
8–7   Adding and Subtracting Fractions   187
8–8   Changing a Mixed Expression to a Fraction   197
8–9   Additional Work with Complex Fractions   198

9   Equations Containing Fractions   201

9–1   Solving Equations Containing Fractions   202
9–2   Solving Fractional Equations   204
9–3   Literal Equations Containing Fractions   207
9–4   Evaluating Formulas   209

10   Applying Fractions to Electrical Circuits   215

10–1   Voltage Division in the Series Circuit   216
10–2   Conductance of the Parallel Circuit   219
10–3   Equivalent Resistance of the Parallel Circuit   222
10–4   Current Division in the Parallel Circuit   225
10–5   Solving Parallel Circuit Problems   228
10–6   Using Network Theorems to Form Equivalent Circuits   232
     Section Challenge for Chapters 8, 9, and 10   239

11   Special Products, Factoring, and Equations   240

11–1   Mentally Multiplying Two Binomials   241
11–2   Product of the Sum and Difference of Two Numbers   244
11–3   Square of a Binomial   245
11–4   Factoring the Difference of Two Squares   247

11–5   Factoring a Perfect Trinomial Square   247
11–6   Factoring by Grouping   249
11–7   Combining Several Types of Factoring   250
11–8   Literal Equations   252

12   Applying Mathematics to Electrical Concepts   256

12–1   Ratio, Percent, and Parts per Million   257
12–2   Accounting for Empirical Error in Calculations   265
12–3   Efficiency   271
12–4   Proportion   278
12–5   Electrical Conductors   279
    Section Challenge for Chapters 11 and 12   290

13   Relations and Functions   292

13–1   Meaning of a Function   293
13–2   Variables and Constants   293
13–3   Functional Notation   295
13–4   Functional Variation   297
13–5   Simplifying Formulas   302

14   Graphs and Graphing Techniques   307

14–1   Rectangular Coordinates   308
14–2   Graphs of Equations   311
14–3   Graphs of Linear Equations   316
14–4   Deriving a Linear Equation from a Graph   322
14–5   Graphing Empirical Data   325

15   Applying Graphs to Electronic Concepts   330

15–1   Graphic Estimation of Static Parameters   331
15–2   Graphic Estimation of Dynamic Parameters   336
15–3   Graphic Analysis of Linear Circuits   342
15–4   Graphic Analysis of Nonlinear Circuits   345
    Section Challenge for Chapters 13, 14, and 15   351

16   Solving Systems of Linear Equations   353

16–1   Addition or Subtraction Method   354
16–2   Substitution Method   357
16–3   Deriving Electrical Formulas   360
16–4   Determinants of the Second Order   361
16–5   Determinants of the Third Order   366

17   **Applying Systems of Linear Equations to Electronic Concepts**   **373**

17–1   Applying Kirchhoff's Voltage Law   374
17–2   Mesh Analysis   381
17–3   Solving Networks by Mesh Analysis   387
     Section Challenge for Chapters 16 and 17   396

18   **Solving Quadratic Equations**   **398**

18–1   Introduction   399
18–2   Solving Incomplete Quadratic Equations   399
18–3   Solving Complete Quadratic Equations   401
18–4   Solving Quadratic Equations by the Quadratic Formula   406
18–5   Graphing the Quadratic Function   409
18–6   Applying the Techniques of Solving Quadratic Equations to Electronic Problems   417

19   **Exponents, Radicals, and Equations**   **423**

19–1   Laws of Exponents   424
19–2   Zero and Negative Integers as Exponents   425
19–3   Fractional Exponents   426
19–4   Laws of Radicals   429
19–5   Simplifying Radicals   430
19–6   Radical Equations   434
     Section Challenge for Chapters 18 and 19   437

20   **Logarithmic and Exponential Functions**   **438**

20–1   Common Logarithms   439
20–2   Common Logarithms and Scientific Notation   441
20–3   Antilogarithms   444
20–4   Logarithms, Products, and Quotients   445
20–5   Logarithms, Powers, and Radicals   448
20–6   Natural Logarithms   451
20–7   Changing Base   454
20–8   Further Properties of Natural Logarithms   455
20–9   Logarithmic Equations   458
20–10   Exponential Equations   459
20–11   Semilog and Log–Log Plots   461
20–12   Nomographs   464

21 Applications of Logarithmic and Exponential Equations to Electronic Concepts 467

21–1 The Decibel 468
21–2 System Calculations 472
21–3 *RC* and *RL* Transient Behavior 478
21–4 Preferred Number Series 485
Section Challenge for Chapters 20 and 21 490

22 Angles and Triangles 492

22–1 Points, Lines, and Angles 493
22–2 Special Angles 497
22–3 Triangles 499
22–4 Right Triangles and the Pythagorean Theorem 501
22–5 Similar Triangles; Trigonometric Functions 503
22–6 Using the Trigonometric Functions to Solve Right Triangles 506
22–7 Inverse Trigonometric Functions 510
22–8 Solving Right Triangles when Two Sides are Known 512

23 Circular Functions 517

23–1 Angles of Any Magnitude 518
23–2 Circular Functions 519
23–3 Graphs of the Circular Functions 521
23–4 Inverse Circular Functions 524
23–5 The Law of Sines and the Law of Cosines 525
23–6 Polar Coordinates 530
23–7 Converting Between Rectangular and Polar Coordinates 532

24 Vectors and Phasors 536

24–1 Scalars and Vectors 537
24–2 Complex Plane 537
24–3 Real and Imaginary Numbers 538
24–4 Complex Numbers 539
24–5 Phasors 542
24–6 Transforming Complex Number Forms 545
24–7 Resolving Systems of Phasors and Vectors 547

25 The Mathematics of Phasors 552

25–1 Addition and Subtraction of Phasor Quantities 553
25–2 Multiplication of Phasor Quantities 556

25–3   Division of Phasor Quantities   560
25–4   Powers and Roots of Phasor Quantities   563
Section Challenge for Chapters 22, 23, 24, and 25   567

26   Fundamentals of Alternating Current   569

26–1   Alternating-Current Terminology   570
26–2   Resistance   573
26–3   Inductance and Inductive Reactance   575
26–4   Capacitance and Capacitive Reactance   578
26–5   Voltage Phasor for Series Circuits   582
26–6   Current Phasor for Parallel Circuits   590

27   Alternating-Current Circuits   596

27–1   Impedance of Series AC Circuits   597
27–2   Solving Series AC Circuits   606
27–3   Admittance Concepts   612
27–4   Admittance of Parallel AC Circuits   616

28   Sinusoidal Alternating Current   624

28–1   Time and Displacement   625
28–2   Power and Power Factor   629
28–3   Instantaneous Equations and the *EI* Phasor Diagram   636
Section Challenge for Chapters 26, 27, and 28   641

29   Additional Trigonometric and Exponential Functions   642

29–1   Auxiliary Trigonometric Functions   643
29–2   Graphs of the Auxiliary Trigonometric Functions   645
29–3   Trigonometric Identities   646
29–4   Hyperbolic Functions   648
29–5   Graphing the Hyperbolic Functions   651
29–6   Hyperbolic Identities   651
29–7   Inverse Hyperbolic Functions   652

30   Mathematical Analysis   656

30–1   Domain and Range   657
30–2   Discontinuities   658
30–3   Functions of Large Numbers   662
30–4   Asymptotes   664
Section Challenge for Chapters 29 and 30   667

31 Computer Number Systems   669

31–1 Decimal Number System   670
31–2 Three Additional Number Systems   671
31–3 Converting Numbers to the Decimal System   676
31–4 Converting Decimal Numbers to Other Systems   678
31–5 Converting Between Binary, Octal, and Hexadecimal   682
31–6 Binary Addition and Subtraction   685
31–7 Octal Addition and Subtraction   687
31–8 Hexadecimal Addition and Subtraction   689
31–9 Complements   691
31–10 Binary Arithmetic with Complements   696
31–11 Review   701

32 Mathematics of Computer Logic   704

32–1 Introductory Concepts   705
32–2 Inversion Operator (NOT)   707
32–3 Conjunction Operator (AND)   708
32–4 Disjunction Operator (OR)   712
32–5 Application of Logic Concepts   715
32–6 Introduction to Karnaugh Maps   720
32–7 DeMorgan's Theorem   727
32–8 Boolean Theorems   733
32–9 Applications   735
Section Challenge for Chapters 31 and 32   739

Glossary of Selected Terms   740

Appendix A: Reference Tables   747
Symbols
Constants
Greek Alphabet
Selected Abbreviations
American Wire Gauge
Preferred Number Series
Color Code
Selected Identities and Conversion Factors
Prefixes and Symbols for Multiples and Submultiples of the SI Units
Development of the International System of Units

Appendix B: Answers to Selected Problems   755

Appendix C: Solutions to Section Challenges   820

Index   830

INTRODUCTION

The Introduction outlines the scope and structure of the text and offers assistance in selecting a calculator or in helping you match the one you have to the requirements of the text. Additionally, it gives several study hints and some general information to aid you in your use of this text.

I–1   SCOPE AND STRUCTURE OF THE TEXT

Scope

The book comprises selected topics from the field of mathematics that support your career goal in technology. The mathematical topics have been tailored and the presentation streamlined so that the needs of the technology you are studying are met and a reasonable level of instruction is maintained.

The text material starts with prealgebra concepts and then moves to a presentation of number notation, measurement, and units. It continues with the principles of algebra and their application to electronics and then goes on to functions and graphing. Systems of simultaneous equations with applications are addressed, followed by the study of quadratics, logarithmic functions, trigonometric functions, circular functions, and mathematics of phasors with applications to ac electric circuits. The text concludes with math analysis, computer number systems, and the mathematics of computer logic.

Structure

A gradual progression from the known to the unknown, from the simple to the complex, is achieved in the sequencing of the chapters and the topics within each chapter. The rigorous, in-depth pursuit of a single topic common to formal courses in mathematics is not found in this textbook. Topics are introduced in an early section of the text and then augmented at a later time in an application chapter. Through this technique, concepts are reinforced.

The text contains hundreds of detailed examples to enable you to "learn on your own" and become an active participant in the process of educating yourself. If the class you are attending is taught in a traditional lecture/demonstration manner, then the self-paced feature of the text may be used to aid in preparing for the next lesson or to assist in doing the assigned out-of-class work. Additionally, if you have been absent, you can study and master the material missed with the aid of the detailed examples.

Using the Structure of the Text

To educate yourself is among life's more challenging tasks. Most of you have major commitments to your work, home, and family while attending classes to better your understanding of technology and ensure your future in a technology-related world. With the myriad of commitments you have, it is important that you make time for daily study, even if it is only for a few minutes at a time.

Have a study plan that includes working through each example with pencil, paper, and calculator. Once you have an understanding of the material in a section, then test your understanding by working through each problem in that section's exercises. Check your work using the answers found in the back of the book. Rework any problems that you did incorrectly. By being active in your education, you will ensure your success.

I–2　SELECTING A CALCULATOR

First and foremost, purchase a *full-function* scientific calculator with display and rounding capability. Use the information in Table I–1 to guide you in your purchase, making sure that the calculator has *all* the functions listed. Double check to be sure that the $\boxed{\rightarrow \text{P}}$ and $\boxed{\rightarrow \text{R}}$ are among the special function keys and that $\boxed{\text{SCI}}$ and $\boxed{\text{ENG}}$ are also on the keyboard. An inadequate calculator will cost you time and energy that would be better spent elsewhere.

Calculators use one of two data entry systems—either the *reverse Polish notation* (RPN) or the *algebraic entry system* (AES). The RPN system, commonly called the *Polish system,* is based on the work of the Polish mathematician Jan Lukasiewicz. Unlike the algebraic entry system, the Polish system does not use parentheses when entering numerical expressions into the calculator.

In calculators with the algebraic entry system (commonly called the *algebraic system*), the operators ($+$, $-$, $\times$, $\div$, etc.) come between the two numbers, as in 7×3. This is not the case with reverse Polish notation. Before you purchase either an "algebraic" or a "Polish" calculator, try out various calculators, talk to other students in your department about their calculators, listen to and talk with your instructor, and look over Table I–1 to become familiar with the functions and operations needed in a calculator.

Today's full-function scientific calculators are usually programmable. You will find that once the program feature is learned and understood, you will have added flexibility in your calculations. Being able to program your calculator to repeatedly solve a function for different values of the variables can be both convenient and timesaving.

TABLE I–1   Summary of Calculator Functions of Operations

Typical Key Symbol	Function/Operation	Comments	Introduced in Chapter	Alternative Key Stroke
$+$	Add		1	
$-$	Subtract	Simple arithmetic operations	1	
$\times$	Multiply		1	
$\div$	Divide		1	
CHS	Change sign		1	$+/-$
( )	Parentheses	Chain calculation w/AES	1	
y^x	Exponential	Raise number to power	2	x^y or $\wedge$
$1/x$	Reciprocal	Calculate the reciprocal	2	x^{-1}
$\sqrt{x}$	Square root	Calculate square root	2	
x^2	Square x	Square a number	2	
FIX	Fix point notation		2	
SCI	Scientific notation	Display and rounding	2	
ENG	Engineering notation		2	
STO	Store	Memory		
RCL	Recall			
$x \gtrless y$	Exchange x and y			EXC
EE	Enter exponent		2	EEX or EXP
log	Common logarithm		20	
10^x	Common antilog	Logarithmic and exponential functions	20	INV LOG
$\ln x$	Natural logarithm		20	
e^x	Natural antilogarithm		20	INV LN
SIN	Sine		22	
COS	Cosine	Trigonometric functions	22	
TAN	Tangent		22	
SIN^{-1}	Arc sine		22	INV SIN
COS^{-1}	Arc cosine	Inverse trigonometric functions	22	INV COS
TAN^{-1}	Arc tangent		22	INV TAN
DEG	Degree	Angular mode selection	22	DRG
RAD	Radian		22	
$\rightarrow$P	Rectangular to polar	Polar/rectangular coordinate conversion	24	
$\rightarrow$R	Polar to rectangular		24	

From time to time you will see in the margin the note (PROGRAMMABLE), which will alert you to topics and problems that are appropriate for the programmable calculator. Although these noted topics may be learned without the programmable calculator, we would encourage you to learn to use your calculator to its fullest.

I–3   ASSUMPTIONS MADE

We have made several assumptions:

- We assume that you have mastered the skills of arithmetic, including adding, subtracting, multiplying, dividing, fractions (both decimal fractions [0.75] and built-up fractions [3/4 or $\frac{3}{4}$]), and percentage.
- We assume that you will have a scientific calculator with you while studying this text.
- We assume that you have an active interest in the field of electronics.
- We assume that you will work through each example. A great deal of information has been included in the examples.
- We assume that you will have your owner's guide for your calculator available with your calculator.

I–4   GENERAL INFORMATION

In addition to the previous assumptions, you need to be aware of the following:

- The leading zero in decimal numbers, as in 0.357, is used to set off the decimal point.
- Numbers are set off in groups of three without commas, as in 5 282 621 rather than 5,282,621.
- Whole numbers such as 50, 18, and 143 are written with an *implied* decimal point to the right of the last digit in the number. Numbers containing decimal fractions such as 1.3, 0.59, and 10.15 are written with an *explicit* decimal point.
- Reference tables, including Symbols, Constants, Greek Alphabet, Selected Abbreviations, American Wire Gauge, Color Code, and Selected Identities and Conversion Factors, are located in Appendix A.
- Answers to selected exercises are given in Appendix B.
- Answers to the section challenges are found in Appendix C.
- No trigonometric or logarithmic tables have been included in this text because the scientific calculator is being used exclusively.
- Chapters, sections, rules, guidelines, figures, tables, examples, and exercises are numbered so that they can be quickly found. Thus, Example 3–7 is the seventh example in Chapter 3.

- Words set in bold blue are defined as *Selected Terms* at the end of the chapter in which they are found. These terms may also be found in the *Glossary of Selected Terms* in the back of the book.
- To ensure alignment of hexadecimal numbers, a fixed-width type font is used to express hexadecimal numbers in this text.
- When the example focuses on using a calculator, calculator key symbols will be used to indicate the operation. Thus, $\boxed{+}$ means addition, $\boxed{-}$ means subtraction, $\boxed{\times}$ means multiplication, and $\boxed{\div}$ means division on your calculator.

We wish you well in your study of this text and know that the field you have selected will bring challenge and reward to you.

The world belongs to the dissatisfied. . . . The great underlying principle of all human progress is that divine discontent which makes men strive for better conditions and improved methods.

Charles Proteus Steinmetz (1865–1923)

1

Selected Prealgebra Topics

1-1 Natural Numbers and Number Systems

1-2 Signed Numbers

1-3 Numerical Expressions and Equations

1-4 Order of Operations

1-5 Symbols of Grouping

1-6 Double Meaning of + and −

1-7 Absolute Value of a Signed Number

1-8 Combining Signed Numbers

1-9 Relational Operators

1-10 Multiplying with Signed Numbers

1-11 Dividing with Signed Numbers

PERFORMANCE OBJECTIVES

- Use order of operations to solve numerical expressions.
- Solve numerical expressions that include symbols of grouping.
- Form signed numbers.
- Take the absolute value of a number.
- Utilize a calculator to combine, multiply, and divide signed numbers.
- Construct compound inequalities using relational operators.

Static electricity from the Van de Graaf generator spikes the hair of the three youngsters. (Courtesy of Pearson Learning)

For some, this chapter serves as a review of prealgebra concepts and principles; for others, this chapter is the beginning of an understanding of the fundamentals of algebra. For all, this chapter is a preparation for the study of algebra.

Whole numbers are used to introduce you to several concepts used in algebra, including signed numbers, symbols of grouping, and operators.

Some abbreviations have been used to add clarity to the examples. The symbols within *boxes* are used to indicate calculator operations. The other symbols are used to indicate manual operations. The following have been used in this chapter and are used throughout the book:

$\boxed{+}$ and **A** add $\boxed{÷}$ and **D** divide

$\boxed{-}$ and **S** subtract $\boxed{\text{CHS}}$ change sign*

$\boxed{×}$ and **M** multiply $\therefore$ therefore

Note: $\boxed{+/-}$ is also used to change sign.

The following symbols are listed here for your reference (a complete list may be found in Appendix A):

$=$ is equal to $\neq$ is not equal to

$>$ is greater than $\geq$ is greater than or equal to

$<$ is less than $\leq$ is less than or equal to

$\Rightarrow$ yields $|\ |$ take absolute value

It is assumed that you have read the introduction and that you will be using a calculator to aid in your calculations. If you have not read the introduction, go back and read it before continuing this chapter.

1–1 NATURAL NUMBERS AND NUMBER SYSTEMS

Our first introduction to mathematics was counting. The numbers used were one, two, three, and so forth, and are called the counting numbers or **natural numbers.** Since zero is not used in counting, zero is not a natural number. This is why our current calendar started with year 1, not year 0.

Number Systems

The system of numbers used by most people in Europe, the Americas, Africa, Australia, and New Zealand is the *decimal system.* This system is a base-ten system; that is, ten symbols (0, 1, 2, 3, 4, 5, 6, 7, 8, 9) are used in this system. These ten symbols are called ***digits.*** Because the decimal system is a *positional number system,* only ten digits are needed to express any number. This means that the value of a digit depends on the position of the digit. In contrast, Roman numerals use new symbols for larger denominations. For example, the number three hundred thirty-three is written as CCCXXXIII in Roman numerals. The same number is written as 333 in the decimal system.

There are many other number systems in use today. Since you are studying electronics, you will probably need to learn about the number systems in use by computers. Table 1–1 is a summary of some of these computer number systems. Notice that the number of symbols used in the number system is the same as the base of the system: base two (binary), 2 symbols; base eight (octal), 8 symbols; base ten (decimal), 10 symbols; base sixteen (hexadecimal), 16 symbols. Notice also that each system starts with zero.

TABLE 1–1 Computer Number Systems

System Name	Number of Symbols	Symbols Used	Name of Symbol
Binary	Two	01	Binary digits or bits
Octal	Eight	01234567	Octal digits
Decimal	Ten	0123456789	Digits
Hexadecimal	Sixteen	0123456789 ABCDEF	Hex digits

While the binary and octal systems use the familiar decimal digits, the hexadecimal system needs six extra digits. Instead of inventing new symbols, the first six capital letters (A, B, C, D, E, and F) of the alphabet are used as the six extra digits. The computer number systems of Table 1–1 are covered in greater detail in Chapter 31.

1–2 SIGNED NUMBERS

To solve the wide variety of problems found in electronics, more is needed than the natural numbers. Signed numbers and zero are needed. Numbers preceded by a + sign are called ***positive numbers;*** those preceded by a − sign are called ***negative numbers.*** A plus sign (+) *may* be written before a number to show that it is a positive number, or the plus sign *may* be left off. A negative sign (−) is **always** written before a number to show that it is a negative number. If a number has no sign before it, it has an implied plus sign and is a positive number.

Look at the thermometer pictured in Figure 1–1(a). Notice that two kinds of numbers are shown. The positive numbers are measuring the distance above the zero. The negative numbers are measuring the distance below the zero. Thus, zero is a reference point from which positive numbers are greater and negative numbers are less. Zero is *neither positive nor negative.*

In the following exercise many of the problems can be done mentally, but some require calculations. Use a calculator when you need it.

FIGURE 1–1 Use of signed numbers: (a) temperature measured above (+) and below (−) zero; (b) voltage measured above (+) and below (−) zero reference voltage.

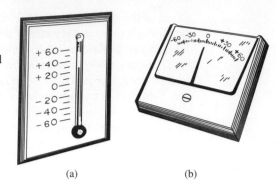

(a)   (b)

EXERCISE 1–1

1. If +3 stands for a gain of $3, what number would be used to stand for a loss of $7?

2. If an increase of 15% in the cost of resistors is shown by +15, how would a decrease of 3% be shown?

3. If −8 means 8 meters below sea level, how can 20 meters above sea level be represented?

4. In the diagram below, distance to the right of the starting point is positive and distance to the left is negative. Complete the following chart for points B, C, E, F, and G.

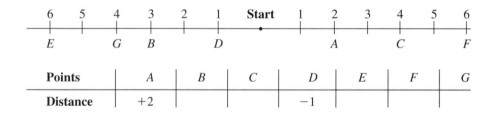

Points	A	B	C	D	E	F	G
Distance	+2			−1			

5. How would you keep score in a game if you were "down 8 points"? Would you use 8, +8, or −8 to stand for "down 8"?

6. Which is colder, −13° or −18°?

7. Below are 12 test scores for a class of students in an electronics math course. The average grade on the test is 65. Show by signed numbers the amount each score is above or below the average score of 65. The first grade is 55, which is 10 below the average. So a −10 is written below the 55. The next grade is 65. The difference between it and 65 is zero, so 0 is written below the 65. Complete the table for the remaining scores.

Test Score	55	65	90	40	80	50	75	95	30	65	100	45
Difference	−10	0										

8. A manager of an electronics store kept a record of the monthly sales. She compared them with the monthly sales of the year before. She used a plus sign to show an increase in sales and a minus sign to show a loss in sales. Complete the following table:

Monthly Sales	Jan.	Feb.	Mar.	Apr.	May
This year	4250	7840	5628	10 112	6339
Last year	3020	8118	3782	9286	7002
Change	+1230				

1–3   NUMERICAL EXPRESSIONS AND EQUATIONS

In arithmetic, symbols or operators are used to indicate addition ($+$), subtraction ($-$), multiplication ($\times$), and division ($\div$). These same symbols are also used in algebra.

To show that two numbers are to be added, such as 2 and 5, write the **numerical expression:** $2 + 5$. This expression is read "the sum of 2 and 5" or simply "2 plus 5." The result of adding 2 and 5 is 7. Seven is the sum of 2 and 5. To state that "$2 + 5$" and "7" are the same number, write the **numerical equation:** $2 + 5 = 7$. This equation is read "the sum of 2 and 5 *is equal to* 7" or simply "2 plus 5 *equals* 7." Table 1–2 shows the forming of numerical expressions and equations for the arithmetic operators of $+$, $-$, $\times$, and $\div$.

TABLE 1–2	Operators	
Operator	**Expression**	**Equation**
$+$	$2 + 6$	$2 + 6 = 8$
$-$	$6 - 2$	$6 - 2 = 4$
$\times$	2×6	$2 \times 6 = 12$
$\div$	$6 \div 2$	$6 \div 2 = 3$

EXAMPLE 1–1   Use the pair of numbers 8 and 4 and the arithmetic operators of Table 1–2 to form numerical expressions and equations for each of the operators.

Solution   Given 8, 4, and $+$, $-$, $\times$, $\div$.
Form expressions:

$$8 + 4, \quad 8 - 4, \quad 8 \times 4, \quad 8 \div 4$$

Solution   The order of operations indicates that the multiplications are performed before the addition.

M: 5×3 $15 + 6 \times 2$
M: 6×2 $15 + 12$
A: $15 + 12$ 27

∴ $5 \times 3 + 6 \times 2 = 27$

Chain Calculations

Scientific calculators have aids for working complex arithmetic problems without having to write down intermediate answers. Check your calculator and owner's guide for the following features used to make **chain calculations:** memory, stack registers, parentheses, and preprogrammed order of operations. Each of these features can, in its own way, *save* an intermediate answer for later use when making a chain calculation. Thus, by chaining from one operation to the next, the need to write down intermediate answers is eliminated.

EXERCISE 1–3

In each of the following numerical equations, state whether the answer (right-hand number) is correct or not. If the answer is correct, write *true*. If the answer is not correct, write *false* and then state the correct answer.

1. $7 + 3 - 2 = 8$ **2.** $14 - 3 \times 2 = 22$
3. $8 \times 2 - 3 = 14$ **4.** $12 - 12 \div 3 = 0$
5. $2 + 6 \div 2 = 5$ **6.** $-2 + 4 \times 5 = 18$
7. $12 \div 3 + 8 = 12$ **8.** $6 \div 2 \times 3 = 1$
9. $-5 - 2 \div 2 = -4$ **10.** $-9 \times 2 \div 3 = -6$

Calculator Drill

Use your calculator with Convention 1–1 (order of operations) to perform the following chain calculations. *Note:* You will find all the answers to the calculator drills in Appendix B. Check each of your answers.

11. $2 + 4 \times 7$ **12.** $3 - 5 \times 3$ **13.** $8 + 9 \div 3$
14. $-6 + 12 \div 4$ **15.** $6 \div 2 \times 3$ **16.** $15 \times 2 \div 5$
17. $-8 \div 2 \times 6$ **18.** $-16 \div 4 \times 3$ **19.** $2 \times 3 + 4 \times 2$
20. $5 \times 3 + 7 \times 2$ **21.** $3 \times 3 - 4 \times 5$ **22.** $8 \times 4 - 9 \times 3$
23. $15 \div 3 - 8 \times 2$ **24.** $18 \div 6 + 4 \times 5$ **25.** $9 \div 3 - 14 \div 7$
26. $-10 \div 2 - 35 \div 7$ **27.** $39 \div 3 + 64 \div 4$ **28.** $27 \div 9 + 20 \div 10$
29. $6 + 12 - 4 \times 2$ **30.** $5 - 2 - 18 \div 6$ **31.** $7 \div 2 \times 4 + 7$

1–5 SYMBOLS OF GROUPING

As we have just learned, there is an agreed-upon order in solving a numerical expression. Suppose, however, that you need to add before multiplying. How can we say, "Add first, then multiply"? By using symbols of grouping, the normal order of operation can be overridden. When a particular operation needs to be performed first, or if a particular operation needs to be emphasized, then parentheses ( ) are used.

EXAMPLE 1–4   Compute the number represented by the expression

$$4 \times (3 + 2)$$

Solution   Since $3 + 2$ is enclosed in parentheses, it is done first.

A: $(3 + 2)$ 4×5
M: 4×5 20

$\therefore$ $4 \times (3 + 2) = 20$

Besides parentheses ( ), several other symbols are used to show grouping. These include brackets [ ] and braces { }. When these signs of grouping are used together, they are used in the following order: parentheses ( ) first, then brackets [ ], and finally braces { }. When one pair of grouping symbols is enclosed in another, the operation in the inner symbol is performed first.

EXAMPLE 1–5   Compute the number represented by the expression

$$[2 + (6 - 3)] \times 5$$

Solution   Work within the parentheses first, and then within the brackets:

S: $(6 - 3)$ $[2 + 3] \times 5$
A: $[2 + 3]$ 5×5
M: 5×5 25

$\therefore$ $[2 + (6 - 3)] \times 5 = 25$

EXAMPLE 1–6   Compute the number represented by the numerical expression

$$3 + [(6 + 1) + (6 + 8 \div 4 \times 2)]$$

Solution   Work within each set of parentheses first and then within the brackets:

A:	$(6 + 1)$	$3 + [7 + (6 + 8 \div 4 \times 2)]$
D:	$8 \div 4$	$3 + [7 + (6 + 2 \times 2)]$
M:	2×2	$3 + [7 + (6 + 4)]$
A:	$(6 + 4)$	$3 + [7 + 10]$
A:	$[7 + 10]$	$3 + 17$
A:	$3 + 17$	20

$$\therefore \quad 3 + [(6 + 1) + (6 + 8 \div 4 \times 2)] = 20$$

EXERCISE 1–4

1. In working with the expression $[(6 + 3) - (2 \times 2)] + 1$, either of two operations may be performed first. What are they?

2. In working with the expression $[(9 - 2) \times (5 + 4)] - (3 + 2)$, any of three operations may be performed first. What are they?

Calculator Drill

Compute the number represented by the following numerical expressions. Use your calculator to chain calculate the answer. Check each of your answers.

3. $(6 + 3) \div 9$ 4. $(17 - 2) \times 2 \div 3$

5. $5 \times (12 - 10) \div 5$ 6. $80 \div (4 + 12)$

7. $(8 + 9 - 5) \div 3$ 8. $(2 + 3) + (5 + 4)$

9. $55 - 6 \times (3 + 4)$ 10. $(7 + 10 + 8) \div 5$

11. $12 + 5 \times (6 + 7)$ 12. $30 - 2 \times (11 - 7)$

13. $6 - [(4 + 1) - (3 + 2)]$ 14. $12 + [6 \div 3 + (7 - 4)]$

15. $21 \times (16 \div 2) + 26$ 16. $1040 \div (47 - 39) - 123$

17. $[17 + (14 \times 3)] - (3 \times 19)$

18. $[(5 + 4) \times (10 \div 5)] \div (54 \div 27)$

19. $[(9 + 3) \div 3 - (18 \div 9)] + 1$

20. $[(6 - 1) \div 5 + (4 \times 4)] - [(3 \times 3) + 1]$

21. $\{[(12 \times 2) \div (8 \times 2 \div 4)] + 3\} - (3 + 3) \div 2$

22. $\{[(9 + 7) \times 2 \div 2 + (8 \div 2)] - 4\} \div 16$

1–6   DOUBLE MEANING OF $+$ AND $-$

Before learning to add algebraically, you must first have additional understanding of the meaning of the two symbols, plus ($+$) and minus ($-$). In arithmetic, $+$ is used as the sign (operator) for addition, while $-$ is used as the sign (operator) for subtraction.

In algebra, each of these symbols has two uses. The $+$ may be used as the operator for addition or as the sign for a positive number. The $-$ may be used as the operator for subtraction or as the sign for a negative number. In the expression -3, the minus indicates a negative number, while the minus in $5 - 3$ indicates subtraction. In the expression $(-3) + (-7)$, the plus means to add and the minuses indicate negative numbers. When double signs occur in an expression, they may be simplified to a single sign. Table 1–4 shows how to combine double signs into a single sign.

TABLE 1–4   Combining Double Signs

Double Sign	Single Sign	Double Sign	Single Sign
$+\ +$	$+$	$-\ +$	$-$
$+\ -$	$-$	$-\ -$	$+$

EXAMPLE 1–7   Combine the double sign in $(-3) + (-7)$ to a single sign.

Solution   Remove the parentheses:

$$-3 + -7$$

Replace $+\ -$ with $-$:

$$-3 - 7$$

$\therefore$ $(-3) + (-7) \Rightarrow -3 - 7$

EXAMPLE 1–8   Simplify $6 - (-8)$.

Solution   Remove the parentheses:

$$6 - -8$$

Replace $-\ -$ with $+$:

$$6 + 8$$

$\therefore$ $6 - (-8) \Rightarrow 6 + 8$

EXERCISE 1–5

Rewrite each of the following expressions with all the double signs combined into single signs. Use Table 1–4.

1. $3 + (-2)$
2. $-16 + (+3)$
3. $-8 - (-2)$
4. $(+4) - (+3)$

5. $2 - 3 + (-7)$

6. $(-9) - (+3) - (-5)$

7. $1 + (+2) - (+6)$

8. $2 + 3 - +8$

9. $7 - -1 + (-4)$

10. $-5 - +7 - -3$

11. $-4 + (+6) - (+7)$

12. $+ (+8) + (-6) - (-5)$

13. $- (+9) - (-3) + (+6)$

14. $+ (-5) + (+3) + (-7)$

15. $- (+7.3) + (+6.1) - -3$

16. $- (-3.1) + (-6.2) + (+2.4)$

17. $+ (-1.4) + (+5.3) - (+6.7)$

18. $- (9.3) + (+18.1) + (-6.4)$

19. $- (-4.3) + (-8.5) - (+27)$

20. $+ (+16) - (-21) - (+13)$

1–7 ABSOLUTE VALUE OF A SIGNED NUMBER

When you work with signed numbers, it is sometimes easier to manipulate the number without the sign. The **absolute value** of a signed number is just the number without the sign. The absolute value of a number is indicated by placing a vertical bar on either side of the number. The following example will show you how to indicate that the absolute value is being taken.

EXAMPLE 1–9	Indicate the absolute value of -5 and $+7$.
Solution	Absolute value of -5 is indicated by $|-5|$.
	Absolute value of $+7$ is indicated by $|+7|$.

As you have seen, the absolute value of a number is indicated by placing a vertical bar on either side of the number. The solution to the operation of taking the absolute value is the number without any sign. Therefore, the absolute value of a number is understood to be positive.

EXAMPLE 1–10	Take the absolute value of -5 and $+7$.
Solution	Indicate the absolute value of -5:
$\therefore$	$|-5| \Rightarrow 5$
	Indicate the absolute value of $+7$:
$\therefore$	$|+7| \Rightarrow 7$

EXERCISE 1–6

Take the absolute value of each of the following signed numbers, combining double signs first.

1. $+3$	**2.** -4	**3.** -9	**4.** $+2$	**5.** -8
6. $+(+3)$	**7.** $-(+7)$	**8.** $-(-6)$	**9.** $-(+1)$	**10.** $+(-5)$

1–8 COMBINING SIGNED NUMBERS

When we add two numbers together, a *sum* is formed. The numbers used to form the sum are called the **terms** of the sum. For example, in the numerical equation $2 + 5 = 7$, the terms of the sum are 2 and 5. The addition and subtraction of signed numbers is done by applying three rules. Each of these rules will be presented with examples to illustrate their application.

Rule 1–1. Adding Like Signed Numbers

To add two numbers with the *same* sign:

1. Add their absolute values.
2. Give the sum the same sign as the terms.

EXAMPLE 1–11   Add -5 and -7.

Solution   Evaluate $-5 + (-7)$:

$|\ |$: $-5, -7$ $5, 7$
A: $5 + 7$ 12

Assign $(-)$ to the sum:

-12

$\therefore$ $-5 + (-7) = -12$

EXAMPLE 1–12   Add 3 and 7.

Solution   Evaluate $3 + 7$:

$|\ |$: $+3, +7$ $3, 7$
A: $3 + 7$ 10

Assign $+$ sign:

$+10$

$\therefore$ $3 + 7 = +10$

Rule 1–2. Adding Opposite Signed Numbers

To add two signed numbers with *different* signs:

1. Subtract the smaller absolute value from the larger.
2. Give the difference the sign of the larger term.

EXAMPLE 1–13	Add 8 and -3.
Solution	Evaluate $8 + (-3)$:
| |: $+8, -3$	$8, 3$
	Subtract the smaller term from the larger:
S: $\quad 8 - 3$	5
	Assign the sign of the larger term $(+)$:
	$+5$
$\therefore$	$8 + (-3) = +5$

EXAMPLE 1–14	Add -9 and 2.
Solution	Evaluate $-9 + 2$.
| |: $-9, +2$	$9, 2$
	Subtract the smaller term from the larger:
S: $\quad 9 - 2$	7
	Assign the sign of the larger term $(-)$:
	-7
$\therefore$	$-9 + 2 = -7$

Rule 1–3.   Subtracting Signed Numbers

To subtract two signed numbers:

1. Change the sign of the number to be subtracted.
2. Add the numbers using either Rule 1–1 or Rule 1–2.

EXAMPLE 1–15	Subtract -3 from 7.
Solution	Use Rule 1–3 to evaluate $7 - (-3)$.
	Change the sign of -3 and restate problem using Rule 1–1:
	$7 + 3$
| |: $+7, +3$	$7, 3$
A: $\quad 7 + 3$	10
	Assign $(+)$ to the sum:
	$+10$
$\therefore$	$7 - (-3) = 10$

Observation	When you see the word *subtract,* think ***sign change.*** In Example 1–15 you may see that the statement "subtract −3 from 7" is translated into the numerical expression $7 - (-3)$. By applying the methods for simplifying double signs found in Table 1–4, this expression becomes $7 + 3$. However, rather than concern yourself with condensing the double sign, it is easier to think, ***sign change and add.***

EXAMPLE 1–16   Subtract 6 from 2.

Solution   To subtract 6 from 2, change the sign of 6 and add. Thus, 2 subtract 6 becomes 2 add −6. Use Rule 1–2 to evaluate

$$2 + (-6)$$

| | : 2, −6   2, 6

Subtract the smaller term from the larger:

S: 6 − 2   4

Assign the sign of the larger (−):

$$-4$$

∴ $2 - 6 = -4$

As in Example 1–16, subtraction can result in a negative number. What would such a result mean? We have seen that negative numbers are used to indicate amounts below a reference value, for instance, the temperature below freezing, the distance below sea level, and the amount by which a checking account is overdrawn.

EXAMPLE 1–17   You have $42.00 in your checking account. You write checks for the following amounts: $6.00, $18.00, $7.00, and $13.00. How much do you have in your account?

Solution
$$42.00 - (6.00 + 18.00 + 7.00 + 13.00)$$
$$42.00 - 44.00$$
$$-2.00$$

∴   Your account is overdrawn by $2.00.

From this example we see that −$2.00 means that you have insufficient funds in your account and will no doubt be charged a service charge for the overdraft. This certainly has real meaning for you.

Combine the following signed numbers by adding or subtracting as indicated. Use the rules of addition and subtraction, and solve without the use of your calculator. Add the following numbers.

1. 7 and 3	**2.** −3 and −5	**3.** 3 and +2
4. −6 and −1	**5.** 8 and 4	**6.** −5 and −7
7. −17 and −6	**8.** 21 and +14	**9.** −13 and −39

Subtract the first number from the second in the following problems.

10. −5 from 3	**11.** 6 from 2	**12.** 8 from −3
13. 12 from −4	**14.** −18 from −3	**15.** 9 from −4
16. 32 from 14	**17.** 12 from −7	**18.** 8 from +16

Using signed numbers, rewrite the following phrases into numerical expressions and then compute the number represented by the expression.

19. Earning $21 and spending $16

20. Climbing up 14 stairs and then climbing up 17 more

21. A temperature rise of 18° and a fall of 26°

22. A gain of $3.00 and a loss of $9.00

23. Going 12 steps backward and then going 16 steps forward

24. A raise of $10 in pay and an increase of $3 in taxes

Using a Calculator with Signed Numbers

Your calculator is designed to work with signed numbers. It uses the operator keys $\boxed{+}$ and $\boxed{-}$ for addition and subtraction. It uses the change sign key $\boxed{+/-}$ or $\boxed{\text{CHS}}$ to enter negative numbers. It has been programmed to perform arithmetic with signed numbers. The following example will help you to become familiar with these features. It is recommended that you consult your owner's guide for additional information on working with signed numbers.

EXAMPLE 1–18   Evaluate $-7 + 8 - 6$.

Solution   Use your calculator. Enter 7; then change the sign.

$\boxed{\text{CHS}}$ 7 -7

$\boxed{+}$ 8   1

$\boxed{-}$ 6 -5

$\therefore$ $-7 + 8 - 6 = -5$

EXERCISE 1–8

Calculator Drill

Evaluate each of the following expressions.

1. $-18 + (-2)$ **2.** $47 + (-17)$ **3.** $12 - (-4)$

4. $-18 - (-14)$ **5.** $-27 + 8$ **6.** $17 - 13$

7. $-19 - 17$ **8.** $22 + (-17)$ **9.** $-13 + 9$

10. $-25 - (-12)$ **11.** $6 + (-3) + 5$ **12.** $120 - 30 - (-10)$

13. $75 - 15 + (-12)$ **14.** $-13 + (-15) + 39$ **15.** $-37 + 14 - (23)$

1–9   RELATIONAL OPERATORS

We have learned that a numerical equation uses the equal sign (=) to say that the numbers or expressions on either side of the equal sign are exactly the same. The equal sign is one of the relational operators.

The relationship of **inequality** may be shown in several ways. Suppose that we wanted to state the relationship between 5 and 7. We cannot say that 5 and 7 are equal, but we can say that they are not equal. This is stated as the mathematical sentence $5 \neq 7$. The not-equal symbol ($\neq$) is a relational operator. We also know that 7 is greater than 5. Using the greater-than relational operator ($>$), we can write $7 > 5$. The statement "$7 > 5$" is read "seven is greater than five." Finally, by using the less-than relational operation ($<$), we can write $5 < 7$. The inequality "$5 < 7$" is read "five is less than seven." Table 1–5 is a summary of these relational operators.

TABLE 1–5   Relational Operators

Symbol	Name	Use to:	Example	Read:
=	Is equal to	Show equality	$3 + 5 = 8$	3 plus 5 equals 8
$\neq$	Is not equal to	Show inequality	$2 \neq 9$	2 is not equal to 9
$>$	Is greater than	Show inequality and relative size	$6 > 1$	6 is greater than 1
$<$	Is less than	Show inequality and relative size	$4 < 5$	4 is less than 5
$\geq$	Is greater than or equal to	Show relative size	$7 \geq 1 + 5$	7 is greater than or equal to 1 plus 5
$\leq$	Is less than or equal to	Show relative size	$0 \leq 3$	0 is less than or equal to 3

EXERCISE 1–9

Use the appropriate relational operator in place of the words to express each statement.

1. 7 is less than 10.   **2.** 5 is smaller than 8.

3. 23 is more than 13.   **4.** -4 is greater than -7.

5. 3 is the same as -10 plus 13.   **6.** 3 plus 8 is not the same as -2.

7. The sum of 2 and 7 is greater than 3.

8. The sum of 8 and -3 is less than the difference between 25 and 7.

Use =, >, or < in place of "?" to make each statement true.

9. $7 ? -12 + 5$ **10.** $-8 ? 3 + 4$

11. $|+6| ? |-3|$ **12.** $-5 ? |-9|$

13. $-3 ? 13 - (10 + 3)$ **14.** $36 + 4 ? 48 - 12$

15. $-3 + 4 ? -7 + (-2)$ **16.** $8 + 2 - 4 ? -8 + 4$

17. $102 - 8 ? 91 - 89$ **18.** $-68 + 14 ? -59 + 2$

Compound Inequalities

In electronic applications of relational operators we often write compound inequality statements. These statements use two inequalities that have a number in common. For example, $5 < 8$ and $8 < 10$. These statements may be combined into a single compound statement of $5 < 8 < 10$. This statement is read "5 is less than 8, which is less than 10." Here is another compound inequality: $63 > 32 > 25$. This statement is read "63 is greater than 32, which is greater than 25."

In a compound inequality, both inequalities must be true in order for the compound inequality to be true. If either inequality is false, the entire statement is false.

EXAMPLE 1–19   Determine if $(15 - 3) < 20 < (14 - 8)$ is a true statement.

Solution   Combine and remove parentheses:

S: $(15 - 3)$ $12 < 20 < (14 - 8)$
S: $(14 - 8)$ $12 < 20 < 6$

Determine if the first inequality is true:

Yes, 12 is less than 20.

Determine if the second inequality is true:

No, 20is not less than 6.

∴   The compound inequality is false.

EXERCISE 1–10

Determine which of the following compound statements are true and which are false.

1. $3 < 6 < 10$ **2.** $2 > 5 > 3$

3. $17 < 28 < 7$ **4.** $11 > 9 > 7$

5. $-7 < 3 < 4$

6. $-4 > -3 > -6$

7. $13 > 18 > -21$

8. $-5 < -4 < -2$

9. $(3 - 7) > 3 > (-9 + 3)$

10. $(-5 - 3) < -6 < (-2 - 3)$

1–10 MULTIPLYING WITH SIGNED NUMBERS

When we multiply two numbers together, we form a *product*. The numbers used to form the product are called the **factors** of the product. For example, in the numerical equation $2 \times 3 = 6$, 2 and 3 are the factors of the product 6. Six is the product of the factors 2 and 3.

The multiplication of signed numbers is done by applying two rules. Each of these rules will be presented with examples to illustrate its application.

Rule 1–4.   *Multiplying Like Signed Numbers*

To multiply two numbers with the *same* sign:

1. Multiply the absolute values.
2. Give the product a plus sign $(+)$.

EXAMPLE 1–20   Multiply -3 times -4.

Solution   Use Rule 1–4 to evaluate $(-3) \times (-4)$:

| |: $-3, -4$ 3, 4

M: 3×4 12

Assign plus sign $(+)$ to product:

$+12$

$\therefore$ $(-3) \times (-4) = 12$

EXAMPLE 1–21   Multiply 3 times 4.

Solution   Use Rule 1–4 to evaluate $(+3) \times (+4)$:

| |: $+3, +4$ 3, 4

M: 3×4 12

Assign plus sign $(+)$ to product:

$+12$

$\therefore$ $(+3) \times (+4) = 12$

> ### Rule 1–5. Multiplying Unlike Signed Numbers
>
> To multiply two numbers with *different* signs:
> 1. Multiply the absolute values.
> 2. Give the product a minus sign $(-)$.

EXAMPLE 1–22   Multiply -8 times 3.

Solution   Use Rule 1–5 to evaluate -8×3:

$| \ |$: $-8, 3$ 8, 3

M: 8×3 24

Assign minus sign $(-)$ to the product:

-24

$\therefore$ $-8 \times 3 = -24$

EXAMPLE 1–23   Multiply 5 times -3.

Solution   Use Rule 1–5 to evaluate $5 \times (-3)$:

$| \ |$: $5, -3$ 5, 3

M: 5×3 15

Assign minus sign $(-)$ to the product:

-15

$\therefore$ $5 \times (-3) = -15$

From the previous examples, observe that the product of two numbers with like signs is positive, while the product of two numbers with unlike signs is negative. Thus, in multiplication, **like signs give a positive result** and **unlike signs give a negative result.** This concept is graphically presented in Table 1–6.

TABLE 1–6   Multiplying Signed Numbers

Signs of Factors	Sign of Product	Signs of Factors	Sign of Product
+ times +	+	+ times −	−
− times −	+	− times +	−

When using your calculator, remember that it is designed to work with signed numbers. It has been programmed to perform multiplication with signed numbers according to the rules you have learned.

EXAMPLE 1–24	Use your calculator to multiply $15 \times (-6)$.
Solution	Enter 15:
$\boxed{\times}$ 6 $\boxed{\text{CHS}}$	-90
$\therefore$	$15 \times (-6) = -90$
Observation	The indicated keystrokes are to remind you of the operations used in the calculation, but they may not be the same strokes and they may not be in the same order that you will use with your calculator.

EXERCISE 1–11

Use the information in Table 1–6 to multiply the following signed numbers.

1. 2 and -3 **2.** $+8$ and 4   **3.** -5 and -2

4. -1 and -2 **5.** -5 and 7   **6.** -8 and 3

7. -6 and -7 **8.** $+4$ and $+3$ **9.** $+2$ and -2

10. -8 and $+5$ **11.** 9 and -10 **12.** -7 and -1

Calculator Drill
Using your calculator, compute the number represented by each of the following numerical expressions.

13. -31×40 **14.** $28 \times (-15)$ **15.** $+16 \times (+12)$

16. $34 \times (-27)$ **17.** -92×102 **18.** $-72 \times (-13)$

19. $+18 \times 13$ **20.** -47×52 **21.** $-37 \times (-53)$

22. $-49 \times (-25)$ **23.** -13×94 **24.** $37 \times (-14)$

1–11 DIVIDING WITH SIGNED NUMBERS

Division is indicated in arithmetic by the division symbol ($\div$), as in $6 \div 2 = 3$. In algebra, division is usually indicated as a fraction. A fraction is made up of three parts, as shown in Figure 1–2: (1) the numerator (the dividend or the number to be divided), (2) the denominator (the divisor or the number to do the dividing), and (3) the bar between the numerator and the denominator. The bar may appear horizontally or diagonally (— or /), as pictured in Figure 1–2(b). Like parentheses, the bar is a symbol of grouping. Thus, the bar controls the order of operations. The following example will demonstrate this important property of the bar.

$$\frac{\text{Dividend}}{\text{Divisor}} = \text{Quotient} \qquad \frac{\text{Numerator}}{\text{Denominator}} = \text{Numerator/Denominator}$$

(a)   (b)

FIGURE 1–2   Parts of a fraction. (a) The result of dividing the dividend by the divisor is the quotient; (b) Another name for dividend is numerator. Another name for divisor is denominator. The bar may be horizontal or diagonal.

EXAMPLE 1–25   Discuss how to evaluate the following expression and then show the solution:

$$\frac{2 + 13}{5}$$

Discussion   Since $2 + 13$ is grouped together by the bar, we evaluate $2 + 13$ first. We then divide by 5. Notice that the usual order of division before addition has been overridden by the bar, a symbol of grouping.

Solution $\dfrac{2 + 13}{5}$

A: $2 + 13$ 15/5

D: 15/5   3

$\therefore$ $\dfrac{2 + 13}{5} = 3$

The following rules are used when dividing signed numbers. Each of these rules will be presented with examples to illustrate its application.

> **Rule 1–6.**   *Dividing Like Signed Numbers*
>
> To divide two numbers with the *same* sign:
> 1. Divide the absolute values.
> 2. Give the result a plus sign ($+$).

EXAMPLE 1–26   Divide -8 by -4.

Solution   Use Rule 1–6 to evaluate $-8/-4$:

| |: $-8, -4$ 8, 4

D: 8/4   2

Assign plus sign $(+)$:

$$+2$$

$$\therefore \quad -8/-4 = 2$$

Rule 1–7. *Dividing Unlike Signed Numbers*

To divide two numbers with *different* signs:
1. Divide the absolute values.
2. Give the result a minus sign $(-)$.

EXAMPLE 1–27 Divide 6 by -2.

Solution Use Rule 1–7 to evaluate $6/-2$:

| |: 6, -2 6, 2

D: 6/2   3

Assign minus sign $(-)$:

$$-3$$

$$\therefore \quad 6/-2 = -3$$

EXAMPLE 1–28 Use your calculator to evaluate $\dfrac{50 - 11}{-13}$.

Solution The bar is a sign of grouping, so combine the numbers in the numerator.
Enter 50:

$\boxed{-}$ 11   39

$\boxed{\div}$ 13 $\boxed{\text{CHS}}$ -3

$$\therefore \quad \frac{50 - 11}{-13} = -3$$

EXERCISE 1–12

Use Rules 1–6 and 1–7 to evaluate the following expressions.

1. 15/3	**2.** $-8/2$	**3.** $6/-2$
4. 12/-6	**5.** $-14/7$	**6.** 10/5

7. $-18/-6$ **8.** $-15/-5$ **9.** $16/-4$

10. $-20/-10$ **11.** $25/5$ **12.** $24/8$

Calculator Drill

In the following, remember that the bar is a sign of grouping.

13. $-27/9$ **14.** $-36/4$ **15.** $-34/-17$

16. $-64/16$ **17.** $42/-14$ **18.** $75/-15$

19. $\dfrac{7 + 2}{3}$ **20.** $\dfrac{8 - 6}{2}$ **21.** $\dfrac{24 + 36}{-12}$

22. $\dfrac{-13 - 7}{-4}$ **23.** $\dfrac{19 - 11}{5 + 3}$ **24.** $\dfrac{-8 + 2}{-3}$

25. $\dfrac{5 + 15}{-15 + 5}$ **26.** $\dfrac{-12}{25 - 13}$ **27.** $\dfrac{16 - 6}{-2 + 7}$

28. $\dfrac{3 + 17}{-5 + 15}$ **29.** $\dfrac{-15 + 15}{20 + 15}$ **30.** $\dfrac{-19 + (-89)}{17 - 29}$

SELECTED TERMS

absolute value   The value of a number without regard to its sign.

chain calculation   The technique of solving arithmetic problems using a calculator without writing out all the intermediate results.

factors   A product is formed when two or more numbers are multiplied; each of the numbers in the product is called a factor of the product.

inequality   A statement that two numerical expressions do not have the same value.

natural numbers   The set of numbers (1, 2, 3, etc.) used in counting; the counting numbers.

numerical equation   A statement that declares two numerical expressions have the same value.

numerical expression   A number or a list of numbers joined by arithmetic operators; e.g., 9, or $25 - 8$.

terms   A sum is formed when two or more numbers are added; each of the numbers in the sum is called a term of the sum.

2

Number Notation and Operation

2-1   Introduction to Exponents

2-2   Number Notation

2-3   Numeric Operations and Rounding

2-4   Operations with Approximate Numbers

2-5   Square Roots, Radicals, and Reciprocals

2-6   Combined Operations

2-7   Powers of Ten and Approximations

PERFORMANCE OBJECTIVES

- Read and write numbers with exponents.

- Express numbers in engineering and scientific notation.

- Recognize the significant figures in an approximate number.

- Use the calculator's rounding and display features.

- Record sums and products with appropriate precision.

- Work with roots, powers, and fractional exponents.

- Apply a reciprocal when carrying out division with a fraction.

- Chain calculate to avoid rounding error.

- Approximate an answer with the aid of powers of ten.

Designed for mounting above the fluid surface, the ultrasonic level sensor offers reliable, high accuracy, non-contact continuous measurement of distance up to 6 m (20 ft). (Courtesy of George Fischer, Inc.)

In the study of technology in general and electronics in particular, you will work with many physical quantities; some will be represented by exact numbers and others will be represented by approximate numbers. When a measured quantity (approximate number) is used in a computation, the result must be stated to a reasonable degree of precision. In order to know how to state the result, an understanding of exponents, number notation, significant figures, and rounding is essential.

In this chapter, you will gain a knowledge of how to write answers to calculations so that the result is in line with the precision of the measured quantity. Also, you will have an opportunity to use your calculator to do basic mathematical operations, including powers, roots, and reciprocals. Finally, the rules of exponents will be presented in conjunction with the technique of approximating an answer.

In addition to the abbreviations used in Chapter 1, the following abbreviations are new to this chapter:

DN	decimal notation	$\boxed{x^2}$	square a number
EN	engineering notation	$\boxed{\text{EE}}$	exponent entry**
R	round	$\boxed{\sqrt{}}$	square root
SF	significant figures	$\boxed{\text{ENG}}$	engineering notation
SN	scientific notation	$\boxed{\text{SCI}}$	scientific notation
$\boxed{1/x}$	reciprocal*	$\boxed{y^x}$	raise to an exponent***

*Note: $\boxed{x^{-1}}$ is also used for taking the reciprocal.

**Note: $\boxed{\text{EXP}}$ and $\boxed{\text{EEX}}$ are also used for exponent entry.

***Note: $\boxed{\wedge}$ is also used to raise to an exponent.

The following symbols are new to this chapter and are listed here for your reference:

$\equiv$ identically equal to $\approx$ approximately equal to

$\pm$ plus or minus

Special attention is given in this chapter to some features your calculator might have. Therefore, it would be helpful to have the owner's guide for your calculator handy as you study.

2–1 INTRODUCTION TO EXPONENTS

In the equation $4 \times 4 = 16$, recall that the number 16 is the **product** and the two 4's are called the **factors** of the product. This equation could be written in a shorter way by using exponent notation. The equation $4 \times 4 = 16$ becomes $4^2 = 16$. The small 2 written above and to the right of the 4 is called an **exponent** and the 4 is called the **base**. The expression 4^2 is read "four squared" or "four to the second power." The exponent tells how many times the base is used as a factor in a product.

EXAMPLE 2–1	Read the following expression and indicate its meaning; then compute the number it represents: 5^4.
Discussion	5^4 is read "five to the fourth power" and means $5 \times 5 \times 5 \times 5$.
Solution	$5^4 = 5 \times 5 \times 5 \times 5$
	$5^4 = 625$
$\therefore$	$5^4 = 625$

EXAMPLE 2–2	Use exponents to simplify the following expression: $8 \times 8 \times 8 \times 8 \times 8 \times 8 \times 8 \times 8 \times 8 \times 8 \times 8$
Solution	Count the number of times 8 is used as a factor:
	11 times
	Rewrite using an exponent:
	8^{11}
$\therefore$	$8 \times 8 \times 8 \times 8 \times 8 \times 8 \times 8 \times 8 \times 8 \times 8 \times 8 = 8^{11}$

EXERCISE 2–1

Read the following expressions and explain what each means.

1. 5^2 **2.** 6^3 **3.** 2^4 **4.** 3^5

5. 4^4 **6.** 7^2 **7.** 1^3 **8.** 8^7

Simplify each of the following expressions by rewriting the expression using an exponent.

9. $3 \times 3 \times 3$ **10.** 4×4

11. $8 \times 8 \times 8 \times 8$ **12.** $2 \times 2 \times 2 \times 2 \times 2$

13. $5 \times 5 \times 5$ **14.** $10 \times 10 \times 10 \times 10$

15. $9 \times 9 \times 9 \times 9$ **16.** $6 \times 6 \times 6 \times 6$

17. $7 \times 7 \times 7 \times 7 \times 7 \times 7 \times 7$ **18.** $13 \times 13 \times 13$

19. 14×14 **20.** $3 \times 3 \times 3 \times 3 \times 3$

Using a Calculator with Exponents

Scientific calculators can evaluate expressions like 2^5 with a special function key such as $\boxed{y^x}$ or $\boxed{x^y}$. Check your calculator and your owner's guide to see how your calculator performs this operation. Pay special attention to the order in which the numbers are to be entered.

EXAMPLE 2–3	Evaluate 5^4 using your calculator.
Solution	Enter 5:

$$\boxed{y^x}\ 4 \qquad 625$$

$$\therefore \qquad 5^4 = 625$$

EXERCISE 2–2

Calculator Drill

Compute the number represented by each of the following expressions.

1. 2^5	**2.** 4^3	**3.** 9^3	**4.** 5^3	**5.** 8^4
6. 7^3	**7.** 6^5	**8.** 3^5	**9.** 10^3	**10.** 4^8
11. 2^7	**12.** 9^5	**13.** 0.50^2	**14.** $32^{0.2}$	**15.** $81^{0.25}$

2–2   NUMBER NOTATION

Powers of Ten Notation

Powers of ten notation provides a simple way to express any number as a decimal number times a power of ten. This notation enables you to conveniently work with both very large and very small numbers. In electronics, it is not uncommon to work with extremely small decimal fractions like 0.000 0073 and 0.000 000 000 435. By applying the concepts of powers of ten notation, even these numbers may be expressed in a simple manner.

In Section 2–1 you learned how to use exponents. You now know that $10 \times 10 \times 10$ can be expressed as 10^3. Since 1000 is $10 \times 10 \times 10$, we may conclude that $1000 = 10^3$. Suppose that we wanted to write 5000 using powers of ten notation; 5000 is 5×1000 or 5×10^3. Therefore, 5000 can be written as 5×10^3 in *powers of ten notation*. Figure 2–1 shows the pattern of a number written in powers of ten notation. Table 2–1 shows several examples of numbers written in powers of ten notation.

Notice in Table 2–1 that the decimal fractions 0.0234, 0.0087, and 0.345 have negative exponents. Also notice that the placement of the decimal point in the powers of

FIGURE 2–1   Powers of ten notation. (a) Pattern for powers of ten notation is a decimal coefficient times 10 to an exponent; (b) example of powers of ten notation.

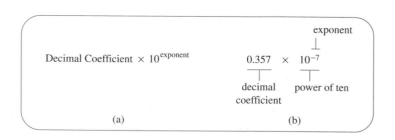

TABLE 2–1   Powers of Ten Notation Examples

Decimal Notation	Powers of Ten Notation
12 430	124.30×10^2
8350	83.5×10^2
0.0235	235×10^{-4}
104 200	10.42×10^4
0.0087	8.7×10^{-3}
192	0.192×10^3
0.345	34.5×10^{-2}

ten notation is not consistent. The decimal point may be placed anywhere you desire, with one restriction: there must be at least one digit to the left of the decimal point. This allows a great deal of flexibility when forming decimal numbers into powers of ten notation.

The steps used to form decimal numbers into powers of ten notation are summarized in Rule 2–1. They are also pictured in Figure 2–2. To explore the use of powers of ten in comparing the size of things, go to the website of the National High Magnetic Field Laboratory (http://micro.magnet.fus.edu/primer/java/scienceopticsu/powersof10/).

RULE 2–1.   *Expressing Decimal Numbers in Powers of Ten Notation*

To express a decimal number in powers of ten notation:

1. Form the decimal coefficient by moving the decimal point to the desired position.

2. Determine the exponent by counting the number of places the decimal point was moved.

3. The proper sign for the exponent is selected as:
   (a) A minus sign if the decimal point is moved to the right.
   (b) A plus sign if the decimal point is moved to the left.

4. Write the number in powers of ten notation as a product of the decimal coefficient and the power of ten.

FIGURE 2–2   Forming powers of ten notation.

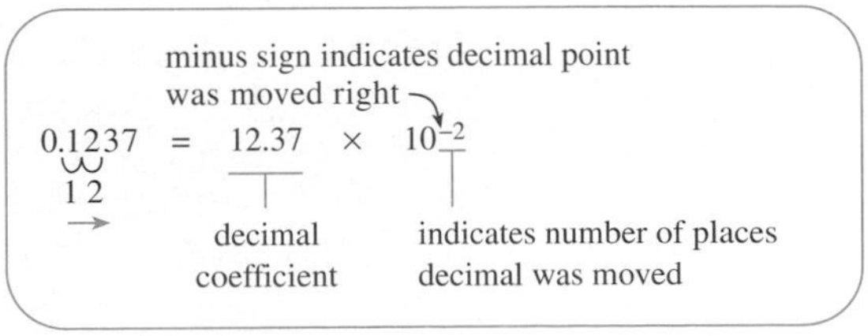

minus sign indicates decimal point was moved right

$$0.1237 = 12.37 \times 10^{-2}$$

decimal coefficient — indicates number of places decimal was moved

EXAMPLE 2–4	Write 0.003 725 in powers of ten notation. Place the decimal between the seven and the two.
Solution	Follow the steps of Rule 2–1.
Step 1:	Move the decimal point:

$$0.003\,725 \Rightarrow 37.25$$

Step 2:	Count places moved:

0.003 725   moved four places

1 2 3 4
$\longrightarrow$

Step 3:	Determine the sign of the exponent. Since the decimal point was moved right, a minus sign is assigned to the exponent: (-4).
Step 4:	Form powers of ten notation:

$$37.25 \times 10^{-4}$$

$\therefore$	$0.003\,725 = 37.25 \times 10^{-4}$

EXAMPLE 2–5	Write 1⌃320 in powers of ten notation. Move the decimal to the place indicated by the caret (⌃).
Solution	Express in powers of ten notation using Rule 2–1.
Step 1:	Move the decimal point:

$$1320 \Rightarrow 1.320$$

Step 2:	Count places:

1320   moved three places

3 2 1
$\longleftarrow$

Step 3:	Determine the sign of the exponent. Since the decimal point was moved left, the exponent is positive: $(+3)$.
Step 4:	Form the powers of ten notation and drop the trailing zero.

$$1.320 \times 10^{3}$$

$\therefore$	$1320 = 1.32 \times 10^{3}$

Express the following decimal numbers in powers of ten notation. Since the decimal point may be placed anywhere, a caret (∧) has been placed under the number to show you where to place the decimal point. Drop the trailing zeros.

1. 75∧4 000 000   **2.** 0.000 8∧83 36   **3.** 0.003∧72

4. 9∧976   **5.** 4∧01   **6.** 0.000 006 0∧9

7. 0.5531∧   **8.** 176∧2   **9.** 0.002∧49

10. 3∧1 982   **11.** 82∧92   **12.** 0.3∧12

13. 28∧6 514   **14.** 7∧21.12   **15.** 1.788∧2

16. 42.06∧8   **17.** 73.5∧0   **18.** 0.610∧

Scientific Notation

Scientific notation is a special form of powers of ten notation. In scientific notation the decimal point is always placed to the right of the leftmost nonzero digit. Table 2–2 shows several examples of scientific notation. You will notice that in every instance the decimal point is located in the same place—to the right of the first nonzero digit.

As can be seen from Table 2–2, scientific notation has a very definite form. Rule 2–2 outlines the procedure for writing a decimal number in scientific notation.

TABLE 2–2   Comparison of Notation

Decimal Notation	Powers of Ten Notation	Scientific Notation
5350	53.5×10^2	5.35×10^3
0.002 87	287×10^{-5}	2.87×10^{-3}
0.033 33	333.3×10^{-4}	3.333×10^{-2}
26 220	26.22×10^3	2.622×10^4
0.0805	80.5×10^{-3}	8.05×10^{-2}

RULE 2–2.   *Expressing Decimal Numbers in Scientific Notation*

To express a decimal number in scientific notation:

1. Form the decimal coefficient by placing the decimal point to the right of the leftmost nonzero digit.

2. Determine the exponent by counting the number of places the decimal point was moved.

(continued)

3. The proper sign for the exponent is selected as:

    (a) A minus sign if the decimal point is moved to the right.

    (b) A plus sign if the decimal point is moved to the left.

4. Write the number in scientific notation as a product of the decimal coefficient and the power of ten.

EXAMPLE 2–6	Express 0.008 72 in scientific notation.
Solution	Use the steps of Rule 2–2.
Step 1:	Form the decimal coefficient.

$$0.008\ 72 \Rightarrow 8.72$$

Steps 2–3:   Count places and determine the sign of the exponent.

$$0.008\ 72 \qquad \text{moved three places to the right } (-3)$$

$$1\ 2\ 3$$

Step 4:   Form the number.

$$8.72 \times 10^{-3}$$

$$\therefore \qquad 0.008\ 72 = 8.72 \times 10^{-3}$$

EXERCISE 2–4

Express the following numbers in scientific notation and drop the trailing zeros.

1. 300	**2.** 2000	**3.** 280	**4.** 36
5. 0.50	**6.** 0.009	**7.** 0.36	**8.** 0.85
9. 7 052	**10.** 0.008 83	**11.** 0.0063	**12.** 992
13. 0.001	**14.** 409	**15.** 578 000	**16.** 0.004 93
17. 5550	**18.** 43 500	**19.** 0.68	**20.** 0.007 04
21. 0.013 13	**22.** 18.52	**23.** −282.3	**24.** 0.151
25. −0.008 25	**26.** 1316.2	**27.** −47.8	**28.** −301 000

Decimal Notation

Numbers that are written without powers of ten are written in decimal notation. For example, 75 824, 0.005 13, and 19.3 are in decimal notation. In electronics, it is sometimes necessary to change numbers in powers of ten notation to decimal notation. Rule 2–3 gives a general procedure for doing that.

> ### RULE 2–3. Converting from Powers of Ten Notation to Decimal Notation
>
> To change from powers of ten notation to decimal notation, form the decimal number from the decimal coefficient and:
>
> 1. If the exponent is positive, then move the decimal point to the right the number of places equal to the exponent. If necessary, add trailing zeros to hold the position of the decimal point.
>
> 2. If the exponent is negative, then move the decimal point to the left the number of places equal to the exponent. If necessary, add leading zeros to hold the position of the decimal point. The number will be in decimal notation.

EXAMPLE 2–7

Write (a) 6.28×10^3 and (b) 5.31×10^{-4} in decimal notation.

Solution

(a) Move the decimal point three places to the right:

DN:
$$6.28 \times 10^3 = 6280. = 6280.$$
$$1\ 2\ 3 \longrightarrow$$

Notice that the zero was added to fill out the number.

$\therefore \qquad 6.28 \times 10^3 = 6280$

(b) Move the decimal point four places to the left:

DN:
$$5.31 \times 10^{-4} = 0.000\ 531 = 0.000531$$
$$4\ 3\ 2\ 1 \longleftarrow$$

Notice that zeros were added to fill out the number.

$\therefore \qquad 5.31 \times 10^{-4} = 0.000\ 531$

Table 2–3 contains more examples that illustrate the use of Rule 2–3.

TABLE 2–3 Application of Rule 2–3

Powers of Ten Notation	Movement of Decimal Point		Decimal Notation
	Direction	No. of Places	
6.38×10^{-3}	Left	3	0.006 38
0.904×10^5	Right	5	90 400
34.16×10^{-4}	Left	4	0.003 416
7.24×10^6	Right	6	7 240 000
5.00×10^3	Right	3	5000

Write the following numbers in decimal notation.

1. 6.35×10^{-2}

2. 1.058×10^6

3. 382.8×10^{-6}

4. 52.42×10^3

5. 9.935×10^{-3}

6. 7015×10^2

7. 462×10^{-2}

8. 14.86×10^{-5}

9. 0.8831×10^2

10. $0.003\ 62 \times 10^4$

11. -1.83×10^5

12. 0.252×10^{-1}

13. 288.2×10^{-3}

14. -192.76×10^2

15. -0.0715×10^{-2}

16. 3652.5×10^{-5}

17. 25.2×10^{-1}

18. 0.082×10^{-2}

Engineering Notation

Engineering notation is a useful form of power of ten notation where the exponent of ten is a multiple of three. Table 2–4 shows several examples of numbers written in engineering notation. You will notice that each number written in engineering notation has an exponent of ten that is a multiple of 3. Decimal numbers written in engineering notation are used with unit prefixes to form multiples and submultiples of the *metric units.* Guidelines 2–1 gives you some insight into forming numbers written in engineering notation.

TABLE 2–4   Comparison of Notation

Decimal Notation	Engineering Notation
3200	3.20×10^3
62 400 000	62.4×10^6
9 880 000 000	9.88×10^9
0.0255	25.5×10^{-3}
0.000 0820	82.0×10^{-6}

GUIDELINES 2–1.   *Expressing Decimal Numbers in Engineering Notation*

1. Adjust the decimal point to express the decimal coefficient as a number with a value from 0.100 to 1000 times an exponent of ten that is a multiple of three.

2. Write the number in engineering notation as a product of the decimal coefficient with a preferred range between 0.100 and 1000 times a power of ten with an exponent that is a multiple of 3.

EXAMPLE 2–8	Write 0.000 72 in engineering notation.
Solution	Use Guidelines 2–1. The decimal point is moved three places right to create a decimal coefficient with a value from 0.100 to 1000.

$$0.000\ 72 \Rightarrow 0.72 \times 10^{-3}$$

∴　　0.000 72 is expressed as 0.72×10^{-3} in engineering notation.

Guidelines 2–1 may be carried out quickly, simply, and automatically with the $\boxed{\text{ENG}}$ stroke of the calculator. Use your calculator to do the following exercise. Consult your calculator owner's guide for the procedure for using this stroke.

EXERCISE 2–6

Calculator Drill
Write each of the following decimal numbers in engineering notation.

1. 1320	**2.** 564 000	**3.** 0.000 085
4. 8648	**5.** 0.0047	**6.** 4 280 000
7. 0.008	**8.** 47 300	**9.** 0.000 572

Change the following numbers written in engineering notation to decimal notation.

10. 27×10^{-3}	**11.** 0.284×10^{3}	**12.** 7.32×10^{6}
13. 420×10^{3}	**14.** 8.20×10^{-9}	**15.** 0.57×10^{6}
16. 1.25×10^{-6}	**17.** 94.5×10^{6}	**18.** 0.39×10^{-3}

2–3   NUMERIC OPERATIONS AND ROUNDING

Exact Numbers

In science and technology, a few of the numbers encountered are totally free from error. These numbers are called **exact numbers** and are obtained by counting objects, by definition, or by computation with exact numbers (they are derived). Some examples of exact numbers include:

12 donuts in a dozen—by definition

9 square feet in a square yard—derived from 3 feet in a yard

4 hubcaps on a car—by counting

Exact numbers occur in formulas as constants. In the formula for the volume of a sphere ($V = 4/3\pi r^{3}$), the constants 4, 3, and π are exact numbers.

Approximate Numbers

Approximate numbers result from measured quantities, and all measured quantities have some error due to the measuring instrument and its use. In the course of measuring, a technician needs to exercise care to ensure that the amount of measurement error is reduced as much as possible.

Measurements are reported as an amount (how much of a quantity), whereas the **precision** of the measurement is indicated by the number of digits in the number used to record the measurement. For example, 12.035 volts is more precise than 12 volts, and 2.368 amperes is more precise than 2 amperes. As the precision of the measuring instrument increases, more digits are reported in the measured value. The **accuracy** (how close the reading is to a recognized standard) of the measurement is limited by the precision of the instrument, the type of instrument (mechanical, electrical, etc.) and its state of calibration, and by the technician's ability (experience and judgment) to make the reading.

EXERCISE 2–7

Determine whether each of the following is an exact or approximate number.

1. 3.92 minutes to run 1.00 kilometer
2. 2 wheels on a motorcycle
3. 1000 millivolts in 1 volt
4. A box wrench having 2 ends
5. A 737 airliner traveling 512 miles per hour
6. A temperature of 24.3° C
7. A gear with 18 teeth
8. An aluminum rod with diameter 28.23 millimeters

Significant Figures

In recording a reading of a measurement, each stated digit in the measurement is a **significant figure.** Suppose the electric current in a circuit is measured with a digital ammeter and found to be 15.35 amperes. In the measured value of 15.35 amperes, there are four significant figures. However, the last digit, 5, is uncertain because of the error inherent in the ammeter and its application to an electric circuit.

From the 15.35-ampere measurement, it is known that the actual value of the electric current is somewhere between 15.30 and 15.40 amperes, or 15.35 ± 0.05 amperes. The *plus-or-minus* value indicates the **tolerance** of the reading.

When measurements are used in calculations, it is very important that the result not be overstated. That is, the act of calculating cannot improve the stated precision of a measurement. For example, suppose you measure the diameter of a pulley with a dial caliper (Figure 2–3) to the nearest tenth of a millimeter as 54.7 millimeters. From this measurement, the circumference is computed ($C = \pi d$) using a scientific calculator. Because the rounding and format feature of the calculator was not set properly, the calculator displayed the answer as 171.845 118 2 millimeters. Obviously the precision of

FIGURE 2–3 Dial caliper measuring the outside diameter of a pulley.

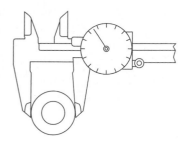

the answer is overstated. It would be absurd to claim precision to ten significant figures when the measurement of the diameter has precision only to three significant figures. With the calculator's rounding and format feature set, the correctly stated answer to the calculation is 172 millimeters.

When recording measured quantities (approximate numbers) or when using a calculator to solve a problem using approximate numbers, use the following guidelines to help you avoid overstating the precision of the number.

GUIDELINES 2–2. Determining the Number of Significant Figures

1. Usually, all figures in a recorded measurement are significant.

2. If zero is used to hold the decimal point in its place, then it is not significant. For example, the zeros in **0.0**24 V (volts) and **0.00**38 A (amperes) *are not* significant. However, the zeros in the recorded measurements 3**0.7** m (meters), 1**0**8.9 W (watts), 8.3**0** Ω (ohms), and 48**0** V (volts) *are all* significant.

3. The placement of the decimal point does not affect the number of significant figures in a number. Thus, each of the numbers 9.360 W (watts), 0.009 360 W (watts), and 936.0 W (watts) has four significant figures.

4. When calculating, use powers of ten, scientific, or engineering notation to indicate the significant digits of very large and very small quantities. For example, writing 470 000 Ω (ohms) to indicate the resistance of a resistor might indicate six-figure significance, but the manufacturer's tolerance of the component indicates only three significant figures. By expressing the number in engineering notation as 470×10^3 Ω, only three significant figures are indicated.

5. All numbers (including zero) in the decimal coefficients of powers of ten notation are significant. Thus, both 2.70×10^4 Ω (ohms) and 480×10^3 V (volts) have three significant figures.

6. Exact numbers can be expressed to any amount of precision with as many significant figures as may be deemed necessary. For example, in multiplying 93.28 (an approximate number) by π, where π is an exact number, π is thought of as being 3.142, making it compatible to 93.28 in the number of significant figures.

(continued)

7. If the number lacks sufficient digits to indicate the desired significance, then append zeros to the number to fill it out. For example, if it is desired to express 80 to three significant figures, then one zero is appended to 80 to make it 80.0.

8. Use judgment in forming the answer to a calculation. Most of the numbers used in electronics are significant only to two or three figures.

EXERCISE 2–8

State the precision of the following measurements by determining the number of significant figures in each recorded number.

Example: 6.65 V **Answer:** Precise to *three* significant figures.

1. 0.345 m (meter)
2. 0.0589 A (amperes)
3. 3.005 Ω (ohms)
4. 5207 W (watts)
5. 0.00032 A (amperes)
6. 40.600×10^3 V (volts)
7. 24.008 kg (kilograms)
8. 9024.06 Ω (ohms)
9. 2000 W (watts)
10. 0.0075 V (volt)
11. 1.30×10^{-3} m (meter)
12. 400×10^2 kg (kilograms)

Rounding

Rounding is a process used to shorten the calculator answer to the desired number of significant figures. Although there are several methods commonly used to round, the following very simple rounding procedure is the one programmed into most calculators with display and rounding options.

RULE 2–4. Rounding

To round to the desired number of significant figures:

1. Decide on the number of significant figures in the answer.

2. Examine the first digit to the right of the last significant figure:
   (a) If it is 5 or more, then add 1 to the last significant figure.
   (b) If it is 4 or less, then leave all figures alone.

3. Drop everything to the right of the last significant figure.

EXAMPLE 2–9   Express 740 800 to three significant figures using scientific notation.

Solution   Follow the steps of Rule 2–4 to round:

SN: $7.408\ 00 \times 10^5$

R to 3 SF:	The fourth digit is 8, which is greater than 5; so add 1:
	7.41×10^5
$\therefore$	$740\ 800 \Rightarrow 7.41 \times 10^5$ to three significant figures.

EXAMPLE 2–10	Express 0.008 895 to three significant figures using engineering notation.
Solution	Round the number using Rule 2–4:
EN:	8.895×10^{-3}
R to 3 SF:	The fourth digit is 5, so 1 is added to 9, which becomes 10; so 89 becomes 90:
	8.90×10^{-3}
$\therefore$	$0.008\ 895 \Rightarrow 8.90 \times 10^{-3}$

As a final observation, be aware that most modern scientific calculators have both *rounding* and *display features* that may be used to set the type of display and the number of digits displayed by the calculator. These features include: **FIX** (*fixed decimal*—decimal notation), **SCI** (*scientific*—scientific notation), **ENG** (*engineering*—engineering notation).

The need for several format options (fixed, scientific, and engineering) results from their typical applications. Engineering notation is used in science and technology in conjunction with the metric system of units and unit prefixes. Scientific notation is used in science and technology in general and with the English system of units. Decimal notation is used in our everyday expression of numbers.

Decimal notation cannot always convey the desired number of significant figures to the reader. For example, 48 000 can be interpreted to have up to five significant figures when all the zeros are seen as significant figures. Suppose, however, that the last two zeros are only placeholders and are not significant figures. To indicate three significant figures, a power of ten notation (scientific or engineering notation) must be used to express the desired amount of significance. It is only through the use of scientific or engineering notation that a number can be consistently expressed to the desired degree of significance.

EXERCISE 2–9

Calculator Drill

Express the following numbers in engineering notation to three significant figures. Use the ENG calculator key to format the numbers. Consult your owner's guide for the procedure for setting up the rounding and engineering notation features of your calculator.

1. 7568.3	**2.** 0.9905	**3.** 1996	**4.** 0.0392
5. 1776	**6.** 726 899	**7.** 0.0593	**8.** 99 999

9. 3 481 500	**10.** 14 080	**11.** 0.04	**12.** 149 200
13. 0.005 206	**14.** 56.56	**15.** 1994	**16.** 0.07
17. 0.000 877 5	**18.** 48 000	**19.** 454.72	**20.** 0.0249

2–4 OPERATIONS WITH APPROXIMATE NUMBERS

Addition and Subtraction

When computing with approximate numbers, some guidelines may be applied to ensure that the precision of the answer is not overstated. The guidelines below are used when adding or subtracting. These guidelines require that you identify the positional value of the least precise, rightmost significant figure in the numbers to be used in the mathematical operation. This is done by inspecting the numbers to determine the positional location of the rightmost significant figure in each of the numbers. For example, 2.5 is to be added to 8.06. Of the two numbers, the 5 in the number 2.5 is the least precise, rightmost significant figure because it is located in the tenths column, whereas the 6 in 8.06 is located in the hundredths column. In this case, the sum, 10.56, is rounded to the nearest tenth of a unit and is stated as 10.6. For your reference, Figure 2–4 lists the positional value names for decimal numbers.

FIGURE 2–4 Positional-value names for the digits in the decimal number 495 017.358 62.

Hundred thousands	Ten thousands	Thousands	Hundreds	Tens	Units		Tenths	Hundredths	Thousandths	Ten-thousandths	Hundred-thousandths
4	9	5	0	1	7	.	3	5	8	6	2

GUIDELINES 2–3. Addition and Subtraction of Approximate Numbers

1. Identify the least precise, rightmost significant figure in the numbers to be added or subtracted.
2. Carry out the calculation.
3. Round the answer (sum or difference) to the same precision as the positional value of the least precise, rightmost significant figure used in the calculation.

EXAMPLE 2–11	Add 55.24 to 60.
Solution	Follow the steps of Guidelines 2–3.
Step 1:	Of the significant figures, the 0 in 60 is the least precise, rightmost significant figure and it is located in the units column of the number.
Step 2:	Add 55.24 to 60.
$+$	115.24
Step 3:	Round the answer to the nearest whole unit.
R to units:	115
$\therefore$	$55.24 + 60 = 115.24 \Rightarrow 115$

EXAMPLE 2–12	Subtract 2.23×10^3 from 55.27×10^2.
Solution	First write each number in decimal notation and then follow the steps of Guidelines 2–3.
DN:	$2.23 \times 10^3 = 2230$ and $55.27 \times 10^2 = 5527$
Observation	In converting from powers of ten notation to decimal notation, a zero was added to 223**0**. This zero is not a significant figure as it is used to fill out the number and hold the decimal point. The original number, 2.23×10^3, has three significant figures, as does the equivalent decimal notation number, 2230.
Step 1:	Of the significant figures, the 3 in 2230 is the least precise, rightmost significant figure, and it is located in the tens column of the number.
Step 2:	Subtract 2230 from 5527.
$-$	3297
Step 1:	Round the answer to the nearest ten units.
R to tens:	3300
$\therefore$	$5524 - 2230 = 3294 \Rightarrow 3300$, or 3.30×10^3
Observation	The zero next to the three (33**0**0) is a significant figure, whereas the rightmost zero is not; it is a placeholder.

EXERCISE 2–10

State the precision of the following measurements (approximate numbers). Use Figure 2–4 to aid in determining the positional value name of the least precise, rightmost significant figure in the number. For example, the measurement 1.08 Ω is precise to the nearest *hundredth unit* because the 8 is located in the hundredths column of the approximate number 1.08.

1. 520 m (meters)
2. 0.32 V (volt)
3. 2.725 W (watts)
4. 26×10^{-3} A (ampere)
5. 33.0×10^3 Ω (ohms)
6. 88.5 m (meters)
7. 0.0442 A (ampere)
8. 0.190 30 V (volt)

Solve the following and express the answer to the appropriate precision. All numbers are approximate.

9. $62.750 + 7.93$
10. $24\,000 + 4875$
11. $3.0505 - 2.2$
12. $0.54 - 0.0728$
13. $248.64 - 72.375$
14. $23.4 + 18.32 + 50$
15. $3.75 - 6.2 - 12$
16. $-67.05 + 5.4 + 54.26$
17. $9.2 \times 10^6 + 2.5 \times 10^7$
18. $22.4 \times 10^3 - 1.72 \times 10^5$
19. $83.00 \times 10^5 + 7.100 \times 10^6$
20. $68.3 \times 10^{-2} - 2.53 \times 10^{-4}$

Multiplication and Division

The following guidelines are used when multiplying or dividing with approximate numbers to ensure that the precision of the answer is not overstated. These guidelines require that you identify the number with the fewest significant figures. This is done by inspecting the numbers to determine which number has the fewest number of significant figures. The answer can contain no more significant figures than are contained in the number with the fewest significant figures. For example, 9.3 is to be divided by 3.106. Of the two numbers, 9.3 has the fewest significant figures with two; 3.106 has four significant figures. In this case, the quotient, 2.994, is rounded to two significant figures and is stated as 3.0.

GUIDELINES 2–4. *Multiplication and Division of Approximate Numbers*

1. Determine (by inspection) which number has the fewest number of significant figures.
2. Carry out the calculation.
3. Round the answer (product or quotient) to the same number of significant figures as contained in the number with the fewest significant figures used in the calculation.

EXAMPLE 2–13	Multiply 38.00 by 7.2.
Solution	Follow the steps of Guidelines 2–4.
Step 1:	Because 7.2 has the fewest number of significant figures, two, the product will be rounded to two significant figures.
Step 2:	Multiply 38.00 by 7.2.
$\boxed{\times}$	273.6
Step 3:	Round the answer to two significant figures.
R to 2 SF:	$273.6 = 2.7 \times 10^2$
$\therefore$	$38.00 \times 7.2 = 273.6 \Rightarrow 270 = 2.7 \times 10^2$
Observation	The zero in 270 is not a significant figure; it is a placeholder.

EXAMPLE 2–14	Solve $53.42 \times (0.291 \times 10^3)/562.0$.
Solution	Follow the steps of Guidelines 2–4.
Step 1:	Because 0.291 has the fewest number of significant figures, three, the quotient will be rounded to three significant figures.
Step 2:	Multiply 53.42 by 0.291×10^3 and divide by 562.0.
$\boxed{\times}\boxed{\div}$	27.66
Step 3:	Round the answer to three significant figures.
R to 3 SF:	27.7
$\therefore$	$53.42 \times (0.291 \times 10^3)/562.0 = 27.66 \Rightarrow 27.7$

Before you start the next exercise, we call your attention to a very useful feature of your calculator. It is the exponent entry key $\boxed{\text{EE}}$, or $\boxed{\text{EXP}}$, or $\boxed{\text{EEX}}$. Learn to use it, because it can greatly simplify your work with powers of ten notation.

EXERCISE 2–11

Calculator Drill
Solve the following and express the answer to the appropriate number of significant figures. All numbers are approximate except where noted.

1. 5×84 **2.** 82 390/22 688
3. 314×2001 **4.** 45×860.0

5. 93.25/26.65

6. 0.08260/1.250

7. 0.078×3.049

8. 13.03×145.0

9. $3.3 \times 17.0 \times 9$

10. $75 \times (30/12)$

11. $0.875 \times 42 \times 10^{-3}$

12. $(15.0 \times 10^2) \times (53.6 \times 10^4)$

13. 300/6.275   (300 is exact)

14. $(0.31 \times 10^{-3})/(0.1536 \times 10^{-2})$

15. $27 \times 14/0.678$

16. $5.94 \times \pi$ (π is exact)

17. $(72.2 \times 10^3)/0.0536$

18. $23.4/(6.42 \times 4.00)$

19. $(86 \times 10^{-5})/(2536 \times 10^{-7})$

20. $(72.2 \times 10^3)/0.536 \times \dfrac{4}{3}$

(3 and 4 are exact)

2–5 SQUARE ROOTS, RADICALS, AND RECIPROCALS

To take the square root of a number, we simply enter the number into our calculator and press the square-root key $\boxed{\sqrt{}}$. What was once a very difficult and time-consuming task is now as simple as adding when we are aided by a calculator. There is, however, more to be known about taking a square root than the mechanics of pressing the square-root key.

Square Root

You are now very familiar with the meaning of 4^2. It, of course, means to use 4 twice as a factor in a product. A **square root** of a product is one of the two equal factors that were used to form the product. So the square root of 16 ($\sqrt{16}$) is 4. This is because the two equal factors of 16 are 4. If $4 \times 4 = 16$, then what does $(-4) \times (-4)$ equal? 16! Then -4 is also a square root of 16.

> **RULE 2–5.   Evaluating Square Roots**
>
> Every positive number has two square roots that have equal absolute values but are opposite in sign.

EXAMPLE 2–15   What are the square roots of 196?

Solution   Use your calculator and Rule 2–5.

$\boxed{\sqrt{}}$ $\sqrt{196} = 14$

So the square roots of 196 are $+14$ and -14.

Check

$\boxed{\times}$ $\qquad$ $(+14) \times (+14) = 196$

$\boxed{\times}$ $\qquad$ $(-14) \times (-14) = 196$

$\therefore$ $\qquad$ The square roots of 196 are ± 14.

Observation $\qquad$ The symbol $\pm$ is read "plus or minus."

If you want only the positive square root of a number, say, 25, called the **principal square root,** then you indicate this as $\sqrt{25}$ without a sign. The negative square root is indicated as $-\sqrt{25}$. If both positive and negative roots are desired, then the $\pm$ symbol is used, as in $\pm\sqrt{25}$. In solving electrical problems, generally we use only the positive or principal square root. In this book, unless otherwise noted, we shall work with the principal root of the number.

EXAMPLE 2–16 $\qquad$ Evaluate $-\sqrt{240}$, expressing the answer in decimal notation to three significant figures.

Solution

$\boxed{\sqrt{}}$ $\qquad$ $-\sqrt{240} = -15.49$

$\therefore$ $\qquad$ $-\sqrt{240} = -15.5$

EXERCISE 2–12

Calculator Drill
Evaluate the following, expressing the answers in decimal notation to four significant figures:

1. $\sqrt{372.0}$ $\qquad$ **2.** $\sqrt{0.008\ 700}$ $\qquad$ **3.** $\pm\sqrt{0.7850}$

4. $\sqrt{99.92}$ $\qquad$ **5.** $\sqrt{1020}$ $\qquad$ **6.** $-\sqrt{65.92}$

7. $\sqrt{0.008\ 825}$ $\qquad$ **8.** $\pm\sqrt{7.285 \times 10^{-1}}$ $\qquad$ **9.** $-\sqrt{192.3 \times 10^4}$

10. $\sqrt{20\ 970 \times 10^{-4}}$ $\qquad$ **11.** $-\sqrt{0.058\ 70 \times 10^3}$ $\qquad$ **12.** $\pm\sqrt{409.5 \times 10^{-3}}$

Radical

The symbol $\sqrt{}$ is a **radical sign.** Figure 2–5(a) shows a radical sign with **index** n. This combination indicates the nth root. When the index is missing, it is an implied index of 2 and indicates the square root. The radical sign combined with the bar, a sign of grouping, is written over the number. This combination of number and symbol is called a **radical.** The number under the radical sign is called the **radicand.** Figure 2–5 summarizes these concepts.

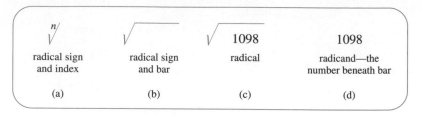

$\overset{n}{\sqrt{}}$	$\sqrt{}$	$\sqrt{}\,1098$	1098
radical sign and index	radical sign and bar	radical	radicand—the number beneath bar
(a)	(b)	(c)	(d)

FIGURE 2–5   Summary of the parts of a radical.

Because the bar is a sign of grouping, a radicand that is an arithmetic expression is computed first before the root is taken.

EXAMPLE 2–17   Solve $\sqrt{16 + 9}$.

Solution   Enter 16 and add 9:

$\boxed{+}$ 25

$\boxed{\sqrt{}}$ 5

$\therefore$ $\sqrt{16 + 9} = 5$

EXERCISE 2–13

Calculator Drill

Evaluate the following, expressing the answers in scientific notation to three significant figures.

1. $\sqrt{82.0 - 18.0}$
2. $\sqrt{-108 + 133}$
3. $\sqrt{8.00 \times 32.0}$
4. $\sqrt{64.0/4.00}$
5. $\sqrt{7.35 + 12.4}$
6. $\sqrt{802 \times 2.39}$
7. $\sqrt{0.950/0.003\ 80}$
8. $\sqrt{1.22 + 56.3}$
9. $\sqrt{72.1 \times 406}$
10. $-\sqrt{0.0580 - 0.248 + 2.03}$
11. $\sqrt{3.56 \times 0.923 \times 10^4}$
12. $\sqrt{15.8 + 8.06 - 7.86}$

Fractional Exponents

Since fractional exponents follow all the rules for powers, it is generally easier to work with fractional exponents rather than with radical signs. The square root of a number may be indicated by a fractional exponent of $\frac{1}{2}$. Thus, $\sqrt{100} = 100^{1/2} = 100^{0.5}$. By raising a number to the $\frac{1}{2}$, or 0.5, power, we mean *compute the principal square root of the number.*

EXAMPLE 2–18	Use your calculator to evaluate $256^{1/2}$.
Solution	Enter 256:

$$\boxed{\sqrt{}} \qquad 16.0$$

$$\therefore \qquad 256^{1/2} = 16.0$$

EXAMPLE 2–19	Evaluate $(732 \times 10^3)^{0.5}$, expressing the answer in scientific notation to three significant figures.

Solution   Enter 732×10^3:

$\boxed{y^x}$ 0.5 $(732 \times 10^3)^{0.5} = 855.57$

R to 3 **SF**: $855.57 \Rightarrow 856$

SN: $856 = 8.56 \times 10^2$

$\therefore$ $(732 \times 10^3)^{0.5} = 8.56 \times 10^2$

EXERCISE 2–14

Calculator Drill

With your calculator formatted to three significant figures and the display set to scientific notation (SCI), evaluate the following, expressing each of your answers in scientific notation to three significant figures.

1. $242^{1/2}$
2. $0.0116^{1/2}$
3. $52.4^{0.5}$
4. $1380^{0.5}$
5. $9.93^{1/2}$
6. $(672 \times 10^5)^{1/2}$
7. $(14.3 \times 10^{-3})^{0.5}$
8. $(7.28 \times 10^4)^{0.5}$
9. $(203 \times 10^5)^{1/2}$
10. $(0.775 \times 10^3)^{1/2}$
11. $(805.3/0.385)^{1/2}$
12. $(43.8 \times 10^2 + 2.01)^{0.5}$
13. $(16.3^2 - \sqrt{720})^{0.667}$
14. $(\pi \times 3.68/0.6910)^{2.718}$
15. $[0.0268/(15.3 \times 10^{-3})]^{3/2}$

Square Root of a Negative Number

Although we will not be working with the square root of a negative number until Chapter 24, we mention it here because of its curious nature. Suppose that we want to determine the square root of -49. Place this in your calculator and press the square-root key. You get a strange response, but not an answer. Your calculator is indicating an improper operation.

From this we conclude that the calculator is no help in our investigation. Exploring, we try $7 \times 7 = 49$, not -49, so 7 is not the square root of -49. We next try $-7 \times -7 = 49$, not -49, so -7 is not the square root. If neither 7 nor -7 is the square root, then what is? Since it is not a number that we know about, it must be some other kind of number that we do not know about. This new kind of number is called an **imaginary number.** This very important kind of number is used to analyze ac circuits, and it is covered in Chapter 24.

Reciprocals

Numbers like $\frac{2}{3}$ and $\frac{3}{2}$ are **reciprocals** of each other because their product equals 1. Thus, $\frac{2}{3} \times \frac{3}{2} = (2 \times 3)/(3 \times 2) = \frac{6}{6} = 1$. For example:

- 2 and $\frac{1}{2}$ are reciprocals because $2 \times \frac{1}{2} = \frac{2}{2} = 1$.
- -1 and -1 are reciprocals because $-1 \times -1 = 1$.
- 0 has no reciprocal because 0 times any number is 0, *not* 1.

To form the reciprocal of a number, simply place one (1) over that number.

EXAMPLE 2–20   Write the reciprocal of 8 and then compute its value. Express the answer in decimal notation rounded to three significant figures.

Solution   Write the reciprocal:

$$\frac{1}{8}$$

$\boxed{\div}$ $\frac{1}{8} = 0.125$

$\therefore$ The reciprocal of 8 is $\frac{1}{8} = 0.125$.

EXAMPLE 2–21   Write the reciprocal of 23.82. Compute its value, expressing the answer in scientific notation to four significant figures.

Solution   Write the reciprocal:

1/23.82

Use the reciprocal key on your calculator. Enter 23.82.

$\boxed{1/x}$ 4.198×10^{-2}

$\therefore$ The reciprocal of 23.82 is $1/23.82 = 4.198 \times 10^{-2}$.

Use the reciprocal key $\boxed{1/x}$ to aid in the solution of the following set of problems.

EXERCISE 2–15

Calculator Drill
Compute the reciprocal of each of the following numbers. Express the answers in engineering notation to three significant figures.

1. 5.00	**2.** 4.00	**3.** 6.00
4. 9.00	**5.** 3.00	**6.** 92.6
7. 0.0835	**8.** 776	**9.** 503
10. 0.001 38	**11.** 6.28×10^3	**12.** -920
13. 4.09×10^{-2}	**14.** 1.86×10^{-3}	**15.** 70.7×10^2
16. -47.3	**17.** 2.60×8.92	**18.** $52.0 - 61.3$

Division by Multiplying

Division can be accomplished by taking the reciprocal of the divisor and then multiplying. For example, $\frac{16}{8} = 16 \times \frac{1}{8} = 2$. This is useful when performing combined operations on your calculator.

EXAMPLE 2–22   Solve $\frac{20}{4}$ by multiplying.

Solution   Use your calculator.

Enter 4:

$\boxed{1/x}$	0.25
$\boxed{\times}$ 20	5.00
$\therefore$	$\frac{20}{4} = 0.25 \times 20 = 5.0$

Division with a fractional divisor is done by reciprocating the divisor and then multiplying. Rule 2–6 summarizes the steps in solving this type of problem.

RULE 2–6. *Dividing by Fractions*

To divide a number by a fraction:
1. Reciprocate the divisor (invert the fraction).
2. Multiply the numerators.
3. Form a new fraction of the product over the denominator.
4. Take the quotient of the resulting fraction.

EXAMPLE 2–23   Find the value of 16/(4/3) by multiplying.

Solution   Use Rule 2–6.

Step 1: Reciprocate the denominator:

$$\frac{3}{4}$$

Step 2: Multiply numerators:

$$16 \times 3 = 48$$

Step 3: Form a new fraction:

$$\frac{48}{4}$$

Step 4: Divide:

$$\frac{48}{4} = 12$$

$$\therefore \qquad 16/(4/3) = (16 \times 3)/4 = 12$$

EXERCISE 2–16

Calculator Drill

Use your calculator to aid in the solution of the following problems. First reciprocate the denominator and then multiply. Express the answers in decimal notation to three significant figures.

1. $\dfrac{48.00}{16.00}$ 2. $\dfrac{84.00}{6.00}$ 3. $\dfrac{96.00}{480.0}$

4. $\dfrac{720.0}{432.0}$ 5. $\dfrac{20.00}{5/4}$ 6. $\dfrac{18.00}{9/2}$

7. $\dfrac{7.09}{1/2}$ 8. $\dfrac{15.4}{3/12}$ 9. $\dfrac{17.36}{0.0820}$

10. $\dfrac{0.0385}{3.23 \times 10}$ 11. $\dfrac{723.6}{0.6923}$ 12. $\dfrac{0.004\ 356}{3/125}$

2–6 COMBINED OPERATIONS

You now have a fundamental understanding of how to use your calculator not only to do calculations, but also to round and display the answer so that the precision of the answer is not overstated.

In this section, the calculator will be used to solve complex arithmetic problems using the technique of *chain calculations*. When calculations are chained, no intermediate answers are recorded. Instead, the full capability of the calculator is used to produce an answer free of intermediate rounding error.

Depending on your calculator's data-entry system, you will either use parentheses in conjunction with a preprogrammed order of operation (Algebraic Entry System) or stack registers (Reverse Polish Notation) to avoid having to store an intermediate result in memory. The chain calculation of a complex problem is carried out one operation after the other in a prescribed order, which results in an error-free, rounded, and formatted answer.

In the solution of most electronic problems, several operations need to be performed to get the answer. For example, the power supplied to an electrical motor may be calculated by multiplying the resistance of the motor by the square of the electrical current. To facilitate entry of the data into the calculator without confusion (multiply, then square; or square, then multiply?), the following convention for combined operations is used.

CONVENTION 2–1.   Combined Operations

It is conventional in problems with several operations to solve the problem from left to right and to:

1. Raise to powers and extract roots first.
2. Multiply and divide next.
3. Add and subtract next.
4. Set priority of operation by interrupting the usual order with parentheses (), the bar (−), or other signs of grouping.

EXAMPLE 2–24   Solve $\dfrac{3 \times \sqrt{18 \times 32} + 18}{9}$.

Solution   The solution begins with the radicand.

$\boxed{\times}$	$18 \times 32 \Rightarrow 576$	$\dfrac{3 \times \sqrt{576} + 18}{9}$
$\boxed{\sqrt{}}$	$\sqrt{576} \Rightarrow 24$	$\dfrac{3 \times 24 + 18}{9}$
$\boxed{\times}$	$3 \times 24 \Rightarrow 72$	$\dfrac{72 + 18}{9}$
$\boxed{+}$	$72 + 18 \Rightarrow 90$	$90/9$
$\boxed{\div}$	$90/9 \Rightarrow 10$	10

$$\therefore \quad \frac{3 \times \sqrt{18 \times 32} + 18}{9} = 10$$

EXAMPLE 2–25   Solve $\dfrac{5.26^2 \times 0.8326 \times 10^2}{37.06}$ using chain calculations. Express the answer in decimal notation to three significant figures.

Solution The solution begins by squaring 5.26.

$\boxed{x^2}$ $5.26^2 = 27.667$ $\dfrac{27.667 \times 0.8326 \times 10^2}{37.06}$

$\boxed{\times}$ $27.667 \times 0.8326 \times 10^2 = 2303.6$ $2303.6/37.06$

$\boxed{\div}$ $2303.6/37.06 = 62.2$ 62.2

$\therefore$ $\dfrac{5.26^2 \times 0.8326 \times 10^2}{37.06} = 62.2$

Observation Because of chain calculations, none of the intermediate results are normally recorded. They are given in this example only to demonstrate the order in which the problem was solved.

EXERCISE 2–17

Calculator Drill

Evaluate each of the following expressions. Express the answers in engineering notation to three significant figures.

1. $\dfrac{7.09}{2.82 \times 6.55}$

2. $\dfrac{12.2 + 9.51}{2.03}$

3. $\dfrac{(75.4)(0.952)}{(-4.05)(12.8)}$

4. $\dfrac{(-8.38)(-5.51)}{-4.22}$

5. $(72.8 + 9.005)45.2$

6. $(10.935/0.081)52.6$

7. $556 \times 21.92 - 30.7^2$

8. $(0.172 \times 0.58)/(2.78 - 5.03 + 3.00)$

9. $(\sqrt{1720} - 3.11 \times 4.32)^3$

10. $\dfrac{11.5 - (18.4)^2 + (22.1)^{0.5}}{0.0821 \times 10^4}$

11. $\dfrac{193^{0.5} + 0.0391^2 \times 16.0}{13.9}$

12. $\dfrac{0.707^2 + 844/12.3}{3.00^{0.5}}$

13. $\left(\dfrac{6.25 \times 21.63^2}{4.27^2}\right)^{0.5}$

14. $\dfrac{\sqrt{18.2} \times 4.17^2 - 18}{8/12}$

15. $\dfrac{(12.8^2 \times 14.3/8.03)^2}{4.81 \times 23.2^2}$

16. $\left(\dfrac{(0.221 \times 10^{-3})\sqrt{0.562}}{3.79^2 - 18.6^{0.5}}\right)^2$

2–7 POWERS OF TEN AND APPROXIMATIONS

This unit on powers of ten and approximations is included so that you will not be a *slave* to your calculator. By approximating the answer to an arithmetic expression before solving with your calculator, you will have a check on your work. Also, the technique of approximations is used daily in electrical and electronic technology by technicians to quickly approximate circuit parameters.

Multiplication with Powers of Ten

This section deals with the use of powers of ten as an aid in making approximations when multiplying. The following rule summarizes the steps used to multiply numbers that are expressed in powers of ten notation.

> **RULE 2–7.   Multiplying with Powers of Ten**
>
> To multiply numbers expressed in powers of ten notation:
> 1. Multiply the decimal coefficients.
> 2. Add the exponents.
> 3. Form the product.

EXAMPLE 2–26   Solve $(3 \times 10^{-7}) \times (2 \times 10^3) \times (1 \times 10^{-2})$.

Solution   Follow the steps of Rule 2–7.

Step 1:   Multiply decimal coefficients: $3 \times 2 \times 1 = 6$

Step 2:   Add exponents:

$$-7 + 3 + (-2) = -6$$

Step 3:   Form the product:

$$6 \times 10^{-6}$$

$\therefore$ $(3 \times 10^{-7}) \times (2 \times 10^3) \times (1 \times 10^{-2}) = 6 \times 10^{-6}$

When working with powers of ten, we sometimes encounter the situation where 10 is raised to the zero power (10^0). By definition, ***any number*** (except zero) ***raised to the zero power is 1.*** So $10^0 = 1$. You should remember this important definition.

EXERCISE 2–18

By inspection, combine the following exponents.

1. $10^2 \times 10^3$
2. $10^8 \times 10^{-9}$
3. $10^{-5} \times 10^{-2}$
4. $10^{-3} \times 10^7$
5. $10^1 \times 10^0 \times 10^{-1}$
6. $10^8 \times 10^{-3} \times 10^{-5}$
7. $10^{-9} \times 10^{-3} \times 10^{-6}$
8. $10^2 \times 10^4 \times 10^5$
9. $10^{-4} \times 10^7 \times 10^3$
10. $10^7 \times 10^0 \times 10^{-9}$

Without the use of a calculator, solve the following. Record the answer to two significant figures.

11. $(5.0 \times 10^3)(6.0 \times 10^{-2})$

12. $(8.0 \times 10^{-3})(4.0 \times 10^5)$

13. $(1.0 \times 10^{-1})(9.0 \times 10^{-1})$

14. $(5.0 \times 10^7)(2.0 \times 10^{-5})$

15. $(8.0 \times 10^4)(4.0 \times 10^2)$

16. $(6.0 \times 10^6)(3.0 \times 10^3)$

17. $(7.0 \times 10^{-6})(6.0 \times 10^6)$

18. $(9.0 \times 10^{-1})(4.0 \times 10^{-9})$

19. $(5.0 \times 10^{-4})(5.0 \times 10^{-2})$

20. $(7.0 \times 10^2)(7.0 \times 10^{-2})$

Division with Powers of Ten

This section deals with the use of powers of ten as an aid in making approximations when dividing. The following rule summarizes the steps used to divide numbers that are expressed in powers of ten notation.

RULE 2–8.   *Dividing with Powers of Ten*

To divide numbers that are expressed in either scientific notation or in powers of ten notation:

1. Divide the decimal coefficients.
2. Determine the exponent of 10 by subtracting the exponent of 10 in the denominator from the exponent of 10 in the numerator.
3. Form the quotient.

EXAMPLE 2–27   Divide 8×10^{-4} by 4×10^{-8}.

Solution   Follow the steps of Rule 2–8.

Step 1:   Divide the decimal coefficients:

$$8/4 = 2$$

Step 2:   Subtract exponents:

$$-4 - (-8) = 4$$

Step 3:   Form the quotient:

$$2 \times 10^4$$

$\therefore$ $\dfrac{8 \times 10^{-4}}{4 \times 10^{-8}} = 2 \times 10^4$

EXAMPLE 2–28   Divide 10^{-6} by 10^3.

Solution   Use the steps of Rule 2–8.

Step 1:	Not used.
Step 2:	Subtract exponents:
	$-6 - (3) = -9$
Step 3:	Form quotient:
∴	$10^{-6}/10^3 = 10^{-9}$

EXERCISE 2–19

Mentally divide the following powers of ten.

1. $10^3/10^2$
2. $10^5/10^7$
3. $10^{-3}/10^{-4}$
4. $10^8/10^{-3}$
5. $10^9/10^4$
6. $(5 \times 10^6)/10^2$
7. $\dfrac{5 \times 10^3}{10^2}$
8. $\dfrac{6 \times 10^{-5}}{2 \times 10^{-9}}$
9. $\dfrac{9 \times 10^{12}}{3 \times 10^3}$
10. $\dfrac{7 \times 10^6}{10^9}$
11. $\dfrac{8 \times 10^0}{4 \times 10^7}$
12. $\dfrac{2 \times 10^2}{10^2}$

Power of a Power

What does the expression $(10^3)^3$ mean? It means to use 10^3 as a factor in a product three times: $10^3 \times 10^3 \times 10^3 = 10^9$. This example shows that a power of ten raised to a power is equal to 10 raised to the product of the two exponents. Thus, $(10^3)^3 = 10^{3 \times 3} = 10^9$.

RULE 2–9. *Powers and Powers of Ten*

To raise a power of ten to a power, multiply the exponents.

EXAMPLE 2–29   Solve $(10^{-5})^2$.

Solution   Multiply the exponents.

$$(10^{-5})^2 = 10^{-5 \times 2} = 10^{-10}$$

∴ $(10^{-5})^2 = 10^{-10}$

EXERCISE 2–20

Mentally solve each of the following.

1. $(10^2)^3$
2. $(10^5)^3$
3. $(10^4)^2$
4. $(10^2)^2$
5. $(10^6)^3$
6. $(10^{-5})^6$
7. $(10^{-7})^2$
8. $(10^{-3})^{-3}$
9. $(10^{-2})^6$
10. $(10^3)^{-8}$
11. $(10^5)^{-2}$
12. $(10^{-9})^3$

Power of a Product

A product raised to a power is called a **power of a product.** This expression is a power of a product: $(3 \times 10^3)^2$. From our understanding of exponents, we know that $(3 \times 10^3)^2 = (3 \times 10^3) \times (3 \times 10^3)$ or $3 \times 10^3 \times 3 \times 10^3$, which may be written $3 \times 3 \times 10^3 \times 10^3$. Multiplying, we get 9×10^6. We arrive at this same answer by applying Rule 2–10.

RULE 2–10.   Powers and Products

To simplify the power of a product:

1. Apply the power to each factor.
2. Form the product.

EXAMPLE 2–30	Simplify $(3 \times 10^3)^2$.
Solution	Use the steps of Rule 2–10.
Step 1:	Apply the power to each factor:

$$3^2 = 9$$
$$(10^3)^2 = 10^6$$

Step 2:	Form the product:

$$9 \times 10^6$$

$$\therefore \quad (3 \times 10^3)^2 = 9 \times 10^6$$

EXAMPLE 2–31	Simplify $(2 \times 10^{-5})^4$.
Solution	Follow the steps of Rule 2–10.
Step 1:	Apply the power to each factor:

$$2^4 = 16$$
$$(10^{-5})^4 = 10^{-20}$$

Step 2:	Form the product:

$$16 \times 10^{-20}$$

$$\therefore \quad (2 \times 10^{-5})^4 = 16 \times 10^{-20}$$

EXERCISE 2–21

Simplify each of the following.

1. $(2 \times 10^3)^2$
2. $(3 \times 10^2)^2$
3. $(5 \times 10^{-4})^2$
4. $(6 \times 10^{-5})^2$
5. $(2 \times 10^4)^3$
6. $(3 \times 10^6)^3$

7. $(8 \times 10^{-6})^2$ **8.** $(4 \times 10^{-7})^2$ **9.** $(9 \times 10^{1})^2$

10. $(2 \times 10^{3})^5$ **11.** $(7 \times 10^{-5})^2$ **12.** $(10 \times 10^{-4})^3$

Power of a Fraction

A fraction raised to a power is called a **power of a fraction.** The expression $\left(\frac{3}{10}\right)^3$ is a power of a fraction. From our understanding of exponents, we know that $\left(\frac{3}{10}\right)^3 = \frac{3}{10} \times \frac{3}{10} \times \frac{3}{10}$, which may be written $(3 \times 3 \times 3)/(10 \times 10 \times 10)$. Simplifying, we get $\frac{27}{1000} = 0.027$. We will arrive at this same answer by applying Rule 2–11.

RULE 2–11.   **Powers and Fractions**

To simplify the power of a fraction:
1. Work within the parentheses first.
2. Raise the resulting quotient to the power.

EXAMPLE 2–32   Simplify $\left(\dfrac{10^3}{10^6}\right)^2$.

       Solution   Use the steps of Rule 2–11.

       Step 1:   Work within parentheses:

$$\left(\frac{10^3}{10^6}\right) = 10^{3-6} = 10^{-3}$$

       Step 2:   Raise to the power:

$$(10^{-3})^2 = 10^{-3 \times 2} = 10^{-6}$$

$$\therefore \quad \left(\frac{10^3}{10^6}\right)^2 = 10^{-6}$$

EXERCISE 2–22

Mentally solve the following.

1. $\left(\dfrac{10^3}{10^5}\right)^2$ **2.** $\left(\dfrac{10^4}{10^2}\right)^3$ **3.** $\left(\dfrac{10^{-2}}{10^3}\right)^3$

4. $\left(\dfrac{10^{-3}}{10^5}\right)^4$ **5.** $\left(\dfrac{10^6}{10^2}\right)^2$ **6.** $\left(\dfrac{10^{-4}}{10^{-3}}\right)^5$

7. $\left(\dfrac{10^{-4}}{10^{-4}}\right)^3$ **8.** $\left(\dfrac{10^1}{10^7}\right)^4$ **9.** $\left(\dfrac{10^{-2}}{10^4}\right)^3$

10. $\left(\dfrac{10^8}{10^5}\right)^2$ **11.** $\left(\dfrac{10^6}{10^5}\right)^3$ **12.** $\left(\dfrac{10^7}{10^{-4}}\right)^2$

13. $\left(\dfrac{6.00 \times 10^3}{3.00 \times 10^{-6}}\right)^2$ **14.** $\left(\dfrac{27.0 \times 10^{-2}}{3.00 \times 10^4}\right)^2$

Approximating Combined Operations

The emphasis in approximating is on speed, not accuracy. The approximation is carried out mentally or with pencil and paper—but not with a calculator!

The following are general guidelines to aid in approximating answers. Note that $\approx$ means **approximately equal to.**

GUIDELINES 2–5.   *Approximating*

1. Express each number in scientific notation to one significant figure. For example, 285 becomes 3×10^2; 0.0835 becomes 8×10^{-2}.

2. When multiplying, express the product to the nearest 10. For example, $6 \times 3 \approx 20$, $6 \times 8 \approx 50$, and $3 \times 4 \approx 10$.

3. Express improper fractions as integers. For example, $\frac{9}{4} \approx 2$, $\frac{7}{5} \approx 1$, and $\frac{8}{3} \approx 3$.

4. When solving proper fractions, add a zero to the numerator and decrease the power of ten by 1. Then, treat as an improper fraction. For example, $\frac{2}{7}$ becomes $20 \times 10^{-1}/7 \approx 3 \times 10^{-1}$, and $\frac{3}{5}$ becomes $30 \times 10^{-1}/5 \approx 6 \times 10^{-1}$.

5. When squaring, express the answer as simply as possible. For example, $4^2 \approx 20$, $3^2 \approx 10$, and $8^2 \approx 60$.

6. When taking the square root of a number, first express the number in powers of ten notation with one or two digits in the decimal coefficient and an exponent of ten divisible by 2. Then divide the exponent by 2 and approximate the square root of the decimal coefficient. For example, $\sqrt{3285} \approx (30 \times 10^2)^{0.5} \approx 30^{0.5} \times 10^{2 \times 0.5} \approx 5 \times 10^1$, and $\sqrt{328} \approx (3^{0.5} \times 10^{2 \times 0.5}) \approx 2 \times 10^1$.

EXAMPLE 2–33   Approximate $\dfrac{0.043 \times 2820}{\sqrt{5}}$.

Solution   Express in scientific notation:

$$\frac{4 \times 10^{-2} \times 3 \times 10^3}{\sqrt{5}}$$

Simplify numerator by multiplying:

$$(10 \times 10^1)/\sqrt{5}$$

Approximate denominator:

$$(10 \times 10^1)/2$$

Divide:

$$5 \times 10^1$$

$$\therefore \quad \frac{0.043 \times 2820}{\sqrt{5}} \approx 5 \times 10^1$$

EXAMPLE 2–34 Approximate $\dfrac{0.0327 \times (542)^{0.5}}{77.4}$.

Solution Express in scientific notation:

$$\frac{3 \times 10^{-2} \times (5 \times 10^2)^{0.5}}{8 \times 10^1}$$

Approximate the square root:

$$\frac{3 \times 10^{-2} \times (2 \times 10^1)}{8 \times 10^1}$$

Multiply numerator:

$$\frac{6 \times 10^{-1}}{8 \times 10^1}$$

Divide; use step 4 of Guidelines 2–5:

$$\frac{60 \times 10^{-1}}{8} \times \frac{10^{-1}}{10^1} \approx 8 \times 10^{-3}$$

$$\therefore \quad \frac{0.0327 \times (542)^{0.5}}{77.4} \approx 8 \times 10^{-3}$$

EXERCISE 2–23

Without the aid of a calculator, approximate each of the following expressions.

1. $\frac{2}{7}$

2. $\frac{2}{9}$

3. $\frac{83}{230}$

4. $0.572/67.4$

5. 528×41

6. 0.37×132

7. 0.752×0.0904

8. $292 \times 0.41 \times 1.43$

9. $\dfrac{0.083 \times 272}{4.32}$

10. $\dfrac{682}{0.047 \times 6.12}$

11. $\dfrac{35^2 \times 8.07}{216}$

12. $\dfrac{12.2 \times \sqrt{92}}{0.0162}$

13. $\dfrac{1776}{582 \times 820}$

14. $\dfrac{8.25 \times 10^4 \times 6.02}{\sqrt{6}}$

15. $\dfrac{0.043 \times 2820}{5}$

16. $\dfrac{10.6 \times 311 \times 5.92}{\sqrt{14.5} \times 32.4}$

17. $\dfrac{4.13^2 + 378 + 15.6^2}{27.6/3.5^2}$

18. $\sqrt{12/3 + 9 \times 4}$

19. $\dfrac{0.083 \times \sqrt{1.97}}{75.2 \times 10^{-3} + 0.0982}$

20. $\dfrac{3.75 \times 18.25 \times 10^{-1}}{\sqrt{16.9}/2.91^2}$

21. $\dfrac{\sqrt{0.82}}{4.9 \times 0.76}$

SELECTED TERMS

accuracy   The degree that a measurement or a calculation conforms to a recognized standard or a specified value.

base   A number raised to a power. In 4^2, the number 4 is the base.

engineering notation   Numbers written in powers of ten notation, having exponents that are multiples of 3 (e.g., 83.7×10^{-3}).

exact number   A number having no error.

exponent   In a power, the number of times the base is used as a factor in a product; e.g., in 2^3, the base is 2 and the exponent is 3; thus, 2 is used 3 times in the product $(2 \times 2 \times 2)$.

precision   In measurement, the degree to which individual measurements agree with each other, i.e., repeatability. In common use, it implies exactness.

principal square root   The positive square root.

radical   An expression of the form $\sqrt[n]{a}$.

scientific notation   Numbers written in powers of ten notation having the decimal point placed after the leftmost nonzero digit (e.g., 5.73×10^2).

significant figure   Any figure (digit) needed to define a value in a calculation or a quantity in a measurement.

square root   One of two equal factors of a product.

3

Quantities and Units of Measurement

3–1 International System of Units

3–2 Selected Physical Quantities

3–3 Forming Decimal Multiples and Submultiples of the SI Units

3–4 Unit Analysis and Conversion Between Systems

3–5 Applying Unit Analysis to Energy Cost

3–6 Units and Exponents

PERFORMANCE OBJECTIVES

- Understand the International System of Units (SI).

- Recognize, name, and apply SI unit symbols.

- Know and use physical symbols to represent electrical quantities in formulas.

- Identify, name, and use SI unit prefixes.

- Employ unit fractions and the techniques of unit analysis.

- Calculate energy cost using unit analysis.

- Carry out calculations containing powers of prefixed units.

The toaster symbolizes the myriad of electrical and electronic household appliances used in our daily lives. (Courtesy of Silver Burdett Ginn)

The application of electronics involves measurements of physical quantities. To give meaning to measurements, you will be introduced to the International System of Units—a modern, complete, and coherent system of units of measure used worldwide.

3–1 INTERNATIONAL SYSTEM OF UNITS

Introduction

When making measurements, it is customary to record both the quantity (how much) and the unit (of what). The term **physical quantity** is used to describe the result of measuring. In science and technology, great importance is placed on the unit of measurement.

Without the unit, the measurement doesn't have much meaning. To say, for example, that a building is 16 high or that you walked 1.5 in the city leaves much to the imagination. Was the building 16 stories, 16 feet, or 16 meters high? Did you walk 1.5 blocks, 1.5 miles, or 1.5 kilometers? Without a clear statement of the unit of measure, no one knows what is meant. From these examples, it is apparent that a system of units is necessary in order to completely specify a physical quantity.

Benefits of SI Units

The **International System of Units (SI)** was named and adopted in 1960 by the eleventh General Conference of Weights and Measures (CGPM, Conférence Générale des Poids et Mesures), which changed and simplified an earlier metric system to form the modern metric system. The International System of Units is indicated worldwide in all languages by the letters **SI** (pronounced ess-eye). For an understanding of the development of the International System of Units, refer to Appendix A.

The reasons that SI has received worldwide acceptance lie in its design. First it is complete—covering all areas of science and technology. Additionally, because of the coherent nature of the SI, numbers are directly substituted into formulas without the addition of conversion factors; thus, a unit length times a unit length results in a unit area—so a meter times a meter results in a square meter.

Besides being coherent, SI makes use of the decimal relation between multiples and submultiples using unit prefixes instead of renaming the unit for each multiple or submultiple. For example, the unit for length is the *meter* (m). Multiple and submultiple units of length incorporate the unit name, meter, in their name. Units such as the centi*meter* (cm), milli*meter* (mm), and kilo*meter* (km) are all obvious units of length in the International System of Units.

Finally, SI has a tremendous advantage over previous systems because it uses a unique unit name for each physical quantity, and it assigns a unique symbol for each name. All these advantages have made SI the system of choice for all the industrial nations of the world.

The SI System of Measurement

The International System of Units has seven base units, several derived units with special names, and many derived units with compound names.

The seven **base units** are the building blocks from which the derived units are constructed. Each base unit is defined by a very precise measurement standard that gives the exact value of the unit. The base units are not related to one another, nor do they depend on each other for their definition. Table 3–1 lists the seven SI base units, along with their unit symbols and their physical symbols (formula symbols). Note that the radian (unit for plane angle) and steradian (unit for solid angle) are now classed as derived units. They are no longer called supplementary SI units. The separate class of supplementary units was eliminated by the 20th CGPM in 1995.

The SI **derived units** are formed from the previously defined SI base units. Many derived units have convenient names and symbols in place of the complicated ones that result from derivation from the base units. The watt (W), a measure of electrical power, is such a unit. Its base units include the meter, kilogram, and the second. Without the special name watt and the symbol W, electrical power would be expressed as $m^2 \cdot kg/s^3$, read "meter squared kilogram per second cubed." Table 3–2 lists many of the SI derived units used in electronics.

Because of the strict nature of SI, great care must be used when writing the symbol (abbreviation) for a particular unit. Uppercase letters must be easily distinguished from the lowercase letters.

In Tables 3–1 and 3–2, you might have noticed that the abbreviations (unit symbols) are made up of both lowercase and uppercase letters. When a unit is named to honor a scientist, an uppercase letter is used. Notice that all the derived units with special names in Table 3–2 are uppercase letters and have been named for a person. The guidelines on page 64 summarize these ideas and will be helpful to you when applying SI units to measurements.

TABLE 3–1 SI Base Units

Physical Quantity	Physical Symbol[a]	Unit Name	Unit Symbol[b]
Length	l	meter	m
Mass	m	kilogram	kg[c]
Time	t	second	s
Electric current	I	ampere	A
Temperature	T	kelvin	K
Luminous intensity	I_L	candela	cd
Amount of substance	M	mole	mol

[a]Standardized letters used in formulas or equations to represent the physical quantity in the calculation.

[b]An abbreviation used to indicate the units in the statement of physical measurement.

[c]Among all the units in SI, the unit for mass (kilogram) is the only one that contains a prefix—a historical remnant of an earlier system.

TABLE 3–2 SI Derived Units with Special Names

Physical Quantity	Physical Symbol	Unit Name	Unit Symbol
Angle (plane)	θ	radian	rad[a]
Capacitance	C	farad	F
Conductance	G	siemens	S
Electric charge	Q	coulomb	C
Electromotive force	E	volt	V
Energy, work	W	joule	J
Force	F	newton	N
Frequency	f	hertz	Hz
Inductance	L	henry	H
Power	P	watt	W
Resistance	R	ohm	Ω

[a]Radian is a dimensionless unit. Its unit symbol may be used or omitted when expressing units of physical quantities.

GUIDELINES 3–1. *Using the International System of Units**

1. When writing the symbol for the unit, do not use a period after the symbol. For example: 25 V not 25 V. is used to express twenty-five volts.

2. The unit symbol is not modified to include plural units. For example: fifteen watts is written 15 W, not 15 Ws. The unit symbol is capitalized when the unit is named for a person. For example: the symbol for the ampere is written as A, the watt as W, and the volt as V; however, the symbol for the meter is written as m and for the kilogram as kg.

3. When the physical quantity is expressed as a numerical value (the usual case), then a unit symbol is used and the unit is not spelled out. For example: 15 W not 15 watts.

4. In writing the numbers that precede the unit in the physical quantity, it is customary to set off the number in groups of three when the number is longer than four digits. For example: 2345200 would be written 2 345 200. Decimal fractions are also set off in groups of three. For example: 0.002253 would be written as 0.002 253. Notice that SI quantities **never** use commas as delineators.

5. Decimal fractions are always written with a zero before the decimal point. For example: 0.25 not .25.

6. The word *per* is used to indicate division. For example: The angular velocity of a sixty-cycle-per-second (60 Hz) alternating current is 377.0 radians per second. This is expressed as 377.0 rad/s.

*These guidelines are based on IEEE/ASTM SI 10-1997, *Standard for Use of the International System of Units (SI): The Modern Metric System.*

Use Guidelines 3–1 for the International System of Units to help you write the following correctly. Do not add unit prefixes at this time, because it is assumed that you have no knowledge of them.

1. .003125 newtons
2. 47 watts of power
3. 12000 seconds
4. 14 thousandths of a meter
5. 234000 joules of energy
6. .525 amperes of current
7. .083265 ohms/meter
8. 20106 ohms
9. 34,500 volts
10. 823 radians per second

3–2 SELECTED PHYSICAL QUANTITIES

Electric Charge

The coulomb (C) expressed in SI units is the *ampere second* (A·s). It is a measure of the quantity of electricity or charge (one coulomb is equal to the charge of 6.242×10^{18} electrons). The name coulomb was selected to honor the French physicist Charles A. de Coulomb (1736–1806). The symbol Q (for quantity) is used to represent charge in electrical formulas. See Figure 3–1.

Electric Current

The ampere (A) is one of the seven base SI units. The name ampere was selected to honor the French physicist André M. Ampère (1775–1836). The symbol I (for intensity) is used to represent current in electrical formulas. See Figure 3–2.

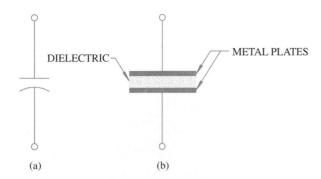

FIGURE 3–1 The capacitor is a device for storing charge. (a) Schematic symbol used for a capacitor; (b) electric charge is stored in the dielectric between the metal plates.

FIGURE 3–2 Schematic symbol used to represent an ammeter. The ammeter is used to measure electric current.

Electromotive Force

The volt (V) expressed in SI units is the *joule per coulomb* (J/C). The volt is a unit for potential difference as well as electromotive force. Voltage is the difference of electric potential between two points in an electric circuit. Both the symbols, E, for a voltage rise (electromotive force), and V, for a voltage drop (potential drop) are used in electrical formulas. The name volt was selected to honor the Italian inventor of the battery, Alessandro Volta (1745–1827). See Figure 3–3.

Resistance

The ohm (Ω) expressed in SI units is *volts per ampere* (V/A). It is a measure of the opposition to the movement of electrical current. One ohm of resistance limits the intensity of electrical current to one ampere when the applied voltage is one volt. The name ohm was selected to honor the German physicist Georg Simon Ohm (1787–1854). The symbol R is used to represent resistance in electrical formulas. See Figure 3–4.

Conductance

The siemens (S) expressed in SI units is *amperes per volt* (A/V). It is a measure of the ease of the movement of electrical current in a conductor. One siemens of conductance permits the electrical current of one ampere to flow when the applied voltage is one volt. The name siemens was selected to honor the German inventor Ernst Werner von Siemens (1816–1892). The symbol G is used to represent conductance in electrical formulas. Observation: although the *mho,* an outdated unit for conductance, still may be found in use in the United States (1 mho = 1 siemens), the use of the mho is now strongly disapproved.

Energy, Work

The joule (J) expressed in SI units is the *newton meter* (N·m). One joule of work is done when an applied force of one newton moves an object one meter in the direction of the

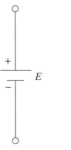

FIGURE 3–3   Schematic symbol for a dc voltage source. The longer line indicates the positive terminal and E represents electromotive force.

FIGURE 3–4   Schematic symbol for any kind of resistance.

FIGURE 3–5 The cape buffalo does work by moving the cart in the direction of the arrow by exerting a horizontal force on the cart. Work = force × distance.

force. The joule is also used as a unit of energy. In an electrical system, work is done in moving an electric charge. The name joule was selected to honor the English physicist James P. Joule (1818–1889). The symbol *W* (for work) is used to represent work and energy in electrical formulas. See Figure 3–5.

Power

The watt (W) expressed in SI units is *joules per second* (J/s). It is a measure of the rate of converting (using or producing) energy or doing work. The watt is both a measure of electrical and mechanical power. When one joule of work is done in one second, the rate of energy conversion is one watt. The name watt was selected to honor the Scottish engineer James Watt (1736–1819). The symbol *P* (for **power**) is used to represent power in electrical formulas. See Figure 3–6.

Force

The newton (N) expressed in SI units is *joules per meter* (J/m). The newton is the amount of force needed to accelerate a mass of one kilogram to one meter per second in one second. The name newton was selected to honor the English scientist, astronomer, and mathematician Sir Isaac Newton (1642–1727). The symbol *F* is used to represent force in formulas. Newton's second law, $F = ma$, is possibly the most important equation in mechanics.

Temperature

Degree Celsius (°C) is commonly used for temperature measurement in electronics. Degree Celsius is the same size as the SI base unit of temperature, the kelvin (K). The

FIGURE 3–6 The electric motor converts electrical power to mechanical power. The electric power (*P*) supplied to the motor is computed by the formula $P = EI$.

FIGURE 3–7 A thermometer illustrating the relationship between the kelvin and the common unit of temperature, the degree Celsius (°C).

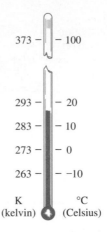

two temperature scales start from different reference points. The same temperature measured in kelvin is 273.15 greater than that measured in degrees Celsius.

$$T \ (°C) = T \ (K) - 273.15$$

This concept is pictured in Figure 3–7. The name Celsius was selected to honor the Swedish astronomer Anders Celsius (1701–1744). See Figure 3–8.

Time

The SI base unit for time is the second (s).

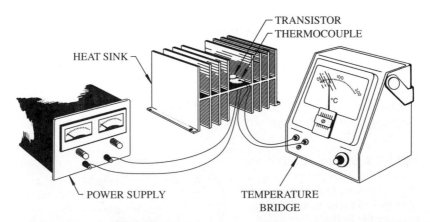

FIGURE 3–8 A temperature bridge and thermocouple are being used to measure the case temperature of a power transistor.

Match one item from the lettered list at the right to one numbered item at the left. Some items in the right list may be used more than once.

1. electrical current	**7.** V/A	**a.** N	**g.** IEEE
2. voltage rise	**8.** A/V	**b.** S	**h.** R
3. J/s	**9.** force	**c.** coulomb	**i.** E
4. $\circ\!\!-\!\!\bigwedge\!\!\!\bigwedge\!\!-\!\!\circ$	**10.** Q	**d.** N·m	**j.** J/m
5. Newton	**11.** joule	**e.** °C	**k.** power
6. SI base unit	**12.** temperature	**f.** I	**l.** ampere

3–3 FORMING DECIMAL MULTIPLES AND SUBMULTIPLES OF THE SI UNITS

When working with the SI units, it often becomes necessary to use larger or smaller units of a particular quantity. Multiples and submultiples of the SI units are formed by using the prefixes in Table 3–3, which lists the SI prefixes used in science and technology. Those prefixes in boldface type are commonly used with electrical units.

TABLE 3–3 Prefixes and Symbols for Multiples and Submultiples of the SI Units

Prefix	Symbol	Phonic	Multiple or Submultiple	Power of Ten
yotta	Y	yott′ə	1 000 000 000 000 000 000 000 000	10^{24}
zetta	Z	zett′ə	1 000 000 000 000 000 000 000	10^{21}
exa	E	ex′ə	1 000 000 000 000 000 000	10^{18}
peta	P	pet′ə	1 000 000 000 000 000	10^{15}
tera	T	ter′ə	1 000 000 000 000	10^{12}
giga	G	gig′ə	1 000 000 000	10^{9}
mega	M	meg′ə	1 000 000	10^{6}
kilo	k	ki′lō	1 000	10^{3}
hecto	h	hec′tō	100	10^{2}
deka	da	dek′ə	10	10^{1}
deci	d	des′ə	0.1	10^{-1}
centi	c	sen′tə	0.01	10^{-2}
milli	m	mil′ē	0.001	10^{-3}
micro	μ	mī′krō	0.000 001	10^{-6}
nano	n	nān′ō	0.000 000 001	10^{-9}
pico	p	pē′kō	0.000 000 000 001	10^{-12}
femto	f	fĕm′tō	0.000 000 000 000 001	10^{-15}
atto	a	ăt′tō	0.000 000 000 000 000 001	10^{-18}
zepto	z	zĕp′tō	0.000 000 000 000 000 000 001	10^{-21}
yocto	y	yŏc′tō	0.000 000 000 000 000 000 000 001	10^{-24}

EXAMPLE 3–4	Add 4.16 kV and 482 V. Express the answer with an appropriate prefix to an appropriate precision.

Solution   Convert 4.16 kV to volts:

$$4.16 \text{ kV} = 4.160 \times 10^3 \text{ V} = 4160 \text{ V}$$

A: $4160 + 482 = 4642$

Express in kV and round to the nearest ten units:

$$4642 \text{ V} = 4.64 \times 10^3 \text{ V} = 4.64 \text{ kV}$$

∴ $4.16 \text{ kV} + 482 \text{ V} = 4642 \text{ V} = 4.64 \times 10^3 \text{ V} = 4.64 \text{ kV}$

EXERCISE 3–3

Express the following measured quantities in the named multiple or submultiple unit.

	Express in:			*Express in:*
1. $282 \ \Omega$	$k\Omega$	**2.** 0.041 A		mA
3. 1.76×10^{-4} H	μH	**4.** 6940 V		kV
5. 19.0×10^5 W	MW	**6.** 0.560 m		cm
7. 1520 Hz	kHz	**8.** 72×10^{-11} F		pF
9. $180\,025 \ \Omega$	$k\Omega$	**10.** 47.2×10^{-10} s		ns

Express the following quantities without unit prefixes in decimal notation.

11. 628 cm	**12.** 2.35 kV	**13.** 17.0 mJ
14. 3.30 kΩ	**15.** 902 μS	**16.** 487 kN
17. 0.854 kW	**18.** 250 mS	**19.** 47.0 μF
20. 702 mg	**21.** 208.7 kJ	**22.** 0.3015 mW

Calculator Drill

Express each of the following as a decimal multiple or submultiple of the stated SI unit. Use the ENG calculator stroke. Round the answer to three significant figures.

23. 7834 V	**24.** 0.000 083 26 s	**25.** 0.065 378 H
26. 17 342 W	**27.** 0.152 S	**28.** 51 621.35 Hz
29. 680 800 Ω	**30.** 0.000 000 916 F	**31.** 0.020 39 A
32. 362.08 N	**33.** 50 286 J	**34.** 1196 V

Using your calculator, combine the following measurements and round the answer to an appropriate precision (as needed, use an appropriate prefix with the SI unit).

35. 10.0 μA + 0.175 mA	**36.** 1.42 S − 572 mS	**37.** 330 kΩ + 1.20 MΩ
38. 250 μs − 0.180 ms	**39.** 1.02 V + 93.7 mV	**40.** 64.0 mg + 0.18 g
41. 982 mm − 37.0 cm	**42.** 230 W + 1.77 kW	**43.** 410 km − 0.250 Mm
44. 44.6 MHz + 770 kHz	**45.** 50.0 μA + 0.25 mA	**46.** 8.45 V + 0.386 V

With your calculator, solve the following and express the answer to an appropriate number of significant figures with a unit and unit prefix (as needed). Round where necessary and express the decimal coefficient of the solution in the range of 0.1 to 1000. All numbers are approximate.

47. $(12.0 \times 10^{-2})^2$ m

48. $(36.00 \times 10^3)^{0.5}$ V

49. $(8.3)^{3/2}$ A

50. 0.572/67.4 W

51. 528×41 A

52. 0.37×132 V

53. 0.752×0.0904 Ω

54. $292 \times 0.41 \times 1.43$ Ω

55. $\dfrac{0.083 \times 272}{4.32}$ H

56. $\dfrac{682}{0.047 \times 6.12}$ J

57. $\dfrac{35^2 \times 8.07}{216}$ W

58. $\dfrac{12.2 \times \sqrt{92}}{0.0163}$ Ω

59. $\dfrac{1776}{582 \times 820}$ V

60. $\dfrac{8.25 \times 10^4 \times 6.02}{\sqrt{6}}$ Hz

61. $\dfrac{0.043 \times 2820}{5.0}$ N

62. $\dfrac{10.6 \times 311 \times 5.92}{\sqrt{14.5} \times 32.4}$ m

63. $\dfrac{4.13^2 + 378 + 15.6^2}{27.6/3.5^2}$ V

64. $\sqrt{12/3 + 9 \times 4}$ Hz

65. $\dfrac{0.083 \times \sqrt{1.97}}{75.2 \times 10^{-3} + 0.0982}$ A

66. $\dfrac{3.75 \times 18.25 \times 10^{-1}}{\sqrt{16.9/2.91^2}}$ J

67. $\dfrac{\sqrt{0.82}}{4.9 \times 0.76}$ S

3–4 UNIT ANALYSIS AND CONVERSION BETWEEN SYSTEMS

When adding or removing unit prefixes or converting between the English and SI systems, you may become confused in knowing when to divide and when to multiply. If you use a technique called **unit analysis,** you are less likely to have this problem. This technique uses the concept that a quantity may be multiplied by 1 without changing its value. Thus, the identity 100 centimeters = 1 meter may be expressed as a **unit fraction**: 100 cm/1 m = 1 or, in its reciprocal form, 1 m/100 cm = 1. In either case, we see the unit fraction is worth 1 and has no dimensions.

Unit Fraction

Unit fractions are formed with the information found in tables of identities and unit conversion factors, such as Table 3–5. Here are some examples of unit fractions that may be constructed using the information found in Table 3–5:

$$\left(\frac{1 \text{ mi}}{5280 \text{ ft}}\right) \text{ or } \left(\frac{5280 \text{ ft}}{1 \text{ mi}}\right), \left(\frac{746 \text{ W}}{1 \text{ hp}}\right) \text{ or } \left(\frac{1 \text{ hp}}{746 \text{ W}}\right), \left(\frac{4.448 \text{ N}}{1 \text{ lbf}}\right) \text{ or } \left(\frac{1 \text{ lbf}}{4.448 \text{ N}}\right)$$

When applying unit fractions to change from one unit to another, multiply or divide the numerical coefficients and strike out the unit names of like units. The following example demonstrates these concepts.

EXAMPLE 3–5   Find the number of kilometers (km) in 21.6 mi (miles).

Solution   From Table 3–5, select the identity 1 km = 0.6214 mi. Form a unit fraction so that miles appear in the denominator of the fraction.

$$21.6 \text{ mi} = 21.6 \text{ mi}\left(\frac{1 \text{ km}}{0.6214 \text{ mi}}\right) = 34.8 \text{ km}$$

$\therefore$ 21.6 mi = 34.8 km

Observation   By placing miles in the denominator of the fraction, miles are canceled, leaving kilometers in the numerator. The numeric coefficients are divided to get the answer.

Applying Unit Analysis

The process of applying unit analysis techniques is summarized in the following guidelines.

GUIDELINES 3–3.   Unit Analysis Technique

Unit analysis is generally applied to a physical quantity by:

1. Expressing an identity (identical equation) that has the desired units.
2. Transforming the identity into a unit fraction.
3. Forming the product of the physical quantity and the unit fraction.
4. Factoring the unit(s), leaving the desired unit(s).
5. Writing the resulting product with the desired units, unit prefix, and appropriate precision.
6. Repeating the preceding steps if more than one conversion is needed.

EXAMPLE 3–6   An automobile is traveling 55.0 mi/h (miles per hour). Determine the speed in meters per second (m/s).

Solution   Because both miles and hours are being changed, select two identities from Table 3–5. Select the identities of 1000 m

TABLE 3–5 Selected Identities and Conversion Factors*

Displacement (length):

1 m = 100 cm = 1000 mm = 3.2808 ft = 39.370 in
1 km = 0.621 37 mi = 3280.8 ft
1 in = 2.54 cm = 25.4 mm
1 mi = 5280 ft
1 yd = 3 ft = 36 in
1 ft = 12 in
1 revolution = 360° = 2π rad = 6.2832 rad

Area:

$1 \text{ m}^2 = 10.7639 \text{ ft}^2 = 1550.0 \text{ in}^2$ **$1 \text{ yd}^2 = 9 \text{ ft}^2$**
$1 \text{ cm}^2 = 0.155\ 00 \text{ in}^2$ **$1 \text{ ft}^2 = 144 \text{ in}^2$**
$1 \text{ m}^2 = 10\ 000 \text{ cm}^2$
$1 \text{ cm}^2 = 100 \text{ mm}^2$

Time:

1 h = 60 min = 3600 s
1 min = 60 s

Force:

1 N = 0.224 81 lbf   **1 lbf = 16 ozf**
1 lbf = 4.4482 N   **1 ton = 2000 lbf**

Mass:

1 kg = 0.068 522 slug   **1 kg = 1000 g**
1 slug = 14.594 kg   **1 g = 1000 mg**
1 kg = 2.2046 lbm

Velocity (speed):

1 m/s = 3.6 km/h = 2.2369 mi/h = 3.2808 ft/s
60 mi/h = 88 ft/s
1 ft/s = 0.3048 m/s
1 rev/min = 0.104 72 rad/s   **60 rev/min = 1 cps = 1 Hz**

Work (energy, torque):

1 J = 0.737 56 ft·lbf   1 ft·lbf = 1.3558 J
1 kWh = 3.6 MJ = 2.655×10^6 ft·lbf   1 N·m = 0.737 56 lbf·ft
1 BTU = 1055.1 J = 778.17 ft·lbf

Power (mechanical):   **Power (electrical):**

1 hp = 550 ft·lbf/s = 33,000 ft·lbf/min = 745.70 W   **1 hp = 746 W**
1 W = 0.737 56 ft·lbf/s   1 kW = 1.3405 hp
1 kW = 1.3410 hp

*__Boldface__ physical quantities are exact.

(1 km) = 0.6214 mi and 1 h = 3600 s. Form the unit fractions. In the first unit fraction, miles appear in the denominator; while in the second, seconds appear in the denominator.

$$55.0 \text{ mi/h} = \left(\frac{55.0 \text{ mi}}{\text{h}}\right)\left(\frac{1000 \text{ m}}{0.6214 \text{ mi}}\right)\left(\frac{1 \text{ h}}{3600 \text{ s}}\right) = 24.6 \text{ m/s}$$

$$\therefore \qquad 55.0 \text{ mi/h} = 24.6 \text{ m/s}$$

Electrical energy is measured in units of joules; however, for some purposes this is a very small unit, so commercial power companies sell energy in a larger unit, the kilowatthour (kWh). Figure 3–9 pictures a kilowatthour meter.

FIGURE 3–9 Kilowatthour meter used to measure electrical energy.

The following example demonstrates the derivation of the conversion factor that relates energy in kilowatthours (kWh) to energy in megajoules (MJ). As you know from the discussion in Section 3–2, the watt is related to the joule by the expression 1 W = 1 J/1 s, or a watt is equal to a joule per second. Solving this expression for energy (J) in terms of power and time yields 1 J = 1 W·s. This identity, a joule is equal to a wattsecond, will be used to develop the conversion factor between kilowatthours and megajoules.

EXAMPLE 3–7   Using the techniques of unit analysis as outlined in Guidelines 3–3, derive the conversion factor that relates energy in kilowatthours (kWh) to energy in megajoules (MJ).

Solution   Since both watts and hours are being changed, two identities are needed. From Table 3–5, select 1 h = 3600 s, and from the preceding discussion, let 1 J = 1 W·s. Form the unit fractions. In the first unit fraction, hour appears in the denominator; while in the second, wattsecond appears in the denominator. Let 1 kWh = 1000 W·h.

$$1 \text{ kWh} = 1000 \text{ W·h} \left(\frac{3600 \text{ s}}{1 \text{ h}}\right)\left(\frac{1 \text{ J}}{1 \text{ W·s}}\right) = 3.6 \times 10^6 \text{ J}$$

$$\therefore \qquad 1 \text{ kWh} = 3.6 \times 10^6 = 3.6 \text{ MJ} \quad (\text{exactly})$$

Observation   Today, the kilowatthour (kWh) is a common unit of electrical energy. However, it is an obsolete unit that will be replaced by the megajoule (MJ) some time in the future.

EXERCISE 3–4

Solve by employing the following aids: Guidelines 3–3, the definitions in Section 3–2, and the identities and conversion factors in Table 3–5. Use rounding to express the answer to the correct precision and, when necessary, use scientific notation with English units and prefixes with SI units.

1. Express 35.0 mi/h (miles per hour) as ft/s (feet per second).
2. Express 4.506×10^3 lbm (pounds mass) as slug (mass).
3. Express 6.20 km/min (kilometers per minute) as m/s (meters per second).
4. Express 790.08 mA·min (milliampere minutes) as C (coulombs).
5. Express 8.57×10^{-3} N·m as J (joules).
6. Express 77.25×10^6 J/m as N (newtons).
7. Express 31.4 rad (radians) as rev (revolutions).
8. Express 8.15×10^4 BTU/h (British thermal units per hour) as ft·lbf/s (foot pounds-force per second).
9. Express 1.08×10^5 in·lbf/min (inch-pounds-force per minute) as hp (horsepower).
10. Express 2880.0° (degrees of angular displacement) as rad (radians).
11. Convert 6.23×10^{-2} ft (feet) to m (meters).
12. Convert 1270 ft·lbf (foot pounds-force) to J (joules).
13. Convert 163.5 m/s (meters per second) to mi/h (miles per hour).
14. Convert 27.30 kWh (kilowatthours) to ft·lbf (foot pounds-force).
15. Convert 0.810 km (kilometers) to in (inches).
16. Convert 0.823 BTU (British thermal units) to J (joules).
17. Convert 3280 ft·lbf/s (foot pounds-force per second) to W (watts).
18. Convert 52.0 lbm/ft (pounds-mass per foot) to kg/m (kilograms per meter).
19. Convert 4.700×10^5 BTU/h (British thermal units per hour) to W (watts).
20. Convert 11 360 rev/min (revolutions per minute) to Hz (hertz).

Express the answers to the following in decimal notation to an appropriate precision.

21. Using 1 in = 2.54 cm (exactly), derive the conversion factor that relates the length of 1 m to inches.

22. Using 1 gal = 4 qt and 1 L = 0.2642 gal, derive the conversion factor that relates the volume of 1 qt to liters (L).

23. Using 1 lbf = 16 ozf and 1 N = 0.224 81 lbf, derive the conversion factor that relates the force of 1 N to ozf (ounce-force).

24. Using 1 in = 2.54 cm (exactly) and 1 N = 0.224 81 lbf, derive the conversion factor that relates the torque of 1 N·m to ozf·in (ounce-force inches).

25. Using 1 kg = 2.2046 lbm, derive the conversion factor that relates the flow rate of 1 lbm/min to kilograms per second.

3–5 APPLYING UNIT ANALYSIS TO ENERGY COST

Practical Unit of Energy

As you now know, electric energy is measured in joules (J). Because of its small size, the joule is not a practical unit for the sale of electric energy. The traditional, but non-SI, unit for the sale of electric energy is the **kilowatthour** (kWh).

In the paragraph leading up to Example 3–7, it was demonstrated that energy in joules is equal to the product of power in watts and time in seconds. That is to say, energy (W) is equal to (=) the product of power (P) and time (t). Stated in a formula, W = Pt. This formula, stated as Equation 3–1, is used to calculate energy use in kilowatthours. To use the formula, power (P) must be expressed in kilowatts (kW) and time must be expressed in hours (h). Example 3–8 demonstrates this concept.

$$W = Pt \quad \textbf{(kWh)} \tag{3–1}$$

where W = energy in kilowatthours (kWh)
  P = power in kilowatts (kW)
  t = time in hours (h)

EXAMPLE 3–8 Determine the amount of energy (in kilowatthours) used to operate a 150-W color television for a month (30 d) if it is in use six hours per day (6 h/d).

Solution Use Equation 3–1 to aid in the calculation:
First, express power in kilowatts:

$$P = 150 \text{ W} = 0.150 \times 10^3 \text{ W} \Rightarrow 0.150 \text{ kW}$$

Next, express time in hours:

$$t = 30 \text{ d} \left(\frac{6 \text{ h}}{\text{d}} \right) = 180 \text{ h}$$

Then, substitute power and time into Equation 3–1 and express the answer in kWh:

$$W = Pt$$
$$W = 0.150 \text{ kW} \times 180 \text{ h}$$
$$W = 27.0 \text{ kWh}$$

Measuring Electrical Energy Use

Electrical energy use is measured by an instrument called a kilowatthour meter. The instrument pictured in Figure 3–10 is used to record the energy use on a series of dials.

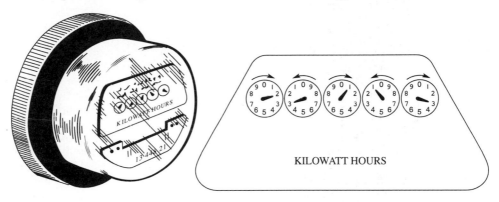

KILOWATT HOURS

FIGURE 3–10 (a) A dial-type kilowatthour meter used in residential or commercial buildings to measure energy supplied. (b) The dials record the use. The dials are read from left to right and rotate in the direction indicated. The reading is 23 113, or 23 113 kWh.

Determining the Cost of Electrical Energy

The cost of electrical energy is given as a rate in cents per kilowatthour (¢/kWh) and varies from utility company to utility company. Cost may be computed with the aid of Equation 3–2.

$$\textbf{cost = rate} \times \textbf{number of kilowatthours} \qquad (3\text{–}2)$$

EXAMPLE 3–9   If the rate is 7.80¢/kWh, then compute the cost of operating the television in Example 3–8.

Solution   The Example 3–8 solution indicates that 27.0 kWh were used in one month of operation.
Use Equation 3–2 and substitute:

$$\text{cost} = \text{rate} \times \text{number of kilowatthours}$$
$$\text{cost} = \frac{7.80¢}{\text{kWh}} \times 27.0 \text{ kWh}$$
$$\text{cost} = 211¢$$
$$\therefore \quad \text{Cost} = \$2.11.$$

TABLE 3–6 Wattage Ratings of Major Household Appliances

Appliance	Wattage Rating (W)	Typical Energy Use per Month (kWh)
Clothes washer	500	10
Clothes dryer	5000	80
Computer monitor	250	15
Dishwasher	1200	30
Heat pump	4000	500
Microwave oven	1500	25
Refrigerator/freezer		
Frost free	450	180
TV (color)	150	30
Toaster oven	1200	18
Water heater	4500	450

Table 3–6 lists the wattage ratings of several major household appliances.

EXAMPLE 3–10 Determine the cost of operating the microwave oven listed in Table 3–6 for 30 d if the oven is used for an average of 30 min/d. Assume an electrical rate of 10.4¢/kWh.

Solution First, determine the number of kilowatthours.

$$W = Pt$$

$$W = 1.50 \text{ kW} \times 30 \text{ d} \left(\frac{30 \text{ min}}{\text{d}}\right)\left(\frac{1 \text{ h}}{60 \text{ min}}\right)$$

$$W = 1.50 \text{ kW} \times 30 \times 30 \times \frac{1 \text{ h}}{60}$$

$$W = 22.5 \text{ kWh}$$

Then, determine the cost using Equation 3–2 when the rate = 10.4¢ /kWh.

$$\text{cost} = \text{rate} \times \text{kilowatthours}$$

$$\text{cost} = \left(\frac{10.4¢}{\text{kWh}}\right) 22.5 \text{ kWh}$$

∴ Cost = 234¢ = $2.34.

The kilowatthour-to-megajoule conversion factor is used in the next example to calculate the cost of energy use. From Example 3–7, you now know that every kilowatthour of electrical energy delivered to your home equates to the use of three million, six hundred thousand joules of energy (1 kWh = 3.6 MJ).

EXAMPLE 3–11   Determine the cost for the use of 5.40 GJ of electric energy when the rate per kilowatthour is $0.0862/kWh.

Solution   Express 5.40 GJ as 5400 MJ and then use the identity of 1 kWh = 3.6 MJ to solve for the number of kilowatthours.

$$\text{cost} = \text{rate} \times \text{kilowatthours}$$

$$\text{cost} = \left(\frac{\$0.0862}{1\ \text{kWh}}\right) \times (5400\ \text{MJ})\left(\frac{1\ \text{kWh}}{3.6\ \text{MJ}}\right) = 129.30$$

∴   Cost for the use of 5.40 GJ of energy is $129.30.

As a final thought, Table 3–5 has only a small sampling of the available identities and conversion factors. Many identities and intersystem conversion factors are listed in engineering, physics, and chemical handbooks.

EXERCISE 3–5

1. Determine the amount of energy, in kilowatthours, used to operate a 3800-W heat pump for a week (7 d) if it is in use 9.50 h/d.

2. Determine the amount of energy, in kilowatthours, used to operate the computer monitor listed in Table 3–6 for 2 weeks if the monitor is on for an average of 5.25 h/d.

3. Determine the amount of energy, in kilowatthours, supplied to a stepper motor during its 72.8 h of operation. The rate of supplying energy to the motor is 82.0 J/s (82.0 W).

4. Determine the amount of energy, in megajoules, supplied to a three-phase induction motor if the motor has been operating for 15.75 h. The rate of supplying energy to the motor is 7460 J/s (7.460 kW).

5. Determine the amount of energy, in megajoules, supplied to a clothes dryer during its 45.5 min of operation. The rate of supplying energy to the dryer is 4650 J/s (4.650 kW).

6. Determine the cost, in dollars, to operate the toaster oven listed in Table 3–6 for 22 d if the oven is used for an average of 15.0 min/d. The electric rate is $0.0685/kWh.

7. Determine the cost, in dollars, to operate the water heater listed in Table 3–6 for 30 d if the heater is operating for an average of 3.75 h/d. The electric rate is 9.60¢/kWh.

8. Determine the cost for 2.875 TJ of electric energy supplied to an electric utility by an energy production company at a rate of $0.004 250/MJ.

9. With an energy rate of 10.0¢/kWh, determine:
   (a) The cost of operating both a 75-W and a 40-W lamp over the 1500-h life of the lamp.
   (b) The savings in dollars by using the 40-W lamp instead of the 75-W lamp.

10. Determine the difference in operating cost between a 450-W frost-free refrigerator/freezer and a 360-W manual-defrost refrigerator/freezer over a 96-month

period (30 d/mo) if each runs 10.0 h/d. Assume that both are equal in size and that the average energy rate is 7.80¢/kWh.

3–6 UNITS AND EXPONENTS

Physical Quantity Raised to a Power

When substituting data into a formula, the units associated with the substituted data may need to be converted from one system to another. This is done to ensure that the desired outcome of the calculation is realized. The following example illustrates a case where the radius is in inches but the answer is needed in square millimeters (mm^2).

In the course of solving this problem, the radius in inches is converted to millimeters, 28.626 mm, which is then squared: $(28.626 \text{ mm})^2$. When a physical quantity is acted upon by a power, as in this case, the power is applied to both the decimal coefficient and the unit. Thus, $(28.626 \text{ mm})^2 \Rightarrow 28.626^2 \text{ mm}^2 = 819.45 \text{ mm}^2$. So, a physical quantity raised to a power is treated as though it were a product raised to a power—i.e., a *power of a product*. This concept was discussed in Section 2–7.

EXAMPLE 3–12	Determine the area, in square millimeters (mm^2), of the end of a round steel bar having a measured radius of 1.127 in. The formula for the area of a circle is $A = \pi r^2$, where A is the area, r is the radius, and π (an exact number) is entered as a calculator keystroke.
Solution	Express the radius in millimeters (mm) using 1 in = 25.4 mm:

$$A = \pi r^2$$

$$r = 1.127 \text{ in} \left(\frac{25.4 \text{ mm}}{1 \text{ in}} \right) = 28.626 \text{ mm}$$

$$A = \pi (28.626 \text{ mm})^2 = 2574 \text{ mm}^2$$

∴	The area of the end of the bar is 2574 mm^2.
Observation	To minimize rounding error in the calculation, the result of the conversion was expressed to more precision than the original data would indicate. Once the calculation was completed, the final solution was rounded to the appropriate precision—in this case four significant figures. Recall that the identity 1 in = 25.4 mm is an exact number by definition.

Powers of Units

A prefixed unit with an exponent attached to the unit (as in squared or cubed units, mm^2 or cm^3) is an indication that the prefix is also raised to the power expressed by the exponent. The prefix can be expressed as a power of ten when the base unit, without the prefix, is needed in a formula. Carefully study each of the following:

$$3 \text{ cm}^2 \Rightarrow 3 \times (10^{-2} \text{ m})^2 = 3 \times 10^{-2 \times 2} \text{ m}^2 = 3 \times 10^{-4} \text{ m}^2$$
$$25 \text{ mm}^3 \Rightarrow 25 \times (10^{-3} \text{ m})^3 = 25 \times 10^{-3 \times 3} \text{ m}^3 = 25 \times 10^{-9} \text{ m}^3$$
$$584 \text{ km}^3 \Rightarrow 584 \times (10^3 \text{ m})^3 = 584 \times 10^{3 \times 3} \text{ m}^3 = 584 \times 10^9 \text{ m}^3$$
$$7385 \text{ cm}^2 \Rightarrow 7385 \times (10^{-2} \text{ m})^2 = 7385 \times 10^{-4} \text{ m}^2 = 0.7385 \text{ m}^2$$
$$22.68 \times 10^6 \text{ cm}^3 \Rightarrow 22.68 \times 10^6 \times 10^{-6} \text{ m}^3 = 22.68 \text{ m}^3$$
$$0.585 \times 10^{-5} \text{ km}^2 \Rightarrow 5.85 \times 10^{-6} \times 10^6 \text{ m}^2 = 5.85 \text{ m}^2$$

RULE 3–1. Powers of Units

An exponent attached to a unit symbol containing a prefixed unit is an indication that the multiple or submultiple (prefix) of the unit is raised to the power expressed by the exponent.

EXAMPLE 3–13   The rectangular cross-sectional area of a relay core is determined to be 6.45 cm². Convert this area to an area in square meters (m²) so it may be used in the formula $F = B^2 A/(2\mu_0)$ to calculate the relay's closure force in newtons.

Conversion
$$A = 6.45 \text{ cm}^2 \Rightarrow 6.45 \times (10^{-2} \text{ m})^2 = 6.45 \times 10^{-2 \times 2} \text{ m}^2$$
$$A = 6.45 \times 10^{-4} \text{ m}^2$$

Solution
$$F = B^2 A/(2\mu_0)$$
$$B = 0.85 \text{ T} \quad \text{(tesla)}$$
$$A = 6.45 \times 10^{-4} \text{ m}^2$$
$$\mu_0 = 4\pi \times 10^{-7} \text{ Wb/A·m (webers per ampere meter)}$$
$$F = 0.85^2 (6.45 \times 10^{-4})/(2 \times 4\pi \times 10^{-7})$$
$$F = 185 \text{ N}$$

Observation   Here we have seen the need to change from square millimeters to the base unit of square meters. This operation is usually carried out mentally, because it involves integer values of both powers of ten and exponents. Similarly, 86 mm³ is expressed as 86×10^{-9} m³ in the base unit of cubic meters. Table 3–7 lists several conversion factors for powers of units.

Because most formulas have been defined in terms of the base and the derived SI units, the unit prefix must be replaced with its equivalent powers of ten value before the quantity is substituted. To avoid errors in calculations, get in the habit of replacing prefixes with powers of ten notation when substituting into an equation.

TABLE 3–7 Conversion Factors for Powers of Units

$1 \text{ m}^2 = 1 \times 10^6 \text{ mm}^2$	$1 \text{ m}^3 = 1 \times 10^9 \text{ mm}^3$
$1 \text{ m}^2 = 1 \times 10^4 \text{ cm}^2$	$1 \text{ m}^3 = 1 \times 10^6 \text{ cm}^3$
$1 \text{ m}^2 = 1 \times 10^{-6} \text{ km}^2$	$1 \text{ m}^3 = 1 \times 10^{-9} \text{ km}^3$

EXERCISE 3–6

By inspection, express each of the following without a unit prefix in powers of ten notation with the base unit squared or cubed.

1. 3.0 km^2 **2.** $6.5 \text{ mm}^2/\text{s}$ **3.** 718 cm^3

4. $26.8 \text{ cm}^2/\text{s}$ **5.** 398.0 km^2 **6.** $2.72 \text{ cm}^3/\text{s}$

7. $63.4 \text{ mm}^3/\text{s}$ **8.** 0.254 mm^3 **9.** $692 \text{ } \mu\text{m}^3$

By inspection, simplify the following. Express each as a squared or cubed base unit without a unit prefix or power of ten.

10. $6\,906\,000 \text{ cm}^3$ **11.** $0.820 \times 10^6 \text{ mm}^2$ **12.** $52 \times 10^9 \text{ mm}^3$

13. $8.25 \times 10^{-7} \text{ km}^2$ **14.** $100 \text{ cm} \times 100 \text{ cm}$ **15.** 150 cm^2

With your calculator, square or cube the decimal coefficient in each of the following physical quantities. Express the answer to an appropriate precision with squared or cubed prefixed units.

16. $(5.28 \text{ } \mu\text{m})^3$ **17.** $(27.08 \text{ km})^2$ **18.** $(0.680 \text{ cm})^3$

19. $(33.60 \text{ km})^2$ **20.** $(10.5 \text{ cm})^2$ **21.** $(9.32 \text{ mm})^3$

Using your calculator, solve the following and express the result with only a squared or cubed base unit (without a unit prefix) to an appropriate precision. Where needed, use powers of ten notation.

22. $(10.28 \text{ mm})^2$ **23.** $(37.18 \text{ mm})^2$ **24.** $(0.570 \text{ cm})^3$

25. $(47.80 \text{ km})^2$ **26.** $(36.50 \text{ cm})^2$ **27.** $(932 \text{ } \mu\text{m})^3$

Solve the following. Round to express the answer to an appropriate precision, with the results expressed in the units specified in the problem.

28. The area of a circle, A, is equal to π, 3.1416, times the radius, r, squared. That is, $A = \pi r^2$. Find the area in m^2 of a circular cover plate used on a cable box. The radius of the cover is 22.86 cm. Use the $\boxed{\pi}$ key to enter π.

29. The area of a trapezoid, A, is equal to one-half the altitude, h, times the sum of the bases, b_1 and b_2, where b_1 and b_2 are the parallel sides. That is, $A = \frac{1}{2}h(b_1 + b_2)$. How many square inches of sheet aluminum are needed to make one trapezoidal plate having an altitude of 20.5 cm and bases of 28.2 cm and 42.6 cm?

30. The volume of a cylinder, V, is equal to π times the radius, r, squared, times the altitude, h. That is, $V = \pi r^2 h$. What is the volume in cm^3 of a cylindrical servo motor shipping canister? The canister has a radius of 0.545 ft and a height of 9.25 in.

31. The volume of a sphere, V, is equal to four-thirds π times the radius, r, cubed. That is, $V = \frac{4}{3}\pi r^3$. Determine the volume of a spherical antenna cover in ft^3 when the radius of the cover is 10.28 m.

Using only the conversion factors found in Tables 3–5 and 3–7, solve the following. More complete tables may be found that would simplify the problems, but this is not intended to be an exercise in looking up conversion factors in a handbook.

32. Express 0.0235 m^2 in square centimeters (cm^2).

33. Express 7.45×10^4 mm^2 in square centimeters (cm^2).

34. Derive the conversion factor that relates the volume of 1 m^3 to cubic yards (yd^3).

35. Derive the conversion factor that relates the flow of 1 m^3/s to cubic feet per minute (ft^3/min).

36. The unit for pressure in SI, the pascal, is defined as one newton per square meter. That is, 1 Pa = 1 N/m^2. Convert 15.0 kPa into pressure expressed in pound-force per square inch (lbf/in^2).

37. As noted in the previous problem, the unit for pressure in SI is the pascal (1 Pa = 1 N/m^2). Using only the conversion factors found in Table 3–5, convert a pressure reading of 62.0 lbf/in^2 to kPa.

38. The loading on a floor is given in units of pounds-mass per square foot (lbm/ft^2). Convert 150 lbm/ft^2 into a load expressed in kg/m^2.

39. A crate with a mass of 356.0 kg is placed on a warehouse floor. If the area of the bottom of the box in contact with the floor is 0.825 m^2, what is the load on the floor in lbm/in^2?

40. A block of plastic measuring 2.83 ft long by 3.50 in wide by 0.875 in high has a mass of 3.80 lbm. Determine its density in grams per cubic centimeter (g/cm^3).

SELECTED TERMS

base unit   The building blocks of the measurement system. A precise standard that gives an exact value of the unit.

derived unit   Formed from base units. Usually given convenient unit name and symbol, for example, volt (V), watt (W), ampere (A), etc.

kilowatthour   Unit of energy used by power companies in the sale of energy to customers. One kilowatthour equals 3.6 megajoules.

power   Rate of converting (producing or using) energy; rate of doing work; J/s, W.

SI   International System of Units. The metric system.

unit analysis   Technique used in reduction within a system of measurement, or in conversion between systems of measurement, in which conversion factors are expressed as fractions having values equal to 1.

SECTION CHALLENGE

WEB CHALLENGE FOR CHAPTERS 1, 2, AND 3

To evaluate your comprehension of Chapters 1, 2, and 3, log on to **www.prenhall.com/ harter** and take the online True/False and Multiple Choice assessments for each of the chapters.

SECTION CHALLENGE FOR CHAPTERS 1, 2, AND 3*

Your challenge is to compute the value of the parameters in the following problems using chain calculation and rounding to express the final answer to an appropriate precision. Where appropriate, express the answer with a prefixed unit when the unit is SI or with scientific notation when the unit is BES (British Engineering System).

A. The force of attraction *(F)* resulting from an air gap in a magnetic circuit is the basis of operation for the solenoid pictured in Figure C–1. With the coil energized, the plunger is drawn into the coil and rests against the stationary pole. Using the following equation, determine the force, in newtons (N), between the faces of the poles (plunger and stationary pole) of the open frame solenoid of Figure C–1

$$F = \frac{B^2 A}{2\mu_0}$$

when the flux density *(B)* in the steel magnetic pathway is 0.82 T (tesla), the cross-sectional area *(A)* of the 10.0-mm diameter steel plunger is 78.5 mm^2, and the permeability of air (μ_0) is $4\pi \times 10^{-7}$ Wb/A·m (webers per amp-meter). Solve

$$F = \frac{0.82^2 \times 78.5 \times 10^{-6}}{2 \times 4\pi \times 10^7}$$

and express the force in newtons (N).

*The solution to this Section Challenge is found in Appendix C.

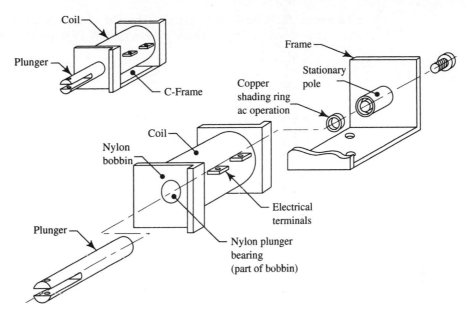

FIGURE C–1 The parts of a C-frame type of solenoid (*Electromechanics: Principles, Concepts, and Devices,* Pearson Education Inc., 2003).

B. When a 3-ϕ electric motor (under load) is initially energized, the rotor shaft is stationary. This condition is referred to as a *locked rotor.* Using the following equation, determine the locked rotor current (I_{LR})

$$I_{LR} = \frac{\text{kVA/hp} \times P \times 1000}{\sqrt{3}E}$$

when the kilovolt-amps per horsepower (kVA/hp) is 6.4, the motor power *(P)* is 15 hp, and the line voltage *(E)* is 460 V. Solve

$$I_{LR} = \frac{6.4 \times 15 \times 1000}{\sqrt{3} \times 460}$$

and express the locked rotor current (I_{LR}) in amperes (A).

C. The torque (τ), in units of pound-feet (lbf·ft), that a machine's drive train will carry is approximated by the following equation. Determine the torque (τ)

$$\tau = \frac{5250P \times \text{SF}}{\omega}$$

when the power (P) in the drive train is 3.60 hp, the service factor (SF) of the load is 1.50, and the speed (ω) of the rotating drive shaft is 1730 rev/min. Solve

$$\tau = \frac{5250 \times 3.60 \times 1.50}{1730}$$

and express the torque in both pound-feet (lbf·ft) and newton-meters (N·m).

D. The dc current (I_B) in the base lead of a bipolar transistor is calculated using the following equation. Determine the base current (I_B)

$$I_B = \frac{V_{CC}}{R_1 + (\beta + 1)R_E}$$

when the collector bias voltage (V_{CC}) is 15.0 V, the bias resistor (R_1) is 390 kΩ, the current transfer ratio (β) is 80.0, and the emitter resistor (R_E) is 470 Ω. Solve

$$I_B = \frac{15.0}{390 \times 10^3 + (80.0 + 1)\,470}$$

and express the base current (I_B) in amperes (A).

4

Algebra Fundamentals I

4–1   Variables, Subscripts, and Primes

4–2   Indicating Multiplication

4–3   General Numbers

4–4   Algebraic Expressions

4–5   Products, Factors, and Coefficients

4–6   Combining Like Terms

4–7   Polynomials

4–8   Adding Polynomials

PERFORMANCE OBJECTIVES

- Read variables represented by English and Greek letters—including those with subscripts and primes.

- Recognize rational, irrational, and fractional algebraic expressions.

- Employ the words variable, coefficient, term, and factor when working with algebraic expressions.

- Combine algebraic expressions having like terms.

- Identify and name polynomials with one, two, and three terms.

- Utilize the associative and the commutative laws of addition to combine polynomials.

An illustrative advertisement in *Life* magazine for the Baker Electric Automobile (1917–1924) circa 1920. The easy-to-operate car was primarily an urban car and, like its modern counterpart, it had a limited range. Its popularity was due in part to the gasoline shortages caused by World War I. (Courtesy of Library of Congress)

In this chapter and the one that follows, the *language of algebra* is introduced. Like any written language, algebra uses many special symbols and technical words. It is very important that you learn the names of these symbols and become familiar with the vocabulary of algebra. Once this is done, you will then be ready to solve equations and work with the *techniques of algebra*.

4–1 VARIABLES, SUBSCRIPTS, AND PRIMES

Variables

As your understanding of algebra grows, so will your appreciation of how algebraic methods are used to complement arithmetic methods for solving problems. The effectiveness of algebra comes from its use of letters to stand in for numbers. These letters are called **variables.** *Variable* is the technical term for a letter which is used in place of a number. The number represented by a letter may be known or unknown, it may be constant or changing. The important idea is that the letter represents a number and obeys the laws of numbers.

The letters most commonly used as variables in electronic problems are taken from the letters of the English and Greek alphabets. Table 4–1 is a partial listing of the Greek alphabet. It contains the letters commonly associated with electronic applications. Since these letters occur in electronics, you should learn their names and how to write them. Appendix A contains a complete listing of the Greek alphabet.

TABLE 4–1 Greek Letters Commonly Associated with Electronic Applications

Name	Capital	Lowercase	Used to Designate:
Alpha		α	Angles, temperature coefficient, current transfer ratio
Beta		β	Angles, current transfer ratio
Delta	Δ		Increment, determinant of coefficients
Eta		η	Efficiency
Theta		θ	Phase angle, thermal impedance
Lambda		λ	Wavelength
Mu		μ	Permeability, micro, amplification factor
Pi		π	3.1416 (ratio of circumference of circle to diameter)
Rho		ρ	Specific resistance
Sigma	Σ	σ	Summation (capital), conductivity (lowercase)
Phi	Φ	ϕ	Magnetic flux (capital), phase angles and angles (lowercase)
Psi		ψ	Electric flux
Omega	Ω	ω	Ohms (capital), angular velocity (lowercase)

Subscripts

In naming variables for electronics use, sometimes there is a need for the same letter to be used more than once. In this case, a subscript or a prime is used. A **subscript** is a number, a letter, or a group of letters written to the right and below a variable.

EXAMPLE 4–1   Using the letter V for voltage, select variables to represent three different voltages.

Solution   Since each voltage is different, a subscript will be used with each variable. The numbers 1, 2, 3, or the letters a, b, c could be used as subscripts. Each is shown:

V_1 read "V one" or "V sub one"

V_2 read "V two" or "V sub two"

V_3 read "V three" or "V sub three"

V_a read "V a" or "V sub a"

V_b read "V b" or "V sub b"

V_c read "V c" or "V sub c"

Primes

A **prime** is an apostrophe mark made to the *right* and *above* a variable. It is placed in the same position as an exponent. It is used to distinguish variables with the same name.

EXAMPLE 4–2   Using the Greek letter β, select variables to represent two corresponding angles in two triangles.

Solution   Use β to represent the angle in the first triangle. Use β', "beta prime," for the angle in the second triangle. See Figure 4–1.

FIGURE 4–1   The two corresponding angles are labeled β and β'.

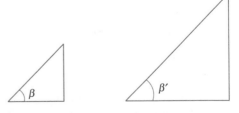

Read each of the following variables.

1. L_1	**2.** I_{max}	**3.** G_m	**4.** k'
5. R_p	**6.** ω_0	**7.** R'	**8.** β_2
9. μ	**10.** m'	**11.** R_{ac}	**12.** i_p
13. e'	**14.** C_{cb}	**15.** λ_t	**16.** A'_v

4–2 INDICATING MULTIPLICATION

In arithmetic, multiplication was indicated by the multiplication sign ($\times$). When we indicate multiplication in algebra, we may use the multiplication sign ($A \times B$), a dot ($A{\cdot}B$), parentheses ($(A)(B)$), or no sign at all (AB). These four ways of indicating multiplication are summarized in Table 4–2. Note that we cannot omit the multiplication sign when multiplying numbers because they would be misunderstood. For example, we would understand 32 (*three two*) as the number thirty-two and not as three times two.

When a multiplication sign ($\times$) is used to indicate multiplication, care must be taken to make it look different from the letter x. When using the dot to indicate multiplication, be careful to place it above the writing line so that it does not appear as a decimal point. For example, $2{\cdot}B$, not $2 \,.\, B$.

TABLE 4–2 Examples Indicating Multiplication

Symbol	Possible Ways of Expression		
$\times$	3×2	$2 \times b$	$\rho \times \omega$
$\cdot$	$3 \cdot 2$	$2 \cdot b$	$\rho \cdot \omega$
$()$	$(3)(2)$	$(2)(b)$	$(\rho)(\omega)$
None	None	$2b$	$\rho\omega$

Write the following in two other ways.

1. 2×3	**2.** $x \cdot z$	**3.** 7×2	**4.** $5b$
5. $(b)c$	**6.** $m \cdot n$	**7.** $\alpha\beta$	**8.** $3h$
9. $\Delta \cdot \Sigma$	**10.** $9 \times R$	**11.** $\lambda\eta$	**12.** $(4)(2)$
13. $(6)(c)$	**14.** $r \cdot s$	**15.** $\theta \times \phi$	**16.** $3h_{oe}$

4–3 GENERAL NUMBERS

Numbers that are represented by letters are called **literal numbers** or **general numbers.** By using general numbers rather than words, electrical formulas can be easily stated. Ohm's law, an important formula of electronics, is easily expressed as $E = IR$. To express this same idea in words, the following statement would be needed. One volt of electromotive force (E) causes one ampere of current (I) to flow through a resistance (R) of one ohm. As you can see, written expressions are cumbersome and more easily misunderstood, whereas formulas written with general numbers are concise and less likely to be misunderstood.

4–4 ALGEBRAIC EXPRESSIONS

An **algebraic expression** is a collection of variables, numbers, and exponents connected by operators $(+, -, \times, /, \sqrt{\ })$ and signs of grouping. Table 4–3 shows several examples of algebraic expressions.

TABLE 4–3   Examples of Algebraic Expressions

(1) $\sqrt{8y}$	(2) $8y + 5$	(3) $(x + 2)(x - 3)$
(4) $x^{3.5}$	(5) $2/y$	(6) $\frac{1}{2}$
(7) $\sqrt{2x + y} + 29$	(8) $3x$	(9) $\dfrac{R_1 + 2}{3}$
(10) $5x^2 - \frac{1}{3}x - 5$	(11) 3	(12) $\dfrac{R_b + R}{\alpha + \beta'} + \theta$

Algebraic expressions are classified by mathematicians in several ways. One way is to describe them as *rational expressions* or *irrational expressions.* A **rational expression** is an algebraic expression that contains no radical signs and has only whole numbers as exponents. An **irrational expression** is an algebraic expression that contains a fractional exponent or a radical sign. In Table 4–3, Examples (1), (4), and (7) are irrational expressions; the rest are rational expressions.

Rational expressions are further classified as either polynomial or fractional expressions. A rational expression that has no variable as a divisor (no division with a variable) is called a **rational integral expression,** or simply a **polynomial.** A rational expression that has a variable as a divisor (division with a variable) is called a **fractional expression.** In Table 4–3, examples (2), (3), (6), and (8) through (11) are polynomials. Examples (5) and (12) are fractional expressions. Figure 4–2 summarizes the classification of algebraic expressions.

FIGURE 4–2 Classification of
algebraic expressions.

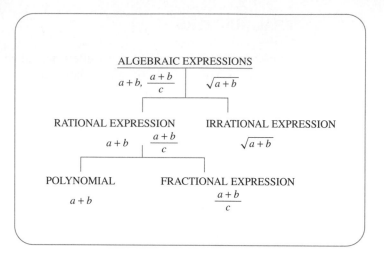

EXAMPLE 4–3   In the following algebraic expressions, list all the terms that apply to each. Are they irrational, rational, polynomial, or fractional expressions?

(a) $3x + y/2$ (b) $\dfrac{1}{x + y}$ (c) $\sqrt{4x - 3}$

Solution   (a) $3x + y/2$ rational, polynomial expression

(b) $\dfrac{1}{x + y}$ rational, fractional expression

(c) $\sqrt{4x - 3}$ irrational expression

Now that you have an understanding of algebraic expressions, the *general number* can be fully defined. Every number, variable, rational expression, and irrational expression is a **general number.** Because each of these (number, variable, rational expression, and irrational expression) is a numerical quantity, each follows all the rules for numbers.

EXAMPLE 4–4   List several general numbers.

Solution   Because every number, variable, rational expression, and irrational expression is a general number, all entries in Table 4–3 are general numbers. Examples would be 6.32, x, $5/x$, and $\sqrt{3/5}$.

In the following algebraic expressions, list all the terms that apply to each. Describe whether each is a rational or an irrational expression. If rational, then further indicate whether each is a polynomial or a fractional expression.

1. 12.7

2. $\sqrt{5x}$

3. $\dfrac{a^2 + 3b}{5y}$

4. $5x + \sqrt{6y}$

5. $1 + \sqrt{y}$

6. $1/x$

7. $\eta_0 + 2$

8. $7\pi + (x/3)$

9. $\sqrt{3x}$

10. $\sqrt{x + y}$

11. $4x^{0.5}$

12. $(3/x) + \sqrt{5y}$

13. $5 + 3\mu$

14. 7.0

15. $5x^2 + \dfrac{1}{3x}$

4–5 PRODUCTS, FACTORS, AND COEFFICIENTS

A *product* is formed when two or more general numbers (numbers, variables, rational expressions) are multiplied together. Thus, $3y$ is a product; it is the result of multiplying 3 (a number) times y (a variable).

The general numbers that are used to form a product are called the *factors* of the product. Since $3 \times y = 3y$, 3 and y are factors of $3y$. Three is called the *numerical factor*; y is called the *literal factor*.

A **coefficient of a product** is any factor of a product. Each factor is the coefficient of the other factor. In the product $3y$, the 3 is the coefficient of y and y is the coefficient of 3. The numerical factor of a product is called the *numerical coefficient*. If a variable has no written numerical coefficient, then it has an **implied coefficient** of 1. For example, T means $1T$ and rs means $1rs$.

EXAMPLE 4–5 In the product $5ab$, state the coefficient of:

(a) ab (b) 5   (c) a

Solution (a) The coefficient of ab is 5.
(b) The coefficient of 5 is ab.
(c) The coefficient of a is $5b$.

State the numerical coefficient of each product.

1. $3x$

2. $\frac{2}{5}\mu$

3. ab

4. $\frac{1}{2}zy$

5. b

6. $0.3c$

7. $2y\alpha$ **8.** 0.8ω **9.** $\frac{1}{3}s$

10. 1.2λ **11.** $\frac{7}{4}\beta$ **12.** 16η

State the coefficient of π in each of the following problems.

13. 2π **14.** πb **15.** πr^2

16. $2\pi fL$ **17.** $5b\pi$ **18.** $\sqrt{\pi S}$

19. $16\pi d^2$ **20.** π **21.** $7.4\pi\omega$

4–6 COMBINING LIKE TERMS

An algebraic expression made up of a number, a variable, a product, or a quotient is called a **term.** The algebraic expression $5y$ is a term, as are 7, $-2cd$, $b(2 + c)$, x^2, β, $3a/5$, $1/x$, and $\sqrt{5x}$. Notice that both rational and irrational algebraic expressions may be terms.

Like Terms

Terms that have the same literal factors, but different numerical coefficients, are called **like terms.** Examples are $5a$ and $2a$, 8μ and 2μ, $2x^2y$ and $0.5x^2y$. Like terms can be added and subtracted. For example, $2R + 5R = 7R$.

> **RULE 4–1.** *Combining Like Terms*
>
> To combine *like terms:*
> 1. Combine the numerical coefficients.
> 2. Form the new term from the combined numerical coefficients and the literal factor.

EXAMPLE 4–6   Simplify $2a + 5a$ by combining like terms.

Solution   Use the steps of Rule 4–1.

Step 1:   Combine numerical coefficients:

$$2 + 5 = 7$$

Step 2:   Form new term:

$$7a$$

$\therefore$ $2a + 5a = 7a$

EXAMPLE 4–7	Simplify $3a - 4a$ by combining like terms.
Solution	Follow the steps of Rule 4–1.
Step 1:	Combine numerical coefficients:
	$3 - 4 = -1$
Step 2:	Form new term:
	$-1a$
$\therefore$	$3a - 4a = -a$

EXERCISE 4–5

Simplify each of the following algebraic expressions by combining like terms.

1. $2a + 3a$
2. $7\omega - 3\omega$
3. $9x + 3x$
4. $-ab - 6ab$
5. $-c - 6c$
6. $7y - (-8y)$
7. $5b - (+7b)$
8. $2\mu - 3\mu + 7\mu$
9. $4h + 3h - 2h$
10. $4k - 9k - k$

11. $4A + 3B - B + 6A$
12. $16x + 5y - 4x$
13. $c^2b - 4c^2b$
14. $x^2y^2 - 3x^2y^2$
15. $5x^2 - 3x - 7 - 2x^2 + 6x + 9$
16. $5y^2 + 3y - 8y^2 + 2y$
17. $2x - 3x^3 + 2x^2 - 4x - x^2 + x^3$
18. $9x^2y - 2xy^2 - 3x^2y$
19. $3xy^2 - 5x^2y - 3xy^2$
20. $2y - 3y^2 - 4y^3 - 2y + 4y^3 - y^2$

4–7 POLYNOMIALS

Polynomials are sums of products. Certain polynomials have special names. If the polynomial has only one term, it is called a **monomial.** If the polynomial has two terms, it is called a **binomial.** A three-term polynomial is called a **trinomial.** Special names are not used for polynomials with more than three terms. Table 4–4 is a list of sample polynomials.

TABLE 4–4 Sample Polynomials	
Special Name	**Example**
Monomial	$3y^2$
Binomial	$7y^2 + 4$
Trinomial	$m^2 - mn + n^2$

Of the following rational expressions, which can be classified as (a) monomials, (b) binomials, (c) trinomials, (d) fractional expressions?

1. $5h$

2. $2a^2 + 3a + 5$

3. $\pi + 2\lambda$

4. $27y^3$

5. $\dfrac{e + ir}{4x}$

6. 28

7. $E_1 + E_2 + IR$

18. $\frac{5}{8}$

9. $x + 5y$

10. $\dfrac{\mu + e_o}{e_i}$

11. $\dfrac{\eta_1 + \alpha_0}{\Sigma}$

12. $3\psi^3 + 2\psi^2 + \psi$

4–8 ADDING POLYNOMIALS

When adding polynomials, there are two important concepts to keep in mind. First, two terms may be added in either order. Thus, the sum $2x + 3x$ is $5x$; also the sum $3x + 2x$ is $5x$. This concept is called the **commutative law of addition.** Second, three terms may be grouped in either manner to indicate which addition is to be performed first. Thus, $(5x + 3x) + 4x = 5x + (3x + 4x)$. This concept is called the **associative law of addition.** In summary, **addition is independent of order and of grouping.**

EXAMPLE 4–8	Simplify $6x^2 + 5 + 3x + 5x + 2x^2 + 4$.
Solution	Apply the commutative and associative laws in order to group like terms:

$$(6x^2 + 2x^2) + (5 + 4) + (3x + 5x)$$

Combine like terms:

$$8x^2 + 9 + 8x$$

$$\therefore \qquad 6x^2 + 5 + 3x + 5x + 2x^2 + 4 = 8x^2 + 9 + 8x$$

> **RULE 4–2.   *Adding Polynomials***
>
> To add polynomials:
> 1. Arrange the terms in both polynomials in the same order.
> 2. Write like terms under each other.
> 3. Add each column, connecting the sums with their correct sign.

EXAMPLE 4–9	Add $6x^2 + 5 + 3x$ and $5x + 2x^2 + 4$.
Solution	Follow the steps of Rule 4–2.
Step 1:	Arrange terms:

$$(6x^2 + 3x + 5) + (2x^2 + 5x + 4)$$

Step 2:   Place like terms in columns:

$$
\begin{array}{r}
6x^2 + 3x + 5 \\
2x^2 + 5x + 4 \\
\hline
\end{array}
$$

Step 3:   Add: $8x^2 + 8x + 9$

$\therefore$ $6x^2 + 5 + 3x + 5x + 2x^2 + 4 = 8x^2 + 8x + 9$

EXAMPLE 4–10	Add $3m + 6n - 4m^2$ and $2 + 6m^2 - 8n$.
Solution	Use the steps of Rule 4–2.
Step 1:	Arrange terms:

$$(-4m^2 + 3m + 6n) + (6m^2 - 8n + 2)$$

Step 2:   Place like terms in columns:

$$
\begin{array}{r}
-4m^2 + 3m + 6n \\
6m^2 \qquad - 8n + 2 \\
\hline
\end{array}
$$

Step 3:   Add: $2m^2 + 3m - 2n + 2$

$\therefore$ $3m + 6n - 4m^2 + 2 + 6m^2 - 8n = 2m^2 + 3m - 2n + 2$

EXAMPLE 4–11	Add $-\theta^3 - 7 + \theta^2$ and $-\theta^2 + \theta^3 + 5$ and $-2 - \theta^3 + \theta^2$.
Solution	Use the steps of Rule 4–2.
Step 1:	Arrange terms:

$$(-\theta^3 + \theta^2 - 7) + (\theta^3 - \theta^2 + 5) + (-\theta^3 + \theta^2 - 2)$$

Step 2:   Place like terms in columns:

$$
\begin{array}{r}
-\theta^3 + \theta^2 - 7 \\
\theta^3 - \theta^2 + 5 \\
-\theta^3 + \theta^2 - 2 \\
\hline
\end{array}
$$

Step 3:   Add: $-\theta^3 + \theta^3 - 4$

$\therefore$ $-\theta^3 - 7 + \theta^2 + -\theta^2 + \theta^3 + 5 + -2 - \theta^3 + \theta^2 = -\theta^3 + \theta^2 - 4$

Read each of the following polynomials, and state the terms in descending powers of the variables.

1. $3x^2 + 2 + 2x$ **2.** $5y^2 + 4 + 8y$

3. $9 + 2x^2 + 4x$ **4.** $6 + 3\phi^2 + 13\phi$

5. $3n + 2n^2 - 11$ **6.** $5\eta - 2 + 2\eta^2$

7. $-8V + 2 - V^2 + 2V^3$ **8.** $-5 + 3\alpha^3 - \alpha^2 + 4\alpha$

9. $x^2 + 3y^2 + 2xy$ **10.** $\omega^3 + 2\omega^2\pi + 2\pi^2 + 3 + \omega\pi^2$

Add the following.

11. $2y + 3$
$\underline{4y + 6}$ **12.** $3x + 5$
$\underline{2x - 4}$

13. $x^2 + 3$
$\underline{x^2 - 3}$ **14.** $\mu - \beta^2$
$\underline{\mu + \beta^2}$

15. $2x + 4y + \ z$
$\underline{\ x - 2y + 3z}$ **16.** $3\psi^2 + 4\psi - 1$
$\underline{\ \ \psi^2 \ \ \ \ \ \ + 1}$

17. $V^3 + 2V^2 - \ V$
$\underline{\ \ \ \ \ \ \ \ \ V^2 + V + 5}$ **18.** $\ \ \ c^3 - 2c^2 + \ c$
$\underline{-2c^3 \ \ \ \ \ \ \ + 3c - 9}$

19. $\ \ 2x + \ y - 3z$
$\ \ \ \ \ \ \ \ - 2y + 4z$
$\underline{-3x \ \ \ \ \ \ \ - \ z}$ **20.** $\ \ -\alpha^3 + \ \alpha^2 - 7\alpha$
$\ \ \ \ 2\alpha^3 - \ \alpha^2 + 4\alpha$
$\underline{-\alpha^3 + 2\alpha^2 - 2\alpha}$

21. $n + 5$ and $2n + 3$ **22.** $2\omega - 7$ and $3\omega + 7$

23. $3x^3 - x^2$ and $x^3 - 4x$ **24.** $5\mu^4 + 3\mu^2$ and $\mu^3 - 8\mu^2$

25. $4m - 2n$ and $3m + 3n$ **26.** $-3t + S$ and $2t + 4S$

27. $3x^2 - 2x$ and $-x^2 - 3x$

28. $5y^2 - 2xy$ and $-10y^2 + 14xy$

29. $7.2\omega - 3.1\rho$ and $0.8\omega + 5.7\rho$

30. $-4.2V + 0.5W$ and $3.9W - 1.8V$

Simplify each expression by adding like terms.

31. $(x^3 - 4x^2 - 7x + 6) + (2x^3 - 5x + 2)$

32. $(m^2 - mn + n^2) + (4n^2 + m^2 - 3mn)$

33. $(6a^3 - 5a) + (7a - 11a^3 + 5b^2)$

34. $(4x^2 - 2x^3 + 3x + 1) + (x^3 + 3x^2 - 1 - 5x)$

35. $(-3t^2 + t^2 - 2 + 5t) + (-5t - t^2 + 2 + 3t^2)$

36. $(2x + 4y - 12) + (-5y + 7x - 2) + (-1 - 3y + 4x)$

37. $(7\beta - 9 + \theta) + (-5\alpha + \theta) + (3 - 2\theta + 8\beta)$

38. $(-3y^2 - 3y + 2) + (-3y^2 + 7) + (-7y - 5y^2)$

39. $(x - y - z) + (2z - 5x + 3y) + (2y - 7x + 2z)$

40. $(7\lambda^2 - 6 + 7\lambda) + (7\lambda - 4 + \lambda^2) + (-1 + 2\lambda^2)$

Calculator Drill

Work the following problems and express the coefficients in the answers in decimal notation to the appropriate precision. All numbers are approximate.

41. $(9.26x^2 - 4.25x + 8.25) + (-3.81x^2 + 5.01x + 9.88)$

42. $(30.1x^2 + 22.8x - 14.3) + (5.11x^2 - 24.3x - 13.4)$

43. $(10.9x^2 - 7.65x + 4.33) + (18.88x^2 - 7.52x + 7.99)$

44. $(52.4x^2 - 14.2x - 46.5) + (-12.3x^2 - 5.53x + 18.3)$

SELECTED TERMS

associative law of addition   Three terms may be grouped in either way to indicate which addition is performed first; thus, $(a + b) + c = a + (b + c)$.

coefficient of a product   Any factor of a product; each factor is the coefficient of the other factor.

commutative law of addition   Two terms may be added in either order; thus, $a + b = b + a$.

fractional expression   Algebraic expression having a variable in the denominator of one or more terms.

irrational expression   Algebraic expression containing a fractional exponent or a radical sign.

like terms   Having the same literal factors but different numerical coefficients.

polynomial   Rational expression having no variable as a divisor.

rational expression   Algebraic expression having whole numbers as exponents and containing no radical signs.

variable   Symbol for a number that may be constant or varying in an algebraic expression or equation.

5

Algebra Fundamentals II

5–1 Multiplying Monomials

5–2 Multiplying a Monomial and a Binomial

5–3 Multiplying a Monomial and a Polynomial

5–4 Subtracting Polynomials

5–5 Additional Work with Polynomials

5–6 Division of Monomials

5–7 Dividing a Polynomial by a Monomial

5–8 Factoring Polynomials with a Common Monomial Factor

5–9 Evaluating Algebraic Expressions

PERFORMANCE OBJECTIVES

- **Utilize the associative and the commutative laws of multiplication to multiply monomials.**

- **Demonstrate proficiency in making sign changes to carry out subtraction of polynomials.**

- **Employ the concepts of implied exponent, zero exponent, and division of powers of a variable to the division of a polynomial by a monomial.**

- **Evaluate by substituting numbers for variables in an algebraic expression.**

The antennas on the range instrumentation ship, USS General HH Arnold, track and gather data on the eastern test range in support of NASA, DOD, and USAF. (Courtesy of U.S. Air Force)

This chapter continues our discussion of the concepts of algebra. By becoming familiar with the topics presented here, you will be ready to solve equations.

5–1 MULTIPLYING MONOMIALS

In multiplication, as in addition, we will use several important concepts. First, two factors may be multiplied together in either order. Thus, $A \times B = B \times A$. This concept is called the **commutative law of multiplication.** Second, three factors may be grouped in either manner to indicate which multiplication is to be performed first. Thus, $(5 \times A) \times D = 5 \times (A \times D)$. This concept is called the **associative law of multiplication.** In summary, **multiplication is independent of order and of grouping.**

In Section 2–1, the use of exponents was introduced. At that time we stated that the exponent tells us how many times a number is used as a factor in a product. Thus, $y^3 = y \cdot y \cdot y$. In Section 2–7 you were shown how to multiply with powers of ten. Recall that powers of ten can be multiplied by adding their exponents. Thus, $10^3 \times 10^4 = 10^7$. In like manner, powers of a single variable may be multiplied by adding their exponents.

RULE 5–1. *Multiplying Powers of a Variable*

To multiply powers of a variable, add the exponents:

$$a^m a^n = a^{m+n}$$

EXAMPLE 5–1	Solve $x^3 \cdot x^4$.
Solution	Add the exponents.
	$$x^3 \cdot x^4 = x^{3+4} = x^7$$
Check	We can check our answer by "longhand." Thus,

$$x^3 = x \cdot x \cdot x$$

$$x^4 = x \cdot x \cdot x \cdot x$$

$$\therefore \quad x^3 \cdot x^4 = (x \cdot x \cdot x)(x \cdot x \cdot x \cdot x) = x^7.$$

When multiplying monomials we form the product by applying Rule 5–2.

> **RULE 5–2. Multiplying Monomials**
>
> To multiply monomials:
> 1. Multiply the numerical coefficients.
> 2. Multiply the literal factors.
> 3. Form the product by multiplying the numerical factor and the literal factor.

EXAMPLE 5–2	Multiply $5a$ by 3.
Solution	Follow the steps of Rule 5–2.
Step 1:	Multiply numerical coefficients:
	$3 \times 5 = 15$
Step 2:	Not needed.
Step 3:	Form the product:
	15a
$\therefore$	$5a \times 3 = 15a$

EXAMPLE 5–3	Multiply $3y^5$ by $2y^4$.
Solution	Use the steps of Rule 5–2.
Step 1:	Multiply numerical coefficients:
	$3 \times 2 = 6$
Step 2:	Multiply literal factors:
	$y^5 y^4 = y^{5+4} = y^9$
Step 3:	Form the product:
	6y^9
$\therefore$	$3y^5(2y^4) = 6y^9$

EXAMPLE 5–4	Multiply $-6c$ by $-3b$.
Solution	Use the steps of Rule 5–2.
Step 1:	Multiply numerical coefficients:
	$(-6)(-3) = 18$
Step 2:	Multiply literal factors:
	$c \cdot b = cb$

CHAPTER 5   Algebra Fundamentals II

Step 3:	Form the product:
	$18bc$
$\therefore$	$(-6c)(-3b) = 18bc$
Observation	It is conventional to put variables in alphabetical order.

EXAMPLE 5–5	Multiply $-7IR$ by $6IZ$.
Solution	Follow the steps of Rule 5–2.
Step 1:	$(-7)(6) = -42$
Step 2:	$(IR)(IZ) = IIRZ = I^2RZ$
Step 3:	$-42I^2RZ$
$\therefore$	$(-7IR)(6IZ) = -42I^2RZ$

EXERCISE 5–1

Multiply each of the following.

1. $x^5 \cdot x^4$	**2.** $y^2 \cdot y^6$
3. $\omega^3 \cdot \omega^{-3}$	**4.** $a^5 \cdot a^{-2}$
5. $b \cdot b$	**6.** $-4(3a)$
7. $-7y(-2)$	**8.** $-6z(2)$
9. $7(-5ab)$	**10.** $4\mu\,(-8\beta)$
11. $10^2 \cdot 10^{-4}$	**12.** $-5c \times -3c$
13. $(-xy)(x^3y^2)$	**14.** $(ab^2)(-7ab)$
15. $(-8\phi)(-2\theta\eta)$	**16.** $(3a^5)(-7a)$
17. $(-8x^3)(-2x^{-3})$	**18.** $(-2ab)(4cb)$
19. $(2y^{-3})(-4yx^2)$	**20.** $(-2x^3)(5y^4)$
21. $2x^{-2} \cdot 3x \cdot x^3$	**22.** $-5(a^2)(ab)$
23. $(-2y)(-2y)(-2y)$	**24.** $(6ab)(-bc)(-3ac)$

5–2   MULTIPLYING A MONOMIAL AND A BINOMIAL

When multiplying a monomial and a binomial, we will use the *distributive law of multiplication over addition,* commonly called the **distributive law.** This law states that each term in the binomial is multiplied by the monomial, and the resulting products are added together as in Figure 5–1.

FIGURE 5–1 Distributive law: a is distributed over b and c.

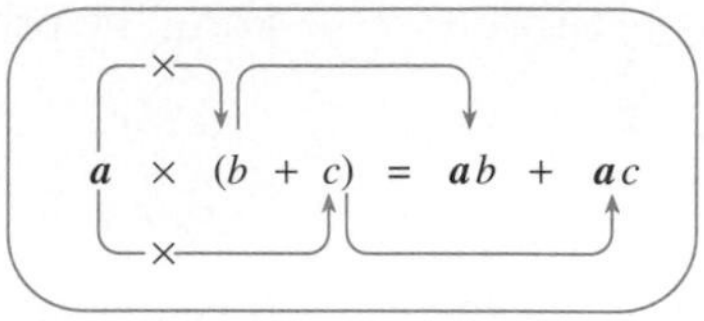

$$a \times (b + c) = ab + ac$$

RULE 5–3. Distributive Law of Multiplication

To multiply a polynomial by a monomial:
1. Multiply each term of the polynomial by the monomial.
2. Add the resulting products.

EXAMPLE 5–6   Multiply $(a - b)$ by 6.

Solution   Use the steps of Rule 5–3 with the expression $6(a - b)$.

Step 1:   Multiply each term by 6:

$$6(a) = 6a, \, 6(-b) = -6b$$

Step 2:   Add the products:

$$6a + -6b \Rightarrow 6a - 6b$$

$\therefore$ $6(a - b) = 6a - 6b$

EXAMPLE 5–7   Multiply $5x + 3$ by $2y$.

Solution $2y(5x + 3)$

$$(2y \cdot 5x) + (2y \cdot 3)$$

$$10xy + 6y$$

$\therefore$ $2y(5x + 3) = 10xy + 6y$

EXAMPLE 5–8   Multiply $-3x^2 - 2x$ by $-x$.

Solution $-x(-3x^2 - 2x)$

$$(-x \cdot -3x^2) + (-x \cdot -2x)$$

$$(3x^3) + (2x^2)$$

$\therefore$ $-x(-3x^2 - 2x) = 3x^3 + 2x^2$

Before you begin work on Exercise 5–2, review Sections 1–6 and 1–10.

EXERCISE 5–2

Apply the distributive law to each of the following.

1. $3(7 + 2)$	**2.** $-2(5 - 3)$	**3.** $2(y - 3)$
4. $-4(x + 3)$	**5.** $3(\mu - 5)$	**6.** $-1(a - 2)$
7. $-Z(3 - Z)$	**8.** $a(-5 - 2a)$	**9.** $\beta(c - 4)$
10. $-d(a + 6)$	**11.** $-3(x - y)$	**12.** $-6(U - V)$
13. $-2a(a - 2b)$	**14.** $3x(x^2 - y)$	**15.** $-7\theta(2\omega + 5\phi)$
16. $-x^3(x^2 + xy)$	**17.** $y^4(-y^3 - 4y^2)$	**18.** $8ab(-a - b)$
19. $-3cd(-d^2 - c^2)$	**20.** $-5xy(x^2 - y^2)$	**21.** $3s(s^2 + 4s)$

5–3 MULTIPLYING A MONOMIAL AND A POLYNOMIAL

In this section we will build on a technique used in the preceding section. To multiply a monomial by a polynomial, we use the distributive law (Rule 5–3).

EXAMPLE 5–9 Multiply $3x^2 - x + 3$ by $-2x$.

Solution $-2x(3x^2 - x + 3)$

Distribute $-2x$ over each term in the trinomial:

$$(-2x \cdot 3x^2) + (-2x \cdot -x) + (-2x \cdot 3)$$

Add the products:

$$-6x^3 + 2x^2 - 6x$$

$\therefore \qquad -2x(3x^2 - x + 3) = -6x^3 + 2x^2 - 6x$

EXAMPLE 5–10 Multiply $3xy^2 - 2x + 3y - 7$ by $-3xy$.

Solution $-3xy(3xy^2 - 2x + 3y - 7)$

Distribute $-3xy$ over each term in the polynomial:

$$(-3xy \cdot 3xy^2) + (-3xy \cdot -2x) + (-3xy \cdot 3y) + (-3xy \cdot -7)$$

Add the products:

$$-9x^2y^3 + 6x^2y - 9xy^2 + 21xy$$

$\therefore \qquad -3xy(3xy^2 - 2x + 3y - 7) = -9x^2y^3 + 6x^2y - 9xy^2 + 21xy$

EXERCISE 5–3

Apply the distributive law to each of the following.

1. $4(3x + 2y + 5)$ **2.** $-3(2a + b - c)$

3. $-7(-3u + 2v - w)$ **4.** $-1(2\rho - 3\lambda - 6)$

5. $5(-8x + y - 3)$ **6.** $2(7 + 2x - 3y)$

7. $-6(-\mu + 2\theta - 3)$ **8.** $-8(-2a + b - 3c)$

9. $-x(2x + 4 - y)$ **10.** $2x(-3 + 4y - z)$

11. $-3h(-4j + 5\theta - 1)$ **12.** $4b(-2b^2 - b + 2)$

13. $-3ab(2a - 3b + 4c)$ **14.** $-a^2c(-a + bc - 1)$

15. $3x^2(2x^2 - 7x - 6)$ **16.** $-2y^2(-y - 2x - 3)$

17. $-2ab(a^2 - b^2 + c - 3)$ **18.** $-x^2yz(z^2 - 2x^2 + xy + 4)$

19. $-1(a^2 + 2b - 3c + 5)$ **20.** $-1(x^2 - 3x + y - 7)$

Calculator Drill

Work each of the following problems, expressing the coefficients in the answers in decimal notation to three significant figures.

21. $3.87y(1.95y^2 - 6.41x + 4.05)$

22. $12.8b(-5.32b + 3.55c - 8.17d)$

23. $-42.8\mu(2.93\mu^2 - 5.25\mu + 7.15)$

24. $-3.81a^2(16.5b^2 + 4.32a - 8.47)$

5–4 SUBTRACTING POLYNOMIALS

Recall from Section 1–8 that we subtract by changing the sign and adding. But a polynomial has more than one sign, so what do we do? Distribute the sign change over each term of the polynomial.

RULE 5–4. *Subtracting Polynomials*

To subtract one polynomial from another:

  1. Change the sign of **each and every** term in the polynomial to be subtracted.

  2. Combine like terms as in addition.

EXAMPLE 5–11   Subtract $2x - 3$ from $5x + 4$.

    Solution   Use the steps of Rule 5–4.

      Step 1:   Change both signs of $2x - 3$:

$$2x - 3 \Rightarrow -2x + 3$$

Step 2: Add like terms:

$$5x + 4$$
$$\underline{-2x + 3}$$
$$3x + 7$$

$$\therefore \quad (5x + 4) - (2x - 3) = 3x + 7$$

Signs of grouping preceded by a plus sign may be removed without affecting the signs of the terms. Signs of grouping preceded by a minus sign may be removed by changing the sign of **each and every** term. These concepts are demonstrated in the following examples.

EXAMPLE 5–12   Simplify $(8a + 3) - (5a - 2)$ by removing parentheses and adding like terms.

Solution   Remove the parentheses.

Observation   There is an implied plus sign in front of $(8a + 3)$.

$$(8a + 3) - (5a - 2) \Rightarrow 8a + 3 - 5a + 2$$

Combine like terms:

$$3a + 5$$

$$\therefore \quad (8a + 3) - (5a - 2) = 3a + 5$$

EXAMPLE 5–13   Simplify $(3x^2 + 3x - 1) - (x^2 - 5x + 4)$.

Solution   Remove the parentheses:

$$(3x^2 + 3x - 1) - (x^2 - 5x + 4) \Rightarrow$$
$$3x^2 + 3x - 1 - x^2 + 5x - 4$$

Combine like terms:

$$2x^2 + 8x - 5$$

$$\therefore \quad (3x^2 + 3x - 1) - (x^2 - 5x + 4) = 2x^2 + 8x - 5$$

EXERCISE 5–4

1. Subtract $2a$ from $3a - 8$.
2. Subtract $-5x$ from $7x + 3$.
3. Subtract $2a - 3c$ from $6c + a$.
4. Subtract $4\eta + 2\omega$ from $-3\eta - 2\omega + 1$.
5. Subtract $5x^2 + 2x - 1$ from $-7x^2 - 4x + 3$.

Simplify each expression by removing parentheses and combining like terms.

6. $5y - (3y - 2)$
7. $(3x + 2) - 3x$
8. $-7\mu - (2\mu + 6)$
9. $7\theta - (4\theta + \beta)$
10. $(a + b) - (a + b)$
11. $(x - y) - (x + y)$
12. $(2x + 3y) - (2y + 3x)$
13. $(3x + 2y) - (x + y)$
14. $(2a + 3b) - (2a - b)$
15. $(-4\mu + \pi) - (-5\mu - \pi)$
16. $(-3x - 7) - (-x + 5)$
17. $(V^2 - 3U + 2) - (-V^2 - 2U + 2)$
18. $(a^3 - 3a^2 + 4) - (2a^3 + 5a^2 - 5a)$
19. $(2x^4 - 7x^2 + 1) - (x^4 - 3x^2 - x + 3)$
20. $(3\theta^3 - \theta^2 + 2\theta) - (5\theta^3 - \theta^2 + \theta - 7)$
21. $(-3y^2 + 4y - 5) - (2y^2 - 7)$

5–5 ADDITIONAL WORK WITH POLYNOMIALS

This section is intended to allow you to improve your skills in working with addition and multiplication as they apply to simple polynomial expressions. Before working through the following examples, review the rules pertaining to signs of grouping in Section 1–5.

EXAMPLE 5–14   Simplify $2a[-3(5a - 4) - 5a]$.

Solution

| M: | -3 | $2a[-15a + 12 - 5a]$ |

| A: | $-15a - 5a$ | $2a[-20a + 12]$ |

| M: | $2a$ | $-40a^2 + 24a$ |

$\therefore$ $2a[-3(5a - 4) - 5a] = -40a^2 + 24a$

EXAMPLE 5–15   Simplify $2x - [4 + 2(x - 2)]$.

Solution

| M: | 2 | $2x - [4 + 2x - 4]$ |

| A: | $4 - 4$ | $2x - [0 + 2x]$ |

Remove brackets:

$2x - 0 - 2x$

A: $2x - 2x$ $2x - 2x = 0$

$\therefore$ $2x - [4 + 2(x - 2)] = 0$

EXAMPLE 5–16   Simplify $-3[-2y(y^2 - 2) - 3y(-2y^2 + 1)]$.

Solution

M: $-2y$ $-3[-2y^3 + 4y - 3y(-2y^2 + 1)]$
M: $-3y$ $-3[2y^3 + 4y + 6y^3 - 3y]$
A: $-2y^3 + 6y^3$ $-3[4y^3 + 4y - 3y]$
A: $4y - 3y$ $-3[4y^3 + y]$
M: -3 $-12y^3 - 3y$

$\therefore$ $-3[-2y(y^2 - 2) - 3y(-2y^2 + 1)] = -12y^3 - 3y$

EXERCISE 5–5

Simplify the following.

1. $x \cdot x \cdot x - (2x^2 + 4)$
2. $-a(a^2 + a - 5) + a \cdot a$
3. $-3y - 4y(2 + y)$
4. $2x + 4x(-3x + 1)$
5. $-(\theta + \eta - 2) + 2(\theta - \eta)$
6. $5x + 3(2 - x)$
7. $3 - 7[2x(x - 1)]$
8. $y^2 - [y \cdot y + 3(-2y + 2)]$
9. $3c[-(2a + 3) - 5c]$
10. $-5a - [3 + 4a(4 - 3a)]$
11. $5\mu + [7 - 2\mu(6 - 3\mu)]$
12. $-3\pi[-2(\pi + 3) - 6\pi]$
13. $-\theta - [3\theta - 2(\theta - 1)]$
14. $-5[4c(c \cdot c - 3) - 2c(c^2 + 1)]$
15. $3[-x^3 + 2x(x^2 + x - 3) - 4]$
16. $2[2t(t - 4) - 2(t^2 + 4t)]$
17. $-2a[-2(a - 2 + 3b) - 3b(a - 2)]$
18. $-3[-2x^2 - 2x - 2(-3x^2 - x)]$

5–6   DIVISION OF MONOMIALS

Dividing Powers of a Variable

In Section 5–1 we multiplied powers of a single variable by adding their exponents. Division of powers of a single variable is just as easy.

RULE 5–5.   *Dividing Powers of a Variable*

To divide powers of a variable, subtract the exponent in the denominator from the exponent in the numerator.

$$a^m/a^n = a^{m-n}$$

| EXAMPLE 5–17 | Solve x^5/x^2. |

Solution   Subtract denominator exponent from numerator exponent.

$$x^5/x^2 = x^{5-2} = x^3$$

$$\therefore \quad x^5/x^2 = x^3$$

Division by Zero

In this book we will not consider division by zero. **_Division by zero is not a permissible operation._** You may wish to try dividing by zero on your calculator.

Implied Exponent

When a variable is written without any exponent, the variable occurs only once as a factor. This means that the exponent is 1 ($x = x^1$). When the exponent of 1 is omitted, it is an **implied exponent.**

Zero Exponent

Any number (other than zero) divided by itself is equal to 1. For example, $3/3 = 1$, $a/a = 1$, $x^3/x^3 = 1$. If we applied Rule 5–5 to x^3/x^3, the result would be x^0. Since there can be only one answer, $x^0 = 1$. **_Any number (except zero) or any variable raised to the zero power is 1 by definition._**

Division of Monomials

In dividing monomials we apply the preceding concepts. In applying the following rule, we will treat each variable separately.

> ### RULE 5–6.   *Dividing One Monomial by Another*
>
> To divide one monomial by another:
>   1. Divide the numerical coefficients.
>   2. Divide the literal factors.
>   3. Form the quotient.

| EXAMPLE 5–18 | Simplify $9y/(3y)$. |

Solution   Follow the steps of Rule 5–6.

Step 1:   Divide numerical coefficients:

$$9/3 = 3$$

Step 2: Divide literal factors:

$$y/y = 1$$

Step 3: Form the quotient:

$$(3)(1) = 3$$

$$\therefore \quad 9y/(3y) = 3$$

EXAMPLE 5–19 Divide $15x^2y$ by $5xy$.

Solution

D: 5
$$\frac{15x^2y}{5xy} = \frac{3x^2y}{xy}$$

D: x
$$= 3xy/y$$

D: y
$$= 3x$$

$$\therefore \quad 15x^2y/(5xy) = 3x$$

EXERCISE 5–6

Find the quotients of the following.

1. $a^5 \div a$
2. $x^8 \div x^2$
3. $c^{10} \div c^8$
4. $-x^4 \div x^3$
5. $\mu^{10} \div \mu^3$
6. $\theta^{12} \div \theta^3$
7. $3y^8/y^8$
8. $12t^4/(3t^2)$
9. $27y^3z^6/(-9yz)$
10. $\dfrac{-32a^5b^4}{4a^5b^2}$
11. $\dfrac{-8c^{10}d^3}{-4c^{12}d^3}$
12. $\dfrac{48\eta^5\lambda^7}{16\lambda^3}$
13. $\dfrac{4^2a^3b^5}{-2ab}$
14. $\dfrac{-27x^5y^2}{3^2x^3y^4}$
15. $\dfrac{-20c^2t}{-5t}$

5–7 DIVIDING A POLYNOMIAL BY A MONOMIAL

Division is the inverse of multiplication. Remember, to multiply a polynomial by a monomial, each term in the polynomial is multiplied by the monomial. Similarly, to divide a polynomial by a monomial, each term in the polynomial is divided by the monomial.

> **RULE 5–7. Dividing a Polynomial by a Monomial**
>
> To divide a polynomial by a monomial, divide each term of the polynomial by the monomial.

EXAMPLE 5–20   Divide $12x^3 + 8x^2 + 4x$ by $4x$.

Solution   Divide each term by $4x$.

$$\frac{12x^3 + 8x^2 + 4x}{4x} = \frac{12x^3}{4x} + \frac{8x^2}{4x} + \frac{4x}{4x}$$

$$= 3x^2 + 2x + 1$$

$$\therefore \quad (12x^3 + 8x^2 + 4x)/(4x) = 3x^2 + 2x + 1$$

EXAMPLE 5–21   Divide $5a^3b + 15a^2b^2 + 10ab^3$ by $5ab$.

Solution   Divide each term by $5ab$.

$$\frac{5a^3b + 15a^2b^2 + 10ab^3}{5ab}$$

$$= \frac{5a^3b}{5ab} + \frac{15a^2b^2}{5ab} + \frac{10ab^3}{5ab}$$

$$= a^2 + 3ab + 2b^2$$

$$\therefore \quad (5a^3b + 15a^2b^2 + 10ab^3)/(5ab) = a^2 + 3ab + 2b^2$$

EXERCISE 5–7

Divide the following.

1. $\dfrac{9x + 12}{3}$

2. $\dfrac{8c + 16}{8}$

3. $\dfrac{21 + 12y}{3}$

4. $\dfrac{10a^2 + 15}{5}$

5. $\dfrac{18U + 6V}{6}$

6. $\dfrac{25\theta^2 + 20}{5}$

7. $\dfrac{6b^3 + 12a^3}{3}$

8. $\dfrac{-18\alpha + 6}{6}$

9. $\dfrac{24x^2 + 12xy}{4x}$

10. $\dfrac{y^3 + y^2}{y^2}$

11. $\dfrac{a^5 + a^3}{a^2}$

12. $\dfrac{27x^3 + 21x^2 + 15x}{3x}$

13. $\dfrac{13c^3 + 9c^2 + 11c^4}{c^2}$

14. $\dfrac{2ax^2 + 4ax - 6a^2x}{2ax}$

15. $\dfrac{5x^2y + 10x^2y^2 + 15xy^3}{5xy}$

16. $\dfrac{3\psi^3\phi + 6\psi\phi^2 + 9\psi^2\phi^2}{3\psi\phi}$

17. $\dfrac{40a^2b^2 + 30a^2b + 20a^2b^2}{10a^2b}$

18. $\dfrac{16U^5 + 32U^4 + 8U^3}{-8U^3}$

19. $\dfrac{28\mu^7 + 16\mu^5 + 20\mu^3}{4\mu^5}$

20. $\dfrac{16y^2z^3 + 8y^3z^4 + 4yz^3}{2y^2z^3}$

114

5–8 FACTORING POLYNOMIALS WITH A COMMON MONOMIAL FACTORS

Factoring

From the distributive law we learned that $a(b + c) = ab + ac$. We know that $ab + ac$ and $a(b + c)$ are two algebraic expressions for the same number. The number expressed as a product is $a(b + c)$. This form is called the **factored form.** Similarly, $3(x + 1)$ is the factored form of $3x + 3$. The process of finding the factors of a sum is called **factoring.** Table 5–1 shows several examples of polynomials that have been factored.

TABLE 5–1   Examples of Factoring

Polynomial	Factors	Factored Form
$5a - 5$	5 and $a - 1$	$5(a - 1)$
$2x^2 - x$	x and $2x - 1$	$x(2x - 1)$
$8y^2 + 4y + 12$	4 and $2y^2 + y + 3$	$4(2y^2 + y + 3)$
$4xy + 8x^2y$	$4xy$ and $1 + 2x$	$4xy(1 + 2x)$

RULE 5–8.   *Factoring Polynomials Containing a Common Monomial Factor*

To factor a polynomial that contains a common monomial factor:

1. By inspection, determine the factors that are common to each term of the polynomial. Form the common factor.
2. Divide the polynomial by the common factor.
3. Write the factored expression as a product of the common factor and the quotient.

EXAMPLE 5–22   Factor $6x + 12$.

Solution   Each term of $6x + 12$ is divisible by both 3 and 6. In factoring, where there is more than one possible common factor, the largest common factor is selected. So 6 is the common monomial factor. The other factor is the quotient of $6x + 12$ and 6; that is, $(6x + 12)/6 = x + 2$. The factored form of $6x + 12$ is expressed as a product of 6 and $x + 2$, which is $6(x + 2)$.

∴ $6x + 12 = 6(x + 2)$

EXAMPLE 5–23   Factor $2a^4 + 6a^3 + 8a^2$.

Solution   Use the steps of Rule 5–8.

Step 1:   By inspection we see that 2 is the largest numerical factor that exactly divides 2, 6, and 8; and a^2 is the highest power of a that divides a^4, a^3, and a^2 exactly. So $2a^2$ is the common monomial factor of $2a^4 + 6a^3 + 8a^2$. It is important that both the numerical and the literal factors be found.

Step 2:   Divide by $2a^2$:

$$\frac{2a^4 + 6a^3 + 8a^2}{2a^2} = a^2 + 3a + 4$$

Step 3:   Write the factored expression as a product:

$$2a^2(a^2 + 3a + 4)$$

$$\therefore \quad 2a^4 + 6a^3 + 8a^2 = 2a^2(a^2 + 3a + 4)$$

EXERCISE 5–8

Write the following polynomials in factored form. Check your solutions by multiplying the factors.

1. $6a + 12$ **2.** $8\mu + 4$ **3.** $6 + 9b$

4. $5x + 25$ **5.** $3m + 3$ **6.** $ax + ay$

7. $\omega u + \omega v$ **8.** $6x + 6y$ **9.** $7\phi + 7\theta$

10. $3c^2 + 7c$ **11.** $\rho^2 + 5\rho$ **12.** $4x^2 + 5x$

13. $7xy + 21y$ **14.** $10h^2 + 5h^3$ **15.** $8a^3 + 16a^2$

16. $7st^3 + 5s^2t$ **17.** $ax + 3a$ **18.** $2\pi r^2 + 2\pi rh$

19. $a^2b^3c + a^4b^5c^3$ **20.** $4d + 6 + 10d^2$ **21.** $5\alpha^2 + 15\alpha + 20$

22. $6 + 3\lambda + 18\lambda^2$ **23.** $ab + a^2b^2 + a^3b^3$

24. $U^4V^2 + U^3V^3 + U^2V^4$ **25.** $16\beta^2\pi^2 + 8\beta^2\pi + 32\beta^2$

26. $42x^2y + 35x^2 + 14x^2y^2$ **27.** $30\eta^2\alpha + 15\eta\alpha - 25\eta\alpha^2$

28. $12a^3b^3 + 6a^2b^3 + 18ab^3$ **29.** $5y^3z^2 + 15y^2z + 10y^4z$

30. $18\theta^3\phi^3 + 12\theta^2\phi^4 + 27\theta\phi^5$ **31.** $21y^4x + 18y^3x^2 + 27xy^3$

5–9   EVALUATING ALGEBRAIC EXPRESSIONS

The process of *finding the value* of an algebraic expression is called **evaluating the expression.** This process is carried out by substituting numbers for the variables in the expression, and then simplifying the result to a single number.

| **EXAMPLE 5–24** | Find the value of $2a^2b$ when $a = 5$ and $b = 3$. |

Solution   Substitute the numbers for the variables and simplify:

$$2a^2b = 2(5)^23 = 2(25)3 = 150$$

∴ $2a^2b = 150$ when $a = 5$ and $b = 3$

| **EXAMPLE 5–25** | Find the value of $\dfrac{x^2 - \sqrt{4y}}{8}$ when $x = 4$ and $y = 16$. |

Solution   Substitute:

$$\frac{x^2 - \sqrt{4y}}{8} \Rightarrow \frac{4^2 - \sqrt{4(16)}}{8}$$

$$= \frac{16 - \sqrt{64}}{8} = \frac{16 - 8}{8} = \frac{8}{8} = 1$$

∴ $\dfrac{x^2 - \sqrt{4y}}{8} = 1$ when $x = 4$ and $y = 16$

| **EXAMPLE 5–26** | Evaluate $\dfrac{3x^3 - (2xy + y^2)}{2x - 7}$ when $x = 2$ and $y = 3$. |

Solution   Substitute:

$$\frac{3x^3 - (2xy + y^2)}{2x - 7} \Rightarrow \frac{3(2)^3 - [(2)(2)(3) + (3)^2]}{2(2) - 7}$$

$$= \frac{3(8) - (12 + 9)}{4 - 7} = \frac{24 - 21}{-3} = \frac{3}{-3} = -1$$

∴ $\dfrac{3x^3 - (2xy + y^2)}{2x - 7} = -1$ when $x = 2$ and $y = 3$

EXERCISE 5–9

Calculator Drill

Evaluate the following expressions for the given values. Express your answers in decimal notation, and round the answer to three significant figures.

If $x = 3$ and $y = 4$, find the value of the following.

1. xy

2. $2x - y$

3. $\dfrac{3y + x}{3}$

4. $x^2 - \sqrt{y}$

5. $5y^2$

6. $y^3 - 10$

7. $-2(y - x)^2$

8. $3x^2 + 5y$

9. $2x^2 - xy$

If $a = 2$, $b = 3$, and $c = 4$, evaluate the following.

10. $2a^2 + 5$

11. $a^3 - b^2$

12. $a^2 + b^2 - c^2$

13. $25 - abc$

14. $c^2 - ab$

15. $(12 - 2c)(2c - b)$

16. $(c - 2)(bc - a)$

17. $(14 - 2c)/b$

18. $c^2 - (ab) + 3$

19. $\dfrac{\sqrt{2c - 2} - (b + c)}{a^2 + cb}$

20. $\dfrac{5(2c - b)}{b(c - b)} - \dfrac{7(b + 1)}{2c + b}$

21. $b^2 - 5/a^2$

If $x = 3.72$ and $y = 1.15$, evaluate the following.

22. $(3xy^2 - 2xy) - (9xy - 3xy^2)$

23. $4x(3x^2 - 2y)$

24. $(6x^2 - 4\sqrt{y} + 8y) \div 2x$

25. $(10x^2 + y - \sqrt{5}) \div (2x - 3y)$

SELECTED TERMS

distributive law   When multiplying a monomial and a polynomial, each term in the polynomial is multiplied by the monomial.

evaluating an expression   Substituting numbers for variables in an algebraic expression.

factoring   The process of finding factors of a sum.

implied exponent   Any number or variable without an exponent has an *implied exponent* of 1.

6

Solving Equations

6-1   **Equations**

6-2   **Finding the Root of an Equation**

6-3   **Using Addition to Transform Equations**

6-4   **Using Multiplication to Transform Equations**

6-5   **Additional Techniques**

6-6   **Equations Containing Parentheses**

6-7   **Solving Formulas**

6-8   **Evaluating Formulas**

6-9   **Forming Equations**

6-10   **Solving Word Problems**

PERFORMANCE OBJECTIVES

- **Utilize the addition, subtraction, multiplication, and division axioms to solve equations.**

- **Employ factoring in the process of solving an equation.**

- **Use the distributive law in the solution of an equation.**

- **Check the solution to an equation.**

- **Evaluate formulas for the desired quantity and represent the result with an SI unit and unit prefix.**

- **Systematically use symbols to solve word problems.**

The formation of the world's first all-nuclear powered surface task force is celebrated by the crew of the USS Enterprise CVAN-65 by spelling out Einstein's equation $E = mc^2$. (Courtesy of U.S. Navy Office of Information)

This chapter will be limited to fairly elementary equations so that you may gain a solid understanding of how to solve algebraic equations by concentrating on the algebraic principles.

6–1 EQUATIONS

A sentence saying that one general number is equal to a second general number is called an **equation.** The expressions that are joined by the equal sign ($=$) to form an equation are called **members** ("sides") **of the equation.** Thus, in the equation $x - 3 = 4$, the left member is "$x - 3$" and "4" is the right member. Equations may be read left to right or right to left. Figure 6–1 shows the structure of an equation.

Equations are divided into two types: identical equations and conditional equations. An **identical equation** is true for all values of the variable contained in the equation. Thus, the identical equation $2(x + 1) = 2x + 2$ is true for any value of x. Identical equations are also called *identities*. A **conditional equation** is only true for particular values of the variable contained in the equation. Thus, the conditional equation $x + 2 = 5$ is true only when $x = 3$. That is, $3 + 2 = 5$. We will be concerned with conditional equations in this chapter. We will use the word *equation* to mean conditional equation.

EXERCISE 6–1

Study each of the following statements to determine if it is true or false.

1. $3 + 2 = 4 + 1$
2. $6 - 2 = 2 + 2$
3. $6(2 + 1) = 15$
4. $8 + 2 \neq 12 - 3$
5. $5^2 = 29 - 4$
6. $9 + 3 \neq 15 - 3$
7. $8(4) = 42 - 10$
8. $3 - 2 > 8 - 11$
9. $-5 - 4 < -7 + 6$
10. $7(2 - 4) = 7(2) - 4(7)$
11. $6(-3) + 4 > 7/(-7) - 14$
12. $(2 \times 3) \times 5 < 3 \times (5 \times 2)$

FIGURE 6–1 Structure of an equation.

In each of the following equations, a value for the variable has been given to the right. Determine if the value given for the variable makes the equation true.

13. $x + 3 = 11$, $x = 8$ **14.** $I - 4 = 7$, $I = 3$

15. $-5\mu = -15$, $\mu = -3$ **16.** $a^2 = 6$, $a = 3$

17. $7 - \beta = 7$, $\beta = 0$ **18.** $-R + 5 = 3^2$, $R = -4$

19. $y^3 + 4 - 6 = 12$, $y = 2$ **20.** $x^{0.5} - 4(2) = -15 + 9$, $x = 4$

6–2 FINDING THE ROOT OF AN EQUATION

The value of the variable that makes the equation true is called the **solution** of the equation or the **root of the equation.** The root is a number that makes the equation a true statement. The equation is said to be *satisfied* by the root of the equation.

EXERCISE 6–2

Select the root from the values at the right of each equation.

1. $x + 3 = 4$, $\{-2, 1, 5\}$

2. $2x + 1 = 5$, $\{1, 2, 3, 4\}$

3. $m + 2 = -6$, $\{-8, -6, -4, 4\}$

4. $2\lambda = -12$, $\{-8, -6, -4\}$

5. $x + 1 = 0$, $\{-1, 0, 1, 2\}$

6. $-5y = -13 + 3$, $\{-3, -2, 2, 3\}$

7. $\alpha - 3 = 6$, $\{5, 7, 9, 11\}$

8. $3(x + 2) = 15$, $\{1, 2, 3\}$

9. $-2(\phi - 4) = 0$, $\{3, 4, 5\}$

10. $-(x + 2) = 1$, $\{-1, -2, -3\}$

6–3 USING ADDITION TO TRANSFORM EQUATIONS

In solving an equation it is usual to isolate the variable in the left member (left side) of the equation. To do this, an *equivalent* equation is formed by *transforming* the original equation. Two equations are equivalent if they have the same roots.

Among the common operations used to transform an equation are:

- Addition (or subtraction) of the same number to (or from) both members of the equation.

- Multiplication (or division) of members of the equation by the same number.

EXAMPLE 6–1	Solve $x - 3 = 5$.
Solution	Add 3 to both members:
A: 3	$x - 3 + 3 = 5 + 3$
	$x + 0 = 8$
	$x = 8$
Check	Is 8 the root of the original equation? Let's check by substituting 8 for x in $x - 3 = 5$.
	$x - 3 = 5$
	$8 - 3 = 5$
	$5 = 5$ Yes!
$\therefore$	$x = 8$ is the root of $x - 3 = 5$.

In the preceding example, 3 was added to both members of the equation so that -3 would be eliminated from the left member. This, then, leaves the *unknown isolated* in the left member. The reason a 3 was added is that it nullifies (makes zero) -3. Because of this property of addition, the following rule is very useful in solving equations.

RULE 6–1.   Addition Axiom

The same number added to both members *(sides)* of an equation results in an equivalent equation.

EXAMPLE 6–2	Solve $y - 5 = 2$.
Solution	Add 5 to both members:
A: 5	$y - 5 + 5 = 2 + 5$
	$y + 0 = 7$
	$y = 7$
Check	Does $y - 5 = 2$ when $y = 7$?
	$7 - 5 = 2$
	$2 = 2$ Yes!
$\therefore$	$y = 7$ is the root.

Checking the Solution

In the preceding examples, we checked the answer by substituting the computed root into the original equation. This is routinely done to catch numerical mistakes that might occur in transforming the equation. It is important that you develop a habit of checking the solution.

Subtraction Axiom

In solving equations, sometimes it is necessary to nullify a positive number. This can be achieved by applying the following rule to subtract the positive number from both members of the equation.

RULE 6–2.   Subtraction Axiom

The same number subtracted from both members (*sides*) of an equation results in an equivalent equation.

EXAMPLE 6–3	Solve $x + 7 = 14$.
Solution	Subtract 7 from both members:
S: 7	$x + 7 - 7 = 14 - 7$
	$x + 0 = 7$
	$x = 7$
Check	Does $x + 7 = 14$ when $x = 7$?
	$7 + 7 = 14$
	$14 = 14$ Yes!
$\therefore$	$x = 7$ is the root.

EXERCISE 6–3

Solve each of the following equations using the addition or subtraction axiom. Check the root (solution) by substituting into the original equation.

1. $x + 6 = 8$ **2.** $x + 3 = 9$ **3.** $x - 5 = 5$

4. $\lambda - 7 = 8$ **5.** $y + 2 = 10$ **6.** $h + 3 = -15$

7. $\theta - 1 = 4$ **8.** $m - 7 = 18$ **9.** $y + 10 = 13$

10. $x + (-3) = 5$ **11.** $n + (-4) = -5$ **12.** $-13 + \beta = 24$

13. $-18 + C = -14$ **14.** $62 + K = 62$ **15.** $\tau + 17 = 0$

16. $K - 4 = -7$ **17.** $22 + m = 9$ **18.** $26 + \omega = -8$

19. $x + 11 = 11$ **20.** $y - 2 = -5$ **21.** $-45 + t = -25$

6–4 USING MULTIPLICATION TO TRANSFORM EQUATIONS

To solve an equation, the variable must have a coefficient of 1. A fractional coefficient can be transformed into 1 by multiplying both members of the equation by the reciprocal of the fraction.

> ### RULE 6–3. Multiplication Axiom
>
> Multiplying both members *(sides)* of an equation by the same ***nonzero*** number results in an equivalent equation.

EXAMPLE 6–4	Solve $\frac{1}{3}Y = 5$.
Solution	Multiply each member by 3, the reciprocal of $\frac{1}{3}$.
M: 3	$3(\frac{1}{3}Y) = 3(5)$
	$1Y = 15$
	$Y = 15$
Check	Does $\frac{1}{3}Y = 5$ when $Y = 15$?
	$(\frac{1}{3})15 = 5$
	$5 = 5$ Yes!
$\therefore$	$Y = 15$ is the root.

EXAMPLE 6–5	Solve $\dfrac{1}{-7}x = -4$.
Solution	Multiply each member by -7, the reciprocal of $1/{-7}$:
M: -7	$-7\dfrac{1}{-7}x = -7(-4)$
	$1x = 28$
	$x = 28$
Check	Does $\dfrac{1}{-7}x = -4$ when $x = 28$?
	$\dfrac{1}{-7}(28) = -4$
	$-4 = -4$ Yes!
$\therefore$	$x = 28$ is the root.

Division Axiom

An equation with a coefficient of the unknown greater than 1 may be transformed to an equation with a coefficient equal to 1 by dividing both members by the coefficient of the unknown.

> ### RULE 6–4. Division Axiom
> Dividing both members *(sides)* of an equation by the same *nonzero* number results in an equivalent equation.

EXAMPLE 6–6	Solve $5x = 15$.
Solution	Divide each member by 5, the coefficient of *x:*
D: 5	$5x/5 = 15/5$
	$x = 3$
Check	Does $5x = 15$ when $x = 3$?
	$5(3) = 15$
	$15 = 15$ Yes!
$\therefore$	$x = 3$ is the root.

EXAMPLE 6–7	Solve $-2x = 14$.
Solution	Divide each member by -2, the coefficient of *x:*
D: -2	$-2x/-2 = 14/-2$
	$x = -7$
Check	Does $-2x = 14$ when $x = -7$?
	$-2(-7) = 14$
	$14 = 14$ Yes!
$\therefore$	$x = -7$ is the root.

EXERCISE 6–4

Apply the multiplication or division axiom to each of the following equations. Check the solutions.

1. $3y = 12$ **2.** $9m = 72$ **3.** $8\mu = 48$

4. $5x = 40$ **5.** $x/3 = 5$ **6.** $h/4 = 9$

7. $y/6 = 10$ **8.** $\psi/8 = 3$ **9.** $6K = -36$

10. $-5t = -15$ **11.** $-4P = 28$ **12.** $-3d = 0$

13. $y/2 = 48$ **14.** $18y = -72$ **15.** $-16\beta = 64$

16. $-2x = 32$ **17.** $x/-9 = 7$ **18.** $K/3 = -13$

19. $-7\phi = -98$ **20.** $13x = 52$ **21.** $5r = -10$

22. $-17x = -85$ **23.** $-y/13 = -4$ **24.** $19w = -57$

Apply the addition or subtraction axiom and then the multiplication or division axiom to each of the following equations. Check the solutions.

25. $x/2 + 6 = 8$ **26.** $3y - 5 = 10$

27. $3 - 4k = 19$ **28.** $28 - 9W = -8$

29. $3y - 6 = -15$ **30.** $h/5 - 12 = -14$

6–5 ADDITIONAL TECHNIQUES

In solving a given equation, several techniques may be needed to transform the equation into the solution. We will explore several additional techniques in this section.

Combining Like Terms in Equations

Complicated looking equations can often be made to look simpler by combining like terms. Like terms are always combined before isolating the unknown.

EXAMPLE 6–8   Solve $8x - 5 + 3x = 12 + 16$.

Solution   Combine like terms in each member:

$$8x - 5 + 3x = 12 + 16$$
$$11x - 5 = 28$$

A: 5 $11x - 5 + 5 = 28 + 5$
$$11x = 33$$

D: 11 $11x/11 = 33/11$
$$x = 3$$

Check   Does $8x - 5 + 3x = 12 + 16$ when $x = 3$?

$$8(3) - 5 + 3(3) = 12 + 16$$
$$24 - 5 + 9 = 12 + 16$$
$$28 = 28 \text{ Yes!}$$

$\therefore$ $x = 3$ is the solution.

EXERCISE 6–5

Solve each equation.

1. $2x + 7x = 18$ **2.** $4x - 2x = 12$

3. $5x + 4x = 35 - 8$ **4.** $5I + 2I - 3I = 22 - 6$

5. $R + 2R - R = 1$ **6.** $7E - 5E + E = 45$

7. $9\phi - 2\phi + 6\phi = 39$ **8.** $3y + 4y + 7y = 70$

9. $0.5m + 0.5m = 1$ **10.** $3\omega + 4\omega = 7 + 35$

Solving Equations with the Unknown in the Right Member

Since the unknown is conventionally isolated in the left member, it then becomes time-consuming to manipulate an equation that has the unknown in the right member. To counter this situation, we may apply Rule 6–5, the **symmetric property of equality.**

> **RULE 6–5.** *Symmetric Property of Equality*
>
> If the members *(sides)* of an equation are interchanged, the resulting equation is equivalent to the original equation. *Example:* If $A = B$, then $B = A$.

EXAMPLE 6–9   Solve $9 = 3y$.

Solution   Use Rule 6–5 to *interchange the members:*

$$3y = 9$$

D: 3 $3y/3 = 9/3$

$$y = 3$$

Check   Does $9 = 3y$ when $y = 3$?

$$9 = 3(3)$$
$$9 = 9 \text{ Yes!}$$

$\therefore$ $y = 3$ is the solution.

EXAMPLE 6–10   Solve $9 + 8 = 4x - x + 5$.

Solution   Combine like terms in each member:

$$17 = 3x + 5$$

S: 5 $17 - 5 = 3x + 5 - 5$

$$12 = 3x$$

Interchange members:

$$3x = 12$$

D: 3 $3x/3 = 12/3$

$$x = 4$$

Check   Does $9 + 8 = 4x - x + 5$ when $x = 4$?

$$9 + 8 = 4(4) - (4) + 5$$
$$9 + 8 = 16 - 4 + 5$$
$$17 = 17 \text{ Yes!}$$

$\therefore$ $x = 4$ is the solution.

EXERCISE 6–6

Solve each equation.

1. $16 = 4y$
2. $-49 = 7x$
3. $3 = \phi - 3 + 5$
4. $-2 = -7 + \alpha - 2\alpha$
5. $4 = -5 - 3R$
6. $-6 = \frac{1}{2}I$
7. $-3 = \frac{1}{4}E$
8. $26 = -13h$
9. $-7 = 2\mu + 3$
10. $13 = 5y - 2 - 2y$
11. $3 + 12 = 6t - 2t - 1$
12. $25 - 13 = 4W + 2 + W$
13. $13 - 3 = 3 + 7u - 5 - 3u$
14. $29 - 6 = 6 - 2v + 2 + 7v$
15. $-42 + 35 = 4\eta - 16 + 3\eta + 9$

Solving Equations with the Unknown in Both Members

So far, the unknown has been in only one member of the equation. We will now work with equations in which the unknown appears in both members.

EXAMPLE 6–11   Solve $5x = 2 + 3x$.

Solution   Collect all the terms containing x in the left member:

S: $3x$ $5x - 3x = 2 + 3x - 3x$

Combine like terms in each member:

$$2x = 2$$
D: 2 $$2x/2 = 2/2$$
$$x = 1$$

Check   Does $5x = 2 + 3x$ when $x = 1$?

$$5(1) = 2 + 3(1)$$
$$5 = 2 + 3$$
$$5 = 5 \text{ Yes!}$$

$\therefore$ $x = 1$ is the solution.

EXAMPLE 6–12   Solve $-7x = 24 - x$.

Solution   Collect all terms containing x in the left member:

A: x $-7x + x = 24 - x + x$

Combine like terms in each member:

$$-6x = 24$$
D: -6 $-6x/-6 = 24/-6$
$$x = -4$$

Check   Does $-7x = 24 - x$ when $x = -4$?

$$-7(-4) = 24 - (-4)$$
$$28 = 24 + 4$$
$$28 = 28 \text{ Yes!}$$

$\therefore$ $x = -4$ is the solution.

EXERCISE 6–7

Solve each equation.

1. $3\Delta = \Delta + 4$ **2.** $6N = N + 4$ **3.** $6x = 20 - 4x$
4. $-10\beta = 24 + \beta$ **5.** $-3y = 15 - 2y$ **6.** $2R - 10 = 12R$
7. $5\pi + 13 = -8\pi$ **8.** $-20 - 6m = 4m$ **9.** $2\theta = -7\theta + 18$
10. $-8 - 5I = 3I$ **11.** $5\eta - 1 = 2\eta + 14$ **12.** $E - 6 = 10 - 3E$
13. $9y - 16 + 6y = 11 + 4y - 5$ **14.** $3a + 5 - 8a = 7 - 9a - 12$
15. $4z + 14 + 2z = 12 - 3z - 8$ **16.** $21 - 3C - 7 = 6C - 2 - C$

6–6   EQUATIONS CONTAINING PARENTHESES

To solve an equation containing parentheses, we must first remove the parentheses. This may require the application of the distributive law of multiplication.

EXAMPLE 6–13   Solve $5(x + 2) = 20$.

Solution   Distribute 5 over $(x + 2)$:

$$5x + 10 = 20$$
S: 10 $5x + 10 - 10 = 20 - 10$
$$5x = 10$$
D: 5 $5x/5 = 10/5$
$$x = 2$$

Check	Does $5(x + 2) = 20$ when $x = 2$?
	$5(2 + 2) = 20$
	$5(4) = 20$
	$20 = 20$ Yes!
$\therefore$	$x = 2$ is the solution.

EXAMPLE 6–14   Solve $10i - 2(i - 8) = (5i + 4)$.

Solution   Distribute -2 over $(i - 8)$ in the left member:

$$10i - 2i + 16 = (5i + 4)$$

Remove the ( ) in the right member:

$$10i - 2i + 16 = 5i + 4$$

Combine like terms:

$$8i + 16 = 5i + 4$$

S: $5i$ $8i + 16 - 5i = 5i + 4 - 5i$

$$3i + 16 = 4$$

S: 16 $3i + 16 - 16 = 4 - 16$

$$3i = -12$$

D: 3 $3i/3 = -12/3$

$$i = -4$$

Check   Does $10i - 2(i - 8) = (5i + 4)$ when $i = -4$?

$$10(-4) - 2(-4 - 8) = (5(-4) + 4)$$
$$-40 - 2(-12) = (-20 + 4)$$
$$-40 + 24 = -16$$
$$-16 = -16 \text{ Yes!}$$

$\therefore$ $i = -4$ is the solution.

EXAMPLE 6–15   Solve $7R - (3 - 2R) + 6 = 21 + 3(R - 4)$.

Solution   Distribute -1 over $(3 - 2R)$ in the left member and distribute 3 over $(R - 4)$ in the right member:

$$7R - 3 + 2R + 6 = 21 + 3R - 12$$

Combine like terms:

$$9R + 3 = 9 + 3R$$

S: 3 $9R + 3 - 3 = 9 + 3R - 3$

$$9R = 6 + 3R$$

S: $3R$ $9R - 3R = 6 + 3R - 3R$

$$6R = 6$$

D: 6 $6R/6 = 6/6$

$$R = 1$$

Check   Does $7R - (3 - 2R) + 6 = 21 + 3(R - 4)$ when $R = 1$?

$$7(1) - (3 - 2 \cdot 1) + 6 = 21 + 3(1 - 4)$$
$$7 - (1) + 6 = 21 + 3(-3)$$
$$12 = 21 - 9$$
$$12 = 12 \text{ Yes!}$$

$\therefore$ $R = 1$ is the solution.

EXERCISE 6–8

Solve each equation.

1. $2(i - 3) = 16$
2. $-3(5 + e) = 3$
3. $4 + 5(\beta - 1) = -6$
4. $3(Z + 6) = 21$
5. $15 - (2x + 10) = -1$
6. $4(-2 + 3\lambda) = -20$
7. $-6 = 3(8 - 2R_1)$
8. $E_1 = 5(E_1 - 4)$
9. $-(-2W + 6) = 3W - 18$
10. $4Q_1 + 6 = -2(Q_1 + 3)$
11. $-(-2L + 13) + 2L = 59$
12. $6\phi + 15 = 7(\phi - 2)$
13. $-4(3 + R_1) + 5(R_1 + 4) = 0$
14. $(7 - K)3 = (-8 + 3K)$
15. $(3\theta - 3) = (\theta + 5)$
16. $9I_1 - (6I_1 - 2) = 8$
17. $13N - (3 + 12N) = -3$
18. $7(C' - 5) = 6 - (C' + 1)$
19. $6V_t - (V_t - 7) = (V_t + 15)$
20. $5A_v - (1 - A_v) + 4 = 9$
21. $5(\alpha + 2) - 4(\alpha + 1) - 3 = 0$
22. $12g_m - (3 - g_m) = 7(5g_m - 1) - 18g_m$
23. $3(8X_C - 2) - 3(1 - X_C) + 8 = 8$
24. $11 + 4(\sigma - 17) = 21(\sigma - 2) - 5(3 - \sigma)$

6–7 SOLVING FORMULAS

In electronics, as in all sciences, there are many laws. These laws are often expressed as equations. This type of equation is called a **formula.** It is through formulas that mathematics is applied to electronics, because formulas express relationships between physical quantities in mathematical language.

Ohm's law is fundamental to all electronics. It describes the relationship among voltage, current, and resistance. As a formula, Ohm's law is written $E = IR$.

In applying formulas, we must sometimes transform the stated formula into an equivalent formula so that a particular quantity may be found. This concept is demonstrated in the following examples.

EXAMPLE 6–16 Solve $I = E/R$ for R.

$\quad$ **M:** R $\qquad$ $IR = ER/R$
$\qquad\qquad\qquad\quad$ $IR = E$
$\quad$ **D:** I $\qquad$ $IR/I = E/I$
$\quad$ $\therefore$ $\qquad\qquad$ $R = E/I$

EXAMPLE 6–17 Solve for R in the formula $P = I^2R$. Express the answer in terms of P and I^2 as a new formula.

$\quad$ **Solution** $\qquad$ $P = I^2R$

$\quad$ **D:** I^2 $\qquad$ $\dfrac{P}{I^2} = \dfrac{I^2R}{I^2}$

$\qquad\qquad\qquad$ $P/I^2 = R$

Interchange members:

$\quad$ $\therefore$ $\qquad\qquad$ $R = P/I^2$

We see in Example 6–17 that the formula $P = I^2R$ was transformed into $R = P/I^2$ using the rules previously learned in this chapter.

EXAMPLE 6–18 Solve $E = IR_1 + IR_2$ for I.

$\quad$ **Solution** $\qquad$ Factor I from the right member:

$\qquad\qquad\qquad\qquad$ $E = I(R_1 + R_2)$
$\quad$ **D:** $R_1 + R_2$ $\quad$ $E/(R_1 + R_2) = I(R_1 + R_2)/(R_1 + R_2)$
$\qquad\qquad\qquad\qquad$ $E/(R_1 + R_2) = I$

Interchange members:

$\quad$ $\therefore$ $\qquad\qquad$ $I = \dfrac{E}{R_1 + R_2}$

EXAMPLE 6–19 Solve $\alpha\beta + \alpha = \beta$ for β.

$\quad$ **Solution**

$\quad$ **S:** α $\qquad$ $\alpha\beta + \alpha - \alpha = \beta - \alpha$
$\qquad\qquad\qquad\quad$ $\alpha\beta = \beta - \alpha$
$\quad$ **S:** β $\qquad$ $\alpha\beta - \beta = \beta - \alpha - \beta$
$\qquad\qquad\qquad\quad$ $\alpha\beta - \beta = -\alpha$
$\qquad\qquad\qquad$ $-\beta + \alpha\beta = -\alpha$

Change the sign of each term by multiplying each term in both sides of the equation by -1:

M: -1 $-1(-\beta) - 1(\alpha\beta) = -1(-\alpha)$

$$\beta - \alpha\beta = \alpha$$

Factor the left member:

$$\beta(1 - \alpha) = \alpha$$

D: $(1 - \alpha)$ $\beta(1 - \alpha)/(1 - \alpha) = \alpha/(1 - \alpha)$

$\therefore$ $$\beta = \frac{\alpha}{1 - \alpha}$$

EXAMPLE 6–20   Solve $I_C = \beta I_B + I_{CBO}(\beta + 1)$ for β.

Observation   This is an actual formula used with transistors. There are three currents denoted by I_C, I_B, and I_{CBO}. Even though the third subscript is composed of three letters, it is only a subscript.

Solution   Distribute I_{CBO} over $(\beta + 1)$ in the right member:

S: I_{CBO} $$I_C = \beta I_B + \beta I_{CBO} + \beta I_{CBO}$$
$$I_C - I_{CBO} = \beta I_B + \beta I_{CBO} + I_{CBO} - I_{CBO}$$
$$I_C - I_{CBO} = \beta I_B + \beta I_{CBO}$$

Factor the right member:

$$I_C - I_{CBO} = \beta(I_B + I_{CBO})$$

D: $I_B + I_{CBO}$ $$\frac{I_C - I_{CBO}}{I_B + I_{CBO}} = \beta\frac{I_B + I_{CBO}}{I_B + I_{CBO}}$$

$$\frac{I_C - I_{CBO}}{I_B + I_{CBO}} = \beta$$

Interchange members:

$\therefore$ $$\beta = \frac{I_C - I_{CBO}}{I_B + I_{CBO}}$$

EXERCISE 6–9

Solve each formula for the indicated variable in terms of the other variables:

Given:	Solve for:
1. $v = f\lambda$	λ
2. $C = Q/V$	V
3. $E = IR$	R
4. $R = E^2/P$	E^2

Given:	Solve for:

5. $r_p = \dfrac{\mu}{g_m}$ g_m

6. $I_C = \beta I_B$ β

7. $Q_0 = \dfrac{BW}{f_0}$ f_0

8. $V_{CC} = I_C R_1 + V_{CE}$ R_1

9. $E_{BB} - E_1 = I_B E_p$ E_1

10. $E_1(R_1 + R_2) = E_T R_1$ E_1

11. $R_O I_L = E_N - E_L$ E_L

12. $X_C = \dfrac{1}{\omega C}$ C

13. $-x = t/RC$ R

14. $Q = \omega_1 L/R$ ω_1

15. $I_2 R_1 = I_1 G - I_2 R_2$ I_2

16. $R = \rho L/A$ ρ

17. $T_J - T_A = P_D \theta_{JA}$ θ_{JA}

18. $cd = 0.2KA(N - 1)$ N

19. $I_1 = \dfrac{E_T - E_1}{R}$ E_1

20. $E_f C_1 + E_f C_2 = E_1 C_1 - E_2 C_2$ C_2

21. $V_1(Y_1 Y_2 - Y^2) = iY_2$ Y_2

22. $R_s Q^2 = R_p - R_s$ R_s

23. $I_T R_{th} = E_{th} - I_T R_L$ I_T

24. $I_E = I_B + \beta I_B$ I_B

25. $A_v R_L + A_v r_p = \mu R_L$ R_L

26. $R_2 = \dfrac{C_3 P_o}{D_2}$ D_2

27. $BU = \dfrac{Y2K}{G}$ Y

28. $\dfrac{HA}{20} = \dfrac{450}{L}$ L

6–8 EVALUATING FORMULAS

When substituting numerical quantities into a formula, the units representing the physical quantity *cannot* be separated from the number, nor can the unit size be neglected. When working with formulas, the usual procedure is to substitute the quantity in base SI units and not in prefixed units. For example, when substituting into the formula for electrical power, $P = EI$, substitute voltage (E) in volts, current (I) in amperes, and calculate power (P) in watts. This is accomplished by converting the prefix into a power of ten.

EXAMPLE 6–21   Compute the power dissipation of an industrial furnace that requires 60.0 A of current at a voltage of 0.480 kV. Use the formula $P = IE$.

Solution   Where needed convert to base units:

$$P = IE$$
$$E = 0.480 \text{ kV} = 0.480 \times 10^3 \text{ V}$$
$$I = 60.0 \text{ A}$$
$$P = 0.480 \times 10^3 \times 60.0$$
$$P = 28\ 800 \text{ W}$$
$$\therefore \quad P = 28.8 \text{ kW}$$

The preceding example shows that unit size must be accounted for in solving a formula. As you become more familiar with the physical quantities used in electronic technology, you may be able to take shortcuts. For now, it is best to convert the unit prefixes into powers of ten before substituting into the formula.

EXAMPLE 6–22   The equation for inductive reactance is $X_L = 2\pi fL$. Compute X_L in ohms when $f = 2.35$ MHz and $L = 100\ \mu$H.

Solution   Where needed convert to base units:

$$X_L = 2\pi fL$$
$$f = 2.35 \text{ MHz} = 2.35 \times 10^6 \text{ Hz}$$
$$L = 100\ \mu\text{H} = 100 \times 10^{-6} \text{ H}$$
$$X_L = 2\pi \times 2.35 \times 10^6 \times 100 \times 10^{-6}$$
$$X_L = 1.48 \times 10^3\ \Omega$$
$$\therefore \quad X_L = 1.48 \text{ k}\Omega$$

Although it is conventional to substitute base units into a formula, some formulas have been derived for special cases. When this is done, a statement is made noting the change from base units to prefix units. Example 6–23 explores such a situation.

EXAMPLE 6–23   A leading semiconductor manufacturer provided a data sheet for the integrated operational amplifier in Figure 6–2. This data sheet presents two special formulas for determining satisfactory values for resistance R_1 (Ω) and for capacitance C_1 (μF).

$$R_1 = 5R_f \ \Omega \qquad C_1 = 0.04/R_f \ \mu\text{F}$$

In these special formulas, R_f is specified in *kilohms* (kΩ), not in ohms, and C_1 is specified in *microfarads*. Determine R_1 and C_1 when R_f is 10 kΩ.

FIGURE 6–2   Schematic for the integrated operational amplifier of Example 6–23. The triangular symbol represents the amplifier.

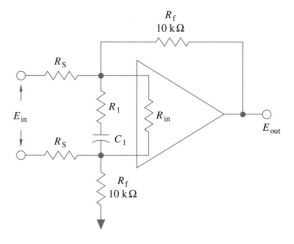

Solution   Summarize the information given:

R_f is in **kilohms** (kΩ). R_f has been selected as 10 kΩ.
R_1 is in ohms (Ω). $R_1 = 5R_f \ \Omega$.
C_1 is in **microfarads** (μF) $C_1 = 0.04/R_f \ \mu\text{F}$.

Determine the values for R_1 and C_1 when R_f is 10 kΩ:

$$R_1 = 5R_f \ \Omega \qquad C_1 = \frac{0.04}{R_f} \ \mu\text{F}$$

Substitute 10 for R_f (not 10 k) in each equation:

$$R_1 = 5 \times 10 \qquad C_1 = \frac{0.04}{10}$$

$$R_1 = 50 \ \Omega \qquad C_1 = 0.004 \ \mu\text{F}$$

$\therefore$ $R_1 = 50 \ \Omega$ and $C_1 = 0.004 \ \mu\text{F} \Rightarrow 4$ nF.

EXERCISE 6–10

Evaluate each of the following formulas for the indicated quantity by first transforming the given equation as needed and then substituting the given values into the formula. With your calculator, compute the answer and express the results to an appropriate number of significant figures with an SI unit and unit prefix (where needed). Round where necessary and express the decimal coefficient of the solution in the range of 0.1 to 1000.

Formula	Solve for:	Given:	Answer Base Unit
1. $E = IR$	I	$E = 52.3$ V $R = 0.470$ kΩ	A
2. $P = I^2R$	R	$P = 52.0$ W $I = 22.0$ mA	Ω
3. $X_C = \dfrac{1}{2\pi fC}$	X_C	$f = 10.0$ kHz $C = 0.100\ \mu$F	Ω
4. $C = \pi d$	d	$C = 3.82$ m	m
5. $T_C = T_K - 273.15$	T_C	$T_K = 872$ K	°C
6. $P = W/t$	t	$P = 15.8$ W $W = 0.273$ kJ	s
7. $R_T = R_1 + R_2$	R_2	$R_1 = 3.30$ kΩ $R_T = 10.1$ kΩ	Ω
8. $G_1 = I/V_1$	V_1	$G_1 = 500\ \mu$S $I = 75.0$ mA	V
9. $\theta = \omega t$	ω	$t = 8.33$ ms $\theta = 3.14$ rad	rad/s
10. $r_p = \mu/g_m$	g_m	$\mu = 80$ (no units) $r_p = 15$ kΩ	S

The following are special formulas. Pay attention to the specified units.

11. $\lambda = \dfrac{300}{f}$ λ is in m f is in **MHz**	λ	$f = 372$ kHz	m
12. $T = \dfrac{I_m R_m}{2}$ I_m is in μA R_m is in kΩ T is in °C	R_m	$T = 50$°C $I_m = 0.152$ mA	Ω

	Formula	Solve for:	Given:	Answer Base Unit
13.	$C = \dfrac{2400}{rR_L}$	r	$C = 4300 \ \mu F$	no unit; a ratio
	C is in μF		$R_L = 8.0 \ \Omega$	
	R_L is in Ω			
	r is a ratio			

6–9 FORMING EQUATIONS

In a word problem, written descriptive information is given about numerical quantities and how they relate to one another. To solve word problems, we must *first* form an equation from the written words, and *second* we must solve the equation.

EXAMPLE 6–24	Express each of the statements using algebraic symbols. Let x represent the unknown number.
STATEMENT A	A number increased by 4 equals 17.
Equation	$x + 4 = 17$
Solution	$x = 13$
STATEMENT B	A number decreased by 8 equals 20.
Equation	$x - 8 = 20$
Solution	$x = 28$
STATEMENT C	A number increased by twice the number equals 33.
Equation	$x + 2x = 33$
Solution	$x = 11$
STATEMENT D	One-third of a number diminished by 2 is 3.
Equation	$\frac{1}{3}x - 2 = 3$
Solution	$\frac{1}{3}x = 3 + 2$
	$x = 3(5)$
$\therefore$	$x = 15$
STATEMENT E	Twice a number plus 3 equals the number minus 6.
Equation	$2x + 3 = x - 6$
Solution	$2x - x = -6 - 3$
$\therefore$	$x = -9$

Use algebraic symbols to form an equation for each of the statements. Let x represent the unknown number. Solve each of the equations.

1. Eight times a number is 40.
2. Four times a number, increased by twice the number, equals 42.
3. Six times a number, diminished by 5, equals 25.
4. A number less 3 is 11.
5. Sixteen diminished by a number is 9.
6. Five less than one-half a number is 6.
7. Add 2 to a number, and double the sum; the result is 18.
8. Add 6 to a number, and decrease the sum by 7; the result is 15.
9. Six times a number diminished by twice the number is 52.
10. Five added to twice a number less 2 equals the number diminished by 3.

6–10 SOLVING WORD PROBLEMS

Solving word problems in mathematics and troubleshooting an electronic system have many similarities. In each case, the problem is verbalized; each may appear complicated, causing confusion in the solution; each requires a systematic approach that includes the use of symbols; a successful conclusion is reached after one or more trials.

Because word problems require that you have *interpretive skills,* you may find them very challenging. We have included the following procedures for solving word problems to aid in developing analytical and interpretive skills. There is no easy way to learn how to troubleshoot or to solve word problems; the skill comes only by a consistent, conscious effort over a relatively long period of time.

GUIDELINES 6–1. General Procedure for Solving Word Problems

1. Carefully read the problem.
2. Select a variable to represent the unknown number.
3. If there is more than one condition described, represent all conditions in terms of the selected variable.
4. Form an equation from the described conditions.
5. Solve the equation for the variable.
6. Check the solution to see that all the described conditions of the problem are satisfied.

EXAMPLE 6–25   Three times the result of subtracting six from an unknown number is equal to minus two times the sum of five times the unknown number and minus four.

Solution   Let x = the unknown number and then write the conditions using mathematical symbols. Thus: "Three times the result of subtracting six from an unknown number" $3(x - 6)$ "is equal to" $(=)$ "minus two times the sum of five times the unknown number and minus four" $-2(5x + -4)$. Stated as an equation:

$$3(x - 6) = -2(5x - 4)$$
$$3x - 18 = -10x + 8$$
$$13x = 26$$
$$\therefore \qquad x = 2$$

Check   Does $3(x - 6) = -2(5x - 4)$ when $x = 2$?

$$3(x - 6) = -2(5x - 4)$$
$$3(2 - 6) = -2(5 \cdot 2 - 4)$$
$$3(-4) = -2(6)$$
$$-12 = -12$$

EXAMPLE 6–26   A test technician tested 714 100-MB/s optical transmitters. Some of the transmitters tested were rejected and sent to *rework*. Fifty times as many systems were accepted as were rejected. How many systems were rejected?

Solution   Let x represent the number of systems rejected, and $50x$ represent the number of systems accepted. The number of systems rejected plus the number of systems accepted equals the number of systems tested.

$$x + 50x = 714$$
$$51x = 714$$
$$x = 14$$

$\therefore$ 14 systems were rejected.

EXAMPLE 6–27   A leading manufacturer of random access memories (RAM) sells three models of its double-in-line memory modules (DIMM). The storage capacity of the largest model is four times the capacity of the intermediate model. The storage capacity of the intermediate model is twice that of the smallest model. If the total combined capacity of all three models is 369 098 752 bytes of storage, find the capacity of each model of memory module. Computer memory storage is measured in units called **bytes.** A one-megabyte (1 MB) memory has 1 048 576 bytes (2^{20} bytes) of storage.

Solution
Let N = the capacity of the smallest memory
$2N$ = the capacity of the intermediate memory
$4(2N)$ = the capacity of the largest memory

Observation
Each of the capacities is related to the smallest memory by using the variable N. No new variable has been introduced. Thus:

$$N + 2N + 4(2N) = 369\ 098\ 752$$
$$N + 2N + 8N = 369\ 098\ 752$$
$$11N = 369\ 098\ 752$$
$$N = \ \ 33\ 554\ 432$$

∴ 33 554 432 bytes (32 MB) is the capacity of the smallest memory *(N)*.
67 108 864 bytes (64 MB) is the capacity of the intermediate memory *(2N)*.
268 435 456 bytes (256 MB) is the capacity of the largest memory *(8N)*.

Observation
A 32-MB memory has 33 554 432 bytes of storage (2^{25}). Although this is more than 33 MB, the computer industry commonly calls this a 32-MB memory, because 32 MB/1 MB = 33 554 432/1 048 576 = 32.

EXERCISE 6–12

Solve each problem using Guidelines 6–1.

1. When five is added to an unknown number, the sum is twice the unknown number. Find the unknown number.

2. Two times the sum of twelve and an unknown number is equal to twenty. Determine the unknown number.

3. When twice the sum of six and an unknown number is added to the unknown number, the result is equal to ten added to the unknown. Find the unknown number.

4. Three times the result of subtracting an unknown number from nine is nine. Determine the unknown number.

5. Twice the sum of four and three times an unknown number is equal to five times the sum of four times the unknown number and ten. Find the unknown number.

6. Three 120-V plug-in transformers purchased from a local supply house cost $16 after a $2 discount. What is the cost of a single transformer without a discount?

7. A large supplier of electronic components utilizes a computerized inventory system. In checking the daily printout of the supply of a particular integrated circuit

(IC), it was found that there were 1210 fewer ICs listed than the amount recorded on the printout of the day before. If the daily printout shows 3126 ICs, what was the number recorded the day before?

8. A special rectangular screen room, which is used to make RF (radio-frequency) measurements, was measured and found to be three times as long as it is wide. If the sum of all the sides is 24 meters, what is the length and width of the room?

9. One resistor and two capacitors cost 82¢. If the capacitor costs 23¢ more than the resistor, find the cost of one capacitor.

10. A 30-m piece of coaxial cable is cut into two pieces. One piece is 2 m shorter than three times the length of the other piece. Find the length of the shorter piece.

11. An 18-min video tape was made to train electronic technicians on the use of the digital voltmeter (DVM) to make voltage and resistance measurements. The part of the tape about voltage measurements lasted 3 min longer than twice the time about resistance measurement. How many minutes of instruction was spent on voltage measurements?

12. A service technician took 35 min to disassemble and reassemble a computer. If it took 7 min less to reassemble than it took to disassemble, how long did it take to disassemble this unit?

13. A resistor costs R cents and a capacitor costs five times more than a resistor. If the combined cost of one resistor and one capacitor is 18¢, what is the cost of a single resistor and a single capacitor?

14. Divide a 96-cm piece of hookup wire into three parts (with no wire left over) so that the first piece is three times as long as the second, and the third piece is as long as the sum of the other two pieces.

15. Two semiconductor diodes, one a rectifier, the other a zener diode, have an average operating temperature of 160°C. What is the operating temperature of each diode if the rectifier diode has a 20°C higher operating temperature than the zener diode? (The average of two values is found by adding the two quantities and dividing by 2.)

SELECTED TERMS

conditional equation   True only for particular values of the variable in the equation.
formula   An equation expressing the relationship between physical quantities.
identical equation   True for all values of the variable in the equation.
members of the equation   The expressions on either side of the equal sign.
root of the equation   A value of the variable that makes the equation true. A solution.
symmetric property of equality   Allows the members of an equation to be interchanged.

7

Applying Mathematics to Electrical Circuits

7–1   Current, Voltage, and Resistance

7–2   Ohm's Law

7–3   Resistance in a Series Circuit

7–4   Applying Ohm's Law

7–5   Summary of the Series Circuit

7–6   Power

PERFORMANCE OBJECTIVES

- Form a series equivalent circuit.

- Apply Kirchhoff's voltage law in conjunction with Ohm's law to aid in the solution of a series circuit.

- Determine the power dissipated in the load of an electric circuit.

An electronic passenger train exiting the Pfaffensprung Tunnel on the Gotthard line in the Swiss Alps. (Courtesy of Switzerland Tourism)

In this chapter, units of measure, numeric notation, and algebraic concepts from previous chapters will be used to solve problems associated with electrical circuits.

The behavior of an electrical circuit is described by Ohm's law. This law relates the three parameters of an electric circuit: the electromotive force, the electric current, and the electrical resistance. Each of these three important parameters will be used along with Ohm's law to compute circuit characteristics and conditions. Additionally, these same three parameters will be used to define and to compute electrical power.

7–1 CURRENT, VOLTAGE, AND RESISTANCE

Current

The purpose of an electric circuit is to move energy (joules) from a source to a load. The energy is moved by charge carriers through conductors (wires) that connect the source of energy to the load. At the load, the energy is converted to light, heat, sound, and so forth. The source of energy in the electric circuit may be as simple as a flashlight battery, and the load may be as simple as a lamp.

The charge carriers of electric energy are the electrons in the wire of the conductors. Since the amount of charge carried by each electron is extremely small and of little consequence, the **coulomb** (C) of charge was defined to have 6.242×10^{18} electrons. This quantity of charge (Q) is sufficient to have a significant effect in providing charge carriers for electric current flow.

As you know from Chapter 3, the ampere (A) is one of the SI base units and as such, it is not a derived unit of measure. However, by using the defined quantity of *electric charge (Q)* and the *time (t)* the charge is allowed to flow, the *intensity (I)* of the electric **current** may be derived (calculated) from the following equation (Equation 7–1).

$$I = Q/t \quad \textbf{(amperes)} \tag{7–1}$$

where I = intensity of the electric current in amperes (A)
Q = quantity of electric charge in coulombs (C)
t = time in seconds (s)

From Equation 7–1 we learn that the intensity (I) of the electrical current is the rate at which electrons move through an electric circuit, and that one ampere of current is equal to the movement of one coulomb of charge per second (1 A = 1 C/s).

EXAMPLE 7–1   What is the current in a circuit where 325 C of charge passes a point in 25 s?

Observation   The phrase *what is the current* is asking for the intensity of the current.

Solution Use $I = Q/t$. Substitute 325 C for Q and 25 s for t:

$$Q = 325 \text{ C}$$
$$t = 25 \text{ s}$$
$$I = 325/25 = 13$$

$$\therefore \qquad I = 13 \text{ A}$$

Electric current is measured by an instrument called an ammeter. See Figure 7–1.

Voltage

The word **voltage** is used to indicate both the *rise* in potential energy due to a generator or a battery, as pictured in Figure 7–2, and the *fall* in potential energy when current flows through the circuit resistance. Since the word voltage is so common in the language of electronics, we will make a conscious effort to indicate a difference between a voltage rise and a voltage drop.

The letter E is used to indicate a **voltage rise.** The rise in potential energy created by the source (battery) is the ***electromotive force*** (emf) that causes current to flow through the resistance of the circuit. The letter E is used to indicate a voltage source.

The letter V is used to indicate a **voltage drop.** A voltage drop occurs in a circuit when current flows through an electrical load, such as the lamp in Figure 7–3(b), and energy is converted to light, heat, or the like.

From Figure 7–3(a) and (b) you may see that in each figure the voltmeter connected across the source voltage registers 9 V. However, the voltmeter across the lamp registers 9 V only when current is flowing in the circuit. **There can be a voltage drop in an electrical circuit only when there is a current flow.**

A voltmeter is used to make voltage measurements. When making voltage measurements, the voltmeter is placed across the load in the circuit. Two voltmeters are pictured in Figure 7–4.

FIGURE 7–1 Measuring current flow: (a) ammeter; (b) digital ammeter; (c) schematic symbol for ammeter.

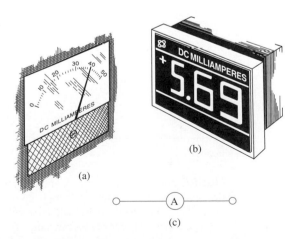

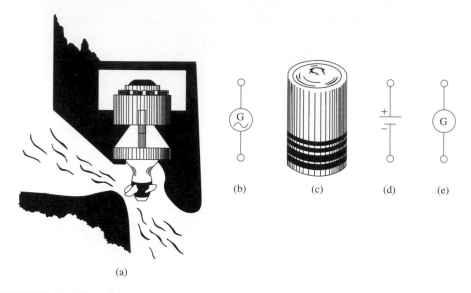

(a)

(b)   (c)   (d)   (e)

FIGURE 7–2   Direct and alternating sources of electromotive force: (a) hydroelectric alternating-current generator; (b) schematic symbol for an alternating-current generator; (c) direct-current battery (cell); (d) schematic symbol for any direct-current source; (e) universal schematic symbol for either an ac or dc generator.

Voltage is expressed as energy per unit charge (joule/coulomb) as stated in Equation 7–2.

$$E \ (\textit{or } V) \ = \ W/Q \quad \textbf{(volts)} \tag{7–2}$$

where E (or V) = potential difference in volts (V)
                 W = energy (work) carried by charge carriers in joules (J)
                 Q = amount of electric charge in coulombs (C)

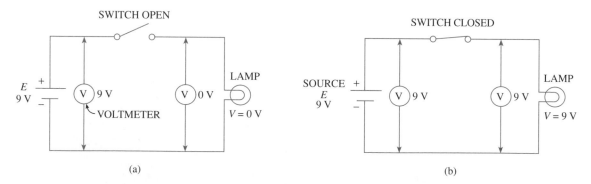

(a)   (b)

FIGURE 7–3   (a) The source is indicated by the letter E. With the switch open, no current flows and the lamp is not lit. There is no potential drop (voltage) seen across the lamp. The meter reads 0 V. (b) With the switch closed, current flows, the lamp lights, and there is potential drop (voltage) seen across the lamp. The meter indicates 9 V.

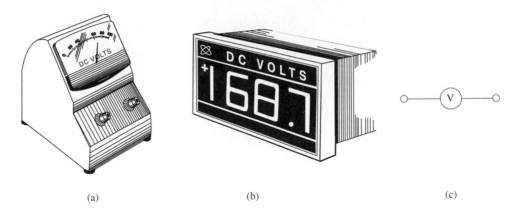

FIGURE 7–4 Measuring voltage: (a) dc voltmeter; (b) digital voltmeter (DVM); (c) schematic symbol for a voltmeter.

EXAMPLE 7–2   What potential difference (voltage) will be measured across the lamp of Figure 7–5 if 150 mC of charge flows through the lamp (load) and converts 0.900 J of energy to heat and light?

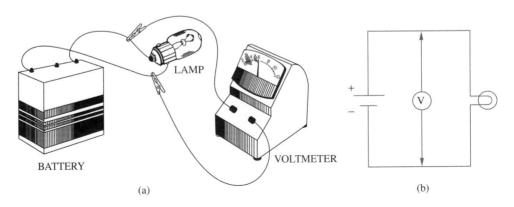

FIGURE 7–5 Circuit for Example 7–2: (a) pictorial diagram; (b) schematic diagram.

Solution   Use $V = W/Q$ and substitute 0.900 J for W and 150 mC for Q:

$$V = W/Q$$
$$W = 0.900 \text{ J}$$
$$Q = 150 \text{ mC} = 150 \times 10^{-3} \text{ C} = 0.150 \text{ C}$$
$$V = 0.900/0.150 = 6.00 \text{ V}$$

∴   The voltage drop across the lamp is 6.00 V.

EXAMPLE 7–3   The current of Figure 7–5 was measured for 2.0 min and found to be a constant 150 mA. What is the voltage of the battery if 108 J of energy was converted to heat and light in the lamp?

Solution   Begin by solving $I = Q/t$ (Equation 7–1) for Q:

$$I = Q/t$$
$$Q = I(t)$$

Substitute 150 mA for I and 2.0 min for t:

$$I = 150 \text{ mA} = 150 \times 10^{-3} \text{ A} = 0.150 \text{ A}$$
$$t = 2.0 \text{ min} \times \frac{60 \text{ s}}{1 \text{ min}} = 120 \text{ s}$$
$$Q = 0.150 \times 120 = 18 \text{ C}$$

Next solve $E = W/Q$ (Equation 7–2) by substituting 108 J for W and 18 C for Q:

$$E = W/Q$$
$$W = 108 \text{ J}$$
$$Q = 18 \text{ C}$$
$$E = 108/18 = 6.0 \text{ V}$$

∴   The source voltage is 6.0 V.

Resistance

The **resistance** of an electrical circuit opposes the flow of electric current. Resistance is related to voltage drop and current as in Equation 7–3.

$$R = V/I \quad \textbf{(ohms)} \tag{7–3}$$

where R = resistance in ohms (Ω)
V = voltage drop across R in volts (V)
I = current through R in amperes (A)

Because all parts of an electrical circuit offer some opposition to current flow, the wires connecting the voltage source to the resistive load are selected to have a very small resistance. These connecting wires are referred to as conductors because of their high *conductance* (or low resistance). Copper is usually used as a conductor in electrical circuits because of its high **conductivity.**

When computing the resistance of a circuit, we often neglect the small resistance of the conductors. We may assume the connecting wires to be perfect conductors. This assumption is allowable because the electrical load (see Figure 7–6) presents a much larger opposition to the flow of current than do the conductors.

FIGURE 7–6 An electrical circuit with several loads shown. $R = V/I$ = 100/5 = 20 Ω. The resistance of the connecting wires has been neglected because it is much smaller than the load resistance.

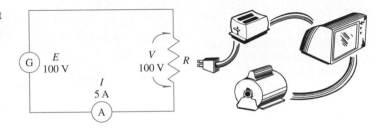

EXAMPLE 7–4   The television set of Figure 7–7 is connected to a 120-V source and the current is 2.50 A. Compute the resistance of the set.

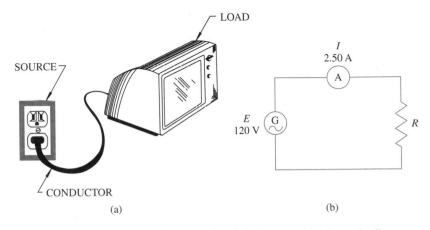

(a)   (b)

FIGURE 7–7 Circuit for Example 7–4: (a) pictorial diagram; (b) schematic diagram.

Solution   Use $R = V/I$ (Equation 7–3). Substitute 120 V for V and 2.50 A for I:

$$R = VI$$
$$V = 120 \text{ V}$$
$$I = 2.50 \text{ A}$$
$$R = 120/2.50 = 48.0 \text{ Ω}$$

∴   The resistance of the load (TV set) is 48.0 Ω.

Because electronics is concerned with amplifying, attenuating, shaping, and generating signals for communication, computation, and control, known amounts of resistance are added to the circuit to control the flow of current. This resistance is in the form of a resistive device called a **resistor.** Figure 7–8 shows two types of fixed resistors used in electronic circuits. The parameters of electrical circuits discussed in this section are summarized in Table 7–1.

FIGURE 7–8 Two common types of fixed resistors used to control current flow: (a) carbon-film resistor; (b) wire-wound resistor; (c) cutaway of a wire-wound resistor.

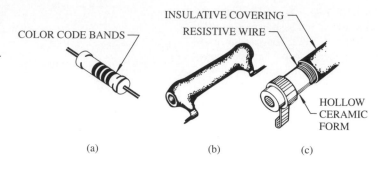

(a) (b) (c)

TABLE 7–1	Parameters of an Electrical Circuit			
Physical Quantity	**Unit**	**Unit Symbol**	**Schematic Symbol**	**Formula**
Current	ampere	A	I	$I = Q/t$
Potential difference rise	volt	V	E	$E = W/Q$
Potential difference drop	volt	V	V	$V = W/Q$
Resistance	ohm	Ω	R	$R = V/I$

EXERCISE 7–1

1. Express the movement of 100 C/h (coulombs per hour) in amperes.

2. Express 8.00 mC/min in amperes.

3. A certain battery is charged with a steady current of 500 mA for 5.00 min. How much charge is accumulated in the battery?

4. How many minutes will it take 1.00 kC to pass through an electrical conductor if the current is 2.50 A?

5. Express 100 J per 8.33 C (100 J/8.33 C) in volts.

6. Express 6.52 kJ/4.35 kC in volts.

7. A certain battery can move 14.0 C of charge through a circuit by supplying 126 J of energy. What is the voltage of the battery?

8. How much voltage drop will be measured across the terminals of a dc motor if 62.2 C of charge is moved to convert 1.49 kJ of energy?

9. Express 440 V/15.0 A in ohms.

10. What is the resistance of an electrical circuit that has a measured source voltage of 120 V and a current of 227 mA?

7–2   OHM'S LAW

Nearly two centuries ago a German physicist, Georg Simon Ohm, observed that the intensity of the current in an electrical conductor was dependent on the electromotive force producing it. He observed that doubling the source voltage doubled the current and that tripling the source voltage tripled the current. From these observations, a formula was written that describes the relationship between the emf and the current flow. This formula is called **Ohm's law.**

$$E = IR \quad \text{(volts)} \tag{7–4}$$

Other forms of Ohm's law are:

$$I = E/R \quad \text{(amperes)} \tag{7–5}$$
$$R = E/I \quad \text{(ohms)} \tag{7–6}$$

where E = voltage in volts (V)
   I = current in amperes (A)
   R = resistance in ohms (Ω)

EXAMPLE 7–5   How much voltage will be measured across the terminals of the battery of Figure 7–9 if 150 mA is passing through the 150-Ω resistor?

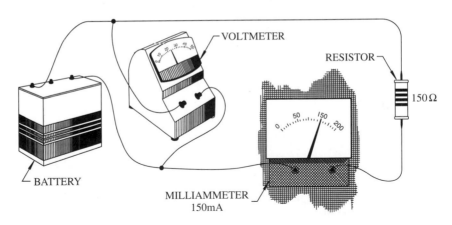

FIGURE 7–9   Pictorial diagram of the connection of the components and meters in the electrical circuit for Example 7–5.

Solution   Draw a schematic diagram and label it with the information given. See Figure 7–10.
Select the appropriate equation: $E = IR$.
Substitute 150 mA for I and 150 Ω for R:

$$E = IR$$
$$E = 150 \times 10^{-3} \times 150$$
$\therefore$ $E = 22.5 \text{ V}$

FIGURE 7–10 Schematic diagram
of the circuit for Example 7–5.

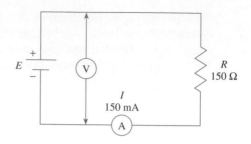

EXAMPLE 7–6

In the circuit pictured by Figure 7–9, the source voltage is 22.5 V. Determine the circuit current when the resistance is increased from 150 Ω to 2.70 kΩ.

Solution

Draw a schematic diagram and label it with the information given. See Figure 7–11.

Select the appropriate equation: $I = E/R$. Substitute 22.5 V for E and 2.70 kΩ for R:

$$I = E/R$$
$$I = \frac{22.5}{2.70 \times 10^3}$$
$$I = 0.008\ 33$$
$$\therefore \quad I = 8.33 \text{ mA}$$

FIGURE 7–11 Schematic diagram
of the circuit for Example 7–6.

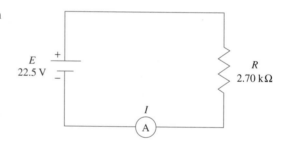

EXAMPLE 7–7

In the circuit pictured by Figure 7–9 the emf is increased to 90 V. What must the circuit resistance be in order to maintain the current at 150 mA?

Solution

Draw a schematic diagram and label it with the information given. See Figure 7–12.

Select the appropriate equation: $R = E/I$.

Substitute 90.0 V for E and 150 mA for I:

$$R = E/I$$
$$R = \frac{90.0}{150 \times 10^{-3}}$$
$$\therefore \quad R = 600 \ \Omega$$

FIGURE 7-12 Schematic diagram for the circuit for Example 7–7.

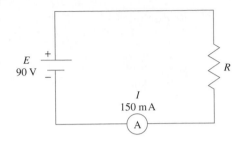

EXERCISE 7-2

In solving each of the following, it is suggested that you (1) draw a schematic diagram and label it with the information given, (2) state the equation being used, and (3) show the solution to the problem. Express the answer to an appropriate precision with unit and unit prefix.

1. Calculate each voltage drop *(V)* given the following information:
   (a) 10 A passing through 100 Ω
   (b) 6.06 mA passing through 3.30 kΩ
   (c) 15 μA passing through 1.9 MΩ
   (d) 0.732 A passing through 320 Ω

2. Calculate each current *(I)* given the following information:
   (a) 100 V across 10 Ω
   (b) 15.0 V across 560 Ω
   (c) 2.2 kV across 4.7 MΩ
   (d) 176 μV across 22.0 Ω

3. A milliammeter connected in a circuit reads 75 mA and a voltmeter connected across the voltage source reads 45 V. What is the resistance of the circuit?

4. A voltmeter is connected across the leads of a 10-kΩ resistor. What should the voltmeter reading be if 9.3 mA of current is flowing through the resistor?

5. The voltage across the heating element of an electric heater is 120 V. If the resistance of the heater is 15 Ω, what should an ammeter indicate when it is placed in the circuit?

6. The terminal of an automobile battery is severely corroded, which causes resistance at the terminals. A voltmeter placed from the terminal post of the battery to the cable (see Figure 7–13) indicates 9.5 V when 15 A of current is flowing through the cable. Compute the resistance of the connection.

7. Two wires were spliced (connected) together by soldering. However, the connection was poorly made and resulted in a resistive connection (a cold solder joint). Compute the resistance of the connection if 2.3 V is measured across the splice when 0.92 A is flowing.

FIGURE 7–13 Measuring the voltage drop across the terminal of a battery.

FIGURE 7–14 Soldering iron for Exercise 7–2, problem 9.

8. A 20-horsepower motor that drives a pump is connected to the source voltage with wiring that has a total resistance of 432 mΩ. Compute the voltage drop across the wiring when 40.0 A of current flows.

9. Compute the *hot* resistance of the soldering iron pictured in Figure 7–14 when it is plugged into a 117-V source. The current rating is 300 mA at 117 V.

10. A digital clock requires 84 mA for operation when plugged into a 120-V outlet. Compute the resistance of the clock.

7–3 RESISTANCE IN A SERIES CIRCUIT

The circuits we have worked with up until now have been concerned with a single resistive load. We will now consider several resistive loads connected so that they form a *series circuit.*

Series Circuit Defined

A series circuit is formed when each component is successively connected so that one end of a component is connected to an end of the next component, and so forth, until a complete path is formed from the beginning to the end. An example of a series circuit is shown in Figure 7–15.

One way to determine if a circuit is connected in series is to *trace* the flow of current from the voltage source through the loads and back to the source. The components are in series if the current flows through each component in turn (one after the other) before returning to the voltage source.

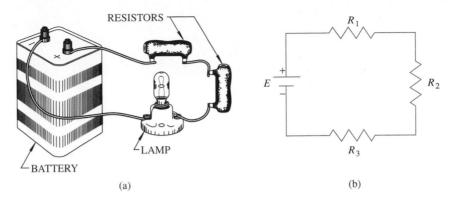

(a)   (b)

FIGURE 7–15 Series circuit consisting of a battery, two wire-wound resistors, and a lamp: (a) pictorial diagram of the connection; (b) schematic diagram with the resistive loads shown and labeled with subscripted variables.

An important characteristic of a series circuit is that the circuit current, I, is the same throughout the entire series circuit at the same time. This is stated mathematically as Equation 7–7.

$$I = I_1 = I_2 = I_3 \quad \textbf{(amperes)} \tag{7–7}$$

Total Resistance in a Series Circuit

The total resistance of a series circuit, R_T, is found by summing (adding up) each of the resistances in the circuit. Thus,

$$R_T = R_1 + R_2 + R_3 \quad \textbf{(ohms)} \tag{7–8}$$
$$R_T = \Sigma R \quad \textbf{(ohms)} \tag{7–9}$$

R_T is the total resistance, and ΣR is read "the sum of the resistances."

EXAMPLE 7–8   Determine the total resistance, R_T, of the circuit pictured in Figure 7–16, where $R_1 = 50 \ \Omega$, $R_2 = 20 \ \Omega$, and $R_3 = 130 \ \Omega$.

FIGURE 7–16 Schematic diagram of the circuit for Example 7–8.

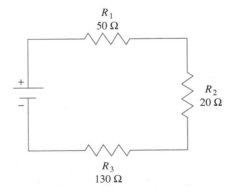

Sum the resistances:

$$R_T = \Sigma R$$
$$R_T = 50 + 20 + 130$$
$$\therefore \quad R_T = 200 \ \Omega$$

Series Equivalent Circuit

A series circuit made up of several resistances may be simplified to a circuit with a single resistance having the value of R_T. This single resistance (R_T) is the **equivalent resistance** of the equivalent circuit, and it is the total resistance of the series circuit. The *equivalent series circuit* has the same source voltage, current, and total resistance as the series circuit from which it was derived.

EXAMPLE 7–9   Draw and label an equivalent series circuit for the circuit of Figure 7–17.

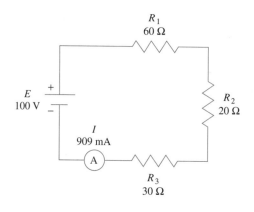

FIGURE 7–17   Circuit for Example 7–9.

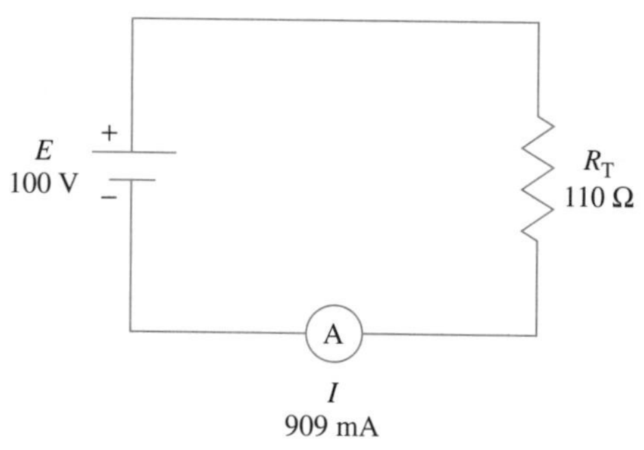

FIGURE 7–18   Equivalent circuit of Figure 7–17.

Solution   Determine the equivalent resistance, R_T:

$$R_T = \Sigma R$$
$$R_T = 60 + 20 + 30$$
$$R_T = 110 \ \Omega$$
$$\therefore \quad R_T = 110 \ \Omega$$

Draw and label the schematic of the equivalent circuit. See Figure 7–18.

1. Three resistances, $R_1 = 4.7$ kΩ, $R_2 = 12$ kΩ, and $R_3 = 6.8$ kΩ, are connected in series across an ac source of voltage. Determine the total resistance, R_T, of the circuit.

2. Find the total resistance, R_T, of the following series-connected carbon resistors: 3.0 kΩ, 0.220 MΩ, 15 kΩ, 180 kΩ, 22 kΩ, and 0.560 MΩ.

3. A wire-wound resistor is connected in series with a lamp. What is the combined resistance (R_T) of the two resistances if the lamp resistance is 2.0 kΩ and the resistance of the resistor is 750 Ω?

4. Draw and label a schematic of an equivalent circuit of a series circuit consisting of a generator of 6.3 V and two resistors, one of 20 Ω and the other of 10 Ω. The circuit current is 210 mA.

5. Three carbon-film resistors ($R_1 = 120$ kΩ, $R_2 = 91$ kΩ, and $R_3 = 62$ kΩ) are connected in series across a 220-V dc source. If $I = 806$ μA:

   (a) Determine the equivalent resistance, R_T.

   (b) Determine current in R_3.

   (c) Draw and label the schematic of the equivalent circuit.

6. To limit the circuit current to 262 mA, a 100-Ω wire-wound resistor is connected in series with a 40-W, 358-Ω soldering iron. If $E = 120$ V:

   (a) Determine the equivalent resistance, R_T.

   (b) Draw and label the schematic of the circuit.

   (c) Draw and label the schematic of the equivalent circuit.

7–4 APPLYING OHM'S LAW

Kirchhoff's Voltage Law

In a series circuit the sum of the voltage drops is equal to the voltage source. Gustav Kirchhoff is credited with the discovery of this important circuit characteristic. Stated mathematically,

$$E = V_1 + V_2 + V_3 \quad \text{(volts)} \qquad \qquad \text{(7–10)}$$
$$E = \Sigma V \quad \text{(volts)} \qquad \qquad \text{(7–11)}$$

where ΣV is read "the sum of the voltage drops."

EXAMPLE 7–10   Verify Kirchhoff's voltage law for the circuit of Figure 7–19, where $V_1 = 0.360$ V, $V_2 = 1.05$ V, $V_3 = 2.00$ V, $V_4 = 2.87$ V, $V_5 = 3.72$ V, and $E = 10.0$ V.

FIGURE 7–19 Circuit for Example 7–10.

FIGURE 7–19 Circuit for Example 7–10.

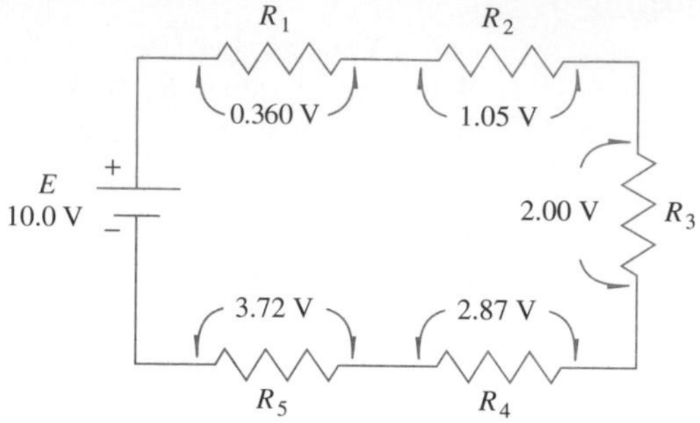

| Solution | Use Equation 7–10. Substitute: |

$$E = V_1 + V_2 + V_3 + V_4 + V_5$$
$$10.0 = 0.360 + 1.05 + 2.00 + 2.87 + 3.72$$
$$10.0 = 10.0$$

∴ The sum of the voltage drops equals the voltage source.

Applying Ohm's Law

When Ohm's law is applied to a series circuit, the symbol for resistance in the Ohm's law formula is R_T, which is determined by adding *all* the resistances in the series circuit ($R_T = \Sigma R$).

$$E = IR_T \quad \text{(volts)} \tag{7–12}$$

where E = source voltage in volts (V)
I = circuit current in amperes (A)
R_T = equivalent resistance in ohms (Ω)

EXAMPLE 7–11   Compute the circuit current (*I*) in a series circuit that has an applied voltage of 22.5 V and resistance of $R_1 = 50\ \Omega$, $R_2 = 150\ \Omega$, and $R_3 = 100\ \Omega$. Draw and label a schematic of the equivalent circuit.

Solution   Draw and label a schematic diagram for the given conditions. See Figure 7–20.
First compute the equivalent resistance, R_T:

$$R_T = \Sigma R$$
$$R_T = 50 + 150 + 100$$
$$R_T = 300\ \Omega$$

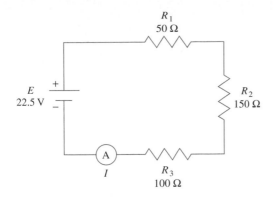

FIGURE 7–20 Circuit for Example 7–11.

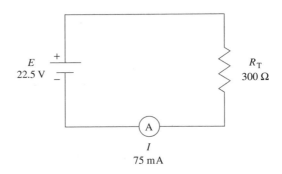

FIGURE 7–21 Equivalent circuit of Figure 7–20.

Next compute the circuit current, I, using $I = E/R_T$:

$$I = E/R_T$$
$$I = 22.5/300 = 0.0750$$
$$\therefore \qquad I = 75.0 \text{ mA}$$

Draw and label the schematic for the equivalent circuit. See Figure 7–21.

EXAMPLE 7–12

A 6.30-V, 150-mA lamp is to be operated from a 117-V source as shown in Figure 7–22. Compute the value of the series resistor (R_1) needed to drop the excess voltage and limit the current to 150 mA.

Solution

To solve for R_1, the values of V_1 and I are needed. Solve for V_1 using $E = V_1 + V_2$ (Equation 7–10):

$$V_1 = E - V_2$$

Substitute $E = 117$ V and $V_2 = 6.30$ V:

$$V_1 = 117 - 6.30$$
$$V_1 = 111 \text{ V}$$

Solve for R_1. I is given as 150 mA:

$$R_1 = V_1/I$$
$$R_1 = 111/(150 \times 10^{-3})$$
$$R_1 = 740 \text{ }\Omega$$

$\therefore \qquad$ 740 Ω is needed to limit the current to 150 mA in the circuit of Figure 7–22.

Observation

A standard-value, 750-Ω, 5%, 25-W, fixed-power resistor is used in this application. When selecting resistor sizes, consult the *Preferred Number Series for Standard Resistance Values*

FIGURE 7–22 Circuit for Example
7–12.

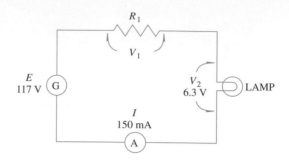

table located in Appendix A. Additionally, take note of the
*Color Code for Specifying Preferred Resistors Values and Tol-
erances* table, also located in Appendix A.

Computing the Voltage Drop Across One Series Resistance

The voltage drop across one resistance in a series circuit is computed by Ohm's law as
Equation 7–13.

$$V_n = IR_n \quad \text{(volts)} \qquad (7\text{–}13)$$

where V_n = voltage in volts (V) across selected resistance, R_n
 I = circuit current in amperes (A)
 R_n = selected resistance in ohms (Ω)
 n = number of the selected resistor

EXAMPLE 7–13   Two carbon-film resistors are connected in series with a generator, as shown in the schematic of Figure 7–23. The current
throughout the circuit is 500 μA.

(a) Compute the voltage drop across each resistor.
(b) Compute E, the generator voltage.

FIGURE 7–23 Circuit for Example
7–13.

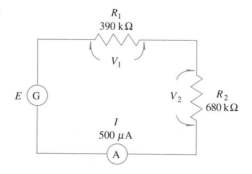

Solution

(a) Use Equation 7–13: $V_1 = IR_1$. Substitute $I = 500 \ \mu A$ and $R_1 = 390 \ k\Omega$:

$$V_1 = IR_1$$
$$V_1 = 500 \times 10^{-6} \times 390 \times 10^3$$
$$V_1 = 195 \ V$$

$V_2 = IR_2$. Substitute $I = 500 \ \mu A$ and $R_2 = 680 \ k\Omega$:

$$V_2 = IR_2$$
$$V_2 = 500 \times 10^{-6} \times 680 \times 10^3$$
$$V_2 = 340 \ V$$

$\therefore \qquad V_1 = 195 \ V, V_2 = 340 \ V$

(b) Use $E = V_1 + V_2$ (Equation 7–10):

$$E = V_1 + V_2$$
$$E = 195 + 340$$
$$E = 535 \ V$$

$\therefore \qquad E = 535 \ V$

EXERCISE 7–4

1. Determine the source voltage of a series circuit having voltage drops of 12, 7.5, and 22.5 V.

2. Determine the source voltage of a series circuit having voltage drops of 238 mV, 686 mV, 52 mV, and 1.024 V.

3. Two resistors are connected in series across a generator; determine the voltage drop across R_2 if the source voltage is 65 V and V_1 is 28 V.

4. Two resistors, $R_1 = 6.0 \ \Omega$ and $R_2 = 24 \ \Omega$, are connected in series with a battery. The circuit current is 400 mA. Determine V_1 and E.

5. Three resistors, $R_1 = 200 \ \Omega$, $R_2 = 500 \ \Omega$, and $R_3 = 800 \ \Omega$, are connected in series across a 120-V source. The circuit current is 80 mA. Show that $E = V_1 + V_2 + V_3$.

6. Two resistors, $R_1 = 100 \ \Omega$ and $R_2 = 50.0 \ \Omega$, are connected across an ac source. The voltage dropped by R_2 is 28.0 V. Determine the current through R_1.

7–5 SUMMARY OF THE SERIES CIRCUIT

The concepts of the preceding sections of this chapter are summarized in Table 7–2. Refer to this table while studying the following examples and working the additional exercises.

TABLE 7–2 Characteristics of a Series Circuit

Equation	Comment
$I = I_1 = I_2 = I_3$	The current (I) is the same throughout the series circuit
$R_T = R_1 + R_2 + R_3$	The equivalent resistance (R_T) is the sum of the individual resistances (ΣR)
$E = V_1 + V_2 + V_3$	Source voltage (E) is the sum of the voltage drops (ΣV)
$E = IR_T$	Ohm's law for a series circuit
$V_n = IR_n$	Voltage across the selected resistance, R_n

EXAMPLE 7–14 Four resistors, 270, 220, 330, and 180 Ω, are connected in series across a 10-V source. Draw a schematic diagram of the series circuit and determine:

(a) The equivalent resistance (R_T)
(b) The circuit current (I)
(c) The voltage drop across each resistor
(d) That the source voltage equals the sum of the voltage drops

Draw and label a schematic of the equivalent circuit.

Solution Draw and label a schematic diagram. See Figure 7–24.
(a) Determine the equivalent resistance:

$$R_T = \Sigma R$$
$$R_T = 270 + 220 + 330 + 180$$
$$R_T = 1000 \ \Omega$$
$$\therefore \qquad R_T = 1.00 \ \text{k}\Omega$$

FIGURE 7–24 Schematic diagram for Example 7–14.

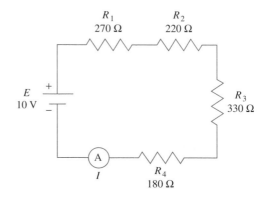

162

(b) Determine the circuit current:

$$I = E/R_T$$
$$I = 10/1.00 \times 10^3 = 10 \times 10^{-3} = 10 \text{ mA}$$
$$\therefore \quad I = 10 \text{ mA}$$

(c) Determine the voltage drop across each resistance.

$$V_1 = IR_1$$
$$V_1 = 10 \times 10^{-3} \times 270$$
$$\therefore \quad V_1 = 2.7 \text{ V}$$

$$V_2 = IR_2$$
$$V_2 = 10 \times 10^{-3} \times 220$$
$$\therefore \quad V_2 = 2.2 \text{ V}$$

$$V_3 = IR_3$$
$$V_3 = 10 \times 10^{-3} \times 330$$
$$\therefore \quad V_3 = 3.3 \text{ V}$$

$$V_4 = IR_4$$
$$V_4 = 10 \times 10^{-3} \times 180$$
$$\therefore \quad V_4 = 1.8 \text{ V}$$

(d) Determine that the source voltage equals the sum of the voltage drops.

$$E = \Sigma V$$
$$10 = 2.7 + 2.2 + 3.3 + 1.8$$
$$10 \text{ V} = 10 \text{ V}$$

$$\therefore$$ They are equal.

Draw and label a schematic diagram of the equivalent series circuit. See Figure 7–25.

FIGURE 7–25 Schematic diagram of the equivalent circuit for Example 7–14.

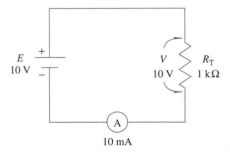

1. Three carbon-film resistors (1.5 kΩ, 3.0 kΩ, and 10.0 kΩ) are connected in series across a 145-V source. Draw a schematic diagram of the series circuit and determine:

   (a) The equivalent resistance

   (b) The current throughout the circuit

   (c) The current through the 10-kΩ resistor

   (d) The voltage drop across the 3.0-kΩ resistor

2. Repeat problem 1 for a source voltage of 1.20 kV.

3. Four identical indicator lights of the type shown in Figure 7–26 are connected in series across a 112-V source. When lit, each lamp has a resistance of 700 Ω. Determine:

   (a) The equivalent resistance

   (b) The circuit current

   (c) The voltage across one of the lamps

4. A lamp similar to the one pictured in Figure 7–26 is to be operated from a 120-V source. If the lamp is specified as 28 V, 40 mA, what resistance is needed to limit the current to the rated value?

FIGURE 7–26 Subminiature indicator light.

5. Two resistors are connected in series across a 90-V source. If a voltmeter connected across R_1 reads 22 V and the current through R_2 is 20 mA, determine:

   (a) The current through R_1

   (b) The voltage across R_2

   (c) The resistance of R_1

   (d) The equivalent resistance, R_T

6. Three resistors are connected in series across a 100.0-V source. If $V_1 = 32.0$ V, $V_2 = 18.0$ V, and $R_3 = 47.0$ kΩ, determine:

   (a) The circuit current

   (b) The value of R_2

   (c) The equivalent resistance, R_T, across the voltage source

7. In the circuit of Figure 7–27, determine:

   (a) The circuit current *(I)*

   (b) The source voltage *(E)*

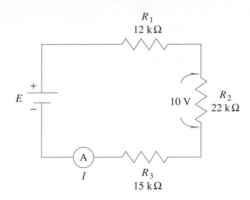

FIGURE 7–27 Circuit for Exercise 7–5, problem 7.

8. Repeat problem 7 for $R_1 = 220\ \Omega$, $R_2 = 620\ \Omega$, and $R_3 = 390\ \Omega$.

9. Four resistors are connected in series. The voltage across R_1 and R_2 is the same. The resistance of R_3 is 15 kΩ, and the voltage drop across R_4 is 20 V. If the source voltage is 150 V and the circuit current is 2.0 mA, determine R_1, R_2, and R_4.

10. Three resistors are connected in series across a generator. It is known that $R_1 = 110\ \Omega$. The resistance of R_3 is twice the resistance of R_1, R_2 has a resistance that is three times that of R_1, and the voltage drop across R_2 is 15 V. Compute the generator voltage.

7–6 POWER

Power was defined in Chapter 3 as the *rate* of converting (producing or using) energy. Power is also the *rate* of doing work. We learned that

$$P = W/t \quad \textbf{(watts)} \tag{7–14}$$

where P = power in watts (W)
W = work in joules (J)
t = time in seconds (s)

Remember that a joule per second (J/s) is a watt (W), and that power is the rate of doing work or converting energy from one form to another. Thus,

$$1 \text{ watt (W)} = 1 \text{ joule per second (J/s)}$$

EXAMPLE 7–15   Five kilojoules (5.00 kJ) of energy was used to move a mass of 170 kg a vertical distance of 3.00 m in 1.25 min. At what rate is energy being used?

Solution   Use $P = W/t$ (Equation 7–14). Substitute 5.00 kJ for work and 1.25 min for time.

$$W = 5.00 \text{ kJ}$$

$$t = 1.25 \text{ min} \times \frac{60 \text{ s}}{1 \text{ min}} = 75.0 \text{ s}$$

$$P = 5.00/75.0 = 66.7 \text{ J/s}$$

$$\therefore \quad P = 66.7 \text{ W}$$

Power in Electrical Circuits

Equation 7–14 is not the best equation to use to compute power in an electrical circuit. A more useful formula is Equation 7–15.

$$P = EI \text{ or } P = VI \quad \text{(watts)} \tag{7–15}$$

where P = power in watts (W)
$\quad E$ = voltage rise in volts (V) or V = voltage drop in volts (V)
$\quad I$ = current in amperes (A)

Since the product of E and I is an expression for power ($P = EI$), it must have the units of power (joules per second). Exploring:

$$E = W/Q \text{ and } I = Q/t$$

Substituting into $P = EI$:

$$P = \frac{W}{Q} \times \frac{Q}{t}$$

Simplifying:

$$P = \frac{W}{\cancel{Q}} \times \frac{\cancel{Q}}{t} = \frac{W}{t}$$

$$P = W/t \quad \text{(joules/second)}$$

$$\therefore \quad P = EI = W/t \quad \text{(watts)}$$

Additional forms of the formula for power may be derived by substituting forms of Ohm's law ($E = IR$ and $I = E/R$) into Equation 7–15. Thus, substituting IR for E in $P = EI$ results in

$$P = IR \times I$$

$$P = I^2R \quad \text{(watts)} \tag{7–16}$$

Substituting E/R for I in $P = EI$ results in

$$P = E \times E/R$$
$$P = E^2/R \quad \text{(watts)} \qquad \text{(7–17)}$$

The power formulas are summarized in Table 7–3.

TABLE 7–3 Power Formulas	
Formula	**Equation Number**
$P = W/t$	(7–14)
$P = EI$ or $P = VI$	(7–15)
$P = I^2R$	(7–16)
$P = E^2/R$ or $P = V^2/R$	(7–17)

EXAMPLE 7–16   An electric iron is attached to a 120-V ac source. Determine the power rating of the iron if 11.0 A of current passes through the iron.

Solution   Use $P = EI$ (Equation 7–15):

$$P = 120 \times 11.0$$
$$P = 1320 \text{ W}$$
$$\therefore \quad P = 1.32 \text{ kW}$$

EXAMPLE 7–17   When electric current passes through a copper conductor, the wire is heated and energy is lost. Determine the rate of the energy loss in watts when 46.0 A is passed through a 6-gauge copper conductor with a resistance of 0.790 Ω.

Solution   Use $P = I^2R$ (Equation 7–16):

$$P = 46.0^2 \times 0.790$$
$$P = 1672 \text{ W}$$
$$\therefore \quad P = 1.67 \text{ kW}$$

Observation   The size and the resistance per unit length of an electrical conductor is specified by the American Wire Gauge (AWG), as shown in Appendix A.

EXAMPLE 7–18   Determine the resistance of a 120-V, 40-W incandescent lamp.

Solution   Use $P = E^2/R$ (Equation 7–17) and solve for R:

$$R = E^2/P$$
$$R = 120^2/40$$
$$\therefore \quad R = 360 = 0.36 \text{ k}\Omega$$

Observation Electrical equipment is commonly specified by the voltage and the energy rate (power) needed for operation.

Power Considerations for Resistors

Resistors are rated by their resistive properties in ohms and by their rate of converting electric energy to heat energy in watts (commonly called power dissipation). Thus, a carbon-film resistor may have a rating of 470 Ω, $\frac{1}{2}$ W, while a wire-wound resistor might be rated as 500 Ω, 10 W.

Although resistors may be operated near their power ratings, it generally is a good practice to operate them below their rated value.

EXAMPLE 7–19 A 1.0-kΩ, $\frac{1}{2}$-W resistor has 25 V dropped across it.

Determine:

(a) The power being dissipated by the resistor
(b) Whether the resistor is being operated within its rating

Solution (a) Select the equation from Table 7–3 that relates power to voltage and resistance:

$$P = V^2/R \quad \text{(Equation 7–17)}$$
$$P = 25^2/(1.0 \times 10^3)$$
$$\therefore \quad P = 0.625 \text{ W}$$

(b) Determine whether the resistor is being operated within its rating:

$$625 \text{ mW} > 500 \text{ mW}$$

$$\therefore \quad \text{No. The wattage rating is exceeded by 125 mW.}$$

EXAMPLE 7–20 Select the wattage rating for an 8.0-kΩ wire-wound resistor that has 39 mA passing through it. This resistor is available as an 8-, 12-, or 20-W unit.

Solution Select the equation from Table 7–3 that relates power to current and resistance:

$$P = I^2R \quad \text{(Equation 7–16)}$$
$$P = (39 \times 10^{-3})^2 \times 8.0 \times 10^3$$
$$P = 12 \text{ W}$$

$$\therefore \quad \text{The 20-W unit is selected.}$$

Total Power of Several Loads

The power dissipated in an electrical circuit may be computed by adding the power dissipation of each of the loads connected to the source. Thus,

$$P_T = P_1 + P_2 + P_3 \quad \text{(watts)} \tag{7–18}$$
$$P_T = \Sigma P \quad \text{(watts)} \tag{7–19}$$

where P_T is the total power dissipation in watts (W).

EXAMPLE 7–21
Two resistances ($R_1 = 200\ \Omega$ and $R_2 = 100\ \Omega$) are connected in series with a 180-V source. Determine:

(a) The power dissipation of each resistance
(b) The power dissipation of the total circuit
(c) That $P_T = P_1 + P_2$

Solution
(a) First solve for I in the series circuit:

$$I = E/R_T$$
$$I = 180/300 = 600\ \text{mA}$$

Then substitute I into Equation 7–16 to solve for power:

$$P_1 = I^2R_1$$
$$P_1 = (600 \times 10^{-3})^2 \times 200$$
∴ $\quad P_1 = 72.0\ \text{W in the 200-}\Omega\text{ resistance}$

$$P_2 = I^2R_2$$
$$P_2 = (600 \times 10^{-3})^2 \times 100$$
∴ $\quad P_2 = 36.0\ \text{W in the 100-}\Omega\text{ resistance}$

(b) Compute the total power (P_T) from the total resistance ($R_T = 300\ \Omega$) and the source voltage, 180 V:

$$P_T = E^2/R_T$$
$$P_T = (180)^2/300$$
∴ $\quad P_T = 108\ \text{W}$

(c) Compute P_T using Equation 7–18:

$$P_T = P_1 + P_2$$
$$P_T = 72 + 36$$
$$P_T = 108\ \text{W}$$

∴ $\quad$ The two methods for calculating P_T give the same answer.

EXERCISE 7–6

1. Find the rate, in watts, at which a space heater converts electric energy into heat if 82.8 kJ of heat is produced for each minute (1.00 min) of operation.

2. Determine the time in seconds needed to deliver 680 J of energy to a 75-W incandescent lamp.

3. Determine the amount of energy, in joules, dissipated by a 1.20-kW toaster oven in 1 week if it is used an average of 18.6 min/d.

4. Calculate the power rating, in watts, of a computer monitor that requires 115 V at 1.37 A for its operation.

5. Determine the amount of energy, in joules, delivered to an external disk drive for 1 month if it is operated for 100 h/month. The drive is supplied from a 5.00-V dc source, and it requires 850 mA for its operation.

6. A 120-V, 1.025-kW electric coffeemaker is operated 25.0 min/d. Determine:

   (a) The resistance of the heating element in ohms

   (b) The quantity of electrical energy, in joules, converted to heat during the 25.0 min of daily operation

7. If a 120-V, 16.0-W DVD player is operated for 2.86 h, determine:

   (a) The electric current in the conductors connecting the player to the source of electrical energy

   (b) The quantity of electrical energy, in joules, delivered to the player during its 2.86 h of operation

8. If a 330-Ω, metal-film resistor is available with 0.5-W, 1.0-W, and 2.0-W ratings:

   (a) Determine the power, in watts, dissipated by the resistor when 65.0 mA of current is passing through it.

   (b) Select one of the three specified power ratings for this resistor.

9. Compute the maximum voltage that can be dropped across an 8.0-W, 1.25-kΩ resistor without exceeding the power rating.

10. Determine the maximum current that can be passed through a 0.50-W, 560-Ω resistor without exceeding the power rating.

11. The service entrance to a home (Figure 7–28) is rated at 240 V, 100 A. Determine:

    (a) How much power is available to the home

    (b) The total power (P_T) of the following loads:

       (1) A 6.90-kW electric range

       (2) A refrigerator/freezer, 120 V, 4.35 A

       (3) A heat pump, 230 V, 17.4 A

       (4) An electric motor, 230 V, 8.20 A

       (5) A 4850-W clothes dryer

    (c) If all the loads in (b) can operate simultaneously without an overload

12. Three resistors, $R_1 = 62.0\ \Omega$, $R_2 = 91.0\ \Omega$, and $R_3 = 51.0\ \Omega$, are connected in series across a 12-V alternator. Draw and label a schematic and determine:

    (a) The circuit current

    (b) The power dissipated by each resistor

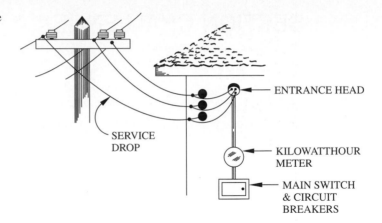

FIGURE 7–28 Electrical service entrance for Exercise 7–6, problem 11.

ENTRANCE HEAD

SERVICE DROP

KILOWATTHOUR METER

MAIN SWITCH & CIRCUIT BREAKERS

(c) The total power provided by the alternator using $P_T = E^2/R_T$

(d) That $P_T = P_1 + P_2 + P_3$

SELECTED TERMS

conductivity   Indication of the ability of a material to pass an electric current.

current   Flow of electric charge along a conductor in a closed circuit.

equivalent resistance   The combining of several resistances in a circuit into a single resistance called the equivalent resistance, R_T.

Ohm's law   Relationship between voltage and current formulated by Georg Simon Ohm; $E = IR$.

resistance   Quotient of voltage divided by current; that is, $R = V/I$. A substance has a resistance of 1 ohm when the application of 1 volt results in a current of 1 ampere.

voltage   Name given to energy added per unit of electric charge as the charge passes through a source.

SECTION CHALLENGE

WEB CHALLENGE FOR CHAPTERS 4, 5, 6, AND 7

To evaluate your comprehension of Chapters 4, 5, 6, and 7, log on to **www.prenhall. com/harter** and take the online True/False and Multiple Choice assessments for each of the chapters.

SECTION CHALLENGE FOR CHAPTERS 4, 5, 6, AND 7*

A. Using the characteristics of the series circuit ($E = V_1 + V_2 + V_3$ and $V = IR$), your challenge is to write an equation for R_E (pictured in Figure C–2) in terms of E, V_{BE}, I_B, and I_E. Start the process by replacing the voltage drops V_1, V_2, and V_3 with V_{60RE}, V_{BE}, and V_{RE}, the voltages across $60R_E$, B-E, and R_E respectively.

FIGURE C–2 A transistor's base-emitter (B-E) equivalent circuit is a series-like circuit.

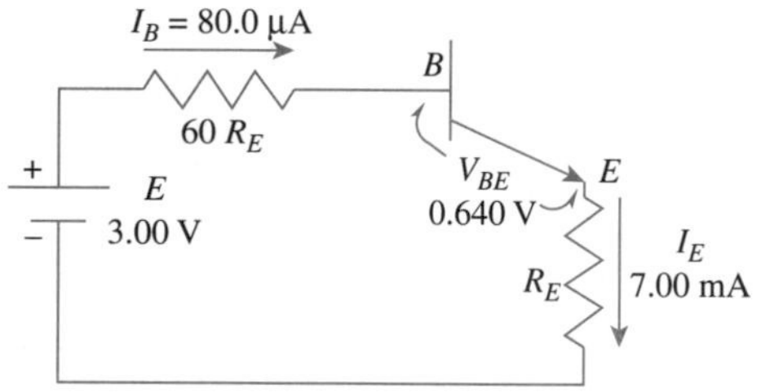

*The solution to this Section Challenge is found in Appendix C.

Once the source voltage, E, is expressed as the sum of the voltage drops, then modify the equation by replacing V_{60RE} with the product of I_B and $60R_E$ and V_{RE} with the product of I_E and R_E. The resulting equation is solved for R_E in terms of E, V_{BE}, I_B, and I_E.

B. Your next challenge is to substitute the values of E, V_{BE}, I_B, and I_E (given in Figure C–2) into the equation for R_E and solve for resistors R_E and $60R_E$ ($60 \times R_E$).

8

Fractions

8-1 Introductory Concepts

8-2 Forming Equivalent Fractions

8-3 Simplifying Fractions

8-4 Multiplying Fractions

8-5 Dividing Fractions

8-6 Complex Fractions

8-7 Adding and Subtracting Fractions

8-8 Changing a Mixed Expression to a Fraction

8-9 Additional Work with Complex Fractions

PERFORMANCE OBJECTIVES

- Identify and name the parts of a fraction.

- Form an equivalent fraction.

- Reduce a fraction to lowest terms by removing common factors.

- Form a common denominator to add fractions.

- Apply multiplication, division, addition, and subtraction to fractions.

- Convert a mixed expression to a fraction.

- Change a complex fraction to a simple fraction.

The F-16 Fighting Falcon fighter jet relies on advanced avionics for its state-of-the-art performance. (Courtesy of U.S. Air Force)

In this chapter we are concerned with both arithmetic fractions, such as $\frac{1}{2}$, $\frac{3}{5}$, and $\frac{5}{8}$, and algebraic fractions, such as 3/*a, x/y,* and (*x* − 2)/4. Fractions are used in many equations that describe electrical and electronic principles and circuits.

8–1 INTRODUCTORY CONCEPTS

The following concepts apply to both arithmetic and algebraic fractions. These concepts are basic to working with fractions.

- As noted in Figure 8–1, a fraction indicates division. Thus, 8 ÷ 4, 8/4, and $\frac{8}{4}$ all indicate division. Furthermore, $\frac{8}{4}$ and 8/4 are the same fraction printed in different styles.

FIGURE 8–1 A fraction indicates division. The bar is a sign of grouping for the numerator and the denominator.

$$\frac{\text{Numerator}}{\text{Denominator}} = \text{Numerator/Denominator}$$
$$= \text{Numerator} \div \text{Denominator}$$

- The fraction 9*y*/3 is read "9*y* divided by 3" or "9*y* over 3."
- The numerator is the quantity above the bar.
- The denominator is the quantity below the bar.
- The numerator and denominator of a fraction may be called the **parts of the fraction.**
- The bar separating the parts of the fraction is called the **vinculum.** The vinculum is a sign of grouping for the numerator and the denominator.
- Fractions may be written in two forms, $\frac{xy}{4}$ and $\frac{a - 2b}{5}$ or $\frac{1}{4}xy$ and $\frac{1}{5}(a - 2b)$. It is important to recognize that the two forms are equal. Thus, $\frac{xy}{4} = \frac{1}{4}xy$ and $\frac{a - 2b}{5} = \frac{1}{5}(a - 2b)$.
- A fraction has three signs: the sign of the fraction, the sign of the numerator, and the sign of the denominator. When the sign is positive, it usually is not written—it is implied. Thus, $\frac{1}{2}$ means $+ \frac{+1}{+2}$.
- Division by zero is not permitted. If the denominator evaluates to zero, then the fraction is not defined. Thus, $\frac{7 + x}{x - 3}$ is an **undefined fraction** when *x* = 3, because the denominator evaluates to zero. $\frac{7 + 3}{3 - 3} = \frac{10}{0}$. Try dividing zero into ten with your calculator.

- The numerator and denominator of a fraction may be multiplied or divided by the same number or expression without changing the value of the fraction. Thus,

$$\frac{1}{4} = \frac{1 \times 3}{4 \times 3} = \frac{3}{12} \quad \therefore \quad \frac{1}{4} \text{ and } \frac{3}{12} \text{ are equivalent fractions}$$

$$\frac{5}{15} = \frac{5 \div 5}{15 \div 5} = \frac{1}{3} \quad \therefore \quad \frac{5}{15} \text{ and } \frac{1}{3} \text{ are equivalent fractions}$$

- A very important set of equivalent fractions has the value 1. Some of its members are

$$1 = \frac{1}{1} = \frac{2}{2} = \frac{3}{3} = \frac{1059}{1059} = \frac{-64.2}{-64.2} = \frac{14a}{14a}, \text{ and so forth}$$

- Any fraction may be multiplied or divided by 1 without changing its value. Thus,

$$\frac{2a}{9} = \frac{2a}{9} \times 1 = \frac{2a}{9}\left(\frac{5\beta}{5\beta}\right)$$

$$\frac{3}{T-5} = \frac{3}{T-5} \times 1 = \frac{3}{T-5}\left(\frac{29}{29}\right)$$

EXAMPLE 8–1 In the expression $\dfrac{x}{21 - 3x}$, determine the value of x for which the fraction is undefined.

Solution Set the denominator equal to zero and solve the resulting equation for x:

$$21 - 3x = 0$$
$$-3x = 0 - 21$$
$$3x = 21$$
$$x = 7$$

$\therefore$ When $x = 7$, the denominator is zero, and the fraction is undefined.

EXERCISE 8–1

Determine the value of the variable in each of the following fractions that causes the fraction to be undefined. Do this by setting the denominator equal to zero and solving the resulting equation.

1. $7/x$ **2.** $c/(5c)$ **3.** $8/(y + 1)$ **4.** $\dfrac{t}{t + 2}$ **5.** $\dfrac{a + 1}{2a - 1}$

6. $\dfrac{2}{2x + 4}$ **7.** $\dfrac{10a}{15 - 3a}$ **8.** $\dfrac{m - 6}{m - 6}$ **9.** $\dfrac{3x + 4}{10 - 2x}$ **10.** $\dfrac{7}{-5\mu - 8}$

8–2 FORMING EQUIVALENT FRACTIONS

A fraction may have its form changed by multiplying both parts of the fraction by the same number. This is done to shape the denominator into a desired form. The value of the fraction that results from multiplying both numerator and denominator by the same number has not been changed—only the form has been changed. Thus, $\frac{1}{3}$, $\frac{2}{6}$, and $\frac{3}{9}$ are **equivalent fractions**; that is, they have the same value.

EXAMPLE 8–2   Change $\frac{2}{5}$ to an equivalent fraction with -15 as the denominator.

Solution   Multiply both 2 and 5 by -3:

M: $$\frac{2}{5}\left(\frac{-3}{-3}\right) = \frac{-6}{-15}$$

∴ $$\frac{2}{5} = \frac{-6}{-15}$$

> **RULE 8–1.   Multiplication Property of a Fraction**
>
> Multiplying both parts of a fraction by the same number (except zero) results in an equivalent fraction. Thus,
>
> $$\frac{a}{b} = \frac{a}{b}\left(\frac{c}{c}\right) = \frac{ac}{bc} \qquad c \neq 0$$

EXAMPLE 8–3   Form $\frac{3}{8}$ into an equivalent fraction having a denominator of $24 - 16x$.

Solution $$\frac{3}{8} = \frac{?}{24 - 16x}$$

Factor a common monomial of 8 out of $24 - 16x$:

$$24 - 16x = 8(3 - 2x)$$

Form the equivalent fraction by multiplying both 3 (the numerator) and 8 (the denominator) by $(3 - 2x)$:

$$\frac{3}{8} = \frac{3}{8}\left(\frac{3 - 2x}{3 - 2x}\right) = \frac{9 - 6x}{24 - 16x}$$

∴ $$\frac{3}{8} = \frac{9 - 6x}{24 - 16x}$$

We see in the preceding examples that multiplying a fraction by one, $(-3/-3)$ in Example 8–2 and $(3 - 2x)/(3 - 2x)$ in Example 8–3, changed the way the fraction looks, but it has not changed the value of the fraction.

EXERCISE 8–2

Form the given fraction into equivalent fraction(s) having the specified denominator(s).

1. $\dfrac{1}{5} = \dfrac{?}{10} = \dfrac{?}{60}$

2. $\dfrac{1}{7} = \dfrac{?}{49} = \dfrac{?}{35}$

3. $\dfrac{1}{13} = \dfrac{?}{-39} = \dfrac{?}{52}$

4. $\dfrac{1}{2} = \dfrac{?}{32} = \dfrac{?}{-16}$

5. $\dfrac{-3}{7} = \dfrac{?}{77} = \dfrac{?}{-21}$

6. $\dfrac{27}{37} = \dfrac{?}{74} = \dfrac{?}{111}$

7. $\dfrac{m}{n} = \dfrac{?}{5n} = \dfrac{?}{-8n}$

8. $\dfrac{x}{y} = \dfrac{?}{cy} = \dfrac{?}{2yb}$

9. $\dfrac{s}{t} = \dfrac{?}{t^2} = \dfrac{?}{-2t^3}$

10. $\dfrac{-2a}{b} = \dfrac{?}{4ab} = \dfrac{?}{-3b^2}$

11. $\dfrac{2y}{-5x} = \dfrac{?}{30x^2} = \dfrac{?}{15xy^2}$

12. $\dfrac{3}{x - y} = \dfrac{?}{-5x + 5y}$

13. $\dfrac{b}{a - b} = \dfrac{?}{2b^2 - 2ab}$

14. $\dfrac{2x + y}{x^2 - y} = \dfrac{?}{3y^2z - 3x^2yz}$

8–3 SIMPLIFYING FRACTIONS

A fraction is in its simplest form when the numerator and denominator have no factors in *common* except the number 1. The process of simplifying fractions is called **reducing the fraction to its lowest terms.** To reduce a fraction to its lowest terms, divide the numerator and denominator by their *common factors.* If the *same* factor occurs in both parts of the fraction, then it is a common factor. Rule 8–2 shows the steps in reducing fractions to their lowest terms.

> **RULE 8–2. Reducing Fractions to Lowest Terms**
>
> To reduce a fraction to lowest terms:
> 1. Factor each part of the fraction.
> 2. Group the common factors.
> 3. Remove the common factors from each part of the fraction.

EXAMPLE 8–4   Reduce $\dfrac{4a + 8}{a + 2}$ to lowest terms.

Solution   Use the steps of Rule 8–2.

Step 1:   Factor:

$$\frac{4(a + 2)}{a + 2}$$

Step 2:   Group the common factors:

$$\frac{4(a + 2)}{1(a + 2)}$$

Step 3:   Remove the common factors:

4/1

$\therefore$ $\dfrac{4a + 8}{a + 2} = \dfrac{4(a + 2)}{a + 2} = 4$

EXAMPLE 8–5   Simplify $\dfrac{3x^2 y}{6xy^2}$.

Solution   Follow the steps of Rule 8–2.

Step 1:   Factor the numerator and denominator:

$$\frac{3 \cdot x \cdot x \cdot y}{2 \cdot 3 \cdot x \cdot y \cdot y}$$

Step 2:   Group factors:

$$\frac{1}{2}\left(\frac{3}{3}\right)\left(\frac{x}{x}\right)\frac{x}{1}\left(\frac{y}{y}\right)\frac{1}{y}$$

Step 3:   Remove the common factors:

$$\frac{1}{2} \cdot \frac{x}{1} \cdot \frac{1}{y} = \frac{x}{2y}$$

$\therefore$ $\dfrac{3x^2 y}{6xy^2} = \dfrac{x}{2y}$

EXERCISE 8–3

Reduce each fraction to its lowest terms.

1. 16/32
2. 12/18
3. 14/35
4. x/x^2
5. $15b/10$
6. $27c^2/18$

7. $\dfrac{26a^2b}{26ab}$

8. $\dfrac{6xy}{15y^3}$

9. $\dfrac{22d^2c}{121d}$

10. $\dfrac{m^2n^2}{mn}$

11. $\dfrac{x^8}{x^3}$

12. $\dfrac{y^5}{y^{13}}$

13. $\dfrac{39a^4b^5}{26a^2b^2}$

14. $\dfrac{72x^2y}{18x^2y}$

15. $\dfrac{(a+1)(a-1)}{a+1}$

16. $\dfrac{8m+24}{7m+21}$

17. $\dfrac{t^2-2t}{4t-8}$

18. $\dfrac{3a+a^2}{15a+5a^2}$

19. $\dfrac{x}{x^2+x}$

20. $\dfrac{(3x-3)(2x+2)}{6x-6}$

21. $\dfrac{\theta\beta^3+\beta^2\theta^2}{\beta+\theta}$

8–4 MULTIPLYING FRACTIONS

When two or more fractions are multiplied, the product that results is formed by multiplying the numerators and dividing by the product of the denominators. Rule 8–3 will guide you in this process.

> **RULE 8–3.   Multiplying Fractions**
>
> The product of two or more fractions results in a new fraction, which is formed by:
> 1. Multiplying the numerators together, forming the numerator of the new fraction.
> 2. Multiplying the denominators together, forming the denominator of the new fraction.
> 3. Multiplying the signs of the fractions together, forming the sign of the new fraction.
> 4. Simplifying the new fraction by reducing to lowest terms.

EXAMPLE 8–6   Multiply -15 by $\frac{3}{5}$.

Solution   Multiply numerators and denominators together:

$$\frac{-15}{1} \times \frac{3}{5} = \frac{-15 \times 3}{1 \times 5}$$

Remove common factors of 5 and multiply:

$$\frac{-3 \cdot 5 \times 3}{1 \times 5} = \frac{-3 \times 3}{1}\left(\frac{5}{5}\right) = -9$$

$$\therefore \quad -15 \times \tfrac{3}{5} = -9$$

EXAMPLE 8–7   Multiply $\frac{4}{5}$ by $-\frac{2}{3}$.

Solution   Multiply numerators, denominators, and signs together:

$$\frac{4}{5}\left(-\frac{2}{3}\right) = (+\ -)\frac{4\times 2}{5\times 3} = -\frac{8}{15}$$

$$\therefore\qquad \frac{4}{5}\left(-\frac{2}{3}\right) = -\frac{8}{15}$$

EXAMPLE 8–8   Multiply $\dfrac{4xy^2}{-6}$ by $\dfrac{18}{8x^2}$.

Solution   Perform the first three steps of Rule 8–3 and factor:

$$\frac{4xy^2}{-6}\times\frac{18}{8x^2} = \frac{4\cdot x\cdot y^2\cdot 6\cdot 3}{-6\cdot 4\cdot 2\cdot x\cdot x}$$

Simplify by removing common factors:

$$\left(\frac{1}{-1}\right)\left(\frac{6}{6}\right)\left(\frac{4}{4}\right)\left(\frac{x}{x}\right)\left(\frac{y^2}{1}\right)\left(\frac{3}{2}\right)\left(\frac{1}{x}\right) = \frac{3y^2}{-2x}$$

$$\therefore\qquad \frac{4xy^2}{-6}\times\frac{18}{8x^2} = \frac{3y^2}{-2x}$$

EXAMPLE 8–9   Multiply $-\dfrac{5x^2-25x}{3x-15}$ by $\dfrac{12xy^2}{10x^3y+10x^2}$.

Solution   Factor both numerator and denominator. Remember that the numerator and the denominator are grouped by the vinculum. Simplify by removing common factors:

$$-\frac{5x(x-5)}{3(x-5)}\times\frac{3\cdot 4(xy^2)}{10x^2(xy+1)}$$

$$= -\left(\frac{5x}{10x^2}\right)\left(\frac{x-5}{x-5}\right)\left(\frac{3}{3}\right)\left(\frac{4}{1}\right)\left(\frac{xy^2}{xy+1}\right)$$

$$= -\frac{20x^2y^2}{10x^2(xy+1)}$$

Remove the common factors of $10x^2$:

$$-\frac{10x^2(2y^2)}{10x^2(xy+1)} = -\frac{2y^2}{xy+1}$$

$$\therefore\qquad -\frac{5x^2-25x}{3x-15}\times\frac{12xy^2}{10x^3y+10x^2} = -\frac{2y^2}{xy+1}$$

Multiply the following fractions.

1. $\dfrac{2}{3} \times \dfrac{6}{8}$ **2.** $\dfrac{5}{9} \times \dfrac{3}{5}$ **3.** $\dfrac{-3}{4} \times \dfrac{8}{15}$

4. $\dfrac{5}{7} \times \dfrac{3}{4} \times \dfrac{14}{15}$ **5.** $\dfrac{-7}{13} \times \dfrac{10}{35} \times \dfrac{39}{40}$ **6.** $\dfrac{-5}{8} \times \dfrac{4}{17} \times \dfrac{51}{60}$

7. $\dfrac{-a}{b} \cdot \dfrac{b}{a}$ **8.** $\dfrac{4m}{9} \cdot \dfrac{1}{2m}$ **9.** $\dfrac{2x}{y} \cdot \dfrac{-3y}{x}$

10. $\dfrac{n}{2m} \cdot \dfrac{6}{p}$ **11.** $\dfrac{2x + y}{6z} \cdot \dfrac{12z}{y}$ **12.** $\dfrac{3a + 2a}{2 + 7} \cdot \dfrac{3a}{5a^2}$

13. $\dfrac{-a}{c + b} \cdot \dfrac{c - b}{a}$ **14.** $\dfrac{x - y}{x + y} \cdot \dfrac{x + y}{x - y}$ **15.** $\dfrac{15}{a^4} \cdot \dfrac{4ab}{9}$

16. $\dfrac{m^3}{12} \cdot \dfrac{1}{m^4} \cdot 3m$ **17.** $\dfrac{2\pi r^2}{1} \cdot \dfrac{1}{8r}$ **18.** $\dfrac{2x - 10}{3x + 9} \cdot \dfrac{3 + x}{x - 5}$

19. $\dfrac{x^2 - x}{2xy - 2y} \cdot \dfrac{2y}{3x}$ **20.** $\dfrac{6a - 12}{4a + 4} \cdot \dfrac{(a^2 + a)(a - 1)}{2a^2 - 4a}$

8–5 DIVIDING FRACTIONS

To divide one fraction by another, ***invert the divisor and multiply.*** That is, multiply the dividend by the reciprocal of the divisor.

EXAMPLE 8–10   Solve $\frac{2}{6} \div \frac{2}{3}$.

Solution   Take the reciprocal of the divisor, $\frac{2}{3}$, and multiply the dividend, $\frac{2}{6}$.

$$\frac{2}{6} \div \frac{2}{3} = \frac{2}{6} \times \frac{3}{2}$$

Multiply and remove the common factor:

$$\frac{6}{6 \times 2} = \frac{1}{2}$$

$$\therefore \quad \frac{2}{6} \div \frac{2}{3} = \frac{1}{2}$$

EXAMPLE 8–11   Divide $\dfrac{2a + 2b}{6}$ by $\dfrac{a + b}{3}$.

Solution   Invert and multiply:

$$\frac{2a + 2b}{6} \div \frac{a + b}{3} = \frac{2a + 2b}{6} \cdot \frac{3}{a + b}$$

Factor and simplify:

$$\frac{2(a + b)}{6} \cdot \frac{3}{a + b} = \frac{6(a + b)}{6(a + b)} = 1$$

$$\therefore \quad \frac{2a + 2b}{6} \div \frac{a + b}{3} = 1$$

EXAMPLE 8–12   Solve $\dfrac{2mn}{(m - n)^2} \div \dfrac{4n}{2m - 2n}$

Solution   Invert and multiply:

$$\frac{2mn}{(m - n)^2} \cdot \frac{2m - 2n}{4n}$$

Factor and simplify:

$$\frac{2mn}{(m - n)(m - n)} \cdot \frac{2(m - n)}{4n}$$

$$= \frac{4mn(m - n)}{4n(m - n)(m - n)} = \frac{m}{m - n}$$

$$\therefore \quad \frac{2mn}{(m - n)^2} \div \frac{4n}{2m - 2n} = \frac{m}{m - n}$$

EXERCISE 8–5

Divide the following fractions.

1. $\dfrac{3}{8} \div \dfrac{1}{4}$

2. $\dfrac{4}{9} \div \dfrac{4}{3}$

3. $\dfrac{3}{10} \div \dfrac{2}{5}$

4. $\dfrac{-3}{16} \div \dfrac{9}{2}$

5. $\dfrac{-3}{4} \div \dfrac{-5}{7}$

6. $\dfrac{x}{y} \div z$

7. $\dfrac{2a}{b} \div \dfrac{a}{2b}$ **8.** $\dfrac{-6m}{7n} \div \dfrac{3m}{-14n}$

9. $\dfrac{3}{2x} \div \dfrac{3y}{2x}$ **10.** $\dfrac{-4ab}{6cb} \div \dfrac{8b}{12ac}$

11. $\dfrac{2x^2}{y^2} \div \dfrac{2xy}{y^2}$ **12.** $\dfrac{b^2}{6a^2} \div \dfrac{3b}{2a}$

13. $\dfrac{-(3a-6)}{9} \div \dfrac{1}{3}$ **14.** $\dfrac{(x-3)(x+3)}{15} \div \dfrac{x-3}{5}$

15. $\dfrac{a^2-ab}{b} \div \dfrac{a}{b}$ **16.** $\dfrac{b}{b-3} \div \dfrac{b^3}{2}$

8–6 COMPLEX FRACTIONS

In the preceding section we learned that division of fractions is carried out by multiplying with the reciprocal of the divisor. The arithmetic operator ÷ was used to indicate division.

In this section we will indicate division of fractions by forming a **complex fraction.** A complex fraction is one that has a fraction in the numerator or the denominator or in both the numerator and the denominator.

$$\frac{\frac{2}{3}}{\frac{3}{4}}, \qquad \frac{\frac{a}{2}}{\frac{b}{4}} \quad \text{and} \quad \frac{a+b}{\frac{a}{b}}$$

are complex fractions.

EXAMPLE 8–13   Write $\dfrac{x-y}{3} \div \dfrac{x-y}{6}$ as a complex fraction.

Solution $\dfrac{x-y}{3} \div \dfrac{x-y}{6} = \dfrac{\dfrac{x-y}{3}}{\dfrac{x-y}{6}}$

Observation   The vinculum of the main fraction is made longer and bolder to distinguish the numerator from the denominator.

To simplify a complex fraction having a simple fraction as the numerator and denominator, divide the denominator into the numerator by:

1. Reciprocating the denominator.
2. Multiplying the numerator by the reciprocal of the denominator.
3. Reducing to lowest terms.

EXAMPLE 8–14   Simplify: $\dfrac{\dfrac{xy}{5}}{\dfrac{2x}{10y}}$.

Solution   Follow the steps of Rule 8–4.

Step 1:   Reciprocate the denominator:

$$\frac{2x}{10y} \Rightarrow \frac{10y}{2x}$$

Step 2:   Multiply by the reciprocal of the denominator:

$$\frac{xy}{5} \cdot \frac{10y}{2x}$$

Step 3:   Remove common factors:

$$\frac{xy}{5} \cdot \frac{10y}{2x} = y^2$$

$$\therefore \quad \frac{xy}{5} \Big/ \frac{2x}{10y} = \frac{xy}{5} \cdot \frac{10y}{2x} = y^2$$

EXAMPLE 8–15   Form $\dfrac{mn + m}{n} \div \dfrac{n + 1}{n^2}$ into a complex fraction, and then simplify.

Solution $\dfrac{mn + m}{n} \div \dfrac{n + 1}{n^2} = \dfrac{\dfrac{mn + m}{n}}{\dfrac{n + 1}{n^2}}$

Take the reciprocal of the denominator $\frac{n + 1}{n^2}$ and multiply:

$$\frac{mn + m}{n} \cdot \frac{n^2}{n + 1}$$

Factor and remove common factors:

$$\frac{m(n + 1)}{n} \cdot \frac{n \cdot n}{n + 1}$$

$$= mn\left(\frac{n}{n}\right)\left(\frac{n + 1}{n + 1}\right) = mn$$

$$\therefore \quad \frac{mn + m}{n} \div \frac{n + 1}{n^2} = mn$$

EXERCISE 8–6

Form each of the following expressions into a complex fraction.

1. $\frac{8}{3} \div \frac{4}{5}$

2. $\frac{3}{10} \div \frac{2}{5}$

3. $\frac{-9}{2} \div \frac{3}{16}$

4. $z \div \frac{x}{y}$

5. $\frac{-6x}{7y} \div \frac{3x}{14y}$

6. $\frac{4ab}{6cb} \div \frac{ab}{cb}$

7. $\frac{R^2}{-t} \div \frac{2tR}{t^2}$

8. $\frac{1}{3} \div \frac{3a - b}{9}$

9. $\frac{b}{b - 3} \div \frac{b^2}{2}$

10. $\frac{3x + 3}{2y} \div \frac{6 + 6x}{y}$

Simplify the following complex fractions.

11. $\dfrac{3}{\frac{3}{10}}$

12. $\dfrac{\frac{-39}{5}}{\frac{13}{10}}$

13. $\dfrac{\frac{18}{20}}{\frac{6}{5}}$

14. $\dfrac{\frac{x + 1}{x^2}}{\frac{yx + y}{x}}$

15. $\dfrac{\frac{3a + 3b}{a - b}}{\frac{6b + 6a}{a - b}}$

16. $\dfrac{\frac{a + b}{b}}{\frac{a - b}{b}}$

17. $\dfrac{\dfrac{x - y}{y}}{\dfrac{x + y}{y}}$ **18.** $\dfrac{a}{\dfrac{a - 1}{a}}$

19. $\dfrac{a + 5}{\dfrac{(a + 5)(a - 5)}{a}}$ **20.** $\dfrac{\dfrac{2a}{x - 3y}}{\dfrac{2ab}{2x - 6y}}$

8–7 ADDING AND SUBTRACTING FRACTIONS

Every fraction has three signs: (1) the sign of the numerator, (2) the sign of the denominator, and (3) the sign of the vinculum (also called the sign of the fraction). When no sign is written, it is understood to be positive. This is an *implied* sign.

It is easier to add fractions when the sign of the vinculum is positive. To change the sign of the vinculum of the fraction, apply this rule:

> **RULE 8–5. Changing Sign**
>
> To change the signs of a fraction, change any two of the three signs: the vinculum and the numerator; or the vinculum and the denominator; or the numerator and the denominator.

We see from Rule 8–5 that the sign of the vinculum and one other sign must be changed. When the sign of the vinculum is changed, either the sign of the numerator or the denominator must also be changed. Thus, $-\dfrac{a}{2b}$ may be written with a positive vinculum if we change either the sign of the numerator, $+\dfrac{-a}{2b}$, or the sign of the denominator, $+\dfrac{a}{-2b}$. So

$$-\frac{a}{2b} = \frac{-a}{2b} = \frac{a}{-2b}$$

EXAMPLE 8–16   Change the sign of the vinculum of $-\dfrac{x - 2y}{5}$.

Solution   Change two of the three signs: the vinculum and the numerator; leave the sign of the denominator unchanged.

$$-\frac{x - 2y}{5} = +\frac{-(x - 2y)}{5}$$

Distribute the sign change through the numerator:

$$\frac{-x + 2y}{5} = \frac{2y - x}{5}$$

$$\therefore \quad -\frac{x - 2y}{5} = \frac{2y - x}{5}$$

In Example 8–16 we chose to change the sign of the numerator along with the sign of the vinculum. The choice was made through the application of the following guidelines.

GUIDELINES 8–1.   *Changing the Sign of the Vinculum*

When changing the sign of the vinculum, select:

1. The numerator for the other sign to change, if both the denominator and the numerator are positive.

2. The denominator for the other sign to change, if both the numerator and the denominator are negative.

3. The one that is negative, if either the numerator or denominator is negative.

EXAMPLE 8–17   Change the sign of the vinculum of the following fractions by applying Guidelines 8–1:

(a) $-\dfrac{4}{9}$ (b) $-\dfrac{1}{-2}$ (c) $-\dfrac{-5}{-8}$

Solution   Using Guidelines 8–1:

$\therefore$ (a) $-\dfrac{4}{9} = +\dfrac{-4}{9} = \dfrac{-4}{9}$

$\therefore$ (b) $-\dfrac{1}{-2} = +\dfrac{1}{+2} = \dfrac{1}{2}$

$\therefore$ (c) $-\dfrac{-5}{-8} = +\dfrac{-5}{+8} = \dfrac{-5}{8}$

EXERCISE 8–7

Change the sign of the vinculum in each of the following fractions by applying Guidelines 8–1.

1. $-\dfrac{1}{4}$ 2. $-\dfrac{-3}{8}$ 3. $-\dfrac{-4}{-5}$

4. $-\dfrac{-7}{-9}$

5. $-\dfrac{x-y}{3a}$

6. $-\dfrac{5t}{-2-s}$

7. $-\dfrac{3y+4}{2x+1}$

8. $-\dfrac{-a^2+a-2}{3b}$

9. $-\dfrac{-2\pi R}{-E-1}$

10. $-\dfrac{-5a-3(b-1)}{a+b}$

11. $-\dfrac{5\omega+\beta}{-6-\beta}$

12. $-\dfrac{(x+y)}{-5}$

Common Denominator

Before fractions may be added, they must have the *same* denominator or a **common denominator.** Each of the fractions to be added is first changed to an equivalent fraction having a common denominator.

EXAMPLE 8–18 Change each fraction to an equivalent fraction having a common denominator of $36a^2$.

(a) $\dfrac{3}{12a}$ (b) $\dfrac{8}{18a^2}$ (c) $\dfrac{5a}{6}$

Solution Form equivalent fractions having $36a^2$ as the denominator.

(a) Multiply $3/12a$ by 1 in the form of $3a/3a$:

$$\therefore \quad \dfrac{3}{12a}\left(\dfrac{3a}{3a}\right)=\dfrac{9a}{36a^2}$$

(b) Multiply $8/18a^2$ by 1 in the form of $2/2$:

$$\therefore \quad \dfrac{8}{18a^2}\left(\dfrac{2}{2}\right)=\dfrac{16}{36a^2}$$

(c) Multiply $5a/6$ by 1 in the form of $6a^2/6a^2$:

$$\therefore \quad \dfrac{5a}{6}\left(\dfrac{6a^2}{6a^2}\right)=\dfrac{30a^3}{36a^2}$$

EXERCISE 8–8

Form each of the following sets of fractions into equivalent fractions having the given common denominators.

Common Denominator

1. $\dfrac{1}{6},\dfrac{2}{3},\dfrac{5}{12}$ 12

2. $\dfrac{7}{15},\dfrac{9}{20},\dfrac{11}{30}$ 60

		Common Denominator
3. $\dfrac{-3}{4}, \dfrac{1}{3}, \dfrac{5}{-9}$		36
4. $\dfrac{2}{5}, -\dfrac{2}{3}, \dfrac{-7}{15}$		15
5. $\dfrac{x}{3}, \dfrac{2b}{7}, \dfrac{5}{6}$		42
6. $\dfrac{2}{13}, \dfrac{5x}{b}, \dfrac{10}{3}$		39b
7. $\dfrac{a}{x}, \dfrac{3}{xy}, ax$		xy
8. $\dfrac{a}{b}, \dfrac{2a+b}{a-b}$		$b(a-b)$
9. $\dfrac{x-b}{a-b}, -\dfrac{x}{b-a}$		$(a-b)$
10. $\dfrac{RE}{Z+1}, \dfrac{E}{R}$		$R(Z+1)$

Adding Fractions Having Common Denominators

To add fractions, each must have the same or a common denominator. Once they have a common denominator, they may be combined by Rule 8–6.

RULE 8–6. Combining Fractions with a Common Denominator

To add or subtract fractions having a common denominator:
1. Make the signs of the vinculums positive.
2. Place the sum (or difference) of the numerators over the common denominator.
3. Simplify by combining like terms and reducing to lowest terms.

EXAMPLE 8–19   Combine $\dfrac{2a}{10a} - \dfrac{7a}{10a}$.

Solution   Use the steps of Rule 8–6.

Step 1:   Make the sign of the vinculum positive:

$$\frac{2a}{10a} + \frac{-7a}{10a}$$

190

Step 2: Place the sum of the numerators over the common denominator:

$$\frac{2a - 7a}{10a}$$

Step 3: Simplify by combining like terms and reducing to lowest terms:

$$\frac{-5a}{10a} = -\frac{1}{2}$$

$$\therefore \quad \frac{2a}{10a} - \frac{7a}{10a} = -\frac{1}{2}$$

EXAMPLE 8–20   Combine $\dfrac{-5a}{2a^2 + 4a} - \dfrac{5 - a}{2a^2 + 4a} - \dfrac{3}{2a^2 + 4a}$.

Solution   Follow the steps of Rule 8–6.

$$\frac{-5a - (5 - a) - 3}{2a^2 + 4a} = \frac{-5a - 5 + a - 3}{2a^2 + 4a}$$

$$= \frac{-4a - 8}{2a^2 + 4a} = \frac{-4(a + 2)}{2a(a + 2)} = \frac{-4}{2a} = \frac{-2}{a}$$

EXERCISE 8–9

Add or subtract each of the following as indicated.

1. $\dfrac{3}{4} + \dfrac{5}{4}$

2. $\dfrac{2}{12} - \dfrac{-8}{12}$

3. $-\dfrac{11}{13} - \dfrac{8}{13} + \dfrac{-7}{13}$

4. $\dfrac{3}{11} - \dfrac{5}{11} + \dfrac{-8}{11}$

5. $\dfrac{3}{a} + \dfrac{7}{a}$

6. $\dfrac{2}{3a} - \dfrac{8}{3a}$

7. $\dfrac{5}{x + 1} - \dfrac{3}{x + 1}$

8. $\dfrac{x}{x - y} - \dfrac{y}{x - y}$

9. $\dfrac{2a}{a + 1} + \dfrac{2}{a + 1}$

10. $\dfrac{3b}{b + 1} + \dfrac{b + 4}{b + 1}$

11. $\dfrac{3x + y}{3xy} - \dfrac{3y - 2x}{3xy} + \dfrac{4x + 2y}{3xy}$

12. $\dfrac{3}{3x^2 - 2x} - \dfrac{2x + 5}{3x^2 - 2x} + \dfrac{5x}{3x^2 - 2x}$

Finding Common Denominators

To add fractions, the numerators must be placed over a common denominator. How do we find a common denominator? Let's explore the process of finding a common denominator through the following examples.

EXAMPLE 8–21	Find some possible common denominators for 1/3 and *a*/5 and select the "simplest."
Solution	Since the common denominator *must* be divisible by both of the denominators, one common denominator could be the product of 3 and 5, or 15. Or any multiple of the product of 3 and 5, such as 30, 45, 60, and 75, could be a common denominator.
∴	By inspection, 15 is the *simplest* common denominator.

EXAMPLE 8–22	Find some possible common denominators for *y*/(2*x*) and 7/(*yz*), and select the *simplest*.
Solution	First select the product of 2*x* and *yz,* which is 2*xyz*. Some other possible denominators are 4*xyz*, 2*x*^{2}*yz*, 8*wxyz*, and 2 "whatever" *xyz;* however, the simplest is 2*xyz*.
∴	2*xyz* is the simplest.

EXAMPLE 8–23	Find some possible common denominators for $\frac{5}{12}$ and $\frac{11}{18}$ and select the simplest.
Solution	First, select the product of 12 and 18, which is 216, as a possible common denominator. Although 216 is a common denominator, it is bigger than we need. Find the common factors of 12 and 18:

$$2, 3, \text{ and } 6$$

Divide 216 by 6, the largest common factor:

$$\frac{216}{6} = 36$$

Observation	36 is a multiple of both 12 and 18.
∴	36 is the simplest common denominator.

EXAMPLE 8–24   Find the simplest common denominator for $\dfrac{2a}{9y}$ and $\dfrac{b}{3y^2}$.

Solution   Find the common factors of $9y$ and $3y^2$.

3, y, $3y$

The simplest common denominator is the product of the two denominators $9y$ and $3y^2$, divided by their largest common factor, $3y$.

$$\frac{9y(3y^2)}{3y} = 9y^2$$

∴ $9y^2$ is the simplest common denominator.

RULE 8–7.   Forming a Common Denominator

To form a common denominator:

1. Find the common factors of each denominator.
2. Form the product of the denominators.
3. Divide this product by the largest common factor. This results in the *simplest* common denominator.

EXAMPLE 8–25   Find the simplest common denominator for $\dfrac{1}{3c-1}$ and $\dfrac{7}{6ac-2a}$.

Solution   Use the steps of Rule 8–7.

Step 1:   Find the common factors of each denominator:

Factors of $3c - 1$ are 1 and $3c - 1$.
Factors of $6ac - 2a$ are 2, a, $3c - 1$, $2a$, $6c - 2$, and $3ac - a$.
The largest common factor is $3c - 1$.

Step 2:   Form the product of $3c - 1$ and $6ac - 2a$:

$(3c - 1)(6ac - 2a)$

Step 3:   Divide the product by the largest common factor, $3c - 1$:

$$\frac{(3c-1)(6ac-2a)}{3c-1} = 6ac - 2a$$

∴ $6ac - 2a$ is the simplest common denominator.

Determine the simplest common denominator for each of the following pairs of fractions:

1. $\dfrac{2}{3}, \dfrac{5}{6}$ **2.** $\dfrac{5}{52}, \dfrac{7}{13}$ **3.** $\dfrac{3}{25}, \dfrac{1}{15}$ **4.** $\dfrac{5}{12}, \dfrac{19}{32}$

5. $\dfrac{B}{2A}, \dfrac{B^2}{A}$ **6.** $\dfrac{4}{3C}, \dfrac{2}{3D}$ **7.** $\dfrac{150}{\pi R}, \dfrac{35}{2\pi R}$ **8.** $\dfrac{7}{3V}, \dfrac{\lambda}{V^2}$

9. $\dfrac{\mu}{4\alpha\beta}, \dfrac{\sigma}{6\beta}$ **10.** $\dfrac{7J}{12K}, \dfrac{9}{4JK}$

11. $\dfrac{7}{8\alpha(\beta + 1)}, \dfrac{3\beta}{\alpha^2(2\beta + 2)}$ **12.** $\dfrac{8}{21W^2}, \dfrac{3}{63W}$

13. $\dfrac{AB}{49K - 7}, \dfrac{NO}{14K - 2}$ **14.** $\dfrac{CB^2}{3A(B - 6)}, \dfrac{13B}{3B - 18}$

15. $\dfrac{3}{st + s}, \dfrac{a}{2t + 2}$ **16.** $\dfrac{P}{15uV - 5u}, \dfrac{17R}{10u^2}$

Adding Fractions with Unequal Denominators

When adding fractions, remember that each fraction must have a common denominator before it can be added. Rule 8–8 shows how to combine fractions with unequal denominators.

RULE 8–8. *Combining Fractions with Unequal Denominators*

To add fractions having unequal denominators:

1. Write each fraction as an equivalent fraction having the same denominator and a positive vinculum.
2. Place the sum (or difference) of the numerators over the common denominator.
3. Simplify by combining like terms and reducing to lowest terms.

EXAMPLE 8–26   Combine $\dfrac{a}{3} - \dfrac{3a}{6}$.

Solution   The common denominator of 3 and 6 is 6. Follow the steps of Rule 8–8.

Step 1:   Write equivalent fractions:

$$\frac{a}{3} \cdot \frac{2}{2} + \frac{-3a}{6} = \frac{2a}{6} + \frac{-3a}{6}$$

Step 2: Place the difference over the common denominator:

$$\frac{2a - 3a}{6}$$

Step 3: Combine like terms:

$$\frac{-a}{6}$$

$$\therefore \quad \frac{a}{3} - \frac{3a}{6} = \frac{-a}{6}$$

EXAMPLE 8–27 Combine $\dfrac{x-1}{3} + \dfrac{2x+1}{2x} - \dfrac{3x+8}{9}$.

Solution The common denominator is $18x$. Use the steps of Rule 8–8.

Step 1: Modify each fraction:

$$\left(\frac{6x}{6x}\right)\frac{x-1}{3} + \left(\frac{9}{9}\right)\frac{2x+1}{2x} + \left(\frac{2x}{2x}\right)\frac{-3x-8}{9}$$

Step 2: Perform multiplication:

$$\frac{6x(x-1)}{18x} + \frac{9(2x+1)}{18x} + \frac{2x(-3x-8)}{18x}$$

$$= \frac{6x^2 - 6x + 18x + 9 - 6x^2 - 16x}{18x}$$

Step 3: Combine like terms:

$$\frac{-4x + 9}{18x}$$

$$\therefore \quad \frac{x-1}{3} + \frac{2x+1}{2x} - \frac{3x+8}{9} = \frac{-4x+9}{18x}$$

EXAMPLE 8–28 Combine $\dfrac{3x+2}{x^2-x} + \dfrac{3}{x} - \dfrac{5}{x-1}$.

Solution The common denominator is $x(x-1)$. Modify each fraction.

$$\frac{3x+2}{x(x-1)} + \left(\frac{x-1}{x-1}\right)\frac{3}{x} + \left(\frac{x}{x}\right)\frac{-5}{x-1}$$

$$= \frac{3x+2}{x(x-1)} + \frac{3x-3}{x(x-1)} + \frac{-5x}{x(x-1)}$$

$$= \frac{3x+2+3x-3+(-5x)}{x(x-1)}$$

Combine like terms and simplify:

$$\frac{x - 1}{x(x - 1)} = \frac{1}{x}$$

$$\therefore \qquad \frac{3x + 2}{x^2 - x} + \frac{3}{x} - \frac{5}{x - 1} = \frac{1}{x}$$

EXERCISE 8–11

Combine and simplify the following fractions.

1. $2 + \dfrac{c}{d}$

2. $\dfrac{x}{y} - 3$

3. $\dfrac{1}{2} + \dfrac{5}{6x}$

4. $8y + \dfrac{1}{3}$

5. $\dfrac{1}{x} + \dfrac{1}{y}$

6. $\dfrac{a}{b} + \dfrac{1}{c}$

7. $n + \dfrac{1}{m}$

8. $\dfrac{4x}{3} + \dfrac{5x}{4} - \dfrac{x}{5}$

9. $\dfrac{3ab}{4} - \dfrac{2ab}{3} + \dfrac{5ab}{8}$

10. $\dfrac{3x}{a} - \dfrac{y}{b}$

11. $\dfrac{m}{x} + \dfrac{2n}{y}$

12. $\dfrac{a}{x} - \dfrac{5}{x^2}$

13. $\dfrac{5t}{4} - \dfrac{3t}{12} - \dfrac{t}{3}$

14. $\dfrac{7}{n^2} + \dfrac{5}{n} + \dfrac{9}{n^3}$

15. $\dfrac{7a - 3b}{3} + \dfrac{2a + 5b}{2}$

16. $\dfrac{5x - 4}{12} - \dfrac{3x - 8}{18}$

17. $\dfrac{3x - 1}{3x} - \dfrac{2x - 4}{2}$

18. $\dfrac{m + n - 5}{m} - \dfrac{2m - n + 1}{2m}$

19. $\dfrac{2a - b}{4b} - \dfrac{a - 3b}{6a}$

20. $\dfrac{x - y}{y} + \dfrac{y^2 - x}{y^2}$

21. $\dfrac{6}{6x - 12} - \dfrac{4}{x - 2}$

22. $\dfrac{5}{6a + 6} - \dfrac{3}{2a + 2}$

23. $\dfrac{2b - 4}{9b^2 - 9b} - \dfrac{5}{6b - 6}$

24. $\dfrac{2a - 3}{16a^2} - \dfrac{2 - a}{8a} + \dfrac{3}{4a}$

25. $\dfrac{3}{m^2 + 5m} - \dfrac{3}{m} - \dfrac{3}{m - 5}$

26. $\dfrac{-3a}{-x + y} + \dfrac{2}{x - y} + \dfrac{1}{x + y}$

8–8 CHANGING A MIXED EXPRESSION TO A FRACTION

A **mixed expression** is formed by adding (or subtracting) a polynomial and a fraction. Thus, $3x - \dfrac{2}{5x}$ is a mixed expression. A mixed expression may be changed to a fraction by adding both terms, as demonstrated in the following examples.

EXAMPLE 8–29   Write $3x - \dfrac{2}{5x}$ as a fraction.

Solution $3x - \dfrac{2}{5x} = \dfrac{3x}{1}\left(\dfrac{5x}{5x}\right) + \dfrac{-2}{5x} = \dfrac{15x^2 - 2}{5x}$

$\therefore \quad 3x - \dfrac{2}{5x} = \dfrac{15x^2 - 2}{5x}$

EXAMPLE 8–30   Write $2a + b - \dfrac{2}{a+1}$ as a fraction.

Solution   The common denominator is $a + 1$.

$$\dfrac{2a}{1} \cdot \dfrac{a+1}{a+1} + \dfrac{b}{1} \cdot \dfrac{a+1}{a+1} - \dfrac{2}{a+1}$$

$$= \dfrac{2a^2 + 2a + ab + b - 2}{a+1}$$

$\therefore \quad 2a + b - \dfrac{2}{a+1} = \dfrac{2a^2 + 2a + ab + b - 2}{a+1}$

EXERCISE 8–12

Change each of the following mixed expressions into fractions.

1. $5\frac{3}{4}$ **2.** $6\frac{2}{5}$ **3.** $2 + \dfrac{1}{3}$ **4.** $7 + \dfrac{3}{5}$

5. $3 - \dfrac{2}{m}$ **6.** $n - \dfrac{5}{n}$ **7.** $3a + \dfrac{2}{3}$ **8.** $\dfrac{a}{b} + 5$

9. $4x - \dfrac{7}{2x}$ **10.** $2\beta - \dfrac{\pi}{2a}$ **11.** $\omega C + 3 - \dfrac{Z}{L}$ **12.** $2b + 1 - \dfrac{5}{b}$

13. $x + \dfrac{2}{x+3}$ **14.** $\dfrac{a+2b}{a-b} + 7$ **15.** $x + 5 + \dfrac{3x-1}{x+2}$ **16.** $y + x + c - \dfrac{3}{yx}$

8–9 ADDITIONAL WORK WITH COMPLEX FRACTIONS

In this section we will work with simplifying complex fractions. This section uses many of the concepts already presented in this chapter.

> ### RULE 8–9. *Simplifying Complex Fractions*
>
> To change a complex fraction to a simple fraction:
> 1. Simplify the numerator to a fraction.
> 2. Simplify the denominator to a fraction.
> 3. Divide the resulting numerator by the resulting denominator.
> 4. Reduce to lowest terms.

EXAMPLE 8–31 Change the complex fraction to a simple fraction:

$$\frac{3 - \dfrac{1}{x + 2}}{5 + \dfrac{1}{x + 2}}$$

Solution Use the steps of Rule 8–9 to simplify the complex fraction.

Step 1: Simplify the numerator:

$$\frac{3(x + 2) - 1}{x + 2} = \frac{3x + 6 - 1}{x + 2} = \frac{3x + 5}{x + 2}$$

Step 2: Simplify the denominator:

$$\frac{5(x + 2) + 1}{x + 2} = \frac{5x + 10 + 1}{x + 2} = \frac{5x + 11}{x + 2}$$

Step 3: Divide:

$$\frac{\dfrac{3x + 5}{x + 2}}{\dfrac{5x + 11}{x + 2}}$$

Invert and multiply:

$$\frac{3x + 5}{x + 2} \cdot \frac{x + 2}{5x + 11}$$

Step 4:   Reduce to lowest terms:

$$\frac{3x + 5}{5x + 11}$$

$$\therefore \quad \left(3 - \frac{1}{x + 2}\right) \Big/ \left(5 + \frac{1}{x + 2}\right) = \frac{3x + 5}{5x + 11}$$

EXAMPLE 8–32   Change the complex fraction to a simple fraction:

$$\frac{2 - \dfrac{2m}{n}}{1 - \dfrac{m}{n}}$$

Solution   Follow the steps of Rule 8–9 to simplify. Simplify the numerator and denominator:

$$\frac{\dfrac{2n - 2m}{n}}{\dfrac{n - m}{n}}$$

Invert and multiply, then reduce to lowest terms:

$$\frac{2(n - m)}{n} \cdot \frac{n}{n - m} = \frac{2}{1}$$

$$\therefore \quad \frac{2 - \dfrac{2m}{n}}{1 - \dfrac{m}{n}} = 2$$

EXERCISE 8–13

Change each of the complex fractions to a simple fraction.

1. $\dfrac{7 + \dfrac{4}{5}}{\dfrac{1}{5} + 5}$

2. $\dfrac{6 - \dfrac{1}{3}}{2 + \dfrac{5}{6}}$

3. $\dfrac{a - \dfrac{1}{b}}{\dfrac{1}{b} + a}$

4. $\dfrac{ax + \dfrac{x}{b}}{ax - \dfrac{x}{b}}$

5. $\dfrac{3 - \dfrac{2}{a}}{6a - 4}$

6. $\dfrac{a + \dfrac{a}{b}}{\dfrac{1}{b} + \dfrac{1}{b^2}}$

7. $\dfrac{2 + \dfrac{a+b}{a-b}}{1 - \dfrac{a+b}{a-b}}$

8. $\dfrac{x - 2 + \dfrac{3}{x}}{1 + \dfrac{1}{x}}$

9. $3 - \dfrac{1}{1 - \dfrac{1}{x-1}}$

10. $5 - \dfrac{2a}{3 + \dfrac{a}{2}}$

11. $\dfrac{\dfrac{1}{a+b} - \dfrac{1}{a-b}}{\dfrac{1}{a-b} + \dfrac{1}{a+b}}$

12. $\dfrac{\dfrac{x}{x+y} - \dfrac{y}{x-y}}{\dfrac{x}{x-y} + \dfrac{y}{x+y}}$

13. $\dfrac{\dfrac{2}{b+3} + \dfrac{1}{b-2}}{\dfrac{2}{b-2} - \dfrac{1}{b+3}}$

14. $\dfrac{\dfrac{3}{y-4} - \dfrac{5}{y+3}}{\dfrac{2}{y+3} + \dfrac{7}{y-4}}$

15. $\dfrac{2 - \dfrac{2}{x+4}}{\dfrac{3}{x+4} - \dfrac{1}{x-1}}$

16. $\dfrac{\dfrac{4}{y-3} + 5}{\dfrac{4}{y-3} - \dfrac{7}{y+3}}$

SELECTED TERMS

common denominator   Equal denominators of several fractions.

complex fraction   Fraction with a numerator or denominator (or both) that contains a fraction.

equivalent fractions   Fractions that have different forms but have the same value.

mixed expression   Sum (difference) of a fraction and a polynomial.

undefined fraction   If the denominator evaluates to zero in a fraction, then the fraction is undefined.

vinculum   Sign of grouping; a bar drawn over terms to show they are treated as a unit.

9 Equations Containing Fractions

9–1 Solving Equations Containing Fractions

9–2 Solving Fractional Equations

9–3 Literal Equations Containing Fractions

9–4 Evaluating Formulas

PERFORMANCE OBJECTIVES

- Employ the common denominator in the solution of an equation containing a fraction.

- Solve both fractional equations and literal equations containing fractions.

- Check fractional equations for an extraneous root.

- Identify conventional variables used in literal equations.

- Evaluate electrical formulas.

Laser light passing through the finger in the finger clamp is detected by the microprocessor-controlled Pulse Oximeter to determine both the oxygen level in arterial blood and the patient's pulse rate. (Courtesy of Pearson Education/PH College)

Many equations used by electronic technicians to solve problems involve fractions. In this chapter you will have an opportunity to expand your understanding of algebra by working with fractions, fractional equations, and literal equations containing fractions and by evaluating formulas.

9–1 SOLVING EQUATIONS CONTAINING FRACTIONS

Equations having numerical coefficients that are fractions are transformed and solved by the methods in earlier chapters. Equations with fractional coefficients are best solved by transforming them into equations without fractions. This is done by multiplying each member of the equation by the common denominator.

EXAMPLE 9–1   Solve $\frac{3}{8}y = \frac{3}{4}$.

Solution   Multiply each member by the common denominator, 8; then simplify:

$$(8)\frac{3}{8}y = \frac{3}{4}(8)$$

$$3y = 6$$

$$y = 2$$

Check   Does $\frac{3}{8}y = \frac{3}{4}$ when $y = 2$?

$$\frac{3}{8}(2) = \frac{3}{4}$$

$$\frac{3}{4} = \frac{3}{4} \text{ Yes!}$$

∴ $y = 2$ is the solution.

EXAMPLE 9–2   Solve $\dfrac{2x}{5} - 3 = \dfrac{5 - 3x}{10}$.

Solution   Multiply each member by the common denominator of 10:

$$10\left(\frac{2x}{5} - 3\right) = 10\left(\frac{5 - 3x}{10}\right)$$

$$10\left(\frac{2x}{5}\right) + 10(-3) = 10\left(\frac{5 - 3x}{10}\right)$$

$$2(2x) - 30 = 5 - 3x$$

$$4x + 3x = 5 + 30$$

$$7x = 35$$

$$x = 5$$

Check   Does $\dfrac{2x}{5} - 3 = \dfrac{5 - 3x}{10}$ when $x = 5$?

$$\frac{2(5)}{5} - 3 = \frac{5 - 3(5)}{10}$$

$$2 - 3 = \frac{5 - 15}{10}$$

$$-1 = -1 \ \text{ Yes!}$$

∴ $x = 5$ is the solution.

EXAMPLE 9–3   Solve $\dfrac{3x + 3}{12} = \dfrac{x - 3}{18} + \dfrac{x}{6}$.

Solution   Multiply each member by the common denominator of 36:

$$36\left(\frac{3x + 3}{12}\right) = 36\left(\frac{x - 3}{18} + \frac{x}{6}\right)$$

$$36\left(\frac{3x + 3}{12}\right) = 36\left(\frac{x - 3}{18}\right) + 36\left(\frac{x}{6}\right)$$

$$3(3x + 3) = 2(x - 3) + 6x$$

$$9x + 9 = 2x - 6 + 6x$$

$$9x - 8x = -15$$

$$x = -15$$

Check   Does $\dfrac{3x + 3}{12} = \dfrac{x - 3}{18} + \dfrac{x}{6}$ when $x = -15$?

$$\frac{3(-15) + 3}{12} = \frac{-15 - 3}{18} + \frac{-15}{6}$$

$$\frac{-45 + 3}{12} = -1 + \frac{-15}{6}$$

$$\frac{-42}{12} = \frac{-6 - 15}{6}$$

$$\frac{-7}{2} = \frac{-21}{6}$$

$$\frac{-7}{2} = \frac{-7}{2} \ \text{ Yes!}$$

∴ $x = -15$ is the solution.

Solve the following equations and check your solutions.

1. $\dfrac{1}{8}x = \dfrac{1}{4}$ **2.** $\dfrac{-x}{2} = \dfrac{1}{6}$ **3.** $\dfrac{a}{15} = \dfrac{1}{10}$

4. $\dfrac{5}{4} = \dfrac{4}{5}y$ **5.** $\dfrac{1}{3}R = -\dfrac{15}{7}$ **6.** $\dfrac{x}{3} = \dfrac{4}{5}$

7. $\dfrac{m-7}{4} = 2m$ **8.** $\dfrac{2b-5}{3} = \dfrac{45}{3}$ **9.** $\dfrac{4}{7} = \dfrac{12V}{16}$

10. $\dfrac{a}{3} - \dfrac{a}{2} = 5$ **11.** $\dfrac{3I+4}{7} = I+2$ **12.** $\dfrac{3}{4} + E = \dfrac{5}{4}E$

13. $5 - \dfrac{2}{3}x = \dfrac{-x}{2}$ **14.** $\dfrac{V}{2} = \dfrac{4V}{5} - 3$

15. $\dfrac{1}{6} = \dfrac{a-15}{3} + \dfrac{7-a}{2}$ **16.** $\dfrac{m-3}{3} = \dfrac{m+4}{5} - 3$

17. $\dfrac{3x}{10} - \dfrac{2}{10} = \dfrac{2x-7}{5} + \dfrac{1}{5}$ **18.** $\dfrac{2R}{5} = \dfrac{4R-1}{4} - \dfrac{3}{16}$

19. $\dfrac{2I-3}{3} - 3 = -\dfrac{3}{6} - \dfrac{I}{2}$ **20.** $\dfrac{8y-1}{3} = 3 - \dfrac{10y}{6} - \dfrac{7}{6}$

21. $\dfrac{3x-3}{4} = \dfrac{2x}{3} - \dfrac{x-5}{2}$ **22.** $\dfrac{4a-3}{2} = \dfrac{4a-3}{5}$

9–2   SOLVING FRACTIONAL EQUATIONS

A **fractional equation** is an equation that has a variable in the denominator of one or more terms of the equation. The solution of fractional equations is carried out in the same way as the solution of equations containing fractions: by multiplying both members of the equation by the common denominator.

EXAMPLE 9–4   Solve $\dfrac{2}{x} = \dfrac{5}{x} - \dfrac{3}{5}$.

     Solution   Multiply each member by the common denominator of $5x$:

$$5x\left(\dfrac{2}{x}\right) = 5x\left(\dfrac{5}{x}\right) + 5x\left(-\dfrac{3}{5}\right)$$
$$10 = 25 - 3x$$
$$3x = 15$$
$$x = 5$$

Check   Does $\dfrac{2}{x} = \dfrac{5}{x} - \dfrac{3}{5}$ when $x = 5$?

$$\frac{2}{5} = \frac{5}{5} - \frac{3}{5}$$

$$\frac{2}{5} = \frac{2}{5} \quad \text{Yes!}$$

$\therefore$ $x = 5$ is the solution.

It is important to check the roots of fractional equations. When we multiply both members of an equation by an expression containing the unknown, we run the risk of introducing an **extraneous root.** An extraneous root will *check* in the derived equation, but **not** in the original equation. Extraneous roots result from multiplying by zero.

EXAMPLE 9–5   Solve $\dfrac{5x}{x - 1} = \dfrac{5}{x - 1}$.

Solution   Multiply each member by $(x - 1)$, the common denominator:

$$(x - 1)\frac{5x}{x - 1} = (x - 1)\frac{5}{x - 1}$$
$$5x = 5$$
$$x = 1$$

Check   Does $\dfrac{5x}{x - 1} = \dfrac{5}{x - 1}$ when $x = 1$?

$$\frac{5(\mathbf{1})}{1 - 1} = \frac{5}{1 - 1}$$

$$\frac{5}{0} = \frac{5}{0}$$

No! Both members of the equation are undefined.
Since division by zero is not permitted, this equation has no root.

$\therefore$ $x = 1$ does not satisfy the equation.

EXAMPLE 9–6   Solve $\dfrac{3x}{x + 1} = 3 - \dfrac{4}{2x}$.

Solution　Multiply each member by the common denominator $2x(x + 1)$ and then factor:

$$\frac{2x(\cancel{x + 1})(3x)}{\cancel{x + 1}} = 2x(x + 1)(3) - \frac{\cancel{2x}(x + 1)(4)}{\cancel{2x}}$$

$$6x^2 = 6x^2 + 6x - 4x - 4$$
$$6x^2 - 6x^2 = 6x - 4x - 4$$
$$0 = 2x - 4$$
$$2x = 4$$
$$x = 2$$

Check　Does $\dfrac{3x}{x + 1} = 3 - \dfrac{4}{2x}$ when $x = 2$?

$$\frac{3(2)}{2 + 1} = 3 - \frac{4}{2(2)}$$

$$\frac{6}{3} = 3 - 1$$

$$2 = 2 \quad \text{Yes!}$$

$\therefore$ 　　$x = 2$ is the solution.

EXERCISE 9–2

Solve the following equations and check your solutions.

1. $\dfrac{3}{x} + \dfrac{5}{x} = 2$

2. $\dfrac{8}{x} = \dfrac{4}{x} - \dfrac{2}{3}$

3. $\dfrac{8}{y} - 1 = \dfrac{7}{y}$

4. $\dfrac{3}{x} - 4 = \dfrac{7}{2}$

5. $\dfrac{8}{x + 1} = 4$

6. $\dfrac{x + 3}{x - 3} = 4$

7. $\dfrac{3y}{4y - 5} = 2$

8. $\dfrac{2x + 5}{x + 4} = 1$

9. $\dfrac{4}{y - 3} = \dfrac{2}{y}$

10. $\dfrac{3x - 2}{5x - 10} = \dfrac{4}{10}$

11. $\dfrac{4x - 2}{4x} - \dfrac{1}{2} = \dfrac{3}{x}$

12. $\dfrac{3}{2} - \dfrac{5}{2y} = 2 - \dfrac{3}{y}$

13. $\dfrac{x - 2}{5x} = \dfrac{4x - 8}{30x}$

14. $\dfrac{1}{y - 1} = \dfrac{3}{y - 3}$

15. $\dfrac{12}{x - 6} = \dfrac{2x}{x - 6}$

16. $5 - \dfrac{5x}{x + 1} = \dfrac{15}{4x}$

17. $\dfrac{2y}{y + 1} = 2 - \dfrac{8}{2y}$ **18.** $\dfrac{3}{y} = \dfrac{3y}{y - 4} - 3$

19. $\dfrac{5}{3 + y} = \dfrac{3y}{y + 3} - \dfrac{3y - 4}{2y + 6}$ **20.** $\dfrac{3}{x} + \dfrac{6}{x^2 + 5x} = \dfrac{6x}{x^2 + 5x}$

9–3 LITERAL EQUATIONS CONTAINING FRACTIONS

Equations that have one or more numbers represented by letters are **literal equations.** The formulas used to solve electrical problems are literal equations.

Variables and Constants

In literal equations, such as $\dfrac{x}{a} - 2 = b$, $\dfrac{c}{3} = y$, and $\dfrac{x}{m} - n = a$, the letters a, b, c, m, and n are *standing in* for numbers that have a fixed or constant value. The letters x and y are *standing in* for numbers that may change (or could be made to change) or vary.

In literal equations, it is conventional to represent constants by the letters a, b, c, g, h, k, l, m, and n. It is conventional to represent variables by the letters s, t, u, v, w, x, y, and z.

Constants are divided into two types. Those represented by letters (a, b, c, etc.) are called **arbitrary constants,** whereas numbers and symbols representing unvarying values are called **numerical constants.** Numerical constants include integers (1, 2, 3, etc.), fractions ($\frac{1}{2}, \frac{2}{3}$, etc.), and irrational numbers ($\pi = 3.14159 \ldots$, $e = 2.71828 \ldots, \sqrt{2} = 1.414 \ldots$, etc.).

EXAMPLE 9–7   Solve the literal equation $\dfrac{x}{a} + 2 = b$ for x.

Solution $\dfrac{x}{a} + 2 = b$

$$\dfrac{x}{a} = b - 2$$

$\therefore$ $x = a(b - 2)$

EXAMPLE 9–8   Solve $\dfrac{7}{b} - \dfrac{y}{a} = 4$ for y.

Solution $\dfrac{7}{b} - \dfrac{y}{a} = 4$

$$7a - yb = 4ab$$

$$-yb = 4ab - 7a$$

Change sign:

$$yb = 7a - 4ab$$

$$\therefore \qquad y = \frac{7a - 4ab}{b}$$

EXAMPLE 9–9   Solve $\dfrac{xm}{2} - \dfrac{m}{3} = x$ for x.

Solution $$\frac{xm}{2} - \frac{m}{3} = x$$

$$\frac{xm}{2} = x + \frac{m}{3}$$

$$\frac{xm}{2} - x = \frac{m}{3}$$

$$6\left(\frac{xm}{2} - x\right) = 6\left(\frac{m}{3}\right)$$

$$3xm - 6x = 2m$$

Factor the left member:

$$x(3m - 6) = 2m$$

$$\therefore \qquad x = \frac{2m}{3m - 6}$$

EXERCISE 9–3

Solve each equation for the *conventional* variable s, t, u, v, w, x, y, or z.

1. $\dfrac{a}{x} = b$

2. $\dfrac{3}{\pi} = \dfrac{a}{W}$

3. $\dfrac{1}{z} - \dfrac{1}{a} = -\dfrac{2}{z}$

4. $\dfrac{hu}{m} + \dfrac{ku}{m} = 5$

5. $\dfrac{y}{2m} + \dfrac{y}{n} = 1$

6. $2g = \dfrac{hke + V}{aV}$

7. $c = \dfrac{b - W}{\pi^2 kW}$

8. $z - 1 - m = \dfrac{5z - 8}{m}$

9. $\dfrac{c}{s} = \dfrac{c}{3}(D - c)$

10. $\dfrac{\pi x + b}{\pi x - b} = 3$

208

11. $\dfrac{m - n}{ay} = b + 1$

12. $\dfrac{2u + 2}{au} - \dfrac{3}{u} = -\dfrac{7 - u}{u}$

13. $\dfrac{5x}{g - h} - 3 = \dfrac{3x}{k}$

14. $\dfrac{1}{4(a - b)} = \dfrac{1}{5(x - a)}$

15. $\dfrac{1}{u} - \dfrac{1}{m} = \dfrac{1}{n} - \dfrac{1}{u}$

16. $\dfrac{1}{w} = \dfrac{1}{a} + \dfrac{1}{b} + \dfrac{1}{c}$

17. $\dfrac{b}{a} = \dfrac{x - b}{x} + \dfrac{2b}{ax}$

18. $\dfrac{4k + 1}{4k} = \dfrac{1}{2z} + \dfrac{2k}{z}$

19. $\dfrac{4g}{3g} - \dfrac{6}{6V} = \dfrac{4g}{V} - \dfrac{1}{3g}$

20. $\dfrac{z}{ab} - \dfrac{1}{ac} = \dfrac{z}{bc}$

9–4 EVALUATING FORMULAS

Evaluating a formula is the process of finding the value of one variable in a formula when all the other variables are known. When evaluating a formula for only one set of values, it is simpler first to substitute the known values into the formula and then solve for the variable representing the desired quantity. This technique is demonstrated in Example 9–10.

EXAMPLE 9–10   Using the formula $F = \frac{9}{5}C + 32$, determine the temperature in Celsius (C) when the temperature in Fahrenheit (F) is 77°F.

Solution   First substitute 77 for F and then solve for C:

$$F = \tfrac{9}{5}C + 32$$

$$77 = \tfrac{9}{5}C + 32$$

$$\tfrac{9}{5}C = 45$$

$$C = 45\left(\tfrac{5}{9}\right)$$

$$\therefore \quad C = 25°C$$

In Example 9–10, we first substituted into the formula, and then solved the problem. When we wish to find several values for a particular variable, it is best first to solve the formula for the desired variable and then substitute into the formula. This concept is demonstrated in Example 9–11.

EXAMPLE 9–11   Using $F = \frac{9}{5}C + 32$, determine the temperature in Celsius (C) for temperatures in Fahrenheit (F) of: (a) 105°F (b) 93°F (c) 85°F.

Solution   First solve for C:

$$F = \frac{9}{5}C + 32$$

$$\frac{9}{5}C = F - 32$$

$$C = \frac{5(F - 32)}{9} \; °C$$

Now, substitute each of the given temperatures.

(a) $F = 105°F$

$$\therefore \qquad C = \frac{5(\mathbf{105} - 32)}{9} = 40.6°C$$

(b) $F = 93°F$

$$\therefore \qquad C = \frac{5(93 - 32)}{9} = 33.9°C$$

(c) $F = 85°F$

$$\therefore \qquad C = \frac{5(85 - 32)}{9} = 29.4°C$$

In the study of electronics, formulas are routinely evaluated. Because of this, it is very important that you develop the skill of successfully working with formulas. The following exercise will give you some practice.

EXERCISE 9–4

In the following formulas, solve for the indicated variable.

Formula:	*Solve for:*
1. $C = \frac{5}{9}(F - 32)$	F
2. $A = \frac{1}{2}h(b + c)$	c
3. $X_C = \dfrac{1}{2\pi f C}$	f

4. $C = \dfrac{1}{C_1} + \dfrac{1}{C_2}$ $\qquad\qquad\qquad\qquad C_2$

5. $k = \dfrac{wv^2}{2g}$ $\qquad\qquad\qquad\qquad w$

6. $\beta = \dfrac{\alpha}{1 - \alpha}$ $\qquad\qquad\qquad\qquad \alpha$

7. $\theta_{JA} = \dfrac{T_J - T_A}{P_D}$ $\qquad\qquad\qquad\qquad T_A$

8. $A'_v = \dfrac{A_v}{1 + \beta A_v}$ $\qquad\qquad\qquad\qquad A_v$

9. $C = \dfrac{E}{R_1 + R_2}$ $\qquad\qquad\qquad\qquad R_2$

10. $\dfrac{V_1}{V_2} = \dfrac{P_2}{P_1}$ $\qquad\qquad\qquad\qquad P_1$

11. $\theta = \dfrac{108(n - 2)}{n}$ $\qquad\qquad\qquad\qquad n$

12. $s - t\left(\dfrac{V_0 + V_1}{2}\right)$ $\qquad\qquad\qquad\qquad V_1$

13. $\eta = \dfrac{A_2 S_2}{F_1 S_1}$ $\qquad\qquad\qquad\qquad S_2$

14. $Q = \dfrac{tKA(T_0 - T_1)}{L}$ $\qquad\qquad\qquad\qquad T_0$

15. $F = \dfrac{q_1 q_2}{4\pi\epsilon_0 d^2}$ $\qquad\qquad\qquad\qquad d^2$

16. $V = \dfrac{R_2(V_{CC} - R_1 I_{\text{sat}} - 1)}{R_1 + R_2}$ $\qquad\qquad\qquad\qquad R_1$

17. $R_T = \dfrac{R_1 R_2}{R_1 + R_2}$ $\qquad\qquad\qquad\qquad R_2$

18. $I_1 = \dfrac{I R_2}{R_1 + R_2}$ $\qquad\qquad\qquad\qquad R_2$

19. $S = \dfrac{R_1 + R_2}{R_E + R_\beta(1 - \alpha)}$ $\qquad\qquad\qquad\qquad \alpha$

20. $R_2 = \dfrac{Z_Z(k + 1)}{k - 1} - R_3$ $\qquad\qquad\qquad\qquad k$

Evaluate each of the following formulas. First assign the indicated values to the variables and then solve. Express the solution to an appropriate precision with a unit and unit prefix (as needed).

21. Determine the resistance (R) of a coil in ohms from

$$Q_0 = \frac{2\pi f_{ar} L}{R}$$

when $f_{ar} = 310$ kHz, $L = 5.00$ mH, $Q_0 = 200$, and $\pi \approx 3.14$.

22. Determine the collector current (I_c) in amperes from

$$I_{CBO} = \frac{I_c - \beta I_B}{\beta + 1}$$

when $I_{CBO} = 5.00\ \mu A$, $I_B = 60.0\ \mu A$, and $\beta = 100$.

23. Determine the total resistance of a parallel network (R_T) in ohms from

$$\frac{1}{R_T} = \frac{1}{R_1} + \frac{1}{R_2}$$

when $R_1 = 2.20$ kΩ and $R_2 = 1.50$ kΩ.

24. Determine the impedance of the load (Z_2) in ohms from

$$\frac{V_1^2}{V_2^2} = \frac{Z_1}{Z_2}$$

when $V_1 = 60$ V, $V_2 = 2$ V, and $Z_1 = 7.2$ kΩ. Refer to Figure 9–1.

FIGURE 9–1 Circuit for Exercise 9–4, problem 24: (a) pictorial diagram showing an output transformer being used to match the load impedance to the amplifier; (b) schematic diagram of the circuit.

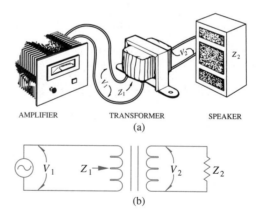

AMPLIFIER   TRANSFORMER   SPEAKER

(a)

(b)

25. Determine the frequency (f) in hertz from

$$X_C = \frac{1}{2\pi f C}$$

when $X_C = 410\ \Omega$, $C = 50$ nF, and $\pi \approx 3.14$.

26. Determine the output voltage (V'_o) in volts from

$$\frac{V'_o}{V_s} = \frac{A_v}{1 - \beta A_v}$$

when $\beta = 0.050$, $A_v = 60$, and $V_s = 1.0$ V. Refer to Figure 9–2.

FIGURE 9–2 Circuit for Exercise 9-4, problem 26. Flow diagram of the signal through an amplifier with feedback.

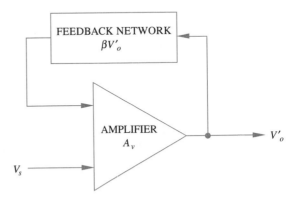

27. Determine the forward current transfer ratio of a transistor (β) from

$$\alpha = \frac{\beta}{\beta + 1}$$

when $\alpha = 0.985$.

28. Determine the load resistance (R_L) in ohms from

$$A_v = \frac{\mu R_L}{r_p + R_L}$$

when $\mu = 18$, $r_p = 0.20$ kΩ, and $A_v = 12$.

29. Determine the value of the bias resistance (R_1) in ohms from

$$I_B = \frac{V_{cc}}{R_1 + (\beta + 1)R_E}$$

when $I_B = 40.0$ μA, $V_{cc} = 18.0$ V, $\beta = 100$, and $R_E = 470$ Ω.

30. Determine the temperature of the case of a power transistor (T_c) in degrees Celsius from

$$P_D = \frac{T_c - T_s}{\theta_{cs}}$$

when $P_D = 75$ W, $T_s = 55°$C, and $\theta_{cs} = 0.35°$C/W.

SELECTED TERMS

extraneous root   A solution of a mathematical equation that is not a solution to the physical problem.

fractional equation   Equation that has a variable in the denominator of one or more terms.

literal equation   Equation with one or more numbers represented by letters.

10

Applying Fractions to Electrical Circuits

10–1 Voltage Division in the Series Circuit

10–2 Conductance of the Parallel Circuit

10–3 Equivalent Resistance of the Parallel Circuit

10–4 Current Division in the Parallel Circuit

10–5 Solving Parallel Circuit Problems

10–6 Using Network Theorems to Form Equivalent Circuits

PERFORMANCE OBJECTIVES

- Apply the principles of voltage division to the series circuit.

- Use the concept of conductance to form a parallel equivalent circuit.

- Employ Kirchhoff's current law in conjunction with the principles of current division and Ohm's law to solve parallel circuits.

- Use network theorems to aid in forming equivalent circuits.

Through combined efforts of scientists, engineers, and technicians, the Space Shuttle Discovery is successfully launched from Cape Canaveral, Florida.   (Courtesy of NASA)

Many of the formulas used with electrical circuits are fractional in form. We will look at voltage division in series circuits and several concepts associated with parallel circuits. These topics involve both fractions and electronics.

10–1 VOLTAGE DIVISION IN THE SERIES CIRCUIT

In Chapter 7 we learned about the characteristics of the series circuit. One important characteristic is that the circuit current, I, is the same throughout the entire series circuit. In the series circuit of Figure 10–1, the circuit current may be determined by Ohm's law as

$$I = \frac{E}{R_T} \quad \text{or} \quad I = \frac{V_1}{R_1} \quad \text{or} \quad I = \frac{V_2}{R_2} \quad \text{(amperes)}$$

Since the current is constant in a series circuit, an equation may be written using the fractions of E/R_T and V_1/R_1. This equation states that the applied voltage, E, is to the equivalent resistance, R_T, as the voltage drop across resistance one, V_1, is to the resistance one, R_1. Thus,

$$\frac{E}{R_T} = \frac{V_1}{R_1} \tag{10–1}$$

Equation 10–1 may be stated more generally as Equation 10–2.

$$\frac{E}{\Sigma R} = \frac{V_n}{R_n} \tag{10–2}$$

where E = applied voltage (V)
 ΣR = summation of all the series resistances ($R_T = \Sigma R$) (Ω)
 V_n = voltage drop across R_n (V)
 R_n = one resistance in the series circuit (Ω)
 n = subscript of the selected resistor

FIGURE 10–1 The source voltage, E, divides between resistance R_1 and R_2. The larger resistance has the greater voltage drop.

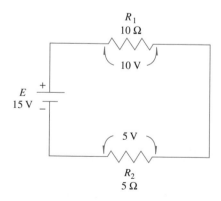

EXAMPLE 10–1 Determine the voltage drop, V_2, across resistor R_2 in Figure 10–1 when $R_2 = 2.2$ kΩ, $R_T = 2.7$ kΩ, and $E = 40$ V.

Solution Use Equation 10–2 and substitute 2.2 kΩ for R_2, 2.7 kΩ for ΣR, and 40 V for E. Solve for V_2:

$$\frac{E}{\Sigma R} = \frac{V_2}{R_2}$$

$$\frac{40}{2.7 \times 10^3} = \frac{V_2}{2.2 \times 10^3}$$

$$V_2 = \frac{40 \times 2.2 \times 10^3}{2.7 \times 10^3}$$

$$\therefore \qquad V_2 = 33 \text{ V}$$

EXERCISE 10–1

Use Equation 10–2 and Figure 10–1 to determine the indicated quantity in each of the following.

1. Find V_1 when $R_1 = 1.4$ kΩ, $\Sigma R = 22.5$ kΩ, and $E = 14.5$ V.
2. Find V_2 when $R_2 = 860$ Ω, $\Sigma R = 2.2$ kΩ, and $E = 16$ V.
3. Find V_2 when $R_2 = 1.5$ MΩ, $\Sigma R = 3.6$ MΩ, and $E = 20$ V.
4. Find E when $\Sigma R = 860$ kΩ, $R_2 = 560$ kΩ, and $V_2 = 12.5$ V.
5. Find E when $\Sigma R = 775$ kΩ, $R_1 = 7.00$ kΩ, and $V_1 = 0.600$ V.
6. Find E when $\Sigma R = 560$ Ω, $R_1 = 270$ Ω, and $V_1 = 375$ mV.
7. Find ΣR when $E = 46$ V, $R_1 = 375$ kΩ, and $V_1 = 12.4$ V.
8. Find ΣR when $E = 28$ V, $R_2 = 47$ kΩ, and $V_2 = 16.5$ V.
9. Find R_1 when $E = 15.0$ V, $\Sigma R = 3.23$ kΩ, and $V_1 = 4.04$ V.
10. Find R_2 when $E = 28$ V, $\Sigma R = 2.95$ kΩ, and $V_2 = 11.4$ V.

Voltage Divider

Since the applied voltage is divided between the resistances in a series circuit, a series circuit may be called a **voltage divider.** The voltage drop across any resistance in a series circuit may be determined by Equation 10–3, which results from solving for V_n in Equation 10–2.

$$V_n = \frac{ER_n}{\Sigma R} \quad \textbf{(volts)} \tag{10–3}$$

where V_n = voltage drop across R_n (V)
$\quad E$ = applied voltage (V)
$\quad R_n$ = one resistance in the voltage divider (Ω)
$\quad \Sigma R$ = summation of all the resistances in the divider (Ω)

EXAMPLE 10–2   Compute the voltage drop across each of the resistances of Figure 10–2. Use Equation 10–3 and show that $E = V_1 + V_2 + V_3$.

FIGURE 10–2   Series voltage divider. Each voltage drop is directly proportional to the resistance it is across. Larger resistances have larger voltage drops than smaller resistances.

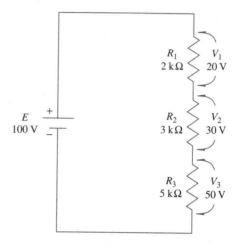

Solution   Note that E is 100 V and ΣR is 10 kΩ in the solution for V_1, V_2, and V_3. Solve for V_1; substitute 2 kΩ for R_1:

$$V_1 = \frac{ER_1}{\Sigma R}$$

$$V_1 = \frac{100 \times 2 \times 10^3}{10 \times 10^3}$$

$\therefore$ $V_1 = 20$ V

Solve for V_2; substitute 3 kΩ for R_2:

$$V_2 = \frac{ER_2}{\Sigma R}$$

$$V_2 = \frac{100 \times 3 \times 10^3}{10 \times 10^3}$$

$\therefore$ $V_2 = 30$ V

Solve for V_3; substitute 5 kΩ for R_3:

$$V_3 = \frac{ER_3}{\Sigma R}$$

$$V_3 = \frac{100 \times 5 \times 10^3}{10 \times 10^3}$$

$\therefore$ $V_3 = 50$ V

Show that the summation of the voltage drops equals the voltage rise:

$$E = V_1 + V_2 + V_3$$
$$100 = 20 + 30 + 50$$
$$100 \text{ V} = 100 \text{ V}$$

∴   The voltage rise equals the sum of the voltage drops.

EXERCISE 10-2

Use Equation 10–3 to determine the voltage drop across the indicated resistor of Figure 10–2, given the following information.

1. Find V_2 when $E = 700$ V, $R_1 = 2.2$ kΩ, $R_2 = 3.0$ kΩ, $R_3 = 1.8$ kΩ.
2. Find V_3 when $E = 36$ V, $R_1 = 390$ Ω, $R_2 = 110$ Ω, $R_3 = 500$ Ω.
3. Find V_2 when $E = 50$ V, $R_1 = 82$ kΩ, $R_2 = 300$ kΩ, $R_3 = 120$ kΩ.
4. Find V_1 when $E = 9.0$ V, $R_1 = 10$ kΩ, $R_2 = 27$ kΩ, $R_3 = 15$ kΩ.
5. Find V_3 when $E = 80$ V, $R_1 = 4.7$ kΩ, $R_2 = 5.6$ kΩ, $R_3 = 6.2$ kΩ.
6. Find V_1 when $E = 220$ V, $R_1 = 750$ kΩ, $R_2 = 1.2$ MΩ, $R_3 = 2.0$ MΩ.
7. Find V_2 when $E = 50$ V, $R_1 = 12$ kΩ, $R_2 = 22$ kΩ, $R_3 = 9.1$ kΩ.
8. Find V_1 when $E = 480$ V, $R_1 = 39$ Ω, $R_2 = 56$ Ω, $R_3 = 20$ Ω.
9. Find V_3 when $E = 1.20$ kV, $R_1 = 1.80$ MΩ, $R_2 = 2.70$ MΩ, $R_3 = 820$ kΩ.
10. Find V_3 when $E = 6.0$ V, $R_1 = 30$ kΩ, $R_2 = 56$ kΩ, $R_3 = 75$ kΩ.

10-2 CONDUCTANCE OF THE PARALLEL CIRCUIT

We have previously learned that resistance, measured in ohms (Ω), indicates the amount of *opposition* offered by a circuit or component to current flow. **Conductance,** measured in siemens (S), indicates the ease with which current *passes* through a circuit or component. Conductance is inversely proportional to resistance. This is stated as Equation 10–4.

$$G = 1/R \quad \textbf{(siemens)} \qquad \text{(10–4)}$$

where G = conductance in siemens (S)
       R = resistance in ohms (Ω)

EXAMPLE 10-3   Determine the conductance of a 500-Ω resistance.
      Solution   Use Equation 10–4 and your calculator.

$$G = 1/R$$
1/x $\quad G = 1/500$
∴ $\quad G = 2.00 \text{ mS}$

EXERCISE 10–3

Calculator Exercise

Use the reciprocal key on your calculator to compute the conductance of the following resistances.

1. $56.0\ \Omega$ 2. $1.80\ k\Omega$ 3. $910\ \Omega$
4. $390\ k\Omega$ 5. $6.20\ k\Omega$ 6. $3.30\ M\Omega$
7. $2.70\ k\Omega$ 8. $22.0\ \Omega$ 9. $8.20\ M\Omega$
10. $750\ \Omega$ 11. $560\ m\Omega$ 12. $3.70\ m\Omega$

Parallel Circuit

When two or more resistances are connected in parallel, as shown in Figure 10–3, the voltage drop across each branch is the same, and it is equal to the source voltage. This is expressed as Equation 10–5.

$$E = V_1 = V_2 = V_3 \qquad \textbf{(volts)} \qquad \textbf{(10–5)}$$

In the parallel circuit, the total conductance of all the branches is determined by summing the conductance of each branch. Thus,

$$G_T = G_1 + G_2 + G_3 \qquad \textbf{(siemens)} \qquad \textbf{(10–6)}$$

where G_T = total conductance of the entire circuit in siemens (S)
         G_1 = conductance of the first branch (S)
         G_2 = conductance of the second branch (S)
         G_3 = conductance of the third branch (S)

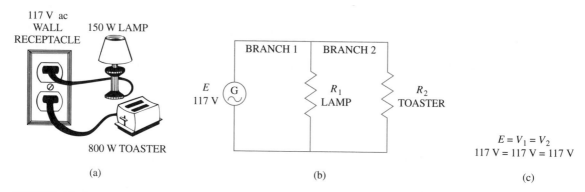

FIGURE 10–3   Parallel circuit consisting of a lamp and a toaster connected to a 117-V wall receptacle: (a) pictorial diagram of the two-branch parallel circuit; (b) schematic diagram of the two-branch parallel circuit; (c) the voltages (V_1 and V_2) across each branch are the same in a parallel circuit and they are equal to the applied voltage, E.

Equation 10–6 may be stated more generally as Equation 10–7:

$$G_T = \Sigma G \qquad \text{(siemens)} \qquad \textbf{(10–7)}$$

where ΣG = summation of the branch conductances in siemens (S).

EXAMPLE 10–4   Compute the total conductance of the parallel circuit pictured by Figure 10–4.

FIGURE 10–4   Circuit for Example 10–4.

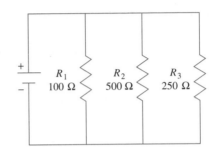

Solution   Compute the conductances of each branch by Equation 10–4:

$1/x$ $G_1 = 1/R_1 = 1/100 = 0.010$ S

$1/x$ $G_2 = 1/R_2 = 1/500 = 0.002$ S

$1/x$ $G_3 = 1/R_3 = 1/250 = 0.004$ S

Use Equation 10–7 to compute the total circuit conductance:

$G_T = \Sigma G$

$+$ $G_T = 0.010 + 0.002 + 0.004$

$G_T = 0.016$ S

$\therefore$ $G_T = 16$ mS

Observation   These computations can be performed in the chain calculation mode on your calculator by recognizing that $G_T = \Sigma 1/R$.

EXERCISE 10–4

Determine the total conductances of each parallel circuit given the following information.

1. $G_1 = 3.0$ mS $G_2 = 15$ mS   2. $G_1 = 0.30$ S $G_2 = 0.80$ S
3. $G_1 = 22$ μS $R_2 = 12$ kΩ 4. $R_1 = 820$ Ω $G_2 = 1.80$ mS
5. $R_1 = 75$ kΩ $R_2 = 180$ kΩ
6. $R_1 = 5.0$ Ω $R_2 = 10$ Ω $R_3 = 20$ Ω

7. $R_1 = 150\ \Omega$	$R_2 = 300\ \Omega$	$R_3 = 100\ \Omega$
8. $R_1 = 2\ k\Omega$	$R_2 = 5\ k\Omega$	$R_3 = 3\ k\Omega$
9. $R_1 = 750\ \Omega$	$R_2 = 680\ \Omega$	$R_3 = 420\ \Omega$
10. $R_1 = 15\ \Omega$	$R_2 = 4.7\ \Omega$	$R_3 = 33\ \Omega$
11. $G_1 = 9.0\ \mu S$	$R_2 = 91\ k\Omega$	$G_3 = 12\ \mu S$
12. $G_1 = 18\ mS$	$R_2 = 27\ \Omega$	$G_3 = 68\ mS$

10–3 EQUIVALENT RESISTANCE OF THE PARALLEL CIRCUIT

A parallel circuit of two or more branches may be simplified to an *equivalent series circuit,* as pictured in Figure 10–5, by computing the equivalent resistance (R_T) of the parallel circuit. The resulting equivalent circuit has the same source voltage, total current, and total resistance as the parallel circuit from which it was derived.

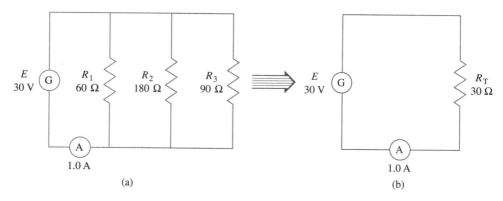

FIGURE 10–5   Equivalent series circuit of a parallel circuit: (a) parallel circuit with three branches (R_1, R_2, R_3); (b) equivalent series circuit with the loads (R_1, R_2, R_3) reduced to R_T.

Computing the Equivalent Resistance

The equivalent resistance (R_T) of a parallel circuit is determined by first adding the conductance of each branch to get the total conductance, and then reciprocating the total conductance to get the equivalent resistance. This is stated as Equation 10–8.

$$R_T = \frac{1}{\Sigma G}\quad \text{(ohms)} \tag{10–8}$$

where R_T = equivalent resistance of the parallel resistances in ohms
ΣG = summation of the branch conductances in siemens

Equation 10–8 is used when the conductances are known. When the resistances are known, Equation 10–9, and Equation 10–10 that will be covered later, are more convenient.

$$R_T = \cfrac{1}{\cfrac{1}{R_1} + \cfrac{1}{R_2} + \cfrac{1}{R_3}} \quad \text{(ohms)} \qquad \text{(10–9)}$$

where R_T = equivalent resistance of the parallel resistance (Ω)

R_1, R_2, R_3 = resistances of the branches of the parallel circuit (Ω)

Equation 10–9 is most often used with a calculator to compute the equivalent resistance of a multi-branch parallel circuit. The following guidelines are presented to aid in evaluating Equation 10–9.

GUIDELINES 10–1. *Computing Equivalent Resistance*

To compute R_T using Equation 10–9 and a calculator:

1. Take the reciprocal of R_1.
2. Add in the reciprocal of R_2 and the reciprocal of R_3.
3. Reciprocate the sum of the reciprocals.

EXAMPLE 10–5   Determine the equivalent resistance (R_T) of the circuit of Figure 10–6.

FIGURE 10–6 Circuit for Example 10–5.

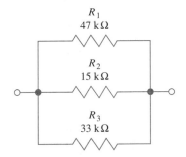

R_1
47 kΩ

R_2
15 kΩ

R_3
33 kΩ

Solution   Use Equation 10–9 and the previous guidelines:

$$R_T = \cfrac{1}{\cfrac{1}{R_1} + \cfrac{1}{R_2} + \cfrac{1}{R_3}}$$

Substitute 47 kΩ for R_1, 15 kΩ for R_2, and 33 kΩ for R_3:

$$R_T = \cfrac{1}{\cfrac{1}{47 \times 10^3} + \cfrac{1}{15 \times 10^3} + \cfrac{1}{33 \times 10^3}}$$

Use your calculator to evaluate the denominator:

$$\boxed{1/x} \qquad R_T = \frac{1}{118 \times 10^{-6}} = 8.46 \times 10^3$$

$$\therefore \qquad R_T = 8.5 \text{ k}\Omega$$

EXAMPLE 10–6 Determine the equivalent resistance (R_T) of the circuit of Figure 10–7.

FIGURE 10–7 Circuit for Example 10–6.

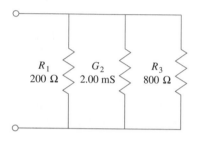

Solution Reciprocate R_1, add in G_2, and then add in the reciprocal of R_3 (G_2 is the reciprocal of R_2):

$$R_T = \frac{1}{\dfrac{1}{200} + 2.00 \times 10^{-3} + \dfrac{1}{800}}$$

$$\boxed{1/x} \qquad R_T = \frac{1}{8.25 \times 10^{-3}} = 121$$

$$\therefore \qquad R_T = 121 \ \Omega$$

Equation 10–10, an important equation used to compute the equivalent resistance (R_T) of a two-branch parallel circuit, is derived from Equation 10–9 and is commonly referred to as the *product over the sum* form of the equivalent resistance equation.

$$R_T = \frac{R_1 R_2}{R_1 + R_2} \quad \text{(ohms)} \tag{10–10}$$

where R_T = equivalent resistance of the parallel resistance (Ω)
R_1 and R_2 = resistances of the branches of the parallel circuit (Ω)

EXAMPLE 10–7 Determine the equivalent resistance (R_T) of the circuit of Figure 10–8.

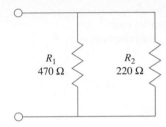

FIGURE 10–8 Circuit for Example 10–7.

Solution   Use Equation 10–10; let $R_1 = 470\ \Omega$ and $R_2 = 220\ \Omega$.

$$R_T = \frac{R_1 R_2}{R_1 + R_2}$$

$$R_T = \frac{470 \times 220}{470 + 220} = 150\ \Omega$$

$$\therefore \qquad R_T = 150$$

EXERCISE 10–5

Compute the equivalent resistance of each parallel circuit given the following information:

1. $R_1 = 100\ \Omega$ $R_2 = 300\ \Omega$ 2. $R_1 = 2.0\ k\Omega$ $R_2 = 1.2\ k\Omega$
3. $R_1 = 750\ \Omega$ $R_2 = 420\ \Omega$ 4. $R_1 = 6.2\ k\Omega$ $R_2 = 3.6\ k\Omega$
5. $R_1 = 33\ \Omega$ $R_2 = 15\ \Omega$
6. $R_1 = 9.1\ k\Omega$ $R_2 = 5.6\ k\Omega$ $R_3 = 3.9\ k\Omega$
7. $R_1 = 7.5\ k\Omega$ $R_2 = 2.2\ k\Omega$ $R_3 = 6.2\ k\Omega$
8. $R_1 = 4.7\ M\Omega$ $R_2 = 2.0\ M\Omega$ $R_3 = 3.0\ M\Omega$
9. $G_1 = 3.00\ \mu S$ $R_2 = 120\ k\Omega$ $R_3 = 91.0\ k\Omega$
10. $R_1 = 4.3\ k\Omega$ $G_2 = 800\ \mu S$ $R_3 = 2.7\ k\Omega$

10–4 CURRENT DIVISION IN THE PARALLEL CIRCUIT

The total circuit current (I_T) in a parallel circuit divides between the various branches of the circuit, as shown in Figure 10–9. The value of each branch current depends directly upon the conductance of that branch. The branch with the largest conductance will have the largest current, while the branch with the smallest conductance will have the smallest current passing through it.

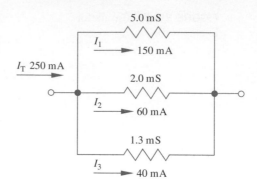

FIGURE 10–9 The division of the total current in a parallel circuit depends directly upon the conductance of each branch.

Kirchhoff's Current Law

The total circuit current (I_T) in a parallel circuit is equal to the summation of the branch currents. Gustav Kirchhoff is credited with the discovery of this important circuit characteristic. Stated mathematically,

$$I_T = I_1 + I_2 + I_3 \quad \textbf{(amperes)} \qquad \textbf{(10–11)}$$

where I_1, I_2, I_3 = branch currents (A)

$$I_T = \Sigma I \quad \textbf{(amperes)} \qquad \textbf{(10–12)}$$

where I_T = total current in the parallel circuit (A)
$\quad\ \Sigma I$ = summation of the branch currents (A)

Current Divider

Since the total current (I_T) is divided between the branches of the parallel circuit, a parallel circuit may be called a **current divider.** The current passing through any branch of a parallel circuit may be determined by Equation 10–13.

$$I_n = \frac{I_T G_n}{\Sigma G} \quad \textbf{(amperes)} \qquad \textbf{(10–13)}$$

where I_n = current through branch n in amperes (A)
$\quad I_T$ = total circuit current in amperes (A)
$\quad G_n$ = conductance of branch n in siemens (S)
$\quad \Sigma G$ = summation of all the branch conductances in siemens (S)

EXAMPLE 10–8   Determine the current in each branch of the circuit of Figure 10–10.

FIGURE 10–10 Circuit for Example 10–8.

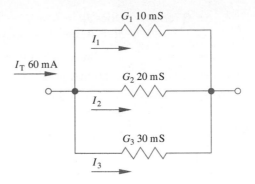

Solution   Compute the total conductance of the circuit pictured in Figure 10–10:

$$\Sigma G = G_1 + G_2 + G_3$$
$$\Sigma G = 10 \times 10^{-3} + 20 \times 10^{-3} + 30 \times 10^{-3}$$

$$\Sigma G = 60 \text{ mS}$$

Note that $I_T = 60$ mA and ΣG is 60 mS in the solution for I_1, I_2, and I_3. Solve for I_1 when $G_1 = 10$ mS:

$$I_1 = \frac{I_T G_1}{\Sigma G}$$

$$I_1 = \frac{60 \times 10^{-3} \times 10 \times 10^{-3}}{60 \times 10^{-3}}$$

∴ $I_1 = 10$ mA

Solve for I_2 when $G_2 = 20$ mS:

$$I_2 = \frac{I_T G_2}{\Sigma G}$$

$$I_2 = \frac{60 \times 10^{-3} \times 20 \times 10^{-3}}{60 \times 10^{-3}}$$

∴ $I_2 = 20$ mA

Solve for I_3 when $G_3 = 30$ mS:

$$I_3 = \frac{I_T G_3}{\Sigma G}$$

$$I_3 = \frac{60 \times 10^{-3} \times 30 \times 10^{-3}}{60 \times 10^{-3}}$$

∴ $I_3 = 30$ mA

Check   Does $I_T = I_1 + I_2 + I_3$ when $I_1 = 10$ mA, $I_2 = 20$ mA, $I_3 = 30$ mA, and $I_T = 60$ mA?

$$60 \text{ mA} = 10 \text{ mA} + 20 \text{ mA} + 30 \text{ mA}$$

$$60 \text{ mA} = 60 \text{ mA} \qquad \text{Yes!}$$

EXERCISE 10–6

Use Equation 10–13 to determine the current through the indicated branch of the parallel circuits given the following information. It is suggested that a schematic be drawn and labeled for each problem.

1. Find I_3 when $I_T = 10$ A, $G_1 = 10$ mS, $G_2 = 25$ mS, $G_3 = 15$ mS.
2. Find I_1 when $I_T = 3.00$ mA, $G_1 = 80.0$ μS, $G_2 = 105$ μS, $G_3 = 150$ μS.
3. Find I_2 when $I_T = 12.0$ mA, $G_1 = 21.3$ μS, $G_2 = 66.7$ μS, $G_3 = 37.0$ μS.
4. Find I_1 when $I_T = 600$ mA, $G_1 = 17.9$ mS, $G_2 = 12.2$ mS, $G_3 = 30.3$ mS.
5. Find I_3 when $I_T = 38.0$ mA, $G_1 = 667$ μS, $G_2 = 370$ μS, $G_3 = 147$ μS.
6. Find I_2 when $I_T = 20.0$ μA, $G_1 = 0.833$ μS, $G_2 = 1.47$ μS, $G_3 = 2.13$ μS.
7. Find I_3 when $I_T = 2.5$ A, $G_1 = 37$ mS, $R_2 = 10$ Ω, $G_3 = 66.7$ mS.
8. Find I_1 when $I_T = 150$ mA, $R_1 = 910$ Ω, $G_2 = 455$ μS, $G_3 = 0.625$ mS.
9. Find I_3 when $I_T = 50$ mA, $R_1 = 2.0$ kΩ, $R_2 = 3.3$ kΩ, $G_3 = 556$ μS.
10. Find I_2 when $I_T = 300$ μA, $R_1 = 910$ kΩ, $R_2 = 680$ kΩ, $R_3 = 1.0$ MΩ.

10–5   SOLVING PARALLEL CIRCUIT PROBLEMS

The solution of parallel circuit problems involves the application of all the concepts covered so far in this chapter, as well as Ohm's law. The current in each branch of a parallel circuit may be computed by Ohm's law when the applied voltage and the branch resistance or conductance are known. Since $I = E/R$, we may write an equation specifically for computing the branch current in a parallel circuit.

$$I_n = \frac{E}{R_n} \quad \text{(amperes)} \tag{10–14}$$

where I_n = current in branch n of the parallel circuit (A)
   E = applied voltage (V)
   R_n = resistance of the branch n (Ω)
   n = branch index

EXAMPLE 10–9   Compute (a) the current in each branch of the circuit of Figure 10–11 and (b) the total current (I_T).

FIGURE 10–11 Circuit for Example 10–9.

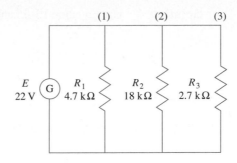

Solution

(a) Use Equation 10–14:

$$I_1 = 22/(4.7 \times 10^3) = 4.7 \text{ mA}$$
$$I_2 = 22/(18 \times 10^3) = 1.2 \text{ mA}$$
$$I_3 = 22/(2.7 \times 10^3) = 8.2 \text{ mA}$$

(b) Use Equation 10–12: $I_T = \Sigma I$.

$$I_T = 4.7 + 1.2 + 8.2$$

$\therefore \qquad I_T = 14.1 \text{ mA}$

Equation 10–14 is used to compute the branch currents when voltage and resistance are known. However, in analyzing solid-state circuits, conductance is often known instead of resistance, so Equation 10–14 is transformed into Equation 10–15, which is used to compute branch current when conductance is known.

$$I_n = EG_n \qquad \textbf{(amperes)} \tag{10–15}$$

where I_n = current in branch n of the parallel circuit (A)
$\quad\quad E$ = applied voltage (V)
$\quad\quad G_n$ = conductance of branch n (S)
$\quad\quad n$ = branch index

EXAMPLE 10–10 Determine the current in branch 2 of the parallel circuit of Figure 10–12.

FIGURE 10–12 Circuit for Example 10–10.

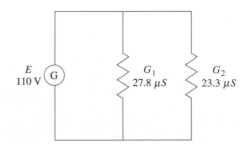

Solution   Use Equation 10–15 and substitute 110 V for E and 23.3 μS for G_2:

$$I_2 = EG_2$$
$$I_2 = 110 \times 23.3 \times 10^{-6}$$
$$\therefore \quad I_2 = 2.56 \text{ mA}$$

Table 10–1 is a summary of the formulas used to solve parallel circuit problems.

TABLE 10–1   Formulas for Parallel Circuits

Equation	Comment
$G = \dfrac{1}{R}$	Conductance G is the reciprocal of resistance R
$G_T = \Sigma G$	Total conductance G_T is the sum of the branch conductances
$E = V_1 = V_2$	Voltage drops across branches V_1 and V_2 are the same and are equal to the source voltage E
$R_T = \dfrac{1}{\Sigma G}$	Equivalent resistance R_T is formed by taking the reciprocal of the total conductance
$R_T = \dfrac{1}{\dfrac{1}{R_1} + \dfrac{1}{R_2} + \dfrac{1}{R_3}}$	Equivalent resistance R_T is the reciprocal of the sum of the reciprocals of the branch resistances
$R_T = \dfrac{R_1 R_2}{R_1 + R_2}$	Equivalent resistance R_T is equal to the product over the sum of two branch resistances
$I_T = \Sigma I$	Total current I_T is the summation of the individual branch currents
$I_n = \dfrac{I_T G_n}{\Sigma G}$	Current in one branch I_n is computed with the current-divider equation
$I_n = \dfrac{E}{R_n}$	Ohm's law is used to find the current in a branch when resistance and voltage are known
$I_n = EG_n$	Current in a branch is found by multiplying voltage and conductance

EXAMPLE 10–11   Two resistances, R_1 and R_2, are connected in parallel across a 5.0-V source. The total circuit current is 0.50 A. The current through R_1 is 200 mA.

(a) Draw a schematic of the circuit.
(b) Determine the value of V_1 and V_2.
(c) Determine the value of R_1 and R_2.
(d) Compute the equivalent resistance, R_T.

Solution   (a) Draw and label a schematic. See Figure 10–13.
(b) V_1 and V_2 are equal to the applied voltage.

$$E = 5.0 \text{ V}$$
$$\therefore \quad V_1 = V_2 = 5.0 \text{ V}$$

FIGURE 10–13 Schematic diagram for the circuit for Example 10–11.

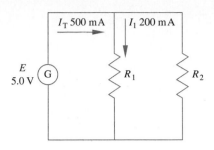

(c) Determine the value of R_1:

$$R_1 = V_1/I_1$$
$$R_1 = 5.0/(200 \times 10^{-3})$$
$$\therefore \quad R_1 = 25 \, \Omega$$

Determine the value of R_2:

$$I_2 = I_T - I_1$$
$$I_2 = 500 \text{ mA} - 200 \text{ mA}$$
$$I_2 = 300 \text{ mA}$$
$$R_2 = V_2/I_2$$
$$R_2 = 5.0/(300 \times 10^{-3})$$
$$\therefore \quad R_2 = 16.7 \, \Omega$$

(d) Compute the equivalent resistance, R_T.

$$R_T = \frac{R_1 R_2}{R_1 + R_2}$$
$$R_T = \frac{25 \times 16.7}{25 + 16.7}$$
$$\therefore \quad R_T = 10 \, \Omega$$

EXERCISE 10–7

Solve the following problems.

1. Determine the current through R_2 of Figure 10–14.
2. Compute the total conductance and the equivalent resistance of the circuit of Figure 10–14.
3. If the source voltage E of Figure 10–14 becomes 12.0 V, determine the current through R_1.
4. Determine the current through resistor R_1 in Figure 10–15 when $I_2 = 80.0$ mA.

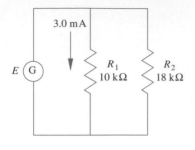

FIGURE 10–14 Circuit for Exercise 10–7, problems 1 through 3.

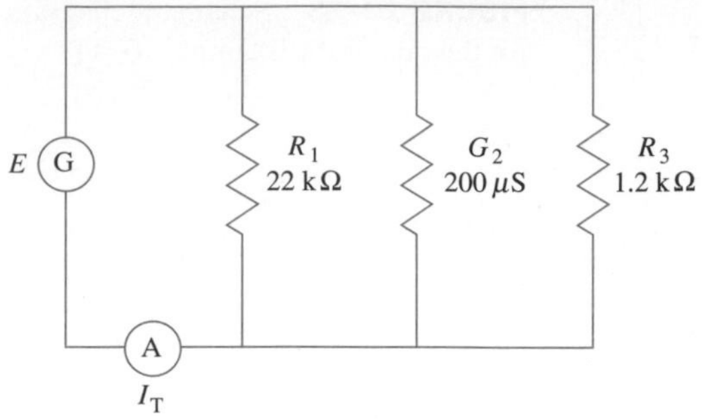

FIGURE 10–15 Circuit for Exercise 10–7, problems 4 through 6.

5. Compute the total circuit current, I_T, for Figure 10–15 when the source voltage, E, is 100 V.

6. Calculate the equivalent resistance, R_T, of the circuit of Figure 10–15.

7. Find I_1 and I_2 in Figure 10–16.

FIGURE 10–16 Circuit for Exercise 10–7, problems 7 through 9.

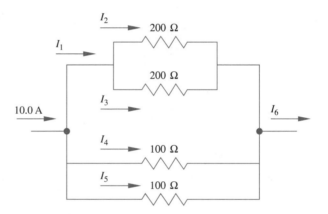

8. Determine I_5 and I_6 in Figure 10–16.

9. Determine I_3 and I_4 in Figure 10–16.

10. Three resistors are connected in parallel. The total circuit current is 300 mA, and R_1 has a conductance of 1.25 mS. The second resistor has a voltage drop of 28.0 V across it, and the third resistor has a current of 112 mA passing through it. Determine the resistance of each of the three resistors.

10–6 USING NETWORK THEOREMS TO FORM EQUIVALENT CIRCUITS

The analysis of an electric circuit is simplified when the *network theorems* are used to form an equivalent circuit. In this section you will be introduced to Thévenin's theorem and Norton's theorem.

In electronics we use mathematical formulation to *model* electrical circuits. When the mathematical model is a good approximation of the actual circuit, then both the original circuit and its model will yield valid results and the two representations are *equivalent*.

Any linear two-terminal circuit may be represented by an electrically equivalent circuit consisting of:

1. A voltage source and a series resistance (Thévenin's)
2. A current source and a parallel resistance (Norton's)

Figure 10–17 pictures each of these equivalent circuits. Figure 10–17(a) shows a **Thévenin's equivalent circuit** consisting of a *Thévenin's equivalent voltage source,* E_{Th}, and a series *Thévenin's equivalent resistance,* R_{Th}. Figure 10–17(b) pictures a **Norton's equivalent circuit** consisting of a *Norton's equivalent current source,* I_N, and a parallel *Norton's equivalent resistance,* R_N.

FIGURE 10–17 Equivalent circuits: (a) Thévenin's equivalent circuit; (b) Norton's equivalent circuit.

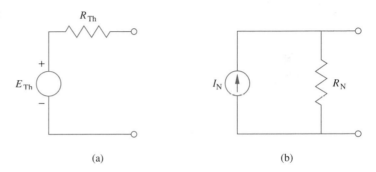

(a)   (b)

Since both Thévenin's and Norton's theorems deal with equivalent models of circuits connected between two terminals, the circuit of Figure 10–18 may be represented by either one of these two *network theorems*. Each of the theorems will be used to model the circuit of Figure 10–18.

FIGURE 10–18 Circuit for Example 10–12.

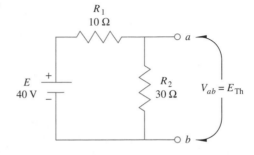

EXAMPLE 10–12   Form the Thévenin's and Norton's equivalent circuits for the circuit of Figure 10–18.

Solution   View the circuit of Figure 10–18 from terminals a–b. Thévenin's equivalent voltage (E_{Th}) is V_{ab} of Figure 10–18. Compute E_{Th} using the voltage-divider equation, Equation 10–3.

$$E_{Th} = V_{ab} = \frac{ER_2}{R_1 + R_2} = \frac{(40)(30)}{40} = 30 \text{ V}$$

Observation V_{ab} is the voltage across R_2.

Norton's equivalent current (I_N) is the current that passes through a short circuit placed across terminals a–b, as pictured in Figure 10–19(a).

$$I_N = E/R_1 = 40/10 = 4.0\text{A}$$

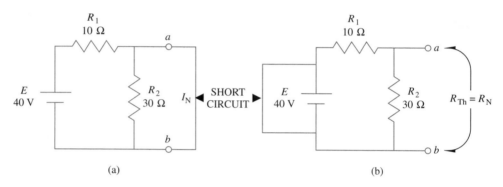

(a)   (b)

FIGURE 10–19   (a) Norton's equivalent current, I_N, is the current passing in the short circuit between terminals a–b. (b) The equivalent resistance ($R_{Th} = R_N$) is calculated by first setting the voltage source to zero (short circuit) and then looking across terminals a–b.

Observation R_2 is bypassed by the short circuit.

Compute Thévenin's equivalent resistance by setting the voltage source to zero (short circuit) as shown in Figure 10–19(b) and calculate the resistance between terminals a–b.

Observation R_1 is now in parallel with R_2. Use Equation 10–10 to compute R_{Th}.

$$R_{Th} = \frac{R_1 R_2}{R_1 + R_2} = \frac{10 \times 30}{10 + 30} = 7.5 \ \Omega$$

Compute Norton's equivalent resistance by setting the voltage source to zero (short circuit) as pictured in Figure 10–19(b) and calculate the resistance between terminals a–b.

Observation   The Norton's equivalent resistance is computed in the same manner as the Thévenin's equivalent resistance. Thus:

$$R_{Th} \equiv R_N \qquad (10\text{--}16)$$

$$R_N = R_{Th} = \frac{R_1 R_2}{R_1 + R_2} = \frac{10 \times 30}{10 + 30} = 7.5 \ \Omega$$

∴   The Thévenin's equivalent circuit is formed as shown in Figure 10–20(b) and the Norton's equivalent circuit is formed as shown in Figure 10–20(c).

Observation   The original circuit is pictured in Figure 10–20(a). The circuit of Figure 10–20(b) is *equivalent* to the circuit of Figure 10–20(c). This is indicated by the equivalent symbol (⇔) between these two figures.

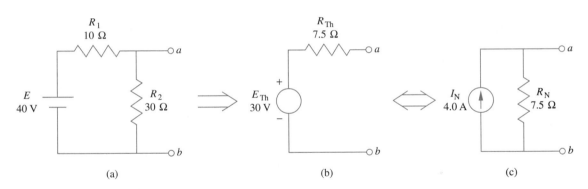

(a)   (b)   (c)

FIGURE 10–20   (a) Circuit for Example 10–12; (b) Thévenin's equivalent circuit of circuit (a); (c) Norton's equivalent circuit of circuit (a).

Before moving on to another example, a word of caution is in order. In forming the Norton's equivalent circuit, the terminals were short circuited to determine the short-circuit current, I_N. You must realize that this is a mathematical technique (pencil-and-paper process) that must not be attempted in the laboratory with real laboratory equipment. This is also true of the technique used to set the voltage source to zero.

EXAMPLE 10–13   Form the Thévenin's and Norton's equivalent circuits for the circuit pictured in Figure 10–21(a).

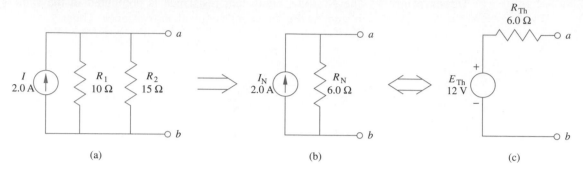

FIGURE 10–21 (a) Circuit for Example 10–13; (b) Norton's equivalent circuit of circuit (a); (c) Thévenin's equivalent circuit of circuit (a).

Solution   Determine the equivalent resistance by setting the current source to zero (open circuit) and calculating the resistance between terminals a–b.

$$R_{ab} = R_{Th} = R_N = \frac{R_1 R_2}{R_1 + R_2} = \frac{10 \times 15}{10 + 15} = 6.0 \ \Omega$$

Norton's equivalent current is the current that passes through a short circuit across terminals a–b. Because the short bypasses R_1 and R_2, the Norton's current is equal to the source current of 2.0 A.

$$I_N = 2.0 \text{ A}$$

Observation   Because the two network theorems are equivalent, the Thévenin's equivalent voltage, E_{Th}, may be determined from the Norton's equivalent current, I_N, and, conversely, the Norton's equivalent current, I_N, may be determined from the Thévenin's equivalent voltage, E_{Th}. Stated mathematically,

$$\begin{array}{lll} I_N = E_{Th}/R_{Th} & \text{where } R_{Th} = R_N & \textbf{(10–17)} \\ E_{Th} = I_N R_N & \text{where } R_N = R_{Th} & \textbf{(10–18)} \end{array}$$

Compute Thévenin's equivalent voltage using Equation 10–17 and Figure 10–21(b).

$$E_{Th} = I_N R_N = (2.0)(6.0) = 12 \text{ V}$$

$\therefore$ The Norton's equivalent is formed as shown in Figure 10–21(b) and the Thévenin's equivalent is formed as shown in Figure 10–21(c).

Observation   The two network theorems are equally valid since both yield the same results. The choice of which theorem to use depends on how you choose to model the circuit.

EXAMPLE 10–14   Demonstrate that the two circuits pictured in Figure 10–22 are equivalent; that is, the parameters of one circuit may be derived from the parameters of the other circuit.

FIGURE 10–22   Conversion from Thévenin's (a) to Norton's (b) equivalent circuit for Example 10–14.

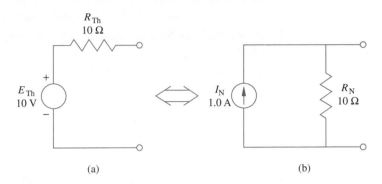

(a)   (b)

Solution   Derive the Norton's equivalent circuit from the Thévenin's equivalent circuit in Figure 10–22(a). Use Equations 10–16 and 10–17.

∴ $R_N = R_{Th} = 10\ \Omega$

and

$I_N = E_{Th}/R_{Th} = 10/10 = 1.0\ A$

Derive the Thévenin's equivalent circuit from the Norton's equivalent circuit of Figure 10–22(b). Use Equations 10–16 and 10–18.

∴ $R_{Th} = R_N = 10\ \Omega$

and

$E_{Th} = I_N R_N = (1.0)(10) = 10\ V$

Summary

Using network theorems to form equivalent circuits is a straightforward procedure. First, calculate the equivalent resistance ($R_N = R_{Th}$) by looking into the two terminals of the selected circuit. Set all sources to zero; that is, short-circuit voltage sources and open-circuit current sources. Second, calculate either the open-circuit voltage (E_{Th}) or the short-circuit current (I_N) at the selected terminals.

EXERCISE 10–8

Solve the following problems using Equations 10–16, 10–17, and 10–18.

1. Determine the parameters of Norton's equivalent circuit of Figure 10–17(b) when $E_{Th} = 25.0\ V$ and $R_{Th} = 150\ \Omega$ in Figure 10–17(a).
2. Repeat problem 1 when $E_{Th} = 16.0\ V$ and $R_{Th} = 200\ \Omega$.

3. Determine the parameters of the Thévenin's equivalent circuit of Figure 10–17(a) when $I_N = 0.30$ A and $R_N = 47\ \Omega$ in Figure 10–17(b).

4. Repeat problem 3 when $I_N = 28.0$ mA and $R_N = 1.20\ k\Omega$.

5. Determine the Thévenin's equivalent circuit of a power supply with a specified open-circuit voltage of 20 V (E_{Th}) and a short-circuit current (I_N) of 2.0 A.

6. Repeat problem 5 for an open-circuit voltage of 12 V and a short-circuit current of 5.0 A.

7. Determine the Norton's equivalent circuit of Figure 10–18 when $E = 15$ V, $R_1 = 68\ \Omega$, and $R_2 = 56\ \Omega$.

8. Determine the Norton's equivalent circuit of Figure 10–18 when $E = 6.30$ V, $R_1 = 180\ \Omega$, and $R_2 = 220\ \Omega$.

9. Determine the Thévenin's equivalent circuit of Figure 10–21(a) when $I = 45$ mA, $R_2 = 2.7\ k\Omega$, and $R_2 = 3.0\ k\Omega$.

10. Determine the Thévenin's equivalent circuit of Figure 10–21(a) when $I = 85$ mA, $R_1 = 3.9\ k\Omega$, and $R_2 = 4.7\ k\Omega$.

SELECTED TERMS

conductance   Measure of the ease with which a conductor carries an electric current; the reciprocal of resistance.

current divider   Two or more resistances (loads) connected in parallel across a source will cause the source current to divide.

voltage divider   Two or more resistances connected in series with a voltage source form a voltage divider.

SECTION CHALLENGE

WEB CHALLENGE FOR CHAPTERS 8, 9, AND 10

To evaluate your comprehension of Chapters 8, 9, and 10, log on to **www.prenhall.com/harter** and take the online True/False and Multiple Choice assessments for each of the chapters.

SECTION CHALLENGE FOR CHAPTERS 8, 9, AND 10*

In Chapter 10, Equations 10–9 and 10–10 are used to determine the equivalent resistance of a parallel circuit. Your challenge is to derive the *product over the sum form* of the equivalent resistance equation (Equation 10–10), from the reciprocal of the *sum of the reciprocals form* of the equation, Equation 10–9.

$$R_T = \frac{1}{\dfrac{1}{R_1} + \dfrac{1}{R_2}} \quad \textbf{(ohms)} \qquad \textbf{(10–9)}$$

$$R_T = \frac{R_1 R_2}{R_1 + R_2} \quad \textbf{(ohms)} \qquad \textbf{(10–10)}$$

Then using the appropriate equation, determine the value of the equivalent resistance of the following parallel connected resistances.

A. $R_1 = 560\ \Omega$ $R_2 = 820\ \Omega$

B. $R_1 = 3.60\ \text{k}\Omega$ $R_2 = 9.10\ \text{k}\Omega$

C. $R_1 = 240\ \text{k}\Omega$ $R_2 = 160\ \text{k}\Omega$

D. $R_1 = 75.0\ \Omega$ $R_2 = 27.0\ \Omega$

E. $R_1 = 390\ \text{k}\Omega$ $R_2 = 82.0\ \text{k}\Omega$

F. $R_1 = 22.0\ \text{k}\Omega$ $R_2 = 120\ \text{k}\Omega$ $R_3 = 43.0\ \text{k}\Omega$

G. $R_1 = 620\ \Omega$ $R_2 = 130\ \Omega$ $R_3 = 300\ \Omega$

H. $R_1 = 330\ \text{k}\Omega$ $R_2 = 91.0\ \text{k}\Omega$ $R_3 = 51.0\ \text{k}\Omega$

I. $R_1 = 0.820\ \text{M}\Omega$ $R_2 = 1.10\ \text{M}\Omega$ $R_3 = 4.70\ \text{M}\Omega$

J. $R_1 = 8.06\ \text{k}\Omega$ $R_2 = 32.4\ \text{k}\Omega$ $R_3 = 18.2\ \text{k}\Omega$

*The solution to this Section Challenge is found in Appendix C.

11

Special Products, Factoring, and Equations

11–1 Mentally Multiplying Two Binomials

11–2 Product of the Sum and Difference of Two Numbers

11–3 Square of a Binomial

11–4 Factoring the Difference of Two Squares

11–5 Factoring a Perfect Trinomial Square

11–6 Factoring by Grouping

11–7 Combining Several Types of Factoring

11–8 Literal Equations

PERFORMANCE OBJECTIVES

- Form a trinomial product by mentally multiplying two binomials.

- Express the product of the sum of two numbers and their difference as the difference of two squares.

- Mentally square a binomial.

- Factor the difference of two squares.

- Apply the rule for factoring a perfect trinomial square.

- Employ factoring in the solution of literal equations.

A seven-story-high vertical axis wind turbine generator located at the Sandia National Laboratories in Albuquerque, New Mexico. The eggbeater-shaped machine has a rotor with a seventeen-meter diameter and is capable of producing 30 kW in a 35 km/h wind and up to 60 kW in a 45 km/h wind. (Courtesy of Sandia National Laboratories)

This chapter will give you the opportunity to increase your ability in using the techniques of algebra to solve equations. In your arithmetic courses you learned ways to make your work easier. In this chapter you will learn ways to make your work with equations easier.

11–1 MENTALLY MULTIPLYING TWO BINOMIALS

Because multiplying is so common in arithmetic, you learned the "times tables." In algebra, multiplying two binomials is a common operation and, like the times tables, it may be done by sight. To learn to write the product of two binomials by sight, study the patterns in Figure 11–1.

FIGURE 11–1
Pattern for multiplying two binomials.

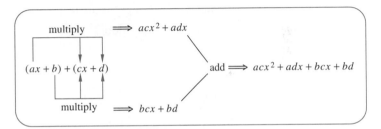

EXAMPLE 11–1 Find the product of $(5x - 1)(2x - 2)$.

Solution

$$\Rightarrow 10x^2 - 10x$$

$$(5x - 1) \quad (2x - 2)$$

$$+ \Rightarrow 10x^2 - 12x + 2$$

$$\Rightarrow -2x + 2$$

$$\therefore \quad (5x - 1)(2x - 2) = 10x^2 - 12x + 2$$

We may see from the preceding example that the process of multiplying two binomials is done by (1) distributing the first binomial (one term at a time) over the second binomial, (2) adding the two products, and (3) combining like terms.

The result of multiplying two binomials having the same variable is a **trinomial product.** The trinomial product is made up of three terms, each having a special name as shown in Table 11–1.

TABLE 11–1   Terms of the Trinomial Product
$$ax^2 + bx + c$$

Term		Degree	Name
First	ax^2	Two	Quadratic term
Second	bx	One	Linear term
Third	c	Zero	Constant term

The procedure for mentally multiplying two binomials is summarized as Rule 11–1.

RULE 11–1.   Multiplying Two Binomials

To multiply two binomials mentally, use the distributive law and:
1. Multiply the second binomial by the first term of the first binomial.
2. Multiply the second binomial by the second term of the first binomial.
3. Add the two resulting products and combine the like terms.
4. Form the trinomial product.

EXAMPLE 11–2   Find the product of $(3x - 4)(x + 1)$.

Solution   To multiply two binomials, use the steps of Rule 11–1.

Step 1:   Distribute (multiply) $3x$ over $(x + 1)$:
$$3x^2 + 3x$$

Step 2:   Distribute (multiply) -4 over $(x + 1)$:
$$-4x - 4$$

Step 3:   Add the products and combine the like terms:
$$3x - 4x = -x$$

Step 4:   Form the trinomial product:
$$3x^2 - x - 4$$

$\therefore$ $(3x - 4)(x + 1) = 3x^2 - x - 4$

EXAMPLE 11–3   Find the product of $(3y - 5)(-y + 3)$.

Solution   Use the steps of Rule 11–1.

Step 1:   Distribute (multiply) $3y$ over $(-y + 3)$:
$$-3y^2 + 9y$$

Step 2: Distribute (multiply) -5 over $(-y + 3)$:

$$5y - 15$$

Step 3: Add the products and combine like terms:

$$9y + 5y = 14y$$

Step 4: Form the trinomial product:

$$-3y^2 + 14y - 15$$

$$\therefore \quad (3y - 5)(-y + 3) = -3y^2 + 14y - 15$$

The following concepts were demonstrated in the preceding example:

- The quadratic term $(-3y^2)$ was formed by multiplying the first term in each binomial: $(3y)(-y) = -3y^2$.
- The constant term (-15) was formed by multiplying the second term in each binomial: $(-5)(3) = -15$.
- The linear term $14y$ was formed by adding the *inner* and *outer* products:

$$\underset{\text{(3y − 5) (−y + 3)}}{} \Rightarrow 5y \atop \Rightarrow 9y \} + \Rightarrow 14y$$

EXERCISE 11–1

Determine the missing terms in the following trinomial products.

1. $(x + 2)(x + 3) = x^2 + 5x + (?)$
2. $(2x + 1)(x - 2) = (?) - 3x - 2$
3. $(5y - 1)(3y - 1) = (?) - 8y + (?)$
4. $(4y - 3)(2y + 2) = 8y^2 + (?) - 6$

Mentally multiply the following pairs of binomials and express the product as a trinomial.

5. $(x + 1)(x + 1)$ 6. $(x + 1)(x + 2)$
7. $(x + 2)(x + 3)$ 8. $(x - 4)(x - 1)$
9. $(x - 5)(x - 3)$ 10. $(x + 5)(x + 3)$
11. $(x + 6)(x + 4)$ 12. $(x - 4)(x - 3)$
13. $(x - 2)(x - 10)$ 14. $(x + 8)(x - 4)$
15. $(x - 5)(x + 2)$ 16. $(2y + 1)(5y - 4)$
17. $(3y - 7)(2y + 3)$ 18. $(4y - 1)(2y - 8)$
19. $(10y - 1)(9y - 2)$ 20. $(2y - 3)(6y - 2)$

21. $(-3x + 4)(2x - 3)$ **22.** $(-2x - 5)(x - 2)$

23. $(6x - 4)(-3x + 4)$ **24.** $(3x - 7)(-5x - 1)$

25. $(-2x - 4)(-2x + 4)$ **26.** $(-5x - 3)(-4x - 6)$

27. $\left(x + \frac{1}{4}\right)\left(x + \frac{1}{2}\right)$ **28.** $\left(x - \frac{1}{3}\right)\left(x + \frac{2}{3}\right)$

29. $\left(y - \frac{2}{5}\right)\left(y + \frac{4}{5}\right)$ **30.** $\left(\frac{3}{8}x + 1\right)\left(\frac{1}{4}x - 1\right)$

11–2   PRODUCT OF THE SUM AND DIFFERENCE OF TWO NUMBERS

A special product is formed when the sum of two numbers is multiplied by their difference. The special product formed by the sum and difference of two numbers is called the **difference of two squares,** which is formed by applying Rule 11–2.

RULE 11–2.   *Multiplying the Sum and Difference of Two Numbers*

The product of the sum of two numbers and their difference equals the difference of their squares.

$$(a + b)(a - b) = a^2 - b^2$$

EXAMPLE 11–4   Solve $(x + 2)(x - 2)$.

Solution   Square x:

$$x^2$$

Square 2:

$$2^2 = 4$$

Form the difference of the two squares:

$$x^2 - 4$$

$\therefore$ $(x + 2)(x - 2) = x^2 - 4$

EXERCISE 11–2

Mentally multiply each of the following and express the product as a polynomial.

1. $(x + 1)(x - 1)$ **2.** $(x + 3)(x - 3)$

3. $(x + 5)(x - 5)$ **4.** $(x + 6)(x - 6)$

5. $(2x + 1)(2x - 1)$ **6.** $(3x + 2)(3x - 2)$

7. $(5x + 5)(5x - 5)$ **8.** $(4x + 3)(4x - 3)$

9. $(-4x + 7)(-4x - 7)$ **10.** $(-2x + 8)(-2x - 8)$

11. $(y + a)(y - a)$ **12.** $(y + b)(y - b)$

13. $(2y + c)(2y - c)$ **14.** $(10y + k)(10y - k)$

15. $(3y + 2t)(3y - 2t)$ **16.** $(-5y + 3a)(-5y - 3a)$

17. $\left(2x + \frac{1}{2}\right)\left(2x - \frac{1}{2}\right)$ **18.** $\left(\frac{1}{4}x + 3\right)\left(\frac{1}{4}x - 3\right)$

19. $\left(\frac{2}{5}y + \frac{1}{8}\right)\left(\frac{2}{5}y - \frac{1}{8}\right)$ **20.** $\left(6x + \frac{1}{3}\right)\left(6x - \frac{1}{3}\right)$

11–3   SQUARE OF A BINOMIAL

The algebraic expression $(a + b)^2$ means to use $(a + b)$ twice as a factor in a product. The expression $(a + b)^2$ is directing us to square the binomial. This may be done by sight by applying the following rule.

> **RULE 11–3.**   *Squaring a Binomial*
>
> To form the square of a binomial:
> 1. Square the first term.
> 2. Add twice the algebraic product of the two terms.
> 3. Add the square of the second term.
> 4. Express the product as a trinomial square with the following pattern:
>
> $$(a + b)^2 = a^2 + 2ab + b^2$$
> $$(a - b)^2 = a^2 - 2ab + b^2$$

EXAMPLE 11–5   Solve $(2x + 3)^2$.

Solution   Use the steps of Rule 11–3 to form the square of the binomial.

Step 1:   Square the first term, $2x$:

$$4x^2$$

Step 2:   Multiply 2 times $2x$ times 3:

$$2(2x)3 = 12x$$

Step 3:   Square the second term, 3:

$$9$$

Step 4:   Express the product as a trinomial square:

$$4x^2 + 12x + 9$$

$\therefore$ $(2x + 3)^2 = 4x^2 + 12x + 9$

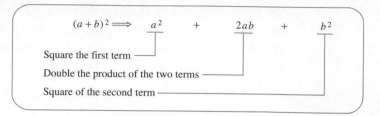

Square the first term ⎯⎯⎯⎯
Double the product of the two terms ⎯⎯⎯⎯
Square of the second term ⎯⎯⎯⎯

The procedure for squaring a binomial is summarized in Figure 11–2. The result is a **perfect trinomial square.**

EXAMPLE 11–6   Solve $(3x - 2)^2$.

Solution   Form the square of the binomial.

Step 1:   Square $3x$:

$9x^2$

Step 2:   Multiply 2 times $3x$ times -2:

$2(3x)(-2) = -12x$

Step 3:   Square -2:

4

Step 4:   Express the product as a trinomial square:

$9x^2 - 12x + 4$

$\therefore$ $(3x - 2)^2 = 9x^2 - 12x + 4$

EXERCISE 11–3

Mentally square each of the following binomials. Express the answer as a trinomial square.

1. $(x + 3)^2$ **2.** $(x + 4)^2$ **3.** $(x - 2)^2$

4. $(x - 3)^2$ **5.** $(x + 5)^2$ **6.** $(x - 7)^2$

7. $(2x + 2)^2$ **8.** $(3x - 1)^2$ **9.** $(4x - 2)^2$

10. $(5x + 4)^2$ **11.** $(-2x - 1)^2$ **12.** $(-x + 6)^2$

13. $(-4x + 3)^2$ **14.** $(-3x - 5)^2$ **15.** $(x + \frac{1}{5})^2$

16. $(-x - \frac{1}{3})^2$ **17.** $(x + \frac{1}{2})^2$ **18.** $(\frac{1}{3}x - 6)^2$

19. $(\frac{1}{4}x + \frac{1}{4})^2$ **20.** $(\frac{3}{4}x - \frac{4}{3})^2$ **21.** $(\frac{4}{5}x + \frac{1}{8})^2$

11–4 FACTORING THE DIFFERENCE OF TWO SQUARES

You know that the product of the sum and difference of two numbers results in the difference of two squares. When the difference of two squares is factored, it factors into the sum and difference of two numbers. Thus, $(a^2 - b^2) = (a + b)(a - b)$. Factoring the difference of two squares is stated as Rule 11–4. **Factoring** is the process of finding the factors of a polynomial.

> **RULE 11–4. Factoring the Difference of Two Squares**
>
> To factor the difference of two squares:
>
> 1. Take the square root of each of the squares.
> 2. Form the factors by writing one of the factors as the sum of the square roots and the other factor as the difference of the square roots.

EXAMPLE 11–7   Factor $4x^2 - 25$.

Solution   Factor the difference of two squares.

Step 1:   Take the square root of $4x^2$ and 25:

$2x$ and 5

Step 2:   Write the factors as the sum and difference of the two numbers:

$(2x + 5)(2x - 5)$

$\therefore$ $(4x^2 - 25) = (2x + 5)(2x - 5)$

EXERCISE 11–4

Factor each of the following.

1. $x^2 - 4$
2. $y^2 - 9$
3. $4x^2 - 64$
4. $x^2 - a^2$
5. $y^2 - c^2$
6. $36x^2 - b^2$
7. $1 - 16c^2$
8. $49x^2 - 4a^2$
9. $4c^2 - 9y^2$
10. $25x^2 - 36$
11. $a^2b^2 - x^2$
12. $m^2n^2 - 4y^2$
13. $\frac{1}{16} - x^2$
14. $9y^2 - \frac{1}{4}$
15. $\frac{1}{9} - 25z^2$

11–5 FACTORING A PERFECT TRINOMIAL SQUARE

We have seen that a perfect trinomial square results when a binomial is squared. Thus,

$$(a + b)^2 = a^2 + 2ab + b^2 \qquad (11\text{–}1)$$
$$(c - d)^2 = c^2 - 2cd + d^2 \qquad (11\text{–}2)$$

If a trinomial follows the pattern of either Equation 11–1 or 11–2, it is a perfect trinomial square and may be factored as the square of a binomial. The procedure for factoring a perfect trinomial square is summarized as Rule 11–5.

RULE 11–5. **Factoring a Perfect Trinomial Square**

To factor a perfect trinomial square:

1. Take the square root of each of the monomial squares.
2. Form the factor by combining the square roots with the sign of the remaining term.
3. Express the trinomial as the square of the binomial factor.

EXAMPLE 11–8   Factor $4x^2 - 12x + 9$.

Solution   Factor the perfect trinomial square.

Step 1:   Take the square root of $4x^2$ and 9:

$2x$ and 3

Step 2:   Form the factor by using the sign of $-12x$:

$2x - 3$

Step 3:   Express the trinomial as the square of the binomial factor:

$(2x - 3)^2$

∴ $4x^2 - 12x + 9 = (2x - 3)^2$

EXAMPLE 11–9   Factor $9y^2 + 30ay + 25a^2$.

Solution   Use the steps of Rule 11–5.

Step 1:   Take the square root of $9y^2$ and $25a^2$:

$3y$ and $5a$

Step 2:   Form the factor by using the sign of $+30ay$:

$3y + 5a$

Step 3:   Express the trinomial as the square of the binomial factor:

$(3y + 5a)^2$

∴ $9y^2 + 30ay + 25a^2 = (3y + 5a)^2$

Factor each of the following.

1. $x^2 - 6x + 9$ 2. $y^2 - 2y + 1$
3. $x^2 - 4x + 4$ 4. $x^2 + 12x + 36$
5. $y^2 + 2y + 1$ 6. $16x^2 + 16x + 4$
7. $9y^2 - 30y + 25$ 8. $9x^2 - 24ax + 16a^2$
9. $4x^2 - 20cx + 25c^2$ 10. $4y^2 + 28yb + 49b^2$
11. $25x^2 - 30xy + 9y^2$ 12. $36y^2 + 24xy + 4x^2$
13. $16y^2 + 24y + 9$ 14. $25z^2 - 20z + 4$
15. $64x^2 + 16xy + y^2$ 16. $81x^2 + 90x + 25$
17. $121z^2 + 88z + 16$ 18. $1.44t^2 - 2.4t + 1$
19. $0.81u^2 - 1.8uv + v^2$ 20. $0.64x^2 + 6.4x + 16$

11–6   FACTORING BY GROUPING

In a polynomial expression, several terms may contain common factors. By grouping the terms of the polynomial, you may recognize the form as a type that is factorable. Study the following examples.

EXAMPLE 11–10   Factor $2y - ax + ay - 2x$.

Solution   Group the terms with common factors y and x:

$$(2y + ay) + (-2x - ax)$$

Factor out the common monomial factors y and x:

$$y(2 + a) - x(2 + a)$$

Factor out the common binomial factor $(2 + a)$:

$$(2 + a)(y - x)$$

$\therefore$ $2y - ax + ay - 2x = (2 + a)(y - x)$

Observation   If we started by grouping on 2 and a, we would arrive at the same factoring.

EXAMPLE 11–11   Factor $3x + by - 3y - bx$.

Solution   Group the terms with common factors x and y:

$$(3x - bx) + (-3y + by)$$

Factor out the common monomial factors x and y:

$$x(3 - b) - y(3 - b)$$

Factor out the common binomial factor $(3 - b)$:

$$(3 - b)(x - y)$$

$$\therefore \quad 3x + by - 3y - bx = (3 - b)(x - y)$$

EXAMPLE 11–12 Factor $z^3 - 3z^2 - z + 3$.

Solution Group terms containing 3 and remaining terms:

$$(3 - 3z^2) + (z^3 - z)$$

Factor out the common monomial factors -3 and z:

$$-3(-1 + z^2) + z(z^2 - 1)$$

Factor out the common binomial factor $(z^2 - 1)$

$$(z^2 - 1)(z - 3)$$

Factor $z^2 - 1$ into two binomial factors:

$$(z + 1)(z - 1)(z - 3)$$

$$\therefore \quad z^3 - 3z^2 - z + 3 = (z + 1)(z - 1)(z - 3)$$

EXERCISE 11–6

Completely factor each of the following polynomials.

1. $3(a + 2) - x(a + 2)$
2. $a^2(x - y) + 3(x - y)$
3. $m(x - 5) - 7(x - 5)$
4. $ax + ay + bx + by$
5. $cx + cy - ax - ay$
6. $x^3 - x - x^2 + 1$
7. $z^2 + 1 + z + z^3$
8. $4x^3 - 8x^2 - 4x + 8$
9. $4b^3 + 5b^2 - 36b - 45$
10. $c^2y - y - 3c^2 + 3$
11. $bx^2 - b - x^2 + 1$
12. $y^4 - 8y + y^3z - 8z$
13. $xy + 2ax - 3a - 1.5y$
14. $8y - x + 4xy - 2$
15. $w^3 + 3w^2 - w - 3$
16. $4z^3 + 4z^2 - z - 1$

11–7 COMBINING SEVERAL TYPES OF FACTORING

When factoring a polynomial, first remove the common monomial factor and then see if the remaining expression may be factored further. In factoring a polynomial, the factoring is continued until all the steps of Rule 11–6 are completed.

> **RULE 11–6.** *Factoring a Polynomial*
>
> To factor a polynomial:
>
> 1. Where possible, remove the common monomial factor from each term.
> 2. Factor a binomial into the sum and difference of two numbers if it is the difference of two squares.
> 3. Factor a perfect trinomial square into the square of a binomial.
> 4. Group the terms with common factors, then factor.

EXAMPLE 11–13 Factor $8y^3 - 2y$.

Solution Factor using the steps of Rule 11–6.

Step 1: Factor out $2y$:

$$2y(4y^2 - 1)$$

Step 2: Factor $4y^2 - 1$ into the sum and difference of two numbers:

$$2y(2y - 1)(2y + 1)$$

$$\therefore \quad 8y^3 - 2y = 2y(2y - 1)(2y + 1)$$

EXAMPLE 11–14 Factor $3x^3y - 6x^2y + 3xy$.

Solution Factor out $3xy$:

$$3xy(x^2 - 2x + 1)$$

Factor the perfect trinomial square into the square of a binomial:

$$3xy(x - 1)^2$$

$$\therefore \quad 3x^3y - 6x^2y + 3xy = 3xy(x - 1)^2$$

EXAMPLE 11–15 Factor $y^3 - 3y^2 - 9y + 27$.

Solution Group the terms containing y^2 and those containing 9:

$$(y^3 - 3y^2) + (-9y + 27)$$

Factor each group:

$$y^2(y - 3) - 9(y - 3)$$

Factor out the binomial $y - 3$:

$$(y - 3)(y^2 - 9)$$

Factor $y^2 - 9$ into the sum and difference:

$$(y - 3)(y + 3)(y - 3)$$

$$\therefore \quad y^3 - 3y^2 - 9y + 27 = (y + 3)(y - 3)^2$$

EXERCISE 11–7

Factor each of the polynomials.

1. $3a + 3$
2. $x^2 - 16x$
3. $y^4 - c^4$
4. $8x^2 - 18$
5. $3a^2 - 3$
6. $9y^2 - 9$
7. $8y^3 - 32y$
8. $3y^2 - 6y + 3$
9. $5y^2 + 20y + 20$
10. $27x^2 - 18x + 3$
11. $8a^2 + 8ab + 2b^2$
12. $2x^2 - 2(a - b)^2$
13. $5y^2 - 5(m + n)^2$
14. $\frac{1}{4}b^2 - \frac{1}{3}bc + \frac{1}{9}c^2$
15. $(x + 3)^2 - 7x - 21$
16. $5ac - 5ad + 5bc - 5bd$
17. $x^3 + x + x^2 + 1$
18. $m^2n^2 - 4m^2 - a^2n^2 + 4a^2$
19. $4y^2 - 4ay + a^2 - c^2$
20. $x^2 + 2xy + y^2 - a^2$

11–8   LITERAL EQUATIONS

When solving literal equations or formulas, it is sometimes necessary to factor one or more expressions within the equation in order to isolate the desired variable as the left member. The following examples illustrate the application of factoring to the solution of literal equations.

EXAMPLE 11–16   Solve $ax + bx = 3a + 3b$ for x.

Solution   Factor both the left and right members:

$$x(a + b) = 3(a + b)$$

D: $(a + b)$ $\qquad\qquad x = 3$

$$\therefore \qquad ax + bx = 3a + 3b \Rightarrow x = 3$$

EXAMPLE 11–17   Solve $bx + b^2 = a^2 - ax$ for x.

Solution   Isolate the terms containing x into the left member:

A: $ax - b^2$ $\qquad bx + ax = a^2 - b^2$

CHAPTER 11   Special Products, Factoring, and Equations

Factor the left and right members:

$$x(a + b) = (a - b)(a + b)$$

Divide by the coefficient of x:

D: $(a + b)$ $\qquad\qquad x = a - b$

$\therefore \qquad bx + b^2 = a^2 - ax \Rightarrow x = a - b$

EXAMPLE 11–18 Solve $\dfrac{1}{R_T} = \dfrac{1}{R_1} + \dfrac{1}{R_2}$ for R_T.

Solution Multiply each member by the common denominator:

M: $\quad R_T R_1 R_2 \qquad R_1 R_2 = R_T R_2 + R_T R_1$

Factor the right member:

$$R_1 R_2 = R_T(R_1 + R_2)$$

Divide by the coefficient of R_T:

D: $(R_1 + R_2)$ $\qquad \dfrac{R_1 R_2}{R_1 + R_2} = R_T$

Interchange the right and left members:

$$R_T = \frac{R_1 R_2}{R_1 + R_2}$$

$\therefore \qquad \dfrac{1}{R_T} = \dfrac{1}{R_1} + \dfrac{1}{R_2} \Rightarrow R_T = \dfrac{R_1 R_2}{R_1 + R_2}$

EXERCISE 11–8

Solve for x in each of the following equations.

1. $x - ax = 4$
2. $bx + 2x = 7$
3. $5x - b = cx$
4. $a - 3x = bx$
5. $bx - 2 = b - 2x$
6. $cx + 5 = 5x - c$
7. $ax - bx = 2a - 2b$
8. $ax - ma = md - dx$
9. $bx - 3x = b^2 - 6b + 9$
10. $3x - 9 = ax - a^2$
11. $bx + 16 = 4x + b^2$
12. $m^2 - mx + 10m = 5x - 25$

Solve for the indicated variable in each of the following.

	Formula	Solve for:		Formula	Solve for:
13.	$R = 3E + 2EI$	E	14.	$CV = T - C$	C
15.	$\beta = \dfrac{\alpha}{1 - \alpha}$	α	16.	$L_1a + L_1b = a - b$	a
17.	$i_1 = \dfrac{\beta iR}{R + Z}$	R	18.	$Z_T = \dfrac{Z_1 Z_2}{Z_1 + Z_2}$	Z_2
19.	$I = \dfrac{nV}{Z + nR}$	n	20.	$C = \dfrac{V_1 - V_2}{\omega V_1}$	V_1
21.	$S = \dfrac{R_E + R_1}{R_E + R_1(1 - \alpha)}$	R_1	22.	$S = \dfrac{R_E(R_1 + R_2) + R_1 R_2}{R_E(R_1 + R_2) + R_1 R_2(1 - \alpha)}$	R_2
23.	Problem 22	R_E	24.	Problem 21	R_E
25.	$R_1 = \dfrac{V_{CC} - I_B R_E(\beta + 1)}{I_B}$	I_B			

Compute the value for each of the following, and express the solution to an appropriate precision with a unit and unit prefix (as needed).

	Formula	Solve for:	When:
26.	$R_T = \dfrac{R_1 R_2}{R_1 + R_2}$	R_2	$R_T = 3.00 \text{ k}\Omega$ $R_1 = 8.20 \text{ k}\Omega$
27.	$I_2 = \dfrac{I_T G_2}{\Sigma G}$	G_2	$I_2 = 2.80 \text{ A}$ $I_T = 5.00 \text{ A}$ $\Sigma G = 42.7 \text{ mS}$
28.	$\dfrac{E}{\Sigma R} = \dfrac{V_3}{R_3}$	R_3	$E = 24.0 \text{ V}$ $V_3 = 8.25 \text{ V}$ $\Sigma R = 378 \ \Omega$
29.	$\dfrac{I_2}{R_1} = \dfrac{I}{R_1 + R_2}$	R_2	$I = 0.835 \text{ A}$ $I_2 = 0.325 \text{ A}$ $R_1 = 470 \ \Omega$

	Formula	Solve for:	When:
30.	$\sqrt{3}E = \dfrac{P_s}{\%Z \times I_{sc}} \times 100$	I_{sc}	$E = 460\ \text{V}$ $\%Z = 2.92$ $P_s = 50.0\ \text{kVA}$

SELECTED TERMS

difference of two squares   A special product resulting from the multiplication of the sum and difference of two numbers.

factoring   The process of finding the factors of a polynomial or other expression.

perfect trinomial square   A special product resulting from the multiplication of a binomial by itself.

trinomial product   The product of two binomials.

12

Applying Mathematics to Electrical Concepts

12–1 Ratio, Percent, and Parts per Million

12–2 Accounting for Empirical Error in Calculations

12–3 Efficiency

12–4 Proportion

12–5 Electrical Conductors

PERFORMANCE OBJECTIVES

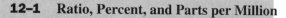

- Apply the concepts of ratio, percent, and parts per million to variation in electrical measurements and circuit components.

- Employ the principles of relative and absolute error to account for deviations in circuit measurements.

- Use per-unit and percent efficiency with electrical energy converting devices.

- Solve electrical conductor problems using proportions.

Integrated into an automated conformal coating system, fan width pattern verification provides consistent, repeatable conformal coating of printed circuit boards (PCBs) and components. (Courtesy of Asymtek, Carlsbad, CA)

This chapter relates the application of fractions to electrical concepts, covering ratio, percent, and parts per million; relative error and absolute error; efficiency; proportion; and electrical conductors.

12–1 RATIO, PERCENT, AND PARTS PER MILLION

We often compare two quantities by dividing one by the other. To compare the cost of two resistors, a ratio is used. If R_1 costs 60¢ and R_2 costs 15¢, we could say that R_1 costs four times as much as R_2. This comparison is made by computing the quotient of 60¢ and 15¢; $60/15 = 4$. This price comparison can be stated as a ratio of "4 to 1." This is written 4:1 using the ratio sign (:).

Ratio

A **ratio** of one number to another is the *quotient* of the *first* number divided by the *second* number. You may express a ratio in several forms, as shown in Table 12–1.

To be a ratio, the two numbers in a ratio (called the *terms*) **must have the same units.** Thus, 3 m and 1 m may be expressed as a ratio of 3:1 with no units (since the units factor); however, the quantities of 60 m and 2 h may be expressed as a *rate* of 60 m per 2 h, or simply 30 m/h. Although a rate may look like a ratio, it is not a ratio since a rate has units and a ratio has no units.

If two numbers express the same quantity, but have different units, then change each to the same unit. So, when computing the ratio of 1 m to 800 cm, first change 1 m to 100 cm, and then find the ratio of 100 cm to 800 cm, shown as 100:800, or 1:8 when reduced to lowest terms.

TABLE 12–1 Expressing Ratios	
The ratio of one number to another may be expressed as a quotient.	
Quotient Indicated by:	**Ratio Expressed as:**
÷ Divide sign	$5 \div 3$
: Ratio sign	5:3
/ Vinculum	$\dfrac{5}{3}$
Decimal fraction	1.667 to 1

RULE 12–1. *Finding the Ratio of Two Like Quantities*

To find the ratio of two like quantities:

1. Express the quantities in the same units.
2. Write the ratio as a fraction.
3. Reduce the fraction to its lowest terms.

EXAMPLE 12–1	Express the ratio of 1.2 m to 80 cm in lowest terms.
Solution	Follow the steps of Rule 12–1.
Step 1:	Change 1.2 m to centimeters:
	$1.2 \text{ m} \Rightarrow 120 \text{ cm}$
Step 2:	Write the ratio as a fraction:
	$\dfrac{120}{80}$
Step 3:	Reduce to lowest terms:
	$\dfrac{120}{80} = \dfrac{3}{2}$
$\therefore$	The ratio of 1.2 m to 80 cm is 3/2.

Because a ratio is a quotient between two numbers, it has all the properties of a fraction and may be formed into an equivalent ratio.

EXAMPLE 12–2	Express the ratio 15:25 as an equivalent ratio.
Solution	

$$15{:}25 = \frac{15}{25}$$

D: 5 $\dfrac{15}{25} = \dfrac{3}{5}$

M: 20 $\dfrac{3}{5} = \dfrac{60}{100}$

$\therefore$ 15:25, 3:5, and 60:100 are equivalent ratios.

EXERCISE 12–1

Reduce each of the following ratios to their lowest terms.

1. 3:12
2. 8:2
3. 14 to 21
4. $7x$ to $13x$
5. 4 cm/6 cm
6. 30 s to 1.5 min
7. 8.42 to 2
8. 0.6 to 6
9. 15 mA to 2.0 A
10. $20k$ to $313.15k$
11. $1.15 to 95¢
12. 440 V ÷ 110 V

Percent

Percent is shorthand for a ratio in which the second number in the ratio is 100. Thus, 8% is the ratio of 8:100 and 58% is the ratio of 58:100.

Errors in electrical measurements are often expressed in percent to indicate the difference between the measured and true values. Equation 12–1 is used to compute the percent error in measurements.

$$\textbf{percent error} = \frac{\textbf{measured value} - \textbf{true value}}{\textbf{true value}} \times \textbf{100\%} \quad \textbf{(12–1)}$$

Percent error is used to indicate the degree of *accuracy* of a measurement. The accuracy of a measurement refers to the amount of difference between the measured value (the instrument reading) and the true value.

EXAMPLE 12–3   Determine the percent error between a digital voltmeter (DVM) and a 50.000-V laboratory standard if the DVM reads 48.3 V when connected to the standard power supply.

Solution   Use Equation 12–1 with your calculator. Substitute 48.3 V for measured and 50.000 V for true:

$$\text{percent error} = \frac{\text{measured value} - \text{true value}}{\text{true value}} \times 100\%$$

$$\text{percent error} = \frac{48.3 - 50.0}{50.0} \times 100$$

$$\text{percent error} = -3.4\%$$

∴   The percent error is −3.4%.

Manufacturers of electronic components and equipment use percent to specify component and equipment tolerances. The percent tolerance of carbon-film resistors (pictured in Figure 12–1) is indicated by the fourth *color band*. The **percent tolerance** is the maximum permitted amount of **deviation** from the *nominal,* or named, value. Percent deviation from the nominal value is used to indicate the deviation between the actual value of an electrical component and the named or nominal value. Equation 12–2 is used to compute percent deviation.

FIGURE 12–1   Color coding of a carbon-film resistor. The fourth color band is the tolerance band. The fourth band colors are red, ±2%; gold, ±5%; silver, ±10%.

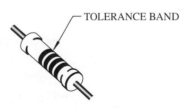

TOLERANCE BAND

$$\text{percent deviation} = \frac{\text{actual value} - \text{nominal value}}{\text{nominal value}} \times 100\% \qquad \textbf{(12–2)}$$

EXAMPLE 12–4 A 180-Ω (nominal value) resistor with a gold fourth band was measured and found to have an actual value of 172.8 Ω. Determine if the resistor is within the specified tolerance.

Solution Use Equation 12–2 and then compare the specified tolerance to the percent deviation. Substitute into Equation (12–2):

$$\text{percent deviation} = \frac{\text{actual value} - \text{nominal value}}{\text{nominal value}} \times 100\%$$

$$\text{percent deviation} = \frac{172.8 - 180}{180} \times 100$$

$$\text{percent deviation} = -4.00\%$$

Observation A gold fourth band indicates a tolerance of $\pm 5\%$.
Does -4% lie between $+5\%$ and -5%? Yes.

$\therefore$ The resistor is within the specified tolerance.

EXAMPLE 12–5 After manufacture but before being labeled, capacitors are measured to determine their actual value. Determine the range of measured values that may be labeled as 15.0 nF if the allowed percent deviation is $\pm 10.0\%$.

Solution Solve Equation 12–2 for the actual value in terms of the percent deviation and the nominal value.

$$\text{percent deviation} = \frac{\text{actual value} - \text{nominal value}}{\text{nominal value}} \times 100\%$$

$$\frac{\text{percent deviation} \times \text{nominal value}}{100} = \text{actual value} - \text{nominal value}$$

$$\text{actual value} = \frac{\text{percent deviation} \times \text{nominal value}}{100} + \text{nominal value}$$

Solve for the maximum actual value by substituting $+10.0$ for the percent deviation and 15.0×10^{-9} (15.0 nF) for the nominal value.

$$\text{actual value} = \frac{10.0 \times 15.0 \times 10^{-9}}{100} + 15.0 \times 10^{-9}$$

$$\text{actual value} = 16.5 \times 10^{-9} = 16.5 \text{ nF}$$

The maximum actual value is 16.5 nF.

Solve for the minimum actual value by substituting -10.0 for the percent deviation and 15.0×10^{-9} (15.0 nF) for the nominal value.

$$\text{actual value} = \frac{-10.0 \times 15.0 \times 10^{-9}}{100} + 15.0 \times 10^{-9}$$

$$\text{actual value} = 13.5 \times 10^{-9} = 13.5 \text{ nF}$$

The minimum actual value is 13.5 nF.

∴   The range of capacitor values labeled as 15.0 nF (nominal value) is made up of actual capacitor values that lie between 13.5 nF and 16.5 nF.

EXERCISE 12–2

1. Determine the percent error between a DVM (digital volt meter) that indicates 46.5 V and a meter calibrator that is set to 50.000 V.

2. A 5-mA panel meter under test indicates 4.20 mA when the laboratory standard is 5.000 mA. Determine the percent error.

3. The dc function of a DVM is being calibrated. The instrument manufacturer has specified the meter to be $\pm 1.0\%$ of reading for all dc range settings. Determine whether the instrument is within the specified tolerance when it indicates 7.2 V. It is connected to an 8.000-V standard source.

4. A radio-frequency (RF) oscillator has a 0.03% error between the true value of 2.5000 MHz and the measured value. Determine the measured value.

5. The operating frequency of a commercial broadcasting station was measured and found to be 800.020 kHz. Determine the assigned frequency if the measured frequency was $+0.0025\%$ in error.

6. A 2.20-kΩ resistor was measured and found to have an actual value of 2862.0 Ω. Determine the percent deviation from the nominal value.

7. A 5000-Ω wire-wound resistor was measured and found to have an actual value of 4732.0 Ω. Determine the percent deviation from the nominal value.

8. The actual value of a 47-Ω carbon-film resistor is 51.43 Ω. What color should the tolerance band be?

9. A voltage standard has a nominal value of 10.000 V. Determine the actual value of voltage from the standard if the deviation is -0.008%.

10. A 1.82 Ω, $\pm 1\%$ carbon-film precision resistor has an actual value of 1.8018 Ω. Determine if the resistor is within the specified tolerance of $\pm 1\%$.

11. A group of metal-film resistors was produced having a targeted nominal value of 220 Ω (color code red-red-brown). As a last step in the production process, each resistor was measured, sorted, and color-banded with an appropriate fourth color band to represent the tolerance (deviation) as $\pm 2.0\%$ (red), $\pm 5.0\%$ (gold), or $\pm 10.0\%$ (silver). Calculate the maximum and minimum actual values for each tolerance range.

12. A batch of high-Q porcelain chip dielectric capacitors for use in a hybrid microwave communication circuit are to be graded before installation on the circuit substrate. If the capacitor nominal value is 68.0 pF, calculate the maximum and minimum actual values for both a $\pm2.0\%$ deviation and a $\pm5.0\%$ deviation.

Parts per Million

Technicians who work in standards laboratories make measurements that result in very high accuracy. Deviations of 0.0001% are frequently encountered. The notation used to show such high accuracy is **parts per million (ppm).** Parts per million is shorthand for a ratio used to express accuracies, errors, deviations from a nominal (named) value, and temperature coefficients in which the second number of the ratio is 1 000 000. The procedure for converting percent to parts per million is outlined in Rule 12–2.

RULE 12–2. *Converting from Percent to Parts per Million*

To convert from percent to parts per million, drop the percent sign and:
1. Move the decimal point four places to the right; ppm = $\% \times 10^4$.
2. Write ppm after the resulting number.

EXAMPLE 12–6	Express 0.0052% in parts per million.
Solution	Use Rule 12–2.
Step 1:	Move the decimal four places to the right:

$$0.0052 \Rightarrow 52.$$
$$\underset{\overrightarrow{1\,2\,3\,4}}{}$$

| **Step 2:** | Write ppm after the number: |

52 ppm

$\therefore \qquad 0.0052\% = 0.0052 \times 10^4 = 52$ ppm

GUIDELINES 12–1. *Applying Percent and Parts per Million*

To Express:	Use:	To Express:	Use:
1.0%	1.0%	0.001%	10 ppm
0.1%	0.1%	0.0001%	1 ppm
0.01%	0.01% or 100 ppm	0.000 01%	0.1 ppm

When to use parts per million or percent will become clear to you with use. Guidelines 12–1 will aid in determining when to use percent and when to use parts per million in expressing the accuracy of a measurement. Guidelines 12–1 are based on the generally accepted and widely used practice in metrology, the art and science of measurement.

From Guidelines 12–1, we can see that *lower accuracy* (larger deviation) is best expressed in percent, such as 1% or 2%. *Higher accuracy* (smaller deviation) is conveniently expressed in parts per million, such as 1 ppm, rather than 0.0001%.

EXAMPLE 12–7   A precision resistor has a nominal value of 1000.000 Ω. If its actual value is 1000.013 Ω, state the deviation from the nominal value in parts per million.

Solution   Use Equation 12–2 with your calculator.

$$\text{percent deviation} = \frac{\text{actual} - \text{nominal}}{\text{nominal}} \times 100\%$$

$$\text{percent deviation} = \frac{1000.013 - 1000.000}{1000.000} \times 100$$

$$\text{percent deviation} = 0.0013\%$$

Use Rule 12–2 to change to parts per million:

$$0.0013\% = 0.0013 \times 10^4 = 13 \text{ ppm}$$

∴   The deviation is 13 ppm.

Frequently, we need to reverse the process and express parts per million as percent. Rule 12–3 explains how this is done.

RULE 12–3.   Converting Parts per Million to Percent

To convert from parts per million to percent, drop the "ppm" and:
1. Move the decimal point four places to the left; $\% = \text{ppm} \times 10^{-4}$.
2. Write % after the resulting number.

EXAMPLE 12–8   Express 180 ppm as a percent.

Solution   Using Rule 12–3, move the decimal point four places to the left and express as a percent:

∴ $180 \text{ ppm} = 180 \times 10^{-4}\% = 0.0180\%$

Precision metal-film resistors ($\leq 1.0\%$) have a *positive temperature coefficient,* a parameter that specifies the effect a change in temperature has on the nominal value of resistance. For example, one manufacturer of 1.00-$M\Omega$, $\pm 1.0\%$, metal-film resistors specifies the temperature coefficient as 50 ppm/°C. That is, for every million ohms of resistance, the nominal value will increase by 50.0 Ω (+50 ppm/°C) or decrease by 50.0 Ω (−50 ppm/°C) for each degree Celsius change in temperature. Thus, a 2°C decrease in temperature will cause the 1.00-$M\Omega$ resistor to decrease in value by 100 Ω, whereas a 10°C increase in temperature results in a 500-Ω increase in its value.

$$R = R_0(1 + \alpha \times 10^{-6}\Delta T) \qquad (12\text{--}3)$$

where R = the resistance at the operating temperature in Ω
R_0 = the resistance at the reference temperature in Ω
α = the temperature coefficient in ppm/°C
ΔT = the difference between the operating and the reference temperature in °C

EXAMPLE 12–9   Determine the resistance of a 71.5-$k\Omega$, $\pm 1.0\%$, metal-film resistor operating at 92.0°C if the temperature coefficient of 100 ppm/°C is specified at 0°C.

Solution   Use Equation 12–3 to compute the resistance at the 92°C operating temperature.

$$R = R_0(1 + \alpha \times 10^{-6}\Delta T)$$
$$R_0 = 71.5 \text{ k}\Omega = 71.5 \times 10^3 \text{ }\Omega$$
$$\alpha = +100 \text{ ppm/°C}$$
$$\Delta T = 92.0°C - 0°C = 92.0°C$$
$$R = 71.5 \times 10^3(1 + 100 \times 10^{-6} \times 92.0)$$
$$R = 72.2 \times 10^3 \text{ }\Omega = 72.2 \text{ k}\Omega$$

∴   The nominal resistor value is 72.2 $k\Omega$ at 92.0°C.

Observation   Because the operating temperature is greater than the reference temperature of 0°C, the nominal resistance will increase with temperature, so ± 100 ppm/°C was used in the calculation.

EXERCISE 12–3

Express each of the following as parts per million (ppm).

1. 0.0015%	**2.** 0.0032%	**3.** 0.0136%
4. 0.0197%	**5.** 0.022%	**6.** 0.0002%

Express each of the following as a percent (%).

7. 1000 ppm	**8.** 318 ppm	**9.** 100 ppm
10. 635 ppm	**11.** 4050 ppm	**12.** 1272 ppm

13. A metal-film resistor has a temperature coefficient of ± 50 ppm/°C. Restate the temperature coefficient as a percent.

14. A digital counter, using a very stable crystal oscillator, has a time-base accuracy of 0.000 025%. Restate the accuracy in parts per million.

15. Determine the resistance of a 100-kΩ, $\pm 1.0\%$, metal-film resistor operating at -20.0°C if the temperature coefficient of 50 ppm/°C is specified at 25°C.

16. Determine the resistance of a 475-kΩ, $\pm 1.0\%$, metal-film resistor operating at either end of its temperature range of -55.0°C to $+175$°C if the temperature coefficient of 80 ppm/°C is specified at 20.0°C.

12–2 ACCOUNTING FOR EMPIRICAL ERROR IN CALCULATIONS

Because manufactured electrical components (resistors, capacitors, etc.) vary in value from the nominal value (named value), they have error and are inaccurate. The source of this error is related to the myriad steps used in the process of manufacturing the component.

As you know from your study of Chapter 2, numbers used to report measurements are approximate. This is so because the very act of measuring alters both the circuit being measured and the instrumentation being used. Also, the state of calibration of an instrument may be a source of error, as are humans. The person operating an instrument may bias the measurement with a tendency to read an **analog instrument** high or low or by using an instrument inappropriately.

Relative Error

An error expressed as a percent, as in 20.00 V $\pm$ 2.0%, is called a **relative error.** Relative to 20.00 V there is an uncertainty (error) in the measurement of $\pm 2.0\%$, or ± 0.40 V ($\pm 2.0\%$ of 20.00 V). Therefore, ± 0.40 V is 2.0% relative to 20.00 V.

Because percentage, called **tolerance**, is used to express error in electrical components, electrical components have relative error. Thus, 470 Ω $\pm$ 5% is an example of a resistor with a tolerance of $\pm 5\%$—a relative error.

Absolute Error

An error expressed as a decimal number, such as ± 0.5V, is called an **absolute error.** An instrument's accuracy statement might report its voltage function having an absolute error of ± 0.030 V in each of its four voltage ranges—that is, 1.000 V $\pm$ 0.030 V, 30.00 V $\pm$ 0.030, 100.0 V $\pm$ 0.030 V, and 300.0 V $\pm$ 0.030 V. As indicated, the variation of ± 0.030 V does not depend on the range—it is the same for all ranges. It is an absolute error.

$$\text{relative error} = \frac{\text{absolute error}}{\text{nominal value}} \times 100\% \qquad (12\text{–}4)$$

where relative error is expressed as a percent (%).

EXAMPLE 12–10 Determine the relative error for the preceding instrument when it is initially set to the 1.000-V range and then to the 30.00-V range; the absolute error is ±0.030 V.

Solution Using Equation 12–4, compute the relative error for both the 1.000-V meter range and the 30.00-V meter range (nominal values) given the absolute error of ±0.030 V.

$$\text{relative error} = \frac{\text{absolute error}}{\text{nominal value}} \times 100\%$$

$$\text{relative error} = \pm(0.030/1.000)100 = \pm 3.0\%$$

∴ The relative error stated as a tolerance for the 1.000-V range is ±3.0%

Observation The word *accuracy* is frequently used in place of the word *tolerance* with relative error in instrumentation.

Solution $$\text{relative error} = \frac{\text{absolute error}}{\text{nominal value}} \times 100\%$$

$$\text{relative error} = \pm(0.030/30.00)100 = \pm 0.10\%$$

∴ The relative error stated as an *accuracy* for the 30.00-V range is ±0.10%

Observation In this example, you saw that the absolute error in the measuring instrument, a fixed decimal amount (±0.030 V), became a variable percentage (depending on the range) when converted into a relative error.

Percentages are used in the design of electrical power circuits, where an allowance up to 4.0% of the specified source voltage is used to determine the voltage drop across the resistance in the circuit's electrical conductors. This drop in voltage, due to conductor resistance, is referred to as the *wiring loss voltage*. A 4.0% allowance with a specified 120-V source results in 115 V at the load (120 − 0.04 × 120 = 115 V), and a 240-V source results in 230 V at the load (240 − 0.04 × 240 = 230 V).

The following example explores the use of percentage in the computation of the load voltage when a voltmeter with an accuracy of ±2.0% is used to measure the source voltage.

EXAMPLE 12–11 The load voltage (V_L) dropped across a heat pump powered by a residential electric circuit is equal to the source voltage (E_S) minus the wiring loss voltage (V_{con}).

$$V_L = E_S - V_{con}$$

Determine the maximum and minimum calculated values of the load voltage (V_L) when the voltage measured at the source (E_S) is 235.2 V $\pm$ 2.0%. The specified source voltage is 240.0 V, and 3.0% of this voltage is allowed for the wire loss voltage (V_{con}).

Solution   Using Equation 12–4, compute the absolute error in the source voltage when the relative error is $\pm$2.0%.

$$\text{absolute error} = (\text{relative error} \times \text{nominal value})/100$$

$$\text{absolute error in } E_S = \pm(2.0 \times 235.2)/100 = \pm 4.7 \text{ V}$$

Observation   The measured source voltage (E_S) is 235.2 V $\pm$ 4.7 V or 239.9 V max, 230.5 V min.

Next, determine the wire loss voltage allowing 3.0% (0.030) of the specified 240.0 V for the voltage drop.

$$V_{con} = E_{S(spec)} \times 0.030 = 240 \times 0.030 = 7.2 \text{ V}$$

Compute the max and min values of the load voltages.

$$V_{L(max)} = E_{S(max)} - V_{con} = 239.9 - 7.2 = 232.7 \text{ V}$$
$$V_{L(min)} = E_{S(min)} - V_{con} = 230.5 - 7.2 = 223.3 \text{ V}$$

$\therefore$ The calculated load voltage is 223.3 V minimum and 232.7 V maximum.

Observation   The heat pump will operate satisfactorily with this range of load voltages, because it is specified to operate with load voltages that range between 197 V and 253 V.

Error in the Product and Quotient of Quantities

When calculations are made with measured quantities that result in a product or a quotient, then the combined *relative error* may be determined from the relative errors of each factor in the calculation by applying Rule 12–4.

RULE 12–4.   The Relative Error in a Product or Quotient

To determine the relative error in a product or quotient:

1. Add the relative error of each factor and assign this sum to the resulting product or quotient.
2. When a factor is squared, multiply the relative error by 2 and then combine (add) it to the relative error of the remaining factor(s). In general, x^m has m times the relative error of x.

EXAMPLE 12–12   Determine the maximum and minimum values of calculated power when a heating element has a relative error of ±8.0% in its resistance of 11.0 Ω (11.0 Ω ± 8.0%) and the measured current of 14.8 A was made with an analog meter with an accuracy of ±5.0% of reading, a relative error of ±5.0% (14.8 A ± 5.0%).

Solution   Using $P = I^2R$, solve for the nominal value of power.

$$P = 14.8^2 \times 11.0 = 2.41 \text{ kW}$$

Combine the relative errors using Rule 12–4.

$$\text{relative error in } P = \pm(2 \times 5.0\% + 8.0\%) = \pm18.0\%$$

Observation I^2 contributed $2 \times \pm5.0\%$ and R, ±8.0%.
Compute the absolute error using Equation 12–4.

$$\text{absolute error} = (\text{relative error} \times \text{nominal value})/100$$
$$\text{absolute error} = \pm(18.0 \times 2.41 \times 10^3)/100 = \pm0.434 \text{ kW}$$

Compute the maximum and minimum values of power by adding and then subtracting the absolute error (±0.434 kW) to or from the nominal value of power, 2.41 kW.

$$2.41 \text{ kW} \pm 0.434 \text{ kW} = 2.84 \text{ kW max and } 1.98 \text{ kW min}$$

∴   Maximum power = 2.82 kW and minimum power = 1.98 kW.

Observation   Although the nominal value (calculated value) of power is 2.41 kW, the true value of power lies somewhere between 1.98 kW and 2.84 kW (2.41 kW ± 0.434 kW).

In summary, when a quantity is calculated as the product or quotient of two or more measurements, the total relative error is the ***sum*** of the *relative errors* in each factor in the calculation.

Error in the Sum and Difference of Quantities

When calculations are made with measured quantities that result in a sum or difference, then the combined ***absolute error*** may be determined from the absolute errors of each term in the calculation by applying the following rule.

RULE 12–5. The Absolute Error in a Sum or Difference

To determine the absolute error in a sum or difference:

1. Add the absolute error of each term and assign this sum to the resulting sum or difference.

2. If a term is squared, first multiply the *relative error* by 2 and then compute the *absolute error* using Equation 12–4. Combine (add) the resulting absolute error with the absolute error of the remaining term(s).

EXAMPLE 12–13   In a two-resistor series circuit, $R_1 = 620\ \Omega \pm 5.0\%$ and $R_2 = 330\ \Omega \pm 10.0\%$. Determine:

(a) The nominal value of the total resistance (R_T)
(b) The maximum and minimum values of the total resistance (R_T)
(c) The relative error in the total resistance

Solution   (a) Compute the nominal value of the total resistance.

$$R_T = R_1 + R_2$$
$$R_T = 620 + 330 = 950\ \Omega$$

∴   The nominal value of total resistance is 950 Ω.

(b) Start the computation for the maximum and minimum values of R_T by first determining the absolute error for each resistor (Equation 12–4); then apply Rule 12–5 to determine the absolute error in R_T.

absolute error = (relative error × nominal value)/100

absolute error in $R_1 = \pm(5.0 \times 620)/100 = \pm 31.0\ \Omega$

absolute error in $R_2 = \pm(10.0 \times 330)/100 = \pm 33.0\ \Omega$

Using Rule 12–5, combine the absolute error in R_1 and R_2.

absolute error in $R_T = \pm(31.0\ \Omega + 33.0\ \Omega) = \pm 64.0\ \Omega$

Compute the maximum and minimum values of the total resistance by adding and then subtracting the absolute error ($\pm 64.0\ \Omega$) to or from the nominal calculated value of total resistance (950 Ω).

$$950\ \Omega \pm 64.0\ \Omega = 1014\ \Omega \text{ max and } 886\ \Omega \text{ min}$$

∴   Maximum $R_T = 1.014\ k\Omega$ and minimum $R_T = 0.886\ k\Omega$.

Observation   Although the nominal value (calculated value) of total resistance for the two resistors connected in series is 950 Ω, the true value of the total resistance (R_T) lies somewhere between 0.886 kΩ and 1.014 kΩ.

(c) Using Equation 12–4, compute the relative error of the total resistance.

$$\text{relative error} = \frac{\text{absolute error}}{\text{nominal value}} \times 100\%$$

$$\text{relative error in } R_T = \pm(64.0/950)100 = \pm 6.74\%$$

∴   The relative error stated as a tolerance for the total resistance is ±6.74%: 950 Ω ± 6.74%.

Observation   The tolerance (relative error) for the 950-Ω total resistance cannot be directly calculated by simply adding the tolerances (relative errors) of the two resistors. The process outlined in this example must be followed.

In conclusion, when a quantity is calculated as the sum or difference of two or more measurements, the total absolute error is the ***sum*** of the *absolute errors* in each term in the calculation.

EXERCISE 12–4

Express each of the following relative errors as an absolute error.

1. 50.0 A ± 8.0%
2. 2.50 V ± 2.0%
3. 560 Ω ± 10.0%
4. 15.0 V ± 3.0%
5. 236.0 W ± 15.0%
6. 22.8 A ± 6.0%

Express each of the following absolute errors as a relative error.

7. 300.0 V ± 0.030 V
8. 3.20 kHz ± 60.0 Hz
9. 50.0 A ± 180 mA
10. 2.00 A ± 50 mA
11. 540 Hz ± 15 Hz
12. 2.50 V ± 80 mV

13. If the specified source voltage for a motor-control center is 480.0 V, calculate the maximum and minimum values of load voltage across the motor when the measured source voltage is 468.8 V ± 5.00%. In the calculation, allow 4.0% of the specified 480.0-V source voltage for the wire loss voltage.

14. If the specified source voltage for a host computer is 120.0 V, calculate the maximum and minimum values of load voltage across the computer when the measured source voltage is 123.6 V ± 2.0%. In the calculation, allow 3.0% of the specified 120.0-V source voltage for the wire loss voltage.

15. A 225-W, 75.0-Ω ± 5.0% power resistor has a measured potential difference (voltage drop) of 92.4 V across its terminals. If the voltage is measured with an accuracy of ±4.0%, determine the power dissipation in the resistor, and specify the tolerance of the result.

16. Three 220-Ω resistors are connected in series with each other. Calculate the maximum and minimum value of the total resistance when one resistor has a tolerance of $\pm 2.0\%$, one resistor has a tolerance of $\pm 5.0\%$, and one resistor has a tolerance of $\pm 10.0\%$.

17. A direct current (dc) power supply is used to power three parallel loads in a computing system. The currents are 18.8 mA, 5.14 mA, and 26.0 mA. The first current (18.8 mA) is measured with an accuracy of $\pm 1.5\%$, whereas the other two currents are measured with an accuracy of $\pm 2.0\%$. Determine the maximum and minimum intensity of the total current coming from the dc power supply.

18. A 0.50-W, 4.30-kΩ $\pm 10.0\%$ metal-film resistor has a measured electric current of 8.53 mA passing into it. If the current is measured with an absolute error of ± 0.60 mA, determine the power dissipation in the resistor, and specify the tolerance of the result.

19. Two resistors are connected in series with each other. Calculate the maximum and the minimum value of the total resistance when one resistor (1.30 kΩ) has a tolerance of $\pm 2.0\%$ and the other resistor (910 Ω) has a tolerance of $\pm 5.0\%$.

20. Determine the maximum and minimum resistance of a submergible heater used to heat chemical reagents during the manufacture of microprocessors. The measured voltage drop across the heater is 228.8 V, whereas the measured current into the heater is 35.4 A. The voltage was measured with an analog instrument ranged to 500.0 V with a full-scale accuracy of $\pm 3.0\%$—a relative error of $\pm 6.6\%$. The intensity of the electric current was measured with a clamp-on digital ammeter with an absolute error of ± 750 mA.

12–3 EFFICIENCY

Evaluating Energy Conversion

Many devices have been invented to convert or transform electrical energy to other forms of energy. In the process of converting energy from one form to another, some energy is wasted, usually as heat. A basic principle of physics is the law of **conservation of energy,** which states that energy may be converted or transformed from form to form, but energy cannot be created or destroyed. Stated mathematically, energy input = energy output + energy wasted. Figure 12–2 illustrates this concept. The motor converts

FIGURE 12–2 Conservation of energy: $W_{in} = W_{out} + W_{wasted}$.

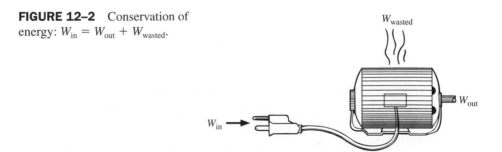

electrical energy (W_{in}) to mechanical energy (W_{out}). However, some energy is converted to heat, in this case a nonuseful form of energy (W_{wasted}), which is *lost* to the output. The **efficiency** of an energy-converting device indicates how well the device performs as an energy converter.

Per-Unit Efficiency

As stated in Equation 12–5, the efficiency of an energy-converting device is expressed as a ratio of the energy out of the device to the energy into the device. Because the terms (numbers) of the efficiency ratio have the same units, the units divide out, resulting in a dimensionless quantity (no units).

$$\eta = \frac{W_{out}}{W_{in}} \qquad (12\text{–}5)$$

where η = efficiency of the device
W_{out} = *useful* output energy (J)
W_{in} = input energy (J)

Because the energy into an electrical machine is always greater than the energy out of the machine, the resulting efficiency, as calculated by Equation 12–5, is always a decimal fraction. When a decimal fraction, such as 0.86, 0.62, etc., is used to represent the efficiency of the energy-converting device, then the efficiency is called a **per-unit efficiency.** That is, for each unit of energy put into an electrical machine (motor, transformer, etc.), a fraction of that unit of energy will come out of the machine.

In summary, a per-unit efficiency (simply called efficiency) results from the ratio of the energy out of a machine to the energy into a machine; efficiency is expressed as a unitless decimal fraction. The power form of the efficiency equation is stated as Equation 12–6.

$$\eta = \frac{P_{out}}{P_{in}} \qquad (12\text{–}6)$$

where η = efficiency of the device
P_{out} = output power (W)
P_{in} = input power (W)

EXAMPLE 12–14   Determine the efficiency of the motor pictured in Figure 12–2 when the work at the output (at the motor shaft) is 2.16 kJ and the work taken in from the power line is 2.63 kJ.

Solution   Substitute into Equation 12–5.

$$\eta = W_{out}/W_{in}$$
$$W_{out} = 2.16 \text{ kJ}$$
$$W_{in} = 2.63 \text{ kJ}$$

$$\eta = 2.16 \times 10^3/2.63 \times 10^3$$

$$\therefore \quad \eta = 0.821$$

Percent Efficiency

Percent efficiency results when the per-unit efficiency is multiplied by 100%. In Equation 12–7, the ratio P_{out}/P_{in} is multiplied by 100%, resulting in percent efficiency ($\eta_\%$).

$$\eta_\% = \frac{P_{out}}{P_{in}} \times 100\% \qquad (12\text{--}7)$$

where $\eta_\%$ = percent efficiency of the device
P_{out} = output power (W)
P_{in} = input power (W)

When percent efficiency is used to express the conversion of power from the input to the output of a machine, the percent efficiency values, such as 72% and 86%, indicate the amount of output power for each one hundred units of input power. For example, a motor with a 68% efficiency would produce 68 units of output power for each 100 units of input power (68/100).

EXAMPLE 12–15   Determine the percent efficiency of a 2400-208Y/120-V service transformer when the output power into the attached resistive load is 22.5 kW. The current and voltage at the input to the transformer were measured and found to be 2.306 kV and 10.2 A, respectively.

Solution   Substitute into Equation 12–7.

$$\eta_\% = P_{out}/P_{in} \ 100\%$$
$$P_{out} = 22.5 \text{ kW} = 22.5 \times 10^3 \text{ W}$$
$$P_{in} = EI = 2.306 \times 10^3 \times 10.2 = 23.52 \times 10^3 \text{ W}$$
$$\eta_\% = (22.5 \times 10^3/23.52 \times 10^3) \times 100$$
$$\therefore \quad \eta_\% = 95.7\%$$

EXAMPLE 12–16   A kettle containing 1 L of water is brought to a boil in 5.0 min by an electric heating element rated at 1200 W. Determine the percent efficiency of the heating element if the water has obtained 300 kJ of heat energy from the heating element.

Solution   First find the value for P_{in}:

$$P_{in} = 1200 \text{ W}$$

Then find the value for P_{out}:

$$P_\text{out} = W/t$$

$$P_\text{out} = 300 \text{ kJ}/5 \text{ min}$$

$$P_\text{out} = \frac{300 \text{ kJ}}{5.0 \text{ min}}\left(\frac{1 \text{ min}}{60 \text{ s}}\right)$$

$$P_\text{out} = 1.0 \text{ kJ/s}$$

$$P_\text{out} = 1.0 \text{ kW} = 1000 \text{ W}$$

Substitute into Equation 12–7:

$$\eta_\% = P_\text{out}/P_\text{in} \times 100\%$$

$$\eta_\% = \frac{1000}{1200} \times 100$$

$$\therefore \quad \eta_\% = 83\%$$

Mechanical Power

The watt is the SI unit of power for both electrical and mechanical power. However, horsepower (hp) is also a commonly used unit for mechanical power. To convert horsepower to watts, use Equation 12–8:

$$\textbf{1 hp} = \textbf{746 W} \qquad \textbf{(exact)} \qquad\qquad \textbf{(12–8)}$$

In the next example, you will be asked to determine the efficiency of an electric motor. The power into the motor (P_in) is equal to the product of the electric current (I) and the voltage (E). The power out of the motor (P_out) is equal to the horsepower developed by the motor in converting the electric power into mechanical power.

EXAMPLE 12–17   Determine both the efficiency and the percent efficiency of a 115-V electric motor that develops 0.500 hp when 4.28 A pass through the motor.

Solution   Compute P_in:

$$P_\text{in} = EI$$
$$P_\text{in} = 115 \times 4.28$$
$$P_\text{in} = 492 \text{ W}$$

Define P_out by converting hp to W:

$$P_\text{out} = 0.500 \text{ hp} \times \frac{746 \text{ W}}{1 \text{ hp}}$$

$$P_\text{out} = 373 \text{ W}$$

Solve for efficiency using Equation 12–6:

$$\eta = P_\text{out}/P_\text{in}$$
$$\eta = 373/492$$
$$\therefore \quad \eta = 0.758$$

Determine $\eta_\%$ by multiplying η by 100%:

$\therefore$ $\eta_\% = 0.758 \times 100 = 75.8\%$

Observation   To convert from percent efficiency to efficiency, simply divide by 100 or move the decimal point two places to the left.

EXAMPLE 12–18   A certain electrical motor delivers 0.250 hp at 1800 revolutions per minute (rpm). The motor is attached to a 120-V source and is 0.785 efficient. What current is used during operation of the motor?

Solution   Use the efficiency equation to find the input power:

$$\eta = \frac{P_{out}}{P_{in}}$$

Solve for P_{in}:

$$P_{in} = \frac{P_{out}}{\eta}$$

Substitute:

$$P_{out} = (0.250 \text{ hp})\left(\frac{746 \text{ W}}{1 \text{ hp}}\right) = 187 \text{ W}$$

$$\eta = 0.785$$

$$P_{in} = \frac{187}{0.785}$$

$$P_{in} = 238 \text{ W}$$

Find current from $P = EI$:

$$I = P/E$$

$$I = 238/120$$

$\therefore$ $I = 1.98 \text{ A}$

Efficiency of a System

When several energy-converting devices are connected together, they form a system. The generating and distribution system of Figure 12–3 is an example of an electrical system. The efficiency of each component in a system contributes to the overall percent of efficiency in the following manner:

$$\eta_{ov} = \eta_1 \times \eta_2 \times \eta_3 \times 100\% \qquad (12\text{–}9)$$

where η_{ov} = overall percent efficiency of the system

η_n = efficiency of component n (*Note:* There may be any number of components in a system.)

In Figure 12–3(a), hydro (water) energy is converted into mechanical energy by a turbine, which then drives the alternator that converts mechanical energy to electrical energy. The overall percent of efficiency of any system decreases each time an additional conversion device is added to the system.

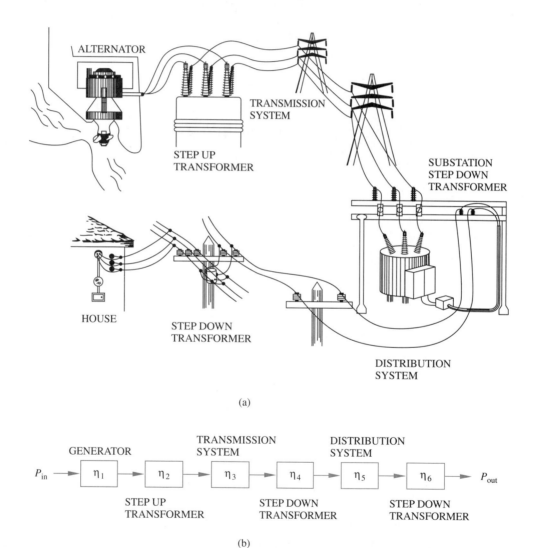

(a)

(b)

FIGURE 12–3 Generation and distribution system: (a) hydroelectric generation and distribution system; (b) cascaded system block diagram of the hydroelectric generation system.

EXAMPLE 12–19   Determine the overall percent of efficiency of the generating system of Figure 12–3(b) if $\eta_1 = 0.89$, $\eta_2 = 0.98$, $\eta_3 = 0.96$, $\eta_4 = 0.98$, $\eta_5 = 0.97$, and $\eta_6 = 0.98$.

Solution
$$\eta_{ov} = \eta_1 \times \eta_2 \times \eta_3 \times \eta_4 \times \eta_5 \times \eta_6 \times 100\%$$
$$\eta_{ov} = 0.89 \times 0.98 \times 0.96 \times 0.98 \times 0.97 \times 0.98 \times 100$$
$$\therefore \quad \eta_{ov} = 78\%$$

Example 12–19 points out how dependent η_{ov} is on each efficiency in the system. We see that the overall efficiency is a product of each efficiency in a system, and it will always be smaller than the smallest efficiency in the system.

EXERCISE 12–5

1. Using the identity 1 hp = 746 W, convert 5.80 hp to **(a)** watts and **(b)** kilowatts.
2. Using the identity 1 hp = 746 W, convert 0.75 hp to **(a)** watts and **(b)** kilowatts.
3. Determine the efficiency of a 1.0-hp electric motor if 850 W of power is required to operate the motor.
4. A power supply for a computer has a 0.76 overall efficiency. Determine the power supplied to the computer by the power supply if 84 W is taken in by the power supply for its operation.
5. How much power, in watts, must be supplied to a home audio system having an overall efficiency of 12.0% if the audio power out of the system is equal to 8.5 W?
6. A battery-charging system is 0.56 efficient when delivering 13.8 V at 23.6 A to the battery being charged. Determine the input current taken from a 120-V ac power line.
7. Two energy-converting devices with $\eta_1 = 0.88$ and $\eta_2 = 0.84$ are cascaded. Determine:
   (a) The output energy if 530 J of energy is put into the system.
   (b) The overall percent efficiency of the cascaded system.
8. The overall efficiency of three energy-converting devices is 58.2%. If $\eta_1 = 0.910$ and $\eta_2 = 0.820$, determine η_3.
9. A motor with an output power of 8.25 hp is operating at 78% efficiency. The electric line into the motor is monitored by a kilowatt meter. Determine:
   (a) The reading of the kilowatt meter.
   (b) The cost at 9.26¢/kWh of wasted energy (lost to heat) for 1 week of continuous operation.
10. A 230-V, 0.75-hp motor has an efficiency of 0.76. Determine:
    (a) The motor's output power in kilowatts.
    (b) The circuit current.
    (c) The cost of operation for 2 weeks at 10¢/kWh if the motor is operated 6 h/d.

11. Determine the power rating, in watts, of an electric-stove heating element if 70.0% of the energy produced in 2.00 min went into heating a kettle of water, and 100.0 kJ of energy was delivered to the water during the 2-min period. One watt is equal to a joule per second.

12. Two differently shaped containers hold equal volumes of water. Each amount of water takes 670 kJ to bring it to a boil. Each container is heated to a boil by identical 1500-W heating elements. Determine:
    (a) The efficiency of each if one takes 9.20 min and the other takes 11.0 min to come to a boil.
    (b) The amount of energy wasted (lost) by each.

13. A 2.00-hp electric motor, 230-V, 60-Hz, having an efficiency of 86% directly drives a 440-V, 400-Hz alternator having an efficiency of 0.83. Draw a cascaded system block diagram and determine:
    (a) The power (in watts) into the motor.
    (b) The overall efficiency of the system.
    (c) The power out of the system.

12–4   PROPORTION

An equation made up of ratios is a **proportion**. The electrical formula $N_1/N_2 = E_1/E_2$ is a proportion. This proportion is read "N_1 is to N_2 as E_1 is to E_2."

EXAMPLE 12–20   Solve the proportion $9/36 = x/16$ for the unknown term x.

Solution
$$\frac{9}{36} = \frac{x}{16}$$

$$x = \frac{9(16)}{36}$$

$$\therefore \quad x = 4$$

EXERCISE 12–6

Solve the following proportions for x.

1. $\dfrac{x}{5} = \dfrac{8}{10}$ 2. $\dfrac{x}{3} = \dfrac{1}{6}$ 3. $\dfrac{9}{x} = \dfrac{12}{4}$

4. $\dfrac{7}{6} = \dfrac{4}{x}$ 5. $\dfrac{1}{x} = \dfrac{7}{13}$ 6. $\dfrac{5}{9} = \dfrac{x}{8}$

Determine which of the following proportions are true. Remember that a proportion is a statement that two ratios (fractions) are equal.

7. $2:3 = 10:15$

8. $\dfrac{3}{8} = \dfrac{5}{9}$

9. $13:11 = 5:4$

10. $\dfrac{5}{7} = \dfrac{7}{5}$

11. $\dfrac{19}{72} = \dfrac{133}{504}$

12. $3:6 = 18:34$

12–5 ELECTRICAL CONDUCTORS

Electrical conductors are manufactured in various forms depending upon the application. Most conductors are made from copper; however, aluminum, nichrome, and tungsten are other metals that are in common use. See Figure 12–4.

The resistance of a conductor depends upon the length, cross-sectional area, temperature, and kind of material. Table 12–2 demonstrates factors governing the resistance of metal conductors.

Length

Two copper wires of equal cross-sectional area and temperature will have different resistances when their **lengths** are different. The longer wire will have more resistance than the shorter wire. We see that the resistance of a conductor increases as the length increases. Mathematically, this is stated as "the resistance of a conductor is **directly proportional** to the length of the conductor." This is written in a **direct proportion** as Equation 12–10.

$$\frac{R_1}{R_2} = \frac{l_1}{l_2} \qquad\qquad \textbf{12–10}$$

where R_1 = resistance of the first conductor (Ω)
R_2 = resistance of the second conductor (Ω)
l_1 = length of the first conductor
l_2 = length of the second conductor

FIGURE 12–4 Metals used in electrical conductors: (a) copper in solid and stranded wire; (b) nichrome in heating elements; (c) tungsten in lamp filaments.

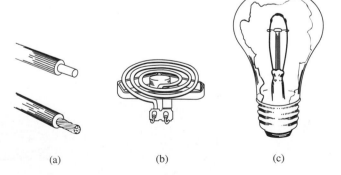

(a)   (b)   (c)

TABLE 12–2 Factors Governing the Resistance of Metal Conductors

Factor	Relative Amount of Resistance			
	More Resistance		**Less Resistance**	
Length		Long		Short
Cross-sectional area		Small		Large
Kind of material		Nichrome		Copper
Temperature		Hot		Cold

EXAMPLE 12–21 The resistance of 10 m of a particular copper wire is 842 mΩ. Determine the resistance of 38 m of the same wire.

Solution Use Equation 12–10. Substitute 842 mΩ for R_1, 10 m for l_1, and 38 m for l_2. Solve for R_2:

$$\frac{R_1}{R_2} = \frac{l_1}{l_2}$$

$$\frac{842 \times 10^{-3}}{R_2} = \frac{10}{38}$$

$$R_2 = \frac{38(842 \times 10^{-3})}{10}$$

$$\therefore \qquad R_2 = 3.2 \ \Omega$$

EXAMPLE 12–22 An RF coil is made from copper wire that has a resistance of 340 Ω/km. Determine the length (in meters) of the wire used to manufacture the coil if the finished coil has a measured resistance of 1.57 Ω. See Figure 12–5.

FIGURE 12–5 Radio-frequency coil for Example 12–22.

Solution Use Equation 12–10. Substitute 340 Ω for R_1, 1 km for l_1, and 1.57 Ω for R_2. Solve for l_2:

$$\frac{R_1}{R_2} = \frac{l_1}{l_2}$$

$$\frac{340}{1.57} = \frac{1 \times 10^3}{l_2}$$

$$l_2 = \frac{1.57(1 \times 10^3)}{340}$$

$$\therefore \qquad l_2 = 4.62 \text{ m}$$

EXERCISE 12–7

In each problem it is assumed that all the conductors are round, solid, copper wire of equal cross-sectional area and temperature:

1. The resistance of 6.0 m of copper wire is 0.52 Ω. Determine the resistance of 17.4 m of the same wire.

2. The radio-frequency coil of Figure 12–5 has a measured resistance of 7.38 Ω. Determine the length of wire in the coil if the wire has a resistance of 213 Ω/km.

3. A 100-m coil of copper wire has a resistance of 2.10 Ω. If 20 m of wire is cut from the coil, what is the resistance of the remaining 80 m of wire?

4. Determine the resistance of the 20 m of wire removed from the coil in problem 3.

5. An unknown amount of wire has been removed from a 1.0-km spool of wire. The remaining wire has a measured resistance of 39.4 Ω. Determine the length of the remaining wire on the spool if a 1.0-km length of wire has a measured resistance of 84.2 Ω.

6. Complete the following table by determining the missing information:

Length	20 cm	100 cm	5.0 m	20 m	100 m
Resistance		0.0910 Ω			

Cross-Sectional Area

Two copper wires of equal length and temperature will have different resistances when their *cross-sectional areas* are different. The wire with the larger cross-sectional area will have a smaller resistance than the wire of smaller cross-sectional area. We see that the resistance of a conductor increases as the wire becomes smaller in cross-sectional area. Mathematically stated, the "resistance is *inversely* proportional to the cross-sectional area." This is written as an **indirect proportion** in Equation 12–11.

$$\frac{R_1}{R_2} = \frac{A_2}{A_1} \qquad\qquad \textbf{(12–11)}$$

where R_1 = resistance of the first conductor (Ω)
R_2 = resistance of the second conductor (Ω)
A_1 = cross-sectional area of the first conductor
A_2 = cross-sectional area of the second conductor

Equation 12–11 may be restated in terms of the conductors' diameters instead of their cross-sectional areas, as in Equation 12–12. Thus,

$$A_1 = \pi \frac{d_1^2}{4}$$

$$A_2 = \pi \frac{d_2^2}{4}$$

$$\frac{R_1}{R_2} = \frac{A_2}{A_1} = \frac{(\pi/4)d_2^2}{(\pi/4)d_1^2}$$

Factor out $\pi/4$ and the result is:

$$\frac{R_1}{R_2} = \frac{d_2^2}{d_1^2} \qquad\qquad \textbf{(12–12)}$$

where R_1 = resistance of the first conductor (Ω)
R_2 = resistance of the second conductor (Ω)
d_1 = diameter of the first conductor
d_2 = diameter of the second conductor

EXAMPLE 12–23   The resistances of two conductors of equal length were measured and found to be 4.14 and 16.6 Ω. Determine the diameter in millimeters of the conductor having a resistance of 4.14 Ω if the other conductor is 1.15 mm in diameter.

Solution   Use Equation 12–12. Substitute 4.14 Ω for R_1, 16.6 Ω for R_2, and 1.5 mm for d_2. Solve for d_1:

$$\frac{R_1}{R_2} = \frac{d_2^2}{d_1^2}$$

$$\frac{4.14}{16.6} = \frac{(1.15)^2}{d_1^2}$$

$$d_1^2 = \frac{16.6(1.15)^2}{4.14}$$

$$d_1 = \sqrt{\frac{16.6(1.15)^2}{4.14}}$$

$$\therefore \qquad d_1 = 2.30 \text{ mm}$$

The American Wire Gauge Table

The **American wire gauge (AWG)** is used to standardize wire sizes. Standard copper wire is designated by one of the 44 standard gauge numbers of the American wire gauge (AWG). The AWG is a sequence of 44 gauge numbers extending from no. 40, the smallest (0.07988 mm), through no. 1 (7.348 mm), to the largest no. 4/0, also written no. 0000 (11.68 mm).

Although the SI system may be used when calculating wire measurements, the wire itself is sold by the foot with the diameter specified in mils (one-thousandth of an inch). For purposes of calculation, mils may be converted to millimeters by the conversion factor

$$1 \text{ mil} = 0.0254 \text{ mm} \quad (\text{exact})$$

TABLE 12–3 Partial List of American Wire Gauge for Solid Annealed Copper Conductors at 20°C

Gauge No.	SI Metric Units		English Units	
	Dia. (mm)	Ω/km	Dia. (mils)*	Ω/1000 ft
10	2.588	3.277	101.9	0.9989
11	2.305	4.132	90.74	1.260
12	2.053	5.211	80.81	1.588
13	1.828	6.571	71.96	2.003
14	1.628	8.285	64.08	2.525
15	1.450	10.45	57.07	3.184
16	1.291	13.17	50.82	4.016
17	1.150	16.61	45.26	5.064
18	1.024	20.95	40.30	6.385
19	0.912	26.42	35.89	8.051
20	0.812	33.31	31.96	10.15
21	0.723	42.00	28.45	12.80
22	0.644	52.96	25.35	16.14
23	0.573	66.79	22.57	20.36
24	0.511	84.21	20.10	25.67
25	0.455	106.2	17.90	32.37
26	0.405	133.9	15.94	40.81
27	0.361	168.9	14.20	51.47
28	0.321	212.9	12.64	64.90
29	0.286	268.5	11.26	81.83
30	0.255	338.6	10.03	103.2

*1 mil = 0.001 in.

For convenience, many recent editions of engineering handbooks include tables for solid, annealed copper wire with the AWG listed in both metric and English units. Table 12–3 is a partial listing of the American wire gauge. The complete table is given in Appendix A.

EXAMPLE 12–24	A certain length of no. 24 gauge copper wire having a diameter of 0.511 mm and a resistance of 12 Ω is replaced with a no. 20 gauge copper wire. Determine the resistance of the 20-gauge wire.
Solution	Use Table 12–3 to determine the diameter of the 20-gauge wire:

$$20 \text{ gauge} \Rightarrow 0.812\text{-mm diameter}$$

Use Equation 12–12 and substitute 12 Ω for R_1, 0.511 mm for d_1, and 0.812 mm for d_2. Solve for R_2:

$$\frac{R_1}{R_2} = \frac{d_2^2}{d_1^2}$$

$$\frac{12}{R_2} = \frac{(0.812)^2}{(0.511)^2}$$

$$R_2 = \frac{12(0.511)^2}{(0.812)^2}$$

$$\therefore \quad R_2 = 4.8 \ \Omega$$

EXAMPLE 12–25	Determine the diameter in millimeters of a conductor needed to replace a no. 30 gauge wire so that the resistance is one-fourth that of the 30-gauge wire. The resistance of the 30-gauge wire is 8.0 Ω.
Solution	Use Table 12–3 to determine the diameter of the 30-gauge wire:

$$30 \text{ gauge} \Rightarrow 0.255 \text{ mm diameter}$$

Use Equation 12–12 and substitute 8.0 Ω for R_1, 0.255 mm for d_1, and 2.0 Ω (8/4 = 2) for R_2. Solve for d_2:

$$\frac{8.0}{2.0} = \frac{d_2^2}{(0.255)^2}$$

$$d_2^2 = \frac{8.0(0.255)^2}{2.0}$$

$$d_2 = \sqrt{\frac{8.0(0.255)^2}{2.0}}$$

$$\therefore \quad d_2 = 0.51 \text{ mm; this is 24-gauge wire.}$$

EXERCISE 12–8

In each problem it is assumed that all the conductors are round, solid, annealed copper wire of equal length and at 20°C.

1. A certain wire has a diameter of 0.0790 mm and a resistance of 10.0 Ω. Determine the resistance when the diameter is increased to 0.254 mm.

2. A no. 32 wire has a diameter of 202 μm and a resistance of 52.7 Ω. Determine the resistance when the diameter is 0.102 mm.

3. One hundred meters of no. 10 wire, which has a diameter of 2.59 mm, has a resistance of 0.3277 Ω. Determine the resistance of 100 m of no. 6 wire, which has a diameter of 4.12 mm.

4. It is desired to reduce the resistance of 0.333 km of no. 18 wire to 3.00 Ω by replacing it with a new wire of a larger diameter.

   (a) Determine the diameter of the new wire.

   (b) Select a wire gauge from Table 12–3 that will meet the requirements of the new wire.

5. Repeat problem 4 for 140 m of no. 22 wire.

6. One hundred meters of no. 10 wire, which has a diameter of 2.59 mm, has a resistance of 0.3277 Ω. Determine the resistance of 100 m of no. 18 wire, which has a diameter of 1.024 mm.

Kind of Material

Conductors made of different metals are compared to one another by their **resistivity** (sometimes called specific resistance). Resistivity is measured in ohm-meters (Ω·m) and is based on a standard conductor, which is 1 m in length and 1 m^2 in cross section at a temperature of 20°C. The Greek letter ρ (rho) is the symbol used to indicate resistivity, as in Equation 12–13.

$$\rho = \frac{RA}{l} \quad \text{(ohm-meters)} \tag{12–13}$$

where ρ = resistivity of the material in ohm-meters (Ω·m)
R = resistance in ohms (Ω)
A = cross-sectional area in square meters (m^2)
l = length in meters (m)

Table 12–4 lists the resistivity of several metals; silver, the best conductor, is listed first.

TABLE 12–4 Resistivity of Some Metals at 20°C

Metal	Resistivity ($\Omega \cdot$m)	Metal	Resistivity ($\Omega \cdot$m)
Silver	0.0164×10^{-6}	Tungsten	0.0552×10^{-6}
Copper	0.0172×10^{-6}	Nichrome	1.00×10^{-6}
Aluminum	0.0283×10^{-6}		

EXAMPLE 12–26 Determine the resistance at 20°C of a nichrome wire of 0.404-mm diameter and 2.0 m long.

Solution Express the diameter in meters:

$$0.404 \text{ mm} = 0.404 \times 10^{-3} \text{ m}$$

Compute the area of the conductor in square meters:

$$A = \frac{\pi}{4}d^2$$

$$A = \frac{3.14}{4}(0.404 \times 10^{-3})^2$$

$$A = 0.128 \times 10^{-6} \text{ m}^2$$

Use Equation 12–13 and substitute 1.00×10^{-6} $\Omega \cdot$m for ρ, 0.128×10^{-6} m^2 for A, and 2.0 m for l. Solve for R:

$$\rho = \frac{RA}{l}$$

$$1.00 \times 10^{-6} = \frac{R(0.128 \times 10^{-6})}{2.0}$$

$$R = \frac{1.00 \times 10^{-6} \times 2.0}{0.128 \times 10^{-6}}$$

$$\therefore \qquad R = 16 \ \Omega$$

EXAMPLE 12–27 Determine the resistance of a solid-copper rectangular conductor that is 10.0 cm long, 2.0 cm wide, and 0.50 cm thick at 20°C.

Solution Compute the cross section in square meters:

$$A = \text{width} \times \text{thickness}$$
$$A = 2.0 \times 10^{-2} \text{ m} \times 0.50 \times 10^{-2} \text{ m}$$
$$A = 1.0 \times 10^{-4} \text{ m}^2$$

Determine ρ for copper from Table 12–4:

$$\rho = 0.0172 \times 10^{-6} \ \Omega\text{·m}$$

Express length in meters:

$$10.0 \text{ cm} = 10.0 \times 10^{-2} \text{ m}$$

Substitute $0.0172 \times 10^{-6} \ \Omega$·m for ρ, 10.0×10^{-2} m for l, and 1.0×10^{-4} m^2 for A in Equation 12–13. Solve for R:

$$\rho = \frac{RA}{l}$$

$$0.0172 \times 10^{-6} = \frac{R(1.0 \times 10^{-4})}{10 \times 10^{-2}}$$

$$R = \frac{(0.0172 \times 10^{-6})(10 \times 10^{-2})}{1.0 \times 10^{-4}}$$

$$R = 1.72 \times 10^{-5} \ \Omega$$

$$\therefore \qquad R = 17 \ \mu\Omega$$

EXERCISE 12–9

1. Determine the resistance of 15.0 m of aluminum wire that has a diameter of 2.588 mm at 20°C.

2. Determine the resistance of 0.200 m of nichrome wire that has a diameter of 0.255 mm at 20°C.

3. What is the length of a 24-gauge tungsten wire that has a resistance of 0.850 Ω at 20°C?

4. How long is an 18-gauge copper wire that has a resistance of 13.0 Ω at 20°C?

5. Determine the resistivity (ρ) of an unknown metal conductor that is 10.0 m long and has a diameter of 0.643 mm. The measured resistance is 1.61 Ω at 20°C.

6. Determine the resistance of the shunt pictured in Figure 12–6. It is made of constantan, which has a $\rho = 0.49 \times 10^{-6} \ \Omega$·m at 20°C. Use the central cross section in Figure 12–6(b) to compute the area.

FIGURE 12–6   Constantan shunt for Exercise 12–9, problem 6.

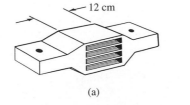

(a)

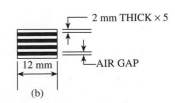

(b)

Temperature

So far, we have investigated the effects that *length, cross-sectional area,* and *kind of material* have on electrical conductors. Now let us consider temperature.

Metals increase in resistance as temperature rises. A copper conductor at 20°C will have a substantial increase in resistance at 80°C. The amount of increase in resistance for each degree Celsius rise in temperature varies from metal to metal. The effect of temperature on the resistance of a copper conductor is computed by Equation 12–14.

$$\frac{R_2}{R_1} = \frac{234.5 + T_2}{234.5 + T_1} \qquad\qquad (12\text{–}14)$$

where R_1 = resistance of the copper conductor at T_1 (Ω)
R_2 = resistance of the copper conductor at T_2 (Ω)
T_1 = condition 1 temperature (°C)
T_2 = condition 2 temperature (°C)
-234.5°C = extrapolated temperature, where copper has a resistance of 0 Ω

EXAMPLE 12–28   Determine the resistance of a copper conductor at 80.0°C if it has a resistance of 10.0 Ω at 20.0°C.

Solution   Solve Equation 12–14 for R_2:

$$R_2 = R_1 \frac{234.5 + T_2}{234.5 + T_1}$$

Substitute 10.0 Ω for R_1, 20.0°C for T_1, and 80.0°C for T_2:

$$R_2 = 10.0 \times \frac{234.5 + 80.0}{234.5 + 20.0}$$

$\therefore$ $R_2 = 12.4 \ \Omega$

EXAMPLE 12–29   Number 28 gauge copper wire has a resistance of 213 Ω/km at 20.0°C. Determine the resistance at -10.0°C.

Solution   Use Equation 12–14 and solve for R_2:

$$R_2 = R_1 \frac{234.5 + T_2}{234.5 + T_1}$$

Substitute 213 Ω for R_1, 20.0°C for T_1, and -10.0°C for T_2:

$$R_2 = 213 \times \frac{234.5 + (-10.0)}{234.5 + 20.0} = \frac{213 \times 224.5}{254.5}$$

$\therefore$ $R_2 = 188 \ \Omega$

---

1. The resistance of 50.0 m of copper wire at 30.0°C is 1.62 Ω. Determine its resistance at 57.0°C.

2. The resistance of an RF coil wound with copper wire is 12.8 Ω at 42.0°C. Determine the resistance at −6.00°C.

3. The copper winding of a motor was measured at a temperature of 22.0°C and found to be 0.48 Ω. While in use, the resistance of the winding was found to be 0.55 Ω. Determine the temperature of the winding.

4. The resistance of the copper wire within an inductor was measured and found to be 2.78 Ω at a temperature of 28.0°C. After several hours of operation, the resistance increased to 3.34 Ω. Determine the temperature of the wire.

SELECTED TERMS

American wire gauge (AWG)   A table of standard wire diameters indicated by gauge numbers.

analog instrument   An instrument with a pointer to indicate values along a graduated scale.

conservation of energy   Energy may be changed in form but energy can neither be created nor destroyed.

deviation   Variation from a specified dimension or design requirement, typically defining upper and lower limits.

direct proportion   Two ratios are in direct proportion when one ratio in the proportion increases and the other ratio increases or one ratio decreases and the other ratio also decreases.

efficiency   Indicates how well an energy converter performs.

indirect proportion   Two ratios are in indirect proportion when one ratio in the proportion increases and the other ratio decreases.

parts per million   Per one million; divided by 1 000 000.

percent   Per one hundred; divided by 100.

proportion   Equation made up of two ratios.

resistivity   Resistance between opposite parallel faces of a 1-meter cube of a material. Resistivity depends only on the properties of the material, whereas resistance depends on the properties of the material as well as length and cross-sectional area.

tolerance   The algebraic difference between the maximum and minimum limits of a specified dimension or value; the total amount of deviation permitted in a quantity.

SECTION CHALLENGE

WEB CHALLENGE FOR CHAPTERS 11 AND 12

To evaluate your comprehension of Chapters 11 and 12, log on to **www.prenhall.com/ harter** and take the online True/False and Multiple Choice assessments for each of the chapters.

SECTION CHALLENGE FOR CHAPTERS 11 AND 12*

Occasionally, a complex circuit configuration may be encountered in which the resistances do not appear to be connected in series or parallel (as in Figure C–3a). However, when examined carefully, a wye (R_1, R_2, and R_3) or a delta configuration (R_A', R_B', and R_C') may be seen. In order to solve the circuit, a set of equations is needed to convert

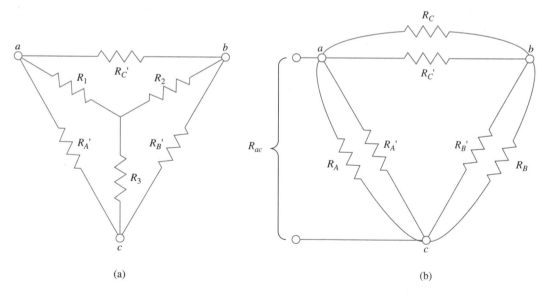

(a) (b)

FIGURE C–3 Complex circuit configuration; (a) resistances don't appear to be connected in series or parallel; (b) converting the Y to Δ, the circuit is transformed so it may be solved.

*The solution to this Section Challenge is found in Appendix C.

from a wye (Y) circuit configuration to a delta (Δ) circuit configuration. In the course of developing the equations, the following three equations were derived.

$$R_1 = \frac{R_A(R_A R_2/R_3)}{R_A + (R_A R_2/R_1) + (R_A R_2/R_3)} \quad \textbf{(ohms)} \qquad \text{(E–1)}$$

$$R_2 = \frac{R_C(R_C R_3/R_1)}{(R_C R_3/R_2) + (R_C R_3/R_1) + R_C} \quad \textbf{(ohms)} \qquad \text{(E–2)}$$

$$R_3 = \frac{R_B(R_B R_1/R_2)}{(R_B R_1/R_2) + R_B + (R_B R_1/R_3)} \quad \textbf{(ohms)} \qquad \text{(E–3)}$$

Your challenge is to solve the first equation (E–1) for R_A in terms of R_1, R_2, and R_3; the second equation (E–2) for R_C in terms of R_1, R_2, and R_3; and the third equation (E–3) for R_B in terms of R_1, R_2, and R_3. Using the resulting equations, determine the resistance values of R_A, R_B, and R_C of the delta configuration of Figure C–4b given the wye configuration resistance values of:

A. $R_1 = 1.00\ \Omega$, $R_2 = 2.00\ \Omega$, and $R_3 = 3.00\ \Omega$
B. $R_1 = 10.0\ \Omega$, $R_2 = 15.0\ \Omega$, and $R_3 = 20.0\ \Omega$
C. $R_1 = 6.67\ \Omega$, $R_2 = 6.67\ \Omega$, and $R_3 = 6.67\ \Omega$

As an added challenge, calculate the equivalent resistance, $R_T = R_{ac}$, of Figure C–3b for each set of values (R_A, R_B, R_C) previously determined in problems A, B, and C. Let $R_A' = R_B' = R_C' = 12.0\ \text{V}$.

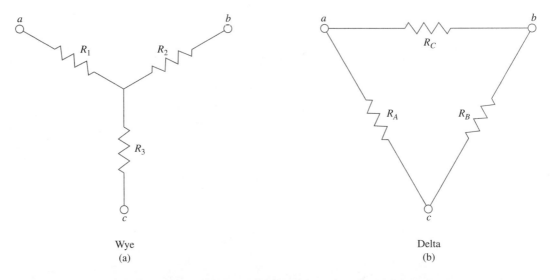

Wye
(a)

Delta
(b)

FIGURE C–4 Two *equivalent* circuit configurations, (a) wye (Y) and (b) delta (Δ), that may be transformed from one form to the other using wye-to-delta or delta-to-wye equations. When *balanced,* the equivalent resistance, R_T, between the same two terminals in either configuration are equal, that is,

$$R_T = R_{ac} = R_1 + R_3 = \frac{R_A(R_B + R_C)}{R_A + R_B + R_C} \text{ when } R_1 = R_2 = R_3 \text{ or } R_A = R_B = R_C.$$

13 Relations and Functions

13–1   Meaning of a Function

13–2   Variables and Constants

13–3   Functional Notation

13–4   Functional Variation

13–5   Simplifying Formulas

PERFORMANCE OBJECTIVES

- Identify dependent and independent variables and the constants in an equation.

- Read functional notation and use it to evaluate a function.

- Recognize direct and inverse variation in formulas.

- Apply the concepts of "much greater than" and "much less than" to the simplification of formulas.

Weather Bureau technicians launch a weather balloon with a radiosonde transmitter attached (the small box in the hand of the technician on the right). The radiometeorgraph (the instrument in the box) transmits weather information back to the ship's radiosonde receiver. (Courtesy of U.S. Coast Guard)

A function is a mathematical statement about the relationship between numerical quantities. Such mathematical statements are used to bring the power of mathematics to bear on physical problems in electronics. The topics in this chapter have been selected to introduce functions and functional notation.

The following symbols are new to this chapter and are listed here for your reference.

$\gg$ Much greater than $f(x)$ "f" function of x

$\ll$ Much less than

13–1 MEANING OF A FUNCTION

By studying the cause and effect relationship between quantities, laws are formulated. Ohm's law was formed by such observations. In the last chapters, many equations were used to solve problems. In applying these equations you may have observed that the quantities represented by the variables are interrelated with one another. A change in the value of one quantity caused a change in another quantity.

When one condition in a relationship depends on one or more other conditions in the relationship, it is said that the first condition is the **function** of the other conditions. In the equation $y = 2x$, y is a function of x. We can say that y is a function of x if for each value of x there is a corresponding value of y.

EXAMPLE 13–1   Determine if the following statement conforms to the definition of a function. The voltage drop across a resistor is a *function* of the current through the resistance.

Solution   Use Ohm's law: $V = IR$. Let R remain constant. If various values of I are substituted into the formula, a value of V results in each case, even when zero is substituted.

$\therefore$ Voltage V is a *function* of current I.

13–2 VARIABLES AND CONSTANTS

Dependent and Independent Variables

A dependence of one variable on another is indicated by the definition of a function. In the equation $y = 5x + 2$, the value of y *depends* upon the selection of the value for x. For example, when $x = 2$, $y = 5(2) + 2 = 12$.

In the equation $y = 5x + 2$, x is the **independent variable.** We may let x assume any value that we wish independent of the equation. In an equation the independent variable appears in the right member.

In the equation $y = 5x + 2$, y is the **dependent variable** because the value of y depends on the value chosen for x. In an equation the dependent variable is the left member.

EXAMPLE 13–2   In the following equations, identify the independent and the dependent variables.

(a) $I_1 = \dfrac{I_T G_1}{\Sigma G}$

(b) $R_T = R_1 + R_2 + R_3$

(c) $E = V_1 + V_2$

Solution   (a) I_1 is the dependent variable.
I_T, G_1, and ΣG are independent variables.

(b) R_T is the dependent variable.
R_1, R_2, and R_3 are the independent variables.

(c) E is the dependent variable.
V_1 and V_2 are the independent variables.

Constants

As was noted previously in Section 9–3, **constants** are divided into two types. First, those represented by numbers and special symbols, for example, 23, 57.9, $\frac{1}{4}$, $\frac{1}{2}$, π (3.14159), e (2.71828), and $\sqrt{3}$ (1.73205). Second, those represented by letters (b, k, etc.). The latter type of constant is a variable (called an *arbitrary constant*) with an arbitrary value that remains fixed during a problem or discussion. For example, when current is determined from the applied voltage and the resistance, Ohm's law is written as $I = E/R$. If R is a fixed resistor of 100 Ω and E is varied, then I is the dependent variable, E is the independent variable, and the resistance, R, is an arbitrary constant.

EXAMPLE 13–3   Identify the symbols representing the constants and variables in the following equations.

(a) $u = 3v + b$, where $b = 16.4$

(b) $A = \pi r^2$

(c) $y = 8x^2 + 6x + e$

Solution   (a) b and 3 are the constants.
u and v are the variables.

(b) π is the constant.
r and A are the variables.

(c) 8, 6, and e are the constants.
x and y are the variables.

In each of the following equations, state which symbols represent constants, dependent variables, and independent variables.

1. The force (F) required to keep motion in a circular path is a function of mass (m), speed (v), and radius (r). The equation for centripetal force is $F = kmv^2/r$, where $k = 1$.

2. The distance traveled by a freely falling body is a function of time (t) and the acceleration due to gravity (g). The equation for displacement of a freely falling body is $s = \frac{1}{2}gt^2$, where g is approximately 9.81 m/s^2 at the surface of the earth.

3. The attractive force (F) between two particles is a function of their masses (m_1 and m_2) and the distance between them (d). The equation for the attractive force is $F = Gm_1m_2/d^2$, where G is the gravitational constant.

4. The brake horsepower (P) developed by a motor is measured by a device called a *Prony brake*. The brake horsepower is a function of the rotational speed (ω) and the difference in force between two balances (ΔF). The equation for brake horsepower is $P = 2\pi r\omega(\Delta F)/33\ 000$, where r is a constant.

5. The speed of sound (v) through a gaseous medium is a function of the absolute temperature (T). The equation for the speed of sound is $v = RT/M$, where R is the gas constant.

6. The electrical resistance (R) of a conductor is a function of the length of the conductor (l) and the cross-sectional area (A). The equation for the resistance is $R = \rho l/A$, where ρ is the resistivity of the conductor—a constant.

13–3 FUNCTIONAL NOTATION

In the equation $y = 5x + 2$, y is a *function of x*. The statement that y is a function of x is written in functional notation as $y = f(x)$, where $f(x)$ is read "function of x," "f of x," or "f function of x." Thus, $y = 5x + 2$ may be written $f(x) = 5x + 2$. In this case, remember that $y = f(x) = 5x + 2$ means that y, $f(x)$, and $5x + 2$ are all the same function of x.

EXAMPLE 13–4   Use functional notation to state that:

(a) e is a function of i.

(b) u is a function of v.

(c) c is a function of r.

Solution   (a) $e = f(i)$

(b) $u = f(v)$

(c) $c = f(r)$

In the preceding example, it is important that you understand that the notation $f(i)$, $f(v)$, and $f(r)$ indicates functional relationship and not that f is multiplied by some variable. The letter f is usually used in functional notation. However, other letters may also be used, as shown in the following example.

EXAMPLE 13–5   Use functional notation to write the following expressions:

(a) y equals f of x.
(b) w is the g function of y.
(c) v is the h function of u.
(d) $s = Q$ of t.

Solution   (a) $y = f(x)$
(b) $w = g(y)$
(c) $v = h(u)$
(d) $s = Q(t)$

The functional notation $y = f(x)$ does not show the specific function of x that y represents; $y = f(x)$ could indicate that y is any function of x. For example,

$$y = f(x) = 5x + 2$$
$$y = f(x) = x^2$$
$$y = f(x) = \sin(x)$$
$$y = f(x) = 3x^3 - 8$$

To indicate a given function of x, such as $5x + 2$, we should simply use $f(x) = 5x + 2$. Functional notation also provides a way to show the specific value of the independent variable to be used. Example 13–6 shows several instances of finding the value of a function for a particular value of the independent variable. This process is called *evaluating a function*.

EXAMPLE 13–6   Given that $f(x) = 3x + 2$, evaluate:

(a) $f(5)$ (b) $f(-3)$ (c) $f(0)$ (d) $f(a)$ (e) $f(m)$

Solution   (a) To find $f(5)$, substitute 5 for x:

$$f(x) = 3x + 2$$
$$f(5) = 3(5) + 2 = 17$$
$$\therefore \quad f(5) = 17$$

(b) To find $f(-3)$, substitute -3 for x:

$$f(x) = 3x + 2$$
$$f(-3) = 3(-3) + 2 = -7$$
$$\therefore \quad f(-3) = -7$$

(c) To find $f(0)$, substitute 0 for x:

$$f(x) = 3x + 2$$
$$f(0) = 3(0) + 2 = 2$$
$$\therefore \quad f(0) = 2$$

(d) To find $f(a)$, substitute the constant a for x:

$$f(x) = 3x + 2$$
$$\therefore \quad f(a) = 3a + 2$$

(e) To find $f(m)$, substitute the constant m for x:

$$f(x) = 3x + 2$$
$$\therefore \quad f(m) = 3m + 2$$

EXAMPLE 13–7   Evaluate $f(x) = 2x^2 - 5x$ for $x = 5$.

Solution
$$f(x) = 2x^2 - 5x$$
$$f(5) = 2(5)^2 - 5(5)$$
$$\therefore \quad f(5) = 25$$

EXERCISE 13–2

Given that $f(e) = 5e - 3$, find:

1. $f(2)$ **2.** $f(-5)$ **3.** $f(4)$ **4.** $f(-1)$
5. $f(-2)$ **6.** $f(0)$ **7.** $f(-b)$ **8.** $f(3a)$

Use functional notation to express each of the following four statements.

9. x is the g function of y.
10. u is the f function of x.
11. z is f of v.
12. y equals g of w.
13. $f(x) = 4x^2 - x + 2$; find $f(2), f(-3), f(0.5)$.
14. $g(r) = \pi r^2$; find $g(3), g(5), g(0.316)$.
15. $Q(e) = \dfrac{1}{e - 2} + \dfrac{1}{e - 3}$; find $Q(1), Q(0.5), Q(5)$.

13–4 FUNCTIONAL VARIATION

When applying a formula to determine the value for a circuit component or a circuit condition, it is often necessary to make several calculations before a satisfactory result is reached. In this process it is very helpful to have an understanding of how the formula affects the selected value in producing the solution.

Suppose that you have constructed a series circuit with a resistive load. You have energized the circuit and measured the source voltage and found it to be too low for your application. You are about to increase the voltage, but you now wonder if the load will be able to dissipate an increase in the power. Your question, then, is, "How does power vary with voltage?" By investigating the functional relationship between power and voltage ($P = E^2/R$), you will have an understanding of how voltage affects power. From the relationship of power and voltage, you may see that an increase by a factor of 10 in voltage produces a *hundredfold* increase in power.

This same basic question can arise in many settings. It will be useful to understand how changing the independent variable will affect the dependent variable without repeating the whole calculation each time.

EXAMPLE 13–8

In the formula $E = IR$, let R be constant, and determine how E changes when I is:

(a) Doubled   (b) Halved   (c) Multiplied by 10
(d) Divided by 10

Solution

(a) $E(I) = IR$
      $E(2I) = 2IR = 2E(I)$

∴ E is doubled when I is doubled.

(b) $E(I) = IR$

$$E(I/2) = \frac{1}{2}IR = \frac{1}{2}E(I)$$

∴ E is halved when I is halved.

(c) $E(I) = IR$
      $E(10I) = 10IR = 10E(I)$

∴ E is multiplied by 10 when I is multiplied by 10.

(d) $E(I) = IR$

$$E(I/10) = \frac{1}{10}IR = \frac{1}{10}E(I)$$

∴ E is divided by 10 when I is divided by 10.

In Example 13–8 we see that the change in I due to multiplying or dividing produces the same change in E. In this example, E *varies directly* as I varies. This behavior is called **direct variation.**

EXAMPLE 13–9   In the formula $X_C = \dfrac{1}{\omega C}$ assume that ω is a constant. Determine how X_C changes when C is:

(a) Multiplied by 5
(b) Divided by 5
(c) Multiplied by 10
(d) Divided by 10

Solution   (a) $X_C(C) = \dfrac{1}{\omega C}$

$$X_C(5C) = \dfrac{1}{\omega 5C} = \dfrac{1}{5}\left(\dfrac{1}{\omega C}\right) = \dfrac{1}{5}X_C(C)$$

$\therefore$ X_C is divided by 5 when C is multiplied by 5.

(b) $X_C(C) = \dfrac{1}{\omega C}$

$$X_C(C/5) = \dfrac{1}{\omega C/5} = 5\left(\dfrac{1}{\omega C}\right) = 5X_C(C)$$

$\therefore$ X_C is multiplied by 5 when C is divided by 5.

(c) $X_C(C) = \dfrac{1}{\omega C}$

$$X_C(10C) = \dfrac{1}{\omega 10C} = \dfrac{1}{10}\left(\dfrac{1}{\omega C}\right) = \dfrac{1}{10}X_C(C)$$

$\therefore$ X_C is divided by 10 when C is multiplied by 10.

(d) $X_C(C) = \dfrac{1}{\omega C}$

$$X_C(C/10) = \dfrac{1}{\omega C/10} = 10\left(\dfrac{1}{\omega C}\right) = 10X_C(C)$$

$\therefore$ X_C is multiplied by 10 when C is divided by 10.

In Example 13–9 we see that the change in C due to multiplying or dividing produces an inverse change in X_C. In this example, X_C *varies inversely* as C varies; this is called an **inverse variation.**

EXAMPLE 13–10   In the formula $P = I^2R$, let R be a constant. Determine how P changes when I is:

(a) Doubled   (c) Multiplied by 10
(b) Halved   (d) Divided by 10

Solution (a) I is used as a factor twice; thus RI^2 is multiplied by 2 twice when I is multiplied by 2 once.

∴ P is multiplied by 4 when I is doubled.

(b) RI^2 is multiplied by $\frac{1}{2}$ twice when I is multiplied by $\frac{1}{2}$ once.

∴ P is multiplied by $\frac{1}{4}$ when I is halved.

(c) RI^2 is multiplied by 10 twice when I is multiplied by 10 once.

∴ P is multiplied by 100 when I is multiplied by 10.

(d) RI^2 is multiplied by $\frac{1}{10}$ twice when I is multiplied by $\frac{1}{10}$ once.

∴ P is multiplied by $\frac{1}{100}$ when I is divided by 10.

In Example 13–10 we see that, when I is multiplied by a number, P is multiplied by the square of the number. Thus, P *varies directly with the square* of I.

EXAMPLE 13–11 In the formula $F = \dfrac{Wv^2}{gr}$, W and g are constant. Determine the effect on F when:

(a) r is halved and v is doubled.
(b) r is doubled and v is tripled.

Solution (a) Since W and g are constant, assume them to be worth 1.

$$F(v,r) = \frac{v^2}{r}$$

Halve r and double v:

$$F(2v,r/2) = \frac{(2v)^2}{1/2\,r}$$

$$F(2v,r/2) = \frac{(2v)^2\,(2)}{r}$$

$$F(2v,r/2) = \frac{(4v^2)(2)}{r}$$

$$F(2v,r/2) = \frac{8v^2}{r} = 8F(v,r)$$

∴ F is multiplied by 8 when r is halved and v is doubled.

(b) Double r and triple v:

$$F(3v,2r) = \frac{(3v)^2}{2r}$$

$$F(3v,2r) = \frac{9v^2}{2r} = \frac{9}{2}F(v,r)$$

∴ F is multiplied by $\frac{9}{2}$ when r is doubled and v is tripled.

EXERCISE 13–3

1. In the conductance formula $G = 1/R$, how does G change when R is:
   (a) Halved?   (b) Tripled?   (c) Divided by 5?
   (d) How does G vary with R?

2. The inductance of a coil (L) is determined by the formula $L = \dfrac{\mu N^2 A}{l}$.
   (a) If μ, N^2, and l are constants, how does L vary with A?
   (b) If μ, N^2, and A are constants, how does L vary with l?
   (c) How does L change when N is tripled and μ, A, and l remain the same?

3. In the Ohm's law formula $I = E/R$:
   (a) How is I changed if E is one quarter as large and R is unchanged?
   (b) How is I changed if R is halved and E is constant?
   (c) How is I changed if both E and R are doubled?

4. The average power dissipated by a resistance is determined by the formula $P = E^2/R$. How is P affected when:
   (a) E is doubled and R remains constant?
   (b) R is one quarter as large and E is the same?
   (c) Both E and R are doubled?

5. The bandwidth (BW) of a resonant circuit is determined by the formula $\text{BW} = \dfrac{f_0 R}{X_L}$. If f_0 is constant, then how does BW change when:
   (a) X_L is doubled and R remains the same?
   (b) R is doubled and X_L remains the same?
   (c) R is halved and X_L is doubled?

6. The resistance of a conductor is determined by the formula $R = \rho\dfrac{l}{A}$. If ρ is constant, then how is R changed when:
   (a) l is divided by 4 and A remains the same?
   (b) Both l and A are divided by 6?
   (c) l is multiplied by 3 and A is divided by 2?

7. The mutual inductance (M) between two coils is determined by the formula $M = k\sqrt{L_1 L_2}$.

   (a) How is M changed if both L_1 and L_2 are doubled and k is constant?

   (b) How is M changed if L_1 is one quarter as large, L_2 is tripled, and k is constant?

8. The closing force developed by the electromagnet of a relay is determined by the formula $F = AB^2/2\mu$. If μ is a constant, then how is F affected when:

   (a) B is halved and A is doubled?

   (b) Both B and A are doubled?

   (c) B is one and one half times as large and A is four times as large?

13–5 SIMPLIFYING FORMULAS

Many formulas used in electronic technology may be simplified by making some assumptions about the value of one or more of the independent variables. These assumptions can only be made with an understanding of the conditions of the circuit under consideration. The formula for finding the equivalent resistance of two resistors in series is such an equation.

EXAMPLE 13–12   Investigate when the formula for the equivalent resistance of a series circuit, $R_T = R_1 + R_2$, may be simplified to $R_T \approx R_1$.

Solution   Select values of 1.0 Ω for R_1 and R_2. Substitute and solve:

$$R_T = 1 + 1 = 2 \ \Omega$$

Observation   The ratio of 2 Ω and 1 Ω is not very close to 1. Increase the value of R_1 an *order of magnitude* from 1.0 Ω to 10 Ω, and let R_2 remain at 1.0 Ω. Substitute and solve:

$$R_T = 10 + 1 = 11 \ \Omega$$

Observation   The ratio of 11 Ω to 10 Ω is much, much closer to 1 than the ratio of 2 Ω to 1 Ω.

∴   When the ratio of R_1 to R_2 is equal to or greater than ($\geq$) 10 to 1, we may simplify the equation $R_T = R_1 + R_2$ to $R_T \approx R_1$.

EXAMPLE 13–13   Investigate when the formula for the equivalent resistance of a parallel circuit, $R_T = \dfrac{R_1 R_2}{R_1 + R_2}$ may be simplified to $R_T \approx R_1$.

Solution   Select values of 1.0 Ω for R_1 and R_2. Substitute and solve $(R_2 : R_1 = 1 : 1)$:

$$R_T = \frac{R_1 R_2}{R_1 + R_2}$$

$$R_T = \frac{1.0 \times 1.0}{1.0 + 1.0} = \frac{1}{2} = 0.5 \ \Omega$$

Increase the value of R_2 an *order of magnitude* from 1.0 Ω to 10 Ω, and let R_1 remain at 1.0 Ω. Substitute and solve $(R_2 : R_1 = 10 : 1)$:

$$R_T = \frac{1.0 \times 10}{1.0 + 10} = \frac{10}{11} = 0.91 \ \Omega$$

Once again increase the value of R_2 an *order of magnitude* from 10 Ω to 100 Ω, and let R_1 remain at 1.0 Ω. Substitute and solve $(R_2 : R_1 = 100 : 1)$:

$$R_T = \frac{1.0 \times 100}{1.0 + 100} = \frac{100}{101} = 0.99 \ \Omega$$

Finally, increase the value of R_2 an *order of magnitude* from 100 Ω to 1000 Ω, and let R_1 remain at 1.0. Substitute and solve $(R_2 : R_1 = 1000 : 1)$:

$$R_T = \frac{1.0 \times 1000}{1.0 + 1000} = \frac{1000}{1001} = 0.999 \ \Omega$$

$\therefore$ $R_T \approx R_1$ when $R_2 : R_1 \geq 10 : 1$

Observation   The approximation is even better when $R_2 : R_1 \geq 100 : 1$. Table 13–1 is a summary of the investigation carried out in this example.

TABLE 13–1   Summary of Example 13–13

Ratio $R_2 : R_1$	$R_T = \dfrac{R_1 R_2}{R_1 + R_2}$	R_1	% error $= \dfrac{R_1 - R_T}{R_T} \times 100\%$
1:1	0.50	1.0	100%
10:1	0.91	1.0	9.9%
100:1	0.99	1.0	1.0%
1000:1	0.999	1.0	0.10%

Ratios of 10:1 or greater are symbolized by $\gg$, which is read "much greater than," as in $R_2 \gg R_1$. Ratios of 1:10 or smaller are symbolized by $\ll$, which is read "much less than," as in $R_1 \ll R_2$.

EXAMPLE 13–14   Use the symbol $\ll$ or $\gg$ to indicate when $R_T = \dfrac{R_1 R_2}{R_1 + R_2}$ may be stated as $R_T \approx R_1$.

Solution   Let R_2 be much greater than R_1. Then $R_1 + R_2 \Rightarrow R_2$. Substitute R_2:

$$R_T = \frac{R_1 R_2}{R_1 + R_2} \approx \frac{R_1 R_2}{R_2}$$

Factor out R_2:

$\therefore$ $R_T \approx R_1$ when $R_2 \gg R_1$ or $R_1 \ll R_2$.

EXAMPLE 13–15   Simplify the formula $C_T = \dfrac{C_1 C_2}{C_1 + C_2}$ for capacitors in series when $C_2 \gg C_1$.

Solution   Looking at the denominator, $C_1 + C_2 \Rightarrow C_2$ when $C_2 \gg C_1$. Substitute C_2 for the denominator:

$$C_T \approx \frac{C_1 C_2}{C_2}$$

Factor out C_2:

$\therefore$ $C_T \approx C_1$ when $C_2 \gg C_1$.

EXAMPLE 13–16   If the low-frequency voltage gain of a junction field-effect transistor (JFET) is given by the formula

$$A_v = \frac{Y_{fs} r_{os} R_L}{r_{os} + R_L}$$

then investigate the condition of r_{os} that will allow the formula to be simplified to $A_v \approx Y_{fs} R_L$. R_L is "on the order of" $10 \text{ k}\Omega$.

Solution   Select r_{os} to be an order of magnitude greater than R_L ($r_{os} \gg R_L$). $R_L = 10 \text{ k}\Omega$ and $r_{os} = 100 \text{ k}\Omega$. Suppose that $Y_{fs} = 5.0 \text{ mS}$. Substitute and solve:

$$A_v = \frac{(5.0 \times 10^{-3})(100 \times 10^3)(10 \times 10^3)}{100 \times 10^3 + 10 \times 10^3}$$

$$A_v = 45.5$$

Substitute and solve in the simplified formula:

$$A_v \approx Y_{fs}R_L$$
$$A_v \approx 5.0 \times 10^{-3} \times 10 \times 10^3 = 50$$

Is $50 \approx 45.5$? Yes, it is within 10%.

$$\therefore \qquad A_v = \frac{Y_{fs}\, r_{os}\, R_L}{r_{os} + R_L} \Rightarrow A_v \approx Y_{fs}R_L \text{ when } r_{os} \gg R_L$$

EXERCISE 13–4

Determine the condition for each formula to yield the simplified formula.

1. $A_v = -\dfrac{g_m r_p R_0}{r_p + R_0} \Rightarrow A_v \approx -g_m R_0$

2. $A_v = \dfrac{g_m R_k}{\dfrac{R_k}{r_p} + g_m R_k + 1} \Rightarrow A_v \approx \dfrac{g_m R_k}{g_m R_k + 1}$

3. $I_1 = \dfrac{IG_1}{G_1 + G_2} \Rightarrow I_1 \approx I$

4. $V_2 = \dfrac{ER_2}{R_1 + R_2} \Rightarrow V_2 \approx E/2$

5. $R_2 = \dfrac{(K + 1)R_{th}}{K} \Rightarrow R_2 \approx R_{th}$

6. $Z_1 = \dfrac{-j}{\omega C_{gp}}\left(\dfrac{k}{k - 1}\right) \Rightarrow Z_1 \approx \dfrac{-j}{\omega C_{gp}}$

7. $I_L = \dfrac{-AE_i}{A'R_f - (r_0 + R_f + R_L)} \Rightarrow I_L \approx -\dfrac{AE_i}{A'R_f}$

8. $I_0 = -\dfrac{Ae_i}{A'r_f} \Rightarrow I_0 \approx -\dfrac{e_i}{r_f}$

9. $y = 1 - \dfrac{4k^2}{4k^2 - 1}\, e^{-x} \Rightarrow y \approx 1 - e^{-x}$

10. $V' = \dfrac{VR_{dc}}{R_{dc} + R} \Rightarrow V' \approx \dfrac{VR_{dc}}{R}$

11. $g = \dfrac{I + I'}{\eta V} \Rightarrow g \approx \dfrac{I}{\eta V}$

12. $\Delta v = (V' - V)\left(\dfrac{R + R'}{R}\right)\left(\dfrac{r_0}{r - r_0}\right) \Rightarrow \Delta v \approx (V' - V)\dfrac{r_0}{r}$

constant   A numeric quantity that has a fixed value for a given problem.

dependent variable   A variable that is the output of a function.

function   A mathematical relationship between a set of input values and a resulting output value.

independent variable   A variable that is the input to a function.

variable   A numeric quantity represented by a letter that can take on more than one value in a problem; also a numeric quantity with an unknown value.

14

Graphs and Graphing Techniques

14–1   Rectangular Coordinates

14–2   Graphs of Equations

14–3   Graphs of Linear Equations

14–4   Deriving a Linear Equation from a Graph

14–5   Graphing Empirical Data

PERFORMANCE OBJECTIVES

- Apply the concepts of the rectangular coordinate system to read a graph.

- Construct graphs of equations and functions.

- Determine the slope of a straight line from its graph.

- Derive the linear equation of a straight line from its graph.

- Graph empirical data.

A technician monitors a storm system on radar. (Courtesy of National Oceanic and Atmospheric Administration)

A **graph** is a visual representation of the relationship between two or more quantities. By the information pictured in the graph, we are able to interpret circuit conditions, values, and trends. Graphs, like pictures, "are worth a thousand words." The topics presented here have been selected to help you gain an understanding of how to construct graphs and how to interpret them.

14–1 RECTANGULAR COORDINATES

On a piece of graph paper, construct both a horizontal and a vertical number line. Draw the lines so that their intersection is at the zero point of each, as shown in Figure 14–1.

FIGURE 14–1 The horizontal and vertical axes of a rectangular coordinate system are formed by two number lines.

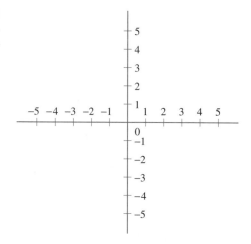

The horizontal and vertical axes of a rectangular coordinate system are formed by two number lines, called **axes.** The horizontal line is called the *x*-**axis;** the vertical line is called the *y*-**axis.** The *x*-axis (horizontal) and the *y*-axis (vertical) divide the graph paper into four parts called *quadrants.* The quadrants are numbered, with Roman numerals, in a counterclockwise (ccw) direction, as shown in Figure 14–2. The intersection of the *x*-axis with the *y*-axis is called the **origin.**

Each point on the graph paper is indicated by a pair of numbers called *coordinates.* The coordinates of the point on the graph paper tell how far the point is from the origin. The first number in the pair of coordinates tells how far the point is along the *x*-axis. This number is called the *x-coordinate* or **abscissa.** The second number, called the *y-coordinate* or **ordinate,** tells how far the point is along the *y*-axis. Together the two coordinates locate the point on the graph paper.

When writing the coordinates of a point, the *x*-coordinate is listed first, and then the *y*-coordinate, as in (*x, y*). Because of this agreed-upon order, the coordinates of a point are called an **ordered pair** of numbers. Figure 14–3 shows several points with their coordinates as ordered pairs of numbers noted.

CHAPTER 14   Graphs and Graphing Techniques

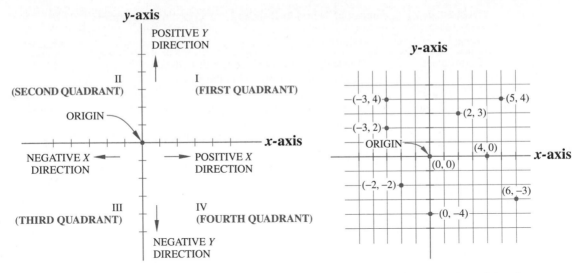

FIGURE 14–2 The graph paper is divided into quadrants.

FIGURE 14–3 The points of the plane are numbered by *rectangular coordinates* with the x-coordinate given first followed by the y-coordinate (x, y). The notation (x, y) is called an *ordered pair.*

When constructing a graph, the independent variable is *usually* assigned to the x-coordinate, and the dependent variable is *usually* assigned to the y-coordinate.

The coordinates of the points in a plane may be negative or positive depending upon the quadrant in which they are located. As seen in Figures 14–1 and 14–2, the x-coordinate is positive to the right of the y-axis and negative to the left of the y-axis. The y-coordinate is positive above the x-axis and negative below the x-axis. Table 14–1 summarizes the signs of the coordinates in each quadrant.

The terms used in the rectangular coordinate system are summarized in Table 14–2. This system of coordinates is sometimes referred to as the Cartesian coordinate system, in honor of its developer René Descartes (1596–1650).

TABLE 14–1   Summary of Signs in Each Quadrant

Quadrant	x-Coordinate (Abscissa)	y-Coordinate (Ordinate)
I	+	+
II	−	+
III	−	−
IV	+	−

TABLE 14–2 Summary of Terms in Rectangular Coordinate System

Term	Definition	Figure
Coordinates	Made up of two numbers that describe the location of each point in the plane	14–3
Ordered pair	Coordinates written as (x, y)	14–3
x-Coordinate	Number in the ordered pair that tells how far to move along the x-axis	14–3
y-Coordinate	Number in the ordered pair that tells how far to move along the y-axis	14–3
x-Axis	Horizontal axis	14–2
y-Axis	Vertical axis	14–2
Abscissa	x-coordinate	
Ordinate	y-coordinate	
Origin	Point where the x- and y-axes cross	14–2
Quadrants	Four quarters of the plane created by the crossing of the x- and y-axes	14–2

EXAMPLE 14–1 Using Figure 14–4, determine the coordinates of points *A*, *B*, and *C* and state in which quadrant each point is located.

FIGURE 14–4 Lettered points for Example 14–1 and Exercise 14–1.

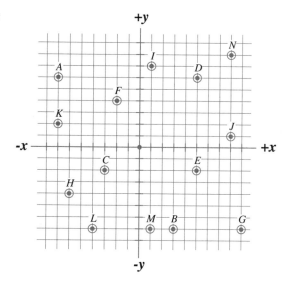

Solution

Point A   Coordinates $(-7, 6)$ located in the second quadrant.
Point B   Coordinates $(3, -7)$ located in the fourth quadrant.
Point C   Coordinates $(-3, -2)$ located in the third quadrant.

Using Figure 14–4, answer the following.

1. In which quadrants are each of the points *A* through *H* located?
2. In which quadrants are each of the points *I* through *N* located?
3. Determine the coordinates of points *A* through *H*.
4. Determine the coordinates of points *I* through *N*.
5. What is the *y*-coordinate of any point on the *x*-axis?
6. What is the *x*-coordinate of any point on the *y*-axis?
7. What is another name for abscissa?
8. What is another name for ordinate?
9. In which quadrant is a point located if its *x*-coordinate is negative and its *y*-coordinate is positive?
10. In which quadrant is a point located if its abscissa is positive and its ordinate is negative?
11. Draw a pair of axes and plot the points $(2, -1)$, $(-3, 4)$, and $(0, -4)$. Label each with its ordered pair.
12. Draw a pair of axes and plot the points $(-5, -2)$, $(6, 0)$, and $(3, -4)$. Label each with its ordered pair.

14–2 GRAPHS OF EQUATIONS

PROGRAMMABLE

In constructing graphs of equations and functions, we must first make a table of ordered pairs. To do this, selected values of the independent variable (*x*) are substituted into the equation, and then values of the dependent variable (*y*) are computed. The ordered pairs are next plotted, and then a smooth curve is passed through the points.

EXAMPLE 14–2   Plot the graph of the equation $y = \frac{1}{2}x^2$.

Solution   Select values of *x* and solve for *y* in the equation given. Some calculations are:

$$x = 0 \qquad y = \frac{1}{2}(0)^2 = 0$$

$$x = 1 \qquad y = \frac{1}{2}(1)^2 = \frac{1}{2}$$

$$x = -1 \qquad y = \frac{1}{2}(-1)^2 = \frac{1}{2}$$

Make a table of ordered pairs:

x	-4	-2	-1	0	1	2	4
y	8	2	$\frac{1}{2}$	0	$\frac{1}{2}$	2	8

∴   Plotting these points results in the *parabola* of Figure 14–5.

FIGURE 14–5   (a) Graph of the equation $y = \frac{1}{2}x^2$ for Example 14–2 correctly drawn with a smooth curve; (b) the graph incorrectly drawn without a smooth curve.

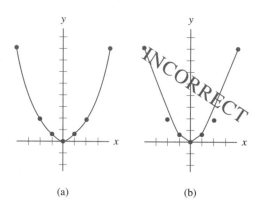

(a)   (b)

Since a graph is a visual representation of all the solutions of an equation, the coordinates of any point on the graph will satisfy the conditions of the equation. The coordinates of any point *not* on the graph will *not* satisfy the conditions of the equation.

EXAMPLE 14–3   Select several coordinates from the graph of $y = x + 2$ as shown in Figure 14–6, and check them by substituting into the equation.

FIGURE 14–6   Graph of the equation $y = x + 2$ for Example 14–3.

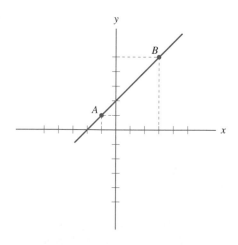

Solution   From the graph of Figure 14–6, select points A $(-1, 1)$ and B $(3, 5)$. Substitute and check:

$$
\begin{array}{cc}
\textit{Point A} & \textit{Point B} \\
y = x + 2 & y = x + 2 \\
1 = -1 + 2 & 5 = 3 + 2 \\
1 = 1 & 5 = 5 \quad \text{Yes!}
\end{array}
$$

∴   Both points satisfy the equation.

The graph of a straight line may be located with only two points as in Figure 14–6. However, when plotting graphs that are not straight lines, more points are needed. It is good to remember that the plot of the graph of an equation is only an approximation of the shape of the "real" graph of the equation. We do not need to spend a great deal of time in computing countless coordinates for the graph. In general, fewer points are needed for the straight portions of the graph than for the curved.

EXAMPLE 14–4   Plot the graph of the equation $y = x^3 - 2x^2 - 5x + 6$.

Solution   From the equation, compute a preliminary table of ordered pairs:

x	-2	-1	0	1	2	3
y	0	8	6	0	-4	0

∴   Plotting these points results in the graph of Figure 14–7(a). Using the graph of Figure 14–7(a), select additional values of x and compute y:

x	-2.5	-1.5	-0.5	0.5	1.5	2.5	3.5
y	-9.6	5.6	7.9	3.1	-2.6	-3.4	9.9

∴   Plotting these points results in the graph of Figure 14–7(b).

FIGURE 14–7 (a) Preliminary graph of the equation $y = x^3 - 2x^2 - 5x + 6$ for Example 14–4; (b) a more complete graph with additional points.

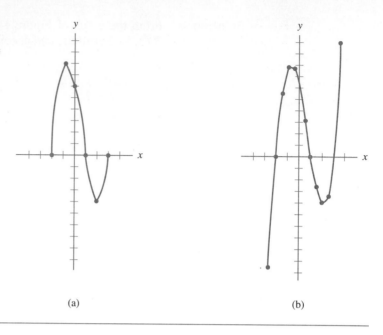

(a)   (b)

How can functions be graphed? A function may be graphed by first forming an equation. Let $y = f(x)$. Then graph the equation, as demonstrated in the following example.

EXAMPLE 14–5   Plot the graph of the function $-x^3$.

Solution $y = f(x) = -x^3$

Select values of x and compute y, and then set up a table of ordered pairs:

x	-2.0	-1.8	-1.5	-1.2	-1.0	-0.5	0.0
y	8.0	5.8	3.4	1.7	1.0	0.1	0.0

x	0.0	0.5	1.0	1.2	1.5	1.8	2.0
y	0.0	-0.1	-1.0	-1.7	-3.4	-5.8	-8.0

∴   Plotting these points results in the graph of Figure 14–8.

FIGURE 14–8 Graph of $y = -x^3$ for Example 14–5.

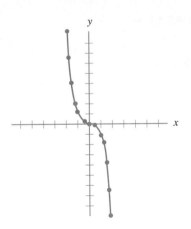

Besides showing the shape of an equation, graphs may be used to solve equations. In most cases the graphical solution is actually easier than the mathematical solution. Graphs, then, provide us with an important alternative method of solving equations.

The **roots** (solutions) of an equation are the abscissas or x-coordinates where the graph **intersects** the x-axis. The roots of an equation may be found by *reading* the graph.

EXAMPLE 14–6 Graph the function $x^2 - 4x - 5$ and determine the roots of the equation $y = f(x) = x^2 - 4x - 5$.

Solution Select values of x and compute y, and then set up a table of ordered pairs:

x	-3	-2	-1	0	1	2	3	5	6	7
y	16	7	0	-5	-8	-9	-8	0	7	16

∴ Plotting these points results in the graph of Figure 14–9. The roots of the equation are the intersection of the x-axis and the curve. Reading from the curve, the intersections are $(-1, 0)$ and $(5, 0)$. The roots are $x = -1$ and $x = 5$.

FIGURE 14–9 Graph of $x = x^2 - 4x - 5$ for Example 14–6. The roots of the equation $y = x^2 - 4x - 5$ are the intersections of the curve and the x-axis; $x = -1$ and $x = 5$ when $y = 0$.

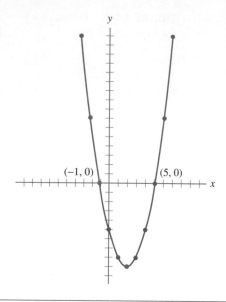

PROGRAMMABLE

Sketch the graph representing each of the following functions. Determine the roots of the functions [when $y = f(x) = 0$].

1. $x + 2$ **2.** $2x - 3$ **3.** $-3x + 1$

4. $-5x - 6$ **5.** $-3x^2$ **6.** x^3

7. $x^3 - 3x^2 - x + 3$ **8.** $-x^2 - 10x - 16$

Sketch the graph representing each of the following equations.

9. $y = x$ **10.** $y = -5x - 6$ **11.** $4x - 2y = 3$

12. $6 = 3y - 5x$ **13.** $y = x^2 + 1$ **14.** $y = x + 3$

15. $x^2 + y^2 = 9$ **16.** $x^2 + y^2 = 25$ **17.** $x^2 + 6y = 0$

18. $2y - 2x^2 = 4x - 6$

14–3 GRAPHS OF LINEAR EQUATIONS

The graph of a linear equation is a straight line. A linear equation has the general form

$$y = mx + b \qquad\qquad (14\text{–}1)$$

where m = coefficient of x
         b = a constant
         x = independent variable
         y = dependent variable

Slope

In Figure 14–10 the straight line has two points, P_1 and P_2, with coordinates (x_1, y_1) and (x_2, y_2). The **slope** of a line is found by taking the difference between the y-coordinates and dividing by the difference between the x-coordinates, as stated in Equation 14–2. The slope is equal to m, the coefficient of x in the general linear equation $y = mx + b$.

$$\text{slope} = m = \frac{y_1 - y_2}{x_1 - x_2}$$

14–2

FIGURE 14–10 The straight line has a slope, m, equal to $(y_1 - y_2)/(x_1 - x_2)$.

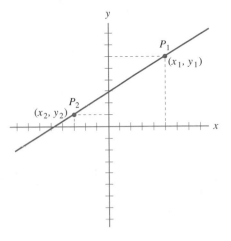

EXAMPLE 14–7 Determine the slope of the lines of Figure 14–11(a) and (b).

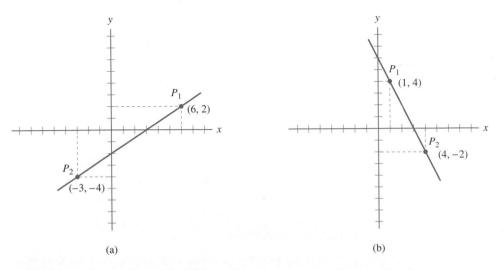

(a) (b)

FIGURE 14–11 Graphs for Example 14–7: (a) positive slope; (b) negative slope.

Solution   Substitute the coordinates of points P_1 and P_2 of Figure 14–11(a) into Equation 14–2:

$$m = \frac{y_1 - y_2}{x_1 - x_2}$$

$$m = \frac{2 - (-4)}{6 - (-3)}$$

$$\therefore \quad m = \frac{2}{3}$$

Substitute the coordinates of points P_1 and P_2 of Figure 14–11(b) into Equation 14–2:

$$m = \frac{y_1 - y_2}{x_1 - x_2}$$

$$m = \frac{4 - (-2)}{1 - 4}$$

$$\therefore \quad m = -2$$

From Example 14–7 we see that the line of Figure 14–11(a) has a *positive* slope and that the line of Figure 14–11(b) has a *negative* slope.

RULE 14–1.   *Slope of a Straight Line*

The slope of a straight line is:

1. Positive—if a point on the lines *rises* as the point moves from left to right along the line.

2. Negative—if a point on the line *falls* as the point moves from left to right along the line.

Intercept

The point where the line intersects the *x*-axis is the **x-intercept.** The *x*-coordinate of the *x*-intercept is the **root** of the equation.

In Equation 14–1, $y = mx + b$, *b* is the **y-intercept.** Thus, *b* is the *y*-coordinate of the point where the line and the *y*-axis intersect. The point where the line intersects the *y*-axis is $(0, b)$.

Graphing Linear Equations

The graph of a straight line can be made with just two points. One of these points can be the *y*-intercept $(x = 0, y = b)$. The other point is determined by selecting an appropriate value for the *x*-coordinate and then computing the *y*-coordinate. This procedure is summarized as Rule 14–2.

RULE 14–2.   Graphing Linear Equations

To graph a straight line:

1. Put the equation in the form of $y = mx + b$.
2. Let $x = 0$ and solve for the *y*-intercept ($y = b$).
3. Select a value for *x* that is away from the origin and solve for *y*.
4. Plot the two points on the graph paper.
5. Draw a straight line through the two points.
6. As a check, select a point from the line and substitute the coordinates into the equation.

EXAMPLE 14–8   Graph $y = 2x + 3$.

Solution   Apply Rule 14–2. Let $x = 0$ and solve for *y*:

$$y = 2x + 3$$
$$y = 2(0) + 3$$
$$y = 3$$

∴   (0, 3) is a point on the line.

Let $x = 3$ and solve for *y*:
$$y = 2(3) + 3$$
$$y = 9$$

∴   (3, 9) is a point on the line.

Construct the graph of $y = 2x + 3$ as shown in Figure 14–12.

FIGURE 14–12   Graph of $y = 2x + 3$ for Example 14–8.

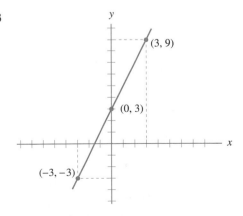

Check   Does the ordered pair $(-3, -3)$ satisfy $y = 2x + 3$?

$$-3 = 2(-3) + 3$$
$$-3 = -6 + 3$$
$$-3 = -3 \qquad \text{Yes!}$$

EXAMPLE 14–9   Graph $-x/2 + y = -5$.

Solution   Apply the steps of Rule 14–2.

Step 1:   Put $-x/2 + y = -5$ into the form $y = mx + b$:

$$y = x/2 - 5$$

Step 2:   Let $x = 0$ and solve for y:

$$y = -5$$

∴ $(0, -5)$ is a point on the line.

Step 3:   Let $x = 6$ and solve for y:

$$y = \frac{6}{2} - 5$$

$$y = 3 - 5$$

$$y = -2$$

∴ $(6, -2)$ is a point on the line.

Steps 4–5:   Construct the graph of $y = x/2 - 5$ as shown in Figure 14–13.

FIGURE 14–13   Graph of $y = (x/2) - 5$ for Example 14–9.

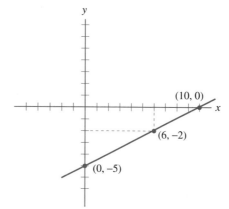

Check   Does the ordered pair $(10, 0)$ satisfy $y = x/2 - 5$?

$$0 = \frac{10}{2} - 5$$

$$0 = 5 - 5$$

$$0 = 0 \qquad \text{Yes!}$$

EXAMPLE 14–10 Graph $E_{th} - I_b(0.2) - 0.7 - 2I_b(0.2) = 0$. Consider $E_{th} = f(I_b)$.

Solution Put the equation in the form $y = mx + b$:

$$E_{th} = (0.2)I_b + (0.4)I_b + 0.7$$
$$E_{th} = (0.6)I_b + 0.7$$

Let $I_b = 0$ and solve for E_{th}:

$$E_{th} = 0.7$$

∴ (0, 0.7) is a point on the line.

Let $I_b = 10$ and solve for E_{th}:

$$E_{th} = (0.6)(10) + 0.7$$
$$E_{th} = 6 + 0.7$$
$$E_{th} = 6.7$$

∴ (10, 6.7) is a point on the line.

Construct the graph of $E_{th} = (0.6)I_b + 0.7$ as shown in Figure 14–14.

FIGURE 14–14 Graph of $E_{th} = (0.6)I_b + 0.7$ for Example 14–10.

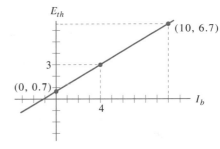

Check Does the ordered pair (4, 3) satisfy the equation?

$$3 = (0.6)(4) + 0.7$$
$$3 = 2.4 + 0.7$$
$$3 = 3.1$$

Observation Notice that the ordered pair did not check exactly. However, it is as close as the graph will allow, and it is quite satisfactory.

EXERCISE 14–3

In each of the following, state (a) whether the slope is positive or negative, (b) the slope, and (c) the coordinates of the y-intercept.

1. $y = -x + 5$

2. $y = x + 3$

3. $y = -2x - 4$

4. $y = x/3 + 1$

5. $y = 4x$

6. $f(x) = -\frac{1}{4}x + 2$

7. $f(x) = -2x/3 + 12$

8. $f(x) = \frac{5}{2}x - 2$

9. $f(x) = x$

10. $f(x) = x - 3$

Graph each of the following equations.

11. $V = 10I + 3$

12. $E = 6R$

13. $A = \frac{1}{2}h$

14. $E_0 = -2I + 3$

15. $I = E/2$

16. $R = E/5 - 4$

17. $R_{th} = -3(E_{th}/6 - 2)$

18. $I_T = 2V + 4$

19. $R_{int} = 5R_s - 2$

20. $Q_T = 10E + 12$

14–4 DERIVING A LINEAR EQUATION FROM A GRAPH

The equation of a straight line may be derived from its graph by determining its y-intercept and the slope. The general equation $y = mx + b$ is made into a specific equation for the graph by substituting values for m and b.

> **RULE 14–3. Deriving an Equation from Its Graph**
>
> To determine an equation of the graph of a straight line:
>
> **1.** Locate the coordinates of the y-intercept.
>
> **2.** Determine the slope of the line.
>
> **3.** Substitute values of m and b into $y = mx + b$.

EXAMPLE 14–11   Derive the equation of the line pictured in Figure 14–15.

Solution   Follow the steps of Rule 14–3.

FIGURE 14–15 Graph for Example 14–11.

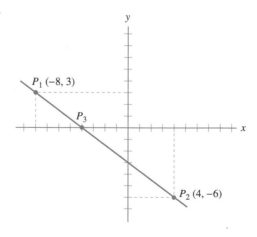

Step 1: Determine the y-intercept:

$$y\text{-intercept} = (0, -3)$$
$$b = -3$$

Step 2: Determine the slope by selecting points P_1 and P_2 as in Figure 14–15. Substitute into Equation 14–2:

$$m = \frac{y_1 - y_2}{x_1 - x_2}$$

$$m = \frac{3 - (-6)}{-8 - 4} = \frac{9}{-12} = -\frac{3}{4}$$

$$\therefore \quad m = -\frac{3}{4}$$

Step 3: Substitute values of m and b into $y = mx + b$:

$$y = -\frac{3}{4}x - 3$$

Check Do the coordinates $(-4, 0)$ of P_3 satisfy the derived equation?

$$y = -\frac{3}{4}x - 3$$

$$0 = -\frac{3}{4}(-4) - 3$$

$$0 = 3 - 3$$

$$0 = 0 \qquad \text{Yes.}$$

$$\therefore \quad y = -\frac{3}{4}x - 3 \text{ is the equation of the line.}$$

EXAMPLE 14–12 Derive the equation of the line shown in Figure 14–16 that has $(4, 2)$ and $(-2, -3)$ as coordinates of points on the line.

FIGURE 14–16 Graph for Example 14–12.

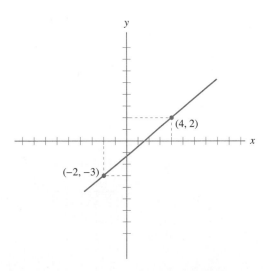

Solution   Determine the slope using Equation 14–2:

$$m = \frac{2 - (-3)}{4 - (-2)} = \frac{5}{6}$$

Determine the y-intercept, b, by substituting the coordinates of one point $(4, 2)$ of the curve of Figure 14–16 along with the slope into the general equation $y = mx + b$.

Thus,

$$2 = \tfrac{5}{6}(4) + b$$
$$12 = 20 + 6b$$
$$6b = -8$$
$$b = \frac{-8}{6} = \frac{-4}{3}$$

Substitute the values for m and b into $y = mx + b$:

$$y = \tfrac{5}{6}x - \tfrac{4}{3}$$

∴   The equation of the line is $y = \tfrac{5}{6}x - \tfrac{4}{3}$.

EXERCISE 14–4

Use Figure 14–17 and derive the equations of the following.

1. The line connecting points A and B
2. The line connecting points C and D
3. The line connecting points E and F

FIGURE 14–17   Graph for Exercise 14–4, problems 1–3.

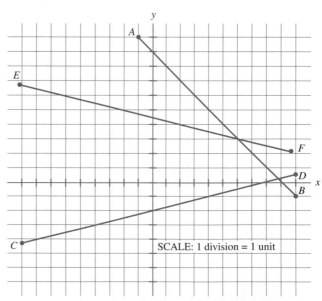

CHAPTER 14   Graphs and Graphing Techniques

Derive the equation of the straight line determined by the following points.

4. (0, 13) and (8, −3)   **5.** (−10, −6) and (10, 2)

6. (0, 6) and (12, 0)   **7.** (−2, 9) and (8, −1)

8. (2, 4) and (−4, −2)   **9.** (−5, 3) and (6, −3)

10. (9, −4) and (0, 0)   **11.** (−4, 5) and (4, 5)

14–5   GRAPHING EMPIRICAL DATA

Data that are gathered by making measurements of circuit conditions are called **empirical data.** As you know, measurements made with test equipment are *not perfect;* instead, a measurement has a range of possible values that depends upon the accuracy statement of the test equipment. Table 14–3 is a table of ordered pairs of measurements for the circuit of Figure 14–18. Notice that each measurement has been given an uncertainty in the form of an *absolute error.*

TABLE 14–3   Empirical Data Taken from the Circuit of Figure 14–18

V(volts)	I(mA)	V(volts)	I(mA)
0.0	0.0	6.0 ± 0.2	75.9 ± 5
1.0 ± 0.2	12.0 ± 5	7.0 ± 0.2	82.0 ± 5
2.0 ± 0.2	25.0 ± 5	8.0 ± 0.2	98.0 ± 5
3.0 ± 0.2	40.0 ± 5	9.0 ± 0.2	105.0 ± 5
4.0 ± 0.2	46.0 ± 5	10.0 ± 0.2	118.0 ± 5
5.0 ± 0.2	60.0 ± 5		

FIGURE 14–18   Resistive load under test.

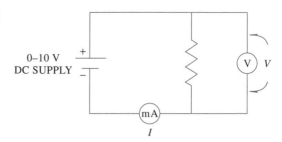

The graph of the data in Table 14–3 is shown in Figure 14–19, with shaded rectangles indicating regions of uncertainty about each point. How do we draw a *smooth curve* through these data points? From the circuit in Figure 14–18 and our understanding of Ohm's law, we expect the *curve* to be a straight line passing through the origin. But which straight line? One that has about as many points above as below and as near

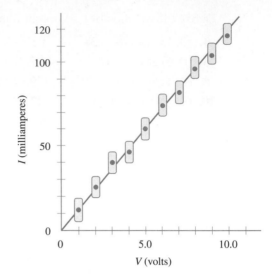

FIGURE 14–19 Graph of the data of Table 14–3 with the uncertainties shown as the shaded area.

all the points as possible. Getting the line as near as possible to all the points is called **averaging.**

When constructing the *smooth* curve, it is important that the curve be *averaged* through the plotted points. This may result in many of the plotted points not lying on the curve, as seen in Figure 14–19. However, the curve does pass through the region of uncertainties of each point. It is because of the uncertainty of the actual points that the smooth curve is averaged through the plotted points.

Notice that the graph in Figure 14–19 shows only the first quadrant. This is because the ordered pairs were all in the first quadrant. In this case, it was not necessary to show any of the other quadrants. In some graphs where the range is very small but the values are large, even the origin is not included! Figure 14–20 is an example of such a case.

FIGURE 14–20 (a) Table of ordered pairs that have a small range and large value; (b) graph of the points on a modified axis. Notice that the origin is not shown.

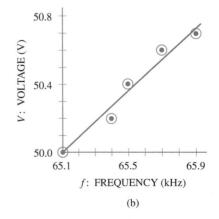

f (kHz)	V (volts)
65.1	50.0
65.4	50.2
65.5	50.4
65.7	50.6
65.9	50.7

(a)

(b)

PROCEDURE 14–1.   Graphing Empirical Data

To graph measured data (empirical data):

1. Make a table of ordered pairs.
2. Determine the range of each variable. From this, decide on the scale for each axis.
3. Construct and label each axis with the scale and units of the variables.
4. Carefully plot the points on the graph. Draw a small circle around each point.
5. Draw a *smooth* curve *averaging* the plotted points. Use as simple a curve as possible; usually a straight line will work.

EXAMPLE 14–13   Using the steps of the procedure, graph the following empirical data:

R	0.5	0.75	1.0	2.0	3.0	4.0
I	10	7.0	5.2	3.0	1.8	1.5

R	5.0	6.0	7.0	8.0	9.0	10
I	1.2	1.0	0.9	0.7	0.5	0.6

Solution   Use the steps of Procedure 14–1.

Step 1:   Let R be the abscissa and I the ordinate.

Step 2:   Select the range of 0 to 10 for each coordinate.

Step 3:   Construct and label each axis with the scale and units of the variables. See Figure 14–21.

Step 4:   Plot the points. See Figure 14–21.

Step 5:   Draw a smooth curve averaging the plotted points. See Figure 14–21.

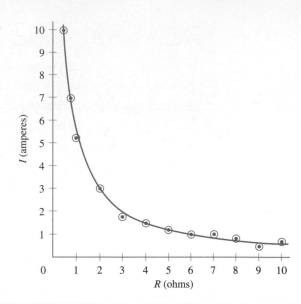

FIGURE 14–21 Graph of the data for Example 14–13.

EXERCISE 14–5

Graph the following ordered pairs of data. Use Procedure 14–1.

1.
E (volts)	0	1	2	3	4	5	6	7	8
I (milliamps)	10	8	7	7	6	4	3	2	0

2.
E (volts)	0	1	2	3	4	5	6	7	8	9	10
I (amps)	0	1	1	2	2	2	3	4	4	4	5

3.
ω (rev/min)	2000	2200	2400	2600	2800	3000
P (watts)	75 000	155 000	185 000	190 000	160 000	75 000

4.
V_F (volts)	0	0.20	0.35	0.50	0.55	0.60	0.64	0.65	0.70	0.75
I_F (milliamps)	0	5	10	20	30	40	50	60	70	100

5.
R (ohms)	0.5	0.75	1.0	2.0	5.0	10.0
I (amps)	3.0	1.9	1.4	0.75	0.30	0.15

6.

V (volts)	0	0.4	1.5	2.0	2.3	2.8	3.3	3.5	3.9	4.2	4.4	4.5	4.6
I (amps)	0	0.2	0.8	1.0	1.2	1.4	1.8	1.8	2.4	2.6	3.0	3.6	3.9

SELECTED TERMS

axis   A number line used as a basis for a coordinate system. The horizontal axis is called the **x-axis,** and the vertical axis is called the **y-axis.**

empirical data   Numerical results from measurements, as opposed to results from calculations.

graph   A visual representation of the relationship between two variables; also the set of coordinates of the points that make up that visual representation.

intercept   When a graph crosses an axis, it is said to intersect the axis. The point at which it intersects the x-axis is called the **x-intercept.** The point at which it intersects the y-axis is called the **y-intercept.**

intersect   Two lines or curves *intersect* where they have a point in common.

roots   The abscissas where a graph intersects the x-axis.

slope   The amount of change produced in the ordinate by a unit change of the abscissa.

15

Applying Graphs to Electronic Concepts

15–1   Graphic Estimation of Static Parameters

15–2   Graphic Estimation of Dynamic Parameters

15–3   Graphic Analysis of Linear Circuits

15–4   Graphic Analysis of Nonlinear Circuits

PERFORMANCE OBJECTIVES

- Approximate static electrical parameters from a graph.

- Estimate dynamic electrical parameters from a graph.

- Determine circuit values from both linear and non-linear circuits by employing graphic analysis.

Cattle grazing in a field of electricity-generating windmills. (Courtesy of U.S. Department of Agriculture)

Because graphs play an important role in describing the behavior of electronic devices, it is important that you have experience in reading and interpreting graphs of technical information. This chapter will explore the use of graphs in determining device parameters and circuit conditions.

15–1   GRAPHIC ESTIMATION OF STATIC PARAMETERS

Before working through the following examples, you must first have an understanding of what is meant by a **static parameter.** Static means stationary, or not moving. So a static parameter is a stationary parameter. In a graph, the static parameter may be the dependent or the independent variable. The value of a static parameter can be read directly from the graph.

EXAMPLE 15–1   A 2-W, carbon-film resistor is operated in an ambient atmosphere of 100°C. What is its *safe power dissipation* at this temperature? Use the temperature derating curve of Figure 15–1.

FIGURE 15–1   Percent power-temperature derating curve for determining the safe power dissipation for a carbon-film resistor operating above 50°C ambient.

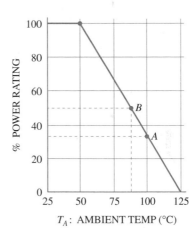

Solution   Enter the graph at a T_A of 100°C. Project vertically upward, strike the derating curve at point A, and then project horizontally across the vertical axis. Read the vertical axis as $\approx$35% of the power rating.
Compute the safe power dissipation of the 2-W resistor:

$$\therefore \quad P_D = 2 \times 0.35 = 0.7 \text{ W}$$

The power derating in the preceding example is a static parameter. The derating curve of Figure 15–1 may be used to compute circuit parameters for a given size and value of resistor, as demonstrated by the following example.

EXAMPLE 15–2   Determine the amount of current that can safely flow through a 2-W, 100-Ω resistor operating at an ambient temperature of 88°C.

Solution   Read the percent power rating from Figure 15–1 (point *B*):

percent power rating = 50%

Compute the safe power dissipation of the 2-W resistor:

$$P_D = 2 \times 0.5 = 1 \text{ W}$$

Compute the current through the 100-Ω resistor:

$$P = I^2R$$
$$I = \sqrt{P/R}$$
$$I = \sqrt{1/100}$$
$$\therefore \qquad I = 0.1 \text{ A}$$

The graph of Figure 15–2 relates voltage drop to the resistance of a 75-W incandescent lamp. Without the curve of Figure 15–2 to assist us, we would be unable to determine the resistance of the filament for a given voltage. This is due to the nonlinear resistance characteristic of the lamp filament.

FIGURE 15–2   Voltage-resistance characteristic of a 75-W incandescent lamp.

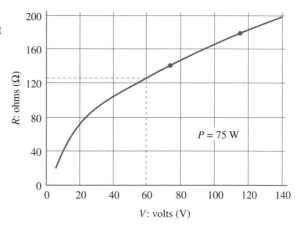

EXAMPLE 15–3   (a) Determine the approximate resistance of a 75-W incandescent lamp when 60 V is dropped across it.
(b) Compute the current passing through the lamp.

Solution   (a) Use the voltage-resistance characteristic of Figure 15–2 to determine the resistance:

$$\therefore \qquad R \approx 125 \text{ Ω}$$

(b) Compute the current through the lamp:

$$I = E/R$$
$$I = 60/125$$
$$\therefore \qquad I = 0.48 \text{ A}$$

Manufacturers of electronic devices may publish a series of curves (called a **family of curves**) on the same axis in order to show the effect of a parameter on the operation of the device. Figure 15–3 is such a graph showing how temperature affects the voltage drop across the device.

FIGURE 15–3 *Family* of forward diode characteristics showing the effect of heat on the parameters of the diode.

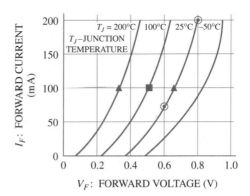

Example 15–4   From the graphs of Figure 15–3, determine:

(a) The voltage drop (V_F) for a current (I_F) of 100 mA at temperatures of 25°C and 200°C

(b) The internal temperature of the device (T_J) when a V_F of 0.5 V is measured at 100 mA I_F

Solution   **(a)** From Figure 15–3 for $T_J = 25°C$:

$$\therefore \qquad V_F = 0.65 \text{ V for an } I_F \text{ of 100 mA at 25°C}$$

From Figure 15–3 for $T_J = 200°C$:

$$\therefore \qquad V_F = 0.35 \text{ V for an } I_F \text{ of 100 mA at 200°C}$$

(b) From Figure 15–3:

$$\therefore \qquad T_J = 100°C \text{ for } V_F = 0.5 \text{ V and } I_F = 100 \text{ mA}$$

The family of curves of Figure 15–4 indicates the effect that the variation of the collector current has on the performance of a family of transistors. In this graph, the transistor's relative dc current gain (h_{FE}), noted on the vertical axis, changes with collector current (I_C).

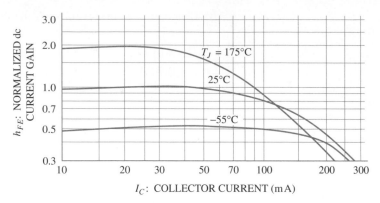

FIGURE 15–4 Normalized dc current gain for a family of small signal transistors.

Because the value of h_{FE} (current gain) varies widely from transistor to transistor type within the family, it is not possible to express h_{FE} as an absolute value on the scale of the graph. Instead, a ***normalized*** per unit multiplier of the actual dc current gain is created by dividing values of h_{FE} obtained at various temperatures and collector currents by a reference h_{FE} (I_C = 20 mA and T_J = 25°C). The current gain, corrected for variations in collector current, is obtained when the known current gain is multiplied by the normalizing factor obtained from the family of curves.

EXAMPLE 15–5 Determine the dc current gain of a transistor with a known h_{FE} of 85 when it is operating with 150 mA of collector current (I_C = 150 mA) at a junction temperature of 25°C (T_J = 25°C).

Solution From Figure 15–4 the approximate per unit multiplier is:

$$\approx 0.6$$

The current gain of the transistor, corrected for variations in collector current, is the product of 0.6 and 85.

$$\therefore \quad h_{FE} = 0.6 \times 85 \approx 50$$

The family of graphs of Figure 15–5 indicates the effect that lead length has on the amount of power that may be dissipated by the semiconductor device. This curve is helpful in determining how long a lead may be to dissipate a specific power when the lead temperature is known.

FIGURE 15–5 Temperature-power derating curve for a solid-state device showing the effect of lead length on the parameters of the device.

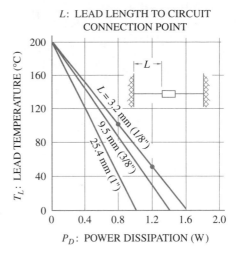

L: LEAD LENGTH TO CIRCUIT CONNECTION POINT

T_L: LEAD TEMPERATURE (°C)

P_D: POWER DISSIPATION (W)

EXAMPLE 15–6   Determine the longest lead length that allows the device to dissipate 1 W when the lead temperature is 60°C.

Solution   From Figure 15–5:

∴ $L = 9.5$ mm for $P_D = 1.0$ W and $T_L = 60°C$

EXERCISE 15–1

1. Using Figure 15–1, determine the *safe power* dissipation of a $\frac{1}{2}$-W carbon-film resistor operating at an ambient temperature of 75°C.

2. Repeat problem 1 for a 2-W carbon-film resistor operating at an ambient temperature of 113°C.

3. The 75-W incandescent lamp of Figure 15–2 is operated from a 40-V source; determine the current flowing in the lamp.

4. Repeat problem 3 for a source voltage of 120 V.

5. The diode of Figure 15–3 has a forward voltage drop (V_F) of 0.7 V when a forward current of 50 mA is flowing. Determine the junction temperature (T_J).

6. Determine the dc current gain of a transistor, represented by Figure 15–4, that is operating with a collector current of 100 mA at 25°C when the manufacturer specifies the current gain (h_{FE}) at 150.

7. Repeat problem 6 for an operating temperature of −55°C.

8. If $h_{FE} = I_C/I_B$, determine the base current (I_B) of the transistor in problem 6 when the collector current is 20 mA and the junction temperature is 175°C.

9. It is desired to dissipate 1.2 W with the device of Figure 15–5 at a maximum lead temperature of 50°C. From the three curves pictured in Figure 15–5, select the most appropriate lead length.

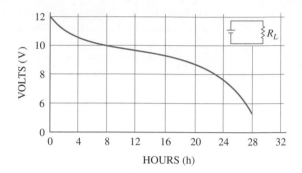

FIGURE 15–6 Service-life curve of a 12-V battery for a load resistance (R_L) of 1000 Ω.

10. Repeat problem 6 for a lead temperature of 120°C and a power dissipation of 0.5 W.

11. The semiconductor of Figure 15–5 is connected to the circuit with a lead length of 10 mm. When the device is operating, the lead temperature was measured at 100°C. Determine the maximum current through the device if the device has a resistance of 775 Ω.

12. The battery of Figure 15–6 is used to power an AM receiver having an equivalent resistance of 1000 Ω. Using the information from the curve of Figure 15–6 and Ohm's law, determine the current through the receiver after 18 h of operation.

13. The receiver of problem 12 is operated for the same amount of time each day for 7 days. The receiver current was measured at the end of the seventh day and was found to be 6.5 mA. Determine how long the receiver was operated each day. Use Figure 15–6.

15–2 GRAPHIC ESTIMATION OF DYNAMIC PARAMETERS

In this section we will learn to determine various **dynamic device parameters** by taking the difference between the coordinates of two points of a curve. We will use special notation to indicate that a difference of two coordinates has been taken. The difference between two ordinates (y-coordinates) is noted as Δy, while the difference between two abscissas (x-coordinates) is noted as Δx.

Approximation of Semiconductor Parameters

Semiconductor parameters are often noted as a small difference or *interval* in one parameter over an interval in another parameter, while a third parameter is held constant. For example, the small-signal, common-emitter, forward-current transfer ratio of a transistor, h_{fe} (also called β), is given by Equation 15–1.

$$h_{fe} = \frac{\Delta i_c}{\Delta i_b}\bigg|_{v_{ce}} \qquad (15\text{–}1)$$

In Equation 15–1, the vertical bar with v_{ce} written to the right of it indicates that v_{ce} is held constant while the intervals Δi_c and Δi_b are selected from the curve.

EXAMPLE 15–7   Determine the forward-current transfer ratio (h_{fe}) of the transistor having an i_c–i_b curve as shown in Figure 15–7 when $v_{ce} = 5$ V.

FIGURE 15–7   Transistor i_c–i_b transfer characteristic curve for Example 15–7.

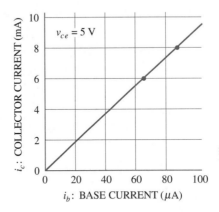

Solution   First, select an interval from the i_c–i_b curve as shown in Figure 15–8. Then, construct a right triangle through the selected points by drawing a horizontal line through point A and a vertical line through point B as illustrated in Figure 15–8.

FIGURE 15–8   Graphic estimation of h_{fe} from a transistor $i_c - i_b$ transfer curve, where
$\Delta i_b = i_{bB} - i_{bA}$
$\Delta i_c = i_{cB} - i_{cA}$

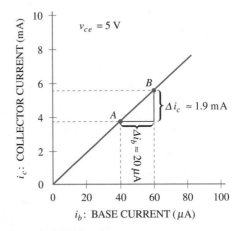

Determine the interval of collector current, Δi_c:

$$\Delta i_c \approx 5.5 - 3.6 \approx 1.9 \text{ mA}$$

Determine the interval of base current, Δi_b:

$$\Delta i_b \approx 60 - 40 \approx 20 \ \mu\text{A}$$

Compute h_{fe} when $v_{ce} = 5$ V:

$$h_{fe} = \frac{\Delta i_c}{\Delta i_b}$$

$$h_{fe} = \frac{1.9 \times 10^{-3}}{20 \times 10^{-6}}$$

$\therefore$ $h_{fe} = 95$

Observation h_{fe} is the slope of the line in Figure 15–7.

EXAMPLE 15–8   Determine the output admittance (h_{oe}) of the transistor having the characteristic curve of Figure 15–9 for $i_b = 60~\mu$A and v_{ce} between 5.0 and 9.5 V, where:

$$h_{oe} = \left. \frac{\Delta i_c}{\Delta v_{ce}} \right|_{i_b} \qquad \text{(siemens)} \qquad \text{(15–2)}$$

FIGURE 15–9   Graphic estimation of h_{oe} from a transistor collector characteristic curve, where
$\Delta v_{ce} = v_B - v_A$
$\Delta i_c = i_B - i_A$

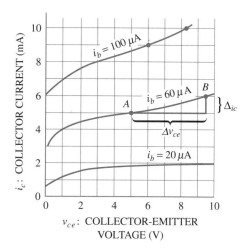

Solution   Hold i_b constant at 60 μA by selecting an interval from the 60-μA curve of Figure 15–9.

Construct a horizontal line through point A and a vertical line through point B. See Figure 15–9.

Determine the vertical interval Δi_c by counting the squares between point B and the horizontal line. Each square equals 1.0 mA.

$$\Delta i_c \approx 6 - 5 \approx 1 \text{ mA}$$

Determine the horizontal interval Δv_{ce} by counting the squares between point A and the vertical line. Each square equals 1.0 V.

$$\Delta v_{ce} \approx 9.5 - 5.0 \approx 4.5 \text{ V}$$

Compute h_{oe} when $i_b = 60 \ \mu A$:

$$h_{oe} = \frac{\Delta i_c}{\Delta v_{ce}}$$

$$h_{oe} = \frac{1 \times 10^{-3}}{4.5}$$

$$\therefore \qquad h_{oe} = 220 \ \mu S$$

In the previous example, h_{oe} is the **local slope** of the curve between the points A and B on the $i_b = 60 \ \mu A$ curve in Figure 15–9. In general when a graph is created, more points are needed where the graph curves and fewer points are needed where the graph is straight. Similarly, to estimate the local slope a smaller Δx is used where the graph curves and a larger Δx is used where the graph is straight. The following guidelines will aid in graphically estimating dynamic parameters.

GUIDELINES 15–1. Graphic Estimation of Dynamic Parameters

1. For graphs with multiple curves, choose the curve that is appropriate for your application.
2. Determine the region of interest for your application and select the operating value (point) of the parameter on the abscissa.
3. Choose two points equally spaced about the operating point.
   a. If the graph is highly curved, make the points closely spaced.
   b. If the graph is very straight, make the points farther apart and easy to read.
4. Calculate the dynamic parameter.

EXAMPLE 15–9

Estimate the *transadmittance* (y_{fs}) of the field-effect transistor (FET) having the transfer characteristic curve of Figure 15–10 for $v_{ds} = 15$ V and $v_{gs} = -3$ V, where:

$$y_{fs} = \frac{\Delta i_d}{\Delta v_{gs}} \bigg|_{v_{ds}} \qquad \text{(siemens)} \qquad \text{(15–3)}$$

Solution

For ease of reading select an interval of -4 to -2 from the 15-V curve of Figure 15–10. Determine Δi_d, the vertical interval:

$$\Delta i_d \approx 5 - 2 \approx 3 \text{ mA}$$

Determine Δv_{gs}, the horizontal interval:

$$\Delta v_{gs} \approx -2 - (-4) \approx 2 \text{ V}$$

FIGURE 15–10 Graphic estimation of y_{fs} from a FET transfer characteristic curve, where
$$\Delta v_{gs} = v_{gsB} - v_{gsA}$$
$$\Delta i_d = i_{dB} - i_{dA}$$

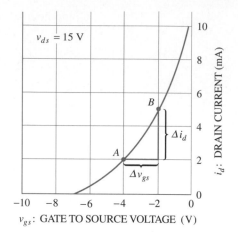

Compute y_{fs} when $v_{ds} = 15$ V:

$$y_{fs} = \frac{\Delta i_d}{\Delta v_{gs}}$$

$$y_{fs} \approx \frac{3 \times 10^{-3}}{2}$$

$\therefore$ $y_{fs} \approx 1.5$ mS

EXAMPLE 15–10   Determine the dynamic resistance (r_{ac}) of the diode having the forward characteristic curve of Figure 15–11. Choose points around $v_f = 0.75$ V:

$$r_{ac} = \frac{\Delta v_f}{\Delta i_f} \quad \text{(ohms)} \tag{15–4}$$

FIGURE 15–11 Graphic estimation of r_{ac} from a diode forward characteristic curve, where
$$\Delta v_f = v_{fB} - v_{fA}$$
$$\Delta i = i_{fB} - i_{fA}$$

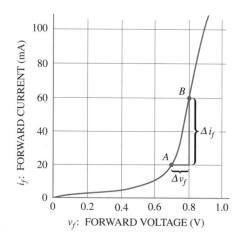

Solution   Select the interval of 0.7 to 0.8 V from the straight portion of the curve of Figure 15–11. Determine Δv_f, the horizontal interval:

$$\Delta v_f \approx 0.8 - 0.7 \approx 0.1 \text{ V}$$

Determine Δi_f, the vertical interval:

$$\Delta i_f \approx 60 - 20 \approx 40 \text{ mA}$$

Compute r_{ac}:

$$r_{ac} = \frac{\Delta v_f}{\Delta i_f}$$

$$r_{ac} \approx \frac{0.1}{40 \times 10^{-3}}$$

$$\therefore \quad r_{ac} \approx 2.5 \ \Omega$$

EXERCISE 15–2

1. Approximate the dynamic current (i) from the interval of the curve of Figure 15–2 delineated by the two noted points, where

$$i = \frac{\Delta V}{\Delta R} \quad \text{(amperes)}$$

2. Approximate the dynamic resistance (r_{ac}) from the region of the curve of Figure 15–3 delineated by the two circled points, where

$$r_{ac} = \left. \frac{\Delta V_F}{\Delta I_F} \right|_{25°C} \quad \text{(ohms)}$$

3. Approximate the thermal resistance (θ_{lead}) of the lead wire from the region of the curve of Figure 15–5 delineated by the two noted points, where

$$\theta_{\text{lead}} = \left. \frac{\Delta T_L}{\Delta P_D} \right|_{3.2 \text{ mm}} \quad °\text{C/W}$$

4. Approximate the h_{fe} or β of a transistor from the region of the curve of Figure 15–7 delineated by the two noted points, where

$$h_{fe} = \left. \frac{\Delta i_c}{\Delta i_b} \right|_{5 \text{ V}}$$

5. Approximate the h_{oe} of a transistor from the region of the curve of Figure 15–9 delineated by the two noted points, where

$$h_{oe} = \left. \frac{\Delta i_c}{\Delta v_{ce}} \right|_{100\,\mu A} \qquad \text{(siemens)}$$

6. Approximate the Z_{zt} of the zener diode from the region of the curve of Figure 15–12 delineated by the two noted points, where

$$Z_{zt} = \frac{\Delta V_z}{\Delta i_z} \qquad \text{(ohms)}$$

FIGURE 15–12 Zener diode reverse characteristic curve for Exercise 15–2, problem 6.

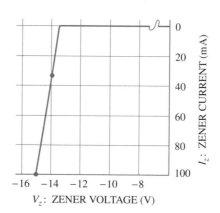

15–3 GRAPHIC ANALYSIS OF LINEAR CIRCUITS

The solution of linear circuits is usually carried out with an algebraic method by writing systems of linear equations and solving them for their common solution (simultaneous equations). This technique will be developed in Chapter 16 and applied to electrical circuits in Chapter 17. This section will give you an opportunity to *see* the solution to a system of linear equations.

When two lines are graphed on the same coordinate system, there are three possible conditions, as shown in Figure 15–13. The three possible conditions are:

- The two lines may have the same slope and not cross because they are **parallel** to each other [Figure 15–13(a)]. This results in *no* points in common and *no* solution to the system of linear equations. This lack of solution results when the equations that were written did not correctly represent the conditions in the electrical circuit.

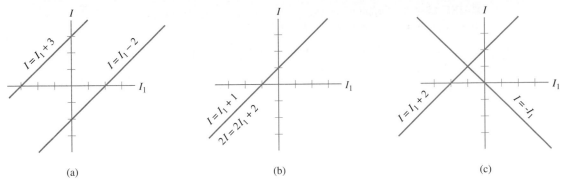

FIGURE 15–13 Graphs of two linear equations may result in one of three conditions: (a) parallel; (b) coincident; (c) intersect.

- The two lines may be *coincident* with each other [Figure 15–13(b)]. This results in *all* points on both lines being in common and *all* points on both lines being solutions to the system of linear equations. This condition is the result of writing two equations that used the same circuit conditions twice and are really only one equation.

- The two lines *intersect* at one point [Figure 15–13(c)]. This results in *one* point in common and *one* solution to the system of linear equations. When well-formulated circuit conditions are represented by two or more equations, then a solution will result.

EXAMPLE 15–11 Graphically solve the following system of *mesh* equations for their intersection.

$$I_1 - 2I_2 = 3$$
$$2I_1 + 3I_2 = -8$$

Solution Write each equation in the form of $y = mx + b$ and then graph as shown in Figure 15–14.

$$I_1 = 2I_2 + 3 \text{ (amperes)}$$
$$I_1 = -\tfrac{3}{2}I_2 - 4 \text{ (amperes)}$$

Read the coordinate of the intersection as (*abscissa, ordinate*).

$$(-2, -1) \quad \text{that is: } I_2 = -2 \text{ A and } I_1 = -1 \text{ A}$$

Check Substitute -2 for I_2 in the first equation and evaluate I_1:

$$I_1 = 2(-2) + 3$$
$$I_1 = -4 + 3$$
$$I_1 = -1$$

This is the value read from the graph. Now check the second equation.

FIGURE 15–14 Graph of the system of equations for Example 15–11.

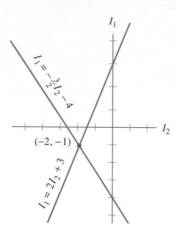

$$I_1 = -\tfrac{3}{2}(-2) - 4$$
$$I_1 = +3 - 4$$
$$I_1 = -1$$

This is also the value read from the graph.

∴ $I_2 = -2$ A and $I_1 = -1$ A is the solution of the system of equations.

Observation   *Mesh analysis* is covered in Section 17–2. *Note:* A negative current indicates that the assumed current direction is opposite to the actual direction.

EXERCISE 15–3

Graph each system of linear equations and determine the solutions of the system.

1. $\begin{cases} y = x \\ y = -x + 2 \end{cases}$ **2.** $\begin{cases} 2x + y = 9 \\ x + 2y = 0 \end{cases}$

3. $\begin{cases} y = 2x \\ y = x - 3 \end{cases}$ **4.** $\begin{cases} x + 3y = 3 \\ x + y = 5 \end{cases}$

5. $\begin{cases} x + y = 4 \\ x - y = 2 \end{cases}$ **6.** $\begin{cases} y = x - 3 \\ -2x - 2y = 6 \end{cases}$

7. $\begin{cases} x + y = 5 \\ x - y = -1 \end{cases}$ **8.** $\begin{cases} x = y + 1 \\ y = 3x + 2 \end{cases}$

9. $\begin{cases} x + y = 0 \\ x - 2y = 6 \end{cases}$ **10.** $\begin{cases} x + 3y = 11 \\ x - 3y = 5 \end{cases}$

15–4 GRAPHIC ANALYSIS OF NONLINEAR CIRCUITS

Many of the devices used in electronics have nonlinear characteristics. These devices are often used with devices having linear characteristics to form a nonlinear series circuit. Figure 15–15, which consists of a resistor in series with a diode, is an example of a nonlinear series circuit. The rules that govern the nonlinear series circuit are the same as those that govern the linear series circuit. They are as follows:

- The total circuit current (I) is the same throughout the circuit.
- The sum of the voltage drops (ΣV) equals the voltage source (E).

The solution of the nonlinear circuit is better done by graphic methods rather than by the methods of algebra. The graphic method is usually faster, easier, and clearer than the methods of algebra.

The graphic solution of a nonlinear series circuit is done by constructing both the curve of the linear device and the curve of the nonlinear device on the same set of coordinates. The point of intersection of the two curves has the circuit current as the ordinate, and the voltage drop across the nonlinear device as the abscissa.

Figures 15–16 and 15–17 are used to develop the mathematical and graphical method of solving a nonlinear series circuit. The nonlinear series circuit of Figure 15–16 has a voltage equation of

$$E = V_F + V_{RL} \qquad (15\text{–}5)$$

Substituting IR_L for V_{RL}:

$$E = V_F + IR_L \qquad (15\text{–}6)$$

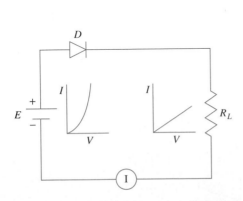

FIGURE 15–15 Nonlinear series circuit with the linear characteristic of the resistor, R_L, and the non-linear characteristic of the diode, D, shown.

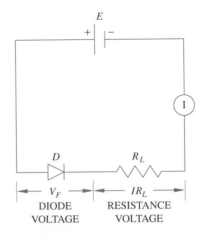

FIGURE 15–16 Simple nonlinear series circuit.

FIGURE 15–17 Graphical solution of the nonlinear circuit of Figure 15–16.

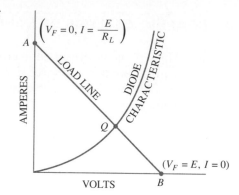

When $V_F = 0$ V, Equation 15–6 becomes $E = 0 + IR_L$ and $I = E/R_L$. The conditions ($V_F = 0$ and $I = E/R_L$) form an ordered pair, which is plotted as point A of Figure 15–17.

When $I = 0$ A, Equation 15–6 becomes $E = V_F + 0$ and $E = V_F$. The conditions ($V_F = E$ and $I = 0$) form an ordered pair, which is plotted as point B of Figure 15–17.

The **load line** (straight line) connecting points A and B intersects the diode characteristic curve at Q. The coordinates of point Q (the operating point) are the operating conditions of the nonlinear circuit of Figure 15–16 and are detailed in Figure 15–18.

The load line of Figure 15–18 has the general equation of $y = mx + b$. In this equation, y is the circuit current *(I)*, x is the voltage across the nonlinear device *(V_F)*, and the slope *(m)* and the y-intercept *(b)* are determined by solving Equation 15–6 for I.

$$I = \frac{E - V_F}{R_L} \qquad \qquad (15\text{–}7)$$

FIGURE 15–18 Graphical solution of the nonlinear circuit of Figure 15–16. The abscissa of the Q point is the voltage across the nonlinear device; the ordinate is the circuit current.

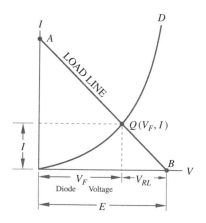

Place each term in the right member over the common denominator (R_L):

$$I = \frac{E}{R_L} - \frac{V_F}{R_L}$$

Then rearrange the terms to form Equation 15–8:

$$I = \frac{-1}{R_L} V_F + \frac{E}{R_L} \qquad (15\text{–}8)$$

Equation 15–8 is the specific equation of the load line where the slope $m = -1/R_L$ and the y-intercept $b = E/R_L$.

The procedure for the graphic solution of a nonlinear series circuit is summarized as Rule 15–1.

RULE 15–1. *Graphical Solution*

To solve a simple nonlinear series circuit:

1. Construct the *load line* on the same coordinates as the curve of the nonlinear device by:

   (a) Establishing two points of the load line:
   Point 1: V of the nonlinear device is zero.

   $$V = 0, \qquad I = E/R_L$$

   Point 2: I of the circuit is zero.

   $$E = V, \qquad I = 0$$

   where V = voltage drop of the nonlinear device
   E = applied voltage of the circuit
   I = circuit current
   R_L = resistance of the linear device
   (b) Constructing the load line as a straight line through the two points.

2. Read the circuit operating condition as the abscissa and the ordinate of the point of intersection of the linear and nonlinear curves.

EXAMPLE 15–12 In the circuit of Figure 15–19, determine the following using the pictured diode characteristics.

(a) The current in the circuit *(I)*
(b) The voltage across the diode
(c) The voltage across the resistance

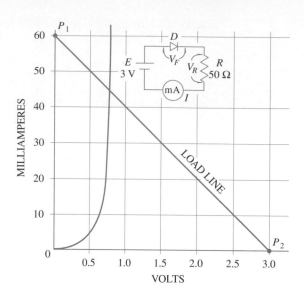

FIGURE 15–19 Circuit and characteristic curve for Example 15–12.

Solution Use the steps of Rule 15–1.

Step 1(a): To construct the load line, use:

Point 1: $V_F = 0$ V, $I = E/R = 3/50 = 60$ mA
Point 2: $I = 0$ mA, $E = V_F = 3$ V

Step 1(b): Plot the two points:

$P_1(0$ V, 60 mA) and $P_2(3$ V, 0 mA)

Draw the load line through these two points. See Figure 15–19.

Step 2: Read the operating condition from the intersection of the two curves:

∴ (a) $I \approx 45$ mA
(b) $V_F \approx 0.8$ V
(c) $V_R = E - V_F = 3 - 0.8 \approx 2.2$ V

EXERCISE 15–4

Using the information given in Figure 15–20:

1. Determine the circuit current *(I)* and the voltage across R when $R = 75\ \Omega$ and $E = 3.0$ V.

2. Determine the circuit current *(I)* and the voltage across R when $R = 40\ \Omega$ and $E = 4.0$ V.

3. Determine the circuit current *(I)* and the voltage across R when $R = 50\ \Omega$ and $E = 6.0$ V.

FIGURE 15–20 Circuit and light-emitting diode (LED) forward characteristic curve for Exercise 15–4, problems 1–3.

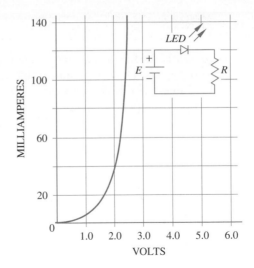

Using the information given in Figure 15–21:

4. Determine the circuit current *(I)*, the voltage across the lamp (nonlinear device), and the voltage across the resistance when $R = 100\ \Omega$ and $E = 100$ V.

5. Determine the circuit current *(I)*, the voltage across the lamp, and the voltage across the resistance when $R = 80\ \Omega$ and $E = 60$ V.

6. Determine the circuit current *(I)*, the voltage across the lamp, and the voltage across the resistance when $R = 250\ \Omega$ and $E = 150$ V.

FIGURE 15–21 Circuit and lamp characteristic curve for Exercise 15–4, problems 4–6.

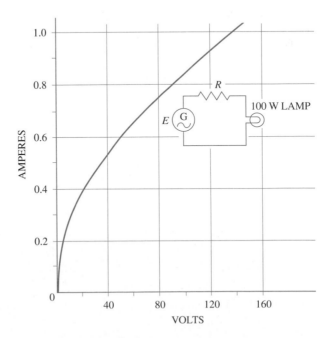

dynamic device parameter   A nonlinear device parameter defined as the local slope of a second device parameter plotted against a third device parameter.

load line   A line drawn on the characteristic curve of a nonlinear device to locate the operating point of the circuit.

static parameter   A parameter that does not change with time.

SECTION CHALLENGE

WEB CHALLENGE FOR CHAPTERS 13, 14, AND 15

To evaluate your comprehension of Chapters 13, 14, and 15, log on to **www.prenhall. com/harter** and take the online True/False and Multiple Choice assessments for each of the chapters.

SECTION CHALLENGE FOR CHAPTERS 13, 14, AND 15*

Your challenge is to determine the value of the gain transfer function *(K)* of a variable flow, pneumatically operated *control valve assembly* like the one pictured in Figure C–5. The gain *(K)* of a control element, in an automatic control system, may be determined from the slope of the curve of the graph of the input/output transfer function of the element (Figure C–6) by selecting two points on the curve (r_1, c_1) and (r_2, c_2) and then taking an interval of the output parameter $(\Delta c = c_1 - c_2)$ and dividing it by an interval of the input parameter $(\Delta r = r_1 - r_2)$. That is:

$$\text{Slope} = m = \text{Gain} = K = \frac{\Delta c}{\Delta r} = \frac{c_1 - c_2}{r_1 - r_2} \left(\frac{\text{gal/min}}{\text{lbf/in}^2} \right) \qquad \textbf{(E–4)}$$

FIGURE C–5 Control valve: (a) General instrumentation symbol with a pneumatic diaphragm pilot assembly (actuator); (b) Variable-flow, pneumatically operated control valve used to vary the flow rate of the product passing through the body of the valve. (*Electromechanics: Principles, Concepts, and Devices.* Pearson Education Inc., 2003).

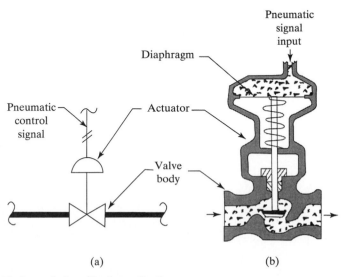

(a)   (b)

*The solution to this Section Challenge is found in Appendix C.

FIGURE C–6 Control valve input/output graph, showing the 3.0–15 lbf/in^2 air pressure that proportionally moves the valve from open to closed (0–250 gal/min). (*Electromechanics: Principles, Concepts, and Devices.* Pearson Education Inc., 2003.)

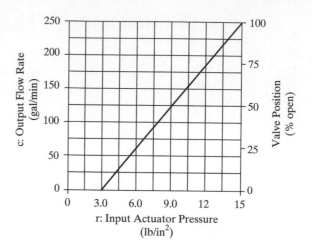

As an added challenge, derive the equation of the output flow rate, *c* (gal/min), in terms of the input pneumatic pressure, *r*, of the pneumatically operated control valve assembly. Use the graph of Figure C–6 and the general equation $y = mx + b$ where $c = y$, $K = m$, $r = x$, and $b = -62.45$ gal/min. Employ the derived equation to determine the output flow rate, *c* (gal/min), of the control valve when a pneumatic control signal, *r* (lb/in^2), is applied to the actuator of the variable-flow control valve. The applied control signals are:

A. $r = 7.5$ lb/in^2

B. $r = 5.0$ lb/in^2

C. $r = 13.0$ lb/in^2

16

Solving Systems of Linear Equations

16–1 Addition or Subtraction Method

16–2 Substitution Method

16–3 Deriving Electrical Formulas

16–4 Determinants of the Second Order

16–5 Determinants of the Third Order

PERFORMANCE OBJECTIVES

- Apply the addition or subtraction method to the solution of a set of linear equations.

- Use the substitution method to solve a system of linear equations.

- Employ the techniques of simultaneous equations to derive a formula from two or more given equations.

- Solve systems of linear equations by applying the rules and techniques of determinants.

Scientists conduct experiments in the environmental biology laboratory of the NASA Ames Research Center in California. (Courtesy of NASA)

This chapter deals with two or more linear equations called a **system of linear equations.** Specifically, we will be interested in the coordinates that satisfy each of the equations in the system *simultaneously.* Two equations that have one common solution are called **simultaneous equations.** Several methods are used to solve systems of linear equations. We will study some of these in this chapter.

In previous chapters we used $|\ |$ to indicate the absolute value of a number. In this chapter we use it to indicate the determinant of a square array of numbers.

16–1 ADDITION OR SUBTRACTION METHOD

This method is an algebraic method that involves eliminating one variable in the system by adding or subtracting equations, as shown in the following three examples.

EXAMPLE 16–1

Solve $\begin{cases} x + 3y = 11 \\ x - 3y = 5 \end{cases}$

Solution

Add the two equations to eliminate the terms containing *y:*

$$x + 3y = 11$$
$$\underline{x - 3y = 5}$$
$$2x + 0 = 16$$

Solve for *x:*

$$2x = 16$$
$$x = 8$$

Solve for *y* by substituting the value of *x* into the first equation:

$$x + 3y = 11$$
$$8 + 3y = 11$$
$$y = 1$$

Check

Substitute $x = 8$ into the second equation:

$$x - 3y = 5$$
$$8 - 3y = 5$$
$$-3y = -3$$
$$y = 1$$

This is the value computed in the solution.

$\therefore$ $x = 8$ and $y = 1$ is the solution of the system of equations.

EXAMPLE 16–2

Solve $\begin{cases} \dfrac{5x}{6} + \dfrac{y}{4} = 7 \\ \dfrac{2x}{3} - \dfrac{y}{8} = 3 \end{cases}$

Solution   Clear the equations of fractions before adding or subtracting:

M: 12

$$\frac{5x}{6} + \frac{y}{4} = 7$$

$$10x + 3y = 84$$

M: 24

$$\frac{2x}{3} - \frac{y}{8} = 3$$

$$16x - 3y = 72$$

Add the two new equations to eliminate the terms containing y and solve for x:

$$
\begin{array}{r}
10x + 3y = 84 \\
\underline{16x - 3y = 72} \\
26x + 0 = 156 \\
x = 6
\end{array}
$$

Solve for y by substituting the value 6 for x into:

$$10x + 3y = 84$$
$$60 + 3y = 84$$
$$y = 8$$

Check   Substitute $x = 6$ and solve for y in each original equation:

$$\frac{5x}{6} + \frac{y}{4} = 7$$

$$\frac{5(6)}{6} + \frac{y}{4} = 7$$

$$\frac{y}{4} = 2$$

$$y = 8$$

This is the value computed in the solution.

$$\frac{2x}{3} - \frac{y}{8} = 3$$

$$\frac{2(6)}{3} - \frac{y}{8} = 3$$

$$4 - y/8 = 3$$

$$-y/8 = -1$$

$$y = 8$$

This is the value computed in the solution.

$\therefore$ $x = 6$ and $y = 8$ is the solution of the system of equations.

EXAMPLE 16–3

Solve $\begin{cases} 6x + 5y = 2 \\ 4x - 2y = 12 \end{cases}$

Solution

Eliminate x by multiplying the first equation by 4 and the second equation by 6:

M: 4
$$6x + 5y = 2$$
$$24x + 20y = 8$$

M: 6
$$4x - 2y = 12$$
$$24x - 12y = 72$$

Subtract the two new equations to eliminate x:

$$24x + 20y = 8$$
$$\underline{-(24x - 12y = 72)}$$
$$0 + 32y = -64$$
$$y = -2$$

Substitute the value of y into the first equation:

$$6x + 5y = 2$$
$$6x + 5(-2) = 2$$
$$6x = 12$$
$$x = 2$$

Check

Substitute $y = -2$ into the second equation:

$$4x - 2y = 12$$
$$4x + 4 = 12$$
$$4x = 8$$
$$x = 2$$

This is the value computed in the solution.

$\therefore$ $\quad x = 2$ and $y = -2$ is the solution.

The techniques for solving a system of linear equations explored in the three previous examples are summarized in Guidelines 16–1.

GUIDELINES 16–1.   Solving a System of Linear Equations

For solving a system of linear equations using addition or subtraction:

1. The goal is to eliminate one of the variables.
2. Clear both equations of any fractions before proceeding.
3. Choose which variable to eliminate.

4. If necessary, form equivalent equations in the following manner. Multiply the first equation by the coefficient of the variable being eliminated in the second equation. Multiply the second equation by the coefficient of the variable being eliminated in the first equation. (Do not multiply by zero in either case!)

5. Combine the equations (by adding or subtracting) to eliminate the chosen variable.

Solve each system of equations by addition or subtraction. Use Guidelines 16–1.

1. $\begin{cases} x + y = 4 \\ x - y = 6 \end{cases}$

2. $\begin{cases} 2x - 3y = -12 \\ x - 3y = 3 \end{cases}$

3. $\begin{cases} x + y = 9 \\ -x + y = 5 \end{cases}$

4. $\begin{cases} x/3 - y = 10 \\ x/5 + 2y/5 = 2 \end{cases}$

5. $\begin{cases} 2x + y = 17 \\ 2x - y = -5 \end{cases}$

6. $\begin{cases} 2x/5 + y = -2 \\ 2x/5 - y/2 = 1 \end{cases}$

7. $\begin{cases} x + 3y = 28 \\ x - 3y = -20 \end{cases}$

8. $\begin{cases} x + y = 3 \\ 3x - 5y = 17 \end{cases}$

9. $\begin{cases} x - 5y = -5 \\ x + 3y = 3 \end{cases}$

10. $\begin{cases} x/6 + y/4 = 3/2 \\ 2x/3 - y/2 = 0 \end{cases}$

16–2 SUBSTITUTION METHOD

As with the addition-subtraction method, this technique of solving the system of equations involves eliminating one of the variables to obtain one equation with one unknown. The approach in substitution is to solve either of the equations for one variable in terms of the other. This expression is then *substituted* into the other equation, resulting in a third equation in one unknown. This principle is shown in the following examples.

EXAMPLE 16–4

Solve $\begin{cases} x + 2y = 1 \\ 2x + 5y = 5 \end{cases}$

Solution

Use substitution. Solve $x + 2y = 1$ for x:

$$x = 1 - 2y$$

Solve for y by substituting $(1 - 2y)$ for x into the second equation:

$$2x + 5y = 5$$
$$2(1 - 2y) + 5y = 5$$
$$2 - 4y + 5y = 5$$
$$y = 3$$

Solve for x by substituting $y = 3$ into $x = 1 - 2y$:

$$x = 1 - 2(3)$$
$$x = -5$$

Check   Substitute $x = -5$ into the second equation and solve for y:

$$2x + 5y = 5$$
$$2(-5) + 5y = 5$$
$$5y = 15$$
$$y = 3$$

This checks the solution.

$\therefore$ $x = -5$ and $y = 3$ is the solution.

EXAMPLE 16–5

Solve $\begin{cases} \dfrac{x}{3} - \dfrac{y}{2} = \dfrac{7}{6} \\ \dfrac{x}{3} - \dfrac{y}{6} = \dfrac{1}{6} \end{cases}$

Solution   Clear the equations of fractions:

M: 6

$$\frac{x}{3} - \frac{y}{2} = \frac{7}{6}$$
$$2x - 3y = 7$$

M: 6

$$\frac{x}{3} - \frac{y}{6} = \frac{1}{6}$$
$$2x - y = 1$$

Solve $2x - 3y = 7$ for x:

$$x = \frac{7 + 3y}{2}$$

Substitute for x in the second equation, $2x - y = 1$:

$$2\left(\frac{7 + 3y}{2}\right) - y = 1$$

Solve for y:

$$7 + 3y - y = 1$$
$$2y = -6$$
$$y = -3$$

Solve for x by substituting $y = -3$ into $x = \dfrac{7 + 3y}{2}$:

$$x = \frac{7 + 3(-3)}{2} = \frac{-2}{2}$$

$$x = -1$$

Check   Substitute $y = -3$ into each equation and solve for x. (We should get -1.)

$$x/3 - y/2 = 7/6$$
$$x/3 + 3/2 = 7/6$$
$$x/3 = -2/6$$
$$x = (-2/6)(3)$$
$$x = -1$$

$$x/3 - y/6 = 1/6$$
$$x/3 + 3/6 = 1/6$$
$$x/3 = -2/6$$
$$x = (-2/6)(3)$$
$$x = -1$$

This completes the check.

$\therefore$ $x = -1$ and $y = -3$ is the solution.

EXERCISE 16–2

Solve each system of equations by substitution.

1. $\begin{cases} x - y = 0 \\ x + y = 2 \end{cases}$

2. $\begin{cases} 3x - 4y = 13 \\ 2x + 3y = 3 \end{cases}$

3. $\begin{cases} -5y = 17 - 3x \\ x + y = 3 \end{cases}$

4. $\begin{cases} 5x + 2y = 36 \\ 8x - 3y = -54 \end{cases}$

5. $\begin{cases} 4x - 3y - 2 = 2x - 7y \\ x + 5y - 2 = y + 4 \end{cases}$

6. $\begin{cases} x + y = 0 \\ x - 2y = 6 \end{cases}$

7. $\begin{cases} x + 3y = 11 \\ x - 3y = 5 \end{cases}$

8. $\begin{cases} 2x + 3(x + y) = 15 \\ 2x - 3y = -1 \end{cases}$

9. $\begin{cases} x - 5y = -5 \\ x + 3y = 3 \end{cases}$

10. $\begin{cases} 8(y + 1) = 2x \\ 3(x - 3y) = 15 \end{cases}$

11. $\begin{cases} x/3 - y/2 = -8/3 \\ x/7 - y/3 = -39/21 \end{cases}$

12. $\begin{cases} x/2 + y/4 = 1 \\ 3x/4 + y = 1/4 \end{cases}$

13. $\begin{cases} \dfrac{x}{3} + \dfrac{y}{6} = \dfrac{1}{2} \\[2mm] \dfrac{x}{2} + \dfrac{3y}{10} = 1 \end{cases}$

14. $\begin{cases} \dfrac{x + y}{2} - \dfrac{x - y}{2} = 2 \\[2mm] \dfrac{x - y}{4} + \dfrac{x + y}{2} = 5 \end{cases}$

16–3 DERIVING ELECTRICAL FORMULAS

The techniques of simultaneous equations may be used to derive a formula from two or more known equations. The following examples demonstrate this concept.

EXAMPLE 16–6   Derive a formula for work (W) in terms of current flow (I), time (t), and voltage (E) from the equations $E = W/Q$ and $Q = It$.

Solution   Eliminate Q in $E = W/Q$ by substituting It for Q:

$$E = W/It$$

Solve for W:

$\therefore$ $W = EIt$

EXAMPLE 16–7   Derive a formula for power (P) in watts (W) from $E = IR$ and $P = IE$ in terms of voltage (E) and resistance (R).

Solution   Eliminate I in $P = IE$ by solving $E = IR$ for I and substituting:

$$I = E/R$$

Substitute E/R for I in $P = IE$:

$$P = \left(\frac{E}{R}\right)E$$

$\therefore$ $P = E^2/R$

EXAMPLE 16–8   Derive a formula for voltage division across a selected resistor in a series circuit (V_n) in terms of the source voltage (E), the selected resistor (R_n), and the total resistance (R_T) from the equations $V_n = IR_n$ and $E = IR_T$.

Solution   Eliminate I in $V_n = IR_n$ by solving $E = IR_T$ for I:

$$I = E/R_T$$

Substituting E/R_T for I in $V_n = IR_n$:

$$V_n = \left(\frac{E}{R_T}\right)R_n$$

$\therefore$ $$V_n = \frac{ER_n}{R_T}$$

---

1. Solve for E in terms of E_{av} from the equations $E = 0.707E_{mx}$ and $E_{av} = 0.637E_{mx}$.

2. Solve for P in terms of E and I from the equations $2P = E_{mx}I_{mx}$, $E = E_{mx}/\sqrt{2}$, and $I = I_{mx}/\sqrt{2}$.

3. Solve for L in terms of ω and X_L from the equations $L/T = 0.1592X_L$, $f = 1/T$, and $\omega = 6.283f$.

4. Solve for P in terms of E, I, R, and Z from the equations $P_A = P/P_F$, $P_A = EI$, and $P_F = R/Z$.

5. Solve for W in terms of L and I from the equations $W = EI/\omega$, $I = E/X_L$, and $X_L = \omega L$.

6. Solve for f^2 in terms of L, C, and the numerical constant of 2π from the equations $X_L = X_C$, $X_L = \omega L$, $X_C = 1/(\omega C)$, and $\omega = 2\pi f$.

7. Solve for Q_0 in terms of R and X_L from the equations $P_q = PQ_0$, $P = I^2R$, and $P_q = I^2X_L$.

8. Solve for BW in terms of f_0 and Q_0 from the equations $2\pi L = R/\text{BW}$, $Q_0 = \omega_0 L/R$, and $\omega_0 = 2\pi f_0$.

9. Solve for R_L in terms of X_L, Q_0, and Q'_0 from the equations $R' = \dfrac{R_L R}{R_L + R}$, $R = \omega L Q_0$, $R' = \omega L Q'_0$, and $X_L = \omega L$.

10. Solve for I_{in} in terms of N and I_{out} from the following equations: $P_{in} = E_{in}I_{in}$, $P_{out} = E_{out}I_{out}$, $N = E_{out}/E_{in}$, and $P_{in} = P_{out}$.

16–4   DETERMINANTS OF THE SECOND ORDER

Meaning of a Determinant

Systems of linear equations may be solved by applying the rules and techniques associated with **determinants.** Programmable calculators have the capacity to evaluate determinants. You should consult your owner's guide to see if your calculator can solve determinants. Because a special symbol is used for the determinant, you may find the idea of what a determinant is and how it is used a new and unusual experience.

A second-order determinant is a tool that can be used to solve a system of two linear equations. A second-order determinant operates on four numbers to produce a new number. A second-order determinant looks like $\begin{vmatrix} a_1 & b_1 \\ a_2 & b_2 \end{vmatrix}$, where the a's and b's are the four numbers on which the determinant operates. They are called the *elements* of the determinant. The numerical value of a determinant is given by Equation 16–1.

$$\begin{vmatrix} a_1 & b_1 \\ a_2 & b_2 \end{vmatrix} = a_1b_2 - a_2b_1 \qquad (16\text{–}1)$$

FIGURE 16–1 Evaluation of a second-order determinant is carried out by multiplying the elements along the diagonals (indicated by the arrows) and then taking the difference of the products.

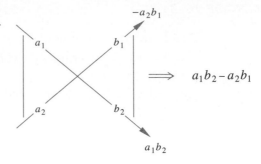

From the definition of a determinant, we see that a number results by first taking the product of the two elements in the *principal* diagonal (from top left to bottom right) as shown in Figure 16–1, and then subtracting the product of the two elements in the other diagonal.

EXAMPLE 16–9   Evaluate $\begin{vmatrix} 3 & 2 \\ -6 & 4 \end{vmatrix}$

Solution   Since

$$\begin{vmatrix} a_1 & b_1 \\ a_2 & b_2 \end{vmatrix} = a_1 b_2 - a_2 b_1$$

$$\begin{vmatrix} 3 & 2 \\ -6 & 4 \end{vmatrix} = 3(4) - (-6)2 = 12 + 12 = 24$$

$$\therefore \qquad \begin{vmatrix} 3 & 2 \\ -6 & 4 \end{vmatrix} = 24$$

This process of evaluating a 2 by 2 determinant is very important and very mechanical (see Figure 16–2). You should practice the process until you master it.

EXAMPLE 16–10   Evaluate $\begin{vmatrix} -5 & 7 \\ 3 & -2 \end{vmatrix}$

Solution   Use Equation 16–1 to evaluate:

$$\begin{vmatrix} a_1 & b_1 \\ a_2 & b_2 \end{vmatrix} = a_1 b_2 - a_2 b_1$$

$$\begin{vmatrix} -5 & 7 \\ 3 & -2 \end{vmatrix} = -5(-2) - (3)7 = 10 - 21 = -11$$

$$\therefore \qquad \begin{vmatrix} -5 & 7 \\ 3 & -2 \end{vmatrix} = -11$$

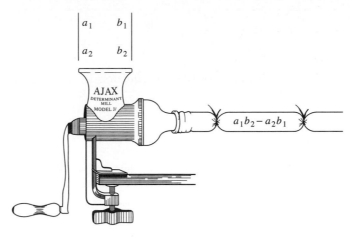

FIGURE 16–2 Turning the crank to expand a determinant.

Solving Systems of Linear Equations

Systems of two linear equations are solved by applying special formulas to the generalized system of linear equations.

$$\begin{cases} a_1x + b_1y = k_1 \\ a_2x + b_2y = k_2 \end{cases} \tag{16–2}$$

where a_1 and a_2 = numerical coefficients of the unknown x
$\qquad$ b_1 and b_2 = numerical coefficients of the unknown y
$\qquad$ k_1 and k_2 = numerical constants
$\qquad\quad$ x and y = coordinates of the common solution to the system of linear equations

The solution of the system of equations (16–2) by determinants is carried out with the following formulas:

$$\Delta = \begin{vmatrix} a_1 & b_1 \\ a_2 & b_2 \end{vmatrix} = a_1b_2 - a_2b_1 \tag{16–3}$$

$$x = \frac{\begin{vmatrix} k_1 & b_1 \\ k_2 & b_2 \end{vmatrix}}{\Delta} = \frac{k_1b_2 - k_2b_1}{\Delta} \tag{16–4}$$

$$y = \frac{\begin{vmatrix} a_1 & k_1 \\ a_2 & k_2 \end{vmatrix}}{\Delta} = \frac{a_1k_2 - a_2k_1}{\Delta} \tag{16–5}$$

EXAMPLE 16–11   Solve $\begin{cases} 4x - 5 = y \\ 12 - 2y = 3x \end{cases}$

Solution   Write the equation in the form of the system of equations (16–2):

$$4x - \ y = 5$$
$$3x + 2y = 12$$

Evaluate the determinant, Δ, using Equation 16–3:

$$\Delta = \begin{vmatrix} a_1 & b_1 \\ a_2 & b_2 \end{vmatrix} = \begin{vmatrix} 4 & -1 \\ 3 & 2 \end{vmatrix}$$

$$\Delta = 4(2) - (3)(-1) = 8 + 3$$

$$\Delta = 11$$

Solve for x using Equation 16–4:

$$x = \frac{\begin{vmatrix} k_1 & b_1 \\ k_2 & b_2 \end{vmatrix}}{\Delta} = \frac{k_1 b_2 - k_2 b_1}{\Delta}$$

$$x = \frac{\begin{vmatrix} 5 & -1 \\ 12 & 2 \end{vmatrix}}{11} = \frac{5(2) - (12)(-1)}{11}$$

$$x = \frac{10 + 12}{11} = \frac{22}{11}$$

$$x = 2$$

Solve for y using Equation 16–5:

$$y = \frac{\begin{vmatrix} a_1 & k_1 \\ a_2 & k_2 \end{vmatrix}}{\Delta} = \frac{a_1 k_2 - a_2 k_1}{\Delta}$$

$$y = \frac{\begin{vmatrix} 4 & 5 \\ 3 & 12 \end{vmatrix}}{11} = \frac{4(12) - (3)(5)}{11}$$

$$y = \frac{48 - 15}{11} = \frac{33}{11}$$

$$y = 3$$

Check Substitute $x = 2$ into each equation and solve for y (we should get 3):

$$4x - y = 5$$
$$4(2) - y = 5$$
$$8 - y = 5$$
$$y = 3$$

$$3x + 2y = 12$$
$$3(2) + 2y = 12$$
$$6 + 2y = 12$$
$$y = 3$$

$\therefore$ $x = 2$ and $y = 3$ is the solution of the system of equations.

EXAMPLE 16–12 Solve $\begin{cases} 3x - y = 10 \\ 4x = 9y - 2 \end{cases}$

Solution Write the equations in the generalized form of the system of equations (16–2):

$$3x - y = 10$$
$$4x - 9y = -2$$

Evaluate the determinant, Δ, using Equation 16–3:

$$\Delta = \begin{vmatrix} 3 & -1 \\ 4 & -9 \end{vmatrix} = 3(-9) - (4)(-1)$$
$$\Delta = -27 + 4$$
$$\Delta = -23$$

Solve for x using Equation 16–4:

$$x = \frac{\begin{vmatrix} 10 & -1 \\ -2 & -9 \end{vmatrix}}{-23} = \frac{10(-9) - (-2)(-1)}{-23}$$
$$x = \frac{-90 - 2}{-23} = \frac{-92}{-23}$$
$$x = 4$$

Solve for y using Equation 16–5:

$$y = \frac{\begin{vmatrix} 3 & 10 \\ 4 & -2 \end{vmatrix}}{-23} = \frac{(3)(-2) - (4)(10)}{-23}$$
$$y = \frac{-6 - 40}{-23} = \frac{-46}{-23}$$
$$y = 2$$

$\therefore$ $x = 4$ and $y = 2$ is the solution of the system of equations.

Evaluate each determinant.

1. $\begin{vmatrix} 1 & 2 \\ -1 & 1 \end{vmatrix}$

2. $\begin{vmatrix} 5 & 0 \\ 3 & 2 \end{vmatrix}$

3. $\begin{vmatrix} -2 & -2 \\ -2 & -2 \end{vmatrix}$

4. $\begin{vmatrix} 3 & 6 \\ 2 & -4 \end{vmatrix}$

PROGRAMMABLE

Solve each system of equations by determinants.

5. $\begin{cases} x + y = 4 \\ x - y = 2 \end{cases}$

6. $\begin{cases} 4x + y = 14 \\ -4x + y = -2 \end{cases}$

7. $\begin{cases} y = -x + 8 \\ x = y + 2 \end{cases}$

8. $\begin{cases} 3x + 2y = 1 \\ 2x = 18 + 3y \end{cases}$

9. $\begin{cases} 3x + y = 11 \\ y = 3x + 5 \end{cases}$

10. $\begin{cases} -3x + 4y = 1 \\ x - y = 0 \end{cases}$

11. $\begin{cases} x + 3y = 1 \\ x - 5 = 3y \end{cases}$

12. $\begin{cases} x/2 + y/3 = 4 \\ x - y = 3 \end{cases}$

13. $\begin{cases} x + y = 0 \\ x - 2y = 6 \end{cases}$

14. $\begin{cases} 2x = y + 4 \\ 3x + y = 5 \end{cases}$

15. $\begin{cases} -x - 5y = 0 \\ -10y - 3x = -10 \end{cases}$

16. $\begin{cases} 3y = 4 + 2x \\ 6x = y + 1 \end{cases}$

17. $\begin{cases} 7x = 56 + 4y \\ 5y - 45 = 3x \end{cases}$

18. $\begin{cases} 2(x + 3y) = 7 \\ 9 = 4x - 3y \end{cases}$

19. $\begin{cases} 9x - 34 = -2y \\ -14 = 6x + 5y \end{cases}$

20. $\begin{cases} 3x/2 - 1 = y/4 \\ x/3 + y/2 = 2 \end{cases}$

16–5 DETERMINANTS OF THE THIRD ORDER

Determinants of the third order have three rows and three columns with nine elements, and have the following form:

$$\begin{vmatrix} a_1 & b_1 & c_1 \\ a_2 & b_2 & c_2 \\ a_3 & b_3 & c_3 \end{vmatrix} = a_1b_2c_3 + b_1c_2a_3 + c_1a_2b_3 - a_3b_2c_1 - b_3c_2a_1 - c_3a_2b_1 \qquad \textbf{(16–6)}$$

PROGRAMMABLE

The expansion of the determinant may be found by copying the first two columns to the right of the determinant, and then forming the downward diagonal products with plus signs and the upward diagonal products with minus signs. Figure 16–3 shows this technique.

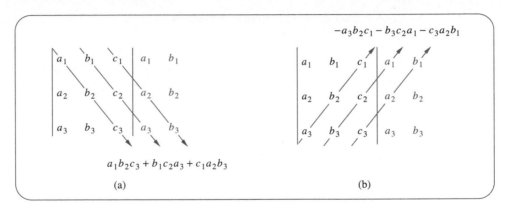

$$-a_3b_2c_1 - b_3c_2a_1 - c_3a_2b_1$$

$$a_1b_2c_3 + b_1c_2a_3 + c_1a_2b_3$$

(a)   (b)

FIGURE 16–3   Forming the expansion of the determinant: (a) downward diagonal products are positive; (b) upward diagonal products are negative.

EXAMPLE 16–13   Evaluate the following determinant:

$$\begin{vmatrix} 1 & 3 & 5 \\ 2 & 1 & 1 \\ 1 & 2 & 4 \end{vmatrix}$$

Solution   Repeat the first two columns:

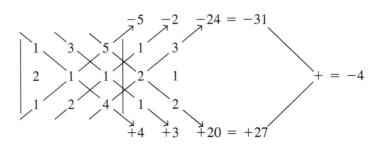

∴   The expansion of the determinant is:
$(4 + 3 + 20) + (-5 - 2 - 24) = 27 - 31 = -4.$

EXAMPLE 16–14   Evaluate the following determinant:

$$\begin{vmatrix} 2 & -1 & 4 \\ -1 & -3 & 2 \\ 2 & 0 & -1 \end{vmatrix}$$

Solution   Repeat the first two columns:

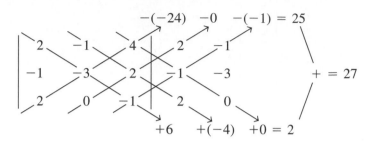

$\therefore$ The expansion of the determinant is:
$(6 - 4 + 0) + (24 - 0 + 1) = 2 + 25 = 27.$

Systems of three linear equations are solved by applying special formulas to the following generalized system of linear equations:

$$\begin{cases} a_1x + b_1y + c_1z = k_1 \\ a_2x + b_2y + c_2z = k_2 \\ a_3x + b_3y + c_3z = k_3 \end{cases} \qquad \text{(16–7)}$$

The solution of the system of equations is carried out with the following formulas:

$$\Delta = \begin{vmatrix} a_1 & b_1 & c_1 \\ a_2 & b_2 & c_2 \\ a_3 & b_3 & c_3 \end{vmatrix} \qquad \text{(16–8)}$$

$$x = \dfrac{\begin{vmatrix} k_1 & b_1 & c_1 \\ k_2 & b_2 & c_2 \\ k_3 & b_3 & c_3 \end{vmatrix}}{\Delta} \qquad \text{(16–9)}$$

$$y = \dfrac{\begin{vmatrix} a_1 & k_1 & c_1 \\ a_2 & k_2 & c_2 \\ a_3 & k_3 & c_3 \end{vmatrix}}{\Delta} \qquad \text{(16–10)}$$

$$z = \dfrac{\begin{vmatrix} a_1 & b_1 & k_1 \\ a_2 & b_2 & k_2 \\ a_3 & b_3 & k_3 \end{vmatrix}}{\Delta} \qquad \text{(16–11)}$$

EXAMPLE 16–15

Solve $\begin{cases} x - y + z = 2 \\ 2x - y + 3z = 6 \\ x + y + z = 6 \end{cases}$

Solution Evaluate the determinant Δ:

$$\Delta = \begin{vmatrix} 1 & -1 & 1 \\ 2 & -1 & 3 \\ 1 & 1 & 1 \end{vmatrix} \begin{matrix} 1 & -1 \\ 2 & -1 \\ 1 & 1 \end{matrix}$$

$$\Delta = -1 - 3 + 2 - (-1) - 3 - (-2)$$

$$\Delta = -2$$

Solve for x using Equation 16–9:

$$x = \frac{\begin{vmatrix} 2 & -1 & 1 \\ 6 & -1 & 3 \\ 6 & 1 & 1 \end{vmatrix} \begin{matrix} 2 & -1 \\ 6 & -1 \\ 6 & 1 \end{matrix}}{-2}$$

$$x = \frac{-2 - 18 + 6 - (-6) - 6 - (-6)}{-2} = \frac{-8}{-2}$$

$$x = 4$$

Solve for y using Equation 16–10:

$$y = \frac{\begin{vmatrix} 1 & 2 & 1 \\ 2 & 6 & 3 \\ 1 & 6 & 1 \end{vmatrix} \begin{matrix} 1 & 2 \\ 2 & 6 \\ 1 & 6 \end{matrix}}{-2}$$

$$y = \frac{6 + 6 + 12 - 6 - 18 - 4}{-2} = \frac{-4}{-2}$$

$$y = 2$$

Solve for z using Equation 16–11:

$$z = \frac{\begin{vmatrix} 1 & -1 & 2 \\ 2 & -1 & 6 \\ 1 & 1 & 6 \end{vmatrix} \begin{matrix} 1 & -1 \\ 2 & -1 \\ 1 & 1 \end{matrix}}{-2}$$

$$z = \frac{-6 - 6 + 4 - (-2) - 6 - (-12)}{-2} = \frac{0}{-2}$$

$$z = 0$$

Check Substitute $x = 4$, $y = 2$, and $z = 0$ into each equation. The left member and the right member should *balance*.

$$x - y + z = 2$$
$$4 - 2 + 0 = 2$$
$$2 = 2$$

$$2x - y + 3z = 6$$
$$2(4) - 2 + 3(0) = 6$$
$$8 - 2 = 6$$
$$6 = 6$$

$$x + y + z = 6$$
$$4 + 2 + 0 = 6$$
$$6 = 6$$

$\therefore$ $x = 4, y = 2$, and $z = 0$ is the solution of the system of equations.

EXAMPLE 16–16

Solve $\begin{cases} y = 4 - x \\ x + 2 = -z \\ z + y = 8 \end{cases}$

Solution

Write the equations in the generalized form as in Equation 16–7. Add zeros where a term is missing.

$$x + y + 0 = 4$$
$$x + 0 + z = -2$$
$$0 + y + z = 8$$

Evaluate the determinant:

$$\Delta = \begin{vmatrix} 1 & 1 & 0 \\ 1 & 0 & 1 \\ 0 & 1 & 1 \end{vmatrix} \begin{matrix} 1 & 1 \\ 1 & 0 \\ 0 & 1 \end{matrix}$$

$$\Delta = 0 + 0 + 0 - 0 - 1 - 1$$

$$\Delta = -2$$

Solve for x using Equation 16–9:

$$x = \frac{\begin{vmatrix} 4 & 1 & 0 \\ -2 & 0 & 1 \\ 8 & 1 & 1 \end{vmatrix} \begin{matrix} 4 & 1 \\ -2 & 0 \\ 8 & 1 \end{matrix}}{-2}$$

$$x = \frac{0 + 8 + 0 - 0 - 4 - (-2)}{-2}$$

$$x = -3$$

Solve for y using Equation 16–10:

$$y = \frac{\begin{vmatrix} 1 & 4 & 0 \\ 1 & -2 & 1 \\ 0 & 8 & 1 \end{vmatrix} \begin{matrix} 1 & 4 \\ 1 & -2 \\ 0 & 8 \end{matrix}}{-2}$$

$$y = \frac{-2 + 0 + 0 - 0 - 8 - 4}{-2}$$

$$y = 7$$

Solve for z using Equation 16–11:

$$z = \frac{\begin{vmatrix} 1 & 1 & 4 \\ 1 & 0 & -2 \\ 0 & 1 & 8 \end{vmatrix}\begin{matrix} 1 & 1 \\ 1 & 0 \\ 0 & 1 \end{matrix}}{-2}$$

$$z = \frac{0 + 0 + 4 - 0 - (-2) - 8}{-2}$$

$$z = 1$$

Check   Substitute $x = -3$, $y = 7$, and $z = 1$ into each equation. The left member should equal the right member.

$$y = 4 - x$$
$$7 = 4 - (-3)$$
$$7 = 7$$

$$x + 2 = -z$$
$$-3 + 2 = -1$$
$$-1 = -1$$

$$z + y = 8$$
$$1 + 7 = 8$$
$$8 = 8$$

$\therefore$ $x = -3$, $y = 7$, and $z = 1$ is the solution of the system of equations.

EXERCISE 16–5

PROGRAMMABLE

Evaluate each determinant.

1. $\begin{vmatrix} 1 & 2 & 1 \\ 2 & -1 & 3 \\ 2 & 0 & 2 \end{vmatrix}$

2. $\begin{vmatrix} 0 & -1 & 1 \\ -3 & 0 & 2 \\ -2 & 1 & 0 \end{vmatrix}$

3. $\begin{vmatrix} 1 & 1 & 1 \\ 1 & 1 & 1 \\ 1 & 1 & 1 \end{vmatrix}$

4. $\begin{vmatrix} 2 & 3 & 4 \\ 0 & 5 & -1 \\ -1 & -2 & -3 \end{vmatrix}$

Solve each system of equations by determinants.

5. $\begin{cases} 3x + 3y - 2z = 2 \\ 2x - 3y + z = -2 \\ x - 6y + 3z = -2 \end{cases}$

6. $\begin{cases} x - 11 + y = -z \\ -11 + z + 3y = -x \\ 3x + z + y = 13 \end{cases}$

7. $\begin{cases} x + y + z = 6 \\ x + y - z = 0 \\ x - y - z = 2 \end{cases}$

8. $\begin{cases} -x + 3z = 7 \\ 4z - 9 + 3x = y \\ 2x + 3y = 15 \end{cases}$

9. $\begin{cases} x + 3y + 4z = 14 \\ z + x - 7 = -2y \\ -2 + 2z + y = -2x \end{cases}$

10. $\begin{cases} 14 = x - z \\ -y + 10 = -x - z \\ y = 21 - z \end{cases}$

11. $\begin{cases} -45I_A + 30I_B + 10I_C = 10 \\ 30I_A - 50I_B + 20I_C = -20 \\ 10I_A + 20I_B - 50I_C = 0 \end{cases}$

12. $\begin{cases} 13I_1 - 8I_2 - 2I_3 = 100 \\ -2I_1 - 4I_2 + 12I_3 = 0 \\ -8I_1 + 22I_2 - 4I_3 = 0 \end{cases}$

SELECTED TERMS

determinant   An operator that combines a square array of numbers into a single number used in solving systems of linear equations.

system of linear equations   Set of linear equations, each involving the same variables.

17

Applying Systems of Linear Equations to Electronic Concepts

17-1 Applying Kirchhoff's Voltage Law

17-2 Mesh Analysis

17-3 Solving Networks by Mesh Analysis

PERFORMANCE OBJECTIVES

- Use Kirchhoff's voltage law to solve for loop currents.
- Employ mesh analysis to solve networks.

A Comstar 2D communications satellite undergoes final preparations before launch aboard an Atlas Centaur launch vehicle. (Courtesy of NASA)

This chapter will deal with the solution of *resistive networks* through the use of a system of linear equations based on Kirchhoff's voltage law.

17–1   APPLYING KIRCHHOFF'S VOLTAGE LAW

Kirchhoff's voltage law (KVL) states that in any complete circuit the algebraic sum of the voltage drops must equal the algebraic sum of the voltage rises. Stated as an equation, Kirchhoff's voltage law becomes:

$$\Sigma V = \Sigma E \qquad (17\text{–}1)$$

or

$$V_1 + V_2 + V_3 = E_1 + E_2 + E_3 \qquad (17\text{–}2)$$

In a complete circuit, as shown in Figure 17–1, we may assume a current (I) passes through the resistive components in the circuit, causing voltage drops. Thus, $V = IR$. Replacing V_1, V_2, and V_3 of Equation 17–2 with their identities, we have a new Kirchhoff's voltage law equation.

$$IR_1 + IR_2 + IR_3 = E_1 + E_2 + E_3 \qquad (17\text{–}3)$$

FIGURE 17–1   A complete electrical circuit is called a *closed loop*. We see that: (a) a closed loop may have zero voltage sources; (b) a closed loop may have one voltage source; (c) a closed loop may have several voltage sources.

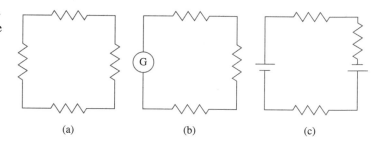

(a)   (b)   (c)

Conventional Direction for Current

In the study of electricity it is *conventional* to think of the current as flowing from positive ($+$) to negative ($-$). This is opposite to *electron flow,* shown in Figure 17–2. For

FIGURE 17–2   The polarity of the voltage drop across a resistance depends upon the choice of current flow—conventional (b) or electron (c). The conventional direction for current flow is used in this chapter.

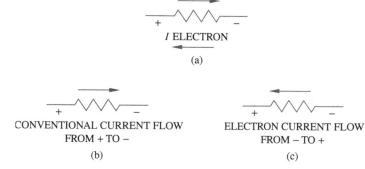

purposes of mathematical discussion, we will use **conventional current** flow. Follow-ing the conventional current direction through Figure 17–3, you will notice that there is a rise in voltage across the source (that is, $-$ to $+$) and a drop in voltage across the resistance (that is, $+$ to $-$).

FIGURE 17–3 When conventional current passes through a resistance, the point of entering is positive and the point of exit is negative.

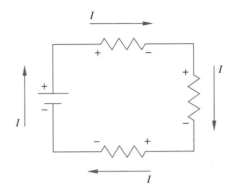

Procedure for Applying KVL Equations

The following series of examples will serve to introduce you to the process of using Kirchhoff's voltage law equations.

EXAMPLE 17–1   Write a KVL equation for the circuit pictured in Figure 17–4 and solve for the loop current (I).

FIGURE 17–4 Circuit for Example 17–1.

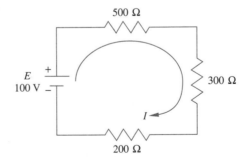

Solution   Label each resistance and polarize the drops across each resis-tance using conventional current as shown in Figure 17–5. Write the KVL equation; use Equation 17–3; start with R_1:

$$IR_1 + IR_2 + IR_3 = E_1 + E_2 + E_3$$

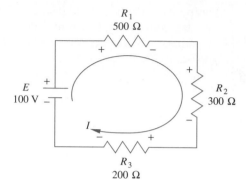

FIGURE 17–5 The voltage drops (across the resistances) are polarized using the conventional direction for current flow.

R_1
500 Ω

E
100 V

R_2
300 Ω

I

R_3
200 Ω

Since there is only one source, set E_2 and E_3 to zero:

$$I(500) + I(300) + I(200) = 100$$

Solve for the loop current (I):

$$1000I = 100$$

$$I = \frac{100}{1000} = 100 \text{ mA}$$

$\therefore$ 　　　　　$$I = 100 \text{ mA}$$

Example 17–2 demonstrates what happens when the assumed loop current is selected opposite to that of the actual loop current. The source is assigned a minus sign and the loop current results in a minus quantity.

EXAMPLE 17–2　　Repeat Example 17–1 for the solution of the loop current (I), only this time assume the loop current to be in the opposite direction.

Solution　　Label each resistance and polarize the voltage drops across each resistance as shown in Figure 17–6.

FIGURE 17–6 The voltage drops (across the resistances) are polarized using the conventional direction of current flow.

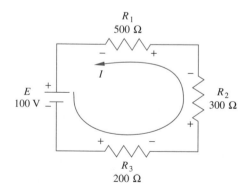

R_1
500 Ω

I

E
100 V

R_2
300 Ω

R_3
200 Ω

Write the KVL equation; use Equation 17–3; start with R_3:

$$IR_3 + IR_2 + IR_1 = E_1 + E_2 + E_3$$

$$I(200) + I(300) + I(500) = -100$$

Observation   Because the assumed loop current passes through the voltage source from $+$ to $-$, a minus sign is assigned to the voltage. Solve for the assumed loop current (I):

$$1000I = -100$$

$$\therefore \qquad I = -100 \text{ mA}$$

Observation   The minus loop current means that the actual direction of the current is *opposite* to that of the assumed direction.

As demonstrated in the preceding examples, the selection of the direction of the loop current is arbitrary and makes no difference, as the direction will be *flagged* by the minus sign. **Remember that a negative loop current indicates that the assumed direction is opposite to the actual conventional current direction.**

Kirchhoff's voltage law may be applied to a simple closed loop through the use of the concepts in the following guidelines.

GUIDELINES 17–1. KVL Loop Procedure

To write a KVL equation for a simple loop, remember that:

1. The direction of the assumed loop current is always positive. Thus, the conventional current enters the resistance from the positive side and leaves from the negative.

2. The polarity of a voltage source is not changed by the direction of the assumed loop current.

3. Assumed loop current flow through a voltage source from $-$ to $+$ is considered positive and is assigned a $+$ sign.

4. Assumed loop current flow through a voltage source from $+$ to $-$ is considered negative and is assigned a $-$ sign.

EXAMPLE 17–3   Apply KVL and Guidelines 17–1 to the circuit of Figure 17–7. Solve for I.

Solution   Assume a loop current direction as shown in Figure 17–8. Label and polarize the resistances and label the voltage sources. Write the KVL equation applying Guidelines 17–1 and Equation 17–3:

$$IR_1 + IR_2 + IR_3 + IR_4 + IR_5 = E_1 + E_2 + E_3$$

FIGURE 17–7 Circuit for Example 17–3.

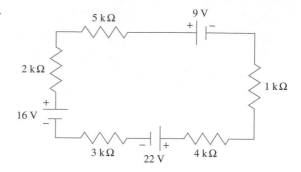

FIGURE 17–8 The resistances of Figure 17–7 are polarized using the conventional direction of current flow.

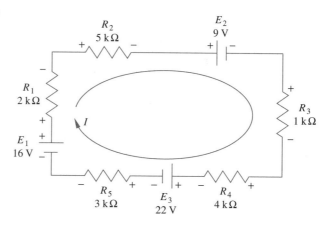

Substitute; pay particular attention to the sign of the voltage sources (Guidelines 17–1, steps 3 and 4); start with R_1:

$$2000I + 5000I + 1000I + 4000I + 3000I = 16 - 9 - 22$$
$$15 \times 10^3 I = -15$$
$$I = -15/(15 \times 10^3)$$
$$I = -1 \text{ mA}$$

∴ The current is -1 mA, and the actual direction of the current flow is *opposite* to the assumed direction.

EXAMPLE 17–4 Apply KVL Equation 17–1 to the circuit of Figure 17–9; solve for V_3.

Solution Since there are two loops that include V_3, we may solve for V_3 with either one. Assume loop current directions as shown in Figure 17–10, and polarize the resistance using the conventional direction of current. Write the KVL equation for loop 1:

$$\Sigma V = \Sigma E$$
$$V_2 + V_3 = E$$

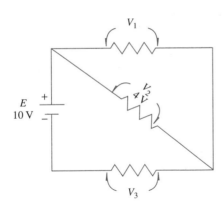

FIGURE 17-9 Circuit for Example 17–4.

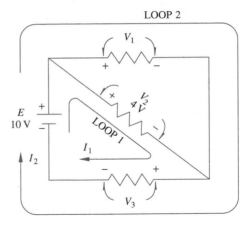

FIGURE 17-10 The resistances of Figure 17–9 are polarized using the conventional direction of current flow.

Substitute and solve for V_3:

$$4 + V_3 = 10$$
$$V_3 = 6 \text{ V}$$

Write the KVL equation for loop 2:

$$\Sigma V = \Sigma E$$
$$V_1 + V_3 = E$$

Substitute and solve for V_3. $V_1 = V_2 = 4$ V because the resistances are connected in parallel:

$$4 + V_3 = 10$$
$$V_3 = 6 \text{ V}$$

∴ V_3 may be found with either loop equation.

EXERCISE 17-1

1. Write a KVL equation for the circuit of Figure 17–11 and solve for the loop current.

2. Write a KVL equation for the circuit of Figure 17–11 when R_1 is 200 Ω. Solve for the loop current.

3. Write a KVL equation for the circuit of Figure 17–11 when E_2 is reversed in polarity. Solve for the loop current.

4. Write a KVL equation for the circuit of Figure 17–11 when R_2 is 600 Ω and E_1 is increased to 15 V. Solve for the voltage drop, V_3.

5. Write a KVL equation for the circuit of Figure 17–12 and solve for the loop current.

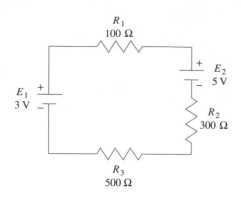

FIGURE 17–11 Circuit for Exercise 17–1, problems 1–4.

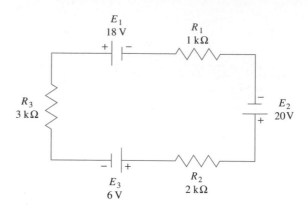

FIGURE 17–12 Circuit for Exercise 17–1, problems 5–8.

6. Write a KVL equation for the circuit of Figure 17–12 when R_1 is 1.8 kΩ. Solve for the voltage drop, V_2.

7. Write a KVL equation for the circuit of Figure 17–12 when E_3 is reversed in polarity. Solve for the loop current.

8. Write a KVL equation for the circuit of Figure 17–12 when R_1 is 4.7 kΩ and E_2 is increased to 36 V. Solve for the voltage drop, V_1.

9. Write a KVL equation and solve for V_3 in Figure 17–13.

10. Write a KVL equation and solve for V_1 in Figure 17–14.

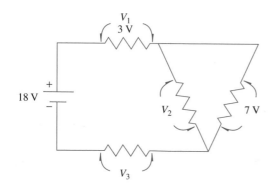

FIGURE 17–13 Circuit for Exercise 17–1, problem 9.

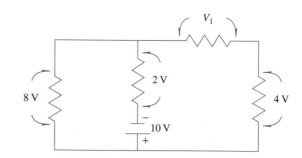

FIGURE 17–14 Circuit for Exercise 17–1, problem 10.

17–2 MESH ANALYSIS

Through the use of mesh analysis, we may solve a *network* (circuit) that requires more than just the rules of series or parallel circuits. *Mesh analysis* is one of several methods used to analyze a network.

A **mesh** is a simple closed loop having no branches. Even a very complicated network can be represented by two or more meshes. Figure 17–15 is an example of a network with two meshes. In solving the network, an equation (similar to a Kirchhoff's voltage law equation) is written for each mesh. The resulting system of equations is solved for each of the **mesh currents.** Once the mesh currents are determined, then the actual branch currents and voltage drops may be determined.

Mutual Resistance

An important idea in mesh analysis is that of **mutual resistance.** The mutual resistance is the resistance *shared* by two mesh currents. In Figure 17–15, R_3 is the mutual resistance.

The mesh currents passing through the mutual resistance may be *aiding* or *opposing,* as shown in Figure 17–16. When we write the mesh equation, we must pay close attention to the mesh currents passing through the mutual resistance. **Since each mesh current of the two meshes passes through the mutual resistance, both must be included in each of the mesh equations.** Examples 17–5 and 17–6 will help to clarify this important concept.

FIGURE 17–15 Network with two meshes. The mesh currents I_1 and I_2 are assumed to flow in a clockwise (cw) direction.

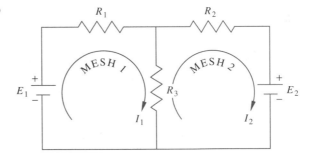

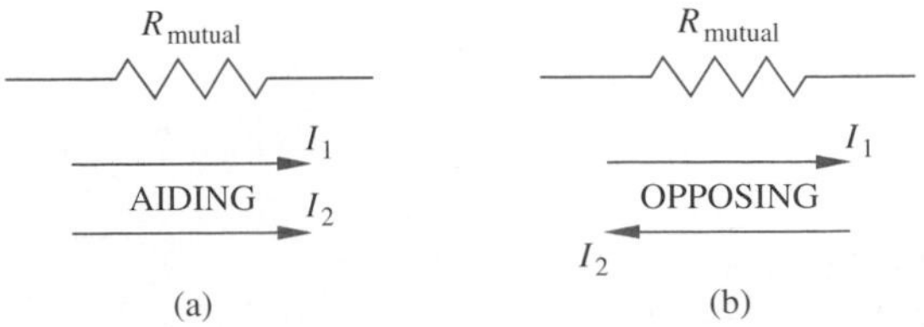

FIGURE 17–16 Mesh currents in the mutual resistance: (a) aiding currents are added; (b) opposing currents are subtracted.

EXAMPLE 17–5   Write the mesh equations for the network of Figure 17–17 and solve for the mesh currents, I_1 and I_2. Arbitrarily assume a clockwise (cw) direction for each current.

FIGURE 17–17   Network for Examples 17–5 and 17–6.

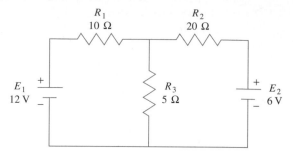

Solution   Draw in mesh currents I_1 and I_2 in a cw direction as shown in Figure 17–18.
Polarize the resistances as was done in the previous KVL section. This is shown in Figure 17–18.

FIGURE 17–18   Network for Figure 17–17 with two meshes. Notice that the mutual resistance R_3 has two polarizations. One polarization is determined by the direction of mesh current I_1; the other polarization is determined by the direction of mesh current I_2.

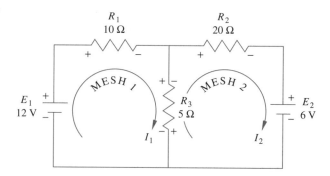

Write the mesh equation for each mesh. Use KVL Equation 17–3 and Guidelines 17–1.
Mesh 1, Figure 17–19:

$$10I_1 + \underbrace{5I_1 - 5I_2}_{\text{current through } R_3} = 12$$

FIGURE 17–19   Mesh 1 of the network for Figure 17–18.

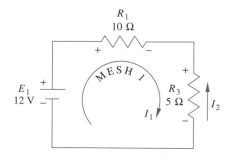

Notice that the mutual resistance ($R_3 = 5\ \Omega$ of Figure 17–18) has opposing currents passing through it. For mesh 1, mesh current I_2 is negative, as emphasized in Figure 17–19.
Mesh 2, Figure 17–20:

$$\underbrace{5I_2 - 5I_1}_{\text{current through } R_3} + 20I_2 = -6$$

current through R_3

Notice that the mutual resistance ($R_3 = 5\ \Omega$ of Figure 17–18) has opposing currents passing through it. For mesh 2, mesh current I_1 is negative, as emphasized in Figure 17–20.

FIGURE 17–20   Mesh 2 of the network for Figure 17–18.

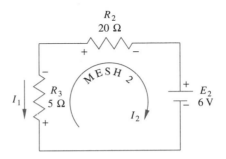

mesh 1: $15I_1 - 5I_2 = 12$
mesh 2: $-5I_1 + 25I_2 = -6$

Solve for I_1 and I_2 using determinants:

$$\Delta = \begin{vmatrix} 15 & -5 \\ -5 & 25 \end{vmatrix} = 375 - 25 = 350$$

$$I_1 = \frac{\begin{vmatrix} 12 & -5 \\ -6 & 25 \end{vmatrix}}{\Delta} = \frac{300 - 30}{350} = \frac{270}{350} = 0.77\ \text{A}$$

$$I_2 = \frac{\begin{vmatrix} 15 & 12 \\ -5 & -6 \end{vmatrix}}{\Delta} = \frac{-90 + 60}{350} = \frac{-30}{350} = -86\ \text{mA}$$

∴ $I_1 = 0.77$ A in the ***assumed*** direction.
$I_2 = 86$ mA in the direction ***opposite*** to that assumed.

Observation   Remember that a negative current indicates that the assumed direction is opposite to the actual direction.

In Example 17–5 the directions of the mesh currents were selected so that the currents passing through the mutual resistance were opposing. In Example 17–6 the directions of the mesh currents are selected so that the currents passing through the mutual resistance are aiding.

EXAMPLE 17–6   Write the mesh equations for the network of Figure 17–21 and solve for the mesh currents, I_1 and I_2.

FIGURE 17–21   Network for Example 17–6. Notice that the mutual resistance R_3 has *aiding currents* passing through it.

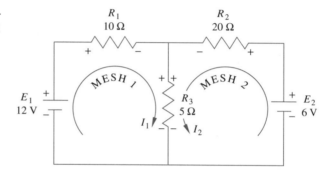

Solution

mesh 1: $10I_1 + \underbrace{5I_1 + 5I_2}_{current\ through\ R_3} = 12$

mesh 2: $20I_2 + \underbrace{5I_2 + 5I_1}_{current\ through\ R_3} = 6$

Since the mutual resistance (R_3 of Figure 17–21) has aiding currents passing through it, I_2 in mesh 1 and I_1 in mesh 2 are both positive. Simplify each mesh equation:

mesh 1: $15I_1 + 5I_2 = 12$
mesh 2: $5I_1 + 25I_2 = 6$

Solve for I_1 and I_2 using determinants:

$$\Delta = \begin{vmatrix} 15 & 5 \\ 5 & 25 \end{vmatrix} = 375 - 25 = 350$$

$$I_1 = \frac{\begin{vmatrix} 12 & 5 \\ 6 & 25 \end{vmatrix}}{\Delta} = \frac{300 - 30}{350} = \frac{270}{350} = 0.77 \text{ A}$$

$$I_2 = \frac{\begin{vmatrix} 15 & 12 \\ 5 & 6 \end{vmatrix}}{\Delta} = \frac{90 - 60}{350} = \frac{30}{350} = 86 \text{ mA}$$

$$\therefore \qquad I_1 = 0.77 \text{ A in the direction assumed.}$$
$$I_2 = 86 \text{ mA in the direction assumed.}$$

Once the mesh currents are computed, the actual branch currents may be determined. Figure 17–22 shows the actual branch currents for the network of Examples 17–5 and 17–6. Notice that the mutual resistance has an actual branch current (I_C) that is equal to the algebraic sum of the mesh currents (I_1 and I_2). Also notice that the actual branch current I_A is equal to the mesh current I_1, and that the branch current I_B is equal to the mesh current I_2.

FIGURE 17–22   Network for Examples 17–5 and 17–6 with the actual branch currents I_A, I_B, and I_C shown and the assumed direction of the mesh currents I_1 and I_2 selected to correspond to the actual direction. $I_A = I_1$; $I_B = I_2$; $I_C = I_1 + I_2$.

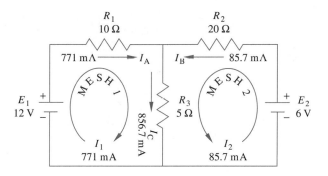

EXAMPLE 17–7   From the following sets of mesh currents, determine:

(a) The actual branch current passing through the mutual resistance, R_3 of Figure 17–23

(b) The actual direction (up or down) when:

Case 1: $I_1 = 15$ A $I_2 = 5$ A
Case 2: $I_1 = -15$ A $I_2 = 5$ A
Case 3: $I_1 = 15$ A $I_2 = -5$ A
Case 4: $I_1 = 5$ A $I_2 = 15$ A

FIGURE 17–23   Network for Example 17–7 and Exercise 17–2.

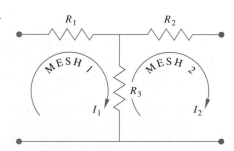

Solution

case 1: $\begin{cases} I_1 = 15 \text{ A} \\ I_2 = 5 \text{ A} \end{cases}$

Because both mesh currents I_1 and I_2 are positive, the actual direction of the currents through the mutual resistance, R_3, is as pictured in Figure 17–23. The currents are *opposing* one another.

$\therefore$ case 1: $I_{R_3} = 15 - 5 = 10$ A downward.

Notice, when two currents are opposing one another, that the direction of the difference is determined by the direction of the larger current.

case 2: $\begin{cases} I_1 = -15 \text{ A} \\ I_2 = 5 \text{ A} \end{cases}$

Because I_1 is negative, its direction is opposite to that pictured in Figure 17–23. The direction of I_2 is as pictured in Figure 17–23. Thus, both I_1 and I_2 are actually flowing upward through R_3 as *aiding* currents.

$\therefore$ case 2: $I_{R_3} = -(-15) + 5 = 20$ A upward.

case 3: $\begin{cases} I_1 = 15 \text{ A} \\ I_2 = -5 \text{ A} \end{cases}$

Because I_2 is negative, its direction is opposite to that pictured in Figure 17–23. The direction of I_1 is as pictured in Figure 17–23. Thus, both I_1 and I_2 are actually flowing downward through R_3 as *aiding* currents.

$\therefore$ case 3: $I_{R_3} = 15 - (-5) = 20$ A downward.

case 4: $\begin{cases} I_1 = 5 \text{ A} \\ I_2 = 15 \text{ A} \end{cases}$

As in case 1, both mesh currents are positive. So the actual direction through the mutual resistance, R_3, is as pictured in Figure 17–23, and the currents are *opposing* one another.

$\therefore$ case 4: $I_{R_3} = 15 - 5 = 10$ A upward.

Observation

When two currents oppose one another, the direction of the difference is determined by the direction of the larger current.

From the following sets of mesh currents, determine (a) the actual branch current passing through the mutual resistance, R_3 of Figure 17–23, and (b) the actual direction—upward or downward.

1. $\begin{cases} I_1 = 3\text{ A} \\ I_2 = 2\text{ A} \end{cases}$ 2. $\begin{cases} I_1 = -10\text{ A} \\ I_2 = 6\text{ A} \end{cases}$

3. $\begin{cases} I_1 = 1\text{ A} \\ I_2 = 3\text{ A} \end{cases}$ 4. $\begin{cases} I_1 = -6\text{ A} \\ I_2 = 8\text{ A} \end{cases}$

5. $\begin{cases} I_1 = -7\text{ A} \\ I_2 = 2\text{ A} \end{cases}$ 6. $\begin{cases} I_1 = -12\text{ A} \\ I_2 = -5\text{ A} \end{cases}$

7. $\begin{cases} I_1 = -3\text{ A} \\ I_2 = -9\text{ A} \end{cases}$ 8. $\begin{cases} I_1 = 7\text{ A} \\ I_2 = -4\text{ A} \end{cases}$

9. $\begin{cases} I_1 = 1\text{ A} \\ I_2 = -1\text{ A} \end{cases}$ 10. $\begin{cases} I_1 = 5\text{ A} \\ I_2 = 5\text{ A} \end{cases}$

17–3   SOLVING NETWORKS BY MESH ANALYSIS

The concepts of mesh currents, Kirchhoff's voltage law, and mutual resistance are brought together as mesh analysis to solve networks.

Guidelines 17–2 are a summary of the techniques used to solve networks with mesh analysis.

GUIDELINES 17–2.   Mesh Analysis

To solve a network by mesh analysis:

1. Draw a mesh current within each mesh in a cw direction and polarize the resistance.
2. Write as many mesh equations as there are meshes, using KVL equations and the concepts of mutual resistance.
3. Solve for the mesh currents with determinants.

Guidelines 17–2 will be applied in the following examples. It is suggested that you work along with pencil, paper, and calculator.

EXAMPLE 17–8   Write the mesh equations for the network of Figure 17–24 and determine the mesh currents.

FIGURE 17–24 Network for Example 17–8.

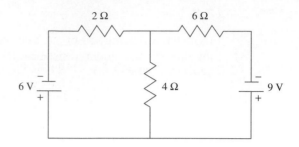

Solution   Use the steps of Guidelines 17–2.

Step 1:   Draw two cw mesh currents within the network as shown in Figure 17–25. Polarize the resistances.

FIGURE 17–25 Network for Example 17–8 with mesh currents (cw) shown and resistance polarized.

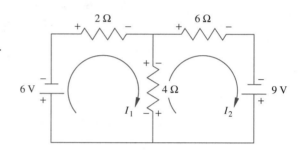

Step 2:   Write two mesh equations:

mesh 1: $2I_1 + 4I_1 - 4I_2 = -6$
$$6I_1 - 4I_2 = -6$$

mesh 2: $4I_2 - 4I_1 + 6I_2 = 9$
$$-4I_1 + 10I_2 = 9$$

Step 3:   Solve for I_1 and I_2 with determinants:

$$\text{mesh 1:}\quad 6I_1 - 4I_2 = -6$$
$$\text{mesh 2:} -4I_1 + 10I_2 = 9$$

$$\Delta = \begin{vmatrix} 6 & -4 \\ -4 & 10 \end{vmatrix} = 60 - 16 = 44$$

$$I_1 = \frac{\begin{vmatrix} -6 & -4 \\ 9 & 10 \end{vmatrix}}{\Delta} = \frac{-60 + 36}{44} = \frac{-24}{44} = -0.55 \text{ A}$$

$$I_2 = \frac{\begin{vmatrix} 6 & -6 \\ -4 & 9 \end{vmatrix}}{\Delta} = \frac{54 - 24}{44} = \frac{30}{44} = 0.68 \text{ A}$$

$$\therefore \quad I_1 = 0.55 \text{ A in the direction } \textit{opposite} \text{ to that assumed.}$$
$$I_2 = 0.68 \text{ A in the } \textit{assumed} \text{ direction.}$$

EXAMPLE 17–9   Write the mesh equations for the network of Figure 17–26 and determine the mesh currents.

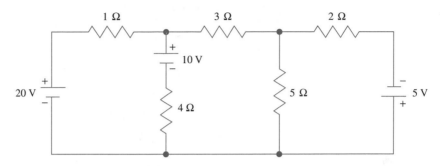

FIGURE 17–26   Network for Example 17–9.

Solution   Use Guidelines 17–2.

Step 1:   Draw three mesh currents within the network as shown in Figure 17–27. Polarize the resistances.

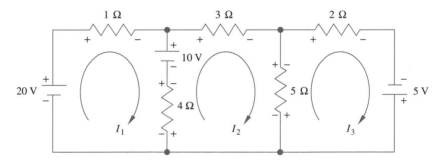

FIGURE 17–27   Network for Example 17–9 with mesh currents (cw) shown and resistances polarized.

Step 2:   Write three mesh equations:

$$\text{mesh 1: } 1I_1 + 4I_1 - 4I_2 = 20 - 10$$
$$5I_1 - 4I_2 = 10$$
$$\text{mesh 2: } 4I_2 - 4I_1 + 3I_2 + 5I_2 - 5I_3 = 10$$
$$-4I_1 + 12I_2 - 5I_3 = 10$$
$$\text{mesh 3: } 5I_3 - 5I_2 + 2I_3 = 5$$
$$-5I_2 + 7I_3 = 5$$

Step 3:   Solve for I_1, I_2, and I_3 using determinants:

mesh 1: $5I_1 - 4I_2 + 0I_3 = 10$
mesh 2: $-4I_1 + 12I_2 - 5I_3 = 10$
mesh 3: $0I_1 - 5I_2 + 7I_3 = 5$

$$\Delta = \begin{vmatrix} 5 & -4 & 0 \\ -4 & 12 & -5 \\ 0 & -5 & 7 \end{vmatrix} = 183$$

$$I_1 = \dfrac{\begin{vmatrix} 10 & -4 & 0 \\ 10 & 12 & -5 \\ 5 & -5 & 7 \end{vmatrix}}{\Delta} = \dfrac{970}{183} = 5.3 \text{ A}$$

$$I_2 = \dfrac{\begin{vmatrix} 5 & 10 & 0 \\ -4 & 10 & -5 \\ 0 & 5 & 7 \end{vmatrix}}{\Delta} = \dfrac{755}{183} = 4.1 \text{ A}$$

$$I_3 = \dfrac{\begin{vmatrix} 5 & -4 & 10 \\ -4 & 12 & 10 \\ 0 & -5 & 5 \end{vmatrix}}{\Delta} = \dfrac{670}{183} = 3.7 \text{ A}$$

$\therefore$ $I_1 = 5.3$ A, $I_2 = 4.1$ A, $I_3 = 3.7$ A

EXAMPLE 17–10   Write the mesh equations for the network of Figure 17–28 and determine the mesh currents I_1, I_2, and I_3.

FIGURE 17–28 Network for Example 17–10.

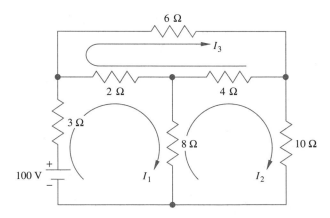

Solution   Write three mesh equations:

mesh 1: $3I_1 + 2I_1 - 2I_3 + 8I_1 - 8I_2 = 100$
$$13I_1 - 8I_2 - 2I_3 = 100$$

mesh 2: $8I_2 - 8I_1 + 4I_2 - 4I_3 + 10I_2 = 0$
$$- 8I_1 + 22I_2 - 4I_3 = 0$$

mesh 3: $4I_3 - 4I_2 + 2I_3 - 2I_1 + 6I_3 = 0$
$$-2I_1 - 4I_2 + 12I_3 = 0$$

Solve for I_1, I_2, and I_3 with determinants:

mesh 1: $13I_1 - 8I_2 - 2I_3 = 100$
mesh 2: $-8I_1 + 22I_2 - 4I_3 = 0$
mesh 3: $-2I_1 - 4I_2 + 12I_3 = 0$

$$\Delta = \begin{vmatrix} 13 & -8 & -2 \\ -8 & 22 & -4 \\ -2 & -4 & 12 \end{vmatrix} = 2240$$

$$I_1 = \frac{\begin{vmatrix} 100 & -8 & -2 \\ 0 & 22 & -4 \\ 0 & -4 & 12 \end{vmatrix}}{\Delta} = \frac{24\,800}{2240} = 11.07 \text{ A}$$

$$I_2 = \frac{\begin{vmatrix} 13 & 100 & -2 \\ -8 & 0 & -4 \\ -2 & 0 & 12 \end{vmatrix}}{\Delta} = \frac{10\,400}{2240} = 4.64 \text{ A}$$

$$I_3 = \frac{\begin{vmatrix} 13 & -8 & 100 \\ -8 & 22 & 0 \\ -2 & -4 & 0 \end{vmatrix}}{\Delta} = \frac{7600}{2240} = 3.39 \text{ A}$$

$\therefore$ $I_1 = 11.1 \text{ A}, I_2 = 4.64 \text{ A}, I_3 = 3.39 \text{ A}$

EXAMPLE 17–11   Use the mesh currents of Example 17–10 and Figure 17–28 to determine the actual branch currents through the 2-, 4-, and 8-Ω resistances.

Solution   Draw the portion of Figure 17–28 in question with the mesh currents shown. See Figure 17–29.
The branch current through the 2-Ω resistance:

$$I_{2\Omega} = I_1 + I_3$$
$$I_{2\Omega} = 11.07 - 3.39 = 7.68 \text{ A}$$

$\therefore$ $I_{2\Omega} = 7.68$ to the right

The branch current through the 4-Ω resistance:

$$I_{4\Omega} = I_3 + I_2$$
$$I_{4\Omega} = -3.39 + 4.64 = 1.25 \text{ A}$$

∴ $I_{4\Omega} = 1.25$ A to the right

The branch current through the 8-Ω resistance:

$$I_{8\Omega} = I_1 + I_2$$
$$I_{8\Omega} = 11.07 - 4.64 = 6.43 \text{ A}$$

∴ $I_{8\Omega} = 6.43$ A downward

Observation The branch currents through the 2-, 4-, and 8-Ω resistances are shown in Figure 17–30.

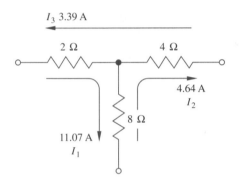

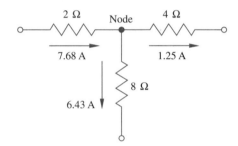

FIGURE 17–29 Network for Example 17–11.

FIGURE 17–30 Branch currents for Example 17–11. Notice that the current entering the node (7.68 A) is equal to the current leaving the node (7.68 = 1.25 + 6.43).

EXERCISE 17–3

Write the mesh equations of Figure 17–31 and solve for I_1 and I_2 when:

1. $R_1 = 2\ \Omega$ $R_2 = 4\ \Omega$ $R_3 = 3\ \Omega$
   $E_1 = 10\ \text{V}$ $E_2 = 20\ \text{V}$

2. $R_1 = 4\ \Omega$ $R_2 = 2\ \Omega$ $R_3 = 1\ \Omega$
   $E_1 = 20\ \text{V}$ $E_2 = 30\ \text{V}$

3. $R_1 = 5\ \Omega$ $R_2 = 10\ \Omega$ $R_3 = 5\ \Omega$
   $E_1 = 9\ \text{V}$ $E_2 = 30\ \text{V}$

4. $R_1 = 500\ \Omega$ $R_2 = 1.0\ \text{k}\Omega$ $R_3 = 300\ \Omega$
   $E_1 = 30\ \text{V}$ $E_2 = 20\ \text{V}$

5. $R_1 = 200\ \Omega$ $R_2 = 700\ \Omega$ $R_3 = 600\ \Omega$
   $E_1 = 15\ \text{V}$ $E_2 = 20\ \text{V}$

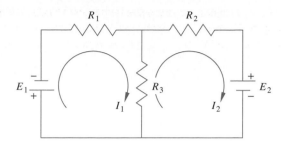

FIGURE 17–31 Network for Exercise 17–3, problems 1–5.

Write the mesh equations of Figure 17–32 and solve for I_1, I_2, and I_3 when:

6. $R_1 = 2\,\Omega$ $R_2 = 4\,\Omega$ $R_3 = 8\,\Omega$ $R_4 = 1\,\Omega$ $R_5 = 5\,\Omega$
  $E_1 = 10\,\text{V}$ $E_2 = 10\,\text{V}$ $E_3 = 10\,\text{V}$

7. $R_1 = 10\,\Omega$ $R_2 = 2\,\Omega$ $R_3 = 4\,\Omega$ $R_4 = 1\,\Omega$ $R_5 = 3\,\Omega$
  $E_1 = 6\,\text{V}$ $E_2 = 9\,\text{V}$ $E_3 = 3\,\text{V}$

8. $R_1 = 20\,\Omega$ $R_2 = 60\,\Omega$ $R_3 = 10\,\Omega$ $R_4 = 30\,\Omega$ $R_5 = 50\,\Omega$
  $E_1 = 15\,\text{V}$ $E_2 = 27\,\text{V}$ $E_3 = 12\,\text{V}$

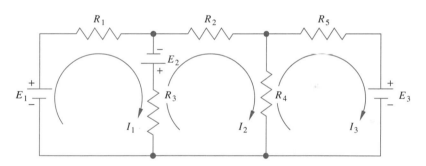

FIGURE 17–32 Network for Exercise 17–3, problems 6–8.

Write the mesh equations of Figure 17–33 and solve for I_1, I_2, and I_3 when:

9. $R_1 = 2\,\Omega$ $R_2 = 7\,\Omega$ $R_3 = 5\,\Omega$ $R_4 = 3\,\Omega$ $R_5 = 6\,\Omega$ $E_1 = 10\,\text{V}$

10. $R_1 = 9\,\Omega$ $R_2 = 1\,\Omega$ $R_3 = 2\,\Omega$ $R_4 = 8\,\Omega$ $R_5 = 5\,\Omega$ $E_1 = 6\,\text{V}$

11. Using the values computed in problem 9, determine the branch current through R_2 of Figure 17–33.

12. Using the values computed in problem 10, determine the branch current through R_1 of Figure 17–33.

13. The circuit of Figure 17–34 is a simplified diagram of an automotive electrical circuit. Using the information given in Figure 17–34, determine if the battery is being charged or discharged, and find the current through R_L.

FIGURE 17–33 Network for Exercise 17–3, problems 9–12.

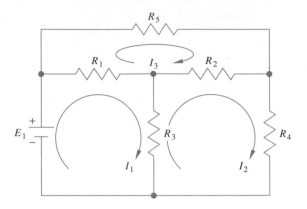

FIGURE 17–34 Network for Exercise 17–3, problem 13.

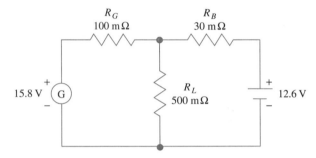

14. The circuit of Figure 17–35 represents a three-wire distribution system in which R_1, R_2, and R_3 represent the wire resistances and R_{L_1} and R_{L_2} represent the loads. Find the current through R_3 using the information given in Figure 17–35.

15. Change R_{L_1} to 25 Ω and leave R_{L_2} at 10 Ω. Repeat problem 14.

FIGURE 17–35 Network for Exercise 17–3, problem 14.

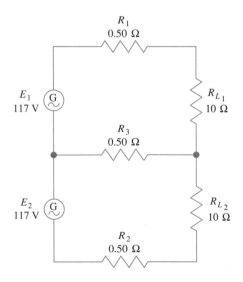

SELECTED TERMS

conventional current A way of thinking of the direction of current flow in an electric circuit. Current is thought of as flowing from positive to negative.

mesh A simple closed loop having no branches.

mesh currents Currents with an assumed direction used to solve a network using mesh analysis. Once the mesh currents are computed, then the actual branch currents in the network may be computed.

mutual resistance The resistance shared by two mesh currents.

SECTION CHALLENGE

WEB CHALLENGE FOR CHAPTERS 16 AND 17

To evaluate your comprehension of Chapters 16 and 17, log on to **www.prenhall. com/harter** and take the online True/False and Multiple Choice assessments for each of the chapters.

SECTION CHALLENGE FOR CHAPTERS 16 AND 17*

You will have several challenges that will require you to use systems of linear equations to solve the problems. You may use any of the several methods for solving systems of equations to arrive at your solution.

A. The following measurements were obtained when an instrument was interconnected to various terminals of a device and capacitance readings were made. Determine the values of capacitance C_1, C_2, and C_3 when:

$$C_1 + C_3 = 11.00 \text{ pF}$$
$$C_3 + C_2 = 3.83 \text{ pF}$$
$$C_2 + C_1 = 9.05 \text{ pF}$$

B. Solve for the branch currents I_{R1}, I_{R2}, and I_{RE}, the currents in resistors R_1, R_2, and R_E respectively, of Figure C–7 when $V_{BB} = 12.0$ V, $R_1 = 39.0$ kΩ, $R_2 = 5.60$ kΩ, $R_E = 910$ Ω, and $V_{BE} = 0.650$ V. Start your solution by assuming a clockwise direction for the mesh currents, I_1 and I_2, as shown in Figure C–8.

C. A tuned circuit is designed to resonate at a certain frequency, f_0, with a specified quality factor, Q_0. The resonant frequency depends on the product of inductance, L, and capacitance, C, and the quality factor depends on the quotient of L and C. Determine the value of inductance, in henrys (H), and capacitance, in farads (F), when the product of L and C is 4.620×10^{-14} and the quotient of L and C is 1.155×10^6.

D. Three resistances, connected in series, have an equivalent resistance, R_T, of 68.0 Ω. Two times the smallest resistance is equal to 29.0 Ω less than the sum of the other two resistances. Three times the mid-value resistance is 5/4 of the sum of the largest and smallest resistance. Determine the values of the three resistances.

*The solution to this Section Challenge is found in Appendix C.

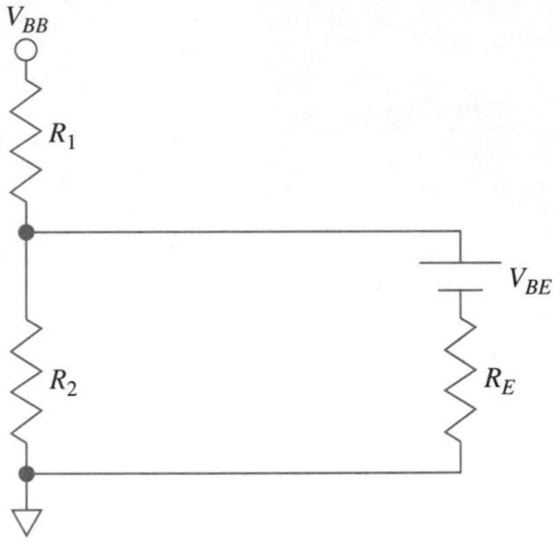

FIGURE C–7 Schematic of a transistor *bias* circuit with labeled components.

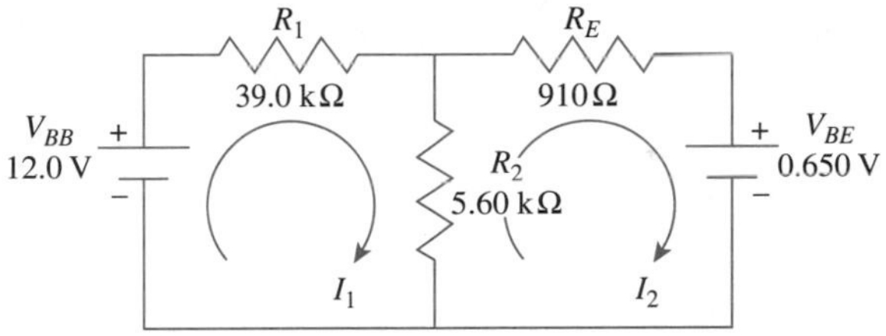

FIGURE C–8 Network drawn with mesh currents and component values indicated.

18

Solving Quadratic Equations

18–1   Introduction

18–2   Solving Incomplete Quadratic Equations

18–3   Solving Complete Quadratic Equations

18–4   Solving Quadratic Equations by the Quadratic Formula

18–5   Graphing the Quadratic Function

18–6   Applying the Techniques of Solving Quadratic Equations to Electronic Problems

PERFORMANCE OBJECTIVES

- Solve incomplete quadratic equations.

- Solve complete quadratic equations.

- Determine the number of roots.

- Determine the location of the vertex.

- Graph a quadratic function.

The 1000-foot diameter parabolic radio/radar antenna of the Arecibo (Puerto Rico) Observatory of the National Astronomy and Ionosphere Center can be used to observe radio and radar signals with wavelengths as short as 6.0 cm. (Courtesy of NASA)

Recall that equations containing a variable to the first power are first-degree, or linear, equations. In this chapter, you will study second-degree, or **quadratic equations.** The topics included in this chapter have been selected to introduce you to the solution and application of quadratic equations.

18–1 INTRODUCTION

A quadratic equation may be solved by applying one of the following four methods: (1) square root, (2) factoring, (3) quadratic formula, or (4) graphing.

Square Root: *Incomplete* or *pure quadratic* equations are best solved with the square-root method. An incomplete quadratic contains only the second power of the variable. Thus, $x^2 = 25$, $4x^2 = 36$, and $x^2 - 9 = 0$ are incomplete or pure quadratic equations.

Factoring: *Complete* quadratic equations may be solved by this method when the roots are rational. A complete quadratic contains terms with both the first and second powers of the variable. Thus, $3x^2 + 5x = 2$ and $6x^2 + 4x - 2 = 0$ are complete quadratic equations.

Quadratic Formula: The quadratic formula may be used to solve any quadratic equation. This method is easily used with a calculator.

Graphing: Any quadratic equation can be solved by graphing when the roots of the equation are real. This method provides a visual understanding of why a quadratic may have *two* real roots, *one* real root, or *no* real roots.

Each of these four methods is explored in the following four sections.

18–2 SOLVING INCOMPLETE QUADRATIC EQUATIONS

Quadratic equations that do not contain a linear term are **incomplete,** or **pure, quadratic equations.** Quadratics of this type have the general form $ax^2 + b = 0$.
Incomplete quadratics are solved by applying Rule 18–1.

RULE 18–1.   *Solving Incomplete Quadratic Equations*

To solve an incomplete quadratic equation:
1. Isolate the squared variable in the left member.
2. Take the square root of each member of the equation.
3. For the first root, assign a plus sign. For the second root, assign a minus sign.
4. Check each root.

EXAMPLE 18–1   Solve $3x^2 - 108 = 0$.

Solution   Solve the equation for x^2:

$$3x^2 - 108 = 0$$
$$x^2 = \frac{108}{3}$$
$$x^2 = 36$$

Take the square root of both members:

$$\sqrt{x^2} = \pm\sqrt{36}$$
$$x = \pm 6$$

Check   Does $3x^2 - 108 = 0$ when $x = 6$?

$$3(36) - 108 = 0$$
$$108 - 108 = 0 \quad \text{Yes!}$$

Check   Does $3x^2 - 108 = 0$ when $x = -6$?

$$3(36) - 108 = 0$$
$$108 - 108 = 0 \quad \text{Yes!}$$

$\therefore$ $x = 6$ and $x = -6$ are the roots of $3x^2 - 108 = 0$.

EXAMPLE 18–2   Solve $4x^2 - 7 = 3x^2 + 9$.

Solution   Rearrange terms:

$$4x^2 - 3x^2 = 9 + 7$$
$$x^2 = 16$$
$$x = \pm\sqrt{16} = \pm 4$$

Check   Does $4x^2 - 7 = 3x^2 + 9$ when $x = 4$?

$$4(16) - 7 = 3(16) + 9$$
$$64 - 7 = 48 + 9$$
$$57 = 57 \quad \text{Yes!}$$

Check   Does $4x^2 - 7 = 3x^2 + 9$ when $x = -4$?

$$4(16) - 7 = 3(16) + 9$$
$$64 - 7 = 48 + 9$$
$$57 = 57 \quad \text{Yes!}$$

$\therefore$ $x = 4$ and $x = -4$ are the roots of $4x^2 - 7 = 3x^2 + 9$.

Solve the following incomplete quadratic equations. Express radical roots as decimal numbers with three significant figures.

1. $x^2 = 25$

2. $x^2 = 49$

3. $x^2 = 8$

4. $x^2 = 13$

5. $y^2 - 49 = 0$

6. $y^2 - 144 = 0$

7. $x^2 - 18 = 0$

8. $x^2 - 32 = 0$

9. $3x^2 = 300$

10. $4y^2 = 100$

11. $16y^2 = 64$

12. $5x^2 = 100$

13. $72y^2 - 2 = 0$

14. $5x^2 - \frac{1}{20} = 0$

15. $4x^2 + 9 = 45$

16. $3y^2 + 9 = 57$

17. $3x^2 - 15.6 = 0$

18. $3x^2 = 48 - 5x^2$

19. $\frac{x^2}{3} + x = \frac{2x + 12}{2}$

20. $\frac{2x + 1}{3} = \frac{3}{2x - 1}$

18–3 SOLVING COMPLETE QUADRATIC EQUATIONS

Quadratic equations having both linear and quadratic terms of the variable are **complete quadratic equations.** Quadratics of this type have the general form $ax^2 + bx + c = 0$. Complete quadratic equations having rational roots may be solved by factoring.

Factoring

To solve complete quadratic equations by factoring, you must first learn how to factor *general* trinomials. The result of factoring trinomials of the form $ax^2 + bx + c$ is the product of two binomials.

Because there is no general rule for factoring this type of trinomial, you will learn this method by *trial and error.* Carefully study the following examples.

EXAMPLE 18–3   Factor $x^2 + 3x + 2$.

Solution   The factors of this trinomial are two binomials whose product is $x^2 + 3x + 2$:

$$(?)(?) = x^2 + 3x + 2$$

Because both 3 and 2 are positive, the signs of the binomial factors are both +:

$$(? + ?)(? + ?) = x^2 + 3x + 2$$

Factor x^2 into x and x; x is the first term of each of the binomials:

$$(x + ?)(x + ?) = x^2 + 3x + 2$$

Factor 2 into 1 and 2; these are the last terms of the binomials:

$$(x + 1)(x + 2) = x^2 + 3x + 2$$

Test

Does the left member of the equation equal the right?

$$x^2 + 3x + 2 = x^2 + 3x + 2 \quad \text{Yes!}$$

$$\therefore \quad x^2 + 3x + 2 = (x + 1)(x + 2)$$

EXAMPLE 18–4

Factor $5x^2 - 17x + 14$.

Solution

Because 14 is positive and 17 is negative, the signs of the binomial factors are both negative:

$$(? - ?)(? - ?) = 5x^2 - 17x + 14$$

Factor $5x^2$ into x and $5x$:

$$(x - ?)(5x - ?) = 5x^2 - 17x + 14$$

Test

Factor 14 into 7 and 2 and **test the product:**

$$(x - 7)(5x - 2) \overset{?}{=} 5x^2 - 17x + 14$$
$$5x^2 - 37x + 14 \neq 5x^2 - 17x + 14$$

This yields the wrong product; interchange the 7 and the 2:

$$(x - 2)(5x - 7) = 5x^2 - 17x + 14$$

Test the product:

$$5x^2 - 17x + 14 = 5x^2 - 17x + 14 \quad \text{Correct.}$$

$$\therefore \quad 5x^2 - 17x + 14 = (x - 2)(5x - 7)$$

EXAMPLE 18–5

Factor $6x^2 - 13x - 5$.

Solution

Because 5 is negative, the signs of the binomial factors are not the same:

$$(? + ?)(? - ?)$$

Set up a series of binomial products made up of the factors of $6x^2$ and 5 and then expand the products:

$$(6x + 1)(x - 5) = 6x^2 - 29x - 5$$
$$(x + 1)(6x - 5) = 6x^2 + x - 5$$
$$(3x + 1)(2x - 5) = 6x^2 - 13x - 5$$
$$(2x + 1)(3x - 5) = 6x^2 - 7x - 5$$
$$(6x - 1)(x + 5) = 6x^2 + 29x - 5$$
$$(x - 1)(6x + 5) = 6x^2 - x - 5$$
$$(3x - 1)(2x + 5) = 6x^2 + 13x - 5$$
$$(2x - 1)(3x + 5) = 6x^2 + 7x - 5$$

By inspection, the third product results in the correct trinomial.

$$\therefore \quad 6x^2 - 13x - 5 = (3x + 1)(2x - 5)$$

Table 18–1 summarizes the sign pattern of the binomial factors implied by the sign pattern of the trinomial to be factored.

TABLE 18–1   Sign Patterns of Binomial Factors

Trinomial	Sign Pattern		Binomial	Factor Sign Pattern
$ax^2 + bx + c$	plus plus	$\Rightarrow$	plus plus	$(\,+\,)(\,+\,)$
$ax^2 - bx + c$	minus plus	$\Rightarrow$	minus minus	$(\,-\,)(\,-\,)$
$ax^2 - bx - c$	minus minus	$\Rightarrow$	minus plus	$(\,-\,)(\,+\,)$
$ax^2 + bx - c$	plus minus	$\Rightarrow$	minus plus	$(\,-\,)(\,+\,)$

EXERCISE 18–2

Factor the following trinomials.

1. $x^2 + 7x + 10$
2. $y^2 + 6y + 8$
3. $a^2 + 8a + 7$
4. $b^2 + 3b + 2$
5. $x^2 + 2x + 1$
6. $y^2 + 6y + 5$
7. $a^2 - 5a + 4$
8. $x^2 - 6x + 5$
9. $y^2 - 7y + 12$
10. $c^2 - 11c + 18$
11. $x^2 - 10x + 24$
12. $y^2 - 2y - 3$
13. $a^2 - 7a - 18$
14. $b^2 - 5b - 14$
15. $x^2 + 3x - 40$
16. $y^2 + 3y - 4$
17. $2x^2 + 5x + 3$
18. $5y^2 - 7y + 2$
19. $3x^2 + 7x - 6$
20. $3x^2 - 14x - 5$
21. $4b^2 + 4b - 15$
22. $5y^2 - 2y - 7$
23. $5a^2 - 17a + 14$
24. $12x^2 + 11x - 15$
25. $6x^2 + 25x + 14$
26. $9y^2 + 3y - 2$
27. $18a^2 - 19a - 12$
28. $24a^2 + 5a - 36$
29. $4b^2 - 11b + 6$
30. $10x^2 + 11x - 18$

Solving Quadratic Equations

Now that you know how to factor trinomials, you can solve complete quadratic equations. The general procedure for solving complete quadratic equations is summarized as Rule 18–2. Rule 18–2 depends on the following observation: If the product of two factors is zero, then one or the other factor is zero. In symbols: if $A \cdot B = 0$, then $A = 0$ or $B = 0$.

> **RULE 18–2.   Solving Complete Quadratic Equations**
>
> To solve a complete quadratic equation with rational roots:
> 1. Express the equation in the form of $ax^2 + bx + c = 0$.
> 2. Factor the left member into two binomial factors.
> 3. Set each factor equal to zero.
> 4. Solve for each root.
> 5. Check the roots.

EXAMPLE 18–6   Solve $6x^2 - 13x = 5$ by factoring.

Solution   Use the steps of Rule 18–2 to solve.

Step 1:   Express the equation in the correct form:

$$6x^2 - 13x - 5 = 0$$

Step 2:   Factor the left member into two binomial factors:

$$(3x + 1)(2x - 5) = 0$$

Steps 3–4:   Set each factor to zero and solve for x:

$$3x + 1 = 0 \qquad 2x - 5 = 0$$
$$3x = -1 \qquad 2x = 5$$
$$x = -\tfrac{1}{3} \qquad x = \tfrac{5}{2}$$

Step 5:   Check.

Check   Does $6x^2 - 13x = 5$ when $x = -\tfrac{1}{3}$?

$$6(\tfrac{1}{9}) - 13(-\tfrac{1}{3}) = 5$$
$$\tfrac{2}{3} + \tfrac{13}{3} = 5$$
$$\tfrac{15}{3} = 5$$
$$5 = 5 \quad \text{Yes!}$$

Check   Does $6x^2 - 13x = 5$ when $x = \tfrac{5}{2}$?

$$6(\tfrac{25}{4}) - 13(\tfrac{5}{2}) = 5$$
$$\tfrac{75}{2} - \tfrac{65}{2} = 5$$
$$\tfrac{10}{2} = 5$$
$$5 = 5 \quad \text{Yes!}$$

$\therefore$ The roots of $6x^2 - 13x = 5$ are $x = -\tfrac{1}{3}$ and $x = \tfrac{5}{2}$.

EXAMPLE 18–7   Solve $x(5x - 17) = -14$.

Solution   To solve, follow the steps of Rule 18–2.

Step 1:   Express the equation in the correct form:

$$5x^2 - 17x + 14 = 0$$

Step 2:   Factor the left member into two binomial factors:

$$(x - 2)(5x - 7) = 0$$

Steps 3–4:   Set each factor to zero and solve for x:

$$x - 2 = 0 \qquad 5x - 7 = 0$$
$$x = 2 \qquad\qquad x = \tfrac{7}{5}$$

Step 5:   Check.

Check   Does $x(5x - 17) = -14$ when $x = 2$?

$$2[5(2) - 17] = -14$$
$$20 - 34 = -14$$
$$-14 = -14 \quad \text{Yes!}$$

Check   Does $x(5x - 17) = -14$ when $x = \tfrac{7}{5}$?

$$\tfrac{7}{5}[5(\tfrac{7}{5}) - 17] = -14$$
$$\tfrac{49}{5} - \tfrac{119}{5} = -14$$
$$-\tfrac{70}{5} = -14$$
$$-14 = -14 \quad \text{Yes!}$$

$\therefore$ The roots of $x(5x - 17) = -14$ are $x = 2$ and $x = \tfrac{7}{5}$.

EXERCISE 18–3

Use factoring to solve each of the following for its roots.

1. $y^2 + 6y + 8 = 0$ **2.** $x^2 + 3x + 2 = 0$

3. $x^2 + 6x + 5 = 0$ **4.** $y^2 - 5y + 4 = 0$

5. $y^2 - 6y = -5$ **6.** $x^2 - 11x = -18$

7. $x(x + 2) + 1 = 0$ **8.** $x(x - 10) = -24$

9. $y^2 - 2y = 3$ **10.** $x(2x + 5) = -3$

11. $x^2 - 5x = 14$ **12.** $y(y + 3) = 4$

13. $y(5y - 7) = -2$ **14.** $4x(x + 1) = 15$

15. $3x^2 - 14x - 5 = 0$ **16.** $5y^2 - 2y - 7 = 0$

17. $12x^2 + 11x - 15 = 0$ **18.** $3y(3y + 1) = 2$

19. $x(24x + 5) = 36$ **20.** $x(10x + 11) = 18$

The general form of a quadratic equation is

$$ax^2 + bx + c = 0 \qquad \text{(18–1)}$$

This general equation for a quadratic may be solved for x. The result of this solution is called the **quadratic formula.** Developing the formula:

$$ax^2 + bx + c = 0$$

$$\text{S: } c \qquad ax^2 + bx = -c$$

$$\text{D: } a \qquad x^2 + \frac{b}{a}x = \frac{-c}{a}$$

Form the left member into a perfect trinomial square by adding $(b/2a)^2$ to each member. (This technique is called **completing the square.**)

$$x^2 + \frac{b}{a}x + \left(\frac{b}{2a}\right)^2 = \left(\frac{b}{2a}\right)^2 - \frac{c}{a}$$

Factor the left member into a binomial square:

$$\left(x + \frac{b}{2a}\right)^2 = \left(\frac{b}{2a}\right)^2 - \frac{c}{a}$$

Take the square root of each member:

$$x + \frac{b}{2a} = \pm\sqrt{\left(\frac{b}{2a}\right)^2 - \frac{c}{a}}$$

Solve for x:

$$x = -\frac{b}{2a} \pm \sqrt{\left(\frac{b}{2a}\right)^2 - \frac{c}{a}} \quad \text{or} \quad x = \frac{-b}{2a} \pm \sqrt{\left(\frac{-b}{2a}\right)^2 - \frac{c}{a}} \qquad \text{(18–2)}$$

Since $b^2 = (-b)^2$, this is the form best used with the calculator. Simplify the radicand:

$$x = -\frac{b}{2a} \pm \sqrt{\frac{b^2}{4a^2} - \frac{4ac}{4a^2}}$$

$$x = -\frac{b}{2a} \pm \sqrt{\frac{b^2 - 4ac}{4a^2}}$$

$$x = -\frac{b}{2a} \pm \frac{\sqrt{b^2 - 4ac}}{2a}$$

$$x = \frac{-b \pm \sqrt{b^2 - 4ac}}{2a} \tag{18-3}$$

This is the traditional form of the quadratic formula. The quadratic equation and the quadratic formulas are summarized in Table 18–2.

TABLE 18–2 The Quadratic Equation and Formula

Quadratic equation	$ax^2 + bx + c = 0$	(18–1)
Quadratic formula (calculator)	$x = \dfrac{-b}{2a} \pm \sqrt{\left(\dfrac{-b}{2a}\right)^2 - \dfrac{c}{a}}$	(18–2)
Quadratic formula (traditional)	$x = \dfrac{-b \pm \sqrt{b^2 - 4ac}}{2a}$	(18–3)

where a = numerical coefficient of x^2
    b = numerical coefficient of x
    c = numerical constant
    x = unknown quantity

EXAMPLE 18–8 Solve $x^2 + 3x + 2 = 0$ by the quadratic formula.

Solution Using Equation 18–3:

$$x = \frac{-b \pm \sqrt{b^2 - 4ac}}{2a}$$

Substitute $a = 1, b = 3, c = 2$:

$$x = \frac{-3 \pm \sqrt{9 - 4(1)(2)}}{2(1)}$$

$$x = \frac{-3 \pm \sqrt{9 - 8}}{2}$$

$$x = \frac{-3 \pm 1}{2}$$

$$x = \frac{-3 + 1}{2} \quad \text{or} \quad x = \frac{-3 - 1}{2}$$

$$x = -1 \text{ or } -2$$

Check Does $x^2 + 3x + 2 = 0$ when $x = -1$?

$$1 - 3 + 2 = 0$$
$$0 = 0 \quad \text{Yes!}$$

Check   Does $x^2 + 3x + 2 = 0$ when $x = -2$?
$$4 - 6 + 2 = 0$$
$$0 = 0 \quad \text{Yes!}$$

∴   The roots of $x^2 + 3x + 2 = 0$ are $x = -1$ and $x = -2$.

EXAMPLE 18–9   Solve $3x^2 - x = 2$ by the quadratic formula.

Solution   Put in correct form:
$$3x^2 - x - 2 = 0$$

Substitute $a = 3$, $b = -1$, and $c = -2$ into the calculator form of the quadratic formula:

$$x = \frac{-b}{2a} \pm \sqrt{\left(\frac{-b}{2a}\right)^2 - \frac{c}{a}}$$

$$x = \frac{1}{6} \pm \sqrt{\left(\frac{1}{6}\right)^2 - \frac{-2}{3}}$$

$$x = 0.1667 \pm 0.8333$$

$$x = 1.00 \text{ or } -0.667$$

Check   When $x = 1.00$, does $3x^2 - x = 2$?
$$3 - 1 = 2$$
$$2 = 2 \quad \text{Yes!}$$

Check   When $x = -0.667$, does $3x^2 - x = 2$?
$$3(-0.667)^2 - (-0.667) = 2$$
$$1.33 + 0.667 = 2$$
$$2 = 2 \quad \text{Yes!}$$

∴   The roots of $3x^2 - x = 2$ are $x = 1.00$ and $x = -0.667$

EXAMPLE 18–10   Solve $31.5x^2 - 52x = 23.6$ by the quadratic formula.

Solution   Put into the correct form:
$$31.5x^2 - 52x - 23.6 = 0$$

Substitute $a = 31.5$, $b = -52$, and $c = -23.6$ into the calculator form of the quadratic formula:

$$x = \frac{-b}{2a} \pm \sqrt{\left(\frac{-b}{2a}\right)^2 - \frac{c}{a}}$$

$$x = \frac{52}{2(31.5)} \pm \sqrt{\left(\frac{52}{2(31.5)}\right)^2 + \frac{23.6}{31.5}}$$

$$x = 2.02 \text{ or } -0.371$$

Check   Does $31.5x^2 - 52x = 23.6$ when $x = 2.02$?

$$31.5(2.02)^2 - 52(2.02) = 23.6$$
$$23.6 = 23.6 \quad \text{Yes!}$$

Check   Does $31.5x^2 - 52x = 23.6$ when $x = -0.371$?

$$31.5(-0.371)^2 - 52(-0.371) = 23.6$$
$$23.6 = 23.6 \quad \text{Yes!}$$

$\therefore$ The roots of $31.5x^2 - 52x = 23.6$ are $x = 2.02$ and $x = -0.371$.

EXERCISE 18-4

PROGRAMMABLE

Use the quadratic formula to solve each of the following for its roots:

1. $2x^2 + 5x = 3$ **2.** $4(y^2 + y) = 15$

3. $3x^2 + 7x = 6$ **4.** $6y^2 + 25y = -14$

5. $18y^2 - 19y - 12 = 0$ **6.** $4x^2 - 11x + 6 = 0$

7. $12x^2 + 11x = 15$ **8.** $24x^2 + 5x - 36 = 0$

9. $5.2y^2 - 7.3y = -2.2$ **10.** $11.3x^2 + 12.1x = 19$

18-5 GRAPHING THE QUADRATIC FUNCTION

Quadratic equations may be graphed and the roots found (assuming that they are real) by constructing a table of values and plotting the resulting points. Besides solving the quadratic equation, graphs provide a means of understanding the behavior of the roots of the quadratic equation. In this section, graphing will be used to explore the shape of the curve and the location of the vertex of the curve, and to determine the nature of the roots of a quadratic equation.

Determining the Shape of the Curve

The general quadratic equation $ax^2 + bx + c = 0$ is a function of x. For purposes of graphing, let $y = f(x)$. Thus,

PROGRAMMABLE

$$y = ax^2 + bx + c \tag{18-4}$$

The effect of a, the coefficient of x^2, will be investigated in the following example.

EXAMPLE 18–11   Determine the effect of a on the shape of the curve of $y = ax^2$ ($b = 0$, $c = 0$).

Solution   Plot $y = ax^2$ for the following values of a:

$$a = 1$$

x	-4	-3	-2	-1	0	1	2	3	4
y	16	9	4	1	0	1	4	9	16

FIGURE 18–1   Effect of a, the coefficient of x^2, on the shape of the curve of the quadratic equation $y = ax^2$.

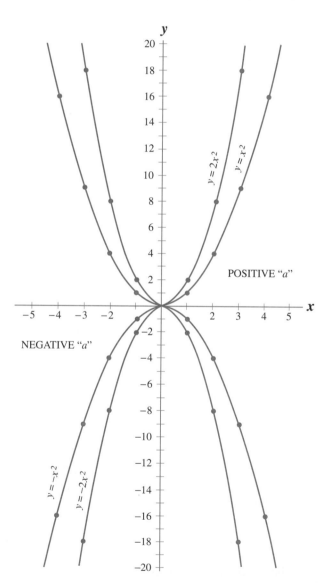

$a = 2$

x	-3	-2	-1	0	1	2	3
y	18	8	2	0	2	8	18

$a = -1$

x	-4	-3	-2	-1	0	1	2	3	4
y	-16	-9	-4	-1	0	-1	-4	-9	-16

$a = -2$

x	-3	-2	-1	0	1	2	3
y	-18	-8	-2	0	-2	-8	-18

Observation　From the curves of Figure 18–1, the following can be concluded:

1. A positive a $(a > 0)$ results in an *opening upward* parabola.
2. A negative a $(a < 0)$ results in an *opening downward* parabola.
3. As a gets larger, the curve gets *narrower* and *steeper*.

The effects of c on the position of the curve are investigated in the following example.

EXAMPLE 18–12　Determine the effects of c on the position of the curves of the form $y = x^2 - 4x + c$.

Solution　Plot $y = x^2 - 4x + c$ for the following values of c:

$c = 0$

x	-2	-1	0	1	2	3	4	5	6
y	12	5	0	-3	-4	-3	0	5	12

$c = 4$

x	-2	-1	0	1	2	3	4	5	6
y	16	9	4	1	0	1	4	9	16

$c = 8$

x	-2	-1	0	1	2	3	4	5	6
y	20	13	8	5	4	5	8	13	20

FIGURE 18–2 Effect of c on the position of the curve of the quadratic equation $y = x^2 - 4x + c$.

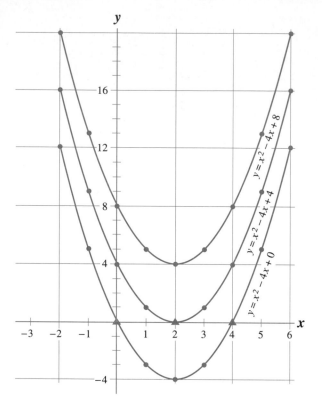

Observation From the *family* of curves of Figure 18–2, we see that c determines the vertical position of the curve. As c gets smaller (decreases in size), the curve moves down.

Determining the Vertex of the Curve

The x-coordinate of the **vertex** (the turning point) of the curve is found by letting $x = -b/2a$. The y-coordinate is then found by substituting the value of x into the general graphic quadratic equation, $y = ax^2 + bx + c$. The ordered pair (x_0, y_0) of the vertex is

$$x_0 = -\frac{b}{2a} \tag{18--5}$$

$$y_0 = ax_0^2 + bx_0 + c = c - \frac{b^2}{4a} \tag{18--6}$$

EXAMPLE 18–13 Compute the coordinates of the vertex of the curve $y = 2x^2 - 8x - 5$.

Solution Find the coordinates using Equations 18–5 and 18–6:

$$x_0 = -\frac{b}{2a}$$

$$x_0 = -\frac{(-8)}{2(2)}$$

$$x_0 = 2 \qquad\qquad y_0 = c - \frac{b^2}{4a}$$

$$y_0 = 2x_0^2 - 8x_0 - 5 \qquad y_0 = -5 - \frac{64}{8}$$

$$y_0 = 2(2)^2 - 8(2) - 5$$

$$y_0 = -13 \qquad\qquad y_0 = -13$$

∴ The vertex coordinates are $(2, -13)$.

EXAMPLE 18–14 Compute the coordinates of the vertex of each of the curves in Figure 18–2.

Solution (a) $y = x^2 - 4x + 0$

$$x_0 = -b/2a \qquad y_0 = (2)^2 - 4(2) + 0$$
$$x_0 = 4/2 = 2 \qquad y_0 = -4$$

∴ The vertex coordinates are $(2, -4)$.

(b) $y = x^2 - 4x + 4$

$$x_0 = \frac{-b}{2a} \qquad\qquad y_0 = (2)^2 - 4(2) + 4$$

$$x_0 = 4/2 = 2 \qquad y_0 = 0$$

∴ The vertex coordinates are $(2, 0)$.

(c) $y = x^2 - 4x + 8$

$$x_0 = \frac{-b}{2a} \qquad\qquad y_0 = (2)^2 - 4(2) + 8$$

$$x_0 = 4/2 = 2 \qquad y_0 = 4$$

∴ The vertex coordinates are $(2, 4)$.

Determining the Characteristics of the Roots

There are three possible cases for the number and kinds of roots of the quadratic equation. Referring to the curves of Figure 18–2, notice each of the three cases.

Case 1: $y = x^2 - 4x + 0$. When the curve passes through the x-axis, the two x-intercepts are the **two real roots** of the quadratic equation. In this curve, the roots are $x = 0$ and $x = 4$.

Case 2: $y = x^2 - 4x + 4$. When the vertex of the curve just touches the x-axis, the two real roots occur together (called a **double root**). In this curve both roots are $x = 2$.

Case 3: $y = x^2 - 4x + 8$. When the entire curve is above (or below) the x-axis as in this case, there are **no real roots.** The roots are **complex.** Complex numbers are defined in Chapter 24.

By "reading" the radicand in the quadratic formula ($b^2 - 4ac$), we can tell which of the three cases will result. Because the quantity represented by $b^2 - 4ac$ tells which one of the three cases holds, it is called the **discriminant** of the quadratic equation. The kind and number of roots are determined by the size of the discriminant in the following manner.

$$b^2 - 4ac > 0 \quad \text{two different real roots}$$
$$b^2 - 4ac = 0 \quad \text{one double real root}$$
$$b^2 - 4ac < 0 \quad \text{no real roots}$$

EXAMPLE 18–15 Compute the discriminant of each of the quadratic equations of Figure 18–2.

Solution (a) $y = x^2 - 4x + 0$

$$b^2 - 4ac = 16 - 4(1)(0) = 16$$

∴ Because the discriminant is greater than zero, there are two different real roots.

(b) $y = x^2 - 4x + 4$

$$b^2 - 4ac = 16 - 4(1)(4) = 0$$

∴ Because the discriminant is equal to zero, there is one double real root.

(c) $y = x^2 - 4x + 8$

$$b^2 - 4ac = 16 - 4(1)(8) = -16$$

∴ Because the discriminant is less than zero, there are no real roots.

Table 18–3 is a summary of the possible number and kinds of roots of a quadratic equation.

TABLE 18–3   Three Cases of the Discriminant $b^2 - 4ac$

Case	Value of Discriminant	Number of Real Roots	Position of Curve
1	Positive	2	Passes through x-axis
2	Zero	1	Vertex touches x-axis
3	Negative	0	No x-intercept; entire curve above or below the x-axis

EXAMPLE 18–16   Solve $x^2 - 10x + 30$ by the quadratic formula.

Solution   Compute the discriminant of the quadratic formula first:

$$b^2 - 4ac = (-10)^2 - 4(1)(30)$$
$$b^2 - 4ac = 100 - 120$$
$$b^2 - 4ac = -20$$

∴   There are no real roots, because the value of the discriminant is less than zero ($-20 < 0$). The solution is terminated at this point.

EXAMPLE 18–17   Graph $y = -2x^2 + 4x + 6$.

Solution   Because a is negative, the parabola will open downward. Compute the discriminant to determine the nature of the roots:

$$b^2 - 4ac = 16 - 4(-2)(6)$$
$$= 16 + 48$$
$$b^2 - 4ac = 64 > 0$$

∴   Two different real roots.

Compute the roots:

$$x = \frac{-b \pm \sqrt{b^2 - 4ac}}{2a}$$

$$x = \frac{-4 \pm \sqrt{64}}{-4}$$

$$x = \frac{-4 \pm 8}{-4}$$

∴ $x = 3 \qquad x = -1$

Compute the coordinates of the vertex:

$$x_0 = -b/2a \qquad y_0 = -2x_0^2 + 4x_0 + 6$$
$$x_0 = -4/-4 \qquad y_0 = -2 + 4 + 6$$
$$x_0 = 1 \qquad y_0 = 8$$

∴   The vertex coordinates are (1, 8).

Construct a table of values that includes the coordinates of the roots and the vertex:

x	-2	-1	0	1	2	3	4
y	-10	0	6	8	6	0	-10

           root   vertex   root

Plot the ordered pairs and construct the curve as shown in Figure 18–3.

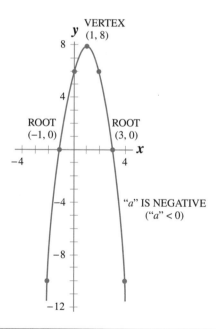

FIGURE 18–3   Graph of the quadratic equation $y = -2x^2 + 4x + 6$, used in Example 18–17.

EXERCISE 18–5

PROGRAMMABLE

In each of the following quadratic equations, determine (a) the shape of the curve (opening up or down), (b) the characteristic of the roots, (c) the coordinates of the vertex, and (d) the coordinates of the real roots.

Graph each of the functions of x by setting $y = f(x)$.

1. $2x^2$
2. $x^2 + 1$
3. $2x^2 + 2$
4. $x^2 + 2x + 1$
5. $-x^2 + 4x$
6. $-x^2 - 2x + 3$
7. $3x^2 + 5x + 3$
8. $x^2 - 3x - 4$
9. $-x^2 - 2x - 2$
10. $-3x^2 + 4x + 1$
11. $\frac{1}{2}x^2 - 6x + 5$
12. $-\frac{1}{3}x^2 - 3x + 4$

18–6 APPLYING THE TECHNIQUES OF SOLVING QUADRATIC EQUATIONS TO ELECTRONIC PROBLEMS

When the conditions of a problem lead to a quadratic equation, two answers will be obtained for the variable. In many problems, one of the answers can be discarded because it is physically impossible or inappropriate. For example, negative number of turns in a coil, negative dimension of an object, and positive values that seem inappropriate are all discarded.

EXAMPLE 18–18   The power dissipated in a 7.00-Ω load is 21.0 W. Determine the current passing through the load.

Solution $P = I^2R$ is an incomplete quadratic equation. Solve for I:

$$I = \pm\sqrt{P/R}$$

Substitute 7.00 Ω for R and 21.0 W for P:

$$I = \pm\sqrt{21.0/7.00}$$
$$I = \pm 1.73 \text{ A}$$

Select the positive root:

$\therefore$ $I = 1.73$ A

EXAMPLE 18–19   Determine the voltage (V_2) of the circuit shown in Figure 18–4.

Solution   Write a voltage loop equation around the path containing I_1, I_2, and E:

(1) $E = 2.0I_1 + V_2$

Write the current equation for node a:

(2) $I_1 = I_2 + I_3$

Substitute equation (2) into equation (1):

(3) $E = 2.0(I_2 + I_3) + V_2$

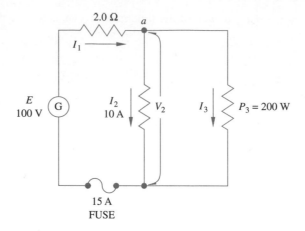

FIGURE 18–4 Circuit for Example 18–19.

Express I_3 in terms of P_3 (200 W) and V_2:

$$P_3 = V_2 I_3$$
$$I_3 = P_3/V_2 = 200/V_2$$

Substitute $200/V_2$ for I_3, 10 for I_2, and 100 for E in equation 3:

$$100 = 2.0(10 + 200/V_2) + V_2$$

Expand and clear fractions:

$$100 = 20 + 400/V_2 + V_2$$
$$100V_2 = 20V_2 + 400 + V_2^2$$

Place into the general quadratic form ($ax^2 + bx + c = 0$):

$$V_2^2 - 80V_2 + 400 = 0$$

Solve for V_2:

$$V_2 = \frac{-b \pm \sqrt{b^2 - 4ac}}{2a}$$

$$V_2 = \frac{80 \pm \sqrt{(-80)^2 - 4(1)(400)}}{2}$$

$$V_2 = 75 \text{ V} \quad \text{or} \quad V_2 = 5.4 \text{ V}$$

Because the 15A fuse will open *(blow)*, 5.4 V is inappropriate.

$$\therefore \quad V_2 = 75 \text{ V}$$

In the previous example, 75 V was selected as the appropriate value of voltage for V_2. This value was selected over 5.4 V based on the analysis of the circuit of Figure 18–4. Using 5.4 V across R_2, the current passing through the fuse is approximately 47 A, which would cause the fuse to open and the circuit to malfunction—an inappropriate condition.

EXAMPLE 18–20 Graphically determine the value of the load resistance that will cause maximum power to be developed in the load of Figure 18–5.

FIGURE 18–5 Circuit for Example 18–20.

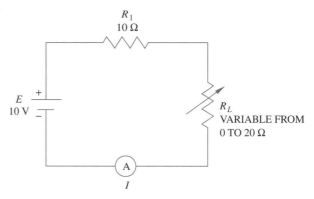

Solution Derive an equation that relates load power (P_L) to the resistance of the load (R_L).

$$(1) \ \ P_L = I^2 R_L$$

also

$$(2) \ \ I = \frac{E}{R_1 + R_L}$$

Substitute $E/(R_1 + R_L)$ for I in equation (1):

$$P_L = \left(\frac{E}{R_1 + R_L}\right)^2 R_L$$

Substitute $E = 10$ V and $R_1 = 10 \ \Omega$:

$$P_L = \left(\frac{10}{10 + R_L}\right)^2 R_L$$

Load power (P_L) is now a function of load resistance (R_L):

$$P_L = f(R_L) = \left(\frac{10}{10 + R_L}\right)^2 R_L$$

Construct a table of values:

R_L	4	6	8	9	10	11	12	16	20
$f(R_L)$	2.0	2.3	2.47	2.49	2.5	2.49	2.48	2.4	2.2

Plot the ordered pairs and construct the curve as shown in Figure 18–6.

FIGURE 18–6 Power transfer curve for the load resistance of Figure 18–5 and Example 18–20.

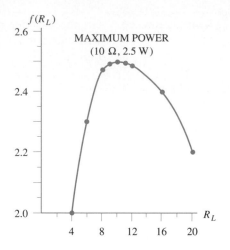

$\therefore$ The maximum power of 2.5 W is developed for $R_L = 10\ \Omega$.

EXERCISE 18–6

1. Determine the value of two resistors (R_1 and R_2) whose product is 440 if $R_2 = R_1 + 2.0\ \Omega$.

2. Determine the value of a resistor whose value squared is 30 more than its value.

3. The value of the voltage V (measured across a circuit load) when added to its square results in 12. What is the measured value of the voltage?

4. Two currents in a parallel circuit have a difference of 14 A and a product of 51. What are the numbers?

5. Determine the voltage across a 12.0-Ω load that is dissipating 500 mW.

6. Determine the current through a 47.0-Ω resistor that is dissipating 0.625 W.

7. Determine the value of two resistances (R_1 and R_2) connected in parallel. The total equivalent resistance is 200 Ω and $R_2 = R_1 + 300\ \Omega$.

8. Determine the value of R_1 and R_3 in the circuit of Figure 18–7.

FIGURE 18–7 Circuit for Exercise 18–6, problem 8.

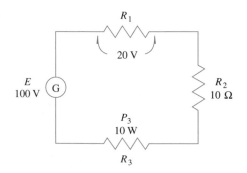

FIGURE 18–8 Circuit for Exercise
18–6, problem 9.

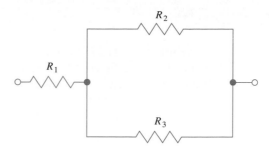

9. Three resistors are connected to form a series-parallel circuit as shown in Figure 18–8. The total equivalent resistance is 60.0 Ω, $R_1 = R_2 + 10.0$ Ω, and $R_3 = 2R_1$. What are the values of R_1, R_2, and R_3?

10. Determine the current (I_3) through R_3, given the information of Figure 18–9 and $R_1 + R_3 = 5.6$ kΩ.

FIGURE 18–9 Circuit for Exercise
18–6, problem 10.

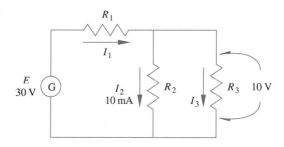

11. Graphically determine the value of load resistance that will cause maximum power to be developed in the load of Figure 18–10.

12. Repeat problem 11 when $R_1 = 1$ kΩ and R_L varies between 500 Ω and 3 kΩ.

FIGURE 18–10 Circuit for Exercise
18–6, problem 11.

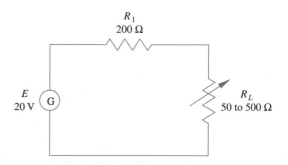

SELECTED TERMS

discriminant   The radicand in the quadratic equation. The number of real roots depends on the sign of the radicand.

quadratic equation   An equation containing the square of the independent variable but no higher order terms.

vertex   The turning point of the graph of a quadratic equation.

19

Exponents, Radicals, and Equations

19–1   Laws of Exponents

19–2   Zero and Negative Integers as Exponents

19–3   Fractional Exponents

19–4   Laws of Radicals

19–5   Simplifying Radicals

19–6   Radical Equations

PERFORMANCE OBJECTIVES

- Apply the laws of exponents to simplify fractional exponents.

- Apply the laws of radicals to simplify radical expressions.

- Employ the technique of rationalizing the denominator to remove the radical from the denominator.

- Solve radical equations.

The Sojourner Rover on the rocky surface of Mars.   (Courtesy of NASA/Jet Propulsion Laboratory)

In this chapter the laws of exponents are applied to literal expressions and to expressions with zero and negative integers as exponents. The laws for radicals are used to simplify expressions with radicals. The chapter concludes with the solution of radical equations.

19–1 LAWS OF EXPONENTS

The fundamental laws for working with exponents, along with an example of the application of each, are given below. Exponents follow the five stated laws:

RULE 19–1.   Five Laws of Exponents

1. $x^m x^n = x^{m+n}$ (19–1)

Example: $x^4 x^3 = x^{4+3} = x^7$

2. $(x^m)^n = x^{mn}$ (19–2)

Example: $(x^7)^3 = x^{7\times 3} = x^{21}$

3. $(xy)^m = x^m y^m$ (19–3)

Example: $(xy)^3 = x^3 y^3$

4. $\left(\dfrac{x}{y}\right)^m = \dfrac{x^m}{y^m}, y \neq 0$ (19–4)

Example: $\left(\dfrac{x}{y}\right)^3 = \dfrac{x^3}{y^3}$

5. $\dfrac{x^m}{x^n} = x^{m-n}, x \neq 0$ (19–5)

Example: $\dfrac{x^5}{x^3} = x^{5-3} = x^2$

EXERCISE 19–1

Use the five laws of exponents to simplify each of the following.

1. $a^3 a^5$ **2.** $(x^2)^3$ **3.** $(2b)^3$

4. $(x/y)^2$ **5.** $\dfrac{c^3}{c^2}$ **6.** $b^x b^{5x}$

7. $-(7c)^2$ **8.** $c^{3a} c^{-2a}$ **9.** $(y^{2x})^2$

10. $\dfrac{8x^3}{x^2}$ **11.** $\left(\dfrac{2x}{3a}\right)^3$ **12.** $(x^m y^n)^2$

13. $(a^b)^c$

14. $\dfrac{24x^5y^4}{-6x^3y^3}$

15. $-2x(xy)^3$

16. $(x^{3a})^a$

17. $y^{a+b}y^{2-b}$

18. $(2ab^2)^3$

19. $\left(\dfrac{2x^5y}{x^2}\right)^3$

20. $\left(\dfrac{2a^2}{b^3}\right)^3$

21. $\left(\dfrac{5S^2}{T}\right)^2$

19–2 ZERO AND NEGATIVE INTEGERS AS EXPONENTS

Zero Exponent

When evaluating the expression a^m/a^m, the solution is 1. Applying the laws of exponents to this expression results in $a^m/a^m = a^{m-m} = a^0$. However, we already know that $a^m/a^m = 1$, so a^0 must also be 1. Rule 19–2 states this.

RULE 19–2. Zero Exponent

Any number, except zero, raised to the zero power is 1. Thus $a^0 = 1$ $(a \neq 0)$.

Negative Exponents

What does the expression a^{-3} mean? Let's investigate the meaning of this result: $a^4/a^7 = a^{4-7} = a^{-3}$. However, a^4/a^7 and a^7/a^4 are reciprocals. Using the laws of exponents, $a^4/a^7 = a^{-3}$, and $a^7/a^4 = a^3$. Then a^{-3} and a^3 are also reciprocals. Therefore, $a^3 = 1/a^{-3}$, and $a^{-3} = 1/a^3$. We see that a^{-3} is the same as $1/a^3$.

RULE 19–3. Negative Exponents

Numbers written with negative exponents may be written with positive exponents by transferring the factor from one term of the fraction to the other term. *Examples:* a^{-m} may be written as $1/a^m$ and $1/b^{-n}$ may be written as b^n.

EXAMPLE 19–1   Express $\dfrac{3}{2x^{-2}}$ with positive exponents.

Solution $\dfrac{3}{2x^{-2}} = \dfrac{3x^2}{2}$

EXAMPLE 19–2   Express $\dfrac{x^5}{3^{-1}x}$ with positive exponents and simplify.

       Solution $\dfrac{x^{-5}}{3^{-1}x} = \dfrac{3}{x^5 \cdot x} = \dfrac{3}{x^6}$

EXERCISE 19–2

Write the value of each of the following.

1. 5^0
         2. $3^0(2)$
         3. $4a^0$

4. $(4a)^0$
       5. $(-4)^0$
       6. $6/(3x^0)$

7. $(2 + 2a^0)/4$
    8. 2^{-1}
          9. 2^{-2}

10. $(-3)^{-2}$
      11. $1/3^{-2}$
        12. 5×3^{-1}

13. $32(4^{-2})$
      14. $1/(2 + a^0)^{-2}$
    15. $1/(2^0 + a^0)$

Express each of the following with positive exponents.

16. a^{-3}
          17. xy^{-2}
        18. $-5b^{-2}$

19. $\dfrac{2a^{-3}}{b}$
         20. $\dfrac{a^{-1}b^{-1}}{c^{-1}}$
     21. $\dfrac{3a}{b^{-2}}$

22. $(x^{-1}y)^{-2}$
     23. $(x^2y^{-3})^{-1}$
     24. $(x^{-2}/y^{-3})^{-2}$

25. $(x/y)^{-2}$
        26. $\left(\dfrac{2ab}{a^2}\right)^{-2}$
     27. $(3b^2c)^{-3}$

19–3   FRACTIONAL EXPONENTS

Fractional exponents follow the laws of exponents where the numerator indicates the power of the base and the denominator indicates the root of the base. Thus, $a^{m/n} = \sqrt[n]{a^m}$, read as "the nth root of a to the mth power."

> ### RULE 19–4.   Fractional Exponents
>
> Fractional exponents follow the laws of integer exponents:
> 1. The numerator of the fraction is the power of the base.
> 2. The denominator of the fraction is the root of the base.

Fractional exponents are used to simplify radicals, as is shown in the following examples. They are the basis of the system of logarithms studied in Chapter 20.

EXAMPLE 19–3   Write $8^{2/3}$ with a radical and evaluate.

Solution $8^{2/3} = \sqrt[3]{8^2}$

Square 8: $8^{2/3} = \sqrt[3]{64}$

Extract root: $\sqrt[3]{64} = 4$

$\therefore$ $8^{2/3} = \sqrt[3]{8^2} = 4$

EXAMPLE 19–4   Evaluate $\sqrt[4]{3^6}$ with your calculator.

Solution   Express $\sqrt[4]{3^6}$ with fractional exponents:

$3^{6/4}$

Express the fraction as a decimal:

$3^{6/4} = 3^{1.5}$

Evaluate $3^{1.5}$ using $\boxed{y^x}$ on your calculator:

$3 \; \boxed{y^x} \; 1.5 \Rightarrow 5.196$

$\therefore$ $\sqrt[4]{3^6} = 5.2$

EXAMPLE 19–5   Simplify $\left(\dfrac{a^2 b^3}{a^3}\right)^{1/6}$.

Solution   Distribute the 1/6 to the other exponents:

$$\left(\frac{a^2 b^3}{a^3}\right)^{1/6} = \frac{a^{2/6} b^{3/6}}{a^{3/6}}$$

Clear the denominator:

$$\frac{a^{2/6} b^{3/6}}{a^{3/6}} = a^{2/6} a^{-3/6} b^{3/6}$$

Simplify:

$$a^{-1/6} b^{1/2} = b^{1/2}/a^{1/6}$$

$\therefore$ $\left(\dfrac{a^2 b^3}{a^3}\right)^{1/6} = b^{1/2}/a^{1/6}$

EXAMPLE 19–6   Simplify $(\sqrt[3]{x^2})(\sqrt[2]{x})^3$.

Solution   Express with fractional exponents:

$$x^{2/3}(x^{1/2})^3 = x^{2/3}x^{3/2}$$

Find the common denominator and add exponents:

$$x^{2/3}x^{3/2} = x^{4/6}x^{9/6} = x^{13/6}$$

Simplify:

$$x^{13/6} = x^2 \cdot x^{1/6} = x^2\sqrt[6]{x}$$

$$\therefore \quad (\sqrt[3]{x^2})(\sqrt[2]{x})^3 = x^2\sqrt[6]{x}$$

EXERCISE 19–3

Calculator Drill
Evaluate each of the following.

1. $25^{1/2}$ 2. $36^{1/2}$ 3. $144^{1/2}$

4. $27^{1/3}$ 5. $64^{1/3}$ 6. $32^{1/5}$

7. $4^{3/2}$ 8. $125^{2/3}$ 9. $32^{2/5}$

10. $48^{5/6}$ 11. $96^{3/8}$ 12. $50^{7/5}$

13. $(\frac{1}{16})^{1/2}$ 14. $(\frac{4}{9})^{3/2}$ 15. $(0.062)^{1/3}$

16. $(0.125)^{3/2}$ 17. $(6^2)^{1/4}$ 18. $(5^4)^{3/4}$

19. $(7^3)^{3/10}$ 20. $(3^5)^{2/5}$ 21. $(4)^{5/6}$

Written Exercise
Express each of the following as a radical.

22. $5^{1/2}$ 23. $a^{1/3}$ 24. $b^{2/3}$

25. $x^{5/2}$ 26. $5a^{1/2}$ 27. $7a^{4/3}$

28. $-(2a^3)^{1/2}$ 29. $-(7b^5)^{1/3}$ 30. $(5c^2)^{3/2}$

Express the following without a radical sign.

31. $\sqrt{2}$ 32. $\sqrt{5}$ 33. $-2\sqrt{a}$

34. $\sqrt[3]{b^2}$ 35. $-5b\sqrt[5]{c^2d}$ 36. $-7c\sqrt{a+b}$

37. $\sqrt[3]{T+s}$ 38. $\sqrt[4]{4\pi + \theta}$ 39. $-\sqrt[5]{R-t}$

Simplify.

40. $\left(\dfrac{a^{1/3}}{b^{2/3}}\right)^6$ 41. $(a^{2/3})^{1/2}$ 42. $\left(\dfrac{a^{1/2}b^{4/3}}{a^{-2}b}\right)^3$

43. $\sqrt[7]{x^4}(\sqrt[7]{x})^3$ 44. $\dfrac{(\sqrt[3]{y})^5}{\sqrt[3]{y^2}}$ 45. $\dfrac{\sqrt[3]{a^4}}{\sqrt[3]{a^3}}$

19–4 LAWS OF RADICALS

The following five basic laws for radicals are used when operating on expressions containing radicals.

RULE 19–5. Five Laws for Radicals

1. $(\sqrt[n]{x})^n = x$ (19–6)

   Example: $(\sqrt[5]{x})^5 = (x^{1/5})^5 = x^{5/5} = x$

2. $\sqrt[n]{xy} = \sqrt[n]{x}\,\sqrt[n]{y}$ (19–7)

   Example: $\sqrt[3]{48} = \sqrt[3]{8 \cdot 6} = \sqrt[3]{8}\,\sqrt[3]{6} = 2\sqrt[3]{6}$

3. $\sqrt[n]{\dfrac{x}{y}} = \dfrac{\sqrt[n]{x}}{\sqrt[n]{y}}$ (19–8)

   Example: $\sqrt[3]{\dfrac{27}{8}} = \dfrac{\sqrt[3]{27}}{\sqrt[3]{8}} = \dfrac{3}{2}$

4. $\sqrt[m]{\sqrt[n]{x}} = \sqrt[mn]{x}$ (19–9)

   Example: $\sqrt[5]{\sqrt[3]{x}} = \sqrt[15]{x} = x^{1/15}$

5. $\sqrt[m]{x^n} = x^{n/m}$ (19–10)

   Example: $\sqrt[3]{x^5} = x^{5/3}$

EXERCISE 19–4

Apply the indicated law for radicals to the following.
Equation 19–6:

1. $(\sqrt[3]{7})^3$ **2.** $(\sqrt[3]{bc})^3$ **3.** $-(\sqrt[3]{6^3})^3$

4. $-(\sqrt{a^3})^2$ **5.** $(\sqrt[7]{x^{14}})^7$ **6.** $-(\sqrt[3]{y^9})^3$

Equation 19–7:

7. $\sqrt[3]{ab}$ **8.** $\sqrt[5]{5x}$ **9.** $-\sqrt[4]{cd}$

10. $-\sqrt{13xy}$ **11.** $\sqrt[7]{4mn}$ **12.** $-\sqrt[3]{11bc}$

Equation 19–8:

13. $\sqrt{\dfrac{2}{3}}$ **14.** $\sqrt{\dfrac{7}{13}}$ **15.** $\sqrt{\dfrac{a}{b}}$

16. $-\sqrt{\dfrac{2b}{c}}$ **17.** $-\sqrt{\dfrac{3x}{5y}}$ **18.** $-\sqrt{\dfrac{ax}{by}}$

Equation 19–9:

19. $\sqrt{\sqrt[3]{7}}$ **20.** $\sqrt{\sqrt[5]{5}}$ **21.** $\sqrt[3]{\sqrt{a}}$

22. $\sqrt[3]{\sqrt[4]{ax}}$ **23.** $-\sqrt[5]{\sqrt{aby}}$ **24.** $-\sqrt[3]{\sqrt[3]{3x}}$

Equation 19–10:

25. $\sqrt[3]{3^2}$ **26.** $\sqrt{2^3}$ **27.** $\sqrt{6^5}$

28. $\sqrt[5]{c^3}$ **29.** $-\sqrt[9]{a^7}$ **30.** $-\sqrt[7]{b^3}$

19–5 SIMPLIFYING RADICALS

A radical may be simplified by the following means: (1) removing factors from the radicand, (2) lowering the index of the radical, or (3) rationalizing the denominator. Each of these means of simplifying a radical is investigated separately.

Simplifying by Removing Factors

Remove factors from the radicand by expressing the radicand as a product of factors using Equation 19–7. Select the factors so that one factor contains all the perfect nth powers, where n is the index of the radical. This is demonstrated in the following examples.

EXAMPLE 19–7   Simplify $\sqrt{27a^3b}$.

     Solution   Factor into two radicals, one containing perfect squares:

$$\sqrt{27a^3b} = \sqrt{9a^2} \ \sqrt{3ab}$$

$$\therefore \quad \sqrt{27a^3b} = 3a\sqrt{3ab}$$

EXAMPLE 19–8   Simplify $\sqrt[3]{-x^7}$.

     Solution   Factor into two radicals, one containing perfect cubes:

$$\sqrt[3]{-x^7} = \sqrt[3]{(-1)^3 x^6} \ \sqrt[3]{x}$$

$$\therefore \quad \sqrt[3]{-x^7} = -x^2 \sqrt[3]{x}$$

Simplifying by Lowering the Index

Lower the index by first expressing the radicand as a power of some quantity, and then apply Equation 19–10. This procedure is shown in Example 19–9.

EXAMPLE 19–9   Simplify $\sqrt[6]{27}$.

Solution   Express 27 as a power of 3:

$$\sqrt[6]{27} = \sqrt[6]{3^3}$$

Express with fractional exponents:

$$\sqrt[6]{3^3} = 3^{3/6} = 3^{1/2}$$

$$\therefore \quad \sqrt[6]{27} = 3^{1/2} = \sqrt{3}$$

Simplifying by Rationalizing the Denominator

The process of removing the radical from the denominator of a fraction is called **rationalizing the denominator.** When the denominator is rationalized, it is changed from an irrational to a rational form. The procedure for rationalizing the denominator is given in Rule 19–6 and demonstrated in the following examples.

RULE 19–6.   *Rationalizing the Denominator*

To rationalize the denominator, multiply the numerator and denominator of the radicand by a factor that will make the denominator a perfect nth power, where n is the index of the radical.

EXAMPLE 19–10   Rationalize $\sqrt{\dfrac{1}{3}}$.

Solution   Multiply numerator and denominator by 3, thus making the denominator a perfect square:

$$\sqrt{\frac{1 \cdot 3}{3 \cdot 3}} = \sqrt{\frac{3}{9}}$$

Apply Equation 19–8:

$$\sqrt{\frac{3}{9}} = \frac{\sqrt{3}}{\sqrt{9}} = \frac{\sqrt{3}}{3} = \frac{1}{3}\sqrt{3}$$

$$\therefore \quad \sqrt{\frac{1}{3}} = \frac{1}{3}\sqrt{3}$$

EXAMPLE 19–11   Rationalize $\sqrt[3]{\dfrac{3}{2x}}$.

Solution   Multiply numerator and denominator by $4x^2$ to make the denominator a perfect cube:

$$\sqrt[3]{\frac{3}{2x}} = \sqrt[3]{\frac{12x^2}{8x^3}}$$

Apply Equation 19–8:

$$\sqrt[3]{\frac{12x^2}{8x^3}} = \frac{\sqrt[3]{12x^2}}{\sqrt[3]{8x^3}} = \frac{\sqrt[3]{12x^2}}{2x}$$

$$\therefore \quad \sqrt[3]{\frac{3}{2x}} = \frac{1}{2x}\sqrt[3]{12x^2}$$

To rationalize the denominator when the radicals are of different *orders,* multiply the numerator and denominator by a factor that will make the denominator a rational form.

EXAMPLE 19–12   Rationalize $\dfrac{\sqrt[3]{3}}{\sqrt{2}}$.

Solution   Multiply numerator and denominator by $\sqrt{2}$:

$$\frac{\sqrt[3]{3}}{\sqrt{2}} = \frac{\sqrt[3]{3}\sqrt{2}}{\sqrt{2}\sqrt{2}} = \frac{\sqrt[3]{3}\sqrt{2}}{2}$$

Combine the radicals; express them using fractional exponents:

$$\frac{\sqrt[3]{3}\sqrt{2}}{2} = \frac{3^{1/3} \cdot 2^{1/2}}{2}$$

Express the exponents with common denominators:

$$\frac{3^{2/6} \cdot 2^{3/6}}{2} = \frac{(3^2 \cdot 2^3)^{1/6}}{2}$$

Simplify and write as a radical:

$$\frac{(3^2 \cdot 2^3)^{1/6}}{2} = \frac{\sqrt[6]{72}}{2}$$

$$\therefore \quad \frac{\sqrt[3]{3}}{\sqrt{2}} = \frac{\sqrt[6]{72}}{2}$$

Simplify each of the following radicals, expressing the answer as a radical.

1. $\sqrt{64x^3y^5}$

2. $\sqrt[3]{8a^2y^6}$

3. $\sqrt[3]{54x^{11}y^8}$

4. $\sqrt{32x^4y^3}$

5. $-\sqrt{18a^6b^8}$

6. $\sqrt{\dfrac{16x^7}{a^2y^6}}$

7. $\sqrt{\dfrac{12a^5}{9b^{10}}}$

8. $\sqrt[8]{16}$

9. $\sqrt[9]{64}$

10. $\sqrt[6]{125}$

11. $\sqrt[10]{32}$

12. $\sqrt[6]{25x^4}$

13. $\sqrt[8]{81a^2}$

14. $\sqrt[10]{x^4y^8}$

15. $\sqrt{\dfrac{3}{5}}$

16. $\sqrt{\dfrac{3}{7}}$

17. $\sqrt{\dfrac{5}{3x^2}}$

18. $\dfrac{\sqrt[3]{4}}{\sqrt{2}}$

19. $\dfrac{\sqrt[3]{2a}}{\sqrt{3a}}$

20. $\dfrac{-\sqrt{6x^2}}{\sqrt[3]{9}}$

21. $\sqrt{\dfrac{1 - \frac{1}{2}}{2}}$

22. $\sqrt{2 - (\frac{1}{5})^2}$

23. $\sqrt{1 - \left(\dfrac{\sqrt{3}}{2}\right)^2}$

24. $\sqrt{1 - (\frac{3}{5})^2}$

25. $\sqrt{4y^2 + 36}$

26. $\sqrt[3]{\dfrac{16x^7}{y^4}}$

27. $\sqrt{\left(\dfrac{b^{2/3}c^{1/2}}{2^{-1}a^{-2}}\right)^6}$

28. $\sqrt[3]{\left(\dfrac{a^2b}{125a^{-3}}\right)^{-1}}$

29. $\sqrt{2x + 6 + \dfrac{9}{2x}}$

30. $\sqrt{\sqrt[5]{32} \cdot \sqrt{4}}$

31. $\sqrt[3]{\dfrac{a}{8} + \dfrac{-3}{y^3}}$

32. $\sqrt{\dfrac{4b^2}{9} - \dfrac{16}{x^2}}$

19–6 RADICAL EQUATIONS

An equation in which a variable appears as part of the radicand (within the radical) or has a fractional exponent is called a **radical equation.** Thus, $\sqrt{y} = 2$ and $\sqrt{x + 1} = 5$ are radical equations. The equation $x + \sqrt{3} = 0$ is not a radical equation because x is not part of the radicand.

RULE 19–7.   *Solving Radical Equations*

To solve a radical equation:

1. Isolate the radical term in one member of the equation.
2. Raise each member of the equation to a power equal to the index of the radical.
3. Solve the resulting equation for the root.
4. Check the root, as the preceding process may introduce roots into the derived equation that the original equation did not have.

EXAMPLE 19–13   Solve $5 + \sqrt{x} = 10$ for x.

Solution   To solve, follow the steps of Rule 19–7.

Step 1:   Isolate $\sqrt{x}$ in the left member:

$$\sqrt{x} = 5$$

Step 2:   Square both members:

$$(\sqrt{x})^2 = 5^2$$

Step 3:   Solve:

$$x = 25$$

Step 4:   Check.

Check   Does $5 + \sqrt{x} = 10$ when $x = 25$?

$$5 + \sqrt{25} = 10$$
$$5 + 5 = 10 \quad \text{Yes.}$$

$\therefore \quad x = 25$ is the root.

EXAMPLE 19–14   Solve $2\sqrt{y - 2} + 4 = 8$.

Solution   Use the steps of Rule 19–7.

Step 1: Isolate $\sqrt{y-2}$ in the left member:

$$\sqrt{y-2} = \frac{8-4}{2} = 2$$

Step 2: Square both members:

$$y - 2 = 4$$

Step 3: Solve:

$$y = 6$$

Check Does $2\sqrt{y-2} + 4 = 8$ when $y = 6$?

$$2\sqrt{6-2} + 4 = 8$$
$$2\sqrt{4} + 4 = 8$$
$$4 + 4 = 8$$
$$8 = 8 \quad \text{Yes.}$$

$\therefore \quad y = 6$ is the solution.

EXAMPLE 19–15   Solve $-4 = x - \sqrt{3x + 10}$ for x.

Solution   Follow the steps of Rule 19–7.

Step 1: Isolate $\sqrt{3x+10}$ in the left member:

$$\sqrt{3x+10} = x + 4$$

Step 2: Square both members:

$$(\sqrt{3x+10})^2 = (x+4)^2$$
$$3x + 10 = x^2 + 8x + 16$$

Step 3: Combine like terms and equate to zero:

$$x^2 + 5x + 6 = 0$$

Solve the quadratic by factoring:

$$(x+3)(x+2) = 0$$
$$x + 3 = 0 \qquad\qquad x + 2 = 0$$
$$x = -3 \qquad\qquad x = -2$$

Step 4: There are two possible roots; check both.

Check   Does $-4 = x - \sqrt{3x+10}$ when $x = -3$?

$$-4 = -3 - \sqrt{-9 + 10}$$
$$-4 = -4 \quad \text{Yes.}$$

Check   Does $-4 = x - \sqrt{3x + 10}$ when $x = -2$?

$$-4 = -2 - \sqrt{-6 + 10}$$
$$-4 = -2 - 2$$
$$-4 = -4 \quad \text{Yes.}$$

$\therefore \quad x = -2$ and $x = -3$ are both roots.

EXERCISE 19–6

Solve and check each of the following.

1. $\sqrt{x} = 7$

2. $\sqrt{x} = 5$

3. $\sqrt{4x} = 8$

4. $\sqrt{3a} = 6$

5. $\sqrt{6x + 1} = 5$

6. $\sqrt{b - 1} = 2$

7. $\sqrt{a + 2} = 8$

8. $5 = \sqrt{1 + 4x}$

9. $2\sqrt{x + 3} = 6$

10. $4\sqrt{2a} = 8$

11. $5\sqrt{3y} = 15$

12. $3\sqrt{x + 1} = 9$

13. $2\sqrt{a - 1} = 2$

14. $\sqrt{2y} + 3 = 1$

15. $\sqrt{6a} - 2 = 5$

16. $3 - \sqrt{2x - 7} = 0$

17. $\sqrt{5c - 2} + 3 = 6$

18. $\sqrt{x^2 - 8} + 4 = x$

19. $\sqrt{5 + a^2} - 5 = -a$

20. $x + 3\sqrt{x} = 10$

21. $\sqrt{y + 4} = 2 + y$

22. $4 + \sqrt{x + 2} = x$

23. $3 - \sqrt{7x - 3} = 2x$

24. $\sqrt{\dfrac{6 + 2x}{5}} = 4$

25. $\sqrt{\dfrac{4y}{3}} - 6 = 2$

26. Solve $4a = c\sqrt{3}$ for c.

27. Solve $t = \sqrt{\dfrac{2S}{g}}$ for g.

28. Solve $t = \pi\sqrt{\dfrac{l}{g}}$ for l.

29. Solve $r = \sqrt[3]{\dfrac{3V}{4\pi}}$ for V.

30. Solve $E = \sqrt{V^2 + 2IR}$ for R.

SELECTED TERMS

fractional exponent   An exponent that is a fraction; an exponent of $1/n$ indicates the nth root; thus $a^{1/n}$ is the nth root of a.

SECTION CHALLENGE

WEB CHALLENGE FOR CHAPTERS 18 AND 19

To evaluate your comprehension of Chapters 18 and 19, log on to **www.prenhall. com/harter** and take the online True/False and Multiple Choice assessments for each of the chapters.

SECTION CHALLENGE FOR CHAPTERS 18 AND 19*

Your challenge is to develop an equation to determine the value of the load resistance (R_L) in the equivalent dc circuit of Figure C–9. Because the equation will be based on the I^2R form of power, the equation will be expressed in a quadratic form.

Start your solution by writing an equation for the current (I) in the load (R_L) in terms of E_{Th}, R_{Th}, and R_L. Substitute this equation into the power equation, $P_L = I^2R_L$, so that the load power (P_L) is stated in terms of E_{Th}, R_{Th}, and R_L.

Solve the power equation for R_L by expressing the equation in the form of a complete quadratic equation, $ax^2 + bx + c = 0$, that is, $aR_L^2 + bR_L + c = 0$. Then, using the quadratic formula, solve for R_L when:

A. $P_L = 4.734$ W, $E_{Th} = 10.00$ V, and $R_{Th} = 5.00 \ \Omega$

B. $P_L = 5.000$ W, $E_{Th} = 10.00$ V, and $R_{Th} = 5.00 \ \Omega$

Check the roots.

FIGURE C–9   Power is transferred from the source (E_{Th}) to the load (R_L) of the equivalent circuit. Maximum power is transferred to the load of a dc equivalent circuit when the load resistance is equal to the equivalent source resistance $(R_L = R_{Th})$.

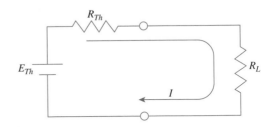

*The solution to this Section Challenge is found in Appendix C.

20

Logarithmic and Exponential Functions

20–1 Common Logarithms

20–2 Common Logarithms and Scientific Notation

20–3 Antilogarithms

20–4 Logarithms, Products, and Quotients

20–5 Logarithms, Powers, and Radicals

20–6 Natural Logarithms

20–7 Changing Base

20–8 Further Properties of Natural Logarithms

20–9 Logarithmic Equations

20–10 Exponential Equations

20–11 Semilog and Log–Log Plots

20–12 Nomographs

PERFORMANCE OBJECTIVES

- Calculate the logarithms and antilogarithms.

- Change the base of a logarithm.

- Solve logarithmic and exponential equations.

- Interpret semilog and log–log graphs.

- Use nomographs to solve problems.

The Hubble Space Telescope (with a new set of solar panels) sits in the cargo bay of the Space Shuttle Endeavour while in orbit above the horizon of Earth after the historic capture of the malfunctioning instrument for repairs. (Courtesy of NASA/Johnson Space Center)

In the preceding chapters, functions involving addition, subtraction, multiplication, division, and raising variables to constant powers have been discussed. In this chapter, we look at raising constants to *variable* powers. We have examined $y = x^2$; now we are going to examine $y = 10^x$.

The following symbols are new to this chapter:

| $\boxed{\log}$ | common logarithm | $\boxed{10^x}$ | common antilog |
| $\boxed{\ln}$ | natural logarithm | $\boxed{e^x}$ | natural antilog |

20–1 COMMON LOGARITHMS

Suppose we know that $a = 10^b$, and we want to solve for b. What do we do? Read $a = 10^b$ as "b is the exponent to which 10 must be raised to yield a." In mathematics, we define common logarithm so that the following statement means the same as the preceding: b is the common logarithm of a. This can be written in the following manner:

If $a = 10^b$, then $b = \log(a)$

The symbol log (a) is read "log of a" and stands for the common logarithm function of a. Notice that the parentheses here indicate "function of" and not multiplication. The parentheses are frequently omitted.

The **common logarithm** of a number is defined as the exponent to which 10 must be raised to yield the number. Thus, 10 is the base of the common logarithms. In general, a **logarithm** is the exponent to which the base must be raised to equal the original number. In the following sections you will see that logarithms follow the rules of exponents.

EXAMPLE 20–1	Find the common logarithm of 1, 10, and 100.
Observation	$1 = 10^0$, $10 = 10^1$, and $100 = 10^2$.
Solution	Use the definition of common logarithms and the observation:

$$\log(1) = \log(10^0) = 0$$
$$\log(10) = \log(10^1) = 1$$
$$\log(100) = \log(10^2) = 2$$

$\therefore$ Log $(1) = 0$, log $(10) = 1$, and log $(100) = 2$.

Now, turn to your calculator and your owner's guide. Locate the common logarithm function in the guide and the function key $\boxed{\log x}$ or $\boxed{\log}$ on your calculator. This function will calculate the common logarithm of any positive number for you. Use your calculator to work the following examples.

EXAMPLE 20–2	Find the common logarithm of 5, 6.5, and 2.98.
Observation	When writing logarithms, we generally write four digits to the right of the decimal point. This corresponds to three place accuracy in the number.
Solution	Use your calculator:

$$5 \boxed{\log} \Rightarrow 0.6990$$

$$6.5 \boxed{\log} \Rightarrow 0.8129$$

$$2.98 \boxed{\log} \Rightarrow 0.4742$$

EXAMPLE 20–3	Find the common logarithm of $10^{0.5}$.
Solution 1	Use the definition of common logarithm:

$$\log (10^{0.5}) = 0.5$$

Observation	Recall that $10^{0.5}$ means the square root of 10.
Solution 2	Use your calculator:

$$\log (10^{0.5}) = \log (\sqrt{10})$$

$\boxed{\sqrt{x}}$ $\log (10^{0.5}) = \log (\mathbf{3.162})$

$\boxed{\log}$ $\log (10^{0.5}) = \mathbf{0.5}$

$\therefore$ Log $(10^{0.5}) = 0.5$ by definition and by calculator.

EXERCISE 20–1

Use the definition to find the common logarithms of the following numbers.

1. 10^3	**2.** 10^1	**3.** 10^4
4. $10^{1.5}$	**5.** $10^{2.5}$	**6.** $10^{0.25}$

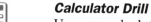

Calculator Drill
Use your calculator to find the common logarithms of the following numbers.

7. 2	**8.** 4	**9.** 8
10. 16	**11.** 32	**12.** 2.1
13. 3.1	**14.** 4.1	**15.** 5.1
16. 6.1	**17.** 7.1	**18.** 6.2
19. 3.41	**20.** 3.78	**21.** 5.92
22. 1.05	**23.** 2.95	**24.** 8.01
25. 7.62	**26.** 9.15	**27.** 4.61

20–2 COMMON LOGARITHMS AND SCIENTIFIC NOTATION

In the preceding section we learned the definition of common logarithms. We also learned how to use our calculators to find the common logarithm of a number. We are going to perform an experiment with our calculators on a series of numbers in scientific notation.

EXAMPLE 20–4 Find the common logarithms of 3×10^0, 3×10^1, 3×10^2, and 3×10^9.

Solution Use your calculator:

$$3 \boxed{EE} 0 \boxed{\log} \Rightarrow 0.4771$$

$$3 \boxed{EE} 1 \boxed{\log} \Rightarrow 1.4771$$

$$3 \boxed{EE} 2 \boxed{\log} \Rightarrow 2.4771$$

$$3 \boxed{EE} 9 \boxed{\log} \Rightarrow 9.4771$$

Observation Notice that the fractional part of each answer is the same number (.4771) and that the integer part of each answer is the same as the exponent of 10.

In Example 20–4, notice that common logarithms of the numbers 3×10^0, 3×10^1, 3×10^2, and 3×10^9 all had the same fractional part. It is a fact that the fractional part of a logarithm depends only on the digits of a number and not on the location of the decimal point nor on the exponent of 10. The fractional part of a logarithm is called the **mantissa.** Notice also that the integer part of the logarithm depends only on the exponent of 10. The integer part of the logarithm is called the **characteristic.** If we have a table of common logarithms for numbers between 1 and 10, or if we have a calculator that will find the common logarithm for a number, we can use Rule 20–1 to find the common logarithm of any positive number.

RULE 20–1. *Finding the Common Logarithm of a Number*

To find the logarithm of a number:
1. Write the number in scientific notation.
2. Determine the mantissa as the common logarithm of the decimal coefficient. (This will be a fraction between zero and one.)
3. Determine the characteristic as the exponent of 10.
4. Form the common logarithm by adding the characteristic and the mantissa.

EXAMPLE 20–5	Find the mantissa, the characteristic, and the common logarithm of 654, 7.83, and 0.0491.
Solution 1	Use the steps of Rule 20–1 and your calculator.
Step 1:	Write the number in scientific notation:
SN:	$654 = 6.54 \times 10^2$
Step 2:	Determine the mantissa:
$\boxed{\log}$	mantissa = log (6.54) = **0.8156**
Step 3:	Determine the characteristic:
	characteristic = 2
Step 4:	Form the common logarithm:
$\therefore$	log (654) = 2.8156
Solution 2	Use Rule 20–1 and your calculator.
Step 1:	Write the number in scientific notation:
SN:	$7.83 = 7.83 \times 10^0$
Step 2:	Determine the mantissa:
$\boxed{\log}$	mantissa = log (7.83) = **0.8938**
Step 3:	Determine the characteristic:
	characteristic = 0
Step 4:	Form the common logarithm:
$\therefore$	log (7.83) = 0.8938
Solution 3	Use Rule 20–1 and your calculator.
Step 1:	Write the number in scientific notation:
SN:	$0.0491 = 4.91 \times 10^{-2}$
Step 2:	Determine the mantissa:
$\boxed{\log}$	mantissa = log (4.91) = **0.6911**
Step 3:	Determine the characteristic:
	characteristic = -2
Step 4:	Form the common logarithm:
$\therefore$	log (0.0491) = $-2 + 0.6911 = -1.3089$

Summary

$$\log(654) = 2 + 0.8156 = 2.8156$$

$$\log(7.83) = 0 + 0.8938 = 0.8938$$

$$\log(0.0491) = -2 + 0.6911 = -1.3089$$

Although the mantissa is positive, the characteristic may be positive or negative. The characteristic is always negative for fractions. In the case of fractions, we combine the characteristic and the mantissa by subtraction. The integer and fractional parts of the resulting logarithm are both negative. This is the answer your calculator will give.

The graph of Figure 20–1 is the plot of the common logarithm function, which reveals several properties of the common logarithm function. For $x > 1$, $\log(x) > 0$; for $1 > x > 0$, $\log(x) < 0$; and for $x = 1$, $\log(x) = 0$.

FIGURE 20–1   Plot of the common logarithm function.

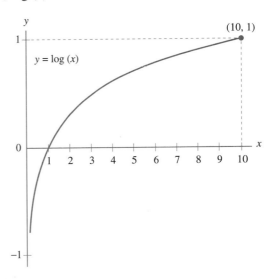

EXERCISE 20–2

Use Rule 20–1 and Example 20–5 to find the characteristics, mantissas, and common logarithms of the following numbers.

1. (a) 7	**(b)** 70	**(c)** 700	**(d)** 7000
2. (a) 2	**(b)** 20	**(c)** 200	**(d)** 2000
3. (a) 0.075	**(b)** 0.75	**(c)** 7.5	**(d)** 75
4. (a) 0.008	**(b)** 0.8	**(c)** 80	**(d)** 8000
5. (a) 1.62	**(b)** 16.2	**(c)** 162	**(d)** 1620
6. (a) 99.1	**(b)** 9.91	**(c)** 0.991	**(d)** 0.0991
7. (a) 5.55	**(b)** 55.5	**(c)** 555	**(d)** 5550
8. (a) 3.61	**(b)** 0.361	**(c)** 0.0361	**(d)** 0.003 61
9. (a) 0.001 05	**(b)** 0.0105	**(c)** 0.105	**(d)** 1.05

20–3 ANTILOGARITHMS

The common logarithm of a number is the exponent to which 10 must be raised to give the original number. Knowing this definition, how can we find the original number if we know the common logarithm? By raising 10 to the common logarithm power! This operation is the definition of the **common antilogarithm** function, which is the **inverse** of the common logarithm function. All of this can be written in symbols:

$$\text{If } b = \log (a), \text{ then } a = \text{antilog } (b) = 10^b$$

Check your calculator and your owner's guide to find out how to evaluate the antilogarithm of a number. Your calculator may use $\boxed{10^x}$ or $\boxed{y^x}$ where you supply the 10 for y. Or your calculator may use two keystrokes, as in $\boxed{\text{INV}}\ \boxed{\log}$ or $\boxed{f^{-1}}\ \boxed{\log}$.

EXAMPLE 20–6   If the common logarithm of A is 2, what is the value of A?

Solution   Use the antilogarithm to find A:

$$2 = \log (A)$$

Apply antilog; see Figure 20–2:

FIGURE 20–2   Effects of antilog.

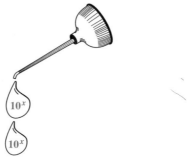

$$b = \log (a) \implies a = \text{antilog } (b) = 10^b$$

$$A = \text{antilog } (2) = 10^2$$

$$\therefore \quad A = 100$$

EXAMPLE 20–7   If the common logarithm of Z is 1.3324, what is the value of Z?

Solution   Use the antilogarithm and your calculator:

$$1.3324 = \log (Z)$$

Apply antilog:

$$Z = \text{antilog} (1.3324) = 10^{1.3324}$$

$$1.3324 \boxed{10^x} \Rightarrow 21.5$$

$$\therefore \qquad Z = 21.5$$

EXERCISE 20–3

Calculator Drill
Find the antilogarithm of the following numbers.

1. 0.3522	**2.** 0.9706
3. 0.1592	**4.** 0.7041
5. 0.5223	**6.** -0.3871
7. -1.0526	**8.** 1.3952
9. 1.4671	**10.** 2.8352
11. 0.3010, 1.3010, 2.3010	**12.** 0.4771, 1.4771, 2.4771
13. 0.6021, 2.6021, 4.6021	**14.** 0.6990, 2.6990, 4.6990
15. -0.1549, -1.1549, -2.1549	

20–4   LOGARITHMS, PRODUCTS, AND QUOTIENTS

In this section we will explore two important properties of logarithms. They will each be stated as rules. These rules will be useful in solving logarithmic equations.

> **RULE 20–2.   Logarithm of a Product**
>
> The logarithm of a product is equal to the sum of the logarithms of the factors.
>
> $$\log (AB) = \log (A) + \log (B)$$

EXAMPLE 20–8   Check the equation $\log (20 \cdot 15) = \log (20) + \log (15)$.

Solution   Evaluate the left and right members separately: Left member:

$\boxed{\times}$

$\boxed{\log}$

$$\log (20 \cdot 15) = \log (\mathbf{300})$$
$$= 2.4771$$

Right member:

$$\text{log } (20) + \text{log } (15) = \mathbf{1.3010 + 1.1761}$$
$$= 2.4771$$

∴ $\quad \text{log } (20 \cdot 15) = \text{log } (20) + \text{log } (15)$

EXAMPLE 20–9   Solve $P = 3 \times 6$ using logarithms and Rule 20–2.

Solution $P = 3 \times 6$
Take the log of each member.

$$\text{log } (P) = \text{log } (3 \times 6)$$

Apply Rule 20–2:

$$\text{log } (P) = \text{log } (3) + \text{log } (6)$$
$$= 0.4771 + 0.7782$$
$$= 1.2553$$
$$\text{log } (P) = 1.2553$$

Take the antilog of each member:

$$P = \text{antilog } (1.2553) = 10^{1.2553} = 18$$

∴ $\quad P = 18$

> **RULE 20–3.   Logarithm of a Quotient**
>
> The logarithm of a quotient is equal to the difference between the logarithm of the dividend and the logarithm of the divisor.
>
> $$\text{log } (A/B) = \text{log } (A) - \text{log } (B)$$

EXAMPLE 20–10   Check the equation $\text{log } \left(\frac{75}{15}\right) = \text{log } 75 - \text{log } 15$.

Solution   Evaluate each member of the equation separately.
Left member:

$$\text{log } \left(\tfrac{75}{15}\right) = \text{log } (5)$$
$$= 0.6990$$

Right member:

$$\log{(75)} - \log{(15)} = \mathbf{1.8751} - \mathbf{1.1761}$$
$$= 0.6990$$

$$\therefore \quad \log{\left(\tfrac{75}{15}\right)} = \log{(75)} - \log{(15)}$$

EXAMPLE 20–11   Solve $Q = \tfrac{20}{4}$ using logarithms and Rule 20–3.

Solution
$$Q = \tfrac{20}{4}$$

Take the log of each member:

$$\log{(Q)} = \log{\left(\tfrac{20}{4}\right)}$$

Apply Rule 20–3:

$$\log{(Q)} = \log{(20)} - \log{(4)}$$
$$= 1.3010 - 0.6021$$
$$= 0.6989$$
$$\log{(Q)} = 0.6990$$

Take the antilog of each member:

$$Q = \text{antilog}\,(0.6990) = 10^{0.6990} = 5$$

$$\therefore \qquad Q = 5$$

EXERCISE 20–4

Check the following equations as in Examples 20–8 and 20–10:

1. $\log{(3 \times 5)} = \log{(3)} + \log{(5)}$
2. $\log{(2 \times 7)} = \log{(2)} + \log{(7)}$
3. $\log{(17 \times 3.5)} = \log{(17)} + \log{(3.5)}$
4. $\log{\left(\tfrac{625}{25}\right)} = \log{(625)} - \log{(25)}$
5. $\log{\left(\tfrac{18}{2}\right)} = \log{(18)} - \log{(2)}$

Determine whether the following statements are true or false:

6. $\log{(75)} = \log{(5)} + \log{(15)}$
7. $\log{(14)} = \log{(7 + 2)}$
8. $\log{(7)} = \log{(35)} - 5$
9. $\log{(20)} = \log{(4)}\log{(5)}$
10. $\log{(9)} = \log{(72)} - \log{(8)}$

Use logarithms to solve the following as in Examples 20–9 and 20–11:

11. $P = 7 \times 4$

12. $P = 23.4 \times 0.0574$

13. $Q = \frac{16}{4}$

14. $Q = 0.213/0.0790$

15. $Q = 1220/203$

20–5 LOGARITHMS, POWERS, AND RADICALS

In the preceding section we learned that through logarithms multiplication becomes addition and division becomes subtraction. In a similar fashion, raising to powers becomes multiplication.

RULE 20–4. *Logarithm of a Power*

The logarithm of a power is equal to the product of the exponent and the logarithm of the base.

$$\log (a^n) = n \log (a)$$

EXAMPLE 20–12   Check the equation $\log (5^4) = 4 \log (5)$.

Solution   Evaluate each member of the equation.
Left member:

$\boxed{y^x}$ $\log (5^4) = \log (\mathbf{625})$

$\boxed{\log}$ $= 2.7959$

Right member:

$\boxed{\log}$ $4 \log (5) = 4 \,(\mathbf{0.6990})$

$\boxed{\times}$ $= 2.7959$

$\therefore$ $\log (5^4) = 4 \log (5)$

Square roots can be indicated with a radical sign and with an exponent of one-half. This combined with Rule 20–4 gives a new rule for the logarithm of the square root of a number.

The logarithm of the square root of a number is equal to one-half the logarithm of the number.

$$\log (\sqrt{A}) = \frac{1}{2} \log (A)$$

EXAMPLE 20–13 Check the equation $\log (\sqrt{49}) = \frac{1}{2} \log (49)$.

Solution Evaluate each member of the equation.
Left member:

| $\boxed{\sqrt{x}}$ | $\log (\sqrt{49}) = \log (7)$ |
| $\boxed{\log}$ | $= 0.8451$ |

Right member:

$\boxed{\log}$	$\frac{1}{2} \log (49) = \frac{1}{2} (\mathbf{1.6902})$
$\boxed{\div}$	$= 0.8451$
$\therefore$	$\log (\sqrt{49}) = \frac{1}{2} \log (49)$

EXAMPLE 20–14 Calculate $\log (\sqrt{72.5})$ without using square root.

Solution Use Rule 20–5.

$$\log (\sqrt{72.5}) = \frac{1}{2} \log (72.5)$$

| $\boxed{\log}$ | $= \frac{1}{2} (\mathbf{1.8603}) = 0.9302$ |
| $\therefore$ | $\log (\sqrt{72.5}) = 0.9302$ |

In the following example we use Rule 20–5 to evaluate the square root of a number. The same steps can be used with Rule 20–4 to evaluate any power of a number.

EXAMPLE 20–15 Calculate $\sqrt{172}$ without using square root.

Solution Use Rule 20–5 and logarithms:

$$\log(\sqrt{172}) = \frac{1}{2} \log (172)$$

| $\boxed{\log}$ | $= \frac{1}{2} (\mathbf{2.2355})$ |
| $\boxed{\div}$ | $= 1.1178$ |

Take the antilog of each member:

$\boxed{10^x}$ $\qquad \sqrt{172} = $ antilog $(1.1178) = \mathbf{10^{1.1178}} = 13.1$

$\therefore \quad \sqrt{172} = 13.1$

EXAMPLE 20–16   Calculate $17^{3.49}$.

Solution   Use Rule 20–4:

$$\log (17^{3.49}) = 3.49 \log (17)$$

$\boxed{\log}$ $\qquad\qquad = 3.49(\mathbf{1.2304})$

$\boxed{\times}$ $\qquad\qquad = 4.2943$

Take the antilog of each member:

$\boxed{10^x}$ $\qquad 17^{3.49} = $ antilog $(4.2943) = \mathbf{10^{4.2943}} = 19\,700$

$\therefore \quad 17^{3.49} = 19\,700$

EXAMPLE 20–17   Solve $N = 4^3$ using logarithms and Rule 20–4.

Solution   Take the log of each member:

$$\log (N) = \log (4^3)$$

Apply Rule 20–4:

$$\log (N) = 3 \log (4)$$

$\boxed{\log}$ $\qquad\quad = 3(\mathbf{0.6021})$

$\boxed{\times}$ $\qquad\quad = 1.8062$

Take the antilog of each member:

$\boxed{10^x}$ $\qquad N = $ antilog $(1.8062) = \mathbf{10^{1.8062}} = 64$

$\therefore \quad N = 64$

EXERCISE 20–5

Check the following equations as in Examples 20–12 and 20–13.

1. $\log (3^2) = 2 \log (3)$
2. $\log (12^2) = 2 \log (12)$
3. $\log (29^2) = 2 \log (29)$
4. $\log (5^3) = 3 \log (5)$
5. $\log (\sqrt{81}) = \frac{1}{2} \log (81)$

6. $\log(\sqrt{961}) = \frac{1}{2}\log(961)$

7. $\log(\sqrt{2^3}) = \frac{3}{2}\log(2)$

Use Rules 20–4 and 20–5 in evaluating the following.

8. $\sqrt{39}$ **9.** $4^{7.5}$ **10.** $\sqrt{409}$

11. $\sqrt{3.5}$ **12.** $\sqrt{11}$ **13.** $(5.2)^{0.28}$

14. $(1.01)^{120}$ **15.** $(87)^{0.314}$ **16.** $\sqrt[3]{6.01 \times 10^{-5}}$

20–6 NATURAL LOGARITHMS

In electronics there are two important logarithm functions. The first is the common logarithm based on the number 10. The second is the Napierian logarithm based on the number e. The value of e to five digits is 2.7183. Napierian logarithms are named for John Napier, the Scottish mathematician who invented logarithms.

Naperian logarithms are also called natural logarithms because the number e arises naturally as the base in exponential equations describing growth or change (see the equations of Tables 21–3 and 21–4 in the next chapter). So Naperian logarithms arise naturally in the solution of such equations.

To distinguish these two logarithm functions, we will continue to use log ( ) for common logarithms. And we introduce a new symbol ln ( ) for natural logarithms. The following is the definition of the **natural logarithm** function:

$$\text{If } A = e^B, \text{ then } B = \text{natural logarithm of } A = \ln(A)$$

The symbol $\ln(A)$ is read "ell-en of a" or "lin of a." It may also be read "natural log of a." As with common logarithms, the parentheses are frequently omitted. Remember that they are used here to denote *function*, not multiplication.

Check your calculator and operating instructions for natural logarithms. Locate the function key $\boxed{\ln x}$ or $\boxed{\ln}$. This function will calculate the natural logarithm of any positive number for you.

EXAMPLE 20–18   Find the natural logarithms of 5, 10, and 15.

Solution   Use your calculator:

$$5\,\boxed{\ln} \Rightarrow 1.6094$$
$$10\,\boxed{\ln} \Rightarrow 2.3026$$
$$15\,\boxed{\ln} \Rightarrow 2.7081$$

Figure 20–3 presents a graph of both logarithm functions. Notice that both graphs pass through the point (1, 0). Also, both have the same *general* behavior. There is no graph for $x \leq 0$. For $0 < x < 1$, both graphs are below the x-axis. This means both functions are negative in this region. For $x > 1$, both graphs are shown above the x-axis and both functions are positive. Finally, as x increases, both graphs increase.

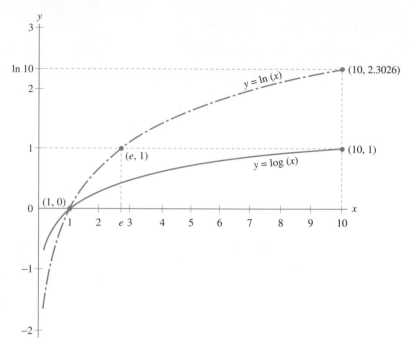

FIGURE 20–3 Comparison of log (x) and ln (x).

Antilogarithm

Recall the definition of the natural logarithm of a number as the exponent to which e must be raised to give the number. Knowing the natural logarithm of a number, how can we find the number? By raising e to the natural logarithm power. This is the definition of the **natural antilogarithm** function, which is the *inverse* of the natural logarithm function. The following symbols express this definition:

$$\text{If } a = \ln (b), \text{ then } b = \text{antiln} (a) = e^a$$

The antiln function is also known as the *exponential function*. The graph of the exponential function is shown in Figure 20–4.

Check your calculator and owner's guide on how to calculate antiln (x). It may use $\boxed{e^x}$ or a two-keystroke sequence, as in $\boxed{f^{-1}}\boxed{\ln}$ or $\boxed{\text{INV}}\boxed{\ln}$. Figure 20–5 shows the effects of the antiln function.

EXAMPLE 20–19   Find y if ln $(y) = 1.9782$.

Solution   Use antiln and your calculator:

$$\ln (y) = 1.9782$$

$$y = \text{antiln} (1.9782) = e^{1.9782}$$

$\boxed{e^x}$ 　　　　　$y = 7.23$

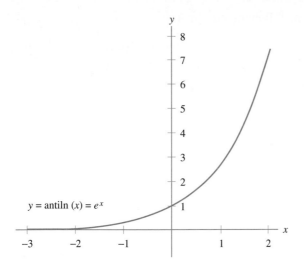

$y = \text{antiln}\,(x) = e^x$

FIGURE 20–4 Graph of the exponential function, e^x.

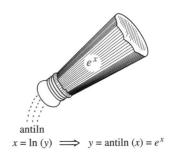

e^x

antiln
$x = \ln\,(y) \implies y = \text{antiln}\,(x) = e^x$

FIGURE 20–5 Effects of antiln.

Check Does $\ln\,(y) = 1.9782$ when $y = 7.23$?

$\boxed{\ln}$ $\ln\,(7.23) = 1.9782$ Yes!

$\therefore$ $y = 7.23$

EXERCISE 20–6

Use the definition to find the natural logarithm of the following terms.

1. $e^{1.2}$ **2.** $e^{3.0}$ **3.** $e^{0.75}$ **4.** $e^{0.12}$ **5.** $e^{-4.1}$

6. $e^{-0.51}$ **7.** e^{a} **8.** e^{-Rt} **9.** e^{-z} **10.** e^{2n}

Calculator Drill
Calculate the natural logarithms of the following numbers.

11. 1.25   **12.** 3.16

13. 4.0   **14.** 5.0

15. 6.0   **16.** 7.0

17. 0.156   **18.** 0.029

19. 0.015   **20.** 18.4

21. 927   **22.** 7040

23. 143.2   **24.** 40.94

25. 6.35×10^4 **26.** 1045

27. 0.831   **28.** 0.0392

29. 79.3   **30.** 2001

Find the natural antilogarithms of the following numbers.

31. 1.0   **32.** 0.0
33. 2.3026   **34.** 2.9444
35. 2.1983   **36.** −5.2785
37. −3.6045   **38.** −1.3863
39. 0.5241   **40.** −99.0919
41. −0.0590   **42.** 4.7205
43. −7.0646   **44.** 0.8751
45. −4.6702   **46.** −3.0011
47. 14.3285   **48.** −4.4273
49. −16.2345   **50.** 0.5623

20–7 CHANGING BASE

Sometimes the question arises of converting a logarithm from one base to another. With the aid of a calculator this is a simple task when the value of the logarithm is known. Take the antilogarithm and then the second logarithm of the number. A more direct method is provided by the following rule:

RULE 20–6. *Changing Base*

1. To change a common logarithm to a natural logarithm, multiply by 2.3026, the ln of 10:

$$\ln (x) = 2.3026 \log (x)$$

2. To change a natural logarithm to a common logarithm, divide by 2.3026, the ln of 10:

$$\log (x) = \ln (x)/2.3026$$

EXAMPLE 20–20 Check Rule 20–6 when $x = 15$.

Solution Substitute for x; then evaluate each member of the equation:

$$\ln (15) = 2.3026 \log (15)$$

Left member:

$\boxed{\ln}$ $\ln (15) = 2.7081$

Right member:

|log| $2.3026 \log (15) = 2.3026 \times \mathbf{1.1761}$

|×| $= 2.7081$

∴ $\ln (15) = 2.3026 \log (15)$

The number 2.3026 may be remembered as ln (10) or 1/log (*e*). In fact, for any positive number *n*, where $n \neq 1$, ln (*n*)/log (*n*) = 2.3026.

EXERCISE 20–7

Use common logarithms to calculate the following.

1. ln (4)	**2.** ln (5.1)	**3.** ln (0.49)
4. ln (8.2)	**5.** ln (0.79)	**6.** ln (10.9)

Use natural logarithms to calculate the following.

7. log (60)	**8.** log (21)	**9.** log (5.7)
10. log (0.015)	**11.** log (0.29)	**12.** log (0.091)

20–8 FURTHER PROPERTIES OF NATURAL LOGARITHMS

As we noted earlier, the natural logarithm function and the common logarithm function have the same general behavior. In fact, the rules for logarithms and products, quotients, powers, and radicals are the same for both logarithm functions.

EXAMPLE 20–21   Check Rule 20–2 for natural logarithms:
$\ln (AB) = \ln (A) + \ln (B)$ when $A = 16$ and $B = 3.5$.

Solution   Substitute for *A* and *B*; then evaluate each member of the equation:

$$\ln (16 \times 3.5) = \ln (16) + \ln (3.5)$$

Left member:

|×| $\ln (16 \times 3.5) = \ln (\mathbf{56})$

|ln| $= 4.0254$

Right member:

|ln| $\ln (16) + \ln (3.5) = \mathbf{2.7726} + \mathbf{1.2528}$

|+| $= 4.0254$

∴ $\ln(16 \times 3.5) = \ln (16) + \ln (3.5)$

EXAMPLE 20–22   Check Rule 20–3 for natural logarithms:
$\ln (A/B) = \ln (A) - \ln (B)$ when $A = 54$ and $B = 60$.

Solution   Substitute for A and B; then evaluate each member of the equation:

$$\ln \left(\tfrac{54}{60}\right) = \ln (54) - \ln (60)$$

Left member:

$\boxed{\div}$

$\boxed{\ln}$

$$\ln \left(\tfrac{54}{60}\right) = \ln (\mathbf{0.9})$$
$$= -0.1054$$

Right member (use chain operations):

$\boxed{\ln}$

$\boxed{-}$

$$\ln (54) - \ln (60) = \mathbf{3.9890 - 4.0943}$$
$$= -0.1054$$

$\therefore$ $\ln \left(\tfrac{54}{60}\right) = \ln (54) - \ln (60)$

EXAMPLE 20–23   Check Rule 20–4 for natural logarithms:
$\ln (a^n) = n \ln (a)$ when $a = 1.67$ and $n = 3$.

Solution   Substitute for a and n and then evaluate each member of the equation:

$$\ln (1.67^3) = 3 \ln (1.67)$$

Left member (perform chain operations):

$\boxed{y^x}$

$\boxed{\ln}$

$$\ln (1.67^3) = \ln (\mathbf{4.66})$$
$$= 1.5385$$

Right member:

$\boxed{\ln}$

$\boxed{\times}$

$$3 \ln (1.67) = 3(\mathbf{0.5128})$$
$$= 1.5385$$

$\therefore$ $\ln (1.67^3) = 3 \ln (1.67)$

EXAMPLE 20–24   Check Rule 20–5 for natural logarithms:

$\ln (\sqrt{A}) = \tfrac{1}{2} \ln (A)$ when $A = 79.6$ (use chain operations).

Solution   Substitute for A and then evaluate each member of the equation:

$$\ln (\sqrt{79.6}) = \tfrac{1}{2} \ln (79.6)$$

Left member:

$$79.6 \ \boxed{\sqrt{x}} \ \boxed{\ln} \ \Rightarrow \ 2.1885$$

Right member:

$$79.6 \boxed{\ln} 2 \boxed{\div} \Rightarrow 2.1885$$

$$\therefore \quad \ln(\sqrt{79.6}) = \tfrac{1}{2}\ln(79.6)$$

In Table 20–1 you will find a summary of the properties of logarithms that we have studied. This table will help in solving logarithmic equations.

TABLE 20–1 Properties of Logarithms

Property	Common	Natural
Base	10	$e = 2.7183$
Logarithm of 1	$\log(1) = 0$	$\ln(1) = 0$
Logarithm of e	$\log(e) = 0.4343$	$\ln(e) = 1$
Logarithm of 10	$\log(10) = 1$	$\ln(10) = 2.3026$
Product	$\log(A \cdot B) = \log(A) + \log(B)$	$\ln(A \cdot B) = \ln(A) + \ln(B)$
Quotient	$\log(A/B) = \log(A) - \log(B)$	$\ln(A/B) = \ln(A) - \ln(B)$
Power	$\log(A^n) = n \log(A)$	$\ln(A^n) = n \ln(A)$
Radical	$\log(\sqrt{A}) = \tfrac{1}{2}\log(A)$	$\ln(\sqrt{A}) = \tfrac{1}{2}\ln(A)$
Antilogarithm	antilog $(x) = 10^x$	antiln $(x) = e^x$
Change of base	$\log(x) = \ln(x)/2.3026$	$\ln(x) = 2.3026 \log(x)$

EXERCISE 20–8

Check the following equations:

1. $\ln(17 \cdot 3) = \ln(17) + \ln(3)$

2. $\ln(9 \cdot 5) = \ln(9) + \ln(5)$

3. $\ln\left(\tfrac{65}{4}\right) = \ln(65) - \ln(4)$

4. $\ln\left(\dfrac{0.695}{0.421}\right) = \ln(0.695) - \ln(0.421)$

5. $\ln(15^2) = 2\ln(15)$

6. $\ln(0.71^2) = 2\ln(0.71)$

7. $\ln(\sqrt{64}) = \tfrac{1}{2}\ln(64)$

8. $\ln(\sqrt{3.06}) = \tfrac{1}{2}\ln(3.06)$

Use natural logarithms, Rules 20–4 and 20–5, and Example 20–15 to calculate:

9. $\sqrt{5.41}$ **10.** $\sqrt{0.0792}$ **11.** 17.1^2 **12.** $(6.45)^{0.295}$

13. $(18.7)^{1/3}$ **14.** $(1.51)^{17}$ **15.** $\sqrt{71.3}$ **16.** $(1.01)^{144}$

20–9 LOGARITHMIC EQUATIONS

An equation involving the logarithm of the unknown is referred to as a **logarithmic equation.** Thus, $16 = 4 \log (x)$ is a logarithmic equation. On the other hand, $25x = x \log (14)$ is not a logarithmic equation. The key to solving a logarithmic equation is to treat the logarithm of the unknown as a new variable. Solve for this new variable and apply the antilogarithm to find the unknown.

EXAMPLE 20–25   Solve $16 = 4 \log (x) + 8$.

Solution   First solve for $\log (x)$:

$$8 = 4 \log (x)$$
$$\log (x) = 2$$

Apply antilog to find x:

$$x = \text{antilog } (2) = 10^2$$

$\therefore$ $$x = 100$$

EXAMPLE 20–26   Solve $4.5 + \ln (x^2) = \ln (x)$.

Solution   Use Table 20–1 to simplify $\ln (x^2)$:

$$4.5 + 2 \ln (x) = \ln (x)$$

Solve for $\ln (x)$:

$$\ln (x) = -4.5$$

Apply antiln to find x:

$\boxed{e^x}$ $$x = \text{antiln } (-4.5) = e^{-4.5}$$
$$x = 1.11 \times 10^{-2}$$

$\therefore$ $$x = 1.11 \times 10^{-2}$$

EXAMPLE 20–27   Solve $\ln (x) - 1 = \log (x)$.

Solution   Convert $\ln (x)$ to $\log (x)$; then proceed:

$$2.3026 \log (x) - 1 = \log (x)$$
$$1.3026 \log (x) = 1$$
$$\log (x) = 1/1.3026$$

$\boxed{1/x}$ $$\log (x) = 0.7677$$

Apply antilog to find x:

$$x = \text{antilog } (0.7677) = 10^{0.7677}$$

$\boxed{10^x}$ $\qquad$ $x = 5.86$

$\therefore$ $\qquad$ $x = 5.86$

EXERCISE 20–9

Solve the following equations.

1. $1.05 = \log (x^2)$

2. $17 + 2 \log (x) = 6$

3. $5 \log (x) - 3 = 2$

4. $7 \ln (x) = 2 \ln (x^3) + 2.71$

5. $4 \ln (\sqrt{x}) = -1.53$

6. $\ln (x/6) = 0.15$

7. $\log (4x) + \log (x) = 2.459$

8. $4 \log (x) - 3 = \ln (x^3)$

9. $5 \log (\sqrt{x}) = \ln (x) + 1.59$

10. $\log (x/4) + \log (x/7) = 0.361$

11. $\ln (17/V) = 4.5$

12. $\ln (13z) + \ln (z^2) = 9.021$

13. $\ln (V^2) - 3.29 = \ln (V)$

14. $\ln (Y^3) - 2 \ln (Y) = 2.3026$

15. $2.3026 \log (Z) = 1$

16. $\ln (t^2) + \ln (1/t) - 4.39$

17. The strength of the signal to the strength of the noise measured in decibels is defined by the equation $N_{dB} = 10 \log (P_S/P_N)$. Solve for P_N.

18. In a particular circuit, the signal-to-noise ratio was found to be 15. How many decibels is this? (See problem 17.)

19. The number of octaves between two frequencies is given by the formula $N = \log (f_2/f_1)/\log (2)$. Solve for f_2.

20. Find the frequency (f_2) 4.5 octaves above 60 Hz (see problem 19).

20–10 EXPONENTIAL EQUATIONS

An equation with the unknown appearing in an exponent is called an **exponential equation.** Thus, $90 = 120 \, e^{-0.01/t}$ is an exponential equation. Exponential equations may be solved by first solving for the term containing the unknown exponent, and then using logarithms to find the unknown.

EXAMPLE 20–28 $\qquad$ Solve $10^{2x} - 4 = 16$ for x.

$\qquad$ **Solution** $\qquad$ Solve for 10^{2x}; then use common logarithms:

$$10^{2x} = 20$$

Take the common log of both members of the equation:

$$2x = \log (20)$$
$$x = \log (20)/2$$
$$x = 1.3010/2$$
$$\therefore \qquad x = 0.6505$$

Observation Since x is like a logarithm, we write it with four digits to the right of the decimal point.

EXAMPLE 20–29 Solve $90 = 120\, e^{-0.01/t}$ for t.

Solution Solve for $e^{-0.01/t}$; then use natural logarithms:

$$90 = 120\, e^{-0.01/t}$$
$$e^{-0.01/t} = \frac{90}{120} = 0.75$$

Take the natural logarithm of both members of the equation:

$$-0.01/t = \ln (0.75)$$

Solve for t:

$$t = -0.01/\ln (0.75)$$
$$t = -0.01/(-0.2877)$$
$$\therefore \qquad t = 0.03476$$

EXAMPLE 20–30 Solve $f_2 - f_1 2^n = 0$ for n.

Solution Solve for 2^n; then use common logarithms to find n:

$$f_1 2^n = f_2$$
$$2^n = f_2/f_1$$
$$\log (2^n) = \log (f_2/f_1)$$

Simplify the left member:

$$n \log (2) = \log (f_2/f_1)$$
$$\therefore \qquad n = \log (f_2/f_1)/\log (2)$$

Observation A similar solution could have been arrived at using natural logarithms. Both equations would give the same numerical value for n.

EXAMPLE 20–31 Evaluate n in the following equations when $f_1 = 330$ Hz and $f_2 = 2700$ Hz.

Solution Use common logarithms

$$n = \log(f_2/f_1)/\log(2)$$

$$n = \log(2700/330)/\log(2)$$

$$n = \log(8.18)/\log(2)$$

$$n = 0.9128/0.3010$$

$$n = 3.032$$

Use natural logarithms

$$n = \ln(f_2/f_1)/\ln(2)$$

$$n = \ln(2700/330)/\ln(2)$$

$$n = \ln(8.18)/\ln(2)$$

$$n = 2.1017/0.6931$$

$$n = 3.032$$

$\therefore$ The two equations give the same answer: $n = 3.032$

EXERCISE 20–10

Solve the following equations.

1. $14 = 10^x$
2. $3.5 = 10^x - 1$
3. $81 = 0.9(10^y)$
4. $0.92 = 3(10^\alpha) + 0.14$
5. $3(10^s) = 11 - 2(10^s)$
6. $18 = e^T$
7. $e^x - 4 = 0$
8. $1.52\, e^{0.1/y} = 40.3$
9. $3.4\, e^x = 0.015 + 1.5\, e^x$
10. $75\, e^x - 125 = 13 + 6\, e^x$
11. $e^{-4/t} = 0.8$
12. $e^{-15/t} = 0.5$
13. $e^{-1/(6t)} = 0.95$
14. $1 - e^{-7.4/t} = 0.26$
15. $1 + e^{8.6/t} = 3.54$
16. $2^x = 32$
17. $3^y = 243$
18. $5^t = 17.3$
19. $21^{1/a} = 0.21$
20. $15^{-1/b} = 0.95$

20–11 SEMILOG AND LOG–LOG PLOTS

Graphs are tools used to present mathematical data. As such, there is no one *right* way to draw a graph. The scales are dictated by use of the graph. Two scales useful in electronics are **semilog** and **log–log.** A semilog plot has one regular (linear) scale and one with logarithmic spacing. Figure 20–6 is an example of a semilog plot. In a log–log plot both scales have logarithmic spacing. Figure 20–7 is an example of a log–log plot.

In Figure 20–6 there are three curves. On rectangular (linear) graph paper, $y = 2x$ would be a straight line, but on semilog paper it is a curve. The graph of $y = x^2$ is a curve on both types of paper, but the curves do not look the same. The graph of $y = 2^x$ is a straight line on semilog paper. These last two curves show the usefulness of semilog plots. First, they compress large values (and expand values close to zero) so that more data may be presented on a single graph. Second, semilog graphs convert exponential curves into straight lines.

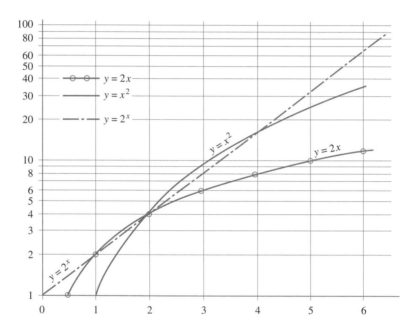

FIGURE 20–6   Examples of semilog plots.

In Figure 20–7, the same three curves are plotted on log–log paper. Here, $y = 2x$ and $y = x^2$ are straight lines, but $y = 2^x$ is a curve. Log–log plots are used to present wide ranges of data, especially data that have the form $y = ax^n$.

EXERCISE 20–11

1. Create a semilog plot of $y = e^x$.
2. Create a semilog plot of $y = 10^x$.
3. Create a log–log plot of $y = x^3$.
4. Create a log–log plot of $y = x^{1.5}$.

FIGURE 20–7 Examples of log–log plots.

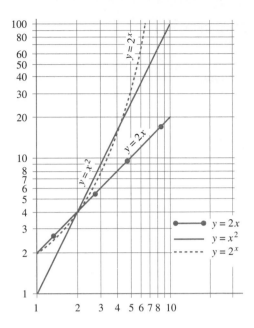

5. On log–log paper, plot $y = \sqrt{x}$ and $y = \ln(x)$.

6. Plot the following points on rectangular, semilog, and log–log paper:

x	0.1	0.2	0.3	0.4	0.5	0.6	0.7
y	0.1	0.203	0.309	0.423	0.546	0.684	0.842

x	0.8	0.9	1.0	1.1	1.2	1.3	1.4
y	1.029	1.26	1.56	1.96	2.57	3.60	5.80

7. Plot the following data on semilog and log–log paper. Do they plot as a straight line?

x	0.5	1	1.5	2	2.5	3
y	0.5	2	4.5	8	12.5	18

8. Plot the following data on semilog and log–log paper. Do they plot as a straight line?

x	1	1.5	2	2.5	3	3.5	4	4.5	5
y	1.07	1.10	1.14	1.18	1.21	1.27	1.31	1.36	1.40

x	5.5	6	6.5	7	7.5	8	8.5	9	9.5	10
y	1.45	1.50	1.55	1.61	1.66	1.72	1.78	1.84	1.90	1.97

20–12 NOMOGRAPHS

In electronics and especially communications, it is often convenient to solve problems graphically. There are many graphical aids to solutions presented in handbooks and periodicals. These graphical aids are called **nomographs.** Figure 20–8 presents a simple nomograph consisting of a single straight line with different gradations on either side. To use this nomograph, find the power ratio on the top; directly under it is the corresponding number of decibels (dB). The reverse process will convert decibels to power ratio. For example, a power ratio of 3 corresponds to 4.8 decibels.

A second type of nomograph relates three variables. The formula for inductive reactance is $X_L = 2\pi fL$. A nomograph relating X_L, f, and L is presented in Figure 20–9. When two of the parameters are known, the third is found by placing a straightedge to join the two known values. The value of the third parameter is read from the appropriate scale under the straightedge.

EXAMPLE 20–32   Find the frequency if the inductance is 400 μH and the reactance is 2 kΩ.

Solution   Use the nomograph in Figure 20–9. Place a mark at 400 μH and a second at 2 kΩ. Draw a straight line through these points. Read the frequency as 800 kHz.

Check   Check the graphical solution:

$$X_L = 2\pi fL$$

Solve for f:

$$f = X_L/(2\pi L)$$

Substitute:

$$f = 2(10^3)/(2\pi 400 \times 10^{-6})$$
$$f = 796\ 000 \text{ Hz}$$
$$f = 796 \text{ kHz} \approx 800 \text{ kHz}$$

$\therefore$ The graphical solution checks.

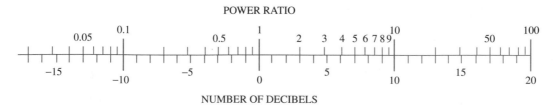

POWER RATIO

FIGURE 20–8   Nomograph for converting from power ratio to number of decibels.

CHAPTER 20   Logarithmic and Exponential Functions

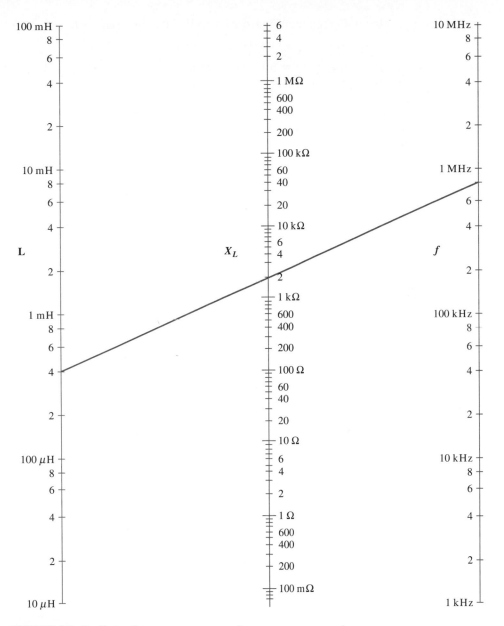

FIGURE 20–9 Inductive reactance versus frequency nomograph.

EXERCISE 20–12

Use the nomograph in Figure 20–8 to convert the following.

1. 6 dB to power ratio
2. 15 dB to power ratio
3. 0.02 to number of decibels
4. 20 to number of decibels

Use the nomograph in Figure 20–9 to find the third parameter.

5. $L = 200 \, \mu\text{H}$ $f = 1$ MHz   Find X_L.
6. $L = 6$ mH $f = 50$ kHz   Find X_L.
7. $X_L = 60 \, \Omega$ $f = 10$ kHz   Find L.
8. $X_L = 10 \, \text{k}\Omega$ $f = 10$ MHz   Find L.
9. $X_L = 1 \, \text{M}\Omega$ $L = 100$ mH   Find f.
10. $L = 30$ mH $X_L = 300 \, \Omega$ Find f.

SELECTED TERMS

characteristic   The integer part of a common logarithm.
common antilogarithm   The inverse of the logarithm function: when $a = \log(b)$ then $b = \text{antilog}\,(a)$.
common logarithm   Base ten logarithm
logarithm   The exponent to which the base must be raised to equal the original number.
mantissa   The fractional part of a common logarithm.
natural logarithm   Base e (2.71828 . . .) logarithm.

21

Applications of Logarithmic and Exponential Equations to Electronic Concepts

21–1   The Decibel

21–2   System Calculations

21–3   *RC* and *RL* Transient Behavior

21–4   Preferred Number Series

PERFORMANCE OBJECTIVES

- Express power gain or loss in decibels.

- Calculate overall system gain or loss using decibels.

- Solve for instantaneous values of current and voltage for series circuits under charge.

- Compute the values used in each of the five preferred number series.

The Viking Lander Project's landing module equipped with an array of instruments.   (Courtesy of NASA)

This chapter applies the techniques of logarithmic and exponential equations to two areas of electronics. The first area is gain and loss in an electronic system. The second is rise time in *RC* and *RL* series circuits. The topics have been arranged to progress from the decibel, through system calculations, to *RC* and *RL* transient behavior.

21–1   THE DECIBEL

Power in Decibels

The human ear hears in a *logarithmic* manner. That is, the multiplication of the acoustic power of any sound by 1.258 is heard as an increment of loudness of one decibel (dB). A second multiplication of the acoustic power gives rise to another increment of one decibel. For example, to increase the loudness by ten decibels of the sound of an audio amplifier operating at four watts, the power must be increased to 40 watts. To increase the loudness another ten decibels, the power must be increased to 400 W!

Because of its logarithmic nature, the **decibel** (dB) has been adopted as the practical unit for measuring what the ear hears. Over the years the decibel has been applied to indicate the performance of all types of electronic circuits and devices, including amplifiers (audio frequency as well as radio frequency), filters, antennas, and microphones.

The decibel is a logarithmic expression that compares two power levels. Expressed mathematically,

$$N_{dB} = 10 \log \left(\frac{P_{out}}{P_{in}} \right) \quad \textbf{dB} \quad\quad\quad \textbf{(21–1)}$$

where N_{dB} = gain or loss in decibels (dB)
   P_{in} = input power (W)
   P_{out} = output power (W)

EXAMPLE 21–1   Determine the power gain of an amplifier (N_{dB}) when an input power of 0.50 W produces an output power of 50 W.

Observation   An **amplifier** is an electronic device that produces an electrical signal at its output that is an enlarged reproduction of the essential features of its input.

Solution   Use Equation 21–1 and substitute:

$$N_{dB} = 10 \log (P_{out}/P_{in})$$
$$N_{dB} = 10 \log (50/0.50)$$
$$\therefore \quad N_{dB} = 20.00 \text{ dB}$$

Observation   The decibel is a logarithm multiplied by 10. If the power ratio is known to two significant figures, then the number of decibels is known to two digits to the right of the decimal point.

Expressing Gain and Loss in Decibels

In the preceding example the larger power of 50 W was placed over the smaller power of 0.5 W. This caused the log of the ratio to be positive. However, when this ratio is inverted, the log of the ratio is negative. A *gain* is indicated with a plus sign: $P_{out}/P_{in} > 1 \Rightarrow \log(P_{out}/P_{in}) > 0\,(+)$. A *loss* is indicated with a minus sign: $P_{out}/P_{in} < 1 \Rightarrow \log(P_{out}/P_{in}) < 0\,(-)$.

EXAMPLE 21–2   Use an appropriate sign to express each of the stated conditions as a gain or a loss: gain of 32 dB and loss of 17 dB.

Solution   gain of 32 dB $= +32$ dB
loss of 17 dB $= -17$ dB

EXAMPLE 21–3   The attenuator network pictured in Figure 21–1 has a loss of 6.00 dB (-6.00 dB). Determine the input power (P_{in}) if the output power (P_{out}) is 300 mW.

FIGURE 21–1   Circuit for Example 21–3 with the attenuator network pictured within the dashed lines.

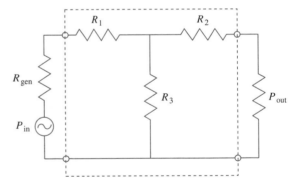

Solution   Substitute into Equation 21–1 and solve for P_{in}:

$$N_{dB} = 10 \log\left(\frac{P_{out}}{P_{in}}\right)$$

$$-6.00 = 10 \log\left(\frac{300 \times 10^{-3}}{P_{in}}\right)$$

$$-0.60 = \log\left(\frac{300 \times 10^{-3}}{P_{in}}\right)$$

Take the antilog of each member:

$$0.251 = \frac{300 \times 10^{-3}}{P_{in}}$$

$$P_{in} = \frac{300 \times 10^{-3}}{0.251}$$

$$\therefore \qquad P_{in} = 1.2 \text{ W}$$

Observation   An **attenuator** is a resistive network that reduces the magnitude of an electric signal.

EXAMPLE 21–4   Determine the voltage (V_o) developed across the load (1.0 kΩ) of the +24.00-dB amplifier shown in Figure 21–2.

FIGURE 21–2 Circuit for Example 21–4. The amplifier has a 24-dB gain.

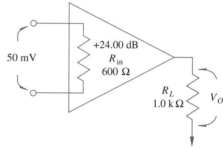

Solution   Determine P_{in}:

$$P_{in} = E^2/R_{in}$$

$$P_{in} = \frac{(50 \times 10^{-3})^2}{600}$$

$$P_{in} = 4.17 \ \mu\text{W}$$

Substitute into Equation 21–1 and solve for P_{out}:

$$N_{dB} = 10 \log \left(\frac{P_{out}}{P_{in}} \right)$$

$$24.00 = 10 \log \left(\frac{P_{out}}{4.17 \times 10^{-6}} \right)$$

$$2.400 = \log \left(\frac{P_{out}}{4.17 \times 10^{-6}} \right)$$

Take the antilog of each member:

$$251 = \frac{P_{out}}{4.17 \times 10^{-6}}$$

$$P_{out} = 1.05 \text{ mW}$$

Solve for V_o:

$$P_{out} = V_o^2/R_L$$
$$V_o = \sqrt{P_{out} R_L}$$
$$V_o = \sqrt{(1.05 \times 10^{-3})(1 \times 10^3)}$$
$$\therefore \qquad V_o = 1.0 \text{ V}$$

Power Gain or Loss in Decibels from Voltage Ratio

The power gain or loss in decibels may be computed from a voltage ratio if the measurements are made across equal resistances. Equation 21–1 is expressed as Equation 21–2 when $R_{in} = R_{out}$.

$$N_{dB} = 10 \log \left(\frac{P_{out}}{P_{in}} \right)$$

However,

$$P_{out} = E_{out}^2/R_{out} \text{ and } P_{in} = E_{in}^2/R_{in}$$

Thus,

$$N_{dB} = 10 \log \left(\frac{E_{out}^2/R_{out}}{E_{in}^2/R_{in}} \right)$$

Substitute $R_{in} = R_{out}$ and simplify:

$$N_{dB} = 10 \log \left(\frac{E_{out}^2/R_{out}}{E_{in}^2/R_{out}} \right)$$

$$N_{dB} = 10 \log \left[\left(\frac{E_{out}}{E_{in}} \right)^2 \right]$$

$$N_{dB} = 2(10) \log \left(\frac{E_{out}}{E_{in}} \right)$$

$$\mathbf{N_{dB} = 20 \log \left(\frac{E_{out}}{E_{in}} \right)} \qquad \mathbf{dB} \qquad\qquad (21\text{–}2)$$

EXAMPLE 21–5   Determine the decibel loss in a 50-Ω coaxial cable that has an input of 12 V. The output is 4.0 V across a 50-Ω termination.

Solution   Use Equation 21–2 because each resistance is 50 Ω:

$$N_{dB} = 20 \log (E_{out}/E_{in})$$
$$N_{dB} = 20 \log (4.0/12)$$
$$\therefore \quad N_{dB} = -9.54 \text{ dB}$$

EXERCISE 21–1

1. An audio amplifier develops 20 W in the output load from an input of 75 mW. Determine the power gain of the amplifier in decibels.

2. An RF linear amplifier requires 2.5-W input to develop 54-W output. Determine the power gain of the amplifier in decibels.

3. An equalizer has an insertion loss of 6.44 dB when placed in an audio line. Determine the output power when the input power is 1.5 W.

4. A low-pass filter has an attenuation of 38.5 dB at 50 MHz. Determine the output power at 50 MHz when the input power is 5.0 W.

5. The input to a preamp is 10 mV and the output is 2.0 V. Assuming that each voltage is developed across the same amount of resistance, determine the power gain of the preamp.

6. Determine the input voltage needed to develop 38 V across a resistance if the power gain of an amplifier is +37.15 dB. Both input and output resistances are equal.

7. Determine the amount of input voltage needed to develop 650 mW in a 16-Ω load when the power gain of the amplifier is +26.99 dB. The input resistance is 10 kΩ.

8. The input resistance to an amplifier is 1.2 kΩ, and the output resistance is 200 Ω. If the amplifier has a +68.45 dB gain, determine what input voltage is needed to develop 28 V in the output.

9. A ceramic transducer develops 180 mV across the 15-kΩ input resistance of an amplifier. The output power developed across an 8.0-Ω load is 24 W. Determine the decibel power gain of the amplifier.

10. Determine the voltage dropped across a 4.0-Ω load when a microphone (at the input) develops 6.0 mV across 600 Ω. The gain of the amplifier is +36.50 dB.

21–2   SYSTEM CALCULATIONS

In the preceding section you learned that decibels (dB) are used to compare two power levels. When working with electronic systems, it is sometimes necessary to know the actual power or voltage of this system in relation to a stated reference. We will consider two ways of using decibels to state absolute values. One term, the decibel milliwatt (dBm), represents the actual power. The other term, the decibel millivolt (dBmV), represents the actual voltage.

Decibel Milliwatt (dBm)

Audio communication systems having input and output resistances of 600 Ω use a reference power of 1 mW as 0 dBm. The unit *decibel milliwatt* (dBm) is used to indicate the actual power of an amplifier, an attenuator, or an entire system. The gain or loss in decibels based on 1 mW is computed by Equation 21–3.

$$N_{dBm} = 10 \log \left(\frac{P}{1 \text{ mW}} \right) \qquad \text{dBm} \qquad (21\text{–}3)$$

EXAMPLE 21–6   Determine the power represented by +20.00 dBm.

Solution   Substitute into Equation 21–3 and solve for P:

$$20.00 = 10 \log \left(\frac{P}{1 \text{ mW}} \right)$$

$$2.00 = \log \left(\frac{P}{1 \times 10^{-3}} \right)$$

Take the antilog:

$$100 = \frac{P}{1 \times 10^{-3}}$$

$$P = 100 \times 10^{-3}$$

$$\therefore \qquad P = 100 \text{ mW}$$

From this example we see that +20 dBm represents 100 mW. Because the decibel is logarithmic, the actual power present in an audio communication system may be determined by first algebraically adding the powers represented in decibel milliwatts and then solving Equation 21–3 for P.

A quick estimate of the power represented by N_{dBm} may be made with the information of Table 21–1.

EXAMPLE 21–7   Determine the actual power in the audio system pictured in Figure 21–3.

FIGURE 21–3   Audio system for Example 21–7.

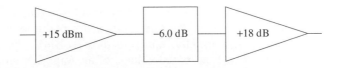

Solution   Algebraically add the gains and losses:

$$N_{dBm} = +15 - 6 + 18$$

$$N_{dBm} = +27 \text{ dBm}$$

Solve Equation 21–3 for P:

$$N_{\text{dBm}} = 10 \log \left(\frac{P}{1 \text{ mW}} \right)$$

$$27 = 10 \log \left(\frac{P}{1 \times 10^{-3}} \right)$$

$$2.7 = \log \left(\frac{P}{1 \times 10^{-3}} \right)$$

Take the antilog:

$$501 = \frac{P}{1 \times 10^{-3}}$$

$$P = 0.50 \text{ W}$$

$\therefore$ The output power is 0.5 W for the 1 mW of input.

TABLE 21–1 Signal Power Compared to Decibels Milliwatt

dBm	P/1 mW	Signal Power
+40	10 000	10 W
+30	1 000	1 W
+20	100	100 mW
+10	10	10 mW
+6	4	4 mW
+3	2	2 mW
0	1	1 mW
−3	0.5	0.5 mW
−6	0.25	0.25 mW
−10	0.10	100 μW
−20	0.01	10 μW
−30	0.001	1 μW
−40	0.000 1	0.1 μW

Decibel Millivolt (dBmV)

Another special application of the decibel is in community antenna television (CATV) for the measurement of the signal intensity. In this system, a signal of 1 mV across 75 Ω is the reference level that corresponds to 0 dBmV. The unit *decibel millivolt* (dBmV) is used to indicate the actual voltage of an amplifier, an attenuator, or an entire system. The actual gain or loss in dBmV is computed by Equation 21–4.

$$N_{\text{dBmV}} = 20 \log \left(\frac{V}{1 \text{ mV}} \right) \qquad \text{dBmV} \qquad (21\text{–}4)$$

EXAMPLE 21–8 Express a signal strength of $100 \ \mu V$ in dBmV.

Solution Use Equation 21–4:

$$N_{dBmV} = 20 \log \left(\frac{V}{1 \ mV} \right)$$

$$N_{dBmV} = 20 \log \left(\frac{100 \times 10^{-6}}{1 \times 10^{-3}} \right)$$

$$N_{dBmV} = -20 \ dBmV$$

∴ A signal strength of $100 \ \mu V$ is noted as -20 dBmV.

Table 21–2 provides information to make a quick estimate of the actual voltage represented by N_{dBmV}.

TABLE 21–2   Signal Voltage Compared to Decibels Millivolt

dBmV	V/1 mV	Signal Voltage	dBmV	V/1 mV	Signal Voltage
+60	1 000	1 V	−3	0.7	0.7 mV
+40	100	100 mV	−6	0.5	0.5 mV
+20	10	10 mV	−20	0.1	100 μV
+6	2	2 mV	−40	0.01	10 μV
+3	1.4	1.4 mV	−60	0.001	1 μV
0	1	1 mV			

EXAMPLE 21–9 Determine the actual signal level across a resistance of 75 Ω for a power of $+25.00$ dBmV.

Solution Substitute the given information into Equation 21–4 and solve for V:

$$N_{dBmV} = 20 \log \left(\frac{V}{1 \ mV} \right)$$

$$25.00 = 20 \log \left(\frac{V}{1 \ mV} \right)$$

$$1.25 = \log \left(\frac{V}{1 \times 10^{-3}} \right)$$

Take the antilog of each member:

$$17.8 = \frac{V}{1 \times 10^{-3}}$$

$$V = 17.8 \text{ mV}$$

∴   The signal level is 18 mV.

EXAMPLE 21–10   Determine the actual signal level in millivolts (mV) across the connecting terminals of the television set of Figure 21–4. The signal at the antenna terminals is 700 μV, and the connecting cable has an attenuation (loss) of 9 dB/100 m. The resistance of the system is the same throughout.

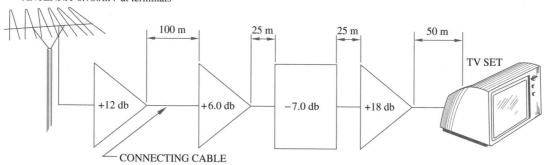

FIGURE 21–4   Simplified distribution system for Example 21–10.

Solution   State the incoming signal of 700 μV in dBmV:

$$N_{\text{dBmV}} = 20 \log\left(\frac{700 \times 10^{-6}}{1 \times 10^{-3}}\right)$$

$$N_{\text{dBmV}} = -3 \text{ dBmV}$$

Compute the cable loss at 9 dB/100 m:

$$100 \text{ m} + 25 \text{ m} + 25 \text{ m} + 50 \text{ m} = 200 \text{ m}$$

$$200 \text{ m} \,(-9 \text{ dB}/100 \text{ m}) = -18 \text{ dB}$$

Compute the system gain in decibel millivolts:

$$-3 \text{ dBmV} - 18 \text{ dB} + 12 \text{ dB} + 6 \text{ dB}$$

$$-7 \text{ dB} + 18 \text{ dB} = +8 \text{ dBmV}$$

Compute the signal level across the connecting terminals of the television. Use Equation 21–4 and solve for V:

$$N_{dBmV} = 20 \log \left(\frac{V}{1\ mV} \right)$$

$$8 = 20 \log \left(\frac{V}{1 \times 10^{-3}} \right)$$

$$0.40 = \log \left(\frac{V}{1 \times 10^{-3}} \right)$$

Take the antilog:

$$2.51 = \frac{V}{1 \times 10^{-3}}$$

$$V = 2.51 \times 10^{-3}$$

$$\therefore \qquad V = 2.5\ mV$$

Summary

- The actual signal level may be determined in an electronic system by referencing decibel (dB) gains and decibel (dB) losses in the system to a definite amount of voltage or power.
- The overall dB gain or loss in a system may be determined by algebraically adding the individual gains and losses.
- Gains or losses in decibel milliwatts (dBm) and decibel millivolts (dBmV) are not used together to describe the same system.

EXERCISE 21–2

1. Express each of the following power levels in decibel milliwatts.
   - **(a)** 15 mW
   - **(b)** 200 mW
   - **(c)** 18 μW
   - **(d)** 2.2 W
   - **(e)** 610 μW
   - **(f)** 87 mW

2. What is the power level in watts (W) represented by the following?
   - **(a)** +17.00 dBm
   - **(b)** −7.00 dBm
   - **(c)** +30.50 dBm
   - **(d)** −5.60 dBm
   - **(e)** −18.00 dBm
   - **(f)** +42.00 dBm

3. Express each of the following voltage levels in decibel millivolts.
   - **(a)** 0.32 V
   - **(b)** 150 μV
   - **(c)** 3.8 V
   - **(d)** 72 mV
   - **(e)** 840 mV
   - **(f)** 27 μV

4. What is the signal level in millivolts represented by the following?
   - **(a)** +23.00 dBmV
   - **(b)** +2.80 dBmV
   - **(c)** −6.90 dBmV
   - **(d)** −10.80 dBmV
   - **(e)** +11.00 dBmV
   - **(f)** −1.40 dBmV

5. A certain 75-Ω antenna develops a signal of 4700 μV. Express this signal level in dBmV.

6. Determine the specifications in decibel milliwatts of an amplifier that has an output power of 8 W.

7. Determine the specifications in decibel millivolts of a CATV system that provides an output of 8.7 mV of signal to the subscriber's set.

8. A certain amplifier has an output of 290 mV across 600 Ω. Express this signal level in dBm.

9. Determine the actual voltage level across a 600-Ω load in an audio system that has a preamp with an output power of +28 dBm connected through a matching transformer of −4 dB to an amplifier with a gain of +18 dB. The output of the amplifier is connected through a −6-dB matching pad (attenuator) to the 600-Ω load.

10. Specify the gain needed in decibels to provide 1500 μV of signal to the 75 Ω input of an FM receiver if the signal at the antenna is 100 μV. The 75 Ω antenna is connected to the receiver by 50 m of 75-Ω coaxial cable having a loss of 4 dB/100 m.

21–3   *RC* AND *RL* TRANSIENT BEHAVIOR

RL Transient Behavior

When the switch of Figure 21–5 is closed, the current through the resistance goes instantly from 0 to 2 A. The graph of current as a function of time verifies this instantaneous change in current. However, when an inductor (coil) is added in series with the resistance of Figure 21–5, the current no longer changes instantaneously. Figure 21–6 shows how the inductance (measured in henries, H) of the inductor changes the characteristic of the current. We see that it takes some time for the current to come to a *steady state*. The increase in the circuit current is a **transient** behavior. The graph of Figure 21–6 is not linear, but is an exponential function. During the transient time, the **instantaneous value** of the current in the circuit is determined with Equation 21–5.

$$i = \frac{E}{R}[1 - e^{-t/(L/R)}] \qquad \text{(amperes)} \qquad (21\text{–}5)$$

FIGURE 21–5   (a) The current in a resistive circuit changes from 0 to 2 A when the switch is closed; (b) the graph of *I* as a function of time.

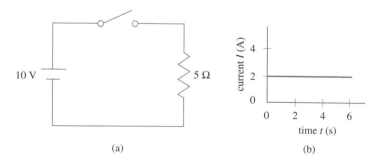

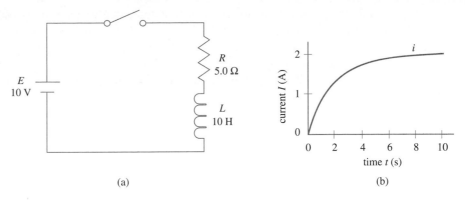

(a)   (b)

FIGURE 21–6   Current in an *RL* circuit: circuit (a) has a transient behavior as shown in (b).

where i = circuit instantaneous current in amperes at time t
        t = time in seconds, after switch is closed
        e = basc of the natural logarithms
        E = source voltage in volts
        R = circuit resistance in ohms
        L = circuit inductance in henries

EXAMPLE 21–11   Determine the instantaneous current in the circuit of Figure 21–6 two seconds after the switch is closed.

Solution   Substitute $R = 5.0\ \Omega$, $L = 10$ H, $t = 2.0$ s, and $E = 10$ V into Equation 21–5:

$$i = \frac{E}{R}\,[1 - e^{-t/(L/R)}]$$

$$i = \frac{10}{5.0}\,[1 - e^{-2.0/(10/5.0)}]$$

$\boxed{e^x}$ $i = 2.0(1 - \mathbf{0.368})$

$\therefore$ $i = 1.3$ A

EXAMPLE 21–12   In the circuit of Figure 21–6, determine the instantaneous current 3.5 s after the switch is closed.

Solution   Substitute into Equation 21–5:

$$i = \frac{E}{R}\,[1 - e^{-t/(L/R)}]$$

$$i = \frac{10}{5.0}\,[1 - e^{-3.5/(10/5.0)}]$$

$\boxed{e^x}$ $i = 2.0(1 - \mathbf{0.174})$

$\therefore$ $i = 1.7$ A

Table 21–3 is a summary of the exponential equations for the *charging* of an *RL* series circuit. The equations in Table 21–3 refer to the circuit in Figure 21–7.

TABLE 21–3   The *RL* Exponential Equations for Series Circuits Under Charge

Instantaneous circuit current

$$i = \frac{E}{R}[1 - e^{-t/(L/R)}] \qquad (21\text{–}5)$$

Instantaneous voltage across inductance

$$v_L = Ee^{-t/(L/R)} \qquad (21\text{–}6)$$

Instantaneous voltage across resistance

$$v_R = E[1 - e^{-t/(L/R)}] \qquad (21\text{–}7)$$

where i = circuit instantaneous current in amperes at time t
   t = time in seconds, after switch is closed
   e = base of the natural logarithms
   E = source voltage in volts
   R = circuit resistance in ohms
   L = circuit inductance in henries
   v_L = instantaneous voltage across the inductor
   v_R = instantaneous voltage across the resistance

FIGURE 21–7   Series *RL* circuit for Equations 21–5 through 21–7 in Table 21–3.

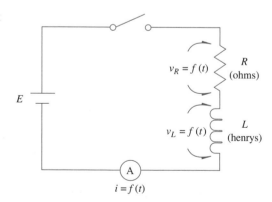

EXAMPLE 21–13   Determine the inductance in henries (H) of the inductor of Figure 21–7 if the instantaneous current is 500 mA 3.0 s after the closing of the switch. $E = 12$ V and $R = 8.0$ Ω.

Solution   Let $x = -t/(L/R)$ in Equation 21–5:

$$i = \frac{E}{R}(1 - e^x)$$

Solve for x:

$$\frac{iR}{E} = 1 - e^x$$

$$e^x = 1 - \frac{iR}{E}$$

$$\ln(e^x) = \ln\left(1 - \frac{iR}{E}\right)$$

$$x = \ln\left(1 - \frac{0.5(8.0)}{12}\right)$$

$$x = -0.405$$

Substitute $-t/(L/R)$ for x:

$$-t/(L/R) = -0.405$$

Solve for L:

$$t = 0.405(L/R)$$

$$\frac{tR}{0.405} = L$$

$$L = \frac{3.0(8.0)}{0.405}$$

$$\therefore \qquad L = 59 \text{ H}$$

EXERCISE 21–3

1. Determine the instantaneous current in the circuit of Figure 21–7 two seconds after the switch is closed. $R = 2.0$ Ω, $L = 5.0$ H, and $E = 30$ V.

2. Repeat problem 1 for $R = 1.0$ Ω.

3. For the circuit of Figure 21–7:

   (a) Find the instantaneous current (i) in the circuit 5.0 ms after the switch is closed. $R = 200$ Ω, $L = 500$ mH, and $E = 50$ V.

   (b) Repeat part **(a)** for v_L.

   (c) Repeat part **(a)** for v_R.

4. An inductor of 12 H is connected in series with a 90-V dc source. What is the instantaneous current in the inductor 0.20 s after the source is connected? The internal resistance of the inductor is 50 Ω.

Determine the values missing in the following table. Use Equations 21–5 through 21–7 and Figure 21–7.

	E	R	L	t	i	v_R	v_L
5.	18 V	56 Ω	1 H	35 ms	?	?	?
6.	36 V	150 Ω	250 mH	800 μs	?	?	?
7.	10 V	8.0 Ω	?	2.5 s	?	?	3.0 V
8.	20 V	4.0 Ω	?	1.0 s	?	12 V	?

RC Transient Behavior

When a capacitance is connected in series with a resistance and a dc source of voltage (as shown in Figure 21–8), the instantaneous current is not linear, but is an exponential function. The *transient behavior* of the circuit of Figure 21–8 is described mathematically by Equations 21–8 through 21–10 in Table 21–4.

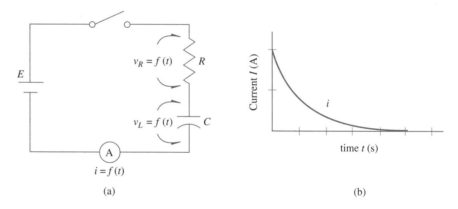

(a)   (b)

FIGURE 21–8   Current in an *RC* circuit: circuit (a) has a transient behavior as shown in (b).

EXAMPLE 21–14   In the circuit of Figure 21–8(a), determine the instantaneous voltage across the capacitor 10 s after the switch is closed. $R = 100$ kΩ, $C = 50$ μF, and $E = 100$ V.

Solution   Use Equation 21–9 from Table 21–4:

$$v_C = E(1 - e^{-t/RC})$$

$$v_C = 100\left[1 - e^{-10/(100 \times 10^3 \times 50 \times 10^{-6})}\right]$$

$$\therefore \qquad v_C = 86 \text{ V}$$

TABLE 21–4 The *RC* Exponential Equations for Series Circuits Under Charge

Instantaneous circuit current	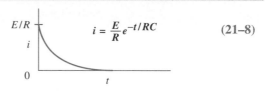	(21–8)
Instantaneous voltage across capacitance	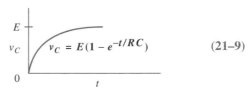	(21–9)
Instantaneous voltage across resistance	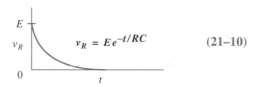	(21–10)

where i = circuit instantaneous current in amperes at time t

t = time in seconds, after switch is closed

e = base of the natural logarithms

E = source voltage in volts

R = circuit resistance in ohms

C = circuit capacitance in farads

v_C = instantaneous voltage across the capacitance

v_R = instantaneous voltage across the resistance

EXAMPLE 21–15

Determine how long it will take for the voltage across the capacitor of Figure 21–8(a) to reach 90% of the source voltage once the switch is closed. $R = 15$ kΩ, $C = 10$ μF, and $E = 30$ V.

Solution

Let $x = -t/RC$ in Equation 21–9:

$$v_C = E(1 - e^x)$$
$$v_C = 0.9E$$

Solve for x by eliminating v_C:

$$0.9E = E(1 - e^x)$$
$$e^x = 1 - 0.9$$

Take the ln:

$$x = \ln(0.1)$$
$$x = -2.30$$

Substitute $-t/RC$ for x:

$$-t/RC = -2.30$$

Solve for t:

$$t = 2.30\,RC$$
$$t = 2.30 \times 15 \times 10^3 \times 10 \times 10^{-6}$$
$$\therefore \qquad t = 0.35 \text{ s}$$

Summary

Before going on to the next exercise, take a moment to look over Tables 21–3 and 21–4. Compare the equation forms to their exponential curves. Here you will see that quantities (voltage or current) that **rise,** as in Equations 21–5, 21–7, and 21–9, have a general equation form of

$$y = Y_{\text{mx}}\,(1 - e^{-t/\tau}) \qquad\qquad \textbf{(21–11)}$$

Quantities that **fall,** as in Equations 21–6, 21–8, and 21–10, have a general equation form of

$$y = Y_{\text{mx}}e^{-t/\tau} \qquad\qquad \textbf{(21–12)}$$

In each of these general equations, y represents the instantaneous quantity (i if current or e if voltage). Y_{mx} represents the maximum value of source voltage (E) or source current $(I = E/R)$. The t represents time in seconds after the start of charge or discharge. The Greek letter tau, τ, is used to represent the time constant RC ($\tau = RC$) in a capacitive circuit or L/R ($\tau = L/R$) in an inductive circuit. By memorizing these general forms, you will be able to recall the appropriate equation for rise or fall of a given quantity in a series RC or RL circuit.

EXERCISE 21–4

1. A 10.0-μF capacitor is charged through a 47.0-kΩ resistor by a 36.0-V source. Determine the instantaneous current in the circuit after 2.00 s.

2. Determine the voltage across the capacitor (v_C) of problem 1 after 0.800 s.

3. Determine the voltage across the resistor of problem 1 after 1.50 s.

4. How long will it be after the switch of Figure 21–9 is closed before the neon lamp will flash? The lamp flashes when there is 80 V across it.

FIGURE 21–9 Circuit for Exercise 21–4, problem 4.

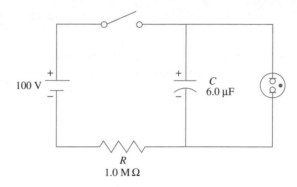

Determine the values missing in the following table. Use Equations 21–8 through 21–10 and Figure 21–8.

	E	R	C	t	i	v_R	v_C
5.	25 V	10 kΩ	470 μF	?	?	18 V	?
6.	10 V	68 kΩ	?	30 ms	?	?	4.0 V
7.	100 V	100 kΩ	0.010 μF	1.80 ms	?	?	?
8.	36 V	820 kΩ	1.0 μF	?	23.9 μA	?	?

21–4 PREFERRED NUMBER SERIES

Military specification (MIL-spec) and industrial-consumer fixed resistor, capacitor, and inductor sizes are designated with any one of the five *preferred number series.* The number series are

E6: 6 numbers used to specify components with ±20% tolerance.

E12: 12 numbers used to specify components with ±10% tolerance.

E24: 24 numbers used to specify components with ±5% tolerance.

E96: 96 numbers used to specify components with ±1% tolerance.

E192: 192 numbers used to specify components with ±0.1% tolerance.

Table 21–5 lists the numbers in one decade for each of the five preferred number series. These numbers serve as the base numbers for a wide range of standard values; for example, 2.2 is one of the E6, E12, and E24 numbers. It may be used to specify a 220-μF ± 20% capacitor (2.2 × 100 μF) or a 220-kΩ ± 5% resistor (2.2 × 10^5 Ω) or a 22-mH ± 10% inductor (2.2 × 0.01 H).

When you first look at the numbers in the preferred number series (Table 21–5), there seems to be no logical relationship among them. However, they are related by a simple equation (Equation 21–13), which is used to calculate the sequential numbers for each of the five preferred number series.

$$N_{x+1} = 10^{x/n} \qquad \text{(21–13)}$$

where $\quad N_{x+1}$ = a value in the preferred number series

$\qquad n$ = number of values in one of the five preferred number series $(n \neq 0)$

$\qquad x$ = whole numbers ranging from 0 to $n - 1$ in value

As noted in Table 21–5, the E6 preferred number series is written with two significant figures in each number. This is also true for the E12 and E24 number series; however, the E96 and E192 series have three significant figures in each number.

TABLE 21–5 Preferred Number Series for Standard Component Values for One Decade*

E6 ± 20%	E12 ± 10% (bold) E24 ± 5% (all)	E96 ± 1% (bold) and E192 ± 0.1% (all numbers)						
1.0	**1.0**	**1.00**	1.01	**1.02**	1.04	**1.05**	1.06	**1.07**	1.09
	1.1	**1.10**	1.11	**1.13**	1.14	**1.15**	1.17	**1.18**	1.20
	1.2	**1.21**	1.23	**1.24**	1.26	**1.27**	1.29	**1.30**	1.32
	1.3	**1.33**	1.35	**1.37**	1.38	**1.40**	1.42	**1.43**	1.45
1.5	**1.5**	**1.47**	1.49	**1.50**	1.52	**1.54**	1.56	**1.58**	1.60
	1.6	**1.62**	1.64	**1.65**	1.67	**1.69**	1.72	**1.74**	1.76
	1.8	**1.78**	1.80	**1.82**	1.84	**1.87**	1.89	**1.91**	1.93
	2.0	**1.96**	1.98	**2.00**	2.03	**2.05**	2.08	**2.10**	2.13
2.2	**2.2**	**2.15**	2.18	**2.21**	2.23	**2.26**	2.29	**2.32**	2.34
	2.4	**2.37**	2.40	**2.43**	2.46	**2.49**	2.52	**2.55**	2.58
	2.7	**2.61**	2.64	**2.67**	2.71	**2.74**	2.77	**2.80**	2.84
	3.0	**2.87**	2.91	**2.94**	2.98	**3.01**	3.05	**3.09**	3.12
3.3	**3.3**	**3.16**	3.20	**3.24**	3.28	**3.32**	3.36	**3.40**	3.44
	3.6	**3.48**	3.52	**3.57**	3.61	**3.65**	3.70	**3.74**	3.79
	3.9	**3.83**	3.88	**3.92**	3.97	**4.02**	4.07	**4.12**	4.17
	4.3	**4.22**	4.27	**4.32**	4.37	**4.42**	4.48	**4.53**	4.59
4.7	**4.7**	**4.64**	4.70	**4.75**	4.81	**4.87**	4.93	**4.99**	5.05
	5.1	**5.11**	5.17	**5.23**	5.30	**5.36**	5.42	**5.49**	5.56
	5.6	**5.62**	5.69	**5.76**	5.83	**5.90**	5.97	**6.04**	6.12
	6.2	**6.19**	6.26	**6.34**	6.42	**6.49**	6.57	**6.65**	6.73
6.8	**6.8**	**6.81**	6.90	**6.98**	7.06	**7.15**	7.23	**7.32**	7.41
	7.5	**7.50**	7.59	**7.68**	7.77	**7.87**	7.96	**8.06**	8.16
	8.2	**8.25**	8.35	**8.45**	8.56	**8.66**	8.76	**8.87**	8.98
	9.1	**9.09**	9.20	**9.31**	9.42	**9.53**	9.65	**9.76**	9.88

*Note: Once a particular number series is selected, then the standard values within a specified range are determined from the 1 to 10 decade by multiplying by 0.01, 0.1, 10, 100, 1000, etc.

EXAMPLE 21–16   Using Equation 21–13, determine the six base numbers of the E6 number series.

Solution   Compute the E6 base numbers using Equation 21–13 and either the $\boxed{y^x}$ or the $\boxed{10^x}$ calculator operation.

$$N_{x+1} = 10^{x/n}$$
$$n = 6$$
$$x = 0, 1, 2, 3, 4, 5$$

Set n equal to 6 and substitute each value of x into the equation. Solve using the $\boxed{y^x}$ or the $\boxed{10^x}$ calculator operation. Calculate the answer to three significant figures and then write the answer to two significant figures.

$\boxed{y^x}$

$N_1 = 10^{0/6}$	$N_2 = 10^{1/6}$	$N_3 = 10^{2/6}$
$N_1 = 10^0$	$N_2 = 10^{0.1667}$	$N_3 = 10^{0.3333}$
$N_1 = 1.00$	$N_2 = 1.47$	$N_3 = 2.15$
$N_1 = 1.0$	$N_2 = 1.5$	$N_3 = 2.2$

$\boxed{y^x}$

$N_4 = 10^{3/6}$	$N_5 = 10^{4/6}$	$N_6 = 10^{5/6}$
$N_4 = 10^{0.5000}$	$N_5 = 10^{0.6667}$	$N_6 = 10^{0.8333}$
$N_4 = 3.16$	$N_5 = 4.64$	$N_6 = 6.81$
$N_4 = 3.2$	$N_5 = 4.6$	$N_6 = 6.8$

∴   The calculated base numbers for the E6 number series are 1.0, 1.5, 2.2, 3.2, 4.6, and 6.8.

Observation   Notice that the calculated values for N_4 (3.2) and N_5 (4.6) differ from the traditional values of 3.3 for N_4 and 4.7 for N_5 shown in Table 21–5. This practice predates the Radio-Electronics-Television Manufacturers' Association (RETMA) standards of the 1940s.

In summary, the preferred number series is widely used to specify electronics components. These number series are established to support component tolerances of $\pm 20\%$, $\pm 10\%$, $\pm 5\%$, $\pm 1\%$, and $\pm 0.1\%$. The step multiplier for a given number series is based on a root of ten where the index of the radical is the number of values in one of the five preferred number series. Thus, the step multipliers for tolerances of $\pm 20\%$, $\pm 10\%$, $\pm 5\%$, $\pm 1\%$, and $\pm 0.1\%$ are $\sqrt[6]{10} = 1.4678$, $\sqrt[12]{10} = 1.2115$, $\sqrt[24]{10} = 1.1007$, $\sqrt[96]{10} = 1.0243$, and $\sqrt[192]{10} = 1.0121$, respectively.

EXAMPLE 21–17   Using the step multiplier for the ±5%, E24 number series, calculate the next three values in the series starting at 5.1.

Solution   The step multiplier for the ±5%, E24 number series is $\sqrt[24]{10} = 1.1007$. The first value after 5.1 is

$$5.1 \times 1.1007 = 5.6136 = 5.6$$

The second value after 5.1 is

$$5.6 \times 1.1007 = 6.1639 = 6.2$$

The third value after 5.1 is

$$6.2 \times 1.1007 = 6.8243 = 6.8$$

∴   The next three values in the ±5%, E24 number series after 5.1 are 5.6, 6.2, and 6.8.

Observation   Consult Appendix A, "Reference Tables," for the color code for specifying preferred resistor values and tolerances.

EXERCISE 21–5

Specify the appropriate preferred number series for the following electronic components with a specified tolerance range.

1. Metalized polyester-film tubular capacitors with a ±5% tolerance
2. Low-noise metal-film MIL-R-10509 fixed resistors with a ±1% tolerance
3. Epoxy-molded MIL-C-15305 shielded inductors with a ±10% tolerance
4. Deposited carbon-film fixed resistors with a ±5% tolerance
5. Miniature aluminum electrolytic capacitors with a ±20% tolerance
6. Subminiature axial-leaded tubular solid-electrolyte tantalex capacitors with a ±10% tolerance
7. Low-power ultra-precision wirewound resistors with a ±0.1% tolerance
8. Subminiature RF filter chokes with pi-windings and a ±20% tolerance

Solve the following and express the answers to the specified number of significant figures.

9. Using Equation 21–13, verify the twelfth entry in the E24, ±5% preferred number series (Table 21–5) by computing its value to three significant figures.
10. Using Equation 21–13, verify the twenty-second entry in the E96, ±1% preferred number series (Table 21–5) by computing its value to four significant figures.
11. Using Equation 21–13, compute the first eight numbers (N_1 through N_8) in the E192, ±0.1% preferred number series (Table 21–5) by computing each of the values to four significant figures.

12. Using Equation 21–13, compute the last six numbers (N_{19} through N_{24}) in the E24, ±5% preferred number series (Table 21–5) by computing each of the values to three significant figures.

13. Using the step multiplier for the ±10%, E12 number series, calculate the next four values in the series starting at 1.5. Express the answer to three significant figures and then round to two significant figures to compute the next number.

14. Using the step multiplier for the ±5%, E24 number series, calculate the next four values in the series starting at 2.2. Express the answer to three significant figures and then round to two significant figures to compute the next number.

15. Using the step multiplier for the ±1%, E96 number series, calculate the next four values in the series starting at 8.66. Express the answer to four significant figures and then round to three significant figures to compute the next number.

SELECTED TERMS

amplifier   A device used to increase the magnitude of an electric signal.

attenuator   A resistive network that reduces the magnitude of an electric signal.

decibel   A logarithmic expression that compares two power levels. The unit used to express gain, loss, and relative power levels.

instantaneous value   The value of current or voltage at a particular instant of time.

transient   A changing action occurring in an electric circuit during the time between the initial application of power and the settling to a steady-state condition.

SECTION CHALLENGE

WEB CHALLENGE FOR CHAPTERS 20 AND 21

To evaluate your comprehension of Chapters 20 and 21, log on to **www.prenhall.com/ harter** and take the online True/False and Multiple Choice assessments for each of the chapters.

SECTION CHALLENGE FOR CHAPTERS 20 AND 21*

A. Your first challenge is to determine the time constant of a small bead-type thermistor used in making temperature measurements. The equation needed to calculate the time constant is given as equation E–5 where the time constant is indicated by the greek letter tau, τ.

One time constant, as pictured in the response curve of Figure C–10, is the time required for the thermistor to indicate 63.2% of a newly impressed temperature. To obtain the necessary parameters for the calculation, an electric current is passed through the thermistor resulting in the self-heating effect, ΔT, an increase in the thermistor's temperature. The length of time, t, that the current is passed through the thermistor is measured as is the power dissipation, P. The dissipation constant, δ, (the power in milliwatts required to raise the thermistor 1.0°C above the surrounding temperature) is specified by the thermistor manufacturer as 8.00 mW/°C.

$$\Delta T = \frac{P}{\delta}\left(1 - e^{-t/\tau}\right) \tag{E–5}$$

where $P = 182$ mW $\delta = 8.00$ mW/°C
       $\Delta T = 12.0$°C $t = 15.0$ s

FIGURE C–10 Response of a thermistor to a change in temperature. One time constant ($1\ \tau$) is the time required for the thermistor to indicate 63.2% of a newly impressed temperature.

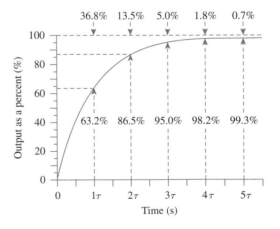

*The solution to this Section Challenge is found in Appendix C.

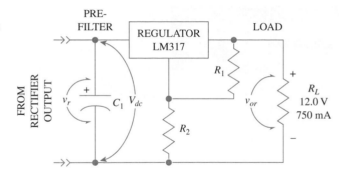

FIGURE C–11 A 12.0-V, 0.750-A regulated power supply with an output ripple, v_{or}, of ≤ 2.50 mV.

B. Your second challenge is to determine the value of the pre-filter capacitor, C_1, of Figure C–11. The series of equations needed to calculate the capacitor's value are given as equations E–6, E–7, and E–8.

The regulated power supply is designed to have an output of 12.0 V (V_0) and a current capability (I_0) of 750 mA with no more than 2.50 mV of output ripple, v_{or}. The regulator (LM317), an integrated circuit, provides a reduction in the ripple voltage (v_r) from the pre-filter (N_{dB}) of -50.0 dB. The dc voltage, from the rectifiers (V_{dc}), across the pre-filter capacitor (C_1) is 14.5 V.

Start your solution by using Equation E–6 to determine the ripple voltage, v_r, across the pre-filter capacitor; next solve for the ripple factor *(r)* at C_1 using Equation E–7; then conclude the solution for C_1 by solving for C in Equation E–8 when R (the resistance seen from the capacitor looking toward the load) equals 19.33 Ω (14.5 V/750 mA). Since Equation E–8 has been derived with R specified in ohms and C specified in microfarads (not farads), the value of C_1 is expressed in units of microfarads, μF.

$$N_{dB} = 20 \log\left(\frac{v_{or}}{v_r}\right) \qquad\qquad \textbf{(E–6)}$$

$$r = \frac{v_r}{v_{dc}} \qquad\qquad \textbf{(E–7)}$$

$$r = \frac{2400}{RC} \qquad\qquad \textbf{(E–8)}$$

22

Angles and Triangles

22–1   Points, Lines, and Angles

22–2   Special Angles

22–3   Triangles

22–4   Right Triangles and the Pythagorean Theorem

22–5   Similar Triangles; Trigonometric Functions

22–6   Using the Trigonometric Functions to Solve Right Triangles

22–7   Inverse Trigonometric Functions

22–8   Solving Right Triangles When Two Sides Are Known

PERFORMANCE OBJECTIVES

- **Convert angles between degrees, radians, and revolutions.**

- **Solve any right triangle given two sides and an acute angle.**

- **Solve any right triangle given one side and one acute angle using trigonometric functions.**

- **Solve any right triangle given two sides using inverse trigonometric functions.**

Apollo 17 Scientist-Astronaut Harrison H. Schmitt collects lunar rock samples on the surface of the moon at the Taurus-Littrow landing site. Notice the backpack life support system worn by the astronaut and the lunar rake being used to collect rock samples (1.3–2.5 cm). (Courtesy of NASA/Johnson Space Center)

This chapter starts a series of chapters that will enable you to understand alternating current (ac) principles and circuits. Once you master this material, you will be able to comprehend advanced topics in electronics, including active devices (amplifiers), filters, and frequency domains.

The key to your understanding of ac circuit principles starts with a knowledge of angles, triangles, and trigonometric functions, which, in turn, allow you to master the concepts of circular functions and phasors.

This chapter uses drawings to introduce many of the geometric concepts. You are encouraged to work along by making your own sketches (drawings) to aid in your understanding and to reinforce your comprehension of the fundamental ideas being presented.

The following symbols are new to this chapter:

$\boxed{\text{SIN}}$	sine	$\boxed{\text{SIN}^{-1}}$	inverse sine
$\boxed{\text{COS}}$	cosine	$\boxed{\text{COS}^{-1}}$	inverse cosine
$\boxed{\text{TAN}}$	tangent	$\boxed{\text{TAN}^{-1}}$	inverse tangent
$\boxed{\text{DEG}}$	degree	$\boxed{\text{RAD}}$	radian
$\angle$	angle		

22–1 POINTS, LINES, AND ANGLES

The first geometric concept that we will consider is the *point*. We have already used points when we were constructing graphs. In this chapter, capital letters are used to designate points. Thus, A and O represent points in Figure 22–1.

The second geometric concept to be considered is the *line segment*. A line segment is the portion of a straight line connecting two points. Figure 22–2 presents a line segment joining the points A and B. The symbol $\overline{AB}$ is used to denote the line segment connecting points A and B.

FIGURE 22–1 Points O and A.   **FIGURE 22–2** Line segment $\overline{AB}$.

Building on the concepts of points and line segments, we can define the concept of an angle. An **angle** is the geometric figure formed when one endpoint of a line segment is held fixed while the second endpoint is moved to a new location without changing the length of the line segment. Thus, in Figure 22–3, the figure represented by $\overline{OA}$ and $\overline{OA'}$ is the angle "swept out" by $\overline{OA}$ as A moves to A', while O is held fixed.

FIGURE 22–3 Sweeping out $\angle AOA'$.

FIGURE 22–4 Sweeping out an angle of one revolution.

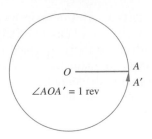

The symbol $\angle AOA'$ is used for the angle AOA'. The fixed endpoint, O, is called the **vertex** of the angle.

Like other mathematical quantities, angles have size and direction. One unit of size is the **revolution.** The angle swept out by moving one endpoint all the way around the fixed endpoint and back to its original position is one revolution in size. This is illustrated in Figure 22–4. The abbreviation for revolution is *rev.*

EXAMPLE 22–1   How many revolutions are swept out by a second hand in 2 min 45 s?

Solution   Each minute, the second hand sweeps out an angle of one revolution. Let n represent the number of revolutions.

$$n = (2 \text{ min } 45 \text{ s})\left(\frac{1 \text{ rev}}{\text{min}}\right)$$

Convert 45 s to 0.75 min:

$$n = (2.75 \text{ min})\left(\frac{1 \text{ rev}}{\text{min}}\right) = 2.75 \text{ rev}$$

∴   The second hand sweeps out 2.75 rev in 2 min 45 s.

EXAMPLE 22–2   Engine speed is measured in revolutions per minute (rev/min). If an automobile engine is operated at 5500 rev/min for 6.0 min, how many revolutions will have been swept out by the crankshaft?

Solution   Let n be the number of revolutions:

$$n = 5500\left(\frac{\text{rev}}{\text{min}}\right) \times 6.0(\text{min})$$

∴ $n = 33\ 000 \text{ rev}$

Often we want to measure fractions of revolutions. For this, a common unit of measure is the degree. One revolution is equal to 360 *degrees,* or 360°, where the superscript ° stands for degree. The following example explores the relation between revolutions and degrees.

EXAMPLE 22–3   What angle size in degrees is swept out by a second hand in 20 s?

Solution   Let n equal the number of revolutions:

$$n = (20 \text{ s})\left(\frac{1 \text{ rev}}{\text{min}}\right)\left(\frac{1 \text{ min}}{60 \text{ s}}\right)$$

$$n = \frac{20}{60} \text{ rev}$$

$$n = \frac{1}{3} \text{ rev}$$

Multiply by 360 to convert to degrees:

$$n = \frac{1}{3} \text{ rev} \frac{360°}{1 \text{ rev}}$$

$$\therefore \quad n = 120°$$

A second unit of size for measuring fractions of revolutions is the **radian.** The distance around a circle is 2π times the length of the radius (see Figure 22–5). In like manner, the angle swept out in one revolution is 2π radians, or 2π.

Although the abbreviation for radians is *rad,* it is not necessary to use any unit indication with angles measured in radians since the radian is not a unit in the same sense as the meter or the ampere. Instead, a radian has no dimension since it is defined as a ratio of the arc length to the radius length of a circle (a ratio of two lengths). Since each is measured in the same unit (m, cm, etc.), the units factor out, leaving the radian as a dimensionless quantity. When an angle is substituted into a formula, the word radian (rad) may be used or it may be omitted in expressing the units of the physical quantities. In this chapter, we introduce the superscript ^{r} as a convenience for indicating radians. When an angular quantity contains π, then no unit indication is used.

Although degrees are "nicer" numbers than radians, advanced mathematics and engineering frequently use radians. So we will practice with both units. Table 22–1 shows conversion factors for revolutions, degrees, and radians. Remember, the number π is approximately 3.14159.

FIGURE 22–5   Geometric properties of radians. (a) The circumference of a circle is related to the radians in one revolution. (b) When the arc length BB' is equal to the sides OB and OB', the angle is one radian.

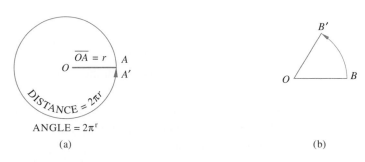

(a)

(b)

TABLE 22–1 Conversion Factors for Revolutions, Degrees, and Radians*

Revolutions	Radians (Using π)	Radians (Decimal)	Degrees
1	**2π**	6.283 19	**360**
0.5	**π**	3.141 59	**180**
0.159 155	**1**	1	57.2958
0.002 777 78	**π/180**	0.017 453 3	**1**

*__Boldface__ entries are exact.

EXAMPLE 22–4   How many degrees are in one radian?

Solution   Start with 1 rev = 360° and 1 rev = 2π; solve for 1^r:

$$2\pi = 360°$$

$$1 = 360/2\pi$$

$\therefore$ $1^r \approx 57.3°$

EXAMPLE 22–5   Convert 75.2° to radians.

Solution   Look up 1° in Table 22–1; multiply 75.2° by the corresponding value under radians (0.0175):

$$75.2 = 75.2(0.0175)$$

$\therefore$ $75.2° = 1.32^r$

EXAMPLE 22–6   Convert 5.25^r to revolutions.

Solution   Look up 1^r in Table 22–1 and multiply 5.25^r by the corresponding value under revolutions (0.159):

$$5.25 = 5.25(0.159)$$

$\therefore$ $5.25^r = 0.835$ rev

Observation   Some calculators have keystrokes that will convert degrees to radians and radians to degrees. Check your owner's guide to see if your calculator has this feature.

EXERCISE 22–1

1. How many revolutions does a second hand sweep out in 1 h?

2. The earth spins on its axis at the rate of 1 rev/day. How many revolutions does it sweep out in 60 h?

3. What angle in revolutions is swept out by a second hand during 48 s? In degrees?

4. How many revolutions does a CD make while playing a 3-min 15-s recording if it rotates 300 rev/min?

Convert the following angles to degrees. (Remember that angles containing π are in radians.)

5. $\frac{1}{4}$ rev   **6.** 2.5^r **7.** 1.76π **8.** 0.15 rev

9. 1.24 rev   **10.** 0.81 rev   **11.** 0.45^r **12.** 5.75^r

Convert the following angles to revolutions.

13. $300°$ **14.** $544°$ **15.** 4.8π **16.** 172π

17. 29.6^r **18.** 31.4^r **19.** 5.27π **20.** $1700°$

Convert the following angles to decimal radians.

21. $114.6°$ **22.** $29.3°$ **23.** 0.75 rev   **24.** 1.54 rev

25. $175°$ **26.** 1.4π **27.** 0.75π **28.** $\pi/3$

22–2 SPECIAL ANGLES

Several angles are so important that they have their own names. The first special angle is $180°$ or $\frac{1}{2}$ rev. This angle, which is a straight line, is called a **straight angle.** Angle β in Figure 22–6 is a straight angle. Two angles that sum to a straight angle are called **supplementary angles.** In Figure 22–7, $\angle\theta$ and $\angle\phi$ are supplementary.

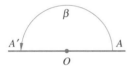

FIGURE 22–6   An angle of $180°$ is a straight angle.   **FIGURE 22–7**   Two supplementary angles, θ and ϕ, sum to a straight angle.

The second special angle is $90°$ or $\frac{1}{4}$ rev. This angle is called a **right angle.** The special symbol $\llcorner$ is used to indicate a right angle, as in Figure 22–8. Two angles that sum to a right angle are called complementary angles. In Figure 22–9, $\angle\alpha$ and $\angle\beta$ are complementary.

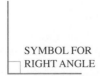

SYMBOL FOR
RIGHT ANGLE

FIGURE 22–8   An angle of $90°$ is a right angle.   **FIGURE 22–9**   The complementary angles α and β sum to a right angle.

Other angles may be classified in terms of straight angles and right angles. If an angle is between a straight angle and a right angle, it is called an *obtuse angle*. If an angle is between zero and a right angle, it is called an *acute angle*. More special angles are presented in Table 22–2. The size of each angle is given in revolutions, degrees, and radians.

TABLE 22–2 Fractional Parts of a Revolution

Revolutions	Radians (Using π)	Radians (Decimal)	Degrees
0	0	0	0
$\frac{1}{12}$	$\pi/6$	0.524	30
$\frac{1}{8}$	$\pi/4$	0.785	45
$\frac{1}{6}$	$\pi/3$	1.047	60
$\frac{1}{4}$	$\pi/2$	1.571	90
$\frac{1}{2}$	π	3.142	180
$\frac{3}{4}$	$3\pi/2$	4.712	270
1	2π	6.283	360

Most devices used to measure angles are similar in operation to the protractor pictured in Figure 22–10. Almost all such devices measure angles in degrees.

FIGURE 22–10 Protractor measuring an angle.

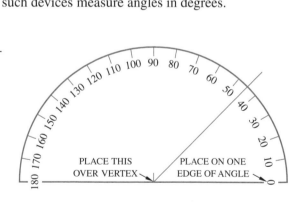

EXERCISE 22–2

Decide whether the following pairs of angles are supplementary, complementary, or neither.

1. 33°, 57°   **2.** 0.374 rev, 0.125 rev

3. 135°, 45°   **4.** 75°, 85°

5. $\pi/3$, $\pi/6$ **6.** 0.125 rev, 0.125 rev

7. $\frac{3}{8}$ rev, $\frac{1}{4}$ rev   **8.** 1.07^r, 0.50^r

9. 2.00^r, 1.14^r **10.** 0.65π, 0.45π

Find the supplement of the following angles.

11. $144°$ **12.** $37.5°$ **13.** $80.7°$

14. $105°$ **15.** $90°$ **16.** $120°$

17. 3.01^r **18.** 1.57^r **19.** 2.61^r

Find the complement of the following angles.

20. $38°$ **21.** $57.3°$ **22.** $22.1°$

23. $45°$ **24.** $83.4°$ **25.** $49.9°$

26. 0.785^r **27.** 1.07^r **28.** 0.52^r

22–3 TRIANGLES

The fourth geometric concept is the triangle. A triangle is a figure formed by joining three line segments endpoint to endpoint as in Figure 22–11. A triangle has three angles, three vertices, and three sides. In Figure 22–11(a), the vertices are labeled A, B, and C. The angles are labeled α, β, and γ. The sides are not labeled in Figure 22–11(a). We can use the line segment notation to refer to them as $\overline{AB}$, $\overline{BC}$, and $\overline{AC}$. This notation is very cumbersome, so we will use the notation of Figure 22–11(b), where the sides are labeled as a, b, and c. Notice that side a is opposite vertex A, side b is opposite vertex B, and side c is opposite vertex C.

FIGURE 22–11 Labeling of the parts of a triangle. The first three letters of the Greek alphabet are α, β, and γ. We use α for the angle opposite side a, β for the angle opposite side b, and γ for the angle opposite side c.

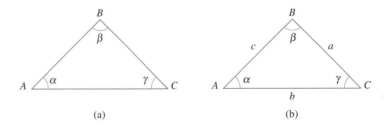

(a)   (b)

A property of triangles that can help you to check your work is stated as Rule 22–1.

> **RULE 22–1.** **Relative Size of the Sides of a Triangle**
>
> The longest side is opposite the largest angle.
> The shortest side is opposite the smallest angle.

EXAMPLE 22–7   Examine the triangle in Figure 22–12 to determine the smallest angle and the shortest side.

FIGURE 22–12   Triangle for Example 22–7.

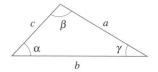

Solution   By inspection, the smallest angle is $\angle\gamma$, and the shortest side is c.

A remarkable property of triangles is that the sum of the angles is a straight angle. This fact can be used to find the size of the third angle if two angles are known. The procedure for finding the third angle is presented as Rule 22–2.

RULE 22–2.   *Finding the Third Angle of a Triangle*

To find the third angle of a triangle:
1. Form the sum of the first two angles.
2. The third angle is the supplement of the sum.

EXAMPLE 22–8   Two angles of a triangle are 35° and 47°; what is the third angle?

Solution   Use the steps of Rule 22–2:

Step 1:   Form the sum:

$$35° + 47° = 82°$$

Step 2:   The supplement of 82° is 180° − 82°, which is 98°.

∴   The third angle is 98°.

EXAMPLE 22–9   In a triangle, $\angle\alpha = 1.20^{\text{r}}$ and $\angle\beta = 0.90^{\text{r}}$; find $\angle\gamma$.

Solution   Use Rule 22–2.

Step 1:   Form the sum:

$$1.20 + 0.90 = 2.10$$

Step 2: Find the supplement:

$$\angle\gamma = 3.14 - 2.10$$

$$\therefore \qquad \angle\gamma = 1.04^{\,r}$$

EXERCISE 22–3

Using the notation of Figure 22–11(b), find the missing angle and determine which is the longest side.

1. $\angle\alpha = 57°$, $\angle\beta = 22°$ 2. $\angle\beta = 51°$, $\angle\gamma = 78°$

3. $\angle\alpha = 101°$, $\angle\gamma = 30°$ 4. $\angle\alpha = 38°$, $\angle\beta = 90°$

5. $\angle\alpha = 29.5°$, $\angle\beta = 50.5°$ 6. $\angle\alpha = 115°$, $\angle\gamma = 15.4°$

7. $\angle\alpha = 1.5^{r}$, $\angle\beta = 0.5^{r}$ 8. $\angle\alpha = 1.57^{r}$, $\angle\beta = 0.52^{r}$

9. $\angle\beta = 1.00^{r}$, $\angle\gamma = 1.00^{r}$ 10. $\angle\alpha = 1.21^{r}$, $\angle\beta = 1.0^{r}$

11. $\angle\alpha = \pi/2$, $\angle\gamma = \pi/4$ 12. $\angle\beta = \pi/3$, $\angle\gamma = \pi/6$

13. $\angle\alpha = \pi/3$, $\angle\beta = \pi/3$ 14. $\angle\alpha = \pi/2$, $\angle\gamma = \pi/6$

15. $\angle\alpha = 57.3°$, $\angle\beta = 57.3°$ 16. $\angle\beta = 90°$, $\angle\gamma = 85°$

17. $\angle\alpha = 79.5°$, $\angle\gamma = 79.6°$ 18. $\angle\alpha = 70°$, $\angle\beta = 42°$

19. $\angle\beta = 45°$, $\angle\gamma = 45°$ 20. $\angle\alpha = 15.2°$, $\angle\gamma = 15.2°$

22–4 RIGHT TRIANGLES AND THE PYTHAGOREAN THEOREM

Triangles containing a right angle make up an important class of triangles. A triangle containing a right angle is called a **right triangle.** The side opposite the right angle is called the **hypotenuse,** as shown in Figure 22–13.

FIGURE 22–13 Parts of a right triangle. The side opposite the right angle is called the hypotenuse.

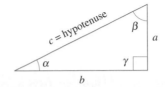

The ancient Greek mathematician Pythagoras is given credit for discovering that, if the lengths of any two sides of a right triangle are known, the length of the third side can be easily determined. This relation, known as the Pythagorean theorem, is stated as Rule 22–3.

RULE 22–3. Pythagorean Theorem

In every right triangle, the square of the hypotenuse equals the sum of the squares of the other two sides. In Figure 22–13,

$$c^2 = a^2 + b^2$$

EXAMPLE 22–10   Find the length of the hypotenuse in a right triangle if the other two sides are 3 m and 4 m.

Solution   Use the Pythagorean theorem, Rule 22–3. Substitute 3 for a and 4 for b, then solve for c:

$$c^2 = a^2 + b^2$$
$$c^2 = 3^2 + 4^2$$
$$c^2 = 9 + 16$$
$$c^2 = 25$$

$\boxed{\sqrt{x}}$

$\therefore \qquad c = 5\text{ m}$

EXAMPLE 22–11   If the hypotenuse is 27 m and a second side is 17 m, how long is the third side in a right triangle?

Solution   Use the Pythagorean theorem, Rule 22–3. Substitute 27 for c and 17 for a; solve for b:

$$c^2 = a^2 + b^2$$
$$27^2 = 17^2 + b^2$$
$$b^2 = 729 - 289$$
$$b^2 = 440$$

$\boxed{\sqrt{x}}$

$\therefore \qquad b = 21\text{ m}$

The angle γ is a right angle in the right triangle of Figure 22–13. We know that the sum of $\angle \alpha$ and $\angle \beta$ is the supplement of $\angle \gamma$. Thus, the sum of $\angle \alpha$ and $\angle \beta$ is a right angle. And $\angle \alpha$ and $\angle \beta$ are complements. This important fact is restated as Rule 22–4.

RULE 22–4. Sum of Two Acute Angles in a Right Triangle

The sum of the two acute angles in a right triangle is a right angle. In Figure 22–13, $\angle \alpha$ and $\angle \beta$ are complements, and

$$\boldsymbol{\alpha + \beta = 90°}$$

EXAMPLE 22–12   Find $\angle\beta$ when $\angle\alpha = 37°$ in a right triangle.

Solution   Use Rule 22–4:

$$\alpha + \beta = 90$$
$$37 + \beta = 90$$
$$\therefore \qquad \beta = 53°$$

EXERCISE 22–4

Find the missing sides and angles in the following right triangles. Use the notation of Figure 22–13.

1.	$a = 1.5$ m	$b = 2.0$ m	$\alpha = 36.9°$
2.	$a = 13$ m	$b = 13$ m	$\alpha = 45°$
3.	$c = 39$ m	$a = 16$ m	$\beta = 65.8°$
4.	$c = 109$ m	$b = 74$ m	$\beta = 43.5°$
5.	$a = 78.5$ m	$b = 44.2$ m	$\beta = 29.4°$
6.	$a = 0.152$ m	$c = 0.305$ m	$\alpha = 29.9°$
7.	$b = 2.59$ m	$c = 4.00$ m	$\alpha = 49.6°$
8.	$a = 7$ m	$b = 24$ m	$\alpha = 16.3°$
9.	$a = 60$ m	$c = 61$ m	$\alpha = 79.6°$
10.	$b = 12$ m	$c = 13$ m	$\beta = 67.4°$

22–5   SIMILAR TRIANGLES; TRIGONOMETRIC FUNCTIONS

Similar Triangles

The two triangles in Figure 22–14 have the same proportions. The ratios of the corresponding sides, a'/a, b'/b, and c'/c, are all equal. When these three ratios are equal, the two triangles are said to be **similar.** If two triangles are similar, their angles are the same size. Thus, in Figure 22–14, $\angle\alpha = \angle\alpha'$, $\angle\beta = \angle\beta'$, and $\angle\gamma = \angle\gamma'$.

FIGURE 22–14   Similar triangles have equal angles: $\angle\alpha = \angle\alpha'$, $\angle\beta = \angle\beta'$, and $\angle\gamma = \angle\gamma'$.

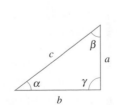
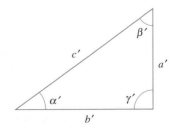

In two similar triangles, the ratio of two sides in one triangle is equal to the ratio of the corresponding sides in the second triangle. Thus, in Figure 22–14, $a/b = a'/b'$, $a/c = a'/c'$, and $b/c = b'/c'$. These ratios become the function of one of the acute angles in a right triangle. This is because in a right triangle we know that one angle is a right angle, and knowing a second angle gives us enough information to determine the third angle. Knowing all three angles, we know the proportions of the sides. Thus, the ratios a/c, b/c, and a/b can be considered functions of the angle.

Trigonometric Functions

In a right triangle we define the **sine function** of an angle as the *ratio* of the length of the **side opposite** the angle **to** the length of the **hypotenuse.** It is customary to use the symbol sin ( ) to indicate the sine function. Thus, in Figure 22–15, sin (α) = side opposite/hypotenuse = a/c. Sin (α) is read "sine of alpha." Remember that the parentheses indicate function, not multiplication.

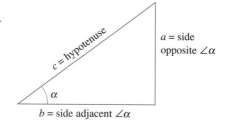

FIGURE 22–15   Right triangle used in definition of the trigonometric functions.

Check your calculator and your owner's guide for calculating the sine function. Look for a key like $\boxed{\sin x}$ or $\boxed{\text{SIN}}$. Pay special attention to how you tell your calculator whether the angle is in degrees or in radians. Two systems are in common use. The first is an angular mode key $\boxed{\text{DRG}}$ with which you can select degrees or radians. The second is a series of keystrokes ($\boxed{\text{RAD}}$ or $\boxed{\text{DEG}}$). The keystrokes allow you to change the system, and an indicator tells you in which system the calculator is working. Consult your owner's guide to determine if there is a restriction on the size of the angle your calculator will accept.

In a right triangle we define the **cosine function** of an angle as the *ratio* of the length of the **side adjacent** to the angle **to** the length of the **hypotenuse.** It is customary to use the symbol cos ( ) to indicate the cosine function. Thus, in Figure 22–15, cos (α) = side adjacent/hypotenuse = b/c. Cos (α) is read "cosine of alpha." Again, the parentheses indicate function and not multiplication.

Check your calculator and your owner's guide for how to calculate the cosine function. Look for a key like $\boxed{\cos x}$ or $\boxed{\text{COS}}$. How do you select degrees or radians? Are there any constraints on the size of the angle?

In a right triangle we define the **tangent function** of an angle as the *ratio* of the length of the **side opposite** the angle **to** the length of the **side adjacent** to the angle. The common symbol for the tangent function is tan ( ). Thus, in Figure 22–15, tan (α) = side opposite/side adjacent = a/b. Tan (α) is read "tangent alpha." The parentheses are used to indicate function, not multiplication.

Check your calculator and your owner's guide for how to calculate the tangent function. Look for a key like $\boxed{\tan x}$ or $\boxed{\text{TAN}}$. How do you select degrees or radians? What limits are placed on the size of the angle?

The definitions of the three functions—sine, cosine, and tangent—are summarized in Table 22–3. Refer also to Figure 22–15.

TABLE 22–3 Definition of the Trigonometric Functions

Function	Symbol	Definition	Figure 22–15
sine (α)	sin (α)	Side opposite/hypotenuse	a/c
cosine (α)	cos (α)	Side adjacent/hypotenuse	b/c
tangent (α)	tan (α)	Side opposite/side adjacent	a/b

EXAMPLE 22–13   Use your calculator to evaluate the sine, cosine, and tangent functions of 35.0°:

Solution   Set your calculator to work in degrees $\boxed{\text{DEG}}$:

$\boxed{\text{SIN}}$ sin (35.0°) = 0.574

$\boxed{\text{COS}}$ cos (35.0°) = 0.819

$\boxed{\text{TAN}}$ tan (35.0°) = 0.700

EXAMPLE 22–14   Use your calculator to evaluate the sine, cosine, and tangent functions of 1.00^r.

Solution   Set your calculator to work in radians $\boxed{\text{RAD}}$:

$\boxed{\text{SIN}}$ sin (1.00^r) = 0.841

$\boxed{\text{COS}}$ cos (1.00^r) = 0.540

$\boxed{\text{TAN}}$ tan (1.00^r) = 1.56

It is helpful to learn the brief list of trigonometric functions in Table 22–4. The list will aid you in finding approximate solutions to problems and in detecting errors in keystrokes while using your calculator.

TABLE 22–4 A Brief Table of Trigonometric Functions

Angle (Degrees)	Angle (Radians)	Sine	Cosine	Tangent
0	0.000	0.000	1.000	0.000
30	0.524	0.500	0.866	0.577
45	0.785	0.707	0.707	1.000
60	1.047	0.866	0.500	1.732
90	1.571	1.000	0.000	—

EXERCISE 22–5

Calculator Drill

Evaluate the sine, cosine, and tangent functions of the following angles in degrees.

1. 1.00°	**2.** 25.0°	**3.** 79.0°
4. 15.0°	**5.** 89.0°	**6.** 0°
7. 51.5°	**8.** 70.9°	**9.** 11.4°
10. 17.3°	**11.** 0.05°	**12.** 5.37°
13. 81.6°	**14.** 77.7°	**15.** 49.2°
16. 63.9°	**17.** 50.2°	**18.** 71.4°
19. 21.3°	**20.** 38.4°	

Evaluate the sine, cosine, and tangent of the following angles in radians.

21. 1.50^r	**22.** 0.250^r	**23.** 0.790^r
24. 0.0170^r	**25.** 0^r	**26.** 0.500^r
27. 0.100^r	**28.** 1.20^r	**29.** 1.450^r
30. 0.670^r	**31.** 0.159^r	**32.** 0.318^r
33. 1.170^r	**34.** 0.200^r	**35.** 1.570^r
36. 1.230^r	**37.** 0.972^r	**38.** 0.808^r
39. 0.355^r	**40.** 0.621^r	

22–6 USING THE TRIGONOMETRIC FUNCTIONS TO SOLVE RIGHT TRIANGLES

In the previous section we learned about three functions of angles in right triangles. These functions, sine, cosine, and tangent, are known as trigonometric functions because they are used in measuring triangles. With the use of these functions we can find the missing parts of a right triangle if we know one angle and one side.

EXAMPLE 22–15   In the right triangle in Figure 22–16, $\angle\alpha = 25.7°$ and $c = 15.5$ m. Find side a, side b, and $\angle\beta$.

FIGURE 22–16 Right triangle for Example 22–15.

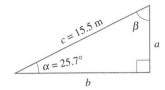

Solution   Solve for side a:

Use the definition of the trigonometric functions to write equations involving the known parts and one unknown side. Which function involves an angle, the hypotenuse, and the side opposite the angle? The sine function:

$$\sin(\alpha) = a/c$$

Solve for the unknown part, a:

$$a = c\sin(\alpha)$$

Substitute:

$\boxed{\text{SIN}}$ $a = 15.5\sin(25.7°)$
$\boxed{\times}$ $a = 15.5(0.434)$
                            $a = 6.72$
$\therefore$ $a = 6.72$ m

Solve for side b:
Which function involves an angle, the hypotenuse, and the side adjacent to the angle? The cosine function:

$$\cos(\alpha) = b/c$$

Solve for the unknown b:

$$b = c\cos(\alpha)$$

Substitute:

$\boxed{\text{COS}}$ $b = 15.5\cos(25.7°)$
$\boxed{\times}$ $b = 15.5(0.901)$
                            $b = 14.0$
$\therefore$ $b = 14.0$ m

Solve for $\angle \beta$:

How do we find the missing angle β? It is the complement of $\angle \alpha$ (25.7°).

$$\therefore \qquad \angle \beta = 64.3°$$

Observation Angles expressed to the nearest tenth of a degree correspond to three significant figures in the length of the sides.

EXAMPLE 22–16 In the right triangle in Figure 22–17, $\angle \alpha = 55.2°$ and $b = 74.5$ mm. Find side c, side a, and $\angle \beta$.

FIGURE 22–17 Right triangle for Example 22–16.

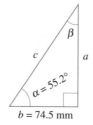

$b = 74.5$ mm

Solution Solve for side c:

Use the definitions of the trigonometric functions to write an equation involving an angle, the side adjacent, and the hypotenuse.

$$\cos (\alpha) = b/c$$

Solve for c:

$$c = b/\cos (\alpha)$$

Substitute:

$\boxed{\text{COS}}$

$\boxed{\div}$

$$c = 74.5/\cos (55.2°)$$
$$c = 74.5/0.571$$
$$c = 131$$

$$\therefore \qquad c = 131 \text{ mm}$$

Solve for side a:

Write an equation involving $\angle \alpha$, b, and a:

$$\tan (\alpha) = a/b$$

Solve for a and substitute:

$$a = b \tan (\alpha)$$

$\boxed{\text{TAN}}$

$$a = 74.5 \tan (55.2°)$$

$$a = 74.5(1.44)$$

$$a = 107$$

∴ $$a = 107 \text{ mm}$$

Solve for ∠β:
Find ∠β as the complement of ∠α (55.2°).

∴ $$∠β = 34.8°$$

EXAMPLE 22–17   In the right triangle in Figure 22–18, ∠β = 47.8° and b = 2.83 m. Find side c, side a, and ∠α.

FIGURE 22–18   Right triangle for Example 22–17.

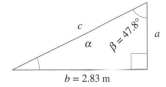

Solution   Solve for side c:
Write an equation involving ∠β, b, and c:

$$\sin(β) = b/c$$

Solve for c and substitute:

$$c = b/\sin(β)$$

$$c = 2.83/\sin(47.8°)$$

$$c = 2.83/(0.741)$$

$$c = 3.82$$

∴ $$c = 3.82 \text{ m}$$

Solve for side a:
Write an equation involving ∠β, b, and a:

$$\tan(β) = b/a$$

Solve for a and substitute:

$$a = b/\tan(β)$$

$$a = 2.83/\tan(47.8°)$$

$$a = 2.83/1.10$$

∴ $$a = 2.57 \text{ m}$$

Solve for ∠α.
Find ∠α as the complement of ∠β = 47.8°:

∴ $$∠α = 42.2°$$

FIGURE 22–19 Right triangle for Exercise 22–6.

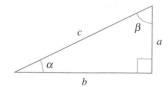

EXERCISE 22–6

Find the missing parts of the right triangle in Figure 22–19 when you are given the following information.

1. $\angle\alpha = 20.0°$, $c = 10$ m
2. $\angle\alpha = 59.0°$, $c = 210$ m
3. $\angle\alpha = 17.5°$, $c = 26.2$ m
4. $\angle\alpha = 83.1°$, $c = 500$ mm
5. $\angle\alpha = 28.4°$, $a = 19.5$ m
6. $\angle\alpha = 52.1°$, $a = 9.05$ m
7. $\angle\alpha = 32.3°$, $a = 2.79$ m
8. $\angle\alpha = 71.5°$, $a = 9.05$ m
9. $\angle\alpha = 14.1°$, $b = 14.1$ m
10. $\angle\alpha = 58.2°$, $b = 351$ mm
11. $\angle\alpha = 45.0°$, $b = 0.70$ km
12. $\angle\alpha = 60.0°$, $b = 1.86$ km
13. $\angle\beta = 15.3°$, $c = 300$ m
14. $\angle\beta = 79.1°$, $c = 121$ km
15. $\angle\beta = 50.8°$, $c = 875$ mm
16. $\angle\beta = 29.7°$, $c = 515$ m
17. $\angle\beta = 36.2°$, $a = 59.4$ m
18. $\angle\beta = 60.5°$, $a = 2.75$ km
19. $\angle\beta = 75.6°$, $b = 1.07$ km
20. $\angle\beta = 43.3°$, $b = 16.9$ mm

22–7 INVERSE TRIGONOMETRIC FUNCTIONS

We have seen that the three trigonometric functions give the value of certain ratios in a right triangle from the knowledge of the size of a particular angle. The inverse functions give the size of the angle from knowledge of the appropriate ratios. Definitions for three inverse trigonometric functions are given below.

In a right triangle we define the **arcsine function** of the ratio of a side to the hypotenuse as *the angle* whose sine is the ratio. There are several symbols in common use for the arcsine: $\sin^{-1}$, Sin^{-1}, arcsin, and Arcsin. For clarity and because we are working with right triangles, we will use Arcsin ( ). This concept is discussed further in Chapter 23. Thus, in Figure 22–20, Arcsin $(a/c) = \angle\alpha$.

FIGURE 22–20 The inverse trigonometric functions:
$\angle\alpha = \text{Arcsin}\ (a/c) = \text{Arccos}\ (b/c)$
$= \text{Arctan}\ (a/b)$

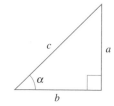

Check your calculator and owner's guide for calculating this function. Look for a single keystroke, $\boxed{\text{SIN}^{-1}}$, or two keystrokes, such as $\boxed{\text{Arc}}$ $\boxed{\text{sin}}$, $\boxed{\text{INV}}$ $\boxed{\text{SIN}}$, or $\boxed{f^{-1}}$ $\boxed{\text{sin}}$. The angle will be returned in the angular unit you have selected, so check whether the calculator is set for degrees or radians! Finally, the arcsine function will work only for numbers between -1 and 1.

In a right triangle we define the **arccosine function** of the ratio of a side to the hypotenuse as *the angle* whose cosine is the ratio. We will use the symbol Arccos ( ) to indicate the arccosine function. Thus, in Figure 22–20, Arccos $(b/c) = \angle\alpha$.

Check your calculator and your owner's guide for calculating this function. The same cautions apply to this function that apply to the arcsine function.

In a right triangle, we define the **arctangent function** of the ratio of the two sides as *the angle* whose tangent is equal to the ratio. We will use the symbol Arctan ( ) to indicate the arctangent function. Thus, in Figure 22–20, Arctan $(a/b) = \angle\alpha$.

Check your calculator and your owner's guide for calculating this function. Notice that there is no limit on the size of the argument for the arctangent function.

EXAMPLE 22–18 Find the angle in degrees whose sine is 0.244. Then find the same angle in radians.

Solution Use your calculator; set it to work in degrees $\boxed{\text{DEG}}$:

$\boxed{\text{SIN}^{-1}}$ Arcsin (0.244) = 14.1°

Set your calculator to work in radians $\boxed{\text{RAD}}$:

$\boxed{\text{SIN}^{-1}}$ Arcsin (0.244) = 0.246^{r}

EXAMPLE 22–19 Find the angle in degrees whose cosine is 0.398. Then find the same angle in radians.

Solution Use your calculator; set it to work in degrees:

$\boxed{\text{COS}^{-1}}$ Arccos (0.398) = 66.5°

Set your calculator to work in radians:

$\boxed{\text{COS}^{-1}}$ Arccos (0.398) = 1.162^{r}

EXAMPLE 22–20 Find the angle in degrees whose tangent is 4.87. Then find the same angle in radians.

Solution Use your calculator; set it to work in degrees:

$\boxed{\text{TAN}^{-1}}$ Arctan (4.87) = 78.4°

Set your calculator to work in radians:

$\boxed{\text{TAN}^{-1}}$ Arctan (4.87) = 1.368^{r}

Calculator Drill

1. Find the arcsine in degrees of the following numbers.

    (a) 0.107   (b) 0.791   (c) 0.255   (d) 0.568

    (e) 0.500   (f) 0.395   (g) 0.642   (h) 0.806

    (i) 0.411   (j) 1.000   (k) 0.043   (l) 0.921

2. Find the arcsine in radians of the numbers in problem 1.

3. Find the arccosine in degrees of the numbers in problem 1.

4. Find the arccosine in radians of the numbers in problem 1.

5. Find the arctangent in degrees of the following numbers.

    (a) 10.0   (b) 0.707   (c) 3.15   (d) 0.444

    (e) 1.57   (f) 1.00   (g) 2.36   (h) 0.290

    (i) 6.03   (j) 0.012   (k) 14.3   (l) 7.08

6. Find the arctangent in radians of the numbers in problem 5.

22–8 SOLVING RIGHT TRIANGLES WHEN TWO SIDES ARE KNOWN

With the aid of the inverse trigonometric functions, we can solve right triangles when two sides are known. The next three examples show how this can be done; it is important that you use your calculator.

EXAMPLE 22–21   In the right triangle in Figure 22–21, the length of the hypotenuse is 200 m and the length of side a is 75.0 m. Find $\angle\alpha$, $\angle\beta$, and side b.

FIGURE 22–21   Right triangle for
Example 22–21.

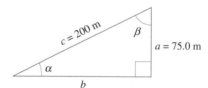

Solution   Find $\angle\alpha$ by writing an equation involving a, c, and $\angle\alpha$:

$$\sin(\alpha) = a/c$$

Apply the arcsine function to both sides to solve for $\angle\alpha$:

$$\text{Arcsin}\,[\sin(\alpha)] = \text{Arcsin}\,(a/c)$$
$$\angle\alpha = \text{Arcsin}\,(a/c)$$

Substitute:

$$\angle\alpha = \text{Arcsin} \,(75.0/200)$$
$$\angle\alpha = \text{Arcsin} \,(0.375)$$
$$\angle\alpha = 22.0°$$

Find $\angle\beta$ by writing an equation involving a, c, and $\angle\beta$:

$$\cos(\beta) = a/c$$

Apply the arccosine function to both sides to solve for $\angle\beta$:

$$\text{Arccos}\,[\cos(\beta)] = \text{Arccos}\,(a/c)$$
$$\angle\beta = \text{Arccos}\,(a/c)$$

Substitute:

$$\angle\beta = \text{Arccos}\,(75.0/200)$$
$$\angle\beta = \text{Arccos}\,(0.375)$$
$$\angle\beta = 68.0°$$

Find side b by the Pythagorean theorem, Rule 22–3:

$$a^2 + b^2 = c^2$$

Solve for b and substitute:

$$b = \sqrt{c^2 - a^2}$$
$$b = \sqrt{200^2 - 75^2}$$
$$b = \sqrt{34\,375}$$
$$b = 185 \text{ m}$$

EXAMPLE 22–22   The length of the hypotenuse is 155 mm and side b is 125 mm. Find $\angle\alpha$, $\angle\beta$, and side a in Figure 22–22.

Solution   Find $\angle\alpha$ by writing an equation involving b, c, and $\angle\alpha$:

$$\cos(\alpha) = b/c$$

FIGURE 22–22   Right triangle for Example 22–22.

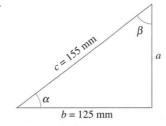

Apply the arccosine function to both sides to solve for $\angle\alpha$:

$$\angle\alpha = \text{Arccos } (b/c)$$

Substitute:

$$\angle\alpha = \text{Arccos } (125/155)$$

$$\angle\alpha = \text{Arccos } (0.806)$$

$$\therefore \quad \angle\alpha = 36.2°$$

Find $\angle\beta$ as the complement of $\angle\alpha$:

$$\angle\beta = 90 - 36.2$$

$$\therefore \quad \angle\beta = 53.8°$$

Find side a by writing an equation involving a, b, and $\angle\alpha$:

$$\tan (\alpha) = a/b$$

Solve for a and substitute:

$$a = b \tan (\alpha)$$

$$a = 125 \tan (36.2°)$$

$$a = 125(0.732)$$

$$\therefore \quad a = 91.5 \text{ mm}$$

EXAMPLE 22–23   In the right triangle in Figure 22–23, the length of side a is 155 km, while side b is 125 km. Find $\angle\alpha$, $\angle\beta$, and the hypotenuse c.

Solution   Find $\angle\alpha$ by writing an equation involving a, b, and $\angle\alpha$:

$$\tan (\alpha) = a/b$$

Apply the arctangent function to both sides to solve for $\angle\alpha$:

$$\angle\alpha = \text{Arctan } (a/b)$$

FIGURE 22–23   Right triangle for Example 22–23.

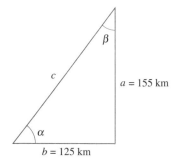

Substitute:

$\div$

$\boxed{\text{TAN}^{-1}}$

$\angle \alpha = \text{Arctan } (155/125)$

$\angle \alpha = \text{Arctan } (1.24)$

$\therefore$　　　　$\angle \alpha = 51.1°$

Find $\angle \beta$ as the complement of $\angle \alpha$:

$\angle \beta = 90 - 51.1$

$\therefore$　　　　$\angle \beta = 38.9°$

Find c from a and $\angle \alpha$:

$\sin (\alpha) = a/c$

$c = a/\sin (\alpha)$

Substitute:

$\boxed{\text{SIN}}$

$\boxed{\div}$

$c = 155/\sin (51.1°)$

$c = 155/0.778$

$\therefore$　　　　$c = 199 \text{ km}$

EXERCISE 22–8

Find the missing parts in the right triangle in Figure 22–24, given the following information.

1. $a = 15.0$ m, $c = 25.0$ m
2. $a = 5.00$ m, $c = 13.0$ m
3. $a = 17.5$ mm, $c = 40.2$ mm
4. $a = 57.4$ m, $c = 77.4$ m
5. $b = 24.0$ m, $c = 25.0$ m
6. $b = 45.0$ m, $c = 100$ m
7. $b = 22.1$ km, $c = 38.3$ km
8. $b = 1.47$ km, $c = 2.00$ km
9. $a = 16.0$ mm, $b = 14.0$ mm
10. $a = 152$ m, $b = 152$ m
11. $a = 500$ m, $b = 866$ m
12. $a = 757$ mm, $b = 602$ mm

FIGURE 22–24　Right triangle for Exercise 22–8.

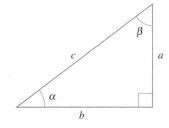

SELECTED TERMS

complementary angles   Two angles that sum to a right angle.

hypotenuse   The side opposite the right angle in a right triangle; also the longest side in a right triangle.

radian   A unit for measuring the size of angles; a complete circle equals 2π (6.283185 . . .) radians.

revolution   A unit for measuring the size of angles; a complete circle equals 1 revolution.

right triangle   A triangle containing a right angle.

supplementary angles   Two angles that sum to a straight angle.

23

Circular Functions

23–1   Angles of Any Magnitude

23–2   Circular Functions

23–3   Graphs of the Circular Functions

23–4   Inverse Circular Functions

23–5   The Law of Sines and the Law of Cosines

23–6   Polar Coordinates

23–7   Converting Between Rectangular and Polar Coordinates

PERFORMANCE OBJECTIVES

- Calculate trigonometric functions for angles of any size.

- Use the law of sines and the law of cosines to solve for the unknown parts of triangles.

- Convert between polar and rectangular coordinates.

The Apollo 15 Lunar Rover on the moon near the west edge of Mount Hadley. The electric-powered Lunar Rover provided a means of transportation to explore the surface of the moon. (Courtesy of NASA)

In the preceding chapter we learned about functions of angles in right triangles. The angles were limited in size to be between 0° and 90°. In this chapter, we will extend these functions to angles of any size.

The following symbols and notation are new to this chapter:

$\boxed{\rightarrow \mathbf{P}}$ rectangular to polar $\boxed{\rightarrow \mathbf{R}}$ polar to rectangular

$\rho \underline{/\theta}$ polar coordinates

23–1   ANGLES OF ANY MAGNITUDE

When the concept of angle was introduced, we said that angles have magnitude and direction. It is now time to consider the direction along with the magnitude. Consider a conventional clock with hands. The hands sweep out angles in clockwise (cw) direction when viewed from the front. This is indicated by the arrow ⌒. The opposite direction is called counterclockwise (ccw). It is indicated by the arrow ⌒.

In mathematics, we call counterclockwise angles positive and clockwise angles negative. Thus, in Figure 23–1, $\angle AOB$ is positive, while $\angle AOC$ is negative.

FIGURE 23–1   Positive and negative angles.

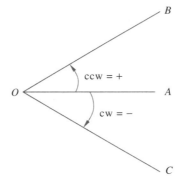

Again, consider a clock showing 2 o'clock. Twelve hours later, the clock will also show 2 o'clock. This property is shared with angles. For any angle, the terminal side is in the same position if the angle is changed by one or more complete revolutions. Any angle can be considered as being made up of two parts: *a fraction of a revolution and an integer number of complete revolutions.* For most applications in electronics, we will be concerned with the fractional part.

EXAMPLE 23–1   Write 451° as the sum of two angles: one angle between −180° and 180°, and the other angle a multiple of 360°.

Solution   Subtract 360° from 451°:

$$451° - 360° = 91°$$

Observation	$-180° < 91° < 180°$
$\therefore$	$451° = 360 + 91°$
Observation	We call 451° and 91° **equivalent angles.**

EXAMPLE 23–2 Find an angle between $-180°$ and $180°$ equivalent to 355°.

Solution Subtract 360° from 355°:

$$355° - 360° = -5°$$

$\therefore$ $-5°$ is equivalent to 355°.

EXAMPLE 23–3 Find an angle between $-180°$ and $180°$ equivalent to $-779°$.

Solution Start by adding 360° to $-779°$:

$$-779° + 360° = -419°$$

Add a second 360° to $-419°$:

$$-419° + 360° = -59°$$

Thus,

$$-779° + 2(360°) = -59°$$

$\therefore$ $-59°$ is equivalent to $-779°$.

EXERCISE 23–1

Find angles between $-180°$ and $180°$ equivalent to the following angles.

1. 365°	**2.** 270°	**3.** 545°
4. 600°	**5.** $-270°$	**6.** $-405°$
7. $-659°$	**8.** $-288°$	**9.** 382°
10. 417°	**11.** $-515°$	**12.** $-323°$

23–2 CIRCULAR FUNCTIONS

We are now ready to extend the trigonometric functions to angles of any size through the use of a circle. First, construct a rectangular coordinate system. Then construct a circle of radius ρ with its center at the origin of the coordinate system. See Figure 23–2. We measure $\angle\theta$ from the positive x-axis. Each point, P, on the circle has four numbers associated with it. The first two, x and y, form the rectangular coordinates of the point. The third is the angle θ, and the fourth is the radius ρ, which is the same for each point on the circle.

FIGURE 23–2 Diagram for defining the circular functions.

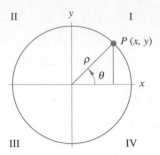

To find the circular functions of any angle, first find the *equivalent angle* between −180° and 180°, and then construct the angle as in Figure 23–2. The definitions for three circular functions are listed in Table 23–1. For angles in the first quadrant (0° ≤ θ ≤ 90°), these definitions coincide with those of the previous chapter. Most calculators will calculate the circular functions for any size of angle.

TABLE 23–1   Definitions for Three Circular Functions

Function	Abbreviation	Definition
sine (θ)	sin (θ)	y/ρ
cosine (θ)	cos (θ)	x/ρ
tangent (θ)	tan (θ)	y/x

Because the length of the radius, ρ, is always positive, the signs of the circular functions depend only on the quadrant in which the terminal side of the angle falls. The signs of the functions are summarized in Table 23–2.

TABLE 23–2   Signs of Circular Functions

Quadrant	Equivalent Angles	x	y	Sine	Cosine	Tangent
I	$0° < \theta < 90°$	+	+	+	+	+
II	$90° < \theta < 180°$	−	+	+	−	−
III	$-180° < \theta < -90°$	−	−	−	−	+
IV	$-90° < \theta < 0°$	+	−	−	+	−

EXERCISE 23–2

In which quadrant do the following angles fall?

1. 75°   **2.** −56°   **3.** −120°

4. 177°   **5.** 220°   **6.** −304°

7. 390°   **8.** −256°   **9.** 15°

10. −100°   **11.** 87°   **12.** 135°

Calculator Drill
Find the sine of the following angles.

13. 75°	**14.** −56°	**15.** −120°	**16.** 179°
17. 220°	**18.** −304°	**19.** 390°	**20.** −256°
21. 15°	**22.** −100°	**23.** 87°	**24.** 135°

Find the cosine of the following angles.

25. 75°	**26.** −56°	**27.** −120°	**28.** 179°
29. 220°	**30.** −304°	**31.** 390°	**32.** −256°
33. 15°	**34.** −100°	**35.** 87°	**36.** 135°

Find the tangent of the following angles.

37. 75°	**38.** −56°	**39.** −120°	**40.** 179°
41. 220°	**42.** −304°	**43.** 390°	**44.** −256°
45. 15°	**46.** −100°	**47.** 87°	**48.** 135°

23–3 GRAPHS OF THE CIRCULAR FUNCTIONS

Figure 23–3 presents the graph of the sine function. Imagine a point starting on the circle at −180° and moving counterclockwise. The radius of the circle is one unit; therefore, the sine function is equal to the height of the point above (or below) the center of the circle. The graph shows how the sine function varies as the angle changes from −180° to 180°. If the angle were to continue to change, the graph would repeat itself. This is called **periodic behavior.** Since the sine function repeats every 360°, the *period* of the sine function is 360°.

Observe that the sine function has a limited range of values. No matter what the angle is, the sine of the angle is between −1 and +1. For any angle θ, $-1 \leq \sin (\theta) \leq 1$. Furthermore, observe from the graph that the sine of an angle, say, 90°, has the same magnitude as the sine of minus the angle (−90°), but it is opposite in sign. Thus, $\sin (-\theta) = -\sin (\theta)$.

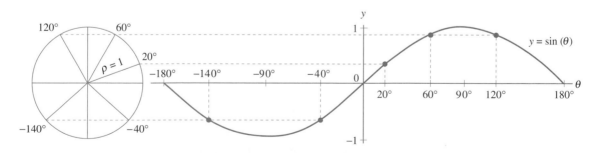

FIGURE 23–3 Circle generating graph of sine function.

EXAMPLE 23–4   Plot $y = \cos(\theta)$ for $-180° \le \theta \le 180°$.

Solution   Build a table of values of θ and y:

θ	$-180°$	$-170°$	$-160°$	$-140°$	$-120°$	$-100°$
y	-1.00	-0.985	-0.940	-0.766	-0.500	-0.174

θ	$-90°$	$-80°$	$-60°$	$-40°$	$-20°$	$-10°$
y	0.000	0.174	0.500	0.766	0.940	0.985

θ	$0°$	$10°$	$20°$	$40°$	$60°$	$80°$
y	1.000	0.985	0.940	0.766	0.500	0.174

θ	$90°$	$100°$	$120°$	$140°$	$160°$
y	0.000	-0.174	-0.500	-0.766	-0.940

θ	$170°$	$180°$
y	-0.985	-1.000

Plot the points from the table (see Figure 23–4). Then draw a smooth curve through the points.

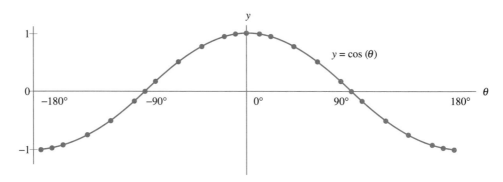

FIGURE 23–4   Graph of the cosine function developed in Example 23–4.

Observation   The graph of the cosine function is similar to the sine function in that the range of values is limited to the range from -1 to $+1$. The cosine function is also a periodic function of period $360°$. However, the cosine function is symmetric about $0°$; that is, $\cos(-\theta) = \cos(\theta)$. Finally, the cosine function looks like a shifted sine function. In fact, $\cos(\theta) = \sin(\theta + 90°)$.

EXAMPLE 23–5   Plot $y = \tan(\theta)$ for $-180° \leqslant \theta \leqslant 180°$.

Solution   Build a table of values of θ and y. Avoid the exact angles of $-90°$ and $90°$.

θ	$-180°$	$-160°$	$-140°$	$-120°$	$-100°$
y	0.00	0.364	0.839	1.73	5.67

θ	$-91°$	$-89°$	$-80°$	$-60°$	$-40°$
y	57.3	-57.3	-5.67	-1.73	-0.839

θ	$-20°$	$0°$	$20°$	$40°$	$60°$	$80°$	$89°$
y	-0.364	0.00	0.364	0.839	1.73	5.67	57.3

θ	$91°$	$100°$	$120°$	$140°$	$160°$	$180°$
y	-57.3	-5.67	-1.73	-0.839	-0.364	0.00

Plot the points. We cannot draw a smooth curve near $-90°$ and $90°$, because the tangent function is *discontinuous* at $-90°$ and $90°$. This is indicated in the graph of Figure 23–5 by the dashed vertical lines.

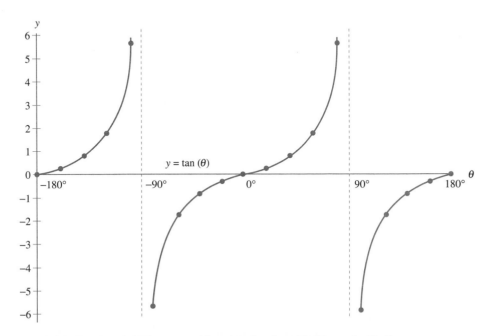

FIGURE 23–5   Graph of the tangent function developed in Example 23–5.

Observation   The graph of the tangent function is unlimited in value. The tangent function can be expressed in terms of the sine and cosine: $\tan(\theta) = \sin(\theta)/\cos(\theta)$. It follows that $\tan(\theta)$ is zero when $\sin(\theta)$ is zero, and that $\tan(\theta)$ is undefined when $\cos(\theta)$ is zero. The graph of the tangent gets closer and closer to the dashed lines as $\angle\theta$ gets closer and closer to $-90°$ or $90°$. But it never reaches the dashed lines.

EXERCISE 23–3

PROGRAMMABLE

Create plots of the following functions for the indicated range of θ.

1. $y = 2\sin(\theta)$ $-180° \leqslant \theta \leqslant 180°$

2. $y = 3\cos(\theta)$ $-180° \leqslant \theta \leqslant 180°$

3. $y = \sin(\theta + 90)$ $-270° \leqslant \theta \leqslant 90°$

4. $y = 1/\tan(\theta)$ $0° \leqslant \theta \leqslant 180°$ set $y = 0$ when $\theta = 90°$

5. $y = \sin^2(\theta)$ $0° \leqslant \theta \leqslant 180°$ ⎤

6. $y = \cos^2(\theta)$ $0° \leqslant \theta \leqslant 180°$ ⎬ **NOTE:** The superscript 2 indicates the square of the function, thus: $\sin^2(\theta) = [\sin(\theta)]^2$.

7. $y = \sin^2(\theta) + \cos^2(\theta)$ $0° \leqslant \theta \leqslant 180°$ ⎦

23–4   INVERSE CIRCULAR FUNCTIONS

The inverse circular functions are extensions of the inverse trigonometric functions. Your calculator will calculate the **principal** value of these functions for you. The sine function takes on all its possible values between $-90°$ and $90°$. The principal value of the arcsine function is an angle between $-90°$ and $90°$. We will capitalize the "a" in arc to indicate principal value. Table 23–3 summarizes the principal values of the inverse circular functions.

TABLE 23–3   Principal Values of Inverse Circular Functions

Function	Symbol	Domain of Argument	Range of Principal Value
Arcsine (x)	Arcsin (x)	$-1 \leqslant x \leqslant 1$	$-90° \leqslant \theta \leqslant 90°$
Arccosine (x)	Arccos (x)	$-1 \leqslant x \leqslant 1$	$0° \leqslant \theta \leqslant 180°$
Arctangent (x)	Arctan (x)	No limit	$-90° \leqslant \theta \leqslant 90°$

CHAPTER 23   Circular Functions

Calculator Drill

Use your calculator to find the principal value of the indicated inverse circular function in degrees:

1. Arcsin (0.707) **2.** Arcsin (0.866) **3.** Arccos (0.500)

4. Arccos (0.866) **5.** Arctan (1.00) **6.** Arctan (0.700)

7. Arcsin (−0.500) **8.** Arcsin (−0.707) **9.** Arcsin (−0.866)

10. Arcsin (−0.342) **11.** Arccos (−0.500) **12.** Arccos (−0.866)

13. Arccos (−0.707) **14.** Arccos (−0.342) **15.** Arctan (−1.00)

16. Arctan (−1.54) **17.** Arctan (−0.700) **18.** Arctan (−0.200)

19. Arctan (−5.67) **20.** Arctan (−57.3) **21.** Arctan (999)

23–5 THE LAW OF SINES AND THE LAW OF COSINES

In the preceding chapter we learned how to solve for the missing parts of a right triangle using the Pythagorean theorem and trigonometric functions of acute angles (sine, cosine, and tangent). In this section we will learn two new laws that hold in every triangle. We will also learn how to use these laws to solve for the missing parts of a triangle. The first law is the **law of cosines,** which is stated as Rule 23–1.

> ### Rule 23–1. The Law of Cosines
>
> For any triangle labeled as in Figure 23–6, the following three equations hold:
>
> $$a^2 = b^2 + c^2 - 2bc \cos (\alpha)$$
> $$b^2 = a^2 + c^2 - 2ac \cos (\beta)$$
> $$c^2 = a^2 + b^2 - 2ab \cos (\gamma)$$

FIGURE 23–6   Triangle for Section 23–5.

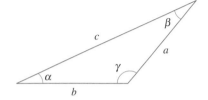

If γ is a right angle, then cos (γ) is zero and the third equation of the law of cosines becomes $c^2 = a^2 + b^2$. For this reason the law of cosines is called an extension of the Pythagorean theorem.

The second law states that within a triangle the length of a side divided by the sine of the opposite angle is a constant. This is known as the **law of sines,** and it is stated in the form of two equations in Rule 23–2.

Rule 23–2. The Law of Sines

For any triangle labeled as in Figure 23–6, the following equations hold:

$$\frac{a}{\sin(\alpha)} = \frac{b}{\sin(\beta)} = \frac{c}{\sin(\gamma)}$$

$$\frac{\sin(\alpha)}{a} = \frac{\sin(\beta)}{b} = \frac{\sin(\gamma)}{c}$$

The law of sines is very easy to use when finding the length of a side; however, caution must be used when using the law of sines to find the size of an angle. This is because the sine of an angle is equal to the sine of its supplement. Thus, the Arcsine cannot return a value greater than 90°. There are three additional facts that we can use with the law of sines to solve triangles. They are:

- The largest angle is opposite the longest side.
- At most, one angle in a triangle can be larger than 90°.
- The sum of the angles of a triangle is 180°.

With the aid of the law of sines and the law of cosines, we can solve three classes of problems involving triangles. The first class of problems is to find the angles of a triangle when the sides are known. This is demonstrated in the following example.

EXAMPLE 23–6   The lengths of the three sides of a triangle are side $a = 37.0$ m, side $b = 43.0$ m, and side $c = 56.0$ m. Find the size of each of the angles. Use the notation of Figure 23–6.

Solution   Use the law of cosines (Rule 23–1) to find the angle opposite the longest side, $\angle\gamma$. Thus:

$$c^2 = a^2 + b^2 - 2ab\cos(\gamma)$$

Solve for $\cos(\gamma)$:

$$\cos(\gamma) = \frac{a^2 + b^2 - c^2}{2ab}$$

Substitute for *a, b,* and *c:*

$$\cos(\gamma) = \frac{37^2 + 43^2 - 56^2}{2 \times 37 \times 43}$$

$$\cos(\gamma) = 0.0258$$

$\boxed{\text{COS}^{-1}}$ Use Arccos to find $\angle\gamma$.

$\therefore$ $\angle\gamma = 88.5°$

Use the law of sines (Rule 23–2) to find $\angle\beta$. Thus:

$$\frac{\sin(\beta)}{b} = \frac{\sin(\gamma)}{c}$$

Solve for $\sin(\beta)$:

$$\sin(\beta) = \frac{b\sin(\gamma)}{c}$$

Substitute for *b, c,* and $\angle\gamma$:

$$\sin(\beta) = \frac{43\sin(88.5)}{56}$$

$$\sin(\beta) = 0.768$$

$\boxed{\text{SIN}^{-1}}$ Use Arcsin to find $\angle\beta$:

$\therefore$ $\angle\beta = 50.1°$

Use the law of sines to find $\angle\alpha$:

$$\frac{\sin(\alpha)}{a} = \frac{\sin(\gamma)}{c}$$

Substitute and solve for $\sin(\alpha)$:

$$\sin(\alpha) = \frac{37\sin(88.5)}{56} = 0.660$$

$\boxed{\text{SIN}^{-1}}$ Use Arcsin to find $\angle\alpha$:

$\therefore$ $\angle\alpha = 41.3°$

Check the solution; the three angles should sum to 180°:

Check $41.3° + 50.1° + 88.5° = 179.9°$

Observation   The sum checks to the accuracy of the solution.

The second class of problems that can be solved with the aid of the law of sines and the law of cosines is finding the missing parts when two sides and the included angle are known, as in the following example.

EXAMPLE 23–7   Two sides and the angle between them are side $a = 74.3$ m, side $b = 37.8$ m, and $\angle \gamma = 52.3°$. Find side c and the other two angles. Use the notation of Figure 23–6.

Solution   Use the law of cosines to find side c:

$$c^2 = a^2 + b^2 - 2ab \cos{(\gamma)}$$

Substitute for a, b, and $\angle\gamma$:

$$c^2 = 74.3^2 + 37.8^2 - 2(74.3)(37.8) \cos{(52.3)}$$

∴
$$c = 59.3 \text{ m}$$

Use the law of sines to find the angle opposite the short side:

$$\frac{\sin{(\beta)}}{b} = \frac{\sin{(\gamma)}}{c}$$

Substitute for b, c, and $\angle\gamma$:

$$\sin{(\beta)} = \frac{37.8 \sin{(52.3)}}{59.3}$$

SIN⁻¹   Use Arcsin to find $\angle\beta$:

∴
$$\angle\beta = 30.3°$$

Use the law of sines to find $\angle\alpha$:

$$\frac{\sin{(\alpha)}}{a} = \frac{\sin{(\gamma)}}{c}$$

Substitute for a, c, and $\angle\gamma$:

$$\sin{(\alpha)} = \frac{74.3 \sin{(52.3)}}{59.3}$$

SIN⁻¹   Use Arcsin to find $\angle\alpha$:

∴
$$\angle\alpha = 82.5°$$

Check the angles: do they sum to 180?

Check $82.5° + 30.3° + 52.3° = 165.1°$

Observation   The sum is wrong; the largest angle, $\angle\alpha$, is too small. Replace $\angle\alpha$ with its supplement. Thus:

$$\angle\alpha = 180° - 82.5° = 97.5°$$

Check the sum of the angles again:

Check $97.5° + 30.3° + 52.3° = 180.1°$

∴ $\angle\alpha = 97.5°$

The third class of problems that can be solved with the aid of the law of sines is finding the missing parts when two angles and one side are known. The following example demonstrates the third class of problems.

EXAMPLE 23–8

One side and two angles in a triangle are side $b = 135$ m, $\angle\alpha = 23.8°$, and $\angle\beta = 44.5°$. Find the other angle and two sides.

Solution

Find $\angle\gamma$ as the supplement of $\angle\alpha + \angle\beta$:

$$\angle\gamma = 180 - (23.8 + 44.5)$$
$$\therefore \qquad \angle\gamma = 111.7°$$

Use the law of sines to find side a:

$$\frac{a}{\sin(23.8)} = \frac{135}{\sin(44.5)}$$

$$\therefore \qquad a = 77.7 \text{ m}$$

Use the law of sines to find side c:

$$\frac{c}{\sin(111.7)} = \frac{135}{\sin(44.5)}$$

$$\therefore \qquad c = 179 \text{ m}$$

EXERCISE 23–5

Solve for the missing parts of the triangle in Figure 23–6 given the following information.

1. $a = 15$ m	$b = 20$ m	$c = 25$ m
2. $a = 40$ m	$b = 20$ m	$c = 30$ m
3. $a = 70.2$ m	$b = 106$ m	$c = 74.6$ m
4. $a = 19.6$ km	$b = 44.4$ km	$c = 48.5$ km
5. $a = 35$ m	$b = 52$ mm	$\angle\gamma = 80°$
6. $a = 156$ m	$c = 76$ m	$\angle\beta = 46°$
7. $b = 79.4$ m	$c = 98.1$ m	$\angle\alpha = 23.3°$
8. $b = 215$ mm	$c = 175$ mm	$\angle\alpha = 31.4°$
9. $a = 40$ m	$\angle\alpha = 40°$	$\angle\beta = 30°$
10. $b = 65$ m	$\angle\alpha = 27°$	$\angle\beta = 45°$
11. $c = 12.5$ m	$\angle\alpha = 10.9°$	$\angle\beta = 53.4°$
12. $c = 58.6$ m	$\angle\alpha = 60°$	$\angle\beta = 60°$

23–6   POLAR COORDINATES

Besides rectangular, semilog, and log–log coordinates, there are many other systems of coordinates. One that is not often used for plots, but which underlies much of electronics, is the system of **polar coordinates.** Construct a skeleton of this new system as in Figure 23–7. Take a piece of paper. Lay out a horizontal and a vertical axis. Mark off 1, 2, and 3 on each axis. Connect the 1s (plus and minus) together with a circle centered at the origin. In like manner, connect the 2s and then the 3s. Label the positive *x*-axis 0°, the positive *y*-axis 90°, the negative *x*-axis 180°, and the negative *y*-axis 270°. The complete drawing in Figure 23–7 illustrates the basic components of the polar-coordinate system. Notice the origin is called the **pole.**

FIGURE 23–7   Skeleton of polar coordinate system.

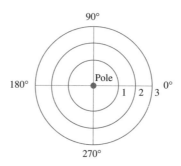

In polar coordinates, each point is associated with two numbers. The first number is the distance out from the pole. We use the symbol ρ for this number. It is called the **magnitude.** The second number is the direction measured from the positive horizontal axis. We use θ for this number. It is called the **argument.** When writing polar coordinates, we write the numbers in this form, $\rho\underline{/\theta}$, read *"rho angle theta."*

In rectangular coordinates, each point has only one set of coordinates. In polar coordinates, each point has many sets of coordinates. When a clock indicates 1 o'clock, the time can be 1 A.M. or 1 P.M. This can also be written 0100 hours and 1300 hours. In the same way, the angles $-360°$, $0°$, $360°$, and $720°$ all determine the same direction on polar graph paper. Look at the polar graph paper in Figure 23–8. Notice that the direction 270° is also labeled $-90°$. Look at the direction labels around the whole graph paper.

EXAMPLE 23–9   Plot the point $2.5\underline{/55°}$ on polar graph paper.

Solution   Refer to Figure 23–8. Begin at the pole (*P*) and move out 2.5 in the 0° direction. Follow the arc to the 55° direction. Label the point *A*.

∴   *A* is the point $2.5\underline{/55°}$

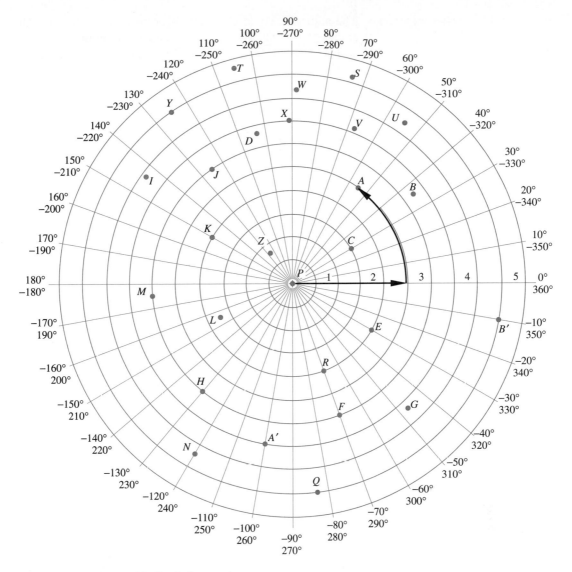

FIGURE 23–8 Polar graph paper.

EXAMPLE 23–10 Determine the coordinates of point B in Figure 23–8.

Solution The argument is between 30° and 40°:

$$\theta = 36°$$

The magnitude is between 3 and 3.5:

$$\rho = 3.25$$

$$\therefore \quad B = 3.25\underline{/36°}$$

EXERCISE 23–6

Determine the polar coordinates of the following points in Figure 23–8.

1. C	**2.** D	**3.** E	**4.** F
5. G	**6.** H	**7.** I	**8.** J
9. K	**10.** L	**11.** M	**12.** N
13. Q	**14.** R	**15.** S	**16.** T
17. U	**18.** V	**19.** W	**20.** X
21. Y	**22.** Z	**23.** A'	**24.** B'

Plot the following points on a piece of polar graph paper as in Figure 23–8.

25. $2\underline{/75°}$ **26.** $3.7\underline{/-90°}$ **27.** $1.2\underline{/-45°}$

28. $2.1\underline{/52°}$ **29.** $0.5\underline{/120°}$ **30.** $2.5\underline{/-145°}$

31. $3.2\underline{/-110°}$ **32.** $3.6\underline{/135°}$ **33.** $4.0\underline{/170°}$

23–7 CONVERTING BETWEEN RECTANGULAR AND POLAR COORDINATES

Both rectangular coordinates (Section 14.1) and polar coordinates represent the location of points on a piece of paper. By inspecting Figure 23–9, we can discover the relationship between (x, y) and $\rho\underline{/\theta}$. The following rules describe this relationship.

FIGURE 23–9   Relationship between rectangular and polar coordinates.

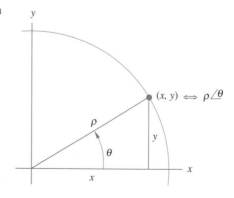

RULE 23–3.   Converting Rectangular to Polar Coordinates

Determine the argument, θ, in two steps:

1. $\theta = \text{Arctan}\,(y/x)$.
2. **(a)** If (x, y) is in the second quadrant, add $180°$.

   (b) If (x, y) is in the third quadrant, subtract $180°$.
3. Determine the magnitude, ρ, by

$$\rho = \sqrt{x^2 + y^2}$$

Observation: ρ, x, and y all have the same level of accuracy. If x and y are accurate to three significant figures, then ρ is accurate to three significant figures and θ is accurate to 0.1 degrees or 0.002 radians.

RULE 23–4.   Converting Polar to Rectangular Coordinates

Calculate the abscissa, x, and the ordinate, y, by:

1. $x = \rho \cos(\theta)$.
2. $y = \rho \sin(\theta)$.

The scientific calculator will perform these conversions for you with one or two keystrokes ($\boxed{\rightarrow P}$, $\boxed{\rightarrow R}$). Check your owner's guide. When using these features, pay close attention to the order of entering the numbers and reading the answers. How are you to input both x and y (or ρ and θ)? How are ρ and θ (or x and y) displayed?

EXAMPLE 23–11   Convert $(-3.5, -2.7)$ to polar coordinates using Rule 23–3.

Solution   Use Rule 23–3.

Step 1: $\theta = \text{Arctan}\,(y/x)$:

$\boxed{\text{TAN}^{-1}}$
$\theta = \text{Arctan}\,(-2.7/-3.5)$
$\theta = \text{Arctan}\,(0.771)$
$\theta = 37.6°$

Step 2: $(-3.5, -2.7)$ is in the third quadrant; therefore, subtract $180°$ from θ.

$$\theta = 37.6° - 180° = -142.4°$$

Step 3: Determine the magnitude:

$$\rho = \sqrt{x^2 + y^2}$$
$$\rho = \sqrt{(-3.5)^2 + (-2.7)^2}$$
$$\rho = \sqrt{12.3 + 7.29}$$

$\boxed{\sqrt{x}}$ 　　$\rho = \sqrt{19.5}$

$$\rho = 4.4$$

∴　$(-3.5, -2.7) \Rightarrow 4.4\underline{/-142°}$, as shown in Figure 23–10.

Observation　Use your calculator with the $\boxed{\rightarrow P}$ keystroke to convert $(-3.5, -2.7)$ to $4.4\underline{/-142°}$.

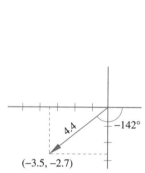

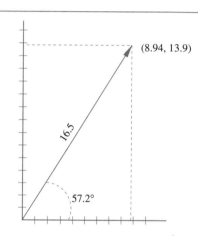

FIGURE 23–10　Diagram for Example 23–11.　　**FIGURE 23–11**　Diagram for Example 23–12.

EXAMPLE 23–12　Convert $16.5\underline{/57.2°}$ to rectangular coordinates using Rule 23–4.

Solution　Use Rule 23–4.

Step 1:　$x = \rho \cos(\theta)$:

$\boxed{\text{COS}}$ 　　$x = 16.5 \cos(57.2°)$
　　　　　$x = 16.5(0.542)$
　　　　　$x = 8.94$

Step 2:　$y = \rho \sin(\theta)$:

$\boxed{\text{SIN}}$ 　　$y = 16.5 \sin(57.2°)$
　　　　　$y = 16.5(0.841)$
　　　　　$y = 13.9$

∴　$16.5\underline{/57.2°} \Rightarrow (8.94, 13.9)$, as shown in Figure 23–11.

Observation　Use your calculator with the $\boxed{\rightarrow R}$ keystroke to convert $16.5\underline{/57.2°}$ to $(8.94, 13.9)$.

Using Rule 23–3, convert from rectangular to polar coordinates.

1. $(15, 0)$	**2.** $(0, -3.57)$	**3.** $(-8.7, 0)$
4. $(1.0, 1.0)$	**5.** $(5.0, 8.7)$	**6.** $(6.34, -4.21)$
7. $(4.15, -5.09)$	**8.** $(174, -108)$	**9.** $(-20.2, 153)$
10. $(-26.9, 12.8)$	**11.** $(-0.866, -0.500)$	**12.** $(15.6, 28.7)$
13. $(39.2, -20.1)$	**14.** $(-15.7, -30.4)$	**15.** $(-4.07, 4.07)$
16. $(15.1, -2.00)$	**17.** $(-13.4, 5.6)$	**18.** $(134, 225)$
19. $(-1.25, 3.45)$	**20.** $(757, -707)$	**21.** $(1.86, -1.41)$

Using Rule 23–4, convert from polar to rectangular coordinates.

22. $17\underline{/90°}$	**23.** $16\underline{/180°}$	**24.** $12.7\underline{/-90°}$
25. $100\underline{/50.0°}$	**26.** $25\underline{/30.0°}$	**27.** $27\underline{/-75.0°}$
28. $56.2\underline{/19.3°}$	**29.** $6.27\underline{/42.3°}$	**30.** $79.3\underline{/-85.1°}$
31. $0.578\underline{/-50.9°}$	**32.** $444\underline{/-135.0°}$	**33.** $74.6\underline{/156.0°}$
34. $12.8\underline{/-85.0°}$	**35.** $17.6\underline{/492.0°}$	**36.** $10.4\underline{/-512.0°}$
37. $22.3\underline{/1.100^r}$	**38.** $6.95\underline{/2.800^r}$	**39.** $30.7\underline{/6.040^r}$
40. $84.7\underline{/5.310^r}$	**41.** $78.3\underline{/4.950^r}$	**42.** $517\underline{/3.890^r}$

Calculator Drill

Repeat problems 1 through 42. Use the $\boxed{\rightarrow P}$ or the $\boxed{\rightarrow R}$ calculator function to solve the problems.

SELECTED TERMS

equivalent angles Angles that differ by an integer number of revolutions.

periodic behavior A behavior or pattern that repeats over and over again, especially in a graph.

polar coordinates A coordinate system based on a set of concentric circles. The coordinates of a point consist of the distance from a reference point (pole) and the direction from a reference line.

24

Vectors and Phasors

24–1   Scalars and Vectors

24–2   Complex Plane

24–3   Real and Imaginary Numbers

24–4   Complex Numbers

24–5   Phasors

24–6   Transforming Complex Number Forms

24–7   Resolving Systems of Phasors and Vectors

PERFORMANCE OBJECTIVES

- Classify numbers as real, imaginary, or complex.

- Plot complex numbers using both rectangular coordinates and polar coordinates.

- Transform complex numbers between rectangular form and polar form.

- Resolve systems of phasors.

An illustration of workers in the dynamo room of the first Edison electric lighting station in New York City.   (Courtesy of Con Edison of New York)

This chapter introduces the application of the circular functions to electronics. To apply these functions, we need to introduce several new concepts. We will look at physical quantities that have only magnitude, and quantities that have both magnitude and direction. For quantities with both magnitude and direction, we will need the concept of vectors.

24–1 SCALARS AND VECTORS

When length is measured, the quantity that results is called a **scalar** quantity, that is, a number and a unit. Quantities such as 6 cm of wire, a speed of 40 km/h, and a volume of 0.2 m^3 are all scalar quantities. Scalar quantities have only magnitude; they do not have direction.

To travel from your house to school, speed alone is not sufficient. Simply driving a car at 40 km/h will not get you there. You need to direct the motion of the car. Directed motion is a *vector quantity,* called *velocity.* In general, quantities expressed with both magnitude and direction are called **vector** quantities.

Vectors are represented by an arrow, as shown in Figure 24–1, in which the magnitude of the vector is the number next to the arrow and the direction of the vector is the direction of the arrow.

FIGURE 24–1 A vector quantity has both magnitude and direction.

50 m/s

It has been traditional to use special notation for vector quantities, such as **V** or $\vec{V}$ for velocity. A contemporary method is to use regular symbols to represent vectors and introduce special symbols only for the magnitude of vectors, such as V for velocity and $|V|$ for the magnitude of velocity. We use the contemporary method of notation. Thus, $E = V_1 + V_2$ is a scalar equation for a dc circuit while it is a vector equation for an ac circuit! Whether a symbol represents a scalar or a vector depends on the context of a problem.

24–2 COMPLEX PLANE

The pilot of a plane flying through the air must keep track of the position of the craft. In order to do this, he must know the compass heading and how the altitude is changing. The position of a plane is seen to be a *three-dimensional* vector (north–south, east–west, up–down).

A concept used with ac circuits is impedance (total opposition to alternating current). Impedance has two components—resistance and reactance. These components do not combine algebraically but combine to form a *two-dimensional vector.* In electronics we use scalars and two-dimensional vectors.

FIGURE 24–2 The complex plane. The complex number $r + jq$ is an alternate form of the rectangular coordinate (r, q). The symbol j is used to delineate the two components of the number.

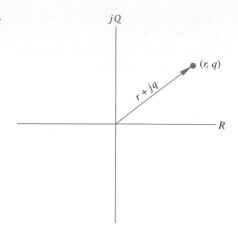

For notational purposes, we will borrow the concept of the complex plane from mathematics. Figure 24–2 shows the axes labeled R and jQ and a vector represented by the *complex number* $r + jq$. In complex numbers, the symbol j is used to indicate the component that is plotted in the vertical direction. This is similar to the ordered pair notation used in rectangular coordinates.

24–3 REAL AND IMAGINARY NUMBERS

We are familiar with real numbers. We have been using them throughout the book. What is an imaginary number? We answer this question with another question. What is the square root of minus one ($\sqrt{-1}$)? We define $\sqrt{-1}$ as the number j! This number is outside our everyday experience. It is so strange that we use the word *imaginary* to describe it. Any number of the form jn is called an **imaginary number.** Mathematicians and physicists use the letter i for $\sqrt{-1}$. However, electrical engineers use j for $\sqrt{-1}$ because i is used for instantaneous current.

EXAMPLE 24–1   Express $\sqrt{-25}$ as an imaginary number.

Solution   Write the radicand as the product of -1 and 25:

$$\sqrt{-25} = \sqrt{(-1)25}$$

Write as two radicals:

$$\sqrt{-25} = \sqrt{-1}\sqrt{25}$$

Simplify; replace $\sqrt{-1}$ with j:

$$\therefore \qquad \sqrt{-25} = j5$$

EXAMPLE 24–2   Express $-\sqrt{-16}$ as an imaginary number.

Solution   Factor:

$$-\sqrt{-16} = -\sqrt{(-1)16}$$

Write as two radicals:

$$-\sqrt{-16} = -\sqrt{-1}\ \sqrt{16}$$

Simplify; replace $\sqrt{-1}$ with j:

$$\therefore \qquad -\sqrt{-16} = -j4$$

EXERCISE 24–1

Express each number as an imaginary number. Use j in forming the imaginary number.

1. $\sqrt{-9}$	**2.** $\sqrt{-64}$	**3.** $\sqrt{-36}$
4. $-\sqrt{-4}$	**5.** $-\sqrt{-25}$	**6.** $-\sqrt{-100}$
7. $\sqrt{-182}$	**8.** $\sqrt{-3.76}$	**9.** $-\sqrt{-144}$
10. $-\sqrt{-90.8}$	**11.** $-\sqrt{-108}$	**12.** $-\sqrt{-31.5}$
13. $\sqrt{-0.875}$	**14.** $-\sqrt{-1286.5}$	**15.** $-\sqrt{-52.82}$

24–4 COMPLEX NUMBERS

A complex number is the sum of a real number and an imaginary number. The **rectangular form** of a complex number is $r + jq$. When $q = 0$, then $r + j0$ is a real number. When $r = 0$ and $q \neq 0$, then $0 + jq$ is an imaginary number. When $r \neq 0$ and $q \neq 0$, then $r + jq$ is a **complex number**.

The *real part* of the complex number $r + jq$ is r. The *imaginary part* of the complex number $r + jq$ is jq. (NOTE: q is a real number while jq is an imaginary number.) Table 24–1 is a summary of the concepts associated with complex numbers. Observe

TABLE 24–1   Summary of Complex Numbers as Related to the Complex Plane

$r + jq$	General form of a complex number
$r + j0$	A real number
$0 + jq, q \neq 0$	An imaginary number
$0 + j0$	Origin in the complex plane
R-axis	Contains all the real numbers
jQ-axis	Contains all the imaginary numbers
$3 + j4$	A complex number located in the first quadrant
$-3 + j4$	A complex number located in the second quadrant
$-3 - j4$	A complex number located in the third quadrant
$3 - j4$	A complex number located in the fourth quadrant

that the real and the imaginary parts of a complex number can be positive or negative *independent of each other.*

EXAMPLE 24–3   Write $8 + \sqrt{-16}$ as a complex number.

Solution   Write $\sqrt{-16}$ as an imaginary number:

$$\sqrt{-16} = \sqrt{-1}\,\sqrt{16}$$
$$\sqrt{-16} = j4$$

Form the complex number:

∴ $8 + \sqrt{-16} = 8 + j4$

EXAMPLE 24–4   Classify the following numbers as real, imaginary, or complex.

(a) 7 (b) $0 - j3$ (c) $-\sqrt{-44}$
(d) $5 + j2$ (e) $6 + j0$ (f) $-8 - j4$

Solution   **(a)** Real   **(b)** Imaginary   **(c)** Imaginary
(d) Complex   **(e)** Real   **(f)** Complex

EXERCISE 24–2

Write each of the following in the form of a complex number.

1. 15 **2.** $-j7$ **3.** $j1.7$
4. $2 - \sqrt{-9}$ **5.** $-5 - \sqrt{-16}$ **6.** $-3 + \sqrt{-36}$
7. $2 - \sqrt{9}$ **8.** -25 **9.** $17 - \sqrt{-1}$

Classify the following numbers as real, imaginary, or complex.

10. 7 **11.** $3 - j6$ **12.** $0 - j0$
13. $-5 + j0$ **14.** 6 **15.** $6 - \sqrt{16}$
16. $-7 - \sqrt{-2}$ **17.** $-\sqrt{18}$ **18.** $\sqrt{-20}$

Plotting Complex Numbers

The *rectangular form* of a complex number, $r + jq$, is very similar to the ordered pair notation, (r, q), used to plot in rectangular coordinates. We will take advantage of this similarity to construct a graphic representation of the complex plane as in Figure 24–3. By using the real part of the complex number as the abscissa and the imaginary part of the complex number as the ordinate, we can directly plot the complex number on the complex plane as in Figure 24–4.

FIGURE 24–3 The coordinates of the horizontal axis (*R*-axis) are real numbers; the coordinates of the vertical axis (*jQ*-axis) are imaginary numbers.

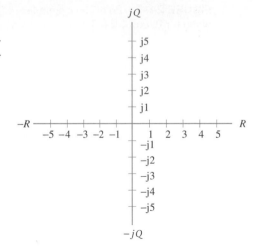

FIGURE 24–4 Points on the complex plane are located with a rectangular coordinate system made up of a real coordinate and an imaginary coordinate.

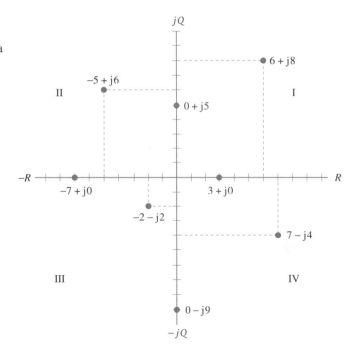

EXERCISE 24–3

Plot the following complex numbers on a complex plane as was done in Figure 24–4.

1. $3 + j2$ 2. $4 + j7$ 3. $-2 - j5$
4. $6 - j3$ 5. $-6 + j4$ 6. $0 + j0$
7. $4 - j6$ 8. $0 - j8$ 9. $1 + j5$

FIGURE 24–5 Points on a complex plane for Exercise 24–3.

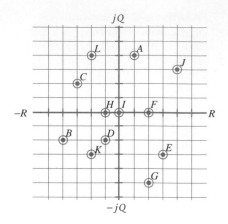

Record the coordinates of the named points of Figure 24–5.

10. A	**11.** I	**12.** K
13. B	**14.** J	**15.** D
16. E	**17.** H	**18.** C
19. F	**20.** G	**21.** L

24–5 PHASORS

In an ac network, the current and voltage are represented by sine waves having magnitudes and directions that are continually changing. A *rotating vector* is used to represent graphically the changing condition of the ac quantities, current, and voltage.

Figure 24–6 shows how ac sinusoidal quantities are represented by the position of a rotating vector. As the vector rotates it generates an angle. The location of the vector on the plane surface is determined by the magnitude (length) of the vector and by the generated angle. This concept of a rotating vector is shown in Figure 24–6(b). The point

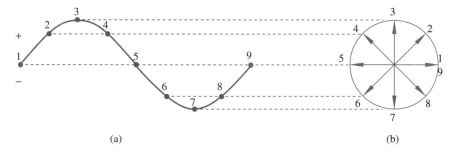

FIGURE 24–6 (a) The magnitude of the sine wave (representing the ac voltage or current) is continually changing. (b) A vector with its end fixed at the origin and rotating in a ccw direction represents this varying condition in ac quantities.

at the top of the arrowhead may be located by the magnitude (the length) and the direction (the angle) of the vector. A rotating vector is represented in general by a function written in **polar form** as

$$\rho \underline{/\alpha(t)} \tag{24–1}$$

where ρ = vector magnitude, which is constant
$\alpha(t)$ = angular displacement from the reference axis, which is a function of time

A **phasor** is a *stop-action photograph* of the changing conditions of the ac quantities, voltage, and current, as in Figure 24–7. A phasor, like a rotating vector, is written in polar form as

$$\rho \underline{/\theta} \tag{24–2}$$

where ρ = phasor magnitude
θ = phase angle, which is equal to $\alpha\,(t_o)$
t_o = time of the "stop-action photograph"

Phasors are graphed on the complex plane using polar coordinates.

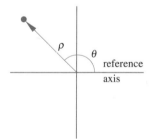

FIGURE 24–7 The value of a phasor is represented by the point at the tip of the arrowhead, which is located by the length of the vector and the size of angle θ.

We previously learned that coordinates of the points of the complex plane may be located by a complex number of the rectangular form $r + jq$. Phasors are also used to locate points of the complex plane by a complex number of the polar form $\rho \underline{/\theta}$. We see then that there are two coordinate systems for representing the points of a complex plane: one is a rectangular coordinate system ($r + jq$) and the other is a polar coordinate system ($\rho \underline{/\theta}$). Because either coordinate system may be used to represent a given point in the plane, the two systems must be equivalent. Thus

$$\rho \underline{/\theta} \Leftrightarrow r + jq \tag{24–3}$$

Figure 24–8 shows that a point of a complex plane located by a phasor may be described in either the polar or the rectangular form of a complex number.

FIGURE 24–8 The point on the complex plane located by the phasor is $4 + j3$ expressed in the rectangular form of a complex number, or $5\underline{/36.9°}$ expressed in the polar form of a complex number.

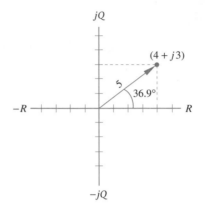

EXAMPLE 24–5   Express the coordinate of point A of Figure 24–9 in polar form.

FIGURE 24–9 Complex plane for Example 24–5 and Exercise 24–4, problems 1–9.

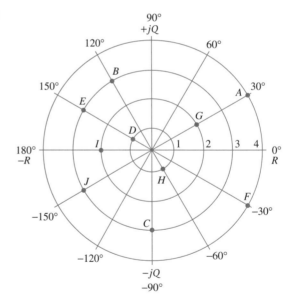

Solution   Draw a phasor from the origin to point A. The magnitude is 4 and the angle is 30°.

∴   Point A is $4\underline{/30°}$

Express the coordinates of the following named points of Figure 24–9 in polar form:

1. *B*	**2.** *F*	**3.** *D*
4. *H*	**5.** *C*	**6.** *E*
7. *G*	**8.** *I*	**9.** *J*

Draw each of the following polar phasors on the complex plane of Figure 24–10:

10. $2/\underline{60°}$	**11.** $3/\underline{-45°}$	**12.** $4/\underline{120°}$
13. $1/\underline{225°}$	**14.** $2/\underline{180°}$	**15.** $3/\underline{-120°}$
16. $1.5/\underline{-135°}$	**17.** $3.5/\underline{150°}$	**18.** $2.5/\underline{-60°}$

FIGURE 24–10 Complex plane for Exercise 24–4, problems 10–18.

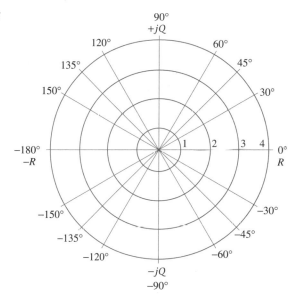

24–6 TRANSFORMING COMPLEX NUMBER FORMS

In working with phasors representing current and voltage, it will be necessary that these complex quantities be expressed in both rectangular and polar form. The process for transforming a complex number from one form to another is carried out with your calculator.

Rectangular to Polar Transformation

To convert a complex number from the rectangular form to the polar form, use the $\boxed{\rightarrow P}$ calculator keystroke as was done in the previous chapter. If needed, check your calculator owner's guide to determine how to use this powerful feature of your calculator. It will be assumed that you have this key on your calculator, as the examples will reflect the use of this keystroke and it will be used exclusively in the remaining chapters.

EXAMPLE 24–6	Express $3.00 + j4.00$ in polar form.
Solution	Use your calculator $\boxed{\rightarrow\textbf{P}}$ key:
$\therefore$	$3.00 + j4.00 \boxed{\rightarrow\textbf{P}} 5.00\underline{/53.1°}$
Observation	Complex numbers expressed in rectangular form to three significant figures result in a polar number with an accuracy of three significant figures in the magnitude and an angle with an accuracy to the nearest tenth of a degree (0.002 rad).

EXAMPLE 24–7	Express $-5.2 - j8.3$ in polar form.
Solution	Use your calculator:
$\therefore$	$-5.2 - j8.3 \boxed{\rightarrow\textbf{P}} 9.8\underline{/-122°}$

Polar to Rectangular Transformation

To convert a complex number from the polar form to the rectangular form, use the $\boxed{\rightarrow\textbf{R}}$ calculator keystroke as was done in the previous chapter. For review, consult your owner's guide for the use of this valuable feature of your calculator. This key will be used exclusively in the remaining chapters.

EXAMPLE 24–8	Express $5.00\underline{/36.9°}$ in rectangular form $(r + jq)$.
Solution	Use your calculator $\boxed{\rightarrow\textbf{R}}$ key:
$\therefore$	$5.00\underline{/36.9°} \boxed{\rightarrow\textbf{R}} 4.00 + j3.00$

EXAMPLE 24–9	Express $12.3\underline{/143.0°}$ in rectangular form.
Solution	Use your calculator:
$\therefore$	$12.3\underline{/143.0°} \boxed{\rightarrow\textbf{R}} -9.82 + j7.40$

Calculator Drill

Express the following complex numbers in polar form with the argument in degrees.

1. $6.00 + j8.00$	**2.** $16.0 - j12.0$	**3.** $-2.14 + j4.28$
4. $5.09 + j3.67$	**5.** $-3.13 - j12.7$	**6.** $6.95 - j7.23$
7. $3.81 + j5.40$	**8.** $7.72 - j9.05$	**9.** $3.15 - j5.52$
10. $11.4 + j17.6$	**11.** $-14.2 - j19.5$	**12.** $10.9 - j5.52$
13. $17.1 - j1.25$	**14.** $23.1 + j303$	**15.** $-0.537 - j4.18$
16. $-12.6 + j10.5$	**17.** $0.225 + j0.903$	**18.** $3.72 - j10.6$

Express the following complex numbers in rectangular form.

19. $5.00\underline{/1.302^{\text{r}}}$	**20.** $3.39\underline{/210.5°}$	**21.** $7.07\underline{/1.234^{\text{r}}}$
22. $8.08\underline{/45.0°}$	**23.** $6.25\underline{/-12.0°}$	**24.** $4.87\underline{/-135.6°}$
25. $12.3\underline{/110.3°}$	**26.** $15.7\underline{/32.7°}$	**27.** $2.19\underline{/-57.3°}$
28. $4.09\underline{/-70.2°}$	**29.** $7.77\underline{/243.5°}$	**30.** $10.3\underline{/140.0°}$
31. $823\underline{/0.784^{\text{r}}}$	**32.** $20.1\underline{/-1.022^{\text{r}}}$	**33.** $0.190\underline{/2.343^{\text{r}}}$
34. $0.587\underline{/0.143^{\text{r}}}$	**35.** $296\underline{/4.684^{\text{r}}}$	**36.** $44.8\underline{/-0.286^{\text{r}}}$
37. $128\underline{/527.3°}$	**38.** $42.3\underline{/9.568^{\text{r}}}$	**39.** $0.358\underline{/7.736^{\text{r}}}$

24–7 RESOLVING SYSTEMS OF PHASORS AND VECTORS

Vectors have the same properties that phasors have—magnitude and direction. Mathematically, vectors and phasors follow the same rules. In this section we chose to use the word *phasor,* but the following concepts also apply to vectors.

A phasor can be thought of graphically as having two components: one component along the R-axis and the second component along the j-axis. This is illustrated in Figure 24–11. Mathematically, this is accomplished by polar to rectangular conversion, as in: $1\underline{/0°}$ $\boxed{\rightarrow R}$ $1 + j0$ and $1\underline{/90°}$ $\boxed{\rightarrow R}$ $0 + j1$.

FIGURE 24–11 The polar phasor $32\underline{/51°}$ of part (a) is equivalent to the two components shown in part (b).

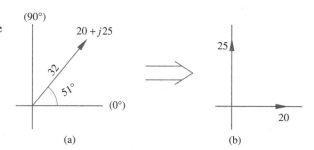

(a)   (b)

The sum of two phasors may be replaced by a single *resultant* phasor in polar form by applying the following guidelines.

> ### GUIDELINES 24–1.   *Resolving Systems of Phasors*
>
> **1.** Two phasors in the same direction are resolved into a single phasor in the same direction by adding their magnitudes.
>
> **2.** Two phasors 180° apart are resolved into a single phasor in the direction of the larger phasor by algebraically adding the magnitudes.
>
> **3.** Two phasors 90° apart, as in Figure 24–12, are resolved into a single phasor by solving for the polar form of a phasor using rectangular to polar conversion.
>
> **4.** Two polar phasors are resolved into a single polar phasor by first expressing each phasor in rectangular form, and then algebraically adding the rectangular phasors (as in cases 1 and 2) so that only two rectangular phasors remain. Finally, the two phasors 90° apart are resolved as in case 3.

FIGURE 24–12   Resolving two phasors 90° apart by rectangular to polar conversion methods as demonstrated in Example 24–12.

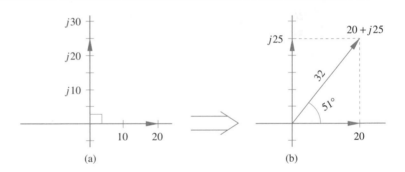

Each of the four cases of Guidelines 24–1 is demonstrated in the following series of examples.

EXAMPLE 24–10   Resolve the two phasors ($4\underline{/0°}$ and $6\underline{/0°}$) of Figure 24–13(a) into a resultant phasor.

FIGURE 24–13   Resolving two phasors in the same direction by adding their magnitudes, as demonstrated in Example 24–10.

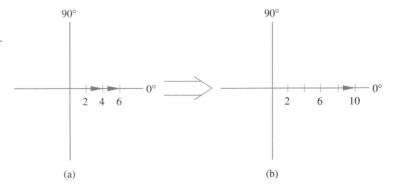

Solution   Use Guidelines 24–1 (**case 1**):

$$4 + 6 = 10$$

$$\therefore \quad 4\underline{/0°} + 6\underline{/0°} = 10\underline{/0°}$$

Observation   The resultant is shown as Figure 24–13(b).

EXAMPLE 24–11   Resolve the two phasors ($j15$ and $-j5$) of Figure 24–14(a) into a resultant phasor.

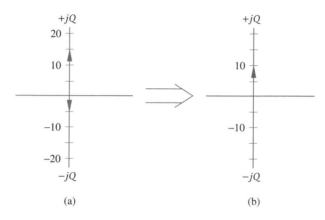

(a)   (b)

FIGURE 24–14   Resolving two phasors 180° apart by algebraically adding the magnitudes, as demonstrated in Example 24–11.

Solution   Use Guidelines 24–1 (**case 2**):

$$\therefore \quad j15 - j5 = j10$$

Observation   The resultant is shown as Figure 24–14(b).

EXAMPLE 24–12   Resolve the two phasors ($20\underline{/0°}$ and $25\underline{/90°}$) of Figure 24–12(a) into a resultant phasor.

Solution   Use Guidelines 24–1 (**case 3**). Start by expressing the phasor in rectangular form:

$$r = 20\underline{/0°} \to 20 + j0$$
$$jq = 25\underline{/90°} \to 0 + j25$$
$$r + jq = 20 + j25$$

$$\therefore \quad 20 + j25 \; \boxed{\to \text{P}} \; 32\underline{/51°}$$

Observation   The resultant is shown as Figure 24–12(b).

EXAMPLE 24–13   Resolve the two phasors ($28.3\underline{/45°}$ and $15\underline{/-30°}$) of Figure 24–15(a) into a resultant phasor.

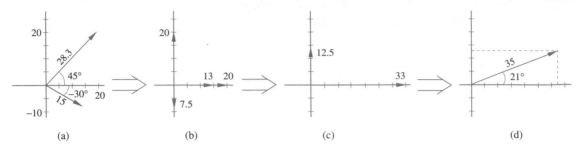

FIGURE 24–15   Resolving two nonrectangular phasors, as demonstrated in Example 24–13.

Solution   Use Guidelines 24–1 (**case 4**):

$$28.3\underline{/45°}\ \boxed{\rightarrow R}\ 20 + j20$$
$$15\underline{/-30°}\ \boxed{\rightarrow R}\ 13 - j7.5$$

Plot each phasor in rectangular form as shown in Figure 24–15(b) and algebraically add:

$$r = 13 + 20 = 33$$
$$jq = j20 - j7.5 = j12.5$$

Plot each component as shown in Figure 24–15(c). Resolve the two 90° phasors into a resultant phasor:

$$r + jq = 33 + j12.5$$

$$\therefore\quad 33 + j12.5\ \boxed{\rightarrow P}\ 35\underline{/21°}$$

Observation   The resultant is shown in Figure 24–15(d).

EXERCISE 24–6

Resolve the following systems of phasors into a single resultant phasor by applying Guidelines 24–1.

1. $\begin{cases} 5\underline{/0°} \\ 3\underline{/0°} \end{cases}$
2. $\begin{cases} 7\underline{/180°} \\ 12\underline{/0°} \end{cases}$
3. $\begin{cases} 2\underline{/0°} \\ 8\underline{/0°} \end{cases}$

4. $\begin{cases} 6.0\underline{/-90°} \\ 3.0\underline{/0°} \end{cases}$
5. $\begin{cases} 5\underline{/90°} \\ 15\underline{/-90°} \end{cases}$
6. $\begin{cases} 4.0\underline{/-180°} \\ 12\underline{/-90°} \end{cases}$

7. $\begin{cases} 11\underline{/90°} \\ 11\underline{/0°} \end{cases}$
8. $\begin{cases} 9.0\underline{/90°} \\ 4.0\underline{/180°} \end{cases}$
9. $\begin{cases} 16\underline{/90°} \\ 14\underline{/90°} \end{cases}$

10. $\begin{cases} 8.00\underline{/20°} \\ 6.00\underline{/-40°} \end{cases}$ **11.** $\begin{cases} 5.0\underline{/50°} \\ 7.0\underline{/120°} \end{cases}$ **12.** $\begin{cases} 4.0\underline{/-60°} \\ 10\underline{/-160°} \end{cases}$

13. $\begin{cases} 2.5\underline{/-20°} \\ 4.7\underline{/120°} \end{cases}$ **14.** $\begin{cases} 7.6\underline{/1.2^r} \\ 2.2\underline{/-2.05^r} \end{cases}$ **15.** $\begin{cases} 6.5\underline{/0.81^r} \\ 8.3\underline{/1.81^r} \end{cases}$

16. $\begin{cases} 15.8\underline{/-3.642^r} \\ 22.7\underline{/0.434^r} \end{cases}$ **17.** $\begin{cases} 9.82\underline{/14.3°} \\ 4.36\underline{/-1.405^r} \end{cases}$ **18.** $\begin{cases} 32.3\underline{/-5.772^r} \\ 51.7\underline{/-0.835^r} \end{cases}$

SELECTED TERMS

complex number   The sum of a real number and an imaginary number.

imaginary number   The square root of a negative quantity.

phasor   A two-dimensional vector used to portray the phase relationship of voltage and current in an electrical circuit.

scalar   A quantity having only magnitude.

vector   A quantity having both magnitude and direction.

25

The Mathematics of Phasors

25-1 Addition and Subtraction of Phasor Quantities

25-2 Multiplication of Phasor Quantities

25-3 Division of Phasor Quantities

25-4 Powers and Roots of Phasor Quantities

PERFORMANCE OBJECTIVES

- Add phasors.
- Subtract phasors.
- Multiply phasors.
- Divide phasors.
- Raise phasors to powers.
- Extract roots of phasors.

The Grand Coulee Dam on the Columbia River in Washington State. The power houses are located at the base of the dam on either side of the spillway. (Courtesy of U.S. Bureau of Reclamation)

In the preceding chapter we introduced the concepts of complex numbers, vectors, and phasors. They can be thought of graphically as arrows or as points. Mathematically, they can be written in the rectangular form or in the polar form of the complex number. Also, complex numbers, vectors, and phasors all follow the same rules.

In this chapter we define phasor operators that are extensions of addition, subtraction, multiplication, and division. Whether quantities are considered as vectors, phasors, or complex numbers, they are combined using the same operators. Thus, the techniques of this chapter can be used with vectors and complex numbers as well as with phasors.

25–1   ADDITION AND SUBTRACTION OF PHASOR QUANTITIES

Addition

A phasor may be written in either the polar or rectangular form as a complex number. To add two phasor quantities, the phasors must be in the **rectangular form.** This important concept is summarized in Rule 25–1.

> **RULE 25–1.   Adding Phasors**
>
> To add phasor quantities, express each in rectangular form and:
> 1. Add the real parts of the phasors.
> 2. Add the imaginary parts of the phasors.
> 3. Form the sum as a phasor written in rectangular form.

The following examples provide you with the opportunity to practice on phasors. It is important that you work along with your calculator. We will be using polar to rectangular $\boxed{\rightarrow R}$ and rectangular to polar $\boxed{\rightarrow P}$ calculator functions throughout this chapter.

EXAMPLE 25–1   Add $3 + j4$ and $5 + j6$.

Solution   Follow the steps of Rule 25–1:

$$\begin{array}{r} 3 + j4 \\ 5 + j6 \\ \hline 8 + j10 \end{array}$$

$\therefore$ $(3 + j4) + (5 + j6) = 8 + j10$

EXAMPLE 25–2   Add $10\underline{/-53°}$ and $-5 - j7$.

Solution   Change $10\underline{/-53°}$ into rectangular form:

$$10\underline{/-53°} \boxed{\rightarrow\text{R}} \; 6 - j8$$

Add:

$$
\begin{array}{r}
6 - j8 \\
\underline{-5 - j7} \\
1 - j15
\end{array}
$$

∴ $(10\underline{/-53°}) + (-5 - j7) = 1 - j15$

EXAMPLE 25–3   Add $18.5\underline{/220.0°}$ and $4.20\underline{/-16.0°}$. Express the answer in polar form.

Solution   Change each phasor to its rectangular form:

$$18.5\underline{/220.0°} \boxed{\rightarrow\text{R}} \; -14.17 - j11.89$$
$$4.20\underline{/-16.0°} \boxed{\rightarrow\text{R}} \; 4.04 - j1.16$$

Add:

$$
\begin{array}{r}
-14.17 - j11.89 \\
\underline{4.04 - \; j1.16} \\
-10.13 - j13.05
\end{array}
$$

Express in polar form:

∴ $-10.13 - j13.05 \boxed{\rightarrow\text{P}} \; 16.5\underline{/-127.8°}$

Subtraction

For phasors to be subtracted, they must be in the **rectangular form.** Rule 25–2 summarizes the steps used in subtracting.

RULE 25–2.   *Subtracting Phasors*

To subtract phasor quantities, express each in rectangular form and:

1. Change the sign of both the real and the imaginary part of the phasor to be subtracted.
2. Add as in Rule 25–1.

EXAMPLE 25–4	Subtract $9 - j4$ from $15 + j12$.
Solution	Change the signs of $9 - j4$:

$$-(9 - j4) = -9 + j4$$

Add:

$$\begin{array}{r} 15 + j12 \\ \underline{-9 + j4} \\ 6 + j16 \end{array}$$

$$\therefore \quad (15 + j12) - (9 - j4) = 6 + j16$$

EXAMPLE 25–5	Subtract $-7 - j3$ from $6 - j5$. Express the difference in polar form.
Solution	Change the signs of $-7 - j3$:

$$-(-7 - j3) = 7 + j3$$

Add:

$$\begin{array}{r} 6 - j5 \\ \underline{7 + j3} \\ 13 - j2 \end{array}$$

Express in polar form:

$$\therefore \quad 13 - j2 \;\boxed{\rightarrow\text{P}}\; 13\underline{/-9°}$$

EXAMPLE 25–6	Subtract $29.3\underline{/142°}$ from $14.3\underline{/-73.5°}$. Express the answer in polar form.
Solution	Change each phasor to its rectangular form:

$$29.3\underline{/142°} \;\boxed{\rightarrow\text{R}}\; -23.09 + j18.04$$

$$14.3\underline{/-73.5°} \;\boxed{\rightarrow\text{R}}\; 4.06 - j13.71$$

Change the signs of $-23.09 + j18.04$:

$$-(-23.09 + j18.04) = 23.09 - j18.04$$

Add:

$$\begin{array}{r} 4.06 - j13.71 \\ \underline{23.09 - j18.04} \\ 27.15 - j31.75 \end{array}$$

Express in polar form:

$$\therefore \quad 27.15 - j31.75 \;\boxed{\rightarrow\text{P}}\; 41.8\underline{/-49.5°}$$

In Examples 25–3 and 25–6, the intermediate steps were stated one place beyond the precision of the original numbers. This was done to minimize error in the calculation due to rounding. Observe that the final answer is expressed to the appropriate precision: in each case, three significant figures in the magnitude and 0.1 degrees in the angle.

EXERCISE 25–1

Perform the indicated addition or subtraction of the given phasor quantities. Express the answer in both rectangular and polar form.

1. $(2.0 + j5.0) + (5.0 - j3.0)$ **2.** $(3.0 + j0) + (-8.0 - j6.0)$

3. $(-4.0 + j7.0) - (8.0 + j6.0)$ **4.** $(9.0 - j9.0) + (0 + j10.0)$

5. $(12\underline{/30°}) + (5.0 + j2.0)$ **6.** $(16\underline{/47°}) - (-12 + j5.0)$

7. $(13 - j4.0) + (5.0\underline{/60°})$ **8.** $(-19 + j11) + (18\underline{/-20°})$

9. $(27 + j16) - (10\underline{/-140°})$ **10.** $(17\underline{/200°}) - (-28 - j12)$

11. $(0.52 + j1.85) + (2.14 - j0.91)$

12. $(-33.7 - j19.5) - (12.5 + j4.14)$

13. $(0.318\underline{/24°}) - (1.5 - j0.85)$

14. $(99 - j10) + (14.1\underline{/-108°})$

15. $(32\underline{/161°}) - (19\underline{/110°})$

16. $(43\underline{/-12°}) + (51\underline{/82°})$

17. $(45\underline{/18°}) + (30\underline{/59°})$

18. $(61\underline{/32°}) - (40\underline{/-72°})$

19. $(127.4\underline{/-2.20^r}) - (9.23\underline{/-0.170^r})$

20. $(0.583\underline{/0.932^r}) + (0.448\underline{/-2.84^r})$

25–2 MULTIPLICATION OF PHASOR QUANTITIES

Phasor quantities may be multiplied when they are expressed in either the rectangular or polar form. Table 25–1 may be used to help multiply phasors in rectangular form. However, the preferred form for multiplying phasors is the **polar form.**

TABLE 25–1 Definitions Used with Imaginary Numbers	
$j = \sqrt{-1}$	$j^3 = -j$
$j^2 = -1$	$j^4 = +1$

Multiplying Phasors in Rectangular Form

Rule 25–3 summarizes the steps used to multiply phasors expressed in rectangular form.

RULE 25–3. Multiplying Phasor Quantities in Rectangular Form

To multiply phasors in rectangular form, multiply the numbers as if they were two binomials by:

1. Distributing the real part of the first complex number over the second complex number.
2. Distributing the imaginary part of the first complex number over the second complex number.
3. Replacing j^2 with -1.
4. Combining like terms.
5. Forming the product as a phasor written in rectangular form.

EXAMPLE 25–7	Multiply $3 + j2$ and $4 - j5$.
Solution	Use the steps of Rule 25–3.
Steps 1–2:	Distribute $(3 + j2)$ over $(4 - j5)$:
	$$(3 + j2)(4 - j5) = 12 - j15 + j8 - j^210$$
Step 3:	Replace j^2 with -1:
	$$12 - j15 + j8 + 10$$
Step 4:	Combine like terms:
	$$22 - j7$$
$\therefore$	$$(3 + j2)(4 - j5) = 22 - j7$$

From the preceding example, we learn that multiplying a real number by an imaginary number results in a product that is an imaginary number, and multiplying two imaginary numbers results in a product that is a real number. These concepts are summarized in Table 25–2.

TABLE 25–2 Summary of Multiplying Complex Numbers*		*NOTE: $j^2 = -1$
Pattern		**Example**
Real number $\times$ real number	$\Rightarrow$ real number	$3 \times 5 = 15$
Real number $\times$ imaginary number	$\Rightarrow$ imaginary number	$4 \times j2 = j8$
Imaginary number $\times$ imaginary number	$\Rightarrow$ real number	$j3 \times j2 = j^26 = -6$

EXAMPLE 25–8   Multiply $7 - j4$ by $-3 - j5$. Express the answer in polar form.

Solution   Use Rule 25–3.

Steps 1–2:   Distribute $7 - j4$ over $-3 - j5$:

$$(7 - j4)(-3 - j5) = -21 - j35 + j12 + j^2 20$$

Step 3:   Replace j^2 with -1:

$$-21 - j35 + j12 - 20$$

Step 4:   Combine like terms:

$$-41 - j23$$

Express $-41 - j23$ in polar form:

$$-41 - j23 \boxed{\rightarrow \text{P}} \; 47\underline{/-150°}$$

$$\therefore \quad (7 - j4)(-3 - j5) = 47\underline{/-150°}$$

Multiplying Phasors in Polar Form

Phasors are easily multiplied when they are expressed in polar form. Rule 25–4 summarizes the steps used to multiply phasors expressed in polar form.

RULE 25–4.   Multiplying Phasor Quantities in Polar Form

To multiply phasors in polar form:
1. Multiply the magnitudes.
2. Add the angles.
3. Form the products as a phasor written in polar form.

EXAMPLE 25–9   Multiply $3\underline{/10°}$ and $5\underline{/20°}$.

Solution   Use Rule 25–4.

Step 1:   Multiply the magnitudes:

$$3 \times 5 = 15$$

Step 2:   Add the angles:

$$10° + 20° = 30°$$

Step 3: Form the product as a polar number:

$$15\underline{/30°}$$

$$\therefore \quad (3\underline{/10°})(5\underline{/20°}) = 15\underline{/30°}$$

EXAMPLE 25–10 Multiply $15.3\underline{/-83°}$ and $26.5\underline{/128°}$.

Solution Use Rule 25–4.

Step 1: Multiply the magnitudes:

$$15.3 \times 26.5 = 405$$

Step 2: Add the angles:

$$-83° + 128° = 45°$$

Step 3: Form the product as a polar number:

$$405\underline{/45°}$$

$$\therefore \quad (15.3\underline{/-83°})(26.5\underline{/128°}) = 405\underline{/45°}$$

EXAMPLE 25–11 Multiply $16.3 - j5.90$ by $22.8\underline{/17.0°}$. Express the answer in polar form.

Solution Convert $16.3 - j5.9$ into polar form:

$$16.3 - j5.90 \boxed{\rightarrow P} 17.3\underline{/-19.9°}$$

Multiply $17.3\underline{/-19.9°}$ and $22.8\underline{/17.0°}$:

$$(17.3\underline{/-19.9°})(22.8\underline{/17.0°}) = 395\underline{/-2.9°}$$

$$\therefore \quad (16.3 - j5.90)(22.8\underline{/17.0°}) = 395\underline{/-2.9°}$$

EXERCISE 25–2

Multiply the following phasors and express the answer in both polar and rectangular form:

1. $(2 + j2)(3 + j3)$ **2.** $(5 + j4)(6 + j2)$

3. $(7 + j2)(4 + j5)$ **4.** $(3 + j5)(6 + j3)$

5. $(3 - j4)(5 + j2)$ **6.** $(-2 + j6)(-2 - j6)$

7. $(5\underline{/15°})(3\underline{/12°})$ **8.** $(4\underline{/9°})(3\underline{/-20°})$

9. $(6\underline{/-40°})(2\underline{/-25°})$ **10.** $(7\underline{/14°})(5\underline{/-14°})$

11. $(38\underline{/115°})(17\underline{/-82°})$ **12.** $(4.5\underline{/-23°})(11.2\underline{/40°})$

13. $(111\underline{/22°})(98\underline{/76°})$ **14.** $(0.593\underline{/-3.6°})(0.218\underline{/8.3°})$

15. $(70.7\underline{/142°})(31.2\underline{/-105°})$ **16.** $(44.3\underline{/29.5°})(14.3\underline{/15.7°})$

17. $(29.1\underline{/-13.4°})(12.5\underline{/9.9°})$

18. $(6.08\underline{/-122.5°})(-9.2 - j3.07)$

19. $(0.358 - j1.12)(2.3 + j0.632)$

20. $(70.4 - j16.9)(-42.1 + j33.0)$

21. $(11.2\underline{/1.5^r})(17.6\underline{/2.2^r})$

22. $(0.34\underline{/-121°})(1.04\underline{/0.94^r})$

23. $(122.6\underline{/78.4°})(96.7\underline{/-0.833^r})$

24. $(16.8\underline{/-1.52^r})(7.65 + j10.8)$

25–3 DIVISION OF PHASOR QUANTITIES

Division of phasor quantities may be carried out in either the polar or rectangular form. However, division with phasors in polar form is *preferred* because of its simplicity.

Dividing Phasors in Rectangular Form

To divide phasors in rectangular form, the divisor (denominator) must be changed to a real number. To change a divisor from a complex number to a real number, we use the **complex conjugate** of the divisor. To form the complex conjugate of a complex number in rectangular coordinates, change the sign of the imaginary part. Thus, $3 + j4$ and $3 - j4$ are complex conjugates. Similarly, to form the complex conjugate of a complex number in polar coordinates, change the sign of the angle. Thus, $5\underline{/53.13}$ and $5\underline{/-53.13}$ are complex conjugates.

EXAMPLE 25–12 Show that the product of $3 + j4$ and its complex conjugate $3 - j4$ is a real number.

Solution The complex conjugate of $3 + j4$ is $3 - j4$. Multiply:

$$(3 + j4)(3 - j4) = 9 - j12 + j12 - j^216$$
$$= 9 - j^216$$
$$= 9 + 16$$
$$= 25$$

∴ $(3 + j4)(3 - j4) = 25$, which is a real number.

Complex conjugates are used in ac electronics to compute the circuit load for maximum power transfer and to determine the value of reactive components needed to correct the load's power factor.

Rule 25–5 summarizes the steps used to divide phasors represented by complex numbers in rectangular form.

RULE 25–5. Dividing Phasors in Rectangular Form

To divide phasors in rectangular form:

1. Multiply the divisor (denominator) and the dividend (numerator) by the complex conjugate of the divisor.

2. Divide the real number and the imaginary number of the dividend by the divisor.

3. Form the quotient as a phasor written in rectangular form.

EXAMPLE 25–13   Divide $(15 + j10)$ by $(2 + j1)$.

Solution   Follow the steps of Rule 25–5.

$$\frac{15 + j10}{2 + j1}$$

Step 1:   Multiply numerator and denominator by $2 - j1$, the complex conjugate of $2 + j1$:

$$\frac{(15 + j10)(2 - j1)}{(2 + j1)(2 - j1)} = \frac{30 - j15 + j20 - j^2 10}{4 - j^2 1}$$

$$= \frac{40 + j5}{5}$$

Step 2:   Divide the real and the imaginary number of the dividend.

$$= \frac{40}{5} + \frac{j5}{5}$$

Step 3:   Write in rectangular form

$$= 8 + j1$$

$$\therefore \qquad (15 + j10)/(2 + j1) = 8 + j1$$

Observation   Division in rectangular form is sometimes used to develop formulas in ac electronics. However, due to its complicated nature, the rectangular form is not usually used for division of phasor quantities.

Dividing Phasors in Polar Form

Phasors are easily divided when they are written in **polar form.** Rule 25–6 summarizes the steps used to divide phasors in polar form.

RULE 25–6.   Dividing Phasors in Polar Form

To divide phasors in polar form:

1. Divide the magnitudes.
2. Subtract the angle of the divisor (denominator) from the angle of the dividend (numerator).
3. Form the quotient as a phasor written in polar form.

EXAMPLE 25–14   Divide $15\underline{/20°}$ by $5\underline{/10°}$.

Solution   Follow the steps of Rule 25–6.

Step 1: Divide the magnitudes:

$$\frac{15}{5} = 3$$

Step 2: Subtract 10° from 20°.

$$20° - 10° = 10°$$

Step 3: Form the quotient:

$$3\angle 10°$$

$$\therefore \quad (15\angle 20°)/(5\angle 10°) = 3\angle 10°$$

EXAMPLE 25–15 Divide $12.5\angle -78°$ by $6.1\angle 19.5°$.

Solution

$$\therefore \quad (12.5\angle -78°)/(6.1\angle 19.5°) = 2.05\angle -97.5°$$

EXAMPLE 25–16 Divide $(13.5 - j3.7)$ by $20.9\angle 42.8°$.

Solution First change $(13.5 - j3.7)$ into polar form:

$$13.5 - j3.7 \boxed{\to P} \, 14.0\angle -15.3°$$

Then divide:

$$\therefore \quad (14.0\angle -15.3°)/(20.9\angle 42.8°) = 0.667\angle -58.1°$$

EXERCISE 25–3

Write the complex conjugates of the following complex numbers.

1. $5 - j3$
3. $12.5 - j4$
5. $-9.4 + j10.6$

2. $3.2 + j8.1$
4. $-7.3 - j12$
6. $14.3 + j15.7$

Divide the following phasors. Express the answer in both polar and rectangular forms.

7. $(2.0 + j3.0)/(3.0 + j4.0)$
9. $(1.0 + j2.0)/(6.0 - j2.0)$
11. $(-10 - j5.0)/(2.0 + j2.0)$
13. $(18\angle 45°)/(6.0\angle -15°)$
15. $(75\angle -80°)/(25\angle -40°)$
17. $(28\angle 112°)/(14\angle -37°)$
19. $(88.9\angle -29.2°)/(215\angle 61.3°)$

8. $(5.0 - j6.0)/(7.0 + j1.0)$
10. $(-4.0 + j8.0)/(2.0 - j5.0)$
12. $(8.0 + j4.0)/(2.0 + j4.0)$
14. $(33\angle -5°)/(3.0\angle -70°)$
16. $(39\angle 52°)/(13\angle -18°)$
18. $(347\angle -161°)/(84\angle 7.5°)$
20. $(0.913\angle 102°)/(0.491\angle 58.8°)$

21. $(18 + j12)/(7.0\underline{/0.23^{\text{r}}})$

22. $(23\underline{/-0.14^{\text{r}}})/(6.0 - j5.0)$

23. $(-0.892 - j1.02)/(2.3\underline{/2.35^{\text{r}}})$

24. $(43.8\underline{/-1.72^{\text{r}}})/(27.2 - j39.5)$

25–4 POWERS AND ROOTS OF PHASOR QUANTITIES

A phasor quantity is most easily raised to a power when it is expressed in **polar form.** Rule 25–7 provides the procedure for raising a phasor to a power.

RULE 25–7. *Power of a Phasor*

To raise a phasor to a power, express the phasor in polar form and:

1. Raise the magnitude to the specified power.
2. Multiply the angle by the exponent.
3. Form the solution as a phasor written in polar form.

EXAMPLE 25–17 Solve $(5\underline{/12°})^3$.

Solution Use the steps of Rule 25–7.

Step 1: Raise 5 to the third power:

$\boxed{y^x}$ $\quad 5^3 = 125$

Step 2: Multiply 12° by 3:

$\boxed{\times}$ $\quad 12° \times 3 = 36°$

Step 3: Form the solution:

$\quad 125\underline{/36°}$

$\therefore \quad (5\underline{/12°})^3 = 125\underline{/36°}$

EXAMPLE 25–18 Solve $(4.0 + j6.0)^2$.

Solution Follow the steps of Rule 25–7. First, express $(4.0 + j6.0)$ in polar form:

$\quad 4.0 + j6.0 \boxed{\rightarrow\text{P}} 7.21\underline{/56.3°}$

Step 1: Raise 7.21 to the second power:

$\boxed{x^2}$ $\quad 7.21^2 = 52$

Step 2: Multiply 56.3° by 2:

$\boxed{\times}$ $\quad 56.3 \times 2 = 113°$

Step 3:   Form the solution:

$$52\underline{/113°}$$

$$\therefore \quad (4.0 + j6.0)^2 = (7.21\underline{/56.3°})^2 = 52\underline{/113°}$$

To take the root of a phasor quantity, express the phasor in **polar form** and apply Rule 25–8.

RULE 25–8.   *Principal Root of a Phasor*

To take the principal root of a phasor, express the phasor in polar form with the angle expressed between $-180°$ and $+180°$ and then:

1. Take the root of the magnitude.
2. Divide the angle by the index of the radical.
3. Form the solution as a phasor written in polar form.

EXAMPLE 25–19   Solve $\sqrt{25\underline{/10°}}$.

Solution   Use the steps of Rule 25–8.

Step 1:   Take the root of 25:

$\boxed{\sqrt{x}}$ $\sqrt{25} = 5$

Step 2:   Divide 10° by the index:

$\boxed{\div}$ $10°/2 = 5°$

Step 3:   Form the solution:

$$5\underline{/5°}$$

$$\therefore \quad \sqrt{25\underline{/10°}} = 5\underline{/5°}$$

EXAMPLE 25–20   Solve $\sqrt[3]{7.20 - j10.9}$.

Solution   Express $7.20 - j10.9$ as a polar number:

$$7.20 - j10.9 \quad \boxed{\rightarrow P} \quad 13.1\underline{/-56.6°}$$

Step 1:   Take the third root of 13.06:

$\boxed{y^x}$ $13.1^{1/3} = 2.36$

Step 2:   Divide $-56.55°$ by the index of the radical:

$\boxed{\div}$ $-56.55°/3 = -18.9°$

Step 3: Form the solution:

$$2.36\underline{/-18.9°}$$

$$\therefore \quad \sqrt[3]{7.20 - j10.9} = 2.36\underline{/-18.9°}$$

EXAMPLE 25–21 Solve $(13.6\underline{/-126°})^{1/2}$. Express the solution in rectangular form.

Solution Take the square root of 13.6 and divide $-126°$ by 2:

$$13.6^{1/2} = 3.69$$

$$-126° \times \tfrac{1}{2} = -63°$$

Form the solution:

$$3.69\underline{/-63°}$$

Change to rectangular form:

$$3.69\underline{/-63°} \;\boxed{\rightarrow R}\; 1.68 - j3.29$$

$$\therefore \quad (13.6\underline{/-126°})^{1/2} = 3.69\underline{/-63°} = 1.68 - j3.29$$

EXERCISE 25–4

Solve each of the following. Express the solution in polar form.

1. $(3\underline{/10°})^2$

2. $(3\underline{/4°})^3$

3. $(2\underline{/10°})^4$

4. $(7\underline{/30°})^2$

5. $(12\underline{/-17°})^2$

6. $(3.25\underline{/-20°})^3$

7. $(14.3\underline{/56.2°})^2$

8. $(410\underline{/26.9°})^2$

9. $(16.8 - j4.2)^2$

10. $(-71.4 - j40.8)^3$

11. $\sqrt{49\underline{/40°}}$

12. $\sqrt{121\underline{/-72°}}$

13. $(125\underline{/-96°})^{1/3}$

14. $(196\underline{/46°})^{1/2}$

15. $(2.72\underline{/17.6°})^{1/2}$

16. $\sqrt[3]{84.3\underline{/-118°}}$

17. $\sqrt{-6.13 + j15.2}$

18. $(70.2 + j64.6)^{1/3}$

19. $(-81.0 - j12.5)^{1/2}$

20. $\sqrt{34.7 + j42.9}$

21. $\sqrt[4]{-143 - j391}$

22. $\sqrt{19.8\underline{/-2.87^r}}$

23. $\sqrt{0.843\underline{/0.571^r}}$

24. $(526.7\underline{/2.95^r})^{1/3}$

complex conjugate   Created by changing the sign of the imaginary component of a complex number in rectangular form. Thus, $2 + j3$ and $2 - j3$ are complex conjugates of one another. In polar form, a change in the sign of the argument results in a complex conjugate. Thus, $5\underline{/53.13}$ and $5\underline{/-53.13}$ are complex conjugates of one another.

SECTION CHALLENGE

WEB CHALLENGE FOR CHAPTERS 22, 23, 24, AND 25

To evaluate your comprehension of Chapters 22, 23, 24, and 25, log on to **www.prenhall. com/harter** and take the online True/False and Multiple Choice assessments for each of the chapters.

SECTION CHALLENGE FOR CHAPTERS 22, 23, 24, AND 25*

A. A *hand bender* is used to make the bends for the offset in the $\frac{3}{4}$-in diameter thin-wall metal conduit pictured in Figure C–12. Your challenge is to determine the finished length, *L,* of the conduit given an offset, side BC of triangle ABC, of 12.0 cm.

B. The chassis support bracket, shown in Figure C–13, is fabricated from 1.50 mm (0.06 in) thick aluminum sheet stock. Using the trigonometric functions, solve triangle ODB for length DB and length OB (the distance from vertex B to the center

FIGURE C–12 The electrical conduit is bent with a 12.0 cm offset at 22.5°.

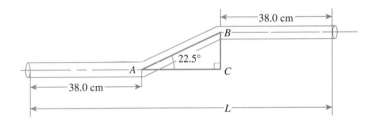

FIGURE C–13 Support bracket where triangle ODB is a right triangle and ∠OBD = ∠OBE.

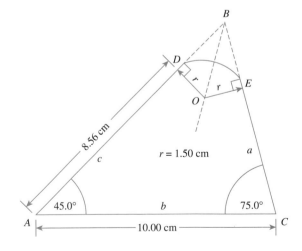

*The solution to this Section Challenge is found in Appendix C.

of radius *r*). Then compute the length of side AB and, with the aid of the law of cosines, determine the length of side BC.

C. As noted in Section 24–1, quantities expressed with both magnitude and direction are called vector quantities. Your last challenge is to resolve the vector quantities pictured in Figure C–14 by using the $\boxed{\rightarrow\text{R}}$ and $\boxed{\rightarrow\text{P}}$ calculator keystrokes. The procedure used to resolve each of the force vectors and combine them into a single resultant force at an angle is:

1. Resolve each vector into its rectangular components using the $\boxed{\rightarrow\text{R}}$ keystroke. Pay attention to the algebraic sign.

2. Algebraically sum the components in each direction to determine the resultant vector *R* in terms of its *x* and *y* components R_x and R_y respectively.

3. Use the $\boxed{\rightarrow\text{P}}$ keystroke to determine the magnitude and the direction of the resultant vector $\rho\underline{/\theta}$.

FIGURE C–14 The resultant of several vectors is found by resolving each vector into its rectangular components and then summing all the *x*-components and all the *y*-components to obtain the *x*- and *y*-components of the resultant.

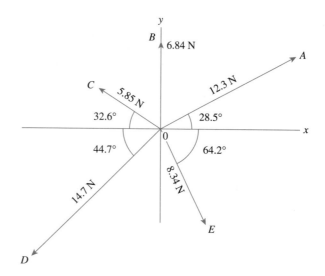

26

Fundamentals of Alternating Current

26–1   Alternating-Current Terminology

26–2   Resistance

26–3   Inductance and Inductive Reactance

26–4   Capacitance and Capacitive Reactance

26–5   Voltage Phasor for Series Circuits

26–6   Current Phasor for Parallel Circuits

PERFORMANCE OBJECTIVES

- Convert between frequency, period, and angular velocity.
- Construct a voltage phasor diagram for an ac LCR series circuit.
- Construct a current phasor diagram for an ac LCR parallel circuit.

View of the left power house of the Columbia Basin Project showing the nine 108 MW generators. (Courtesy of U.S. Bureau of Reclamation)

This chapter introduces you to alternating current (ac) by exploring the sinusoidal (sine wave) properties of alternating current. The chapter begins by introducing the basic names used to describe the sinusoidal nature of alternating current, and then phasors are used to present a picture of the relationship among the current, the voltage, and the components used in circuits.

Starting with this chapter, the vertical axis of the phasor will be labeled with only $+j$ and $-j$ (rather than jQ). This is done to be consistent with the practice in most electric circuit books.

26–1 ALTERNATING-CURRENT TERMINOLOGY

Frequency

When a conductor is rotated through a magnetic flux as pictured in Figure 26–1, a voltage is created. The number of times in a second that the conductor is rotated through a complete revolution is called the **frequency** of the voltage. For example, if a conductor is rotated 20 times in 5 s, the frequency of the resulting sine wave is 4 cycles per second. The unit for frequency is the hertz (Hz). A frequency of 4 cycles per second is written as 4 Hz.

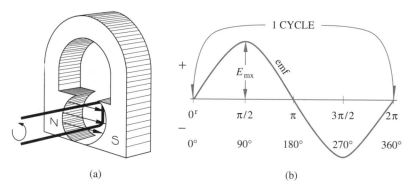

FIGURE 26–1 (a) As the conductor is rotated through the magnetic flux, an emf is created. (b) The emf has a sinusoidal wave shape.

| EXAMPLE 26–1 | Express each of the following frequencies in hertz. |

    (a) 39 cycles in 3.0 s
    (b) 100 cycles in 200 ms
    (c) 1500 revolutions in 0.375 s
    (d) 300 rotations in 1.0 min

Solution
    (a) 39/3.0 = 13 cycles per second = 13 Hz
    (b) 100/0.20 = 500 cycles per second = 500 Hz
    (c) 1500/0.375 = 4000 cycles per second = 4.00 kHz
    (d) 300/60 = 5.0 cycles per second = 5.0 Hz

Period

The time it takes for one complete cycle of the conductor through the magnetic flux is called the **period** of the resulting sine wave. Thus, the period of the voltage created by rotating a conductor 60 times a second (60 Hz) through a flux is $\frac{1}{60}$ s, or 16.67 ms. The relationship between frequency (f) in hertz and period (T) in seconds is expressed as Equation 26–1.

$$T = 1/f \quad \text{s} \tag{26–1}$$

EXAMPLE 26–2 Express the period of each of the following.

(a) 400 Hz   (b) 1.20 MHz   (c) 20.0 Hz   (d) 5.00 kHz

Solution Use Equation 26–1.

(a) $T = 1/400 = 2.50$ ms
(b) $T = 1/(1.20 \times 10^6) = 833$ ns
(c) $T = 1/20.0 = 50.0$ ms
(d) $T = 1/(5.00 \times 10^3) = 200$ μs

Angular Velocity

Each time the conductor of Figure 26–1 rotates one cycle through the flux, it travels 6.28 radians. **Angular velocity** is expressed in radians per second (rad/s). Thus, a wave generated by a conductor traveling 200 rad in 25 s would have an angular velocity of 200 rad/25 s or 8.0 rad/s. Equation 26–2 relates angular velocity (ω) in radians per second to angular displacement (α) in radians and time (t) in seconds.

$$\omega = \alpha/t \quad \text{rad/s} \tag{26–2}$$

EXAMPLE 26–3 Express each of the following as angular velocity in radians per second.

(a) 60 rad in 12 s
(b) 2.4 krad in 0.80 min (0.80 min = 48 s)
(c) 320 rad in 160 s
(d) 45 Mrad in 15 s

Solution Use Equation 26–2.

(a) $\omega = 60/12 = 5.0$ rad/s
(b) $\omega = 2400/48 = 50$ rad/s
(c) $\omega = 320/160 = 2.00$ rad/s
(d) $\omega = (45 \times 10^6)/15 = 3.0$ Mrad/s

Frequency (f) in hertz and angular velocity (ω) in radians per second are related by Equation 26–3, where 2π (6.28) is the number of radians traveled per cycle. The symbol ω (omega) is used exclusively for angular velocity in electronics.

$$\omega = 2\pi f \quad \text{rad/s} \tag{26–3}$$

EXAMPLE 26–4 Express each of the following frequencies as an angular velocity in radians per second (rad/s).

(a) 60.0 Hz (b) 400 Hz (c) 5.00 kHz (d) 200 kHz

Solution Use Equation 26–3.

(a) $\omega = 2\pi60.0 = 377$ rad/s
(b) $\omega = 2\pi400 = 2.51$ krad/s
(c) $\omega = 2\pi5.00 \times 10^3 = 31.4$ krad/s
(d) $\omega = 2\pi200 \times 10^3 = 1.26$ Mrad/s

EXERCISE 26–1

Express the frequency of each of the following waves in hertz.

1. 85 cycles in 5 s
2. 200 revolutions in 3 min
3. 10 rotations in 2 s
4. 120 cycles per second
5. A wave with a period of 28.0 ms
6. A wave with a period of 8.00 μs
7. A wave with $\omega = 377$ rad/s
8. A wave with $\omega = 12.57$ krad/s

Determine the period of each of the following waves in seconds.

9. 60.0 Hz
10. 15.0 Hz
11. 12.0 kHz
12. 5.40 MHz
13. 25 rev/min
14. 4200 rev/min
15. 512 rad/s
16. 212 krad/s

Determine the angular velocity of each of the following waves in radians per second.

17. 3000 rad in 15 s
18. 50 krad in 10 s
19. 1.2 krev in 4 s
20. 900 Hz
21. 6.00 kHz
22. 1.00 MHz
23. Period of 20.0 ms
24. Period of 5.00 μs

26–2 RESISTANCE

Figure 26–2(a) pictures an alternating current passing through the resistance R. The voltage wave of Figure 26–2(a) is the result of the current passing through the resistance. When the curve depicting current is superimposed over the curve depicting voltage, as shown in Figure 26–2(b), you will notice that the two waves start and finish together. Furthermore, both waves reach their maximum positive amplitude and their maximum negative amplitude together. When the current and voltage pass through corresponding points in their cycle at the same time, the two waves are said to be **in phase.** Figure 26–2(c) is the phasor diagram of the current and voltage in phase. The angle between the current and voltage phasor in a phasor diagram is called the **phase angle.** The Greek letter theta (θ) is used to designate the phase angle in this text.

To summarize, when an ac source of voltage is applied to a resistive load, the current and voltage are in phase ($\angle\theta = 0°$), and Ohm's law ($E = IR$) may be applied.

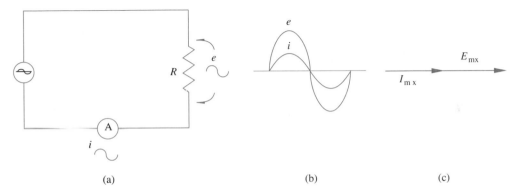

(a)   (b)   (c)

FIGURE 26–2 When an alternating current passes through a resistance (a), the resulting voltage is in phase with the current (b), and the phase angle (the angle between the current and voltage vector) is 0° (c).

EXAMPLE 26–5   A 100-Ω resistor is connected to a 200-V ac source.

(a) Determine the current.
(b) Determine the phase angle.
(c) Draw the phasor diagram representing the current and voltage.

Solution   (a) $I = E/R$:

$\therefore$ $\qquad I = \dfrac{200}{100} = 2.00$ A

(b) Phase angle for a resistive circuit:

$\therefore$ $\qquad \angle\theta = 0°$

(c) See Figure 26–3.

FIGURE 26–3 Phasor diagram of current and voltage for Example 26–5.

$I = 2$ A $E = 200$ V

EXAMPLE 26–6   Determine the value of the resistive load for a circuit having the phasor diagram of Figure 26–4.

FIGURE 26–4   The phasor diagram for Example 26–6.

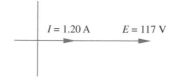

$I = 1.20$ A $E = 117$ V

Solution $R = E/I$

Substitute $E = 117$ V and $I = 1.20$ A:

∴ $R = 117/1.20 = 97.5\ \Omega$

EXERCISE 26–2

Determine the resistive load for each of the phasor diagrams.

1. 5 A ⟶ 50 V ⟶
2. 30 mA ⟶ 90 V ⟶
3. 280 mA ⟶ 30 V ⟶
4. 5.0 V ⟶ 8.0 A ⟶
5. 510 μA ⟶ 18 V ⟶
6. 300 mA ⟶ 12.6 V ⟶
7. 2.00 V ⟶ 12.2 A ⟶
8. 47.0 mA ⟶ 36.0 V ⟶

Determine the source voltage for each of the following.

9. $I = 1.25$ A $R = 50.0\ \Omega$ $\theta = 0°$
10. $I = 78$ mA $R = 2.7$ kΩ $\theta = 0°$
11. $I = 22.0\ \mu$A $R = 1.80$ MΩ $\theta = 0°$

Determine the circuit current for each of the following.

12. $E = 117$ V $R = 35.0\ \Omega$ $\theta = 0°$
13. $E = 2.50$ kV $R = 6.80$ MΩ $\theta = 0°$
14. $E = 14.7$ V $R = 330\ \Omega$ $\theta = 0°$
15. $E = 82.0$ V $R = 47.0\ \Omega$ $\theta = 0°$

26–3 INDUCTANCE AND INDUCTIVE REACTANCE

Figure 26–5 pictures several inductors (coils) and their schematic symbols. The letter *L* is used to designate an inductor in a schematic. The **inductance** of an inductor is measured in henries (H). The inductance in henries of an inductor depends upon the physical makeup of the coil. From Equation 26–4 we see that the inductance (*L*) is dependent on the length (*l*), cross-sectional area (*A*), number of turns of wire (*N*), and the material contained within the core (*μ*).

$$L = \frac{N^2 \mu A}{l} \qquad H \qquad (26\text{–}4)$$

FIGURE 26–5   Inductors: (a) iron core for audio-frequency (AF) applications; (b) air core for radio-frequency (RF) applications.

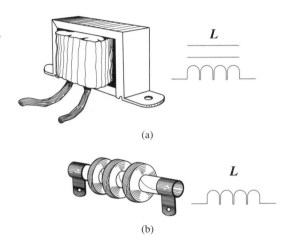

(a)

(b)

When an alternating current passes through an inductor, the inductor "reacts" to the current by producing an opposition to the flow of current. This opposition is the **inductive reactance** of the inductor. The reactance of an inductor (*X_L*) to alternating current is described by Equation 26–5 and is expressed in ohms.

$$X_L = 2\pi f L = \omega L \qquad \Omega \qquad (26\text{–}5)$$

where X_L = inductive reactance in ohms (Ω)
    f = frequency of the current in hertz (Hz) ($\omega = 2\pi f$ rad/s)
    L = inductance in henries (H)

From Equation 26–5 we see that X_L is a direct function of frequency, and that an inductor has zero reactance when $f = 0$ Hz. Thus, direct current ($f = 0$ Hz) will have no opposition owing to the reactive properties of the inductor. Because of the frequency dependence of the inductive reactance, an inductor may be used to *filter out* unwanted frequencies while allowing wanted frequencies to pass. This is demonstrated in the circuit of Figure 26–6.

EXAMPLE 26–7   Determine the reactance of the 100-mH RF *choke coil* of Figure 26–6 to a frequency of:

(a) 0 Hz (dc)
(b) 1.00 MHz (RF)

FIGURE 26–6   The inductor in the dc power-supply circuit blocks the passage of radio-frequency current while allowing the dc and audio-frequency currents to pass.

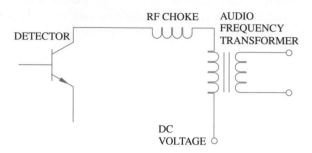

Solution   Use Equation 26–5:

(a) Substitute $f = 0$ Hz and $L = 100$ mH:

∴ $X_L = 2\pi(0)(100 \times 10^{-3}) = 0\ \Omega$

(b) Substitute $f = 1.00$ MHz and $L = 100$ mH:

∴ $X_L = 2\pi(1.00 \times 10^{6})(100 \times 10^{-3}) = 628\ \text{k}\Omega$

The inductor derives its opposition to alternating current from its magnetic properties, and these same properties cause the voltage drop across the inductor to lead the current passing through the inductor by 90°. Thus, $\angle\theta = 90°$. Figure 26–7 shows the relationship of the current and voltage waves of the inductor, as well as the phasor diagram of the current and voltage.

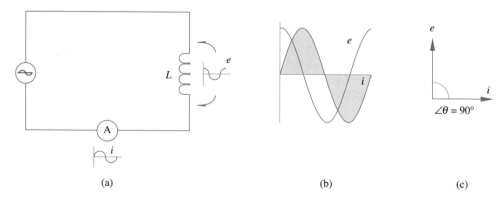

FIGURE 26–7   When ac sinusoidal current passes through an inductor (a), the voltage leads the current (b) by 90°, as shown by the phasor diagram (c).

To summarize, the application of an ac voltage to an inductive load causes an ac current to flow, which causes the inductor to react. This process causes a **phase shift** between the current and voltage of the inductor, resulting in a phase angle (θ) of 90°. Therefore, in a perfect inductor (one without resistance), voltage leads current by 90°. Ohm's law, in the form $E = IX_L$, is used to compute the circuit conditions.

EXAMPLE 26–8   A 2.0-H inductor is connected to a 24-V, 60-Hz ac source.

(a) Determine the circuit current.
(b) Determine the phase angle.
(c) Draw the phasor diagram of the current and voltage phasors.

Solution   (a) $I = E/X_L$; $X_L = 2\pi f L$.
Substitute the second equation into the first:

$$I = E/(2\pi f L)$$
$$I = 24/[2\pi 60(2.0)]$$
$$\therefore \quad I = 32 \text{ mA}$$

(b) Phase angle for a perfectly inductive circuit with current as the reference phasor:

$$\therefore \quad \angle\theta = 90°$$

(c) See Figure 26–8.

FIGURE 26–8   Phasor diagram of the current and voltage for Example 26–8.

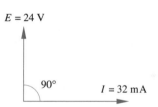

EXAMPLE 26–9   From the phasor diagram of Figure 26–9, determine the inductance of the inductor in henries for a frequency of 2.0 kHz.

FIGURE 26–9   Phasor diagram for Example 26–9.

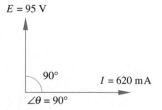

Solution $E = IX_L$
$X_L = E/I$

Substitute $E = 95$ V and $I = 620$ mA:

$X_L = 95/(620 \times 10^{-3}) = 153 \ \Omega$

Solve $X_L = 2\pi fL$ for L:

$L = X_L/2\pi f$

∴ $L = 153/(2\pi 2.0 \times 10^3) = 12$ mH

EXERCISE 26–3

Determine the inductive reactance X_L of each of the following inductors at (a) 60.0 Hz, (b) 5.00 kHz, and (c) 2.00 MHz.

1. 4.00 H	**2.** 10.0 mH	**3.** 250 mH
4. 120 μH	**5.** 12.0 H	**6.** 600 mH

Use the information given in each of the following phasor diagrams to determine (a) the inductive reactance and (b) the inductance of the inductor at a frequency of 1.00 kHz.

7. $E = 100$ V $I = 10$ A   **8.** $E = 35$ V $I = 500$ mA   **9.** $E = 15$ V $I = 6.0$ mA   **10.** $E = 75$ V $I = 150 \ \mu$A

Determine the source voltage for each of the following.

11. $I = 2$ A $X_L = 100 \ \Omega$ $\theta = 90°$
12. $I = 220$ mA $X_L = 1.6$ kΩ $\theta = 90°$

Determine the circuit current for each of the following.

13. $E = 28$ V $L = 2.0$ H $f = 100$ Hz $\theta = 90°$
14. $E = 6.3$ V $L = 570$ mH $f = 2.7$ kHz $\theta = 90°$
15. $E = 132$ V $L = 800 \ \mu$H $f = 3.15$ MHz $\theta = 90°$

26–4 CAPACITANCE AND CAPACITIVE REACTANCE

Figure 26–10 shows several capacitors and their schematic symbols. The letter C is used to designate a capacitor in a circuit diagram. The **capacitance** of a capacitor is measured in farads (F). Capacitance depends upon the construction of the capacitor. From Equation 26–6, the capacitance (C) is dependent on the area of the metallic plates (A), the spacing between the plates (d), and the permittivity of the material used as a dielectric (insulator) between the plates (ϵ).

$$C = \epsilon A/d \quad \text{F} \tag{26–6}$$

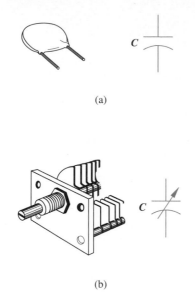

(a)

(b)

FIGURE 26–10 (a) Fixed ceramic disc capacitor; (b) variable capacitor.

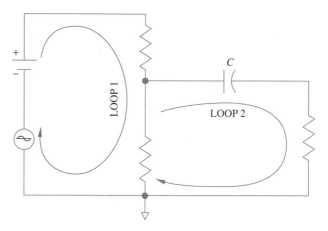

FIGURE 26–11 The current of loop 1 is a mixture of both direct and alternating currents, but the current of loop 2 is only alternating. This is due to the blocking of the direct current by capacitor C.

When an ac voltage is placed across a capacitor, the capacitor "reacts" to the flow of current by producing an opposition to the current. This opposition is the **capacitive reactance** of the capacitor to the alternating current. The reactance of a capacitor (X_C) to alternating current is given by Equation 26–7 and is expressed in ohms.

$$X_C = 1/(2\pi fC) = 1/(\omega C) \qquad \Omega \qquad (26\text{–}7)$$

where X_C = capacitive reactance in ohms (Ω)
$\qquad f$ = frequency of the current in hertz (Hz) ($\omega = 2\pi f$ rad/s)
$\qquad C$ = capacitance in farads (F)

From Equation 26–7 we see that X_C is an inverse function of frequency and that a capacitor has an extremely large reactance when $f = 0$ Hz. Thus, dc current ($f = 0$ Hz) will be blocked due to the reactive properties of the capacitor. Because of the frequency dependence of the capacitive reactance, a capacitor may be used to *filter out* unwanted frequencies while allowing wanted frequencies to pass. This is demonstrated by the circuit of Figure 26–11.

EXAMPLE 26–10 Determine the reactance of the 1.0-μF capacitor of Figure 26–11 to a frequency of

(a) 0 Hz (dc)
(b) 1.0 MHz (RF)

Solution $X_C = 1/(2\pi fC)$.

(a) Substitute $f = 0$ Hz and $C = 1.0 \ \mu F$:

∴ $X_C = \dfrac{1}{2\pi(0)(1.0 \times 10^{-6})} \Rightarrow \infty$ (undefined)

(b) Substitute $f = 1.0$ MHz and $C = 1.0 \ \mu F$:

∴ $X_C = \dfrac{1}{2\pi(1.0 \times 10^6 \times 1.0 \times 10^{-6})} = 0.16 \ \Omega$

The capacitor opposes alternating current because of its inability to take on charge instantly. It is this opposition to the flow of charge that results in the voltage drop across the capacitor lagging the current passing through the capacitor by 90°. Thus, $\angle\theta = -90°$. Figure 26–12 shows the relationship of the current and voltage wave of the capacitor, as well as the phasor diagram of the current and voltage.

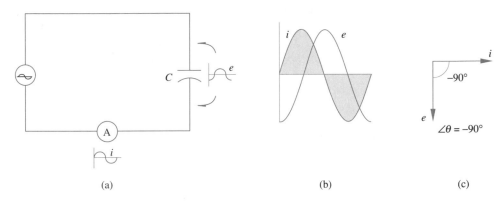

(a)   (b)   (c)

FIGURE 26–12   When alternating sinusoidal current passes through a capacitor (a), the voltage lags the current (b) by 90°, as shown by the phasor diagram (c).

To summarize, the application of an ac voltage to a capacitive load causes the capacitor to react ($X_C = 1/\omega C$). In this process the voltage lags behind the current by 90°, which results in a phase angle (θ) of $-90°$. Ohm's law in the form $E = IX_C$ is used to compute the circuit condition of a perfect capacitive circuit (one without resistance).

EXAMPLE 26–11   A 10-μF capacitor is connected to a 117-V, 60-Hz source.

(a) Determine the circuit current.
(b) Determine the phase angle.
(c) Draw the phasor diagram of the current and voltage phasors.

Solution   (a) $I = E/X_C$, $X_C = 1/(2\pi fC)$.

Substitute the second equation into the first:

$$I = E \bigg/ \left(\frac{1}{2\pi fC}\right)$$

$$I = E2\pi fC$$

$$I = 117(2\pi)(60)(10 \times 10^{-6})$$

$$\therefore \qquad I = 440 \text{ mA}$$

(b) Phase angle for a perfectly capacitive circuit with current as the reference phasor:

$$\therefore \qquad \angle\theta = -90°$$

(c) See Figure 26–13.

FIGURE 26–13   Phasor diagram of the current and voltage for Example 26–11. $\angle\theta = -90°$.

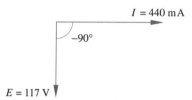

$I = 440$ mA

$-90°$

$E = 117$ V

EXAMPLE 26–12   From the phasor diagram of Figure 26–14, determine the capacitance of the capacitor in farads for a frequency of 100 Hz.

FIGURE 26–14   Phasor diagram for Example 26–12.

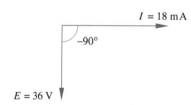

$I = 18$ mA

$-90°$

$E = 36$ V

Solution $E = IX_C$
$X_C = E/I$

Substitute $E = 36$ V and $I = 18$ mA:

$$X_C = 36/(18 \times 10^{-3}) = 2.0 \text{ k}\Omega$$

Solve $X_C = 1/(2\pi fC)$ for C:

$$C = 1/(2\pi fX_C)$$

$$C = \frac{1}{(2\pi \times 100)(2 \times 10^3)}$$

$$\therefore \qquad C = 0.80 \text{ }\mu\text{F}$$

Determine the capacitive reactance X_C of each of the following capacitors at (a) 60.0 Hz, (b) 5.00 kHz, and (c) 2.00 MHz.

1. 15 μF   **2.** 270 pF   **3.** 0.50 μF

4. 3.0 μF   **5.** 1000 μF   **6.** 82 pF

Use the information given in each of the following phasor diagrams to determine the following: (a) the capacitive reactance and (b) the capacitance of the capacitor at a frequency of 500 Hz.

7. I = 620 mA E = 20.0 V   **8.** I = 500 μA E = 2.00 V

9. I = 4.0 A E = 15 V   **10.** I = 12 mA E = 60 V

Determine the source voltage for each of the following.

11. I = 130 mA X_C = 42.5 Ω $\theta = -90°$
12. I = 4.30 A X_C = 10.8 Ω $\theta = -90°$

Determine the circuit current for each of the following.

13. E = 523 V C = 10 μF f = 15 Hz $\theta = -90°$
14. E = 27 V C = 0.10 μF f = 400 Hz $\theta = -90°$
15. E = 440 V C = 720 pF f = 900 kHz $\theta = -90°$

26–5   VOLTAGE PHASOR FOR SERIES CIRCUITS

In the previous sections we learned that the passage of alternating current through a resistance, an inductance, or a capacitance resulted in a phase angle that was different for each of the devices. In this section, we are concerned with the result of passing alternating current through a series circuit made up of a resistance, a capacitance, and an inductance. Table 26–1 summarizes the concepts of the previous three sections of this chapter.

Current in an AC Series Circuit

In an ac series circuit, the current (I) is the same throughout the circuit. Therefore, current is selected as the reference phasor in ac series circuits. A phasor diagram showing the voltage drops in relation to the circuit may be constructed. The following example develops this idea.

TABLE 26–1   Summary of the Effects of Alternating Current on the Components of a Series Circuit

Property				Opposition			
Name	**Symbol**	**Phase (θ)**	**EI Phasor**	**Name**	**Unit**	**Dependence on Frequency**	**Ohm's Law**
Resistance	R	E and I are in phase: $\theta = 0°$		Resistance	Ohm	None	$E = IR$
Inductance	L	E leads I by 90°: $\theta = 90°$		Inductive reactance	Ohm	$X_L = 2\pi f L$	$E = IX_L$
Capacitance	C	E lags I by 90°: $\theta = -90°$		Capacitive reactance	Ohm	$X_C = \dfrac{1}{2\pi f C}$	$E = IX_C$

EXAMPLE 26–13   Construct the phasor diagram for the circuit of Figure 26–15.

FIGURE 26–15   Series circuit for Example 26–13.

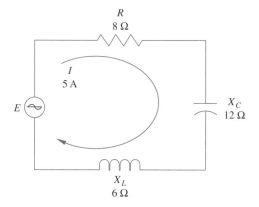

Solution   Determine the voltage drops:

$$|V_R| = IR = 5(8) = 40 \text{ V}$$
$$|V_L| = IX_L = 5(6) = 30 \text{ V}$$
$$|V_C| = IX_C = 5(12) = 60 \text{ V}$$

Use the information of Table 26–1 to determine the phase angle between V and I. Select I as the reference phasor ($0°$).

V_R in phase ($0°$) with I
V_L leads I by $90°$
V_C lags I by $90°$

Construct the circuit phasor diagram as in Figure 26–16.

FIGURE 26–16 Phasor diagram for Example 26–13. Voltage leads current in the inductor, while voltage lags current in the capacitor.

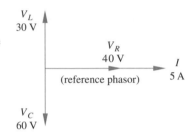

Voltage Phasor

From the preceding example we see that in a series circuit the voltage phasor is constructed so that V_R is along the positive R-axis, V_L along the positive j-axis, and V_C along the negative j-axis. Figure 26–17 pictures the ***voltage phasor diagram*** for a series ac circuit with resistive, capacitive, and inductive components.

FIGURE 26–17 Generalized voltage phasor diagram for a series LCR circuit.

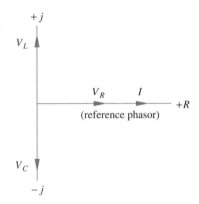

Kirchhoff's Voltage Law (KVL)

Kirchhoff's voltage law (KVL), $E = V_1 + V_2 + V_3$, etc., is valid for ac series circuits. However, it must be remembered that ac quantities are represented by complex numbers and must be operated on with phasor addition. The following example demonstrates this concept.

EXAMPLE 26–14   Verify Kirchhoff's voltage law for the circuit of Figure 26–18.

FIGURE 26–18   Circuit for
Example 26–14.

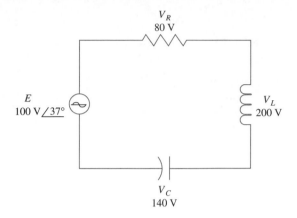

Solution   Construct the voltage phasor diagram as shown in Figure 26–19.

FIGURE 26–19   Voltage phasor diagram for the circuit of Figure 26–18.

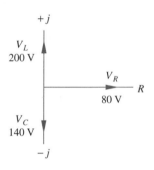

Express each voltage phasor as a complex number in rectangular form:

$$V_R = 80 + j0 \text{ V}$$
$$V_L = 0 + j200 \text{ V}$$
$$V_C = 0 - j140 \text{ V}$$

Add the voltage phasors and express the sum in polar form:

$$
\begin{array}{l}
80 + j0 \\
 0 + j200 \\
\underline{ 0 - j140} \\
80 + j60 \quad \boxed{\rightarrow \text{P}} \ 100\underline{/37°} \text{ V}
\end{array}
$$

∴ $V_R + V_L + V_C = 100\underline{/37°} \text{ V}$

Observation   Kirchhoff's voltage law holds for the circuit of Figure 26–18.
Figure 26–20 shows the sequence to the solution of Example
26–14 through phasor diagrams.

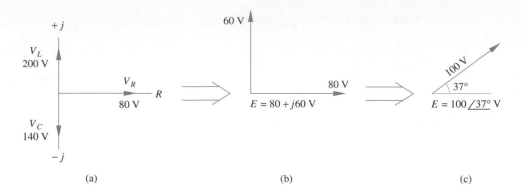

(a)　　　　　　　　　　　　　(b)　　　　　　　　　　　　　(c)

FIGURE 26–20 Solution to Example 26–14: (a) voltage phasor diagram for the voltage drops in the series circuit of Figure 26–18; (b) phasor addition of the voltages; (c) resolution of the rectangular voltage phasors of (b) into a polar phasor.

EXAMPLE 26–15 A 500-Ω wire-wound resistor is connected in series with a 0.5-μF capacitor and a 1.0-H inductor. The circuit is attached to a *test oscillator* operating at 300 Hz. The alternating current is measured as 180 mA.

(a) Compute the voltage drop across each component in the circuit.
(b) Construct the voltage phasor diagram.
(c) Determine the source voltage E.

Solution Draw and label a schematic of the circuit as in Figure 26–21.

(a) Compute the voltage drops:

$$|V_R| = IR = 180 \times 10^{-3} \times 500 = 90 \text{ V}$$
$$|V_L| = IX_L = I(2\pi fL)$$
$$|V_L| = 180 \times 10^{-3} \times 2\pi \times 300 \times 1.0 = 340 \text{ V}$$
$$|V_C| = IX_C = I/(2\pi fC) = \frac{180 \times 10^{-3}}{2\pi \times 300 \times 0.5 \times 10^{-6}}$$
$$|V_C| = 190 \text{ V}$$

(b) Draw the voltage phasor diagram as in Figure 26–22(a).
(c) Determine the source voltage by first expressing each voltage phasor as a complex number in rectangular form and then adding:

$$
\begin{aligned}
V_R &= 90 + j0 \\
V_L &= 0 + j340 \\
V_C &= \underline{0 - j190} \\
E &= 90 + j150 \text{ V}
\end{aligned}
$$

FIGURE 26–21 Circuit for Example 26–15.

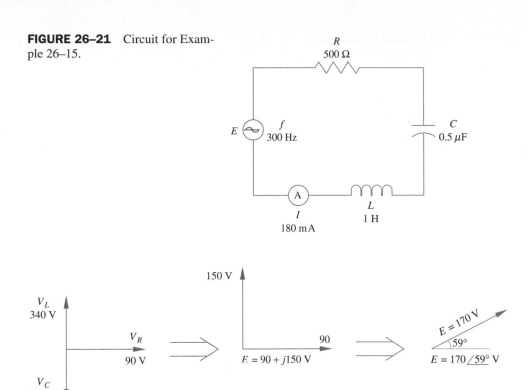

FIGURE 26–22 Voltage phasor diagram for the circuit of Figure 26–21.

Convert the source voltage to polar form:

$\therefore$ $E = 90 + j150 \text{ V} \boxed{\rightarrow P} 170\underline{/59°} \text{ V}$

Observation   Figure 26–22 shows the sequence to the solution for the source voltage through phasor diagrams.

EXAMPLE 26–16   The current passing through a series circuit consisting of a 1.0-μF capacitor and a 700-Ω resistance is $40\underline{/0°}$ mA. If the circuit is powered by a 60-Hz ac generator, first determine:

(a) The voltage across the capacitor
(b) The voltage across the resistor
(c) The source voltage

Then construct:

(d) A voltage phasor diagram
(e) A phasor diagram showing the phase angle between the source voltage and current

FIGURE 26–23 Schematic for Example 26–16.

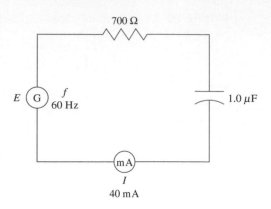

Solution (a) Draw and label a schematic of the circuit as in Figure 26–23 and determine V_C:

$$|V_C| = IX_C = I/(\omega C)$$
$$|V_C| = 40 \times 10^{-3}/[(2\pi60)(1.0 \times 10^{-6})]$$
$$\therefore \quad |V_C| = 106 \text{ V}$$

(b) Determine V_R:

$$\therefore \quad V_R = IR = 40 \times 10^{-3} \times 700 = 28 \text{ V}$$

(c) Determine the source voltage (E) by adding the voltage drops:

$$E = V_R + V_C$$
$$E = (28 + j0) + (0 - j106)$$
$$\therefore \quad E = 28 - j106 \boxed{\rightarrow\text{P}} \ 110\underline{/-75°} \text{ V}$$

(d) Draw the voltage phasor diagram as in Figure 26–24.
(e) Draw the phasor diagram showing the phase angle between source voltage and current as in Figure 26–25.

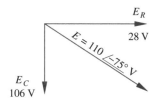

FIGURE 26–24 Voltage phasor diagram for the circuit of Figure 26–23.

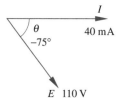

FIGURE 26–25 Phase angle of the current and voltage of the circuit of Figure 26–23.

588

Add each of the following voltage drops and express the sum as a phasor in both rectangular and polar form. We recommend that you construct a voltage phasor diagram to aid your solution.

1. $V_R = 50$ V $\qquad$ $V_L = 50$ V
2. $V_C = 20$ V $\qquad$ $V_R = 10$ V
3. $V_R = 5.0$ V $\qquad$ $V_L = 2.0$ V
4. $V_C = 18$ V $\qquad$ $V_R = 5.0$ V
5. $V_L = 44$ V $\qquad$ $V_C = 14$ V
6. $V_R = 26$ V $\qquad$ $V_C = 9.0$ V $\qquad$ $V_R = 10$ V
7. $V_L = 60$ V $\qquad$ $V_C = 52$ V $\qquad$ $V_L = 12$ V
8. $V_R = 7.0$ V $\qquad$ $V_C = 16$ V $\qquad$ $V_L = 9.0$ V
9. $V_C = 38$ V $\qquad$ $V_R = 75$ V $\qquad$ $V_L = 54$ V
10. $V_L = 81$ V $\qquad$ $V_C = 129$ V $\qquad$ $V_R = 30$ V

Express each of the following voltages in rectangular form as $V_R + jV_L$ or $V_R - jV_C$.

11. $17\underline{/80°}$ V $\qquad\qquad$ 12. $36\underline{/25°}$ V
13. $8.0\underline{/-30°}$ V $\qquad\qquad$ 14. $15\underline{/-72°}$ V
15. $64\underline{/-275°}$ V $\qquad\qquad$ 16. $23\underline{/322°}$ V
17. $29\underline{/288°}$ V $\qquad\qquad$ 18. $40\underline{/-313°}$ V
19. $55.0\underline{/-420°}$ V $\qquad\qquad$ 20. $11\underline{/447°}$ V

For each of the following, construct the voltage phasor and compute the source voltage in rectangular and polar form.

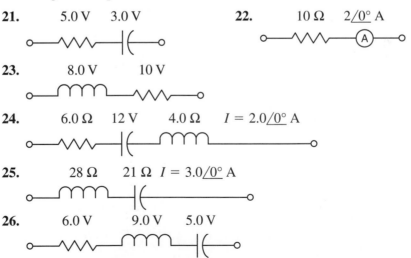

21. $\quad$ 5.0 V $\quad$ 3.0 V $\qquad\qquad$ 22. $\qquad$ 10 Ω $\quad$ $2\underline{/0°}$ A

23. $\qquad$ 8.0 V $\qquad$ 10 V

24. $\quad$ 6.0 Ω $\quad$ 12 V $\quad$ 4.0 Ω $\quad$ $I = 2.0\underline{/0°}$ A

25. $\qquad$ 28 Ω $\quad$ 21 Ω $\;$ $I = 3.0\underline{/0°}$ A

26. $\quad$ 6.0 V $\quad$ 9.0 V $\quad$ 5.0 V

In each of the following, draw a schematic of the circuit and determine (a) the voltage drop across each circuit component and (b) the source voltage E in polar form; then draw (c) the voltage phasor diagram and (d) the phasor diagram of the source voltage and circuit current. Label the phase angle.

27. The 200-Hz current passing through a series circuit consisting of 5.0-kΩ resistance and a 3.0-H inductor is $200\underline{/0°}$ mA.

28. The 8.0-kHz current passing through a series circuit consisting of a 100-μH inductor, a 2.0-μF capacitor, and a 10-Ω resistance is $1.5\underline{/0°}$ A.

29. The 60-Hz current passing through a series circuit consisting of a 10-H inductor and a 1.0-μF capacitor is $90\underline{/0°}$ mA.

30. The 400-Hz current passing through a series circuit consisting of a 75-mH inductor, a 15-μF capacitor, and a 33-Ω resistance is $2.2\underline{/0°}$ A.

26–6 CURRENT PHASOR FOR PARALLEL CIRCUITS

In this section we will investigate the result of passing alternating current through a parallel circuit having branches made up of a resistance, a capacitance, and an inductance. As we have learned, current and voltage are in phase through a resistance, but voltage leads current in an inductor and voltage lags current in a capacitor.

When working with parallel circuits, the voltage is the reference phasor because it is the same across all parts of the parallel circuit. Thus, the current is:

- In phase with voltage across a resistance: $\theta = 0°$
- Leading the voltage across a capacitance: $\theta = 90°$
- Lagging the voltage across an inductance: $\theta = -90°$

These concepts are summarized in Figure 26–26, which shows the ***current phasor diagram*** for a parallel ac circuit with resistive, capacitive, and inductive components.

FIGURE 26–26 Generalized current phasor for a parallel LCR circuit when $E\angle0°$ V is the reference phasor.

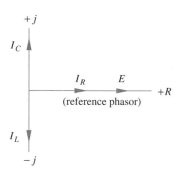

Kirchhoff's Current Law (KCL)

Kirchhoff's current law (KCL), $I = I_1 + I_2 + I_3$, etc., is valid for ac parallel circuits. However, ac quantities are phasor quantities represented by complex numbers and, as such, must be operated on with phasor addition. The following example demonstrates this concept.

EXAMPLE 26–17   Verify Kirchhoff's current law for the circuit of Figure 26–27 $(I = I_R + I_L + I_C)$.

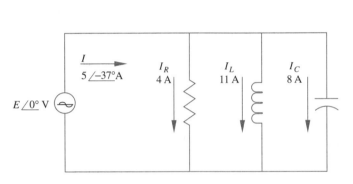

FIGURE 26–27   Circuit for Example 26–17.

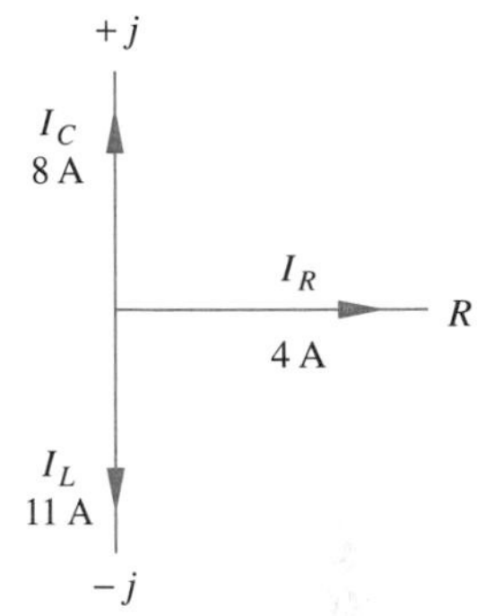

FIGURE 26–28   Current phasor diagram for the circuit of Figure 26–27.

Solution   Construct the current phasor diagram as shown in Figure 26–28. Express each current phasor as a complex number in rectangular form:

$$I_R = 4 + j0 \text{ A}$$
$$I_C = 0 + j8 \text{ A}$$
$$I_L = 0 - j11 \text{ A}$$

Add the current phasors and express the sum in polar form:

$$\begin{array}{c} 4 + j0 \\ 0 + j8 \\ \underline{0 - j11} \\ 4 - j3 \quad \boxed{\rightarrow P} \quad 5\angle{-37°} \text{ A} \end{array}$$

∴   The sum of the branch currents $5\angle{-37°}$ A is equal to the source current $5\angle{-37°}$ A. This verifies Kirchhoff's current law.

Figure 26–29 shows the flow of the solution of Example 26–17. Notice the similarity in the sequence of the solution to that of the series circuit of the previous section.

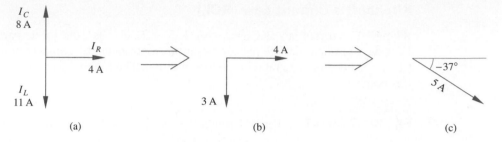

(a)   (b)   (c)

FIGURE 26–29   (a) Current phasor diagram for the current in the branches of the parallel circuit of Figure 26–27; (b) phasor addition of the branch currents; (c) resolution of the rectangular current phasors of (b) into a polar phasor.

EXAMPLE 26–18   A 1.00-kΩ wire-wound resistor is connected in parallel with a 1.00-μF capacitor and an 800-mH inductor. The circuit is attached to a 220-V, 400-Hz alternator.

(a) Compute the current flow through each component in the circuit.
(b) Draw the current phasor diagram.
(c) Determine the source current I.

Solution   Draw and label a schematic of the circuit as in Figure 26–30.

FIGURE 26–30   Circuit for Example 26–18.

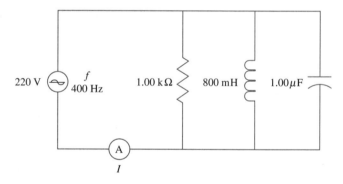

(a) Compute the branch currents:

$$\therefore \quad |I_R| = E/R = 220/(1.00 \times 10^3) = 220 \text{ mA}$$
$$|I_C| = E/X_C = E(2\pi fC) = 220(2\pi 400 \times 1.00 \times 10^{-6})$$
$$\therefore \quad |I_C| = 553 \text{ mA}$$
$$|I_L| = E/X_L = E/(2\pi fL)$$
$$\therefore \quad |I_L| = 220/(2\pi 400 \times 800 \times 10^{-3}) = 109 \text{ mA}$$

(b) Draw the branch-current phasor diagram as in Figure 26–31.

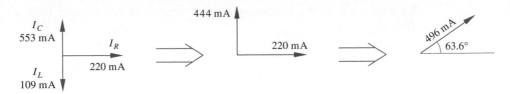

FIGURE 26–31 Current phasor diagram for the circuit of Figure 26–30.

(c) Determine the source current I by first expressing each branch current phasor as a complex number in rectangular form, and then adding:

$$I_R = 0.220 + j0$$
$$I_C = 0 \qquad + j0.553$$
$$\underline{I_L = 0 \qquad - j0.109}$$
$$I = 0.220 + j0.444 \text{ A}$$

Convert the source current to polar form:

$$\therefore \quad I = 0.220 + j0.444 = 496\underline{/64°} \text{ mA}$$

Observation   The rectangular form of the source current may be expressed as $0.220 + j0.444$ A, or $220 + j444$ mA. Whenever unit prefixes are used with phasor quantities, only one prefix is used, and it applies to both parts of the phasor.

EXERCISE 26–6

Add each of the following branch currents, and express the sum as a phasor in both rectangular and polar form. We recommend that you construct a current phasor diagram to aid your solution.

1. $I_R = 2.0$ A	$I_L = 5.0$ A	
2. $I_C = 7.0$ A	$I_R = 3.0$ A	
3. $I_R = 0.300$ A	$I_L = 1.2$ A	
4. $I_C = 700$ μA	$I_R = 0.52$ mA	
5. $I_L = 3.4$ A	$I_C = 1.8$ A	
6. $I_R = 330$ mA	$I_C = 180$ mA	$I_R = 75$ mA
7. $I_L = 8.2$ A	$I_C = 4.4$ A	$I_L = 2.2$ A
8. $I_R = 13$ mA	$I_C = 63$ mA	$I_L = 43$ mA
9. $I_C = 0.732$ A	$I_R = 510$ mA	$I_L = 0.270$ A
10. $I_L = 42$ mA	$I_C = 87$ mA	$I_R = 70$ mA

Express each of the following source currents in rectangular form as $I_R - jI_L$ or $I_R + jI_C$.

11. $12\underline{/72°}$ mA

12. $3.1\underline{/33°}$ A

13. $413\underline{/-19°}$ mA

14. $16\underline{/-63°}$ A

15. $67\underline{/-305°}$ μA

16. $5.4\underline{/282°}$ mA

17. $1.60\underline{/4.5°}$ A

18. $51\underline{/-328°}$ μA

19. $2.7\underline{/443°}$ A

20. $18\underline{/-395°}$ mA

For each of the following, construct the current phasor diagram and compute the source current in rectangular and polar form.

21.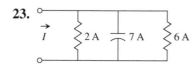

22.

23.

24.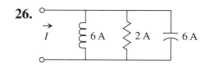

25.

26.

In each of the following, draw a schematic of the circuit and:

   (a) Determine the current through each branch component.

   (b) Determine the source current I in polar form.

   (c) Draw the branch-current phasor diagram.

   (d) Draw the phasor diagram of the source voltage and circuit current, and label the phase angle.

27. The 180-Hz voltage source applied across a parallel circuit consisting of a 4.7-kΩ resistance and a 2.7-H inductor is $100\underline{/0°}$ V.

28. The 10-kHz voltage source applied across a three-branch parallel circuit consisting of a 120-μH inductor, a 2.2-μF capacitor, and a 12-Ω resistance is $10\underline{/0°}$ V.

29. The 60-Hz voltage source applied across a parallel circuit consisting of an 8.0-H inductor and a 2.0-μF capacitor is $160\underline{/0°}$ V.

30. The 800-Hz voltage source applied across a three-branch parallel circuit consisting of a 5.0-mH inductor, a 30-μF capacitor, and a 27-Ω resistance is $12\underline{/0°}$ V.

SELECTED TERMS

angular velocity   The angular rate of travel of a rotating vector expressed in radians per second.

capacitive reactance   The opposition to alternating current by a capacitor; measured in ohms. Its symbol is X_C.

frequency   The number of complete cycles of alternating current in one second; measured in hertz (Hz).

inductive reactance   The opposition to alternating current by an inductor; measured in units of ohms. Its symbol is X_L.

period   The reciprocal of frequency.

phase angle   The angle between the current and voltage phasor in a phasor diagram.

27

Alternating-Current Circuits

27–1   Impedance of Series AC Circuits

27–2   Solving Series AC Circuits

27–3   Admittance Concepts

27–4   Admittance of Parallel AC Circuits

PERFORMANCE OBJECTIVES

- Compute impedance for an ac series circuit.
- Determine the values of each component in an ac series equivalent circuit.
- Solve ac series circuits.
- Compute admittance for an ac parallel circuit.
- Determine the values of each component in an ac parallel equivalent circuit.
- Solve ac parallel circuits.

The 330-kV switching station at the Kariba Hydroelectric works in Zimbabwe, Africa.   (Courtesy of British Information Service)

In the previous chapters you were introduced to the rules for series and parallel circuits. This chapter expands upon these rules to introduce you to *impedance* for series circuits and *admittance* for parallel circuits.

27–1 IMPEDANCE OF SERIES AC CIRCUITS

When alternating current passes through a series *RLC* (resistance, inductance, capacitance) circuit, the intensity of the current is determined by the total opposition offered by the resistance of the resistor and the reactance of the inductor and capacitor. The opposition to the passage of alternating current through a circuit is called **impedance.** Thus, the word impedance may be used to describe the opposition offered by an ac circuit containing only resistance or only capacitance or only inductance. It may also be used to describe ac circuits containing any combination of resistance, capacitance, and inductance. We see, then, that the word impedance is a general label that is used to describe the opposition to the flow of alternating current by any combination of circuit components.

Computing Impedance

Impedance is a *vector quantity* that may be written as a complex number in either the rectangular or polar form. The letter Z is used to represent impedance. Thus, $Z = 5.00\underline{/53.1°}\ \Omega = 3.00 + j4.00\ \Omega$ shows an impedance written in both the polar and rectangular form. Ohm's law may be stated in terms of impedance as Equation 27–1.

$$Z = E/I \qquad \Omega \tag{27–1}$$

EXAMPLE 27–1	Determine the impedance of a series circuit having a current of $2\underline{/0°}$ A and a source voltage of $100\underline{/30°}$ V.
Solution	Substitute $I = 2.0\underline{/0°}$ A and $E = 100\underline{/30°}$ V:

$$Z = E/I$$
$$Z = \frac{100\underline{/30°}}{2.0\underline{/0°}}$$
$$\therefore \qquad Z = 50\underline{/30°}\ \Omega$$

EXAMPLE 27–2	Determine the impedance of a series circuit having a current of $5.0\underline{/0°}$ A and a source voltage of $25\underline{/-20°}$ V.
Solution	Substitute $I = 5.0\underline{/0°}$ A and $E = 25\underline{/-20°}$ V:

$$Z = E/I$$
$$Z = \frac{25\underline{/-20°}}{5.0\underline{/0°}}$$
$$\therefore \qquad Z = 5.0\underline{/-20°}\ \Omega$$

EXAMPLE 27–3   Determine the impedance of a series circuit having a current of $6.0\underline{/0°}$ A and a source voltage of $36\underline{/0°}$ V.

Solution   Substitute $I = 6.0\underline{/0°}$ A and $E = 36\underline{/0°}$ V:

$$Z = E/I$$

$$Z = \frac{36\underline{/0°}}{6.0\underline{/0°}}$$

$$\therefore \quad Z = 6.0\underline{/0°}\ \Omega$$

EXAMPLE 27–4   Determine the impedance of a series circuit having a current of $5.0\underline{/0°}$ A and a source voltage of $15\underline{/90°}$ V.

Solution   Substitute $I = 5.0\underline{/0°}$ A and $E = 15\underline{/90°}$ V.

$$Z = E/I$$

$$Z = \frac{15\underline{/90°}}{5.0\underline{/0°}}$$

$$\therefore \quad Z = 3.0\underline{/90°}\ \Omega$$

EXAMPLE 27–5   Determine the impedance of a series circuit having a current of $4.0\underline{/0°}$ A and a source voltage of $28\underline{/-90°}$ V.

Solution   Substitute $I = 4.0\underline{/0°}$ A and $E = 28\underline{/-90°}$ V:

$$Z = E/I$$

$$Z = \frac{28\underline{/-90°}}{4.0\underline{/0°}}$$

$$\therefore \quad Z = 7.0\underline{/-90°}\ \Omega$$

Each of the five preceding examples used circuit conditions that were different. By relating the phase angle of the applied voltage to the circuit current, we are able to determine the type of components in each of the **series equivalent circuits.** This is demonstrated by the following example. Remember that inductive series circuits have leading phase angles and capacitive series circuits have lagging phase angles.

EXAMPLE 27–6   Determine the circuit components used in each of the preceding five series circuit examples.

Solution   List the impedance $z\underline{/\theta}$ of each circuit, and state the phase relation of the source voltage with respect to the circuit current:

(1) $50\underline{/30°}\ \Omega$ E leads I by $30°$

(2) $5.0\underline{/-20°}\ \Omega$ E lags I by $20°$

(3) $6.0\underline{/0°}\ \Omega$ *E* and *I* are in phase
(4) $3.0\underline{/90°}\ \Omega$ *E* leads *I* by 90°
(5) $7.0\underline{/-90°}\ \Omega$ *E* lags *I* by 90°

∴   Use the phase angle to identify the type of circuit component(s) in the series equivalent circuits.

(1) $\theta = 30°$ resistive and inductive
(2) $\theta = -20°$ resistive and capacitive
(3) $\theta = 0°$ resistive
(4) $\theta = 90°$ inductive
(5) $\theta = -90°$ capacitive

Table 27–1 summarizes the first five examples. Notice that in a *series circuit* a positive phase angle indicates an inductive circuit, whereas a negative phase angle indicates a capacitive circuit.

TABLE 27–1   Summary of Examples 27–1 Through 27–5

Example	$Z\underline{/\theta}$ Ω	Phase Angle θ	Equivalent Circuit Components	*EI* Phasor Diagram
27–1	$50\underline{/30°}$	30°	43 Ω 25 Ω	
27–2	$5.0\underline{/-20°}$	−20°	4.7 Ω 1.7 Ω	
27–3	$6.0\underline{/0°}$	0°	6.0 Ω	
27–4	$3.0\underline{/90°}$	90°	3.0 Ω	
27–5	$7.0\underline{/-90°}$	−90°	7.0 Ω	

Impedance Diagram

In an ac series circuit consisting of a resistor, an inductor, and a capacitor, the following are true statements:

$$(1) \quad I = I_R = I_L = I_C$$
$$(2) \quad E = V_R + j|V_L| - j|V_C|$$

Dividing I of the first equation into each member of the second equation results in

$$\frac{E}{I} = \frac{V_R}{I} + \frac{j|V_L|}{I} - \frac{j|V_C|}{I} \Longrightarrow$$

$$Z = R + jX_L - jX_C \quad \Omega \tag{27–2}$$

When the vector quantities of the right member of Equation 27–2 are graphed, an impedance diagram for the series circuit is formed, as shown in Figure 27–1. It is suggested that the following guidelines be used to aid in the construction of an impedance diagram.

FIGURE 27–1 An impedance diagram has the resistance plotted horizontally, the inductive reactance plotted vertically up, and the capacitive reactance plotted vertically down.

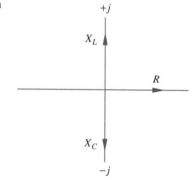

GUIDELINES 27–1. CONSTRUCTING AN IMPEDANCE DIAGRAM

When plotting resistance, inductive reactance, and capacitive reactance on a complex plane:

1. Plot resistance along the positive real axis.
2. Plot inductive reactance along the positive imaginary ($+j$) axis.
3. Plot capacitive reactance along the negative imaginary ($-j$) axis.

EXAMPLE 27–7   Use the information of Table 27–1 to construct an impedance diagram for the circuit of Example 27–2.

Solution   From Table 27–1:

$$R = 4.7\ \Omega, \qquad X_C = 1.7\ \Omega, \qquad Z = 5.0\underline{/-20°}\ \Omega$$

Construct the impedance diagram as shown in Figure 27–2.

FIGURE 27–2 Impedance diagram for Example 27–7. $Z = 4.7 - j1.7 = 5.0\underline{/-20°}\ \Omega$.

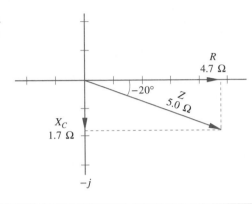

Impedance Forms

The impedance of an ac circuit is written in either the rectangular $(R + jX)$ or the polar $(Z\underline{/\theta})$ form of the complex number. The rectangular form is formatted so that the resistance is written first and then the reactance. Thus, $R + jX$ is the general rectangular form of impedance.

EXAMPLE 27–8   Express each of the following series circuit conditions as (a) an impedance in rectangular form, (b) an impedance in polar form, and (c) an impedance diagram.

(1) Resistance of 6.0 Ω and inductive reactance of 8.0 Ω
(2) Resistance of 20 Ω and capacitive reactance of 15 Ω
(3) Resistance of 6.0 Ω and inductive reactance of 20 Ω

Solution   (a) Fit each condition into the general form of the rectangular impedance, $R + jX$.

(1) $R = 6.0\ \Omega, X_L = 8.0\ \Omega$
  $Z = 6.0 + j8.0\ \Omega$

(2) $R = 20\ \Omega, X_C = 15$
  $Z = 20 - j15\ \Omega$

(3) $R = 6.0\ \Omega, X_L = 20\ \Omega$
  $Z = 6.0 + j20\ \Omega$

(b) Convert each rectangular form to polar form.

(1) $6.0 + j8.0$ →P $10\underline{/53°}\ \Omega$
(2) $20 - j15$ →P $25\underline{/-37°}\ \Omega$
(3) $6.0 + j20$ →P $21\underline{/73°}\ \Omega$

(c) Figure 27–3 is the impedance diagram for each condition.

FIGURE 27–3 Impedance diagram for Example 27–8.

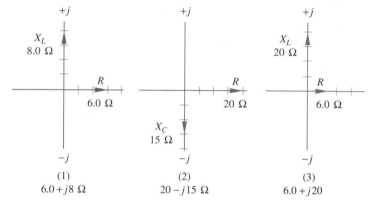

Series Equivalent Circuit

An ac series circuit may have any number of circuit components connected together to form the circuit. The circuit may, however, be expressed at a particular frequency in a simplified form as an equivalent circuit with one resistive component and one reactive component.

The value of each component of the ac *series equivalent circuit* is determined by applying phasor mathematics to the circuit, as demonstrated in the following example.

EXAMPLE 27–9 The following components are connected in series to form the series circuit shown in Figure 27–4(a):

(1) A resistance of 820 Ω
(2) A capacitor having 100 Ω of reactance
(3) An inductor having 200 Ω of reactance
(4) A capacitor having 600 Ω of reactance

For the circuit:

(a) Determine the total impedance in both rectangular and polar form.
(b) Determine the components of the equivalent series circuit when the operating frequency is 5.0 kHz.
(c) Draw the impedance diagram for both the series circuit and its equivalent.

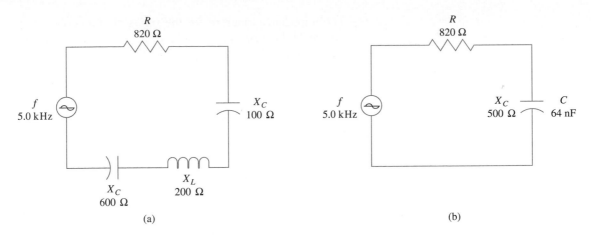

FIGURE 27–4 Schematic of the series circuit for Example 27–9: (a) series ac circuit; (b) equivalent series ac circuit.

Solution   (a) First express each circuit component as an impedance in rectangular form $(R + jX)$, and then add to get the total circuit impedance.

(1) 820-Ω resistance	$820 + j0$
(2) 100-Ω capacitive reactance	$0 - j100$
(3) 200-Ω inductive reactance	$0 + j200$
(4) 600-Ω capacitive reactance	$0 - j600$
	$Z_T = \overline{820 - j500}\ \Omega$

$\therefore \qquad Z_T = 820 - j500\ \Omega\ \boxed{\rightarrow P}\ 960\underline{/-31°}\ \Omega$

(b) The components of the series equivalent circuit are expressed as the rectangular form of the impedance $(R - jX_C)$. Thus, $820 - j500\ \Omega \Rightarrow R = 820\ \Omega$ and $X_C = 500\ \Omega$.

Solving for C when $f = 5.0$ kHz:

$$X_C = 1/(\omega C) \text{ where } \omega = 2\pi f$$
$$C = 1/(\omega X_C)$$
$$C = \frac{1}{2\pi \times 5.0 \times 10^3 \times 500}$$
$$C = 64 \text{ nF}$$

$\therefore$ The equivalent series circuit components for a frequency of 5.0 kHz are a resistor of 820 Ω and a capacitor of 64 nF. Figure 27–4(b) shows the schematic of the equivalent series circuit.

∴ **(c)** The impedance diagram for the series circuit is shown as Figure 27–5(a), and the impedance diagram for the equivalent circuit is shown as Figure 27–5(b).

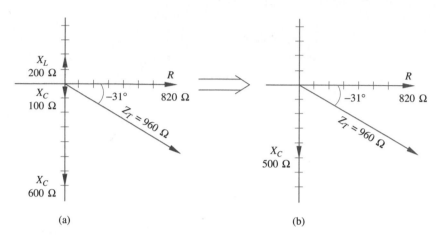

(a)   (b)

FIGURE 27–5 Impedance diagrams: (a) series circuit for Example 27–9; (b) equivalent series circuit for Example 27–9.

The preceding example demonstrates that the rectangular form of the circuit impedance represents the components of the series equivalent circuit. If the circuit impedance is in polar form and the equivalent series circuit is wanted, the polar impedance is converted to its rectangular form.

EXAMPLE 27–10   The circuit impedance of a series circuit is $2.6\underline{/74°}$ kΩ. Determine the value of the components of the equivalent series circuit for a frequency of 400 Hz.

Solution   Convert from polar form to rectangular form:

$$Z = 2.6\underline{/74°}\ \text{k}\Omega\ \boxed{\rightarrow\text{R}}\ 0.72 + j2.5\ \text{k}\Omega$$

Use the rectangular form of Z to find the circuit components:

$$0.72 + j2.5\ \text{k}\Omega \Rightarrow R = 720\ \Omega \text{ and } X_L = 2.5\ \text{k}\Omega$$

Solve for L when $f = 400$ Hz:

$$X_L = \omega L \text{ where } \omega = 2\pi f$$

$$L = X_L/\omega = \frac{2.5 \times 10^3}{2\pi 400} = 0.99\ \text{H}$$

∴   The equivalent series circuit is a resistor of 720 Ω and an inductor of 0.99 H.

604

EXERCISE 27–1

Compute the circuit impedance given the following conditions.

1. $E = 100\underline{/20°}$ V $I = 2.5\underline{/0°}$ A
2. $E = 12\underline{/-90°}$ V $I = 40\underline{/0°}$ mA
3. $E = 18\underline{/0°}$ V $I = 320\underline{/0°}$ mA
4. $E = 320\underline{/33°}$ V $I = 9.4\underline{/0°}$ mA
5. $E = 74\underline{/-8°}$ V $I = 4.3\underline{/0°}$ A
6. $E = 7.5\underline{/-62°}$ V $I = 815\underline{/0°}$ μA

Given the phase angle (θ) between the voltage and the circuit current, determine whether each of the following circuits is resistive, inductive, capacitive, resistive and inductive, or resistive and capacitive. Assume current to be the reference phasor.

7. $\theta = 0°$ 8. $\theta = 90°$ 9. $\theta = 45°$
10. $\theta = -90°$ 11. $\theta = -18°$ 12. $\theta = 60°$

Given the following circuit conditions, express each as an impedance in both the rectangular and polar form of impedance, and construct an impedance diagram for each.

13. A resistance of 10 Ω
14. An inductive reactance of 20 Ω and a resistance of 20 Ω
15. A resistance of 5 Ω and a capacitive reactance of 2 Ω
16. A resistance of 12 Ω and a reactance of $8\underline{/-90°}$ Ω
17. A reactance of $14\underline{/90°}$ Ω
18. A resistance of 25 Ω and a capacitive reactance of 30 Ω

Given the following impedances, determine the resistance and the reactance of the series equivalent circuit for each.

19. $Z = 510\underline{/16°}$ Ω   20. $Z = 4.3\underline{/-50°}$ kΩ
21. $Z = 3.3\underline{/-18°}$ kΩ   22. $Z = 14.3\underline{/5.0°}$ Ω
23. $Z = 83\underline{/67°}$ Ω   24. $Z = 0.635\underline{/-74°}$ MΩ

Given the following circuit conditions, construct an impedance diagram and determine (a) the total impedance in polar form, and (b) the component values for the series equivalent circuit for the stated frequency:

25. $R = 27$ Ω $X_L = 54$ Ω $X_C = 190$ Ω $f = 60$ Hz
26. $X_L = 350$ Ω $R = 95$ Ω $X_L = 65$ Ω $f = 120$ Hz
27. $X_C = 1.5$ kΩ $X_L = 810$ Ω $R = 500$ Ω $f = 400$ Hz
28. $R = 470$ Ω $X_C = 94$ Ω $X_C = 116$ Ω $f = 800$ Hz
29. $R = 2.2$ kΩ $X_L = 1.5$ kΩ $X_C = 3.9$ kΩ $f = 1.0$ kHz
30. $R = 52$ Ω $X_L = 84$ Ω $X_L = 26$ Ω $f = 60$ Hz

27–2 SOLVING SERIES AC CIRCUITS

An ac series circuit may be made up of several resistors, capacitors, and inductors. When this is the case, each kind of element may be combined into a single *equivalent component* by applying one or more of the following equations.

$$R_T = R_1 + R_2 + R_3 + \cdots + R_n \qquad \Omega \qquad\qquad (27\text{–}3)$$

$$L_T = L_1 + L_2 + L_3 + \cdots + L_n \qquad \mathbf{H} \qquad\qquad (27\text{–}4)$$

$$C_T = \cfrac{1}{\cfrac{1}{C_1} + \cfrac{1}{C_2} + \cfrac{1}{C_3} + \cdots + \cfrac{1}{C_n}} \qquad \mathbf{F} \qquad\qquad (27\text{–}5)$$

EXAMPLE 27–11 Compute the equivalent component for each of the following series-circuit conditions.

(a) Three inductors in series:

$$L_1 = 5\ \text{H} \qquad\qquad L_2 = 10\ \text{H} \qquad\qquad L_3 = 2\ \text{H}$$

(b) Three resistors in series:

$$R_1 = 510\ \Omega \qquad\qquad R_2 = 620\ \Omega \qquad\qquad R_3 = 390\ \Omega$$

(c) Three capacitors in series:

$$C_1 = 0.22\ \mu\text{F} \qquad C_2 = 0.50\ \mu\text{F} \qquad C_3 = 0.68\ \mu\text{F}$$

Solution (a) Use Equation 27–4:

$$L_T = L_1 + L_2 + L_3$$
$$L_T = 5 + 10 + 2$$
$$\therefore \qquad L_T = 17\ \text{H}$$

(b) Use Equation 27–3:

$$R_T = R_1 + R_2 + R_3$$
$$R_T = 510 + 620 + 390$$
$$\therefore \qquad R_T = 1.5\ \text{k}\Omega$$

(c) Use Equation 27–5:

$$C_T = \cfrac{1}{\cfrac{1}{C_1} + \cfrac{1}{C_2} + \cfrac{1}{C_3}}$$

$$C_T = \cfrac{1}{\cfrac{1}{0.22 \times 10^{-6}} + \cfrac{1}{0.50 \times 10^{-6}} + \cfrac{1}{0.68 \times 10^{-6}}}$$

$$\therefore \qquad C_T = 0.13\ \mu\text{F}$$

Equations 27–3 through 27–5 are used along with the previously learned rules, concepts, and formulas to solve series ac circuits.

EXAMPLE 27–12   A series circuit made up of the following components is connected to a 120-V, 400-Hz alternator.

$$L_1 = 2\,H \qquad R_1 = 1.8\,k\Omega \qquad L_2 = 4\,H$$
$$C_1 = C_2 = 0.1\,\mu F \qquad R_2 = 750\,\Omega$$

Determine:

(a) The equivalent components R_T, L_T, and C_T
(b) The circuit impedance diagram showing the real component and the imaginary component
(c) The impedance in polar form
(d) The circuit current ($|I|$)
(e) The components of the series equivalent circuit
(f) The equivalent series circuit impedance diagram
(g) The current and voltage phasor diagram

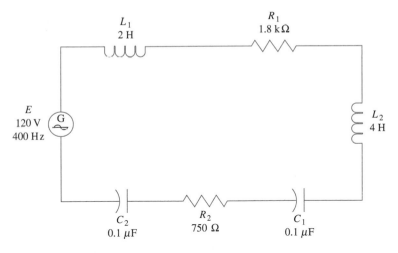

FIGURE 27–6   Schematic diagram for Example 27–12.

Solution   Begin the solution by drawing and labeling a schematic diagram as shown in Figure 27–6.

(a) Use Equation 27–3 to compute R_T:

∴ $R_T = 1800 + 750 = 2.6\,k\Omega$

Use Equation 27–4 to compute L_T:

∴ $L_T = 2 + 4 = 6\,H$

Use Equation 27–5 to compute C_T:

$$\therefore \quad C_T = \frac{1}{\dfrac{1}{0.1 \times 10^{-6}} + \dfrac{1}{0.1 \times 10^{-6}}} = 50 \text{ nF}$$

(b) Compute the value of the inductive and capacitive reactances:

$$\therefore \quad X_{L_T} = \omega L_T = 2\pi 400(6) = 15 \text{ k}\Omega$$

$$\therefore \quad X_{C_T} = 1/(\omega C_T) = \frac{1}{2\pi 400(50 \times 10^{-9})} = 8.0 \text{ k}\Omega$$

Construct the circuit impedance diagram with $R_T = 2.6 \text{ k}\Omega$. Figure 27–7(a) is the impedance diagram for the series circuit.

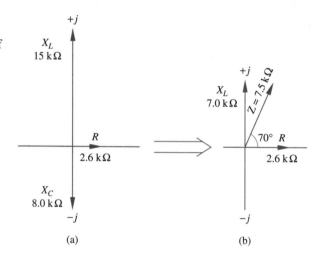

FIGURE 27–7 Impedance diagrams: (a) series circuit of Figure 27–6; (b) series equivalent circuit of Figure 27–8.

(c) Compute the impedance by adding the resistance and reactances of the circuit components:

$$\begin{array}{r} 2600 + j0 \\ 0 + j15\,000 \\ \underline{0 - j8000} \\ Z_T = 2600 + j7000 \ \Omega \end{array}$$

$$\therefore \quad Z_T = 2600 + j7000 \ \boxed{\rightarrow \text{P}} \ 7.5 \underline{/70°} \text{ k}\Omega$$

(d) $|I| = |E|/|Z|$

$$\therefore \quad |I| = 120/(7.5 \times 10^3) = 16 \text{ mA}$$

(e) The components of the series equivalent circuit are expressed as the rectangular form of the impedance. Thus,

$$2.6 + j7.0 \text{ k}\Omega \Rightarrow R = 2.6 \text{ k}\Omega \text{ and } X_L = 7.0 \text{ k}\Omega$$

Compute L from X_L:

$$X_L = 2\pi f L$$
$$L = \frac{X_L}{2\pi f} = \frac{7.0 \times 10^3}{2\pi 400} = 2.8 \text{ H}$$

∴ The series-equivalent circuit of $R = 2.6 \text{ k}\Omega$ and $L = 2.8 \text{ H}$ is shown in Figure 27–8.

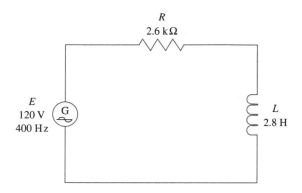

FIGURE 27–8 Series equivalent circuit for Example 27–12.

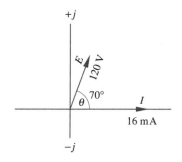

FIGURE 27–9 Phasor diagram of the source emf and circuit current for Example 27–12.

∴ (f) The series-equivalent-circuit impedance diagram is Figure 27–7(b).

(g) The phase angle between the source voltage (120 V) and the circuit current (16 mA) is equal to 70°, the angle of the impedance. In series ac circuits, the current phasor is selected as the reference phasor. In this inductive circuit, the voltage phasor leads the current phasor by 70°. This result may also be determined by phasor mathematics as:

$$E = IZ$$
$$E = (16 \times 10^{-3} \underline{/0°})(7.5 \times 10^3 \underline{/70°})$$
$$E = 120 \underline{/70°} \text{ V}$$

∴ Figure 27–9 is the current and voltage phasor diagram with the phase angle $\theta = 70°$.

EXAMPLE 27–13   The following components are connected in series to a 3.0-V, 10-kHz ac source:

$$R_1 = 5.60 \text{ k}\Omega \qquad C_1 = 820 \text{ pF} \qquad L_1 = 350 \text{ mH}$$
$$C_2 = 1500 \text{ pF}$$

Determine:

(a) The impedance of the circuit
(b) The phase angle between the current and voltage phasors
(c) The circuit current ($|I|$)
(d) The components of the series equivalent circuit

Solution   Begin the solution by drawing and labeling a schematic diagram. This is left for you to do.

(a) Compute the impedance by calculating the capacitive reactance and the inductive reactance:

$$C_T = \cfrac{1}{\cfrac{1}{C_1} + \cfrac{1}{C_2}} = \cfrac{1}{\cfrac{1}{820 \times 10^{-12}} + \cfrac{1}{1500 \times 10^{-12}}}$$

$$C_T = 530 \text{ pF}$$

$$X_C = 1/(2\pi fC)$$

$$X_C = \frac{1}{(2\pi 10 \times 10^3)(530 \times 10^{-12})} = 30 \text{ k}\Omega$$

$$X_L = 2\pi fL$$

$$X_L = (2\pi 10 \times 10^3)(350 \times 10^{-3}) = 22 \text{ k}\Omega$$

$$R = 5.6 \text{ k}\Omega$$

Express the impedance as a sum of the resistance and reactance:

$$\begin{aligned} & 5600 + j0 \\ & 0 - j30\,000 \\ & \underline{0 + j22\,000} \\ Z = & \overline{5600 - j8000} \ \Omega \end{aligned}$$

$$\therefore \quad Z = 5600 - j8000 \boxed{\to \text{P}} \ 9.8\underline{/-55^\circ} \text{ k}\Omega$$

(b) The phase angle (θ) is equal to the angle of the polar form of impedance:

$$\therefore \quad \theta = -55^\circ \text{ (voltage lags current in a}$$
$$\text{capacitive series circuit)}$$

(c) Compute the circuit current:

$$|I| = |E|/|Z|$$
$$|I| = 3.0/(9.8 \times 10^3)$$
$$\therefore \quad |I| = 0.31 \text{ mA}$$

(d) The components of the series equivalent circuit are expressed as the rectangular form of the impedance:

$$5.6 - j8.0 \text{ k}\Omega \Rightarrow R = 5.6 \text{ k}\Omega \text{ and } X_C = 8.0 \text{ k}\Omega$$

Compute C from X_C:

$$C = 1/(2\pi f X_C)$$
$$C = 1/(2\pi \times 10 \times 10^3 \times 8.0 \times 10^3)$$
$$C = 2.0 \text{ nF}$$

∴ The series-equivalent circuit is a resistance of 5.6 kΩ in series with a capacitor of 2.0 nF.

EXERCISE 27–2

Use the given information for each problem to determine (a) the impedance of the circuit (rectangular and polar), (b) the phase angle between the current and voltage phasors, (c) the circuit current ($|I|$), and (d) the components of the series equivalent circuit. It is suggested that you get in the habit of drawing and labeling complete schematics and constructing impedance diagrams and phasor diagrams of current and voltage.

1. $E = 100$ V, 60 Hz $\quad R_1 = 1.2$ kΩ $\quad L_1 = 3$ H $\quad R_2 = 680 \, \Omega$ $\quad C_1 = 1.0 \, \mu$F
2. $E = 1.5$ V, 5.0 MHz $\quad R_1 = 470 \, \Omega$ $\quad L_1 = 4.0 \, \mu$H $\quad C_1 = 160$ pF $\quad L_2 = 10 \, \mu$H
3. $E = 120$ V, 60 Hz $\quad R_1 = 51 \, \Omega$ $\quad C_1 = 10 \, \mu$F $\quad L_1 = 400$ mH $\quad C_2 = 15 \, \mu$F
4. $E = 220$ V, 60 Hz $\quad R_1 = 91 \, \Omega$ $\quad L_1 = 1.8$ H $\quad R_2 = 120 \, \Omega$ $\quad C_1 = 12 \, \mu$F
5. $E = 36$ V, 400 Hz $\quad C_1 = 5.0 \, \mu$F $\quad R_1 = 62 \, \Omega$ $\quad C_2 = 8.0 \, \mu$F $\quad L_1 = 20$ mH
6. $E = 440$ V, 1.0 kHz $\quad L_1 = 500 \, \mu$H $\quad R_1 = 72 \, \Omega$ $\quad L_2 = 3.3$ mH $\quad C_1 = 33 \, \mu$F
7. A 10-V, 1.0-kHz test oscillator (ac sinusoidal) is attached to a series circuit consisting of a 27-kΩ resistor and a 3.5-H inductor. Determine:
   (a) The inductive reactance of the inductor
   (b) The impedance of the circuit
   (c) The phase angle of the voltage with the current as the reference phasor
   (d) The circuit current $|I|$
   (e) The voltage across the resistor
   (f) The voltage across the inductor
   (g) That $E = V_R + V_L$

8. A series circuit is made up of a 220-Ω resistor and a 25-μF capacitor. The circuit is connected to a 120-V, 60-Hz source. Determine:

   (a) The capacitive reactance
   (b) The impedance of the circuit
   (c) The circuit current $|I|$
   (d) The phase angle between the voltage and the circuit current
   (e) That $E = V_R + V_C$

27–3 ADMITTANCE CONCEPTS

Conductance, the reciprocal of resistance, is used in the solution of dc parallel circuits. **Admittance,** the reciprocal of impedance, is used in the solution of ac parallel circuits. Admittance is the vector quantity used to describe the ease of passage of alternating current. It is measured in siemens (S) and is noted by the letter Y. Thus,

$$Y = 1/Z \quad \text{S} \qquad \qquad \textbf{(27–6)}$$

EXAMPLE 27–14   Express an impedance of $10\underline{/30°}$ Ω as an admittance.

Solution   Use Equation 27–6:

$$Y = 1/(10\underline{/30°})$$
$$\therefore \quad Y = 0.10\underline{/-30°} \text{ S}$$

EXAMPLE 27–15   Express an impedance of $4.0 - j3.0$ Ω as an admittance.

Solution   Change Z to polar form:

$$4.0 - j3.0 \boxed{\rightarrow \textbf{P}} 5.0\underline{/-37°} \text{ } \Omega$$

Use Equation 27–6:

$$Y = 1/(5.0\underline{/-37°})$$
$$\therefore \quad Y = 0.20\underline{/37°} \text{ S})$$

Admittance Diagram

Before constructing the generalized admittance diagram, some additional concepts are needed:

1. The reciprocal of the inductive or capacitive reactance is called **susceptance,** noted by the letter B. Susceptance is a measure of the ability of an inductor or a capacitor to pass alternating current. Thus,

$$B_L = 1/X_L \quad \text{S} \tag{27-7}$$
$$B_C = 1/X_C \quad \text{S} \tag{27-8}$$

where B_L = inductive susceptance

B_C = capacitive susceptance

2. In an impedance, diagram X_L has an angle of $+90°$ ($+j$); however, when X_L is reciprocated, the angle becomes $-90°$ ($-j$). Thus,

$$1/(X_L\angle 90°) = B_L\angle -90° = -jB_L \ \text{S}$$

and

$$1/(X_C\angle -90°) = B_C\angle 90° = +jB_C \ \text{S}$$

EXAMPLE 27–16   Express the following reactances as susceptance:

(a) $X_L = 80.0\angle 90°$ (b) $X_C = 4.0\angle -90°$

Solution   (a) Use Equation 27–7:

$$B_L = 1/X_L$$
$$B_L = 1/(80.0\angle 90°)$$
$$\therefore \quad B_L = 12.5\angle -90° \ \text{mS}$$

(b) Use Equation 27–8:

$$B_C = 1/X_C$$
$$B_C = 1/(4.0\angle -90°)$$
$$\therefore \quad B_C = 0.25\angle 90° \ \text{S}$$

Admittance may be expressed in either the polar, $Y\angle\theta$ or the rectangular form, $G + jB$, of the complex number. The rectangular form is written with the conductance (G) first and then the susceptance (B).

EXAMPLE 27–17   Express each of the following admittances in its rectangular form.

(a) $92.0\angle -37.0°$ mS   (b) $13\angle 48°$ μS

Solution   (a) $92.0 \times 10^{-3}\angle -37.0°$ $\boxed{\rightarrow \text{R}}$ $73.5 - j55.4$ mS

$$\therefore \quad 92.0\angle -37.0° \ \text{mS} = 73.5 - j55.4 \ \text{mS}$$

(b) $13 \times 10^{-6}\angle 48°$ $\boxed{\rightarrow \text{R}}$ $8.7 + j9.7\mu$S

$$\therefore \quad 13\angle 48° \ \mu\text{S} = 8.7 + j9.7\mu\text{S}$$

FIGURE 27–10 Admittance diagram for parallel ac circuits.

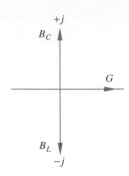

The admittance diagram is constructed with the conductance along the reference axis and the susceptance along the *j*-axis. Figure 27–10 is the general admittance diagram. Notice that the conductance (G) is plotted horizontally, the capacitive susceptance (B_C) is plotted vertically up, and the inductive susceptance (B_L) is plotted vertically down. Some of the concepts presented so far are summarized in Table 27–2.

TABLE 27–2   Admittance Summarized

Property	Noted by:	Equal to:	Units	Schematic Symbol	Vector Diagram	Polar Form	Rectangular Form
Conductance	G	$1/R$	S	G	G	$G\underline{/0°}$	$G + j0$
Susceptance (inductive)	B_L	$1/X_L$	S	B_L	$-90°$ B_L	$B_L\underline{/-90°}$	$0 - jB_L$
Susceptance (capacitive)	B_C	$1/X_C$	S	B_C	B_C $90°$	$B_C\underline{/90°}$	$0 + jB_C$
Admittance	Y	$1/Z$	S	Any one or any combination of the above	θ	$Y\underline{/\theta}$	$G + jB$

Computing Admittance

Admittance may be computed from the source voltage and the circuit current. Thus,

$$Y = I/E \qquad S \tag{27-9}$$

EXAMPLE 27–18   Determine the admittance of a parallel ac circuit having a source voltage of 60 V$\underline{/0°}$ and a circuit current of 180 mA $\underline{/-40°}$. Also construct an admittance diagram.

Solution   Use Equation 27–9:

$$Y = \frac{180 \times 10^{-3} / -40°}{60 / 0°}$$

$\therefore$ $Y = 3.0 / -40°$ mS $\boxed{\rightarrow R}$ $2.3 - j1.9$ mS

Figure 27–11 is the admittance diagram.

FIGURE 27–11   Admittance diagram for Example 27–18.

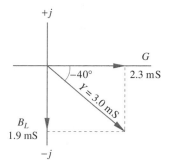

EXERCISE 27–3

Express each of the following impedances as an admittance in polar form:

1. $10 / 15°$ Ω
2. $100 / -55°$ Ω
3. $1000 / 82°$ Ω
4. $750 / -30°$ Ω
5. $56 / -19°$ Ω
6. $270 / 90°$ Ω
7. $120 / -90°$ Ω
8. $300 / 0°$ Ω
9. $910 / -5°$ Ω

Compute the susceptance of each of the following reactances:

10. $80.0 / 90°$ Ω
11. $1.50 / -90°$ kΩ
12. $39.0 / -90°$ Ω
13. $420 / 90°$ Ω
14. $2.70 / 90°$ kΩ
15. $68.0 / -90°$ kΩ
16. $143 / 90°$ kΩ
17. $0.510 / -90°$ MΩ
18. $0.33 / 90°$ kΩ

Compute the admittance of each parallel circuit given the source voltage E and circuit current I. Draw an admittance diagram for each circuit.

19. $E = 36 / 0°$ V $I = 42 / 17°$ mA
20. $E = 50.0 / 0°$ V $I = 7.20 / 48°$ mA
21. $E = 100 / 0°$ V $I = 808 / -61°$ μA
22. $E = 27.0 / 0°$ V $I = 0.618 / -24°$ A
23. $E = 18.0 / 0°$ V $I = 1.43 / 15°$ A
24. $E = 80.0 / 0°$ V $I = 917 / -84°$ mA

27–4 ADMITTANCE OF PARALLEL AC CIRCUITS

Forming Equivalent Components

When several capacitors are connected in parallel, the capacity of a single *equivalent capacitor* may be found by Equation 27–10.

$$C_T = C_1 + C_2 + C_3 + \cdots + C_n \quad \text{F} \qquad (27\text{–}10)$$

Several inductors connected in parallel may be combined into a single *equivalent inductor* by applying Equation 27–11.

$$L_T = \cfrac{1}{\cfrac{1}{L_1} + \cfrac{1}{L_2} + \cfrac{1}{L_3} + \cdots + \cfrac{1}{L_n}} \quad \text{H} \qquad (27\text{–}11)$$

Several resistances in a parallel circuit may be combined into a single *equivalent resistance* by applying Equation 27–12.

$$R_T = \cfrac{1}{\cfrac{1}{R_1} + \cfrac{1}{R_2} + \cfrac{1}{R_3} + \cdots + \cfrac{1}{R_n}} \quad \Omega \qquad (27\text{–}12)$$

EXAMPLE 27–19 Compute the equivalent component for each of the following parallel circuit conditions.

(a) Three resistors in parallel:

$$R_1 = 12 \, \Omega \qquad R_2 = 6.2 \, \Omega \qquad R_3 = 8.2 \, \Omega$$

(b) Three capacitors in parallel:

$$C_1 = 10 \, \mu\text{F} \qquad C_2 = 25 \, \mu\text{F} \qquad C_3 = 5 \, \mu\text{F}$$

(c) Three inductors in parallel:

$$L_1 = 500 \, \text{mH} \qquad L_2 = 80 \, \text{mH} \qquad L_3 = 210 \, \text{mH}$$

Solution (a) Use Equation 27–12 to find the equivalent resistance:

$$R_T = \cfrac{1}{\cfrac{1}{R_1} + \cfrac{1}{R_2} + \cfrac{1}{R_3}}$$

$$R_T = \cfrac{1}{\cfrac{1}{12} + \cfrac{1}{6.2} + \cfrac{1}{8.2}}$$

$$\therefore \qquad R_T = 2.7 \, \Omega$$

(b) Use Equation 27–10 to find the equivalent capacitance:

$$C_T = C_1 + C_2 + C_3$$
$$C_T = 10 + 25 + 5$$
$$\therefore \quad C_T = 40 \ \mu F$$

(c) Use Equation 27–11 to find the equivalent inductance:

$$L_T = \cfrac{1}{\cfrac{1}{L_1} + \cfrac{1}{L_2} + \cfrac{1}{L_3}}$$

$$L_T = \cfrac{1}{\cfrac{1}{500} + \cfrac{1}{80} + \cfrac{1}{210}}$$

$$\therefore \quad L_T = 52 \ \text{mH}$$

Computing Branch Admittance

The admittance of each branch of a parallel circuit is found by first computing the impedance of each branch in polar form, and then taking the reciprocal of the impedance. This procedure is demonstrated in the following example.

EXAMPLE 27–20   Compute the admittance of each branch of the circuit of Figure 27–12.

FIGURE 27–12   Circuit for Examples 27–20 and 27–21.

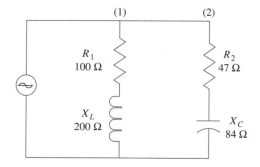

Solution   Determine the impedance of each branch in polar form:

$$Z_1 = 100 + j200 \ \Omega \ \boxed{\rightarrow P} \ 224 \underline{/63.4^\circ} \ \Omega$$
$$Z_2 = 47 - j84 \ \Omega \ \boxed{\rightarrow P} \ 96.3 \underline{/-60.8^\circ} \ \Omega$$

Solve for branch admittances. Use Equation 27–6:

$$Y = 1/Z$$
$$Y_1 = 1/224\underline{/63.4°}$$
$$Y_1 = 4.46\underline{/-63.4°} \text{ mS}$$

$$Y_2 = 1/96.3\underline{/-60.8°}$$
$$Y_2 = 10\underline{/61°} \text{ mS}$$

∴   The admittance of branch (1) is $Y_1 = 4.46\underline{/-63.4°}$ mS.

The admittance of branch (2) is $Y_2 = 10\underline{/61°}$ mS.

Computing Circuit Admittance and Circuit Impedance

Once the branch admittances are known, they are added to obtain the admittance of the entire circuit. Thus,

$$Y_T = Y_1 + Y_2 + Y_3 + \cdots + Y_n \qquad \text{S} \qquad \text{(27–13)}$$

The impedance of the entire circuit may be computed by taking the reciprocal of the admittance. Thus,

$$Z_T = 1/Y_T \qquad \Omega \qquad \text{(27–14)}$$

EXAMPLE 27–21   Use the information of Example 27–20 to compute the admittance and the impedance of the circuit of Figure 27–12.

Solution   Express the branch admittances in rectangular form and add:

$$Y_1 = 4.5\underline{/-63°} \text{ mS} \boxed{\rightarrow\text{R}} \quad 2.0 - j4.0 \text{ mS}$$
$$Y_2 = 10 \quad \underline{/61°} \text{ mS} \boxed{\rightarrow\text{R}} \quad 4.8 + j8.7 \text{ mS}$$
$$Y_T = 6.8 + j4.7 \text{ mS}$$

Convert Y_T to polar form:

$$6.8 + j4.7 \text{ mS} \boxed{\rightarrow\text{P}} \quad 8.3\underline{/35°} \text{ mS}$$

∴ $$Y_T = 8.3\underline{/35°} \text{ mS}$$

Compute Z_T. Use Equation 27–14:

$$Z_T = 1/(8.3 \times 10^{-3}\underline{/35°})$$

∴ $$Z_T = 120\underline{/-35°} \ \Omega$$

EXAMPLE 27–22   Compute the admittance and the impedance of the circuit in Figure 27–13.

FIGURE 27–13 Circuit for Example 27–22.

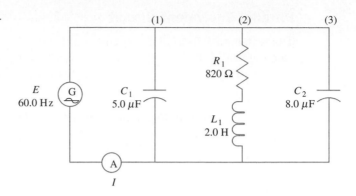

Solution Compute the equivalent component for C_1 and C_2:

$$C_T = C_1 + C_2$$
$$C_T = 13.0 \ \mu F$$

Compute the reactance of L_1 and C_T:

$$\omega = 2\pi 60.0 = 377 \text{ rad/s}$$
$$X_{L1} = \omega L = 377(2.0) = 754 \ \Omega$$
$$X_{CT} = 1/(\omega C) = 1/(377 \times 13.0 \times 10^{-6}) = 204 \ \Omega$$

Determine the impedance of each branch in polar form:

$$Z_{1,3} = 0 - j204 \ \Omega \boxed{\rightarrow P} \ \ 204 \underline{/-90°} \ \Omega$$
$$Z_2 = 820 + j754 \ \Omega \boxed{\rightarrow P} \ \ 1.11 \underline{/42.6°} \ k\Omega$$

Determine the branch admittances:

$$Y_{1,3} = 1/Z_{1,3} = 1/204 \underline{/-90°}$$
$$Y_{1,3} = 4.9 \underline{/90°} \text{ mS}$$

$$Y_2 = 1/Z_2 = 1/1110 \underline{/42.6°}$$
$$Y_2 = 0.90 \underline{/-42.6°} \text{ mS}$$

Express the branch admittances in rectangular form and add:

$$Y_{1,3} = 4.9 \underline{/90°} \text{ mS} \boxed{\rightarrow R} \quad \ \ 0 \ \ \ + j4.9 \text{ mS}$$
$$Y_2 = 0.90 \underline{/-42.6°} \text{ mS} \boxed{\rightarrow R} \ \ 0.66 - j0.61 \text{ mS}$$
$$Y_T = 0.66 + j4.3 \text{ mS}$$

$$\therefore \quad Y_T = 0.66 + j4.3 \text{ mS} \boxed{\rightarrow P} \ \ 4.4 \underline{/81°} \text{ mS}$$

Compute Z_T:

$$Z_T = 1/Y_T = 1/(4.4 \times 10^{-3} \underline{/81°})$$
$$\therefore \quad Z_T = 230 \underline{/-81°} \ \Omega$$

Parallel Equivalent Circuit

The rectangular form of the circuit admittance of an ac parallel circuit represents the components of the *parallel equivalent circuit.* That is, $Y_T = G_{eq} + jB_{eq}$. The parallel equivalent circuit is formed by taking the reciprocal of:

- G_{eq} to get the equivalent parallel resistance R_P
- B_{eq} in polar form to get the equivalent parallel reactance X_P

The value of the reactive component may then be computed if the circuit frequency is known.

EXAMPLE 27–23   Compute the parallel equivalent circuit of Example 27–22 (Figure 27–13) for a frequency of 60 Hz.

Solution   Use the total admittance $Y_T = 0.66 + j4.3$ mS, and express Y_T in terms of G_{eq} and B_{eq}:

$$G_{eq} = 0.66 \text{ mS}$$
$$B_{eq} = 4.3 \text{ mS}$$

Compute the equivalent parallel resistance:

$$R_P = 1/G_{eq} = 1/(0.66 \times 10^{-3}\underline{/0°})$$
$$R_P = 1.5 \text{ k}\Omega$$

Compute the equivalent parallel reactance:

$$X_P = 1/B_{eq} = 1/(4.3 \times 10^{-3}\underline{/90°})$$
$$X_P = 233\underline{/-90°}\ \Omega \text{ (capacitive reactance)}$$

Compute the size of the capacitor:

$$X_C = X_P = 1/(\omega C)$$
$$C = 1/(\omega X_C)$$
$$\omega = 377 \text{ rad/s}$$
$$C = 1/(377 \times 233)$$
$$C = C_P = 11\ \mu\text{F}$$

FIGURE 27–14   Parallel equivalent circuit of Figure 27–13.

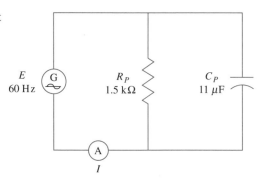

∴ The parallel equivalent circuit is made up of 1.5 kΩ of resistance in parallel with 11 μF of capacitance when the frequency is 60 Hz. Figure 27–14 is the schematic diagram of the parallel equivalent circuit of Figure 27–13.

EXAMPLE 27–24   Determine the components of the parallel equivalent circuit of Figure 27–15 for a frequency of 400.0 Hz if $E = 10.0 \text{ V}\underline{/0°}$ and $I = 602\underline{/-27°}$ mA.

FIGURE 27–15   The box represents an unknown circuit.

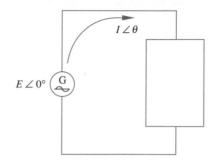

Solution   Compute the circuit impedance:

$$Z = E/I = 10.0\underline{/0°}/(602 \times 10^{-3}\underline{/-27°})$$
$$Z = 16.6\underline{/27°} \, \Omega$$

Compute the circuit admittance:

$$Y = 1/16.6\underline{/27°}$$
$$Y = 60.2\underline{/-27°} \text{ mS } \boxed{\rightarrow\text{R}} \, 53.6 - j27.3 \text{ mS}$$

Solve for R_P and X_P:

$$R_P = 1/(53.6 \times 10^{-3})$$
$$R_P = 18.7 \, \Omega$$
$$X_P = 1/(27.3 \times 10^{-3}\underline{/-90°})$$
$$X_P = 36.6\underline{/90°} \, \Omega \text{ (inductive reactance)}$$

Compute L:

$$X_L = 2\pi fL$$
$$L = X_L/(2\pi f)$$
$$L = 36.6/(2\pi 400.0)$$
$$L = 14.6 \text{ mH}$$

∴ The parallel equivalent circuit is a resistance of 18.7 Ω in parallel with an inductance of 14.6 mH when the frequency is 400.0 Hz.

In each of the following, determine the equivalent component. Assume that the components are connected in parallel.

1. $C_1 = 247$ pF, $C_2 = 88$ pF
2. $R_1 = 82.0$ kΩ, $R_2 = 1.00$ MΩ
3. $L_1 = 50$ μH, $L_2 = 180$ μH
4. $L_1 = 28$ mH, $L_2 = 79$ mH
5. $R_1 = 390$ Ω, $R_2 = 150$ Ω
6. $C_1 = 0.68$ μF, $C_2 = 1200$ pF

Determine the branch admittance for each branch of a two-branch parallel circuit, given the following information.

7. Branch (1) $R = 75$ Ω $\quad X_L = 130$ Ω
   (2) $R = 72$ Ω $\quad X_C = 143$ Ω
8. Branch (1) $R = 560$ Ω $\quad X_C = 318$ Ω
   (2) $R = 2.0$ kΩ
9. Branch (1) $R = 840$ Ω $\quad X_C = 600$ Ω
   (2) $R = 420$ Ω
10. Branch (1) $R = 1.6$ kΩ
    (2) $R = 2.5$ kΩ

In each of the following, determine the circuit admittance and the circuit impedance for a two-branch parallel circuit given the impedance of each branch.

11. Branch (1) $291 + j83$ Ω
    (2) $0 - j153$ Ω
12. Branch (1) $143 + j308$ Ω
    (2) 203 Ω
13. Branch (1) $8.3 + j4.1$ kΩ
    (2) $0 + j3.6$ kΩ
14. Branch (1) $0 - j12$ Ω
    (2) $0 + j48$ Ω
15. Branch (1) $392 - j170$ Ω
    (2) $94 + j568$ Ω

Compute the components of the equivalent circuit contained in the box of Figure 27–15, given the following information. Assume that the components are connected in parallel.

16. $E = 22.0\underline{/0°}$ V $\qquad I = 2.2\underline{/72°}$ A $\qquad f = 60$ Hz
17. $E = 80\underline{/0°}$ V $\qquad I = 380\underline{/-11°}$ mA $\qquad f = 5.0$ kHz
18. $E = 33\underline{/0°}$ V $\qquad I = 92\underline{/31°}$ mA $\qquad f = 400$ Hz
19. $E = 325\underline{/0°}$ V $\qquad I = 0.96\underline{/7°}$ A $\qquad f = 60$ Hz
20. $E = 12\underline{/0°}$ V $\qquad I = 495\underline{/-65°}$ μA $\qquad f = 180$ kHz

Solve the following.

21. Two impedances are connected in parallel: $Z_1 = 210\underline{/81°}\ \Omega$ and $Z_2 = 750\underline{/51°}\ \Omega$. $E = 100\underline{/0°}$ V.

   (a) Determine the admittance of each branch.

   (b) Determine the circuit admittance.

   (c) Determine the circuit impedance.

   (d) Determine the circuit current.

   (e) Construct the circuit current, source voltage, and phasor diagram. Use E as the reference phasor.

   (f) Construct the circuit admittance diagram.

22. Two impedances are connected in parallel: $Z_1 = 920 - j260\ \Omega$, $Z_2 = 382\underline{/-56°}\ \Omega$, and $E = 60\underline{/0°}$ V. Determine:

   (a) The circuit admittance

   (b) The circuit impedance

   (c) The circuit current

   (d) The parallel equivalent circuit

SELECTED TERMS

admittance   The reciprocal of impedance; the vector quantity used to describe the ease of the passage of alternating current; expressed in units of siemens.

impedance   The opposition to the passage of alternating current through an electrical circuit; expressed in units of ohms.

susceptance   The reciprocal of inductive or capacitive reactance; measured in units of siemens.

28

Sinusoidal Alternating Current

28–1   Time and Displacement

28–2   Power and Power Factor

28–3   Instantaneous Equations and the *EI* Phasor Diagram

PERFORMANCE OBJECTIVES

- Calculate instantaneous current and voltage.

- Convert effective values of current and voltage to maximum values.

- Convert maximum values of current and voltage to effective values.

- Determine the power factor.

- Calculate the average power.

Tennessee Valley Authority (TVA) power is carried in these transmission lines near Wilson Dam. The TVA power is purchased by 158 municipal and cooperative power systems for distribution to more than 2,000,000 consumers.   (Courtesy of Tennessee Valley Authority)

An ac generator produces a time-varying voltage. When a load is placed across an ac source, the resulting current is also time varying. In this chapter we study how the varying amplitudes of the voltage and the current are related to the sine wave. We also study how power is determined from a sinusoidal voltage and current.

28–1   TIME AND DISPLACEMENT

If the ac wave produced by a generator has a constant frequency, then a rotating vector representing an alternating current or voltage will also have a constant frequency. When a voltage vector is rotated from the reference axis at a constant frequency, the angle generated will be a function of time. Thus,

$$\alpha = \omega t \quad \text{rad} \tag{28–1}$$

where α = angular displacement from the reference axis in radians
ω = angular velocity in radians per second
t = time in seconds

From Figure 28–1, an equation that relates the instantaneous voltage (e) to the maximum voltage (E_{mx}) and displacement angle (α) may be derived. Applying the concepts of trigonometry, we write a sine function as sin (α) = e/E_{mx}. Solving for e results in Equation 28–2.

$$e = E_{mx} \sin (\alpha) \quad \text{V} \tag{28–2}$$

FIGURE 28–1   The instantaneous voltage e is a function of the angular displacement α from the reference axis.

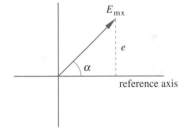

Similarly,

$$i = I_{mx} \sin (\alpha) \quad \text{A} \tag{28–3}$$

Substitute $\alpha = \omega t$:

$$e = E_{mx} \sin (\omega t) \quad \text{V} \tag{28–4}$$
$$i = I_{mx} \sin (\omega t) \quad \text{A} \tag{28–5}$$

where e = instantaneous voltage at time t (V)

$\quad\quad\quad i$ = instantaneous current at time t (A)

$\quad\quad\quad E_{mx}$ = maximum amplitude of the voltage wave in volts (V)

$\quad\quad\quad I_{mx}$ = maximum amplitude of the current wave in amperes (A)

$\quad\quad\quad \omega$ = angular velocity of the generator producing the voltage wave in radians per second (rad/s)

$\quad\quad\quad t$ = time in seconds (s)

Figure 28–2 relates the phasor diagram of voltage to the periodic wave of voltage. As shown in Figure 28–2, the phasor diagram indicates only the condition for a particular time (t_1) or angular displacement (ωt_1).

FIGURE 28–2   The phasor diagram displays the conditions of the periodic wave at time t_1.

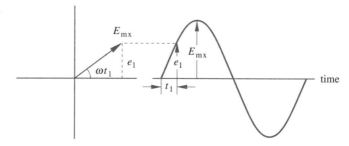

EXAMPLE 28–1

Determine the instantaneous voltage (e) of a 60-Hz ac source that has a maximum amplitude (E_{mx}) of 165 V for the following times.

(a) $t_1 = 3.00$ ms $\quad\quad\quad$ (b) $t_2 = 11.0$ ms

(c) $t_3 = 18.0$ ms $\quad\quad\quad$ (d) $t_4 = 25.0$ ms

Solution

Use Equation 28–4. Remember that $\omega = 2\pi f$.

(a) $e = E_{mx} \sin (\omega t)$

$\boxed{\times}$ $\quad\quad\quad e = 165 \sin (2\pi 60 \times 3.00 \times 10^{-3})$

$\boxed{\text{RAD}}$ $\boxed{\text{SIN}}$ $\quad\quad\quad e = 165 \sin (1.13^r)$

$\boxed{\times}$ $\quad\quad\quad e = 165 \,(0.905)$

$\therefore$ $\quad\quad\quad e = 149$ V at $t = 3.00$ ms

(b) $e = 165 \sin (2\pi 60 \times 11.0 \times 10^{-3})$

Use *chain calculations.*

$\therefore$ $\quad\quad\quad e = -139$ V at $t = 11.0$ ms

(c) $e = 165 \sin (2\pi 60 \times 18.0 \times 10^{-3})$

$\therefore$ $\quad\quad\quad e = 79.5$ V at $t = 18.0$ ms

(d) $e = 165 \sin (2\pi 60 \times 25.0 \times 10^{-3})$

$\therefore \qquad e = 0 \text{ V at } t = 25.0 \text{ ms}$

EXAMPLE 28–2   Compute the angular displacement from the reference axis in radians and in degrees for each of the given times of Example 28–1.

Solution   Use Equation 28–1:

$$\omega = 2\pi 60 = 377 \text{ rad/s}$$

(a) $t_1 = 3.00 \text{ ms}$

$\alpha = \omega t$
$\alpha = 377(3.00 \times 10^{-3})$
$\therefore \qquad \alpha = 1.13^r = 64.8°$

(b) $t_2 = 11.0 \text{ ms}$

$\alpha = 377(11.0 \times 10^{-3})$
$\therefore \qquad \alpha = 4.15^r = 238°$

(c) $t_3 = 18.0 \text{ ms}$

$\alpha = 377(18.0 \times 10^{-3})$
$\therefore \qquad \alpha = 6.79^r = 389°$

(d) $t_4 = 25.0 \text{ ms}$

$\alpha = 377(25.0 \times 10^{-3})$
$\therefore \qquad \alpha - 9.43^r = 540°$

EXAMPLE 28–3   Write the instantaneous equation in terms of time for a current of 400 Hz that has a maximum current (I_{mx}) of 22 A.

Solution   Use Equation 28–5. Compute ω for a frequency of 400 Hz:

$$\omega = 2\pi 400 = 2.51 \text{ krad/s}$$

Substitute $I_{mx} = 22$ A and $\omega = 2.51$ krad/s:

$i = I_{mx} \sin (\omega t)$
$\therefore \qquad i = 22 \sin (2.51 \times 10^3 \, t) \text{ A}$

EXAMPLE 28–4   Compute the instantaneous value for the following phasors.

(a) $E_{mx} = 100 \text{ V}, \; \underline{/\theta} = 140°$
(b) $I_{mx} = 40 \text{ mA}, \; \underline{/\theta} = 0.85^r$

Solution   (a) Use Equation 28–2:

$$e = E_{mx} \sin (\alpha)$$
$$e = 100 \sin (140°)$$
∴ $$e = 64.3 \text{ V}$$

(b) Use Equation 28–3:

$$i = I_{mx} \sin (\alpha)$$
$$i = 40 \times 10^{-3} \sin (0.85^r)$$
∴ $$i = 30 \text{ mA}$$

EXAMPLE 28–5   The circuit current is $i = 50 \times 10^{-3} \sin (500t)$ A.

(a) Determine the instantaneous current when $t = 6.00$ ms.
(b) Draw the phasor diagram for the current.

Solution   (a) Substitute $t = 6.00$ ms into

$$i = 50 \times 10^{-3} \sin (500t):$$
$$i = 50 \times 10^{-3} \sin (500 \times 6.00 \times 10^{-3})$$
∴ $$i = 7.1 \text{ mA when } t = 6.00 \text{ ms}$$

(b) Translate $t = 6.00$ ms into angular displacement in degrees:

$$\alpha = \omega t$$
$$\alpha = 500(6.00 \times 10^{-3})$$
$$\alpha = 3.00^r = 172°$$

∴   The current phasor is $50\underline{/172°}$ mA, as shown in Figure 28–3.

FIGURE 28–3 Current phasor diagram for Example 28–5.

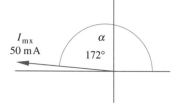

From Example 28–5 it may be seen that the instantaneous equation, $i = 50 \times 10^{-3} \sin (500t)$ A, of the sinusoidal current wave can be translated into the current phasor $50\underline{/172°}$ mA when $t = 6.0$ ms. Thus,

$$i = 50 \times 10^{-3} \sin (172°) \text{ A} \Rightarrow 50 \times 10^{-3}\underline{/172°} \text{ A}$$

With the given information, determine the instantaneous quantity (voltage e or current i) for each of the following.

1. $E_{mx} = 18.0$ V $f = 100$ Hz $t = 2.3$ ms
2. $I_{mx} = 2.0$ A $\alpha = 2.30^r$
3. $I_{mx} = 710$ mA $f = 62.0$ kHz $t = 13.7\ \mu s$
4. $E_{mx} = 120$ V $\alpha = 212°$
5. $I_{mx} = 12.6$ A $\alpha = 4.10^r$
6. $I_{mx} = 4.3$ mA $f = 710$ kHz $t = 500$ ns
7. $E_{mx} = 82$ V $\alpha = 3.10^r$
8. $E_{mx} = 12.6$ V $f = 400$ Hz $t = 1.5$ ms
9. $E_{mx} = 440$ V $\alpha = -36°$
10. $I_{mx} = 1.43$ A $f = 60$ Hz $t = 2.7$ ms

For each of the following instantaneous equations, draw a phasor diagram that expresses the angular displacement (α) in degrees. All angles are in radians.

11. $i = 10.0 \sin (4.50)$
12. $i = 0.370 \sin (3.77)$
13. $e = 24.0 \sin (1.61)$
14. $i = 15 \times 10^{-3} \sin (5.4)$
15. $e = 100 \sin (2.15)$
16. $e = 36.0 \sin (0.53)$

28–2   POWER AND POWER FACTOR

Power in a Resistive Load

When an ac voltage of the general form $e = E_{mx} \sin (\omega t)$ is applied to a resistive load, a circuit current results in the form $i = I_{mx} \sin (\omega t)$, which is in phase with the voltage and the phase angle $\theta = 0$. If the two waves represented by the instantaneous equations are plotted and the instantaneous amplitudes of the waves are multiplied, a third wave representing the instantaneous power ($p = ei$ VA) will result.

Figure 28–4 shows the instantaneous power resulting from multiplying e and i.

FIGURE 28–4   Instantaneous power in a resistive load is always positive.

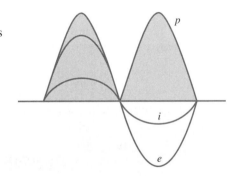

Mathematically, an expression for *average power* may be derived from instantaneous power.

$$p = ei$$
$$(1) \quad e = E_{mx} \sin (\omega t)$$
$$(2) \quad i = I_{mx} \sin (\omega t)$$

Substitute (1) and (2) into $p = ei$:

$$p = [E_{mx} \sin (\omega t)][I_{mx} \sin (\omega t)]$$
$$p = E_{mx}I_{mx} \sin^2 (\omega t)$$

However, $\sin^2(\omega t) = \frac{1}{2}[1 - \cos (2 \omega t)]$:

$$p = \frac{E_{mx}I_{mx}}{2}[1 - \cos (2 \omega t)]$$

$$p = \frac{E_{mx}I_{mx}}{2} - \frac{E_{mx}I_{mx}}{2}\cos (2 \omega t)$$

Notice that the second term is a cosine wave having a frequency twice that of the voltage or current wave. Since the average value of a cosine wave for one cycle is zero, the second term does not contribute to the average. Only the first term remains. This term is called the **average power** and is expressed in a formula as

$$P = \frac{E_{mx}I_{mx}}{2} \quad \text{W} \tag{28–6}$$

Effective Values

Factor the right member of Equation 28–6 using the common denominator $\sqrt{2}$. Thus,

$$P = \frac{E_{mx}}{\sqrt{2}} \cdot \frac{I_{mx}}{\sqrt{2}}$$

Replace $1/\sqrt{2}$ with 0.707:

$$P = (0.707E_{mx})(0.707I_{mx})$$

The factor $0.707E_{mx}$ is given the symbol E, which stands for the **effective voltage;** that is,

$$E = 0.707E_{mx} \quad \text{V} \tag{28–7}$$

And the factor $0.707I_{mx}$ is given the symbol I, which stands for the **effective current;** that is,

$$I = 0.707I_{mx} \quad \text{A} \tag{28-8}$$

Therefore,

$$P = EI \quad \text{W} \tag{28-9}$$

The average power (also called *true power*) dissipated by a resistance in an ac circuit is computed by a formula using the same letters as the formula for power in a dc circuit.

The product of the effective voltage (E) and the effective current (I) results in the same power (watts) as the product of a dc voltage (E) and a dc current (I) of the same magnitudes.

The effective voltage and current are also called root-mean-square (rms) values. This name comes from the mathematical procedure used to derive the effective value. Table 28–1 is a summary of ac voltage and current notation. From the table it may be noted that no subscript notation is used with effective quantities. Example 28–6 demonstrates converting from effective to maximum voltage.

TABLE 28–1   AC Voltage and Current Notation

AC Quantity	Notation	Example	Visual Representation	Comments
Peak-to-peak	E_{pp}, V_{pp}, or I_{pp}	$E_{pp} = 20$ V	Peak-to-peak 20 V	Twice maximum
Maximum or peak	E_{mx} or E_{p} I_{mx} or I_{p}	$E_{p} = 10$ V	Peak 10 V	One half peak to peak
Effective or rms	E or I	$E = 7.07$ V	rms 7.07 V	No subscript $0.707 \times$ peak

EXAMPLE 28–6   Determine the maximum voltage of a 117-V line.

Solution   Remember that 117 V means 117 V effective. Use Equation 28–7:

$$E = 0.707E_{mx}$$

Solve for E_{mx}:

$$E_{mx} = E/0.707$$
$$E_{mx} = 117/0.707$$
$$\therefore \quad E_{mx} = 165 \text{ V}$$

Power in a Reactive Load

When an ac voltage is applied to a reactive load, the voltage is 90° *out of phase* with the current. As with the case of a resistive load, when the instantaneous amplitudes of the voltage and current are multiplied, the resultant wave represents the instantaneous power of the reactor, as shown in Figure 28–5. When the power wave is averaged, the result is zero, which leads to the conclusion that *reactive loads do not dissipate power.*

FIGURE 28–5   Instantaneous power in a reactive load is both positive and negative.

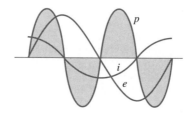

Power Factor

Because only the resistive portion of an ac circuit dissipates power, we are interested in knowing what part of the impedance is resistive. The ratio of the circuit resistance to the magnitude of the circuit impedance is called the **power factor.** Thus,

$$\textbf{power factor} = R/|Z| \qquad \textbf{(28–10)}$$

When the impedance of an ac circuit is resistive ($Z = R$), the power factor is 1. When the impedance is reactive ($Z = jX$), the power factor is zero. The power factor is a positive, unitless number that can be expressed as a percentage or a per unit decimal fraction.

The power factor may be related to the phase angle through the impedance diagram. Looking at Figure 28–6, we may recognize the ratio of $R/|Z|$ as the cosine function of θ. Thus,

$$\textbf{power factor} = \cos(\theta) = R/|Z| \qquad \textbf{(28–11)}$$

FIGURE 28–6   The power factor of
a circuit is cos (θ), which equals R/Z.

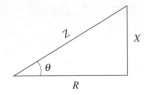

EXAMPLE 28–7   Determine the power factor for each of the following circuits:

(a) $Z = 650\underline{/-34°}\ \Omega$

(b) $Z = 125 + j90.0\ \Omega$

(c) $|Z| = 143\ \Omega,\ R = 18.0\ \Omega$

Solution   (a) $\theta = -34°$.

$$\cos(\theta) = \cos(-34°)$$
$$\therefore \quad \text{power factor} = 0.83$$

(b) Convert $125 + j90.0\ \Omega$ to polar form:

$$125 + j90.0\ \boxed{\rightarrow\text{P}}\ 154\underline{/35.8°}\ \Omega$$
$$\cos(\theta) = \cos(35.8°)$$
$$\therefore \quad \text{power factor} = 0.811$$

(c) Substitute $R = 18.0$ and $|Z| = 143$ into:

$$\text{power factor} = R/|Z|$$
$$\text{power factor} = 18.0/143$$
$$\therefore \quad \text{power factor} = 0.126$$

The word *leading* or the word *lagging* is often used with power factor. The relationship of the current through the load to the voltage as the reference phasor determines *lead* or *lag*. Since current leads voltage in a capacitive load, a **capacitive circuit has a leading power factor.** Since current lags voltage in an inductive load, an **inductive circuit has a lagging power factor.**

The generalized equations for computing the power dissipation in an ac circuit include the power factor as shown in the following:

$$P = |E||I|\cos(\theta) \qquad \text{W} \qquad\qquad (28\text{–}12)$$

$$P = |I|^2|Z|\cos(\theta) \qquad \text{W} \qquad\qquad (28\text{–}13)$$

$$P = \frac{|E|^2}{|Z|}\cos(\theta) \qquad \text{W} \qquad\qquad (28\text{–}14)$$

EXAMPLE 28–8   Determine the average power dissipated in an ac circuit having an impedance of $Z = 750\underline{/-55°}\ \Omega$ and a source voltage of $e = 165\sin(377t)$ V.

Solution   Solve for E in $E = 0.707\ E_{\text{mx}}$:

$$E = 0.707(165)$$
$$E = 117\text{ V}$$

Substitute into Equation 28–14:

$$P = \frac{|E|^2}{|Z|} \cos(\theta)$$

$$P = \frac{117^2}{750} \cos(55°)$$

$\therefore$ $P = 10.5\text{ W}$

EXAMPLE 28–9   A certain ac circuit has a leading power factor of 0.620, an impedance of 310 Ω, and a circuit current $i = 2.30 \sin(754t)$ A.

(a) Determine the power dissipation.
(b) Determine the source voltage E.
(c) Determine the series equivalent components.
(d) Draw the voltage current phasor with the current as the reference phasor.

Solution   (a) $P = |I|^2 |Z| \cos(\theta)$
        $P = [2.30(0.707)]^2 (310)(0.620)$
$\therefore$ $P = 508\text{ W}$

(b) $E = IZ$
        $|E| = (2.30)(0.707)(310)$
$\therefore$ $|E| = 504\text{ V}$

(c) $\angle|\theta| = \text{Arccos (power factor)}$
        $\angle|\theta| = \text{Arccos }(0.620) = 51.7°$

Observation   Because current is the reference phasor and the load is capacitive (leading power factor), the voltage lags and $\theta = -51.7°$.

$$310\underline{/-51.7°} \quad \boxed{\to\text{R}}\quad 192 - j243\ \Omega$$

Compute the capacitance ($\omega = 754$ rad/s):

$$C = 1/(\omega X_C)$$
$$C = 1/[754(243)]$$
$$C = 5.5\ \mu\text{F}$$

$\therefore$ The equivalent series circuit is a resistance of 192 Ω and a capacitance of 5.5 μF at $\omega = 754$ rad/s.

FIGURE 28-7 Phasor diagram for Example 28–9. Notice $i = 0.707I_{mx}$.

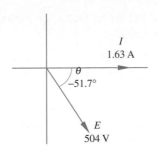

∴ **(d)** The phasor of Figure 28–7 pictures the voltage lagging the current. Remember, in a capacitive circuit (leading power factor) the current leads the voltage and the voltage lags the current.

EXERCISE 28-2

Convert each of the following to its effective value.

1. $E_{mx} = 14.3$ V   **2.** $I_{mx} = 7.4$ A   **3.** $E_{mx} = 17.2$ V

4. $I_{mx} = 10.3$ mA   **5.** $E_{mx} = 180$ V   **6.** $E_{mx} = 4.3$ kV

7. $I_{mx} = 15.7$ A   **8.** $E_{mx} = 62$ V   **9.** $I_{mx} = 8.1$ mA

Convert each of the following to its maximum value.

10. $E = 110$ V   **11.** $I = 1.32$ A   **12.** $I = 81.0$ mA

13. $I = 18.0$ mA   **14.** $E = 2.30$ kV   **15.** $E = 15.0$ V

16. $I = 76.2$ μA   **17.** $E = 44.0$ V   **18.** $I = 604$ μA

Compute the average power in watts given the following phasor diagrams.

19.

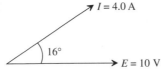

20.

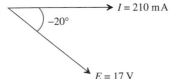

21.

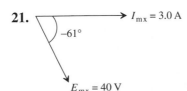

22.

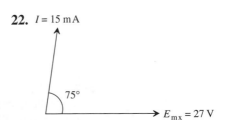

Determine the power factor for each of the following circuit conditions and state if it is leading or lagging.

23. $Z = 12 - j8 \ \Omega$ 　　　　**24.** $Z = 42\underline{/-81°} \ \Omega$

25. $R = 33 \ \Omega, |Z| = 89 \ \Omega$ 　　**26.** $Z = 218 + j143 \ \Omega$

Solve the following.

27. Determine the power dissipated in an ac circuit having a peak source voltage of $E_p = 18.0$ V and an impedance of $Z = 216\underline{/42°} \ \Omega$.

28. Determine the power dissipated in a capacitive load having an impedance of $Z = 717\underline{/-19°} \ \Omega$ and an instantaneous circuit current of $i = 1.25 \sin (377t)$.

29. Determine the equivalent series circuit for an ac circuit having a power dissipation of $P = 30$ W and a lagging power factor of 0.50. The circuit current is $i = 132 \times 10^{-3} \sin (5027t)$ A.

30. Determine the components of the equivalent series circuit for an ac circuit having a power dissipation of $P = 10$ W and a leading power factor of 0.72. The source voltage is $e = 14.3 \sin (377t)$ V.

28–3　INSTANTANEOUS EQUATIONS AND THE *EI* PHASOR DIAGRAM

The graph of the instantaneous equation of voltage and current gives a continuous *picture* of the interrelation between the circuit current and the source voltage. The phasor diagram of current and voltage gives a picture of the circuit conditions for a selected instant of time. Given sufficient information, each may be developed from the other.

The phasor diagram is constructed with one of the quantities acting as the reference phasor and the other phasor located θ radians or degrees from the reference phasor. Thus, if current was the reference phasor, it would be located at $I\underline{/0}$ and the voltage would be located at $E\underline{/\theta}$. We see that the reference phasor assumes 0 as its direction, while the other phasor assumes the phase angle θ as its direction.

When writing the instantaneous equations, one equation is written as the reference equation, while the other contains the phase angle. Either equation may be selected as the reference. The following equations are the generalized equations for the instantaneous voltage and current.

$$e = E_{mx} \sin (\omega t \pm \theta) \qquad \text{V} \qquad\qquad (28\text{–}15)$$
$$i = I_{mx} \sin (\omega t \pm \theta) \qquad \text{A} \qquad\qquad (28\text{–}16)$$

where ωt is in radians and θ is the phase angle in radians or degrees.

Special care must be taken when evaluating the sine of $(\omega t \pm \theta)$. Both ωt and θ must be in the same units (degrees or radians) before adding and evaluating.

EXAMPLE 28–10　　Use the phasor diagram of Figure 28–8 to write the instantaneous equations for voltage and current at a frequency of 60 Hz ($\omega = 377$ rad/s).

FIGURE 28–8 Phasor diagram for Example 28–10.

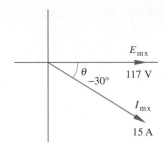

Solution Because voltage is the reference phasor, the instantaneous voltage equation will have a zero phase angle:

$$e = E_{mx} \sin (\omega t \pm \theta)$$
$$\therefore \quad e = 117 \sin (377t) \text{ V}$$

The current equation will include the phase angle:

$$i = I_{mx} \sin (\omega t \pm \theta)$$
$$\theta = -30° \left(\frac{\pi}{180°} \right) = -0.524^r$$
$$\therefore \quad i = 15 \sin (377t - 0.524) \text{ A}$$

Observation Figure 28–9 shows the phase relationship between the instantaneous current and the instantaneous voltage waves.

FIGURE 28–9 The current, i, lags the voltage, e, by $\angle \theta$, the phase angle.

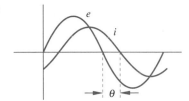

EXAMPLE 28–11 Construct a phasor diagram of the circuit conditions represented by the following equations when $t = 4.0$ ms:

$$e = 10 \sin (200t^r + 20°) \text{ V}$$
$$i = 2 \sin (200t^r) \text{ A}$$

Solution Compute the position in degrees of each phasor at $t = 4$ ms:

$$\alpha = \omega t \text{ (rad)}$$
$$\alpha = 200(4.0 \times 10^{-3})$$
$$\alpha = 0.80^r = 46°$$

$$\therefore \quad E_{mx} \text{ is located at } 66° \ (46° + 20°) \text{ and } I_{mx} \text{ is located at } 46°. \text{ See Figure 28–10.}$$

FIGURE 28–10 Phasor diagram for Example 28–11.

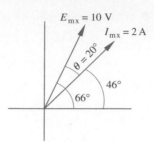

$E_{mx} = 10$ V
$I_{mx} = 2$ A
$\theta = 20°$
$46°$
$66°$

EXAMPLE 28–12 Use the phasor diagram of Figure 28–11 to write the instantaneous equations for voltage and current at a frequency of 400 Hz.

FIGURE 28–11 Phasor diagram for Example 28–12.

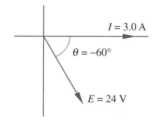

$I = 3.0$ A
$\theta = -60°$
$E = 24$ V

Solution Compute ω using $\omega = 2\pi f$:

$$\omega = 2\pi(400)$$
$$\omega = 2.51 \text{ krad/s}$$

Convert E to E_{mx}:

$$E = 0.707 E_{mx}$$
$$E_{mx} = E/0.707$$
$$E_{mx} = 24/0.707$$
$$E_{mx} = 34 \text{ V}$$

Convert I to I_{mx}:

$$I_{mx} = I/0.707$$
$$I_{mx} = 3.0/0.707$$
$$I_{mx} = 4.2 \text{ A}$$

The current phasor is the reference, so the instantaneous equation for current will have a zero phase angle:

$$i = I_{mx} \sin(\omega t \pm \theta) \text{ A}$$
$$\therefore \quad i = 4.2 \sin(2.51 \times 10^3 t) \text{ A}$$

$$e = E_{mx} \sin(\omega t - \theta) \text{ V}$$
$$e = 34 \sin(2.51 \times 10^3 t^r - 60°) \text{ V}$$

State the phase angle θ in radians. $60° \Rightarrow 1.05^r$:

$$\therefore \quad e = 34 \sin (2.51 \times 10^3 t^r - 1.05^r) \text{ V}$$

EXAMPLE 28–13 The voltage ($E = 36$ V) in an ac circuit leads the current ($I = 0.31$ A) by $27°$. Determine the instantaneous value of current when the voltage has completed 2.00 rad of its cycle.

Solution Because the instantaneous current *tracks* with the voltage, the current has also completed 2.00 rad of its cycle. Assume voltage is the reference phasor. Write the instantaneous equation for current:

$$i = I_{mx} \sin (\omega t - \theta)$$

Substitute $\omega t = 2.00^r$, $\theta = 27°$, and $I_{mx} = 0.31/0.707$ A:

$$i = (0.31/0.707) \sin (2.00^r - 27°)$$

Express $27°$ in radians. $27° \Rightarrow 0.47^r$:

$$i = 0.438 \sin (2.00^r - 0.47^r)$$
$$i = 0.438 \sin (1.53)$$
$$\therefore \quad i = 0.44 \text{ A}$$

EXERCISE 28–3

Write the instantaneous equations for voltage and current from the given phasor diagram and angular velocity in each of the following.

1.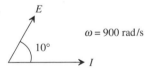

$\omega = 900$ rad/s

2.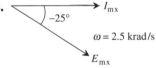

$\omega = 2.5$ krad/s

3.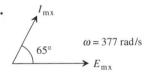

$\omega = 377$ rad/s

4.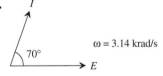

$\omega = 3.14$ krad/s

In the following, identify the phase angle (in degrees) and state which parameter (current or voltage) leads or lags the other.

5. $e = 18 \sin (200t^r)$ V
   $i = 2 \sin (200t^r - 38°)$ A

6. $e = 100 \sin (600t^r + 0.4^r)$ V
   $i = 7.0 \sin (600t^r)$ A

7. $e = 190 \sin (377t^r - 0.7^r)$ V
   $i = 0.62 \sin (377t^r)$ A

8. $e = 143 \sin (700t^r)$ V
   $i = 3.0 \sin (700t^r - 81°)$ A

Solve the following.

9. A 400-Hz, 2.50-V test oscillator delivers 30.0 mA to a reactive load. The voltage leads the current by 48.0°.

   (a) Write the instantaneous equations for voltage and current.

   (b) Determine the value of the instantaneous current at $t = 50.0 \ \mu s$.

   (c) Determine the average power delivered to the load.

10. A 60.0-Hz, 117-V generator delivers 20.0 A to a reactive load. The current lags the voltage by 1.1^r.

   (a) Write the instantaneous equations for voltage and current.

   (b) Determine the value of instantaneous current at $t = 4.00$ ms.

   (c) Determine the average power delivered to the load.

SELECTED TERMS

average power   The power (in watts) dissipated by the resistive component of the impedance in an ac circuit. Sometimes referred to as true power.

power factor   The ratio of resistance to impedance; the cosine of the circuit phase angle.

SECTION CHALLENGE

WEB CHALLENGE FOR CHAPTERS 26, 27, AND 28

To evaluate your comprehension of Chapters 26, 27, and 28, log on to **www.prenhall. com/harter** and take the online True/False and Multiple Choice assessments for each of the chapters.

SECTION CHALLENGE FOR CHAPTERS 26, 27, AND 28*

A. As pictured in Figure C–15(a), a 1.2-kΩ metal-film resistor is connected in parallel with a 0.82-μF capacitor and a 750-mH inductor. The circuit is attached to an airborne 400-Hz, 48.0-V single phase power bus. Your challenge is to (1) determine the branch current in each component and label the current phasor of Figure C–15(b); (2) determine the source current, $I\underline{/\theta}$; (3) label the current-voltage phasor of Figure C–15(c); and (4) write the instantaneous equation for the source voltage and circuit current.

B. Your second challenge uses the parallel circuit of Figure C–16. The loads are connected to a 15.0-V, 1.00-kHz sine wave signal generator. You are to determine (1) the impedance of each branch; (2) the admittance of each branch; (3) the current in each branch; (4) the impedance of the entire circuit; (5) the source current; and (6) the power supplied to the circuit by the source.

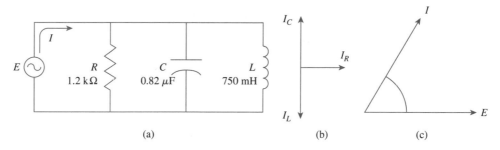

(a)   (b)   (c)

FIGURE C–15 (a) A three-branch parallel circuit; (b) Current phasor diagram; (c) *EI* phasor diagram.

FIGURE C–16 Three-branch parallel circuit powered from a 15.0-V, 1.00-kHz sine wave signal generator.

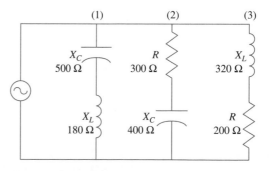

*The solution to this Section Challenge is found in Appendix C.

29

Additional Trigonometric and Exponential Functions

29–1   Auxiliary Trigonometric Functions

29–2   Graphs of the Auxiliary Trigonometric Functions

29–3   Trigonometric Identities

29–4   Hyperbolic Functions

29–5   Graphing the Hyperbolic Functions

29–6   Hyperbolic Identities

29–7   Inverse Hyperbolic Functions

PERFORMANCE OBJECTIVES

- **Evaluate the auxiliary trigonometric functions.**
- **Graph the auxiliary trigonometric functions.**
- **Determine if a trigonometric equation is an identity.**
- **Evaluate the hyperbolic functions.**
- **Graph the hyperbolic functions.**
- **Determine if a hyperbolic equation is an identity.**
- **Evaluate the inverse hyperbolic functions.**

Using triangulation principles, the high-intensity, high-precision analog laser displacement sensor measures distance to a target with accuracies to 4 μm (.00016 in).   (Courtesy of Baumer Electric, Ltd.)

This and the following chapter are designed to help you prepare for a course in calculus. The functions and concepts introduced in this chapter will be encountered again in the following chapter and in most calculus textbooks. As in calculus textbooks, all angles are assumed to be in radians unless specified to be in degrees.

29–1 AUXILIARY TRIGONOMETRIC FUNCTIONS

In this section, three additional trigonometric functions are introduced. These functions are not as common in electronics as the sine, cosine, and tangent functions. However, they do occur in a study of calculus and so we introduce them at this time. These functions are the secant, cosecant, and cotangent.

The first function is the **secant,** which is equal to the reciprocal of the cosine. Thus, for any angle θ, sec $(\theta) = 1/\cos(\theta)$. Many calculators do not have the secant function, but they do have a reciprocal key. The way to calculate the secant function is to calculate the cosine, and then calculate the reciprocal.

EXAMPLE 29–1	Use your calculator to find sec $(45°)$.
Solution	Set your calculator to work in degrees and enter 45:

$$\boxed{\text{COS}} \quad 45 \Rightarrow 0.707$$

Form the reciprocal:

$\boxed{1/x}$ 1.414

$\therefore$ sec $(45°) = 1.414$

The second function is the **cosecant,** which is equal to the reciprocal of the sine. Thus, for any angle θ, csc $(\theta) = 1/\sin(\theta)$. The cosecant function may be calculated as the reciprocal of the sine.

EXAMPLE 29–2	Use your calculator to find csc $(30°)$.
Solution	Set your calculator to work in degrees and enter 30:

$$\boxed{\text{SIN}} \quad 30 \Rightarrow 0.500$$

Form the reciprocal:

$\boxed{1/x}$ 2.0

$\therefore$ csc $(30°) = 2.0$

The third function is the **cotangent,** which is equal to the reciprocal of the tangent. Thus, for any angle θ, cot (θ) = 1/tan (θ). Note that some authors use ctn ( ) to indicate the cotangent. Again, if your calculator does not have this function, use the tangent and the reciprocal to calculate the cotangent.

EXAMPLE 29–3   Use your calculator to find cot (1.42).

Solution   Set your calculator to work in radians and enter 1.42:

$\boxed{\text{TAN}}$ 1.42 $\Rightarrow$ 6.58

Form the reciprocal:

$\boxed{1/x}$ 0.152

$\therefore$ cot (1.42) = 0.152

The definitions of the trigonometric functions are presented in Table 29–1. This table refers to Figure 29–1.

TABLE 29–1 Definitions of Trigonometric Functions

Function	Abbreviation	Definition	Relation to Other Functions
Sine (θ)	sin (θ)	y/z	$1/\csc(\theta)$
Cosine (θ)	cos (θ)	x/z	$1/\sec(\theta)$
Tangent (θ)	tan (θ)	y/x	$\sin(\theta)/\cos(\theta)$
Secant (θ)	sec (θ)	z/x	$1/\cos(\theta)$
Cosecant (θ)	csc (θ)	z/y	$1/\sin(\theta)$
Cotangent (θ)	cot (θ)	x/y	$1/\tan(\theta)$

FIGURE 29–1   Circle used to define the auxiliary trigonometric functions.

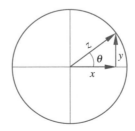

EXERCISE 29–1

Calculator Drill
Evaluate the following functions.

1. sec (29°)	**2.** sec (50°)	**3.** sec (−30°)
4. sec (150°)	**5.** sec (2.3)	**6.** sec (−1.05)
7. sec (0.56)	**8.** sec (6.28)	**9.** csc (82°)
10. csc (98°)	**11.** csc (−25°)	**12.** csc (−135°)
13. csc (1.4)	**14.** csc (6.0)	**15.** csc (−2.0)
16. csc (4.0)	**17.** cot (12.5°)	**18.** cot (37.5°)
19. cot (75°)	**20.** cot (115°)	**21.** cot (−0.32)
22. cot (−1.0)	**23.** cot (−2.0)	**24.** cot (1.05)

29–2 GRAPHS OF THE AUXILIARY TRIGONOMETRIC FUNCTIONS

The secant, cosecant, and cotangent functions are periodic functions, as are their reciprocals. From the previous section we know the definitions of the functions and how to calculate their values. To have a better idea of how a function behaves, it is useful to graph the function.

EXAMPLE 29–4   Create a plot of the secant function for angles from −1.57 to 7.85 rad. On the same graph, plot the cosine function.

Solution   Build a table of θ, cos (θ), and sec (θ) for θ varying from −1.57 to 7.85:

Sample Function Table

θ	cos (θ)	sec (θ)
−1.57	0.0008	1260.0
−1.5	0.071	14.1
−1.4	0.170	5.88
−1.3	0.267	3.74
−1.2	0.362	2.76
−1.0	0.540	1.85

See Figure 29–2 for the graphs.

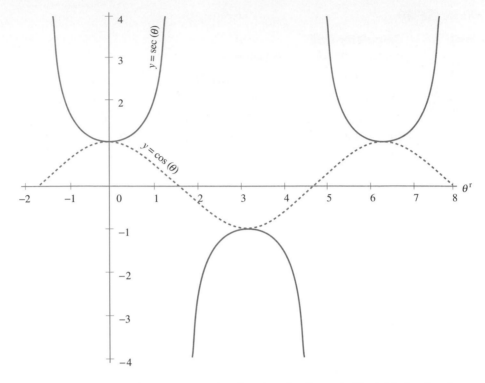

FIGURE 29–2 Graph of secant and cosine functions for Example 29–4. Notice that the secant goes off the graph wherever the cosine is "close" to zero.

EXERCISE 29–2

1. Form a graph of the sine and cosecant functions for angles from 0 to 6.28 rad.
2. Form a graph of the tangent and cotangent functions for angles from -1.57 to 7.85 rad.

29–3 TRIGONOMETRIC IDENTITIES

A trigonometric **identity** is an equation that is true for all angles. For example, the definition of the secant function sec $(\theta) = 1/\cos(\theta)$ is an identity. Many identities can be derived from the definitions of the circular functions. We write these identities using the special symbol $\equiv$ for identically equal.

EXAMPLE 29–5   Show that tan $(\theta) \equiv \sin(\theta)/\cos(\theta)$.

Solution   Substitute the ratio definitions for the circular functions.
Left member:

$$\tan(\theta) = y/x$$

Right member:

$$\sin(\theta)/\cos(\theta) = (y/z)/(x/z)$$
$$= (y/z)(z/x)$$
$$= y/x$$

This is the same value as for the tangent (independent of θ).

∴ $\tan(\theta) \equiv \sin(\theta)/\cos(\theta)$

EXAMPLE 29–6   Check $\cos^2(\theta) + \sin^2(\theta) \equiv 1$. Remember, $\cos^2(\theta)$ means $\cos(\theta)\cos(\theta)$.

Solution   Substitute the ratio definitions. Left member:

$$\cos^2(\theta) + \sin^2(\theta) = (x/z)^2 + (y/z)^2$$
$$= x^2/z^2 + y^2/z^2$$
$$= \frac{x^2 + y^2}{z^2}$$

Apply the Pythagorean theorem ($x^2 + y^2 = z^2$):

$$\cos^2(\theta) + \sin^2(\theta) = z^2/z^2 = 1$$

This is the same value as the right member:

∴ $\cos^2(\theta) + \sin^2(\theta) \equiv 1$

EXERCISE 29–3

Check the following identities.

1. $\cot(\theta) \equiv \cos(\theta)/\sin(\theta)$
2. $\tan(\theta) \equiv \sin(\theta)\sec(\theta)$
3. $\csc(\theta) \equiv \cot(\theta)\sec(\theta)$
4. $\sec^2(\theta) \equiv 1 + \tan^2(\theta)$
5. $\tan(\theta) \equiv \sec(\theta)/\csc(\theta)$
6. $\cot(\theta) \equiv \csc(\theta)/\sec(\theta)$
7. $\tan(\theta)\cot(\theta) \equiv 1$
8. $\sin(\theta)\csc(\theta) \equiv 1$
9. $\tan(\theta) + \cot(\theta) \equiv \sec(\theta)\csc(\theta)$
10. $\cos(\theta)\sec(\theta) \equiv 1$
11. $\sin(\theta)\cos(\theta)\sec(\theta)\csc(\theta) \equiv 1$
12. $1 - 2\sin^2(\theta) \equiv 2\cos^2(\theta) - 1$
13. $\cos(\theta) + \tan(\theta)\sin(\theta) \equiv \sec(\theta)$
14. $\cos^4(\theta) - \sin^4(\theta) \equiv \cos^2(\theta) - \sin^2(\theta)$
15. $\cot(\theta) + \tan(\theta) \equiv \cot(\theta)\sec^2(\theta)$
16. $\csc^2(\theta)\tan^2(\theta) \equiv 1 + \tan^2(\theta)$

17. $\dfrac{1}{1 - \sin(\theta)} + \dfrac{1}{1 + \sin(\theta)} \equiv 2\sec^2(\theta)$

18. $\dfrac{\tan(\theta) - \cot(\theta)}{\tan(\theta) + \cot(\theta)} \equiv 2\sin^2(\theta) - 1$

19. $\dfrac{1 - \cos(\theta)}{\sin(\theta)} \equiv \dfrac{\sin(\theta)}{1 + \cos(\theta)}$

20. $\dfrac{\cos(\theta) - \sin(\theta)}{\cos(\theta) + \sin(\theta)} \equiv \dfrac{\cot(\theta) - 1}{\cot(\theta) + 1}$

21. $\dfrac{1}{\sin(\theta)} - \sin(\theta) \equiv \cot(\theta)\cos(\theta)$

22. $\tan(\theta)\sin(\theta) \equiv \sec(\theta) - \cos(\theta)$

23. $\dfrac{\sin(\theta)}{1 + \cos(\theta)} + \dfrac{1 + \cos(\theta)}{\sin(\theta)} \equiv 2\csc(\theta)$

24. $\dfrac{1 + \tan^2(\theta)}{\tan^2(\theta)} \equiv \csc^2(\theta)$

29–4   HYPERBOLIC FUNCTIONS

These interesting functions occur in the study of the interaction of electric and magnetic fields. They also occur in power transmission applications. These functions can be identified as ratios in geometric figures known as hyperbolas. However, we will present them in terms of the exponential function.

The first hyperbolic function that we introduce is the *hyperbolic sine function,* which is defined by sinh $(x) = (e^x - e^{-x})/2$. The symbol sinh (   ) is read "hyperbolic sine" and is used to indicate the function. Check your calculator and your owner's guide to see if your calculator will calculate this function for you. Look for a key like $\boxed{\text{HYP}}$ used with the sine function key $\boxed{\text{SIN}}$ to calculate the hyperbolic sine. If your calculator does not have this ability, you can use the definition and the exponential function to calculate the hyperbolic sine. Remember that $e^{-x} = 1/e^x$.

EXAMPLE 29–7   Use the definition sinh $(x) = (e^x - e^{-x})/2$ to calculate sinh (1.5).

Solution   Use your calculator; enter 1.5:

$\boxed{e^x}$ $e^{1.5} = 4.48$

Store this value; then change its sign and form its reciprocal:

$\boxed{\text{CHS}}$ $\boxed{1/x}$ $-e^{-1.5} = -0.223$

Form the sum of this and the previous value:

$\boxed{+}$ $\qquad$ $e^{1.5} - e^{-1.5} = 4.26$

Divide by 2:

$\boxed{\div}$ $\qquad$ $\dfrac{e^{1.5} - e^{-1.5}}{2} = 2.13$

$\therefore$ $\qquad$ $\sinh(1.5) = 2.13$

The second hyperbolic function is the *hyperbolic cosine function,* which is defined by $\cosh(x) = (e^x + e^{-x})/2$. The symbol cosh ( ) is read "hyperbolic cosine" and is used to indicate the function. The hyperbolic cosine is calculated in a manner similar to the hyperbolic sine.

EXAMPLE 29–8   Use the definition $\cosh(x) = (e^x + e^{-x})/2$ to calculate the cosh (0.11).

Solution   Using your calculator, enter 0.11 and exponentiate:

$\boxed{e^x}$ $\qquad$ $e^{0.11} = 1.12$

Store this value and form its reciprocal:

$\boxed{1/x}$ $\qquad$ $e^{-0.11} = 0.896$

Form the sum of this and the previous value:

$\boxed{+}$ $\qquad$ $e^{0.11} + e^{-0.11} = 2.01$

Divide by 2:

$\boxed{\div}$ $\qquad$ $\dfrac{e^{0.11} + e^{-0.11}}{2} = 1.01$

$\therefore$ $\qquad$ $\cosh(0.11) = 1.01$

EXAMPLE 29–9   Evaluate $\cosh(-0.21)$.

Solution   Use the hyperbolic function key; enter -0.21.

$\boxed{\text{HYP}}$ $\boxed{\text{COS}}$ $\qquad$ 1.02

$\therefore$ $\qquad$ $\cosh(-0.21) = 1.02$

The third hyperbolic function is the *hyperbolic tangent function,* which is defined by $\tanh(x) = (e^x - e^{-x})/(e^x + e^{-x})$. The symbol tanh ( ) is read "hyperbolic tangent" and is used to indicate the function.

EXAMPLE 29–10   Use the definition tanh $(x) = (e^x - e^{-x})/(e^x + e^{-x})$ to calculate tanh (0.5).

Solution   Using your calculator, enter 0.5 and exponentiate:

$\boxed{e^x}$
$$e^{0.5} = 1.65$$

Store this value, change the sign, and form the reciprocal:

$\boxed{\text{CHS}}$ $\boxed{1/x}$
$$-e^{-0.5} = -0.61$$

Form the sum of this and the previous result:

$\boxed{+}$
$$e^{0.5} - e^{-0.5} = 1.04$$

Save this result as the numerator. To form the denominator recall $e^{0.5}$:

$\boxed{\text{RCL}}$
$$e^{0.5} = 1.65$$

Form the reciprocal:

$\boxed{1/x}$
$$e^{-0.5} = 0.61$$

Sum this and the previous result:

$\boxed{+}$
$$e^{0.5} + e^{-0.5} = 2.26$$

Divide into the numerator:

$\boxed{\div}$
$$(e^{0.5} - e^{-0.5})/(e^{0.5} + e^{-0.5}) = 0.462$$

$\therefore$ tanh $(0.5) = 0.462$

EXAMPLE 29–11   Use the hyperbolic key to calculate tanh (1.0).

Solution   Using your calculator, enter 1.0:

$\boxed{\text{HYP}}$ $\boxed{\text{TAN}}$ 0.762

$\therefore$ tanh $(1.0) = 0.762$

EXERCISE 29–4

Calculator Drill

1. Find the hyperbolic sine of the following numbers:
   (a) -1.0 **(b)** -0.5 **(c)** 0.0
   (d) 0.6 **(e)** 1.2 **(f)** 2.0

2. Find the hyperbolic cosine of the numbers in problem 1.

3. Find the hyperbolic tangent of the numbers in problem 1.

29–5 GRAPHING THE HYPERBOLIC FUNCTIONS

The hyperbolic functions are not periodic functions. This becomes clear when we look at graphs of the functions. In the previous section we developed techniques for using a calculator to evaluate the functions. We can use those techniques to build tables and graphs of the functions.

EXAMPLE 29–12   Graph $y = \tanh(x)$ for values of x between -3 and 3.

Solution   Build a table of function values; then sketch the graph (Figure 29–3):

x	0.0	0.1	0.2	0.5	0.75	1.0
tanh *(x)*	0.0	0.100	0.197	0.462	0.635	0.762

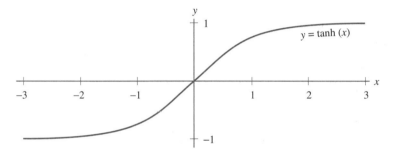

FIGURE 29–3   Graph of hyperbolic tangent function for Example 29–12. Notice that $\tanh(-x) = -\tanh(x)$.

EXERCISE 29–5

Graph the following functions for x between -2.3 and 2.3.

1. $y = e^x/2$ **2.** $y = e^{-x}/2$

3. $y = \cosh(x)$ **4.** $y = \sinh(x)$

29–6 HYPERBOLIC IDENTITIES

A hyperbolic identity is an equation involving the hyperbolic functions that is true for all values of the argument. Again, we will use the symbol $\equiv$ for identically equal.

EXAMPLE 29–13   Check $\cosh^2(x) - \sinh^2(x) \equiv 1$.

Solution   Substitute the exponential definition for $\cosh(x)$ and $\sinh(x)$ in the left member:

$$\cosh^2(x) - \sinh^2(x) = \left(\frac{e^x + e^{-x}}{2}\right)^2 - \left(\frac{e^x - e^{-x}}{2}\right)^2$$

$$= \frac{e^{2x} + 2 + e^{-2x}}{4} - \frac{e^{2x} - 2 + e^{-2x}}{4}$$

$$= 4/4$$

$$= 1$$

This is the value of the right member and it is independent of the value of x.

$$\therefore \quad \cosh^2(x) - \sinh^2(x) \equiv 1$$

EXERCISE 29–6

Check the following identities.

1. $\sinh(x) + \cosh(x) \equiv e^x$
2. $\tanh(-x) \equiv -\tanh(x)$
3. $\tanh(x) \equiv \sinh(x)/\cosh(x)$
4. $\cosh(x) \equiv \cosh(-x)$
5. $\sinh(x) \equiv -\sinh(-x)$
6. $\sinh(2x) \equiv 2\sinh(x)\cosh(x)$
7. $\cosh(2x) \equiv \cosh^2(x) + \sinh^2(x)$
8. $\sinh(2x) + \cosh(2x) \equiv [\sinh(x) + \cosh(x)]^2$
9. $\tanh^2(x) + 1/\cosh^2(x) \equiv 1$
10. $1/\tanh^2(x) - 1/\sinh^2(x) \equiv 1$

29–7 INVERSE HYPERBOLIC FUNCTIONS

If your calculator has the hyperbolic function key, check your owner's guide for the procedure to calculate the inverse hyperbolic functions. The keystroke sequence should look like $\boxed{\text{INV}}\,\boxed{\text{HYP}}\,\boxed{\text{SIN}}$, $\boxed{f^{-1}}\,\boxed{\text{HYP}}\,\boxed{\text{SIN}}$, or $\boxed{\text{HYP}^{-1}}\,\boxed{\text{SIN}}$.

For those calculators without the hyperbolic functions, we can use some of the identities of the previous section to develop formulas that can be used to evaluate the inverse hyperbolic functions.

EXAMPLE 29–14 Develop a formula for the inverse hyperbolic sine. For $S = \sinh(x)$, find x.

Solution We know from the preceding section that $e^x = \sinh(x) + \cosh(x)$. If we can express $\cosh(x)$ in terms of $\sinh(x)$, we can use natural logarithms to find x. We also know that $\cosh^2(x) - \sinh^2(x) = 1$.

Solve for $\cosh(x)$:

$$\cosh^2(x) = \sinh^2(x) + 1$$

$$\cosh(x) = \sqrt{\sinh^2(x) + 1}$$

Substitute for cosh (x) in $e^x = \sinh(x) + \cosh(x)$:

$$e^x = \sinh(x) + \sqrt{\sinh^2(x) + 1}$$

Substitute S for sinh (x):

$$e^x = S + \sqrt{S^2 + 1}$$

Apply natural logarithms to both members:

$$x = \ln(S + \sqrt{S^2 + 1})$$

$$\therefore \quad \sinh^{-1}(S) = \ln(S + \sqrt{S^2 + 1})$$

EXAMPLE 29–15   Check the formula of Example 29–14 for $x = -2$.

Solution   Use your calculator to find sinh (-2) and store the result:

$\boxed{\text{HYP}}\ \boxed{\text{SIN}}$
$$S = \sinh(-2) = -3.6269$$

Use the formula with $S = -3.6269$. Square the result:

$\boxed{x^2}$
$$S^2 = 13.15$$

Add 1:

$\boxed{+}$
$\boxed{\sqrt{x}}$
$$S^2 + 1 = 14.15$$
$$\sqrt{S^2 + 1} = 3.76$$

Recall and add:

$\boxed{+}$
$\boxed{\ln x}$
$$S + \sqrt{S^2 + 1} = 0.135$$
$$\ln(S + \sqrt{S^2 + 1}) = -2.00$$

$\therefore$ The formula checks for this value.

A similar formula for the inverse hyperbolic cosine is given in Equation 29–1.

$$\mathbf{\cosh^{-1}(C) = \ln(C + \sqrt{C^2 - 1})} \tag{29-1}$$

Because cosh $(-x) = \cosh(x)$, there are two possible solutions for $\cosh^{-1}$. Equation 29–1 gives the positive solution, which is called the *principal* solution.

EXAMPLE 29–16   Check Equation 29–1 for cosh (1.75).

Solution   Use your calculator to find cosh (1.75) and store the result:

$\boxed{\text{HYP}}\ \boxed{\text{COS}}$
$$C = \cosh(1.75) = 2.9642$$

Use Equation 29–1 with $C = 2.9642$:

$\boxed{x^2}$
$$C^2 = 8.79$$

Subtract 1:

$\boxed{-}$

$$C^2 - 1 = 7.79$$

$\boxed{\sqrt{x}}$

$$\sqrt{C^2 - 1} = 2.79$$

Recall and add:

$\boxed{+}$

$$C + \sqrt{C^2 - 1} = 5.75$$

$\boxed{\ln x}$

$$\ln (C + \sqrt{C^2 - 1}) = 1.75$$

$\therefore$ The formula checks for this value.

The formula for the inverse hyperbolic tangent is given by Equation 29–2.

$$\tanh^{-1} (T) = \tfrac{1}{2}\ln \left[(1 + T)/(1 - T)\right] \qquad \textbf{(29–2)}$$

EXAMPLE 29–17   Check Equation 29–2 for tanh (-1.00).

Solution   Use your calculator to find tanh (-1.00) and store the result:

$\boxed{\text{HYP}}\ \boxed{\text{TAN}}$ $T = \tanh (-1.00) = -0.7616$

Use Equation 29–2 with $T = -0.7616$. Start by adding 1 to T:

$\boxed{+}$ $1 + T = 0.238$

Store and form $(1 - T)$:

$\boxed{-}$ $1 - T = 1.76$

Recall and divide:

$\boxed{\div}$ $(1 + T)/(1 - T) = 0.135$

$\boxed{\ln x}$ $\ln \left[(1 + T)/(1 - T)\right] = -2.00$

$\tfrac{1}{2}\ln \left[(1 + T)/(1 - T)\right] = -1.00$

$\therefore$ Equation 29–2 checks for tanh (-1.00).

EXERCISE 29–7

Evaluate the following inverse functions.

1. $\sinh^{-1} (0.8223)$ **2.** $\sinh^{-1} (10.018)$

3. $\sinh^{-1} (-6.0502)$ **4.** $\sinh^{-1} (-1.1752)$

5. $\cosh^{-1} (1.5431)$ **6.** $\cosh^{-1} (1.1276)$

7. $\cosh^{-1} (2.1509)$ **8.** $\cosh^{-1} (27.308)$

9. $\tanh^{-1}(0.5005)$ **10.** $\tanh^{-1}(0.9051)$

11. $\tanh^{-1}(-0.9640)$ **12.** $\tanh^{-1}(0.2449)$

SELECTED TERMS

cosecant   The reciprocal of the sine function: $\csc(x) = 1/\sin(x)$.

cotangent   The reciprocal of the tangent function: $\cot(x) = 1/\tan(x)$.

identity   An equation that is true for all values of the variable.

secant   The reciprocal of the cosine function: $\sec(x) = 1/\cos(x)$.

30

Mathematical Analysis

30–1   Domain and Range

30–2   Discontinuities

30–3   Functions of Large Numbers

30–4   Asymptotes

PERFORMANCE OBJECTIVES

- Determine the domain of a function.

- Determine the range of a function.

- Determine if a function has any discontinuities.

- Find the limit of a function as the argument increases to infinity.

A model of a tokamak fusion reactor—a magnetic containment vessel for hydrogen fusion.   (Courtesy of the Department of Energy)

This chapter examines several properties of functions and their behaviors. It is not a complete discussion of analysis; it is only an introduction to the types of questions addressed in a course on calculus.

30–1 DOMAIN AND RANGE

This section presents two useful properties of functions, the domain and the range of the function. The domain is associated with the values of the independent variable; the range is associated with the values of the dependent variable or function.

The **domain** of a function is defined as the *list* of values of the independent variable for which the function is defined. For example, the sine function is defined for all values of the angle while the Arcsine function is defined only for values from -1 to $+1$. Thus, the domain of $Y = \sin(\theta)$ equals all values of θ, sometimes written $-\infty < \theta < \infty$. The domain of $\theta = \text{Arcsin}(Y)$ equals $-1 \leqslant Y \leqslant 1$.

The **range** of a function is defined as the *list* of values that the function can assume over the entire domain. For example, the sine function takes on all values from -1 to 1, while the Arcsine function takes on the values from -1.571^r to 1.571^r. Thus, the range of $Y = \sin(\theta)$ equals $-1 \leqslant Y \leqslant 1$. The range of $\theta = \text{Arcsin}(Y)$ equals $-1.571^r \leqslant \theta \leqslant 1.571^r$. Figure 30–1 illustrates the concepts of range and domain for the Arcsine function.

FIGURE 30–1 The domain and range of the Arcsine function. The domain equals $-1 \leqslant Y \leqslant 1$, and the range equals $-1.571^r \leqslant \theta \leqslant 1.571^r$.

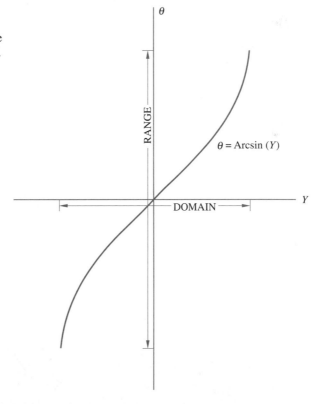

Determine the domain and range of the following functions.

1. $Y = \cos(\theta)$ 2. $\theta = \mathrm{Arccos}(Y)$
3. $Y = \sinh(X)$ 4. $Y = \cosh(X)$
5. $Y = \tanh(X)$ 6. $Y = \ln(R)$
7. $Z = e^u$ 8. $r = \sqrt{S}$
9. $t = 2s$ 10. $t = |2s|$

30–2 DISCONTINUITIES

In the preceding section we looked at concepts of domain and range, which are related to how much of the independent axis and the dependent axis are used to graph the whole function. In this section, we look at the graph itself to see whether there are any "breaks" in it. These breaks are called **discontinuities.**

If there are no breaks or discontinuities in a function, it is said to be continuous. The graph of $s = 3r - 1$ shown in Figure 30–2 is for only a portion of the domain of the function. For any other portion, the graph would *look* the same. The function $s = 3r - 1$ is a continuous function because there are no breaks in the graph.

There are three types of discontinuities. The first type is called a **removable discontinuity,** because if the function is redefined at a single point, it becomes a continuous function. The function $y = (x^2 - 1)/(x - 1)$ is not defined when x is equal to 1. Examination of the graph in Figure 30–3 shows that there is a break in the graph at $x = 1$. Furthermore, to fill the break would require setting the function to 2 at this point. Thus, the function

$$Y = \begin{cases} (x^2 - 1)/(x - 1) & x \neq 1 \\ 2, & x = 1 \end{cases}$$

is a continuous function. The discontinuity has been *removed.*

FIGURE 30–2 Graph of a continuous function.

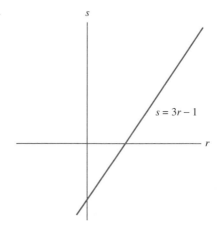

$s = 3r - 1$

FIGURE 30–3 Graph of a removable discontinuity.

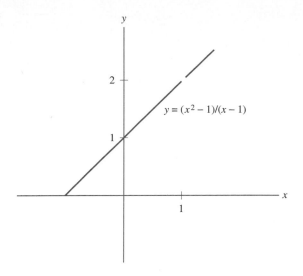

$y = (x^2 - 1)/(x - 1)$

The second type of discontinuity is called a **step discontinuity** because of the appearance of the graph. The trace of a square wave with a rise time of 20 ns on an oscilloscope set at 1 μs per division will look like a function with step discontinuities, as shown in Figure 30–4(a). A faster sweep will produce a continuous trace with ramps as in Figure 30–4(b). A step discontinuity can be replaced with a very steep ramp without totally destroying the function.

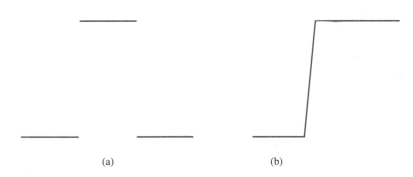

(a)   (b)

FIGURE 30–4 Oscilloscope trace of a square wave: (a) step discontinuity; (b) ramp for faster sweep time.

EXAMPLE 30–1 The greatest integer function, $G(x)$, is defined as the greatest integer less than or equal to x. Replace each step discontinuity with a ramp with a slope of 10.

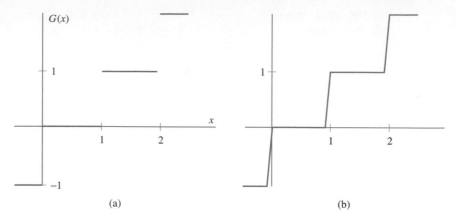

FIGURE 30–5 Graphs for Example 30–1: (a) greatest integer function showing step discontinuities; (b) modified greatest integer function with step discontinuities removed.

Solution   Create a table of function values for $-0.5 \leqslant x \leqslant 2.5$ and plot the function:

x	−0.5	−0.1	0.0	0.5	1.0	1.5	1.99	2.0	2.5
$G(x)$	−1.0	−1.0	0.0	0.0	1.0	1.0	1.0	2.0	2.0

The points are plotted in Figure 30–5(a). At each integer value of x, $G(x)$ has a step increase of 1. To replace the step with a ramp of slope 10, use the definition of slope and solve for Δx:

$$\text{slope} = \Delta G/\Delta x$$
$$10 = 1/\Delta x$$
$$\Delta x = 0.1$$

Redraw the function as in Figure 30–5(b). Begin a ramp at 0.1 ahead of each integer value of x. End the ramp when x is an integer at the proper value of the function.

The third type of discontinuity is called an **essential discontinuity.** It is termed *essential* because it cannot be removed or glossed over, as in the case of removable and step discontinuities. The function $s = 1/r$ is plotted in Figure 30–6. This function has an essential discontinuity at $r = 0$.

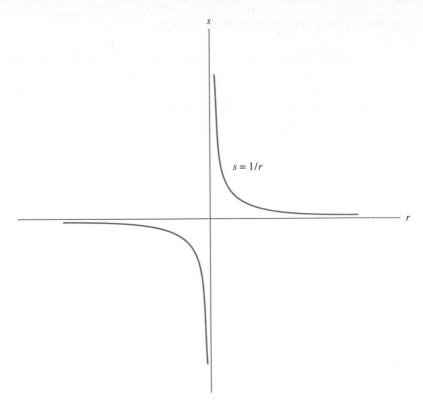

FIGURE 30–6   Graph of $s = 1/r$ showing an essential discontinuity.

Determine whether the following eight functions are continuous or discontinuous, and, if discontinuous, the type of discontinuity.

1. $y = \cos(\theta)$ **2.** $S = \tanh(r)$

3. $Z = \tan(\omega t)$ **4.** $u = (v^2 + v - 2)/(v + 2)$

5. $V = e^t$ **6.** $y = |x|$

7. $w = \sec(\phi)$

8. $y = \text{sign}(x) = \begin{cases} 1 \text{ if } x > 0 \\ 0 \text{ if } x = 0 \\ -1 \text{ if } x < 0 \end{cases}$

9. Define the remainder function $r(x)$ in terms of the greatest integer function $G(x)$ by $r(x) = x - G(x)$. Plot the function for $0 \leqslant x \leqslant 3$. Describe its behavior.

30–3 FUNCTIONS OF LARGE NUMBERS

A concept that underlies much of calculus is the concept of a **limit.** How does a function behave as the independent variable approaches a specific value? In this section, we look at the behavior of several functions as the independent variable is allowed to become *large without bound* or to approach *infinity.*

To aid in our discussions, we introduce several symbols. The first symbol, ∞, is read as "infinity" and is used to mean "larger than any number you can think of." The second symbol, $\lim_{x \to \infty} f(x)$, is read as "limit of $f(x)$ as x approaches infinity."

Now let's consider the behavior of several familiar functions as the independent variable is allowed to approach infinity.

EXAMPLE 30–2   Discuss the behavior of $f(x) = x$ as x approaches infinity.

Solution   Sketch a graph of $f(x)$; see Figure 30–7.
The function is always equal to x; so as x grows, so does the function.

$\therefore \quad \lim_{x \to \infty} x = \infty$

FIGURE 30–7   Graph for Example 30–2.

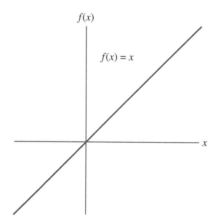

EXAMPLE 30–3   Discuss the behavior of $S(r) = 1/r$ as r approaches infinity.

Solution   Sketch a graph of $S(r)$; refer to Figure 30–6.
As r increases beyond zero, $1/r$ becomes a smaller and smaller positive number. When $r \gg 1$, $0 < 1/r \ll 1$.

$\therefore \quad \lim_{r \to \infty} 1/r = 0$

EXAMPLE 30–4   Find $\lim\limits_{\theta \to \infty} \sin (\theta)$.

Solution   Sketch a graph of sin (θ); see Figure 30–8.

The sine function is a periodic function, and as θ increases, sin (θ) keeps repeating its oscillation between 1 and -1.

∴ $\lim\limits_{\theta \to \infty} \sin (\theta)$ does not exist.

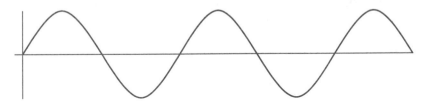

FIGURE 30–8   Sketch of sine function for Example 30–4.

EXAMPLE 30–5   Find $\lim\limits_{x \to \infty}$ Arctan (x) in radians.

Solution   Sketch a graph of Arctan (x); see Figure 30–9.

The Arc tangent function is a continuous function that increases with increasing values of x. Consider the following table:

x	1	10	100	1000	10 000
Arctan (x)	0.785	1.471	1.561	1.570	1.571

∴ $\lim\limits_{x \to \infty}$ Arctan $(x) \approx 1.571$

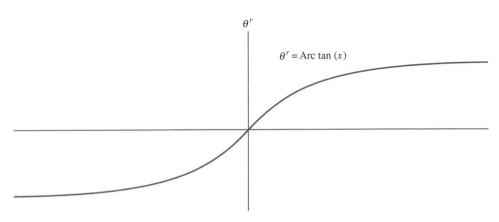

FIGURE 30–9   Sketch of Arctan (x) for Example 30–5.

There are three possible answers to the question of what $\lim\limits_{x\to\infty} f(x)$ is. The limit may be a finite number such as zero or 1.571. The limit may be infinite, either $+\infty$ or $-\infty$. Finally, the limit may not exist because the function keeps oscillating and the amplitude of oscillation is not *tending to zero*.

EXERCISE 30–3

Find the following limits.

1. $\lim\limits_{\theta\to\infty} \cos(\theta)$

2. $\lim\limits_{x\to\infty} e^x$

3. $\lim\limits_{t\to\infty} e^{-t}$

4. $\lim\limits_{Z\to\infty} \sinh(Z)$

5. $\lim\limits_{Z\to\infty} \cosh(Z)$

6. $\lim\limits_{t\to\infty} \tanh(t)$

7. $\lim\limits_{x\to\infty} \sqrt{x}$

8. $\lim\limits_{w\to\infty} \ln(w)$

9. $\lim\limits_{x\to\infty} (2x+3)/x$

10. $\lim\limits_{t\to\infty} e^{1/t}$

30–4 ASYMPTOTES

Some complicated functions behave like much simpler functions for large values of the independent variables. When this happens, the simpler function is referred to as an **asymptote** for the more complicated function. Strictly, an asymptote is a straight line, but we will use this more general definition.

EXAMPLE 30–6   Examine the behavior of tanh (x) as x increases.

Solution   Sketch the graph of $y = \tanh(x)$; see Figure 30–10.
Notice that for $x > 2.5$, $\tanh(x) \approx 1.0$. Also for $x < -2.5$, tanh $x \approx -1.0$.

∴ $y = 1$ and $y = -1$ are asymptotes for $y = \tanh(x)$.

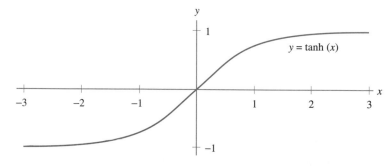

FIGURE 30–10   Sketch of $y = \tanh(x)$ for Example 30–6.

Check Examine the definition of the hyperbolic tangent to see if these results make sense:

$$\tanh (x) = \frac{e^x - e^{-x}}{e^x + e^{-x}}$$

When $x > 2.3$, $e^x > 10$, and $e^{-x} < 0.1$; so $e^x \gg e^{-x}$ and the definition can be simplified:

$$\tanh (x) \approx e^x/e^x$$

∴ $\tanh (x) \approx 1$ for $x > 2.3$

When $x < -2.3$, $e^x < 0.1$, and $e^{-x} > 10$; so $e^x \ll e^{-x}$ and:

$$\tanh (x) \approx -e^{-x}/e^{-x}$$

∴ $\tanh (x) \approx -1$ for $x < -2.3$

EXAMPLE 30–7 Examine $s = \sqrt{t^2 + 4}$ for asymptotic behavior.

Solution When does $t^2 + 4$ look like t^2?

When $t^2 \gg 4$, $t > 7$, or $t < -7$.

$$s \approx \sqrt{t^2} = |t| \text{ for } |t| > 7$$

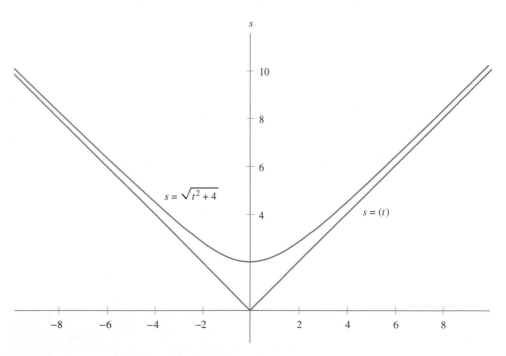

FIGURE 30–11 Graph for Example 30–7.

Plot the two functions $s = \sqrt{t^2 + 4}$ and $s = |t|$ on the same graph; see Figure 30–11.

$$\therefore \qquad s = |t| \text{ is an asymptote.}$$

EXERCISE 30–4

Examine the following functions for asymptotic behavior.

1. $y = e^{-x}$ **2.** $y = 1/x^2$

3. $s = 1 - e^{1/t}$ **4.** $u = \cosh(V) - \sinh(V)$

5. $u = \sqrt{v^2 - 1}$ **6.** $V = 2e^{2/u^2}$

7. $R = \dfrac{2r}{r + 2}$ **8.** $X_C = 1/(\omega C); C = 1 \ \mu F$

SELECTED TERMS

discontinuity A break in the graph of a function.

domain The list of all values of the independent variable for which the function is defined.

range The list of all values that a function assumes over its domain.

SECTION CHALLENGE

WEB CHALLENGE FOR CHAPTERS 29 AND 30

To evaluate your comprehension of Chapters 29 and 30, log on to **www.prenhall.com/ harter** and take the online True/False and Multiple Choice assessments for each of the chapters.

SECTION CHALLENGE FOR CHAPTERS 29 AND 30*

As pictured in Figure C–17, a wooden crate of weight F_W is moved up an inclined plane (a *ramp*), at an *angle of inclination* (θ), by an effort force, F_E. The opposition to the movement of the load is provided by the force of static friction, F_{fs}, resulting from the normal force, F_N, and the coefficient of friction, μ. The efficiency (η) of the ramp is given by Equation E–9:

$$\eta = \frac{\sin(\theta) \sec(\theta) \csc(\theta)}{\mu \csc(\theta) + \sec(\theta)} \qquad \text{(E–9)}$$

A. Using the identities and equivalent forms of Table 29–1 and Exercise 29–3, your challenge is to provide the needed steps to simplify Equation E–9 to:

$$\eta = \frac{\sin(\theta)}{\mu \cos(\theta) + \sin(\theta)} \qquad \text{(E–10)}$$

Then, given the coefficient of friction of wood on wood ($\mu = 0.58$) and the angle of inclination of the ramp ($\theta = 11.0°$), determine the efficiency (η) of the ramp.

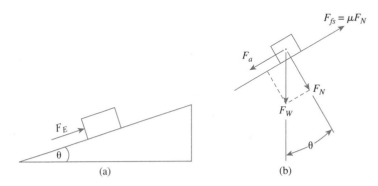

FIGURE C–17   (a) An effort force applied to the crate to move it up the ramp; (b) Free-body diagram of the force components acting on the crate when the crate is at rest on the ramp.

*The solution to this Section Challenge is found in Appendix C.

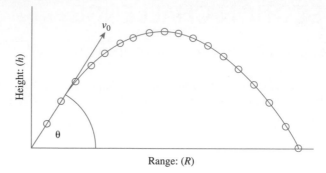

Height: (h)

v_0

θ

Range: (R)

FIGURE C–18   The trajectory of a projectile with an initial velocity, v_o, at an angle of elevation θ.

The motion of a projectile from an artillery shell follows a parabolic trajectory given by Equation E–11 and pictured in Figure C–18.

$$R = \frac{2v_o^2 \sin(\theta) \cos(\theta)}{g} \qquad \text{(E-11)}$$

where v_o = the initial velocity (muzzle velocity)
   θ = the angle of elevation
   g = the gravitational constant, 32.2 ft/s^2 or 9.81 m/s^2
   R = the horizontal range, the horizontal displacement when the projectile returns to its initial elevation

B.   Given the double-angle identity, $\sin(2\theta) = 2 \sin(\theta) \cos(\theta)$, your second challenge is to provide the needed steps to simplify Equation E–11 to:

$$R = \frac{v_o^2 \sin(2\theta)}{g} \qquad \text{(E-12)}$$

When the target and the gun are at the same elevation, it is the angle of elevation (θ) for the artillery that needs to be known since the position of the target (R) can be determined with the range finder and the muzzle velocity (v_o) of the shell is specified by the manufacturer. Solve Equation E–12 for the angle of elevation (θ) given a range of 1650 m, and a muzzle velocity of 320 m/s. Note that with θ between 0° and 90°, there are two solutions, θ or 90° − θ.

31

Computer Number Systems

31–1 Decimal Number System

31–2 Three Additional Number Systems

31–3 Converting Numbers to the Decimal System

31–4 Converting Decimal Numbers to Other Systems

31–5 Converting Between Binary, Octal, and Hexadecimal

31–6 Binary Addition and Subtraction

31–7 Octal Addition and Subtraction

31–8 Hexadecimal Addition and Subtraction

31–9 Complements

31–10 Binary Arithmetic with Complements

31–11 Review

PERFORMANCE OBJECTIVES

- Convert between binary, octal, decimal, and hexadecimal number systems.

- Add and subtract numbers in binary, octal, and hexadecimal.

- Use complements to represent negative numbers.

- Perform addition and subtraction using complement arithmetic.

A portable electronic device used to record medical information manually on the sensitive screen with a light pen—producing a digital record. (Courtesy of Pearson Education/PH College)

More and more computers are coming into use. As a technician you may use a computer in your work, or you may be employed to work on computers. Either way you need knowledge of the number systems in common use with computers. The topics presented here have been selected to help you gain this knowledge.

31–1 DECIMAL NUMBER SYSTEM

The number system we are familiar with is the decimal system. There are three other number systems that are important in electronics and in information technology because they are used with digital equipment. Before we look at these other systems, we need to review the decimal system.

The decimal system uses ten symbols called **digits** to form numbers. The digits are 0, 1, 2, 3, 4, 5, 6, 7, 8, and 9. The number of digits in a system is called the **radix** or *base*. Thus, the radix of the decimal system is 10. Notice that the radix is not one of the digits. Besides the ten digits, the decimal system includes a minus sign ($-$) and a decimal point (.) to indicate negative numbers and fractions.

The reason that the decimal system needs only ten digits is that the system assigns different values to a digit depending on its position relative to the decimal point. Thus, the 9 takes on three different values in 93, 19, and 1.09. The concept of positional value is summarized in Table 31–1.

TABLE 31–1   Positional Value and the Decimal Number System

	To Left of Decimal Point					To Right of Decimal Point				
Position	**5**	**4**	**3**	**2**	**1**	**1**	**2**	**3**	**4**	**5**
Position value	10 000	1000	100	10	1	0.1	0.01	0.001	0.0001	0.000 01
Power of ten	10^4	10^3	10^2	10^1	10^0	10^{-1}	10^{-2}	10^{-3}	10^{-4}	10^{-5}

The leading or leftmost digit is called the **most significant digit,** or **MSD,** because it corresponds to the highest power of ten and contributes most to the value of the number. The trailing or rightmost digit is called the **least significant digit,** or **LSD,** because it corresponds to the lowest power of ten and contributes least to the value of the number.

EXAMPLE 31–1   Find the value of 5, 3, and 1 in 543.21.

Solution   Determine the positions of 5, 3, and 1 and then use Table 31–1:

5 is the third digit left of the decimal point

∴   Its value is 500.

3 is the first digit left of the decimal point

∴   Its value is 3.

1 is the second digit right of the decimal point

∴   Its value is 0.01.

Decimal numbers can extend farther left and right than shown in Table 31–1. Because of this, we need a rule for the general case.

> ### RULE 31–1.   Value of Digits in Decimal Numbers
>
> To determine the value of a digit in a decimal number:
> 1. Determine its position relative to the decimal point.
> 2. For a digit n places to the *left* of the decimal point, the value is given by
>
> $$value = digit \times 10^{n-1}$$
>
> 3. For a digit m places to the *right* of the decimal point, the value is given by
>
> $$value = digit \times 10^{-m}$$

EXERCISE 31–1

1. Find the value of 1 in the following numbers.
   - **(a)** 5107.26
   - **(b)** 21 760.5
   - **(c)** 50.7126
   - **(d)** 6075.12
   - **(e)** 72.0165
   - **(f)** 157.602
2. Find the value of 2 in the numbers in problem 1.
3. Find the value of 5 in the numbers in problem 1.
4. Find the value of 6 in the numbers in problem 1.
5. Find the value of 7 in the numbers in problem 1.
6. Find the most significant digit (MSD) of each number in problem 1.
7. Find the least significant digit (LSD) of each number in problem 1.

31–2   THREE ADDITIONAL NUMBER SYSTEMS

Three additional number systems are becoming more important in electronics. These systems are associated with computers, routers, digital logic trainers, and digital devices. These number systems are the *binary, octal,* and *hexadecimal* systems.

The **binary** system has only two digits, called bits: 0 and 1. These can represent the states of a two-state device. Examples are (1) a switch is on or off, or (2) a diode is conducting or not. Because of this relationship, all computers work in binary *on the device level.*

The radix, or base, of the binary system is 2 because there are two digits. Like the decimal system, the binary system uses positional value. However, the period used to separate the integer and fractional parts of a number is called a **binary point,** not a decimal point.

Since we are working with four different systems, and the period looks and functions the same in each system, we will give it a new name that will be the same in each system. From now on, we will refer to it as the **radix point.**

The **octal** system has eight digits: 0, 1, 2, 3, 4, 5, 6, and 7. The radix of the octal system is 8. The octal system is often used to display binary information from computers.

The third system is the **hexadecimal** system, which has 16 digits: 0, 1, 2, 3, 4, 5, 6, 7, 8, 9, A, B, C, D, E, and F. The radix of the hexadecimal system is 16. The hexadecimal system is also used to display binary information. Table 31–2 lists the first 21 integers in binary, octal, decimal, and hexadecimal. Notice that the binary system needs three digits to represent the decimal values of 1 through 7, whereas the octal system needs only one digit. This reduction in the number of display digits by a factor of 3 for octal and a factor of 4 for hexadecimal is why they are commonly used to display binary data from computers.

TABLE 31–2 Comparison of Four Number Systems

Binary System	Octal System	Decimal System	Hexadecimal System
0	0	0	0
1	1	1	1
10	2	2	2
11	3	3	3
100	4	4	4
101	5	5	5
110	6	6	6
111	7	7	7
1000	10	8	8
1001	11	9	9
1010	12	10	A
1011	13	11	B
1100	14	12	C
1101	15	13	D
1110	16	14	E
1111	17	15	F
10 000	20	16	10
10 001	21	17	11
10 010	22	18	12
10 011	23	19	13
10 100	24	20	14

Notice that the digits in the binary and octal systems are familiar digits, but the hexadecimal system uses six symbols we normally use for letters, not digits. It is sometimes convenient to refer to these digits by their phonetic alphabet names: alpha, bravo, charley, delta, echo, and fox. Take the time now to memorize the decimal values for these special hexadecimal digits given in Table 31–2.

To distinguish the system in which a number is written, we will add the radix as a subscript to the number. Whenever we write the radix, we will write it as a decimal number. For example, 100 in hexadecimal is written 100_{16}.

EXAMPLE 31–2   Indicate that 101 and 110 are binary numbers.

    Solution   The radix of the binary system is 2. Append a subscript of 2 to each number:

$$\therefore \quad 101_2 \text{ and } 110_2$$

All four systems, binary, octal, decimal, and hexadecimal, use positional values. Whole numbers without a written radix point have an implied radix point to the right of the LSD. Some positional values are illustrated in Table 31–3. We can use these facts to find the decimal value of any digit in any number, as outlined in Rule 31–2.

TABLE 31–3 Positional Values

$$1111_2 = 1 \times 2^3 + 1 \times 2^2 + 1 \times 2^1 + 1 \times 2^0$$
$$1001_8 = 1 \times 8^3 + 0 \times 8^2 + 0 \times 8^1 + 1 \times 8^0$$
$$1100_{10} = 1 \times 10^3 + 1 \times 10^2 + 0 \times 10^1 + 0 \times 10^0$$
$$1010_{16} = 1 \times 16^3 + 0 \times 16^2 + 1 \times 16^1 + 0 \times 16^0$$

RULE 31–2.   *Decimal Value of Digits in Numbers*

To determine the value of a digit in a number:

1. Determine its position relative to the radix point.

2. For a digit n places to the left of the radix point, the value is given by

$$\text{value} = \text{digit} \times R^{n-1}$$

where R is the radix.

3. For a digit m places to the right of the radix point, the value is given by

$$\text{value} = \text{digit} \times R^{-m}$$

EXAMPLE 31–3   Find the decimal value of 7 in 70_8.

Solution   Use the steps of Rule 31–2.

Step 1:   7 is two places left of the implied radix point.

Step 2:   value $= 7 \times 8^{(2-1)}$

value $= 7 \times 8$

∴   value $= 56_{10}$

Table 31–4 lists the decimal values for some powers of 2, which will aid in applying Rule 31–2. Similarly, Table 31–5 lists the decimal values for some powers of 8, while Table 31–6 lists the decimal values for some powers of 16. Notice that the decimal values in Tables 31–5 and 31–6 are contained in Table 31–4. Since 8 and 16 are integer powers of 2, every integer power of 8 and every integer power of 16 can be expressed as an integer power of 2.

TABLE 31–4   Powers of 2

2^n	n	2^{-n}
1	0	1
2	1	0.5
4	2	0.25
8	3	0.125
16	4	0.0625
32	5	0.03125
64	6	0.015625
128	7	0.0078125
256	8	0.00390625
512	9	0.001953125
1024	10	0.0009765625
2048	11	0.00048828125
4096	12	0.000244140625

TABLE 31–5   Powers of 8

8^n	n	8^{-n}
1	0	1
8	1	0.125
64	2	0.015625
512	3	0.001953125
4096	4	0.000244140625

TABLE 31-6	Powers of 16	
16^n	n	16^{-n}
1	0	1
16	1	0.0625
256	2	0.00390625
4096	3	0.000244140625

EXAMPLE 31–4 Find the decimal value of B in $1B40_{16}$.

Solution Follow the steps of Rule 31–2.

Step 1: B is three places left of the implied radix point.

Step 2: value = B $\times 16^{(3-1)}$
value = B $\times 16^2$

Convert the hexadecimal digit B to decimal by referring to Table 31–2. Thus, B $\Rightarrow$ 11:

value = 11×16^2

Refer to Table 31–6 to convert 16^2 to 256.

$\therefore$ value = $11 \times 256 = 2816_{10}$

EXAMPLE 31–5 Find the decimal value of 1 in 0.01_2.

Solution Use the steps of Rule 31–2.

Step 1: The number 1 is two places to the right of the radix point.

Step 2: value = 1×2^{-2}

Refer to Table 31–4 to convert 2^{-2} to 0.25.

$\therefore$ value = 0.25_{10}

EXERCISE 31–2

Find the decimal values of the indicated digits.

1. All the 1s in 101.01_2
2. All the 1s in $11\ 010.1_2$
3. The 3 and the 5 in $730\ 51_8$
4. The 1s in 100.1_8
5. The 1s in 100.1_{16}
6. The 7 and the A in $47F.A_{16}$
7. The 4 and the 7 in 7620.04_8
8. The 2 and the 6 in 420.65_8
9. The C and the 9 in $C90_{16}$
10. The 5 and the E in $50.E_{16}$

31–3 CONVERTING NUMBERS TO THE DECIMAL SYSTEM

In the preceding section we learned to calculate the decimal value of any digit in a binary, octal, or hexadecimal number. It is an easy extension to calculate the decimal value of a binary, octal, or hexadecimal number. The procedure is stated in Rule 31–3.

> **RULE 31–3.   Calculating the Decimal Value of a Number**
>
> To calculate the decimal value of a number:
> 1. Calculate the decimal value of each digit.
> 2. The decimal value of the number is the sum of the values of the digits.

EXAMPLE 31–6   Find the decimal value of $10\ 001_2$.

Solution   Follow the steps of Rule 31–3; begin with the LSD.

Step 1:   The position of the LSD is one place to the left of the implied radix point:

$$value = 1 \times 2^0$$
$$value = 1_{10}$$

The next three digits are each zero. Zeros act only as place keepers; their value is always zero. The position of the MSD is five places to the left of the implied radix point:

$$value = 1 \times 2^4$$
$$value = 16_{10}$$

Step 2: $10\ 001_2 = 16_{10} + 0_{10} + 0_{10} + 0_{10} + 1_{10}$

$\therefore$ $10\ 001_2 = 17_{10}$

EXAMPLE 31–7   Find the decimal value of 2001_8.

Solution   Use the steps of Rule 31–3; begin with the LSD.

Step 1:   The LSD is one place to the left of the implied radix point:

$$value = 1 \times 8^0$$
$$value = 1_{10}$$

The MSD is four places to the left of the implied radix point:

$$value = 2 \times 8^3$$
$$value = 1024_{10}$$

Step 2:	$2001_8 = 1024_{10} + 0_{10} + 0_{10} + 1_{10}$
$\therefore$	$2001_8 = 1025_{10}$

EXAMPLE 31–8	Find the decimal value of 2001_{16}.
Solution	Use the steps of Rule 31–3; begin with the LSD.
Step 1:	The LSD is one place to the left of the implied radix point:

$$\text{value} = 1 \times 16^0$$
$$\text{value} = 1_{10}$$

The MSD is four places to the left of the implied radix point:

$$\text{value} = 2 \times 16^3$$
$$\text{value} = 8192_{10}$$

Step 2:	$2001_{16} = 8192_{10} + 0_{10} + 0_{10} + 1_{10}$
$\therefore$	$2001_{16} = 8193_{10}$

EXAMPLE 31–9	Find the decimal value of $C3.0D2_{16}$.
Solution	Follow the steps of Rule 31–3; begin with the LSD.
Step 1:	The 2 is three places to the right of the radix point:

$$\text{value} = 2 \times 16^{-3}$$
$$\text{value} = 0.000\ 488\ 281\ 25_{10}$$

The D is two places to the right of the radix point:

$$\text{value} = 13 \times 16^{-2}$$
$$\text{value} = 0.050\ 781\ 25_{10}$$

The 3 is one place to the left of the radix point:

$$\text{value} = 3_{10}$$

The C is two places to the left of the radix point:

$$\text{value} = 12 \times 16$$
$$\text{value} = 192_{10}$$

Step 2:	$C3.0D2_{16} = 192_{10} + 3_{10} + 0_{10} + 0.050\ 781\ 25_{10}$ $+ 0.000\ 488\ 281\ 25_{10}$
$\therefore$	$C3.0D2_{16} \approx 195.051\ 27_{10}$

Find the decimal value of the following numbers.

1. 110_2	**2.** 1010_2	**3.** $110\ 000_2$
4. $11\ 111_2$	**5.** 0.11_2	**6.** 1.01_2
7. 10.11_2	**8.** 1.011_2	**9.** 11.0101_2
10. $0.111\ 11_2$	**11.** 110_8	**12.** 1010_8
13. 270_8	**14.** $64\ 000_8$	**15.** 0.11_8
16. 1.01_8	**17.** 76.2_8	**18.** 14.7_8
19. 5.34_8	**20.** 0.76_8	**21.** 110_{16}
22. 1010_{16}	**23.** 270_{16}	**24.** $ABCD_{16}$
25. 0.11_{16}	**26.** 1.01_{16}	**27.** 76.2_{16}
28. $0.B1_{16}$	**29.** $F0.CC_{16}$	**30.** $E67.D_{16}$

31–4 CONVERTING DECIMAL NUMBERS TO OTHER SYSTEMS

To convert an octal number to decimal, we use multiplication by powers of eight. To convert a whole number from decimal to octal, we reverse the procedure and use division by eight.

EXAMPLE 31–10   Convert 1024_8 to decimal and back to octal.

Solution   Use Rule 31–2 to indicate the value of each digit:

$$1024_8 = 1 \times 8^3 + 0 \times 8^2 + 2 \times 8^1 + 4$$
$$1024_8 = 512 + 0 + 16 + 4$$
$$\therefore \quad 1024_8 = 532_{10}$$

Reverse the order of the terms, and factor the right member of the first line of the solution:

$$4 + 8^1 \times 2 + 8^2 \times 0 + 8^3 \times 1 = 4 + 8[2 + 8(0 + 8 \times 1)]$$

This factored form shows that the conversion can be considered repeated multiplication by 8. To reverse the process, divide 532 by 8. The remainder is the LSD.

```
8 |532
8 | 66 R 4  ———— LSD ————┐
8 |  8 R 2                │
8 |  1 R 0                │
     0 R 1  ———— MSD ————→ 1024₈
```

$$\therefore \quad 532_{10} = 1024_8$$

The process used in Example 31–10 can be formalized as a rule for converting whole numbers from decimal to any other system.

> ### Rule 31–4.   Converting Decimal Whole Numbers
>
> To convert a whole number from decimal to binary, octal, or hexadecimal, use repeated division by the new radix. The remainder upon the first division is the LSD; the remainder of the last division is the MSD.

EXAMPLE 31–11   Convert 155_{10} to binary.

Solution   Use Rule 31–4 to convert 155_{10} to binary:

$$
\begin{array}{rl}
2\ \underline{|155} & \text{LSD} \\
2\ \underline{|\ \ 77}\ R\ 1 & \\
2\ \underline{|\ \ 38}\ R\ 1 & \\
2\ \underline{|\ \ 19}\ R\ 0 & \\
2\ \underline{|\ \ \ \ 9}\ R\ 1 & \\
2\ \underline{|\ \ \ \ 4}\ R\ 1 & \\
2\ \underline{|\ \ \ \ 2}\ R\ 0 & \\
2\ \underline{|\ \ \ \ 1}\ R\ 0 & \text{MSD} \\
\ \ \ \ \ \ 0\ R\ 1 & \longrightarrow 10\ 011\ 011_2
\end{array}
$$

$\therefore$ $155_{10} = 10\ 011\ 011_2$

EXAMPLE 31–12   Convert 2012_{10} to hexadecimal.

Solution   Use Rule 31–4 to convert 2012_{10} to hex:

$$
\begin{array}{rl}
16\ \underline{|2012} & \\
16\ \underline{|\ \ 125}\ R\ 12 & \text{LSD} \\
16\ \underline{|\ \ \ \ \ 7}\ R\ 13 & \\
\ \ \ \ \ \ \ 0\ R\ \ 7 & \\
& \text{MSD} \longrightarrow \quad 7 \quad D \quad C_{16}
\end{array}
$$

$\therefore$ $2012_{10} = 7DC_{16}$

The process of converting a binary fraction to decimal involves division by 2. To convert a decimal fraction to binary, reverse the process and multiply by 2. The integer result of the first multiplication is the MSD of the binary fraction. The fractional part of the result is again multiplied by 2 to obtain the next digit. The process is terminated when the fractional part is zero or when as many binary digits as desired have been found.

EXAMPLE 31–13   Convert $0.110\ 11_2$ to decimal and back to binary.

Solution

$$0.110\ 11_2 = 1 \times 2^{-1} + 1 \times 2^{-2} + 0 \times 2^{-3}$$
$$+ 1 \times 2^{-4} + 1 \times 2^{-5}$$
$$0.110\ 11_2 = 0.5 + 0.25 + 0 + 0.0625 + 0.03125$$

$\therefore$ $0.110\ 11_2 = 0.843\ 75_{10}$

Use repeated multiplication by 2 to convert $0.843\ 75_{10}$ to binary:

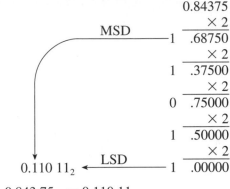

$\therefore$ $0.843\ 75_{10} = 0.110\ 11_2$

The process illustrated in Example 31–13 is summarized in the following rule:

Rule 31–5.   *Converting Decimal Fractions*

To convert a decimal fraction to binary, octal, or hexadecimal, multiply the fraction by the new radix. The integer portion of the product is the MSD. Repeat the process with the fractional remainder for the next digit. The process is terminated when the remaining fraction is zero or when all the desired digits have been determined.

EXAMPLE 31–14   Convert $0.492\ 187\ 5_{10}$ to octal.

Solution   Use Rule 31–5:

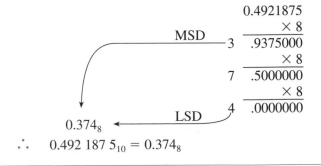

$\therefore$ $0.492\ 187\ 5_{10} = 0.374_8$

EXAMPLE 31–15   Convert $0.492\,187\,5_{10}$ to hexadecimal.

Solution   Use Rule 31–5:

$$
\begin{array}{r}
0.4921875 \\
\times\ 16 \\
\hline
.8750000 \\
\times\ 16 \\
\hline
.0000000
\end{array}
$$

MSD —— 7

LSD —— 14

$0\,.\,7\,E_{16}$

∴ $0.492\,187\,5_{10} = 0\,.\,7\,E_{16}$

In Examples 31–14 and 31–15, notice that the octal and hexadecimal representations require fewer digits than the decimal representation. Some other decimal fractions cannot be represented exactly in binary, octal, or hexadecimal.

EXAMPLE 31–16   Convert 0.1_{10} to octal to nine places.

Solution   Use Rule 31–5:

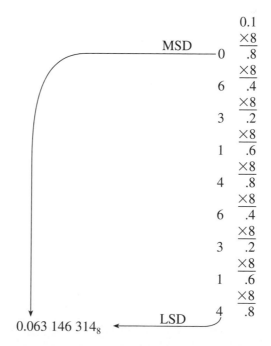

$$
\begin{array}{r}
0.1 \\
\times 8 \\
\hline
.8 \\
\times 8 \\
\hline
.4 \\
\times 8 \\
\hline
.2 \\
\times 8 \\
\hline
.6 \\
\times 8 \\
\hline
.8 \\
\times 8 \\
\hline
.4 \\
\times 8 \\
\hline
.2 \\
\times 8 \\
\hline
.6 \\
\times 8 \\
\hline
.8
\end{array}
$$

MSD —— 0
6
3
1
4
6
3
1
LSD —— 4

$0.063\,146\,314_8$

∴ $0.1_{10} \approx 0.063\,146\,314_8$

To convert a decimal number to binary, octal, or hexadecimal, first convert the integer part and then convert the fractional part.

1. Convert the following decimal numbers to binary, octal, and hexadecimal:
   - **(a)** 19_{10}
   - **(b)** 22_{10}
   - **(c)** 15_{10}
   - **(d)** 36_{10}
   - **(e)** 48_{10}
   - **(f)** 39_{10}
   - **(g)** 64_{10}
   - **(h)** 100_{10}

2. Convert the following decimal numbers to octal and hexadecimal:
   - **(a)** 164_{10}
   - **(b)** 212_{10}
   - **(c)** 408_{10}
   - **(d)** 512_{10}
   - **(e)** 575_{10}
   - **(f)** 650_{10}
   - **(g)** 749_{10}
   - **(h)** 800_{10}

3. Convert the following decimal fractions to binary, octal, and hexadecimal (to five significant digits):
   - **(a)** 0.5_{10}
   - **(b)** 0.25_{10}
   - **(c)** 0.75_{10}
   - **(d)** 0.125_{10}
   - **(e)** 0.0625_{10}
   - **(f)** $0.156\,25_{10}$
   - **(g)** 0.156_{10}
   - **(h)** 0.995_{10}

4. Convert 3.5_{10} to binary.

5. Convert 24.375_{10} to octal.

6. Convert 28.4375_{10} to hexadecimal.

31–5 CONVERTING BETWEEN BINARY, OCTAL, AND HEXADECIMAL

The octal and hexadecimal systems were chosen for the ease of conversion to and from binary. We begin with octal conversion. To convert an octal number to binary, replace each octal digit with the appropriate set of three binary digits from Table 31–7.

TABLE 31–7 Octal Binary Conversion

Octal	Binary	Octal	Binary
0	000	4	100
1	001	5	101
2	010	6	110
3	011	7	111

EXAMPLE 31–17 Convert the octal number 25.24_8 to binary.

Solution Use Table 31–7 to convert each octal digit:

$$25.24_8 = 010\ 101.010\ 100_2$$

Delete leading and trailing zeros:

$$\therefore \quad 25.24_8 = 10\ 101.0101_2$$

___

To convert a binary number to octal, start at the radix point and collect the binary digits into groups of three. If necessary, add *leading zeros* to the integer part and *trailing zeros* to the fractional part. Use Table 31–7 to convert each set of three binary digits to octal digits.

___

EXAMPLE 31–18   Convert the binary number $10\ 110.1001_2$ to octal.

Solution   Group the binary digits into sets of three; start counting at the radix point:

$$10\ 110.1001_2 = 10\ 110.100\ 1_2$$

Add extra leading and trailing zeros to complete the sets of three:

$$10\ 110.100\ 1_2 = 010\ 110.100\ 100_2$$

Use Table 31–7 to convert to octal:

$$\therefore \quad 10\ 110.1001_2 = 010\ 110.100\ 100_2 = 26.44_8$$

___

The conversion of hexadecimal to binary is performed in a manner similar to that of octal to binary conversion. Table 31–8 is used to replace each hexadecimal digit with four binary digits.

TABLE 31–8   Hexadecimal Binary Conversion

Hexadecimal	Binary	Hexadecimal	Binary
0	0000	8	1000
1	0001	9	1001
2	0010	A	1010
3	0011	B	1011
4	0100	C	1100
5	0101	D	1101
6	0110	E	1110
7	0111	F	1111

EXAMPLE 31–19   Convert the hexadecimal number $A3.0F_{16}$ to binary.

Solution   Use Table 31–8 to convert each digit:

$$A3.0F_{16} = 1010\ 0011.0000\ 1111_2$$

$$\therefore \quad A3.0F_{16} = 10\ 100\ 011.000\ 011\ 11_2$$

The conversion of binary to hexadecimal is performed like binary to octal, except the binary digits are grouped in "sets of four" instead of three. Table 31–8 is used to convert the sets of four binary digits to hexadecimal digits.

EXAMPLE 31–20   Convert the binary number $101\ 100\ 101.100\ 111_2$ to hexadecimal.

Solution   Group the binary digits into sets of four; begin at the radix point:

$$101\ 100\ 101.100\ 111_2 = 1\ 0110\ 0101.1001\ 11_2$$

Add leading and trailing zeros to fill out the sets of four:

$$1\ 0110\ 0101.1001\ 11_2 = 0001\ 0110\ 0101.1001\ 1100_2$$

Use Table 31–8 to convert the sets of four to hexadecimal digits:

$$\therefore \quad 0001\ 0110\ 0101.1001\ 1100_2 = 165.9C_{16}$$

A simple way to convert between octal and hexadecimal is to convert first to binary, and then from binary to the desired system.

EXAMPLE 31–21   Convert the octal number 157_8 to hexadecimal.

Solution   First convert to binary:

$$157_8 = 001\ 101\ 111_2$$

Regroup in sets of four:

$$157_8 = 0110\ 1111_2$$

Convert to hexadecimal:

$$\therefore \quad 157_8 = 6F_{16}$$

EXERCISE 31–5

Convert the following octal numbers to binary and hexadecimal.

1. 124_8 **2.** 75_8 **3.** 2001_8 **4.** 25.76_8

Convert the following hexadecimal numbers to binary and octal.

5. 124_{16} **6.** $F2D2_{16}$ **7.** AB_{16} **8.** $A.71_{16}$

Convert the following binary numbers to octal and hexadecimal.

9. 1011.1_2 **10.** $11\ 100.01_2$

11. $101\ 010\ 101_2$ **12.** $111\ 110\ 110\ 101_2$

31–6 BINARY ADDITION AND SUBTRACTION

Before we start binary addition, we will take a small detour to review decimal addition. Generally the process used for adding decimal integers is to add the one's digits and record the answer. Then add the ten's digits and record the answer, and so on. If the sum of the one's digits is larger than nine, then the answer contains a ten's digit. For example, $5 + 7 = 12$. The 2 is the answer's one's digit, and the 1 needs to be included in the sum of the ten's digits. If the original problem is to add 15 to 37, then we add 5 to 7 to get 12. We record the 2 as the one's digit, and we include the 1 in the sum of 1 and 3 to get 5. Thus, $15 + 37 = 52$. The process of including the 1 in the next column's answer is called **carrying,** and the 1 is called the **carry.**

Similarly, when subtracting decimal integers, we begin with the one's digits. If the digit to be subtracted is larger than the other digit, then we borrow ten from the next column's digit. Thus, $24 - 19$ is $(2 - 1)10 + (4 - 9)1$. However, 9 is larger than 4, so we borrow 1 from the 2, which becomes 1 and the 4 becomes 14. Thus, $(1 - 1)10 + (14 - 9)1 = 0 \times 10 + 5 \times 1 = 5$.

Binary addition and subtraction are easy to learn for those who know decimal addition and subtraction. There are only four combinations of digits to learn. The following rule governs binary addition.

> **RULE 31–6. Binary Addition**
>
> $$0 + 0 = 0$$
> $$0 + 1 = 1$$
> $$1 + 0 = 1$$
> $$1 + 1 = 0 \text{ plus a carry (or } 10_2)$$

EXAMPLE 31–22   Add the two binary numbers $10\,110_2$ and $11\,010_2$.

Solution   Use Rule 31–6 and show carries:

$$
\begin{array}{r}
1111 \leftarrow \text{carry} \\
10110 \\
+\ \ 11010 \\
\hline
110000_2
\end{array}
$$

$\therefore$ $10\,110_2 + 11\,010_2 = 110\,000_2$

Binary subtraction is the inverse of addition, and is just as easy to learn. The following rule governs binary subtraction.

RULE 31–7.   Binary Subtraction

$$1 - 0 = 1$$
$$1 - 1 = 0$$
$$0 - 0 = 0$$
$$0 - 1 = 1 \text{ and a borrow } (10_2 - 1 = 1)$$

EXAMPLE 31–23   Subtract 1011_2 from $11\,101_2$.

Solution   Use Rule 31–7:

$$
\begin{array}{r}
0_1 \leftarrow \text{borrow} \\
11\cancel{1}01 \\
-1011 \\
\hline
10010_2
\end{array}
$$

Rules 31–6 and 31–7 are summarized in an addition table, Table 31–9. To add two digits, find the column with the first digit at the top, and find the row with the second digit on the left end; the answer is at the point of intersection of this row and column.

TABLE 31–9   Binary Addition Table

+	0	1
0	0	1
1	1	10

To use Table 31–9 for subtraction, find the column with the digit to be subtracted at the top, and find the row with the digit (or 10_2) to be subtracted from in this column; the answer is at the left end of this row.

EXERCISE 31–6

Perform the indicated binary arithmetic.

1.	101 + 101		**2.**	1010 + 1100
3.	1111 + 1011		**4.**	11 + 11
5.	10111 + 111		**6.**	10101 + 1110
7.	101 110 + 11		**8.**	10 100 + 11
9.	100 − 10		**10.**	101 − 11
11.	1111 − 111		**12.**	1011 − 110
13.	1000 − 1		**14.**	1001 − 11
15.	1010 − 101		**16.**	10101 − 1110

31–7 OCTAL ADDITION AND SUBTRACTION

Frequently, numbers in octal are used for troubleshooting. Skill in adding and subtracting octal numbers is necessary to use these numbers. Table 31–10 is an octal ad-

TABLE 31–10   Octal Addition Table

+	0	1	2	3	4	5	6	7
0	0	1	2	3	4	5	6	7
1	1	2	3	4	5	6	7	10
2	2	3	4	5	6	7	10	11
3	3	4	5	6	7	10	11	12
4	4	5	6	7	10	11	12	13
5	5	6	7	10	11	12	13	14
6	6	7	10	11	12	13	14	15
7	7	10	11	12	13	14	15	16

dition table. It is used the same way that Table 31–9 is used for binary addition and subtraction. As in decimal arithmetic, the path to proficiency is practice.

EXAMPLE 31–24   Add the following pairs of octal numbers.

(a) 2_8 and 4_8 (b) 6_8 and 7_8

Solution   Use Table 31–10.

(a) Look across the top row to find the 2; it is in the fourth column. Look down the left column to find the 4; it is in the sixth row. Look at the point where this row and column intersect; the answer is 6.

∴ $2_8 + 4_8 = 6_8$

(b) Look across the top row to find the 6; it is in the eighth column. Look down the left column to find the 7; it is in the ninth row. Look at the point where the eighth column and the ninth row intersect; the answer is 15.

∴ $6_8 + 7_8 = 15_8$

Observation   When two digits sum to less than 8 in decimal addition, the answer is the same in octal addition. When two digits sum to more than 7 in decimal addition, the answer in octal is two more. The leading "1" in these entries in Table 31–10 indicates a carry.

EXAMPLE 31–25   Add $30\ 477_8$ and $26\ 605_8$.

Solution   Use Table 31–10:

$$111 \leftarrow \text{carry}$$
$$30477$$
$$+\ \underline{26605}$$
$$57304_8$$

EXAMPLE 31–26   Subtract:

(a) 3_8 from 7_8
(b) 6_8 from 13_8.

Solution   Use Table 31–10.

(a) Find the column headed by 3, the digit to be subtracted. Look down the column to find 7, the digit to be subtracted from. Look at the digit at the left end of this row, the answer, 4.

∴ $7_8 - 3_8 = 4_8$

(b) Find the column headed by 6, the digit to be subtracted. Look down the column to find 13, the number to be subtracted from. Look at the digit at the left end of this row, the answer, 5.

$$\therefore \qquad 13_8 - 6_8 = 5_8$$

Observation When subtracting two octal digits, if no borrow is involved, the answer is the same as in decimal. Borrows are indicated in Table 31–10 by two-digit numbers.

EXAMPLE 31–27 Subtract 237_8 from 1052_8.

Solution Use Table 31–10:

$$
\begin{array}{r}
4_1 \leftarrow \text{borrow} \\
10\not{5}2 \\
-\ \ 237 \\
\hline
613_8
\end{array}
$$

EXERCISE 31–7

Perform the indicated octal arithmetic.

1. $3 + 5$	**2.** $2 + 1$	**3.** $6 + 1$
4. $4 + 3$	**5.** $3 + 2$	**6.** $7 + 6$
7. $5 + 4$	**8.** $6 + 3$	**9.** $7 + 5$
10. $2 + 7$	**11.** $32 + 17$	**12.** $124 + 73$

13. $\begin{array}{r} 5605 \\ +2170 \\ \hline \end{array}$ **14.** $\begin{array}{r} 3651 \\ +2026 \\ \hline \end{array}$ **15.** $\begin{array}{r} 50770 \\ +7426 \\ \hline \end{array}$

16. $\begin{array}{r} 35407 \\ +6777 \\ \hline \end{array}$ **17.** $7 - 4$ **18.** $12 - 7$

19. $5 - 3$	**20.** $13 - 5$	**21.** $34 - 7$
22. $63 - 15$	**23.** $135 - 66$	**24.** $215 - 120$

25. $\begin{array}{r} 7365 \\ -1024 \\ \hline \end{array}$ **26.** $\begin{array}{r} 5057 \\ -4011 \\ \hline \end{array}$ **27.** $\begin{array}{r} 70000 \\ -5321 \\ \hline \end{array}$

28. $\begin{array}{r} 25042 \\ -6043 \\ \hline \end{array}$ **29.** $\begin{array}{r} 10000 \\ -\ \ \ \ \ 1 \\ \hline \end{array}$ **30.** $\begin{array}{r} 65432 \\ -56343 \\ \hline \end{array}$

31–8 HEXADECIMAL ADDITION AND SUBTRACTION

More and more computers are using hexadecimal for *troubleshooting* numbers. Just as we took a quick look at octal addition and subtraction, we now look at hexadecimal addition and subtraction using Table 31–11.

TABLE 31–11 Hexadecimal Addition Table

+	0	1	2	3	4	5	6	7	8	9	A	B	C	D	E	F
0	0	1	2	3	4	5	6	7	8	9	A	B	C	D	E	F
1	1	2	3	4	5	6	7	8	9	A	B	C	D	E	F	10
2	2	3	4	5	6	7	8	9	A	B	C	D	E	F	10	11
3	3	4	5	6	7	8	9	A	B	C	D	E	F	10	11	12
4	4	5	6	7	8	9	A	B	C	D	E	F	10	11	12	13
5	5	6	7	8	9	A	B	C	D	E	F	10	11	12	13	14
6	6	7	8	9	A	B	C	D	E	F	10	11	12	13	14	15
7	7	8	9	A	B	C	D	E	F	10	11	12	13	14	15	16
8	8	9	A	B	C	D	E	F	10	11	12	13	14	15	16	17
9	9	A	B	C	D	E	F	10	11	12	13	14	15	16	17	18
A	A	B	C	D	E	F	10	11	12	13	14	15	16	17	18	19
B	B	C	D	E	F	10	11	12	13	14	15	16	17	18	19	1A
C	C	D	E	F	10	11	12	13	14	15	16	17	18	19	1A	1B
D	D	E	F	10	11	12	13	14	15	16	17	18	19	1A	1B	1C
E	E	F	10	11	12	13	14	15	16	17	18	19	1A	1B	1C	1D
F	F	10	11	12	13	14	15	16	17	18	19	1A	1B	1C	1D	1E

EXAMPLE 31–28 Perform the following addition in hexadecimal.

(a) $3 + 6$ (b) $7 + 7$ (c) $6 + D$

Solution Use Table 31–11. Look up the first number at the top of the table, the second at the left. The answer is where the row and column cross.

(a) $\begin{array}{r} 3 \\ +6 \\ \hline 9_{16} \end{array}$ (b) $\begin{array}{r} 7 \\ +7 \\ \hline E_{16} \end{array}$ (c) $\begin{array}{r} 6 \\ +D \\ \hline 13_{16} \end{array}$

EXAMPLE 31–29 Find the sum of $24F_{16}$ and $1AAA_{16}$ in hexadecimal.

Solution Use Table 31–11.

$$\begin{array}{r} 1 \leftarrow \text{carry} \\ 1AAA \\ +\ 24F \\ \hline 1CF9_{16} \end{array}$$

$\therefore$ $1AAA_{16} + 24F_{16} = 1CF9_{16}$

Subtraction is the reverse of addition. With the aid of Table 31–11, subtraction in hexadecimal is not too difficult.

EXAMPLE 31–30   Perform the following subtraction in hexadecimal.

(a) $9 - 6$ (b) $7 - 7$ (c) $F - A$ (d) $19 - B$

Solution   Use Table 31–11. Look up the digit to be subtracted at the top of the table; look down the column for the second number. The digit at the left of this row is the answer.

$$
\begin{array}{cccc}
\textbf{(a)} & \quad 9 & \textbf{(b)} \quad 7 & \textbf{(c)} \quad F & \textbf{(d)} \quad 19 \\
& \underline{-6} & \underline{-7} & \underline{-A} & \underline{-B} \\
& 3_{16} & 0_{16} & 5_{16} & E_{16}
\end{array}
$$

EXAMPLE 31–31   Subtract $2ABC_{16}$ from 3000_{16}.

Solution   Use Table 31–11.

$$
\begin{array}{r}
2FF_1 \leftarrow \text{borrow} \\
\cancel{3}\cancel{0}\cancel{0}0 \\
-2ABC \\
\hline
544_{16}
\end{array}
$$

EXERCISE 31–8

Perform the indicated hexadecimal arithmetic.

1. $7 + 7$	**2.** $8 + 3$	**3.** $3 + A$
4. $6 + 4$	**5.** $1 + 7$	**6.** $9 + 5$
7. $12 + 57$	**8.** $35 + 73$	**9.** $A + B$
10. $F + 6$	**11.** $67 + 67$	**12.** $179 + 4F$

$$
\textbf{13.} \quad \begin{array}{r} 785 \\ +\ BA1 \\ \hline \end{array} \qquad \textbf{14.} \quad \begin{array}{r} 96A \\ +\ C0D \\ \hline \end{array} \qquad \textbf{15.} \quad \begin{array}{r} 2001 \\ +\ FFF \\ \hline \end{array}
$$

$$
\textbf{16.} \quad \begin{array}{r} D2C3 \\ +\ 98A \\ \hline \end{array} \qquad \textbf{17.} \ A - 1 \qquad \textbf{18.} \ F - 4
$$

19. $9 - 6$	**20.** $F - 5$	**21.** $D - A$
22. $E - B$	**23.** $12 - F$	**24.** $A - 6$
25. $B - 4$	**26.** $15 - C$	**27.** $\begin{array}{r} 100 \\ -\ AA \\ \hline \end{array}$

$$
\textbf{28.} \quad \begin{array}{r} F88 \\ -\ A51 \\ \hline \end{array} \qquad \textbf{29.} \quad \begin{array}{r} 2677 \\ -\ 1F0A \\ \hline \end{array} \qquad \textbf{30.} \quad \begin{array}{r} 4A00 \\ -\ 3551 \\ \hline \end{array}
$$

31–9   COMPLEMENTS

When two numbers sum to a specified number, they are called **complements.** In trigonometry, two angles that sum to a right angle are called complements. Computers use complements to perform certain arithmetic functions. There are two systems of complements in use in computers today. The first system is based on the largest digit. The second system is based on the radix.

Largest-Digit Complements

Each number system has a largest digit. The largest digit of the binary system is 1; the largest digit of the octal system is 7. Largest-digit complements (also called **base minus one complements**) for the binary system are based on 1 and are called **one's complements.** The largest-digit complements for the octal system are based on 7 and are called seven's complements. The largest digit and the corresponding name for complements are presented in Table 31–12 for the various number systems.

TABLE 31–12   Largest-Digit Complements

System	Largest Digit	Largest-Digit Complements
Binary	1	One's complements
Octal	7	Seven's complements
Decimal	9	Nine's complements
Hexadecimal	F	F's complements

In the rest of this chapter, we will work with eight-digit numbers, although there is nothing special about eight. The following concepts would work with numbers of any fixed length.

To find the *eight*-digit one's complement of a binary number, simply subtract the number from the *eight*-digit binary number of all ones (1111 1111).

EXAMPLE 31–32   Form the eight-digit one's complements of:

(a) $101\ 0110_2$ (b) $110\ 0000_2$

Solution   (a) $\begin{array}{r} 1111\ 1111 \\ -\ \ \ 101\ 0110 \\ \hline 1010\ 1001_2 \end{array}$ (b) $\begin{array}{r} 1111\ 1111 \\ -\ \ \ 110\ 0000 \\ \hline 1001\ 1111_2 \end{array}$

Observation   To form the one's complement of a binary number, replace each 1 with a 0 and each 0 with a 1.

To find the *eight*-digit seven's complement of an octal number, simply subtract the number from the *eight*-digit octal number of all sevens (7777 7777).

EXAMPLE 31–33   Form the eight-digit seven's complements of:

(a) $545\ 6700_8$ (b) $6321\ 0754_8$

Solution	**(a)**	7777 7777		**(b)**	7777 7777
		$-$ 545 6700			$-6321\ 0754$
		$7232\ 1077_8$			$1456\ 7023_8$

To form the *eight*-digit nine's complement of a decimal number, subtract the number from the *eight*-digit decimal number of all nines (9999 9999).

EXAMPLE 31–34 Find the eight-digit nine's complements of:

(a) $3791\ 0684_{10}$ (b) $5327\ 9926_{10}$

Solution	**(a)**	9999 9999		**(b)**	9999 9999
		$-3791\ 0684$			$-5327\ 9926$
		$6208\ 9315_{10}$			$4672\ 0073_{10}$

To form the *eight*-digit F's complement of a hexadecimal number, subtract the number from the *eight*-digit hexadecimal number of all Fs (FFFF FFFF).

EXAMPLE 31–35 Find the eight-digit F's complements of:

(a) $3791\ 0684_{16}$ (b) $AB25\ CDEF_{16}$

Solution	**(a)**	FFFF FFFF		**(b)**	FFFF FFFF
		$-3791\ 0684$			$-AB25\ CDEF$
		$C86E\ F97B_{16}$			$54DA\ 3210_{16}$

Table 31–13 contains the F's complements for all the hexadecimal digits. Take time to look over the table to see the patterns. It will aid you in forming complements of hexadecimal numbers.

TABLE 31–13 F's Complements

Digit	Complement	Digit	Complement
0	F	8	7
1	E	9	6
2	D	A	5
3	C	B	4
4	B	C	3
5	A	D	2
6	9	E	1
7	8	F	0

Find the eight-digit one's complements of the following binary numbers.

1. $111\ 1111_2$ **2.** $100\ 0110_2$ **3.** $110\ 0110_2$

4. $101\ 0101_2$ **5.** $111\ 0001_2$ **6.** $11\ 0011_2$

Find the eight-digit seven's complements of the following octal numbers.

7. $6420\ 1357_8$ **8.** $777\ 7777_8$ **9.** $111\ 1111_8$

10. $3534\ 6721_8$ **11.** $2451\ 7000_8$ **12.** $14\ 2167_8$

Find the eight-digit nine's complements of the following decimal numbers.

13. $1746\ 2001_{10}$ **14.** $111\ 1111_{10}$ **15.** $9253\ 8679_{10}$

16. $777\ 7777_{10}$ **17.** $9999\ 9999_{10}$ **18.** $3796\ 8421_{10}$

Find the eight-digit F's complements of the following hexadecimal numbers.

19. $111\ 1111_{16}$ **20.** $777\ 7777_{16}$ **21.** $1FAB\ 9824_{16}$

22. $C6A2\ ED57_{16}$ **23.** $3056\ 7CDE_{16}$ **24.** $BF98\ 0134_{16}$

True Complements

The second type of complements (called **true complements**) are named for the radix of each system. Thus, the true complements for the binary system are called **two's complements.** The eight-digit true complement of a number is formed by subtracting the number from 1 0000 0000. This number is 1 larger than the number used to find the largest-digit complement. Thus, the two's complement of a binary number is equal to the one's complement plus 1.

   In like manner, the true complements for octal numbers are called **eight's complements** and are equal to seven's complements plus 1. The true complements for decimal numbers are called **ten's complements** and are equal to nine's complements plus 1. The true complements for hexadecimal numbers are called **sixteen's complements** and are equal to F's complements plus 1.

EXAMPLE 31–36   Find the eight-digit two's complement of the binary number $101\ 0101_2$.

Solution   **(1)** Subtract from 1 0000 0000:

$$\begin{array}{r} 1\ 0000\ 0000 \\ -\ \ \ \ \ 101\ 0101 \\ \hline 1010\ 1011_2 \end{array}$$ (two's complement)

(2) Calculate the two's complement from the one's complement:

$$
\begin{array}{rl}
1111\ 1111 & \\
-\quad 101\ 0101 & \\
\hline
1010\ 1010 & \text{(one's complement)} \\
+\qquad\quad 1 & \text{(add 1)} \\
\hline
1010\ 1011_2 & \text{(two's complement)}
\end{array}
$$

∴   Two's complements can be calculated in two ways.

EXAMPLE 31–37   Find the eight-digit eight's complement of the octal number $3721\ 5640_8$.

Solution   Calculate the seven's complement and add 1:

$$
\begin{array}{rl}
7777\ 7777 & \\
-\quad 3721\ 5640 & \\
\hline
4056\ 2137 & \text{(seven's complement)} \\
+\qquad\quad 1 & \text{(add 1)} \\
\hline
4056\ 2140_8 & \text{(eight's complement)}
\end{array}
$$

EXAMPLE 31–38   Find the eight-digit ten's complement of the decimal number $3721\ 5640_{10}$.

Solution   Calculate the nine's complement and add 1:

$$
\begin{array}{rl}
9999\ 9999 & \\
-\quad 3721\ 5640 & \\
\hline
6278\ 4359 & \text{(nine's complement)} \\
+\qquad\quad 1 & \text{(add 1)} \\
\hline
6278\ 4360_{10} & \text{(ten's complement)}
\end{array}
$$

EXAMPLE 31–39   Find the eight-digit sixteen's complement of the hexadecimal number $A375\ B1EF_{16}$.

Solution   Calculate the F's complement and add 1:

$$
\begin{array}{rl}
FFFF\ FFFF & \\
-\quad A375\ B1EF & \\
\hline
5C8A\ 4E10 & \text{(F's complement)} \\
+\qquad\qquad 1 & \text{(add 1)} \\
\hline
5C8A\ 4E11_{16} & \text{(sixteen's complement)}
\end{array}
$$

Find the eight-digit two's complements of the following binary numbers:

1. $111\ 1111_2$ **2.** $1100\ 1110_2$ **3.** $101\ 0111_2$

4. $100\ 1100_2$ **5.** $110\ 1100_2$ **6.** $111\ 0001_2$

Find the eight-digit eight's complements of the following octal numbers:

7. $2651\ 0777_8$ **8.** $72\ 1654_8$ **9.** $1743\ 1111_8$

10. $1000\ 0000_8$ **11.** $111\ 1111_8$ **12.** $3421\ 5720_8$

Find the eight-digit ten's complements of the following decimal numbers:

13. $2561\ 0777_{10}$ **14.** $94\ 3876_{10}$ **15.** $1000\ 0000_{10}$

16. $4716\ 9582_{10}$ **17.** $3490\ 1627_{10}$ **18.** $2815\ 9376_{10}$

Find the eight-digit sixteen's complements of the following hexadecimal numbers:

19. $\mathtt{71AE\ 642F}_{16}$ **20.** $1000\ 0000_{16}$ **21.** $\mathtt{3D8B\ C905}_{16}$

22. $\mathtt{FA\ 9EDC}_{16}$ **23.** $2561\ 0777_{16}$ **24.** $111\ 1111_{16}$

31–10 BINARY ARITHMETIC WITH COMPLEMENTS

To work with signed numbers, we had to learn several rules for adding and subtracting that depend on the signs of the numbers and their relative sizes. For example, to add two numbers with opposite signs, subtract the smaller from the larger and retain the sign of the larger number. All the rules we learned must be "taught" to a computer's arithmetic unit. This is done through the design of the logic circuits. A common technique used by computer manufacturers is to use complements to represent negative numbers.

Most computers work with only fixed-length words. In such a system there is only a finite number of possible binary numbers. Suppose that we want to be able to count up to 127. If we will settle for the positive integers 1 through 127, we can use a "word" of seven binary digits (called **bits**). Now, if we want to include negative integers, we need an extra bit for the sign (0 for positive and 1 for negative).

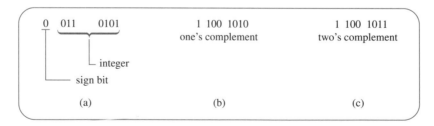

FIGURE 31–1 Bit patterns in computer words: (a) word containing 53_{10}; (b) word containing -53_{10} in one's complement; (c) word containing -53_{10} in two's complement.

For negative numbers, we could simply turn the sign bit "on." Instead, computer manufacturers have chosen to complement the number. Figure 31–1 shows the bit patterns of 53_{10} and -53_{10} in an 8-bit word.

The advantage of using complements to store negative numbers lies in simplifying the computer's arithmetic circuits. Addition of signed numbers, when negative numbers are stored as complements, is a simple extension of binary addition. Subtraction is performed by complementing the number to be subtracted and adding. Some computers have subtractive adders; that is, they know only how to subtract, not how to add.

Addition using one's complements is governed by the following rule:

RULE 31–8. Adding Numbers in One's Complement

To add two numbers in one's complement:

1. Use binary addition.
2. Check for a carry beyond the most significant digit (MSD):
   (a) If there is no carry beyond the MSD, the result of step 1 is the answer.
   (b) If there is a carry beyond the MSD, add 1 to the result obtained in step 1.

EXAMPLE 31–40   Use one's complement to add 75_{10} and -16_{10}. Leave a space between the sign bit and the integer.

Solution   Convert to eight-digit binary and use the steps of Rule 31–8.

$$75_{10} = 0\ 100\ 1011_2$$
$$-16_{10} = -0\ 001\ 0000_2$$

Form one's complement:

$$-16_{10} = 1\ 110\ 1111_2$$

Step 1:   Add:

$$
\begin{array}{r}
0\ 100\ 1011 \\
+\ 1\ 110\ 1111 \\
\hline
1\ 0\ 011\ 1010
\end{array}
$$

Step 2:   Yes, there is a carry beyond the eighth digit, so add 1 to the result obtained in step 1:

$$
\begin{array}{r}
0\ 011\ 1010 \\
+\qquad\quad 1 \\
\hline
0\ 011\ 1011_2
\end{array}
$$

Check   Convert the answer to decimal; it should be 59:

$$0\ 011\ 1011_2 = 1 + 2 + 0 + 8 + 16 + 32$$
$$0\ 011\ 1011_2 = 59_{10}$$

Observation   The process of checking for a carry beyond the most significant digit and conditionally adding 1 to the result is called **end-around carry.**

RULE 31–9. Subtracting Numbers in One's Complement

To subtract binary numbers using one's complements:
1. Form the one's complement of the number to be subtracted.
2. Add, using one's complement arithmetic.

EXAMPLE 31–41   Subtract $0\ 101\ 1000_2$ from $0\ 011\ 0111_2$ using one's complements. Express the answer as a signed binary number.

Solution   Follow the steps of Rule 31–9.

Step 1:   Form the one's complement:

$$\begin{array}{r} 1\ 111\ 1111 \\ -0\ 101\ 1000 \\ \hline 1\ 010\ 0111 \end{array}$$

Step 2:   Add:

$$\begin{array}{r} 0\ 011\ 0111 \\ +1\ 010\ 0111 \\ \hline 1\ 101\ 1110 \end{array}$$

Observation   The sign bit (MSD) is 1; this indicates a negative result, which is in one's complement. To interpret the result, complement the number.

$$\begin{array}{r} 1\ 111\ 1111 \\ -1\ 101\ 1110 \\ \hline 0\ 010\ 0001 \end{array}$$

Delete the leading zeros and prefix the answer with a minus sign.

$$-10\ 0001_2$$

$\therefore$ $0\ 011\ 0111_2 - 0\ 101\ 1000_2 = -10\ 0001_2$

The rule for two's complement addition is slightly different than the rule for one's complement addition.

To add two numbers in two's complement:

1. Use binary addition.
2. Any carry beyond the most significant digit (MSD) is discarded *(end-off carry)*.

EXAMPLE 31–42   Use two's complements to add 75_{10} and -16_{10}.

Solution   Convert to eight-digit binary numbers:

$$75_{10} = 0\ 100\ 1011_2$$
$$-16_{10} = -0\ 001\ 0000_2$$

Form two's complement:

$$-16_{10} = 1\ 111\ 0000_2$$

Apply Rule 31–10:

Step 1:   Add:

$$\begin{array}{r} 0\ 100\ 1011 \\ +\quad 1\ 111\ 0000 \\ \hline 1\ 0\ 011\ 1011 \end{array}$$

Step 2:   End-off carry:

$$0\ 011\ 1011_2 = 59_{10}$$

Observation   The sign bit is off (zero), so the result is positive.

The procedure for subtraction with two's complement numbers is similar to subtraction with one's complement numbers.

To subtract binary numbers using two's complements:

1. Form the two's complement of the number to be subtracted.
2. Add, using two's complement arithmetic.

EXAMPLE 31–43   Subtract $0\ 101\ 1000_2$ from $0\ 011\ 0111_2$ using two's complements. Express the answer as a signed binary number.

Solution   Use Rule 31–11.

Step 1:   Convert to two's complement form:

$$
\begin{array}{ll}
1\ 111\ 1111 & \\
-\ 0\ 101\ 1000 & \\
\hline
1\ 010\ 0111 & \text{(one's complement)} \\
+1 & \text{(add 1)} \\
\hline
1\ 010\ 1000 & \text{(two's complement)}
\end{array}
$$

Step 2:   Add, using two's complement arithmetic.

$$
\begin{array}{l}
0\ 011\ 0111 \\
+\ 1\ 010\ 1000 \\
\hline
1\ 101\ 1111_2
\end{array}
$$

Observation   The sign bit indicates this is a negative number that is in two's complement. To interpret the answer, complement the number.

$$
\begin{array}{ll}
1\ 111\ 1111 & \\
-\ 1\ 101\ 1111 & \\
\hline
0\ 010\ 0000 & \text{(one's complement)} \\
1 & \text{(add 1)} \\
\hline
0\ 010\ 0001 &
\end{array}
$$

Delete the leading zeros and prefix the answer with a minus sign.

$$-10\ 0001_2$$

$\therefore$ $0\ 011\ 0111_2 - 0\ 101\ 1000_2 = -10\ 0001_2$

EXAMPLE 31–44   Subtract $1\ 011\ 0110_2$ from $1\ 110\ 1100_2$ using two's complements.

Solution   Use Rule 31–11.

Step 1:   Form two's complement:

$$
\begin{array}{ll}
1\ 111\ 1111 & \\
-\ 1\ 011\ 0110 & \\
\hline
0\ 100\ 1001 & \text{(one's complement)} \\
1 & \text{(add 1)} \\
\hline
0\ 100\ 1010 & \text{(two's complement)}
\end{array}
$$

Step 2:   Add, using two's complement arithmetic:

$$
\begin{array}{l}
1\ 110\ 1100 \\
+\ 0\ 100\ 1010 \\
\hline
1\ 0\ 011\ 0110
\end{array}
$$

End-off carry:

$$0\ 011\ 0110_2$$

$$\therefore \quad 1\ 110\ 1100_2 - 1\ 011\ 0110_2 = 0\ 011\ 0110_2$$

Observation   The sign bit is off; therefore, the result is positive.

EXERCISE 31–11

Use one's complement to perform the following eight-digit binary arithmetic (if the sign bit is already on, the number is in one's complement).

1.	0 110 1010 +0 000 1111	**2.**	0 011 1111 +0 011 1111
3.	1 100 1010 +0 100 1010	**4.**	0 011 0110 +1 011 0111
5.	0 101 1101 −0 110 0110	**6.**	0 011 1011 −0 001 0100
7.	1 100 1000 −1 010 0110	**8.**	1 011 1011 −0 110 0101
9.	0 011 0101 −1 100 1010	**10.**	1 100 0000 +1 100 0000

11–20. Repeat problems 1–10 using two's complements. (If the sign bit is already on, the number is in two's complement form.)–

31–11 REVIEW

This section will provide a review of the previous material by asking you to solve arithmetic problems for numbers with mixed bases. The proper procedure is to convert the numbers to a common base, evaluate the answer, and convert the answer to the desired base. This procedure will be illustrated in the following examples.

EXAMPLE 31–45   Add 150_{10} to 333_8; express the answer as a hexadecimal number.

Solution   Convert 150_{10} to octal:

$$150_{10} = 222_8$$

Perform octal addition:

$$222_8 + 333_8 = 555_8$$

Convert to binary:

$$555_8 = 101\ 101\ 101_2$$

Convert to hexadecimal:

$$0001\ 0110\ 1101_2 = 16D_{16}$$

$$\therefore \qquad 150_{10} + 333_8 = 16D_{16}$$

EXAMPLE 31–46　Add 256_8 and -127_{16}; express the answer as a decimal number.

Solution　Convert both numbers to binary:

$$256_8 = 010\ 101\ 110_2$$
$$-127_{16} = -0001\ 0010\ 0111_2$$

Convert to 10-digit one's complements and add:

$$\begin{array}{r} 0\ 0\ 1010\ 1110 \\ +1\ 0\ 1101\ 1000 \\ \hline 1\ 1\ 1000\ 0110_2 \end{array}$$

The sign bit is on, so the answer is negative. Complement to obtain:

$$-111\ 1001_2$$

Convert to decimal:

$$-(64 + 32 + 16 + 8 + 1) = -121_{10}$$

$$\therefore \qquad 256_8 - 127_{16} = -121_{10}$$

EXERCISE 31–12

Express the answers to the following problems as decimal numbers.

1.　35_8
 $+67_8$

2.　57_8
 -29_8

3.　43_8
 -72_8

4.　146_8
 $+35_8$

5.　$1A_{16}$
 $+F_{16}$

6.　29_{16}
 $+100_{16}$

7.　40_{16}
 -22_{16}

8.　99_{16}
 $-8E_{16}$

9.　10_{16}
 $+\ 4_8$

10.　29_{16}
 $+21_8$

11.　306_8
 $-\ FA_{16}$

12.　100_{16}
 $-\ 77_8$

13.　$1F9_{16}$
 -77_8

14.　587_8
 $-\ AB_{16}$

15.　11001_2
 $+\ \ 43_8$

16. $\quad 37_8$
$\quad +110101_2$

17. $\qquad D6_{16}$
$\quad -110011_2$

18. 1010110_2
$\qquad 8E_{16}$

19. $\quad 100_8$
$\quad -\ \ 25_{10}$

20. $\quad 263_{10}$
$\quad -342_8$

21. $\quad 79_{10}$
$\quad -1E_{16}$

22. $\quad 308_{16}$
$\quad -517_{10}$

23. $\quad 1A1B_{16}$
$\quad +315_{10}$

24. $\quad F3C_{16}$
$\quad +429_{10}$

SELECTED TERMS

binary   The number system based on the number 2.

complements   A method used to store negative numbers and to perform subtraction in computers.

hexadecimal   The number system based on the number 16.

octal   The number system based on the number 8.

radix   The base of a number system: thus, 2 is the radix of the binary system.

32

Mathematics of Computer Logic

32–1   Introductory Concepts

32–2   Inversion Operator (NOT)

32–3   Conjunction Operator (AND)

32–4   Disjunction Operator (OR)

32–5   Application of Logic Concepts

32–6   Introduction to Karnaugh Maps

32–7   DeMorgan's Theorem

32–8   Boolean Theorems

32–9   Applications

PERFORMANCE OBJECTIVES

- Develop the logic diagram for any logic expression.

- Determine the logic expression represented by any logic diagram.

- Construct a truth table for any logic expression.

- Use a Karnaugh map to minimize any logic expression of four or fewer variables.

The application of computer-controlled welding robots removes workers from a dangerous and repetitious manufacturing operation on this Nissan Motors assembly line.   (Courtesy of Pearson Education/PH College)

The ideas, rules, and theorems of Boolean algebra (in conjunction with other techniques and methods) are used to solve logic problems. Some of the logic problems solved with Boolean concepts have only two possible values for the logical variables. These *two-state* types of logic functions occur in the switching circuitry of digital computers. The output of these logic circuits is either on or off.

Since digital circuits have only two states (on or off, conducting or nonconducting, high or low), it is natural that the binary number system be used to describe these two conditions. The *on* state is indicated by a one (1) while the *off* state is indicated by a zero (0).

32–1 INTRODUCTORY CONCEPTS

When evaluating a logic expression (also referred to as a Boolean expression), the variables and constants of the expression may be assigned only one of two possible values, either a 1 or a 0. A 1 in a logic expression indicates the presence of a certain voltage level (e.g., 5 V) at the input or output of the computer's switching circuitry. A 0 in a logic expression indicates the presence of a voltage level (e.g., 0 V) that is different from that represented by a 1. When analyzing a digital logic system, it is simpler to think in terms of 1s and 0s rather than in terms of voltage levels. Because of the discrete nature of the voltage levels at the inputs and outputs of the logic circuitry, it has become a common practice to let 1 stand for the presence of a specified voltage level and 0 the absence of that level. Thus, a logic 0 might represent a voltage level of 0 V, while a logic 1 might represent a voltage level of 5 V.

In the course of this chapter, we will use switch symbols to represent the *two-state* logic conditions. An open switch will be indicated by a 0, while a closed switch will be indicated by a 1. Figure 32–1 shows a single-pole single-throw (SPST) switch. The condition of the switch is represented by the logic variable A. When the switch is open A = 0; when the switch is closed A = 1.

FIGURE 32–1 The condition of the switch is represented by the logic variable A.

OPEN A = 0

CLOSED A = 1

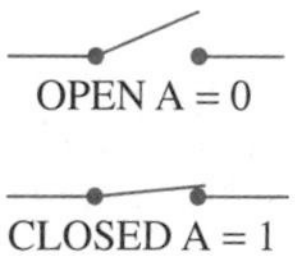

Logic Expressions

Consider the series circuit of Figure 32–2. The presence or absence of light depends on the condition of the switch. We will let A represent the condition of the switch; closed = 1, open = 0. The condition of the light will be represented by f_o (standing for output function); light = 1, dark = 0. The table next to the circuit, Figure 32–2(b), defines the operation of the circuit. This type of table is called a **truth table.**

The lamp will not produce light until the switch is closed. A logical expression representing the direct relationship that exists between the switch and the lamp is

$$f_o = A$$

FIGURE 32–2 (a) Light circuit controlled by a switch; (b) truth table defining the circuit operation.

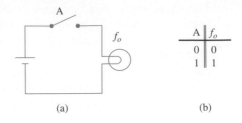

A	f_o
0	0
1	1

(a) (b)

We will use uppercase letters to represent logical conditions and f_o to represent the output of a logical equation.

Logic Operators

Only three basic logic operations are used to design all the logic circuits of a digital computer. These logic operations are:

- The *inversion operator*, commonly called the NOT operator. The NOT operator is indicated by an overbar ($\overline{}$) as in $\overline{A}$; read "NOT A."
- The *disjunction operator*, commonly called the OR operator. The OR operator is indicated by a plus sign (+) as in A + B; read "A OR B."
- The *conjunction operator*, commonly called the AND operator. The AND operator is indicated by any of the usual signs of multiplication (·, ×, etc.), as in A · B or AB; read "A AND B."

EXAMPLE 32–1   Read each of the following logical expressions.

(a) $A + B + \overline{C}$ (b) $\overline{A}B$ (c) $\overline{A} + BC$

Solution   Use the words OR, AND, and NOT:

(a) $A + B + \overline{C}$ is read A OR B OR NOT C.
(b) $\overline{A}B$ is read NOT A AND B.
(c) $\overline{A} + BC$ is read NOT A OR B AND C.

EXERCISE 32–1

Using the words of the logical operators, NOT, OR, and AND, write out the following logical expressions.

1. $\overline{B}$ **2.** $A + B$ **3.** AB

4. $AB + C$ **5.** $C + AB$ **6.** ABC

7. $A\overline{B}C$ **8.** $A + \overline{B}$ **9.** $\overline{A} + \overline{B}$

10. $\overline{A}B + C\overline{D}$ **11.** $A + BC + A\overline{B}$ **12.** $\overline{A}\overline{B} + \overline{C}\overline{D} + E$

32–2 INVERSION OPERATOR (NOT)

We will begin our study of the NOT operator by studying the behavior of the switch pictured in Figure 32–3(a). The switch is a two-state device and, as such, can be in only one state (position) at a time. If setting the logic variable A to 1 represents the switch in the closed position, then the open position (not closed) is indicated by setting the logic variable NOT A to 1 ($\overline{A} = 1$).

FIGURE 32–3 (a) Schematic symbol of a single-pole, double-throw switch; (b) circuit used for Example 32–2; (c) truth table for the circuit of part (b).

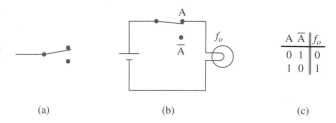

A	$\overline{A}$	f_o
0	1	0
1	0	1

(a)   (b)   (c)

EXAMPLE 32–2   If the closed switch of Figure 32–3(b) is indicated by A = 1, then what is the condition of the lamp (light or dark) when $\overline{A} = 1$?

Solution   Since $\overline{A} = 1$ indicates the switch is in the $\overline{A}$ position, then the switch is open and A = 0. The lamp is dark.

Observation   The truth table for the circuit conditions of Figure 32–3(b) is noted in Figure 32–3(c). In this truth table, $f_o = 1$ means "lamp lit" and $f_o = 0$ means "lamp dark."

From the truth table of Figure 32–3(c), we learn that the lamp is lit when A = 1. Another way of putting this statement would be, the lamp is lit when $\overline{A} = 0$. Furthermore, the lamp is dark when A = 0; or, putting this statement another way, the lamp is dark when $\overline{A} = 1$. Thus, A and $\overline{A}$ are the inverse of one another.

Inverter

The design of digital circuits requires the inversion of logic levels. The digital device used for this operation is called the **inverter,** which performs the NOT operation. The logic diagram symbol for the inverter is shown in Figure 32–4.

FIGURE 32–4 (a) Inverter symbol used to indicate the NOT operation; (b) inverter truth table.

A	$\overline{A}$
0	1
1	0

(a)   (b)

When talking about the NOT operator in relation to logic levels of 1 and 0, the term *complement* is used. For example, $\overline{1} = 0$ and $\overline{0} = 1$. The complement or NOT of 1 is 0, and the complement or NOT of 0 is 1.

The recomplement of a complemented variable or logic level results in the original uncomplemented variable or logic level. Thus, 1 complemented is 0, and 0 recomplemented is 1. Similarly, A inverted is $\overline{A}$, and $\overline{A}$ reinverted is $\overline{\overline{A}}$ or A. Figure 32–5 illustrates this concept.

The words NOT, *complement,* and *inverse* indicate that the opposite condition of the logic variable or logic level is indicated. Thus $\overline{1} = 0$ and $\overline{0} = 1$; NOT A $= \overline{A}$ and NOT $\overline{A} = A$.

FIGURE 32–5 (a) The output logic level of an inverter is the complement of the input logic level. (b) Logic variables are NOTed by an inverter.

(a)   (b)

EXERCISE 32–2

Solve the following.

1. $\overline{1} =$ ___?___

2. $\overline{0} =$ ___?___

3. $\overline{\overline{0}} =$ ___?___

4. $\overline{\overline{1}} =$ ___?___

5. NOT A $=$ ___?___

6. NOT $\overline{A} =$ ___?___

7. $1 \!\!\rhd\!\!\circ$?

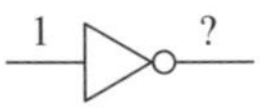

8. $0 \!\!\rhd\!\!\circ$?

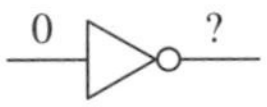

9. $\overline{B} \!\!\rhd\!\!\circ$?

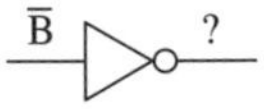

10. ? $\!\!\rhd\!\!\circ$ A

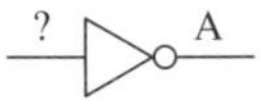

11. ? $\!\!\rhd\!\!\circ \rhd\!\!\circ \rhd\!\!\circ$ 0

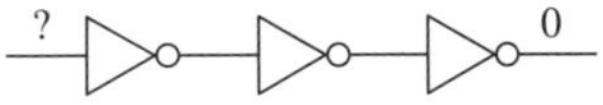

32–3 CONJUNCTION OPERATOR (AND)

The conjunction operator (commonly called the AND operator) depends upon two or more events happening at the same time. When events occur in *conjunction,* a desired function will result. For example, an automobile will start if certain conditions are present in conjunction with one another. Let f_o = car starts, B = charged battery, G = gasoline in tank, K = key in ignition, S = shift lever in park. Then

$$f_o = \text{B AND G AND K AND S}$$
$$f_o = \text{B} \cdot \text{G} \cdot \text{K} \cdot \text{S}$$
$$f_o = \text{BGKS}$$

That is, the car will start "if the battery is charged *and* the tank has gasoline *and* the ignition is unlocked with the key *and* the shift lever is in park."

The AND operator may be visualized as a circuit with two or more series switches.

FIGURE 32–6 (a) AND operation may be thought of as two or more series switches; (b) truth table for the two-variable AND circuit of part (a).

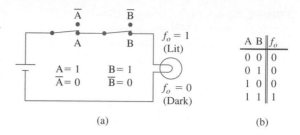

A	B	f_o
0	0	0
0	1	0
1	0	0
1	1	1

(a)   (b)

In Figure 32–6(a), we see that the lamp will light when both switches are closed at the same time. From the truth table of Figure 32–6(b), we also see that both switches must be closed before the lamp will light. When A = 1 (closed) and B = 1 (closed), then f_o = 1 (lit). Notice that when either one or both of the switches are open, the lamp is dark. Thus:

$$f_o = 0 \text{ (dark)} \quad \text{when} \quad A = 0 \text{ (open)} \quad \text{and} \quad B = 0 \text{ (open)}$$
$$f_o = 0 \text{ (dark)} \quad \text{when} \quad A = 0 \text{ (open)} \quad \text{and} \quad B = 1 \text{ (closed)}$$
$$f_o = 0 \text{ (dark)} \quad \text{when} \quad A = 1 \text{ (closed)} \quad \text{and} \quad B = 0 \text{ (open)}$$

In the AND operation, all input logic levels must be 1 if the logic level of the output function is to be 1.

AND Gate

The operation of digital computers uses the conjunction operation to combine several logic levels at the same time. The digital device used for this operation is called the AND gate. The logic diagram symbol for the AND gate is shown in Figure 32–7.

FIGURE 32–7 (a) Two-input AND gate symbol used to indicate the AND operation; (b) two-variable AND gate truth table.

A ———⌐
         $f_o = AB$
         (OUTPUT)
B ———⌐
(INPUTS)

A	B	f_o
0	0	0
0	1	0
1	0	0
1	1	1

(a)   (b)

For a two-input AND gate, there are 2^2 or 4 possible combinations of the variables because each variable has 2 possible values. The general rule is

$$\textbf{number of possible combinations} = 2^n \qquad \textbf{(32–1)}$$

where n is the number of input variables.

EXAMPLE 32–3   (a) Construct the logic diagram for the logic expression $f_o = ABC$.

                    (b) With the aid of Equation 32–1, determine the number of combinations needed to form the truth table.

                    (c) Write the truth table for this three-input AND gate.

	A	B	C	$f_o = ABC$
0	0	0	0	0
1	0	0	1	0
2	0	1	0	0
3	0	1	1	0
4	1	0	0	0
5	1	0	1	0
6	1	1	0	0
7	1	1	1	1

(a)   (b)

Solution

(a) Figure 32–8(a) pictures a three-input AND gate with the inputs noted with the logical variables A, B, and C.

(b) The number of combinations needed to describe a three-variable gate is 2^3 or 8 combinations. These eight combinations are written as the binary equivalent of the decimal numbers 0 through 7.

(c) The truth table is formed as pictured in Figure 32–8(b). Here you see the eight possible combinations for the input variables (A, B, and C) noted by the binary numbers 000 (0_{10}) through 111 (7_{10}). The output AND function (f_o) is determined for each of the listed input combinations by applying the concepts of the AND operation. Thus:

The combination of line 0 yields an output function of zero:

$$0 \text{ AND } 0 \text{ AND } 0 \Rightarrow 0$$

The combination of line 4 yields an output function of zero:

$$1 \text{ AND } 0 \text{ AND } 0 \Rightarrow 0$$

The combination of line 7 yields an output function of one:

$$1 \text{ AND } 1 \text{ AND } 1 \Rightarrow 1$$

In summary, the output function of an AND gate has a logic level of 1 only when all the input logic levels are 1 at the same time. If one or more input logic levels of the AND gate are 0, then the output function of the AND gate is 0. In actual logic circuits, the timing of the logic levels is very important. Figure 32–9 emphasizes this concept. Here we see the important role that timing plays in the operation of the AND gate. Between times t_1 and t_2 of Figure 32–9(b), we see that both input A and input B are *high* (logic level 1), resulting in the output also going *high* (logic level 1). This same condition occurs between time t_5 and t_6. The remainder of the time intervals results in the output function being *low* (logic level 0), because either or both inputs A and B are *low* (logic level 0).

FIGURE 32–9 The output logic level (f_o) of the AND gate (a) is 1 only when the input logic levels of A and B are 1 at the same time as shown in part (b).

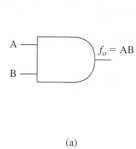

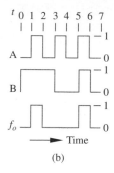

(a)

(b)

EXAMPLE 32–4   (a) Using the logic diagram symbols for NOT and AND, draw the logic diagram for $f_o = \overline{A}B$.

(b) Write the truth table for the two-variable logic expression.

FIGURE 32–10 (a) Logic diagram for the function $f_o = \overline{A}B$; (b) truth table for the logic expression $f_o = \overline{A}B$.

A —▷o— $\overline{A}$

$f_o = \overline{A}B$

B ———

(a)

	A B	$\overline{A}$	$f_o = \overline{A}B$
0	0 0	1 0	
1	0 1	1 1	
2	1 0	0 0	
3	1 1	0 0	

(b)

Solution   (a) Figure 32–10 pictures the implementation of the logic function $f_o = \overline{A}B$. Note the use of the inverter to produce NOT A.

(b) The truth table has 2^2, or 4, possible conditions. In constructing the truth table, an additional column is added for the $\overline{A}$ combination. Figure 32–10(b) shows the completed truth table.

Observation   The method for forming the truth table of Figure 32–10(b) proceeds from left to right. For example, the combination of line 1 results in 0 1 for columns (A, B) and 1 1 for columns (B, $\overline{A}$). 1 AND 1 for $\overline{A}$ AND B results in 1 for f_o.

EXERCISE 32–3

Determine the output logic level of f_o in each of the following expressions.

1. $f_o = ABC$ when A = 1, B = 1, and C = 1

2. $f_o = A\overline{B}C$ when A = 1, B = 0, and C = 1

3. $f_o = ABC$ when $\overline{A} = 0, B = 1$, and $\overline{C} = 1$

4. $f_o = \overline{A}BC$ when $A = 0, B = 0$, and $C = 1$

5. $f_o = \overline{A}BC$ when $A = 1, B = 1$, and $C = 1$

6. $f_o = AB\,\overline{C}$ when $\overline{A} = 0, B = 0$, and $C = 0$

7. $f_o = \overline{A}\,\overline{B}\,\overline{C}$ when $\overline{A} = 1, \overline{B} = 1$, and $C = 1$

8. $f_o = \overline{A}\,B\overline{C}$ when $A = 1, B = 0$, and $C = 1$

Using the logic diagram symbols for NOT and AND, draw the logic diagram for each of the following expressions.

9. $f_o = AB$ 　　10. $f_o = ABC$ 　　11. $f_o = ABCD$

12. $f_o = A\overline{B}$ 　　13. $f_o = \overline{A}\,\overline{B}$ 　　14. $f_o = AB\overline{C}$

Complete the truth table for each of the following logic diagrams.

15.

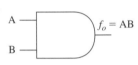

	A	B	$f_o = AB$
0	0	0	0
1			
2			
3			

16.

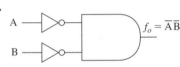

	A	B	$\overline{A}$	$\overline{B}$	$f_o = \overline{A}\,\overline{B}$
0	0	0	1	1	
1					
2					
3					

17. For the logic expression $f_o = \overline{A}BC$:

   (a) Determine the number of logic combinations needed to construct the truth table.

   (b) Construct the truth table.

   (c) Draw the logic diagram.

18. For the logic expression $f_o = \overline{A}B\overline{C}$:

   (a) Determine the number of logic combinations needed to construct the truth table.

   (b) Construct the truth table.

   (c) Draw the logic diagram.

32–4　DISJUNCTION OPERATOR (OR)

The disjunction operator (commonly called the OR operator) allows for a choice between two or more events occurring at the same time. For example, an automobile with an automatic transmission may not be started with the shift lever in drive or if the battery is dead. Let f_o = car won't start, D = shift lever in drive, and B = dead battery. Then

$$f_o = D \text{ OR } B$$
$$f_o = D + B$$

That is, "the car will not start if the shift lever is in drive *or* the battery is dead."

The OR operator may be visualized as a circuit with two or more parallel switches. In Figure 32–11(a), we see that the lamp will light when either of the switches is closed. From the truth table of Figure 32–11(b), we learn that the OR operation includes the possibility of both variables being 1. This type of OR operator is called an *inclusive* OR because of its inclusive nature. We will use the term OR to mean *inclusive* OR. From the truth table, you may notice that when either or both of the switches are closed the lamp is lit, and the lamp is dark only when both switches are open. Thus,

$$f_o = 0 \text{ (dark)} \quad \text{when} \quad A = 0 \text{ (open)} \quad \text{and} \quad B = 0 \text{ (open)}$$
$$f_o = 1 \text{ (lit)} \quad \text{when} \quad A = 0 \text{ (open)} \quad \text{and} \quad B = 1 \text{ (closed)}$$
$$f_o = 1 \text{ (lit)} \quad \text{when} \quad A = 1 \text{ (closed)} \quad \text{and} \quad B = 0 \text{ (open)}$$
$$f_o = 1 \text{ (lit)} \quad \text{when} \quad A = 1 \text{ (closed)} \quad \text{and} \quad B = 1 \text{ (closed)}$$

In the OR operation, if any of the input logic levels is 1, then the output logic level of the output function will be 1.

OR Gate

The operation of digital computers uses the disjunction operation to produce a high output (logic level 1) when any one of several inputs is high (logic level 1). The digital device used for this operation is called the OR gate. The logic diagram symbol for the OR gate is shown in Figure 32–12.

FIGURE 32–11 (a) OR operation may be thought of as two or more parallel switches; (b) truth table for the two-variable OR circuit of part (a).

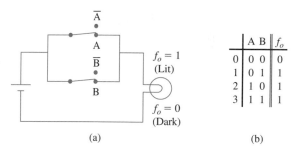

	A	B	f_o
0	0	0	0
1	0	1	1
2	1	0	1
3	1	1	1

(a)   (b)

FIGURE 32–12 (a) Two-input OR gate symbol used to indicate the OR operation; (b) two-variable OR gate truth table.

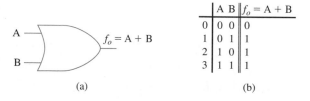

$f_o = A + B$

	A	B	$f_o = A + B$
0	0	0	0
1	0	1	1
2	1	0	1
3	1	1	1

(a)   (b)

Like the AND gate, the OR gate has 2^n possible combinations needed to describe its operation (where n is the number of inputs).

EXAMPLE 32–5

(a) Construct the logic diagram for the logic expression $f_o = \overline{A} + B + \overline{C}$.

(b) With the aid of Equation 32–1, determine the number of combinations needed to form the truth table.

(c) Write the truth table for this three-input OR gate.

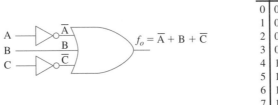

	A B C	$\overline{A}$	B	$\overline{C}$	$f_o = \overline{A} + B + \overline{C}$
0	0 0 0	1	0	1	1
1	0 0 1	1	0	0	1
2	0 1 0	1	1	1	1
3	0 1 1	1	1	0	1
4	1 0 0	0	0	1	1
5	1 0 1	0	0	0	0
6	1 1 0	0	1	1	1
7	1 1 1	0	1	0	1

(a)   (b)

FIGURE 32–13   (a) Three-input OR gate with inverters for Example 34–5; (b) truth table for the three-variable logic expression $f_o = \overline{A} + B + \overline{C}$.

Solution

(a) Figure 32–13(a) shows a three-input OR gate with inverters in the A and C inputs to produce $\overline{A}$ and $\overline{C}$.

(b) The number of combinations needed to describe a three-variable gate is 2^3, or 8, combinations.

(c) The truth table is formed as pictured in Figure 32–13(b). Here you see the eight possible combinations for the input variables (A, B, and C) noted in the binary numbers 000 through 111. In the next column, you see the $\overline{A}$, B, and $\overline{C}$ logic levels listed. Note the combination of line 5, where $\overline{A} = 0$, $B = 0$, and $\overline{C} = 0$ is the only combination that produces an output function of 0 ($f_o = 0$ OR 0 OR 0 = 0).

In summary, the output function of an OR gate has a logic level of 0 only when all the input logic levels are 0 at the same time. If one or more input logic levels of the OR gate is 1, then the output function of the OR gate is 1.

EXERCISE 32–4

Determine the output logic level of f_o in each of the following expressions.

1. $f_o = A + B + C$ when $A = 1$, $B = 1$, and $C = 1$
2. $f_o = A + \overline{B} + C$ when $\overline{A} = 1$, $B = 1$, and $C = 0$

3. $f_o = A + B + C$ when $\overline{A} = 0$, $B = 0$, and $\overline{C} = 1$

4. $f_o = \overline{A} + \overline{B} + C$ when $A = 0$, $B = 0$, and $C = 1$

5. $f_o = \overline{A} + B + C$ when $A = 1$, $B = 1$, and $C = 1$

6. $f_o = A + \overline{B} + \overline{C}$ when $\overline{A} = 1$, $B = 1$, and $C = 1$

7. $f_o = \overline{A} + \overline{B} + \overline{C}$ when $\overline{A} = 1$, $\overline{B} = 1$, and $C = 1$

8. $f_o = \overline{A} + B + \overline{C}$ when $A = 1$, $B = 0$, and $C = 1$

Using the logic diagram symbols for NOT and OR, draw the logic diagram for each of the following expressions.

9. $f_o = A + B$ 10. $f_o = A + B + C$

11. $f_o = A + B + C + D$ 12. $f_o = A + \overline{B}$

13. $f_o = \overline{A} + \overline{B}$ 14. $f_o = A + B + \overline{C}$

Complete the truth table for each of the following logic diagrams.

15.

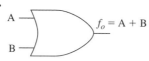

$f_o = A + B$

A	B	$f_o = A + B$
0	0	0	0
1		
2		
3		

16.

$f_o = \overline{A} + \overline{B}$

A	B	$\overline{A}$	$\overline{B}$	$f_o = \overline{A} + \overline{B}$
0	0	0	1	1
1				
2				
3				

17. For the logic expression $f_o = \overline{A} + B + C$:

   (a) Determine the number of logic combinations needed to construct the truth table.

   (b) Construct the truth table.

   (c) Draw the logic diagram.

18. For the logic expression $f_o = B + \overline{C} + \overline{D}$:

   (a) Determine the number of logic combinations needed to construct the truth table.

   (b) Construct the truth table.

   (c) Draw the logic diagram.

32–5 APPLICATION OF LOGIC CONCEPTS

In this section, you will have an opportunity to expand your knowledge of the logic operators as well as the use of truth tables. As you have already learned, logic diagrams are used to symbolize logic expressions. No matter how complex the logic expression may be, it can be symbolized with the three logic operations. By using the AND, OR, and NOT logic symbols independently or collectively, any logic expression may be symbolized. The following examples will demonstrate further use of these *building blocks* in the construction of logic diagrams and truth tables.

EXAMPLE 32–6   Develop the logic diagram for the logic expression $f_o = A + \overline{B}C$.

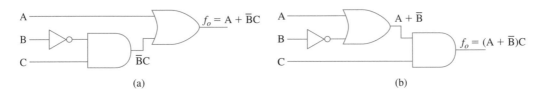

(a)   (b)

FIGURE 32–14   (a) Logic diagram for the expression $f_o = A + \overline{B}C$ for Example 32–6. (b) Logic diagram for the expression $f_o = (A + \overline{B}) C$. This expression demonstrates the use of parentheses to indicate the desired order of operation.

Solution   Figure 32–14(a) is the logic diagram for the logic expression $f_o = A + \overline{B}C$.

Observation   When working with logic expressions, the AND operation has precedence over the OR operation. Thus, in the expression $A + \overline{B}C$, NOT B AND C is ORed with A, as noted in Figure 32–14(a). The defined order of operation may be superseded by parentheses. In the expression $(A + \overline{B})C$, A OR NOT B is ANDed with C. Figure 32–14(b) is the logic diagram for this logic expression.

In the preceding example, you saw the importance of a defined *order of operation* for logic expressions. Convention 32–1 strengthens this concept.

CONVENTION 32–1.   Order of Logic Operators

It is conventional that:

1. The NOT operator has precedence over both the AND and the OR operator.
2. The AND operator has precedence over the OR operator.
3. Signs of grouping, including the bar, may be used to override the conventional order.

The next example demonstrates the use of truth tables to determine if two logic expressions are equivalent or are *logically identical*.

EXAMPLE 32–7   Determine if the logic expressions represented by the logic diagrams of Figure 32–14(a) and (b) are equivalent.

Solution   Form the identity $A + \overline{B}C \equiv (A + \overline{B})C$. Construct a truth table for each member of the equation using 2^3 or 8 lines to represent the possible combinations. Figure 32–15(a) is the truth

	A B C	$\overline{B}$	$\overline{B}C$	$f_o = A + B\overline{C}$
0	0 0 0	1	0	0
1	0 0 1	1	1	1
2	0 1 0	0	0	0
3	0 1 1	0	0	0
4	1 0 0	1	0	1
5	1 0 1	1	1	1
6	1 1 0	0	0	1
7	1 1 1	0	0	1

	A B C	$\overline{B}$	$A + \overline{B}$	$f_o = (A + \overline{B})C$
0	0 0 0	1	1	0
1	0 0 1	1	1	1
2	0 1 0	0	0	0
3	0 1 1	0	0	0
4	1 0 0	1	1	0
5	1 0 1	1	1	1
6	1 1 0	0	1	0
7	1 1 1	0	1	1

(a)   (b)

FIGURE 32–15   Truth tables for Example 32–7.

table of the left member, $A + \overline{B}C$. Figure 32–15(b) is the truth table of the right member, $(A + \overline{B})C$. To determine if they are logically identical, the output function of each member is compared for each of the 8 possible combinations. Make a line-by-line comparison between the output functions of the two truth tables. Notice that lines 4 and 6 do not match. This leads us to conclude that these two logic expressions and their corresponding logic diagrams are not equivalent.

$$\therefore \quad A + \overline{B}C \neq (A + \overline{B})C$$

Observation   In Figure 32–15(a), notice how the term $\overline{B}C$ (NOT B AND C) was evaluated as an intermediate step by ANDing the combinations listed in the C and $\overline{B}$ columns. The use of this intermediate step reduces the possibility of error in the formation of the output function. The output function is formed by ORing the combination of columns A and $\overline{B}C$.

In the preceding example, you saw how two logic diagrams may be examined to see if they are logically identical by comparing the output functions in the corresponding truth tables. This technique is used in the next example to verify that two logic expressions are equivalent.

EXAMPLE 32–8   Determine if $(A + B)(B + C) \equiv AC + B$ is a true statement.

	A B C	A + B	B + C	$f_o = (A + B)(B + C)$
0	0 0 0	0	0	0
1	0 0 1	0	1	0
2	0 1 0	1	1	1
3	0 1 1	1	1	1
4	1 0 0	1	0	0
5	1 0 1	1	1	1
6	1 1 0	1	1	1
7	1 1 1	1	1	1

(a)

	A B C	AC	$f_o = AC + B$
0	0 0 0	0	0
1	0 0 1	0	0
2	0 1 0	0	1
3	0 1 1	0	1
4	1 0 0	0	0
5	1 0 1	1	1
6	1 1 0	0	1
7	1 1 1	1	1

(b)

FIGURE 32–16   Truth tables for Example 32–8.

Solution   Construct a truth table for each member of the identity using 2^3, or 8, lines for each. Figure 32–16(a) is the truth table for the left member, while Figure 32–16(b) is the truth table for the right member. Making a line-by-line comparison between the output functions of the two truth tables leads us to conclude that these two logic expressions and their corresponding logic diagrams are equivalent.

$\therefore$ $(A + B)(B + C) \equiv AC + B$

From the preceding example, we learn that the same output may result from several different logic expressions. Although each expression may produce the desired output, only one expression is the simplest and least expensive to implement. Figure 32–17(a) and (b) are the two equivalent logic diagrams of the previous example. The one on the right is simpler and less costly to implement. The next example provides you with a step-by-step development of the truth table for a logic diagram.

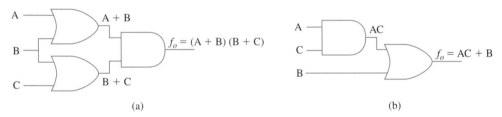

(a)   (b)

FIGURE 32–17   Logic diagrams for the equivalent expressions used in Example 32–8. Part (b) is simpler than part (a) because it has fewer logic gates.

EXAMPLE 32–9   Construct the truth table for the logic diagram of Figure 32–18.

FIGURE 32–18   Logic diagram for Example 32–9.

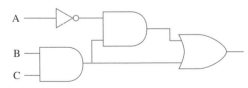

Solution   Because of the three inputs, select three variables (A, B, and C) to represent them. Use 2^3, or 8, lines in the truth table. Start the truth table by listing the 8 combinations in the left side of the truth table, as in Figure 32–19(a). Next, form a column for the NOT operation; label this column $\overline{A}$. Then, form a column for the AND operation, BC, as indicated in Figure 32–19(b). Finally, form columns for the other AND operation, $\overline{A}BC$, and the output of the OR gate, $BC + \overline{A}BC$, as indicated in Figure 32–19(c).

Observation   The logic levels of the output (f_o) are identical to those of column BC. This means that $BC + \overline{A}BC = BC$ and the logic diagram of Figure 32–18 may be simplified to a single AND gate with two inputs, B and C.

LINE	A B C
0	0 0 0
1	0 0 1
2	0 1 0
3	0 1 1
4	1 0 0
5	1 0 1
6	1 1 0
7	1 1 1

(a)

LINE	A B C	$\overline{A}$	BC
0	0 0 0	1	0
1	0 0 1	1	0
2	0 1 0	1	0
3	0 1 1	1	1
4	1 0 0	0	0
5	1 0 1	0	0
6	1 1 0	0	0
7	1 1 1	0	1

(b)

LINE	A B C	$\overline{A}$	BC	$\overline{A}BC$	$f_o = BC + \overline{A}BC$
0	0 0 0	1	0	0	0
1	0 0 1	1	0	0	0
2	0 1 0	1	0	0	0
3	0 1 1	1	1	1	1
4	1 0 0	0	0	0	0
5	1 0 1	0	0	0	0
6	1 1 0	0	0	0	0
7	1 1 1	0	1	0	1

(c)

FIGURE 32–19   Truth table developed for the logic diagram of Figure 32–18.

EXERCISE 32–5

Draw the logic diagram for each of the following logic expressions.

1. $\overline{A}B + C$ **2.** $AC + AB$ **3.** $\overline{A} + ABC$

4. $\overline{A}B + \overline{A}C$ **5.** $A(B + C)$ **6.** $AB(A + C)$

Construct a truth table for the following logic expressions.

7. $A\overline{C} + B$ **8.** $AB\overline{C} + \overline{A}B + C$

9. $A(C + B\overline{C})$ **10.** $C(\overline{B} + \overline{A})$

Using truth tables, determine which of the following equations are true.

11. $A(A + B) = A$ **12.** $A + \overline{A}B = A + B$

13. $A(B + C) = AB + AC$ **14.** $(A + B)(A + C) = A + BC$

15. $AC + \overline{A}\,\overline{C} = A\overline{C} + \overline{A}C$ **16.** $A\overline{B} + C(A + \overline{B}) = AC$

17. Construct the truth table for the logic diagram of Figure 32–20.

18. Construct the truth table for the logic diagram of Figure 32–21.

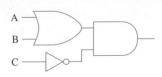

FIGURE 32–20 Logic diagram for Exercise 32–5, problem 17.

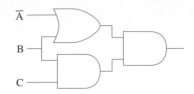

FIGURE 32–21 Logic diagram for Exercise 32–5, problem 18.

32–6 INTRODUCTION TO KARNAUGH MAPS

One goal for the logic circuit designer is to design circuits with as few parts as possible, because a reduction in parts will result in an increase in circuit reliability and a reduction in circuit cost. The **Karnaugh map,** or *K-map,* as it will be called, is a tool for simplifying logic circuits. It is similar in concept to the truth table, but by its organization it allows the designer to simplify a circuit through pattern recognition.

Recall that a truth table contains a line for each combination of values for the input variables. Thus, a truth table for two variables contains four lines, and a truth table for three variables contains eight lines. Furthermore, each line is identified by the decimal number equal in value to the binary number representing the values of the input variables. Similarly, a K-map contains a square for each combination of values for the input variables. Thus, a K-map for two variables contains four squares, as shown in Figure 32–22(b), and a K-map for three variables contains eight squares, as shown in Figure 32–23(b). Furthermore, each square in a K-map is numbered. These numbers correspond to the line or combination numbers in the truth table. The usefulness of a K-map comes from the organization of the data. This means that the numbering system of the squares for the K-maps in Figures 32–22(b), 32–23(b), and 32–24 must be carefully followed.

Examine the K-map for the two variables A and B in Figure 32–22(b). The four small squares are organized as a larger square with two rows and two columns. One column is for A, which is labeled; the other column is for $\overline{A}$, which is usually not labeled. Similarly, one row is for B, which is labeled; the other row is for $\overline{B}$, which is usu-

FIGURE 32–22 Truth table and Karnaugh map for the two-input variables A and B.

LINE	A B	f_o
0	0 0	
1	0 1	
2	1 0	
3	1 1	

(a)

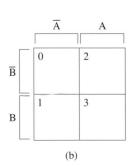

(b)

FIGURE 32–23 Truth table and Karnaugh map for the three-input variables A, B, and C.

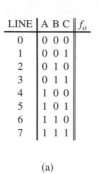

LINE	A B C	f_o
0	0 0 0	
1	0 0 1	
2	0 1 0	
3	0 1 1	
4	1 0 0	
5	1 0 1	
6	1 1 0	
7	1 1 1	

(a)

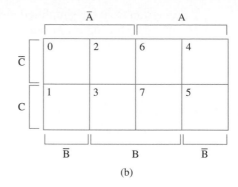

(b)

ally not labeled. The number in the upper left corner of each square is the same as the combination number for the corresponding line in the truth table. Since the upper left square represents the term $\overline{A}\,\overline{B}$, the number in the corner of the upper left square is zero. Similarly, the first line of the truth table in Figure 32–22(a) is for the combination 0_{10} or 00_2—that is, $\overline{A}$ and $\overline{B}$.

The K-map for the three variables A, B, and C of Figure 32–23(b) is slightly more complicated than the two-variable K-map. Two columns are for A and two for $\overline{A}$. In like manner there are two columns for B and two for $\overline{B}$. Notice how the two columns for $\overline{B}$ are split between A and $\overline{A}$. Finally, one row is for C and the other row is for $\overline{C}$. Since the upper left square represents the term $\overline{A}\,\overline{B}\,\overline{C}$, the number in the corner of the upper left square is zero. Figure 32–24 shows a four-variable K-map.

The use of a K-map for logic circuit simplification is part of a systematic process. First, build a truth table from the logic diagram. Second, map the truth table by entering a 1 in each square of the K-map corresponding to a combination having a 1 for the output function in the truth table. (All the other squares are left blank.) Third, group adjacent squares that contain 1s to simplify the logic expression. Fourth, read the simplified logic expression for the circuit from the K-map. Fifth, create a simplified logic

FIGURE 32–24 Pattern for four-variable Karnaugh maps.

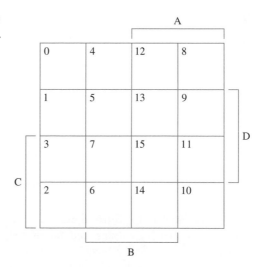

diagram from the simplified logic expression. The preceding concepts will be explored in the following examples.

EXAMPLE 32–10   Create a truth table and a K-map for $f_o = AB + \overline{A}B$ and simplify.

FIGURE 32–25 Truth table and Karnaugh map for Example 32–10, $f_o = AB + \overline{A}B$

LINE	A B	AB	$\overline{A}B$	f_o
0	0 0	0	0	0
1	0 1	0	1	1
2	1 0	0	0	0
3	1 1	1	0	1

(a)

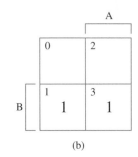

(b)

Solution   Since there are two variables, the truth table will contain four lines and the K-map will contain four squares. See Figure 32–25 for the truth table and the K-map.

Notice that the two 1s in the K-map are adjacent. Because both 1s are in the B row, they can be represented by a single term in the single variable B. Thus, $AB + \overline{A}B$ simplifies to B.

∴ $f_o = B$

EXAMPLE 32–11   Create a truth table and a K-map for $f_o = AB + AC + \overline{A}B\overline{C}$ and simplify.

LINE	A B C	AB	AC	$\overline{A}B\overline{C}$	f_o
0	0 0 0	0	0	0	0
1	0 0 1	0	0	0	0
2	0 1 0	0	0	1	1
3	0 1 1	0	0	0	0
4	1 0 0	0	0	0	0
5	1 0 1	0	1	0	1
6	1 1 0	1	0	0	1
7	1 1 1	1	1	0	1

(a)

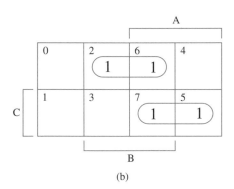

(b)

FIGURE 32–26 Truth table and Karnaugh map for Example 32–11, $f_o = AB + AC + \overline{A}B\overline{C}$.

Solution   Since there are three variables, the truth table will contain eight lines and the K-map will contain eight squares. See Figure 32–26 for the truth table and K-map.

Notice that the three-variable term, $\overline{A}B\overline{C}$, is represented by one square (number 2), while each two-variable term occupies two squares (6, 7) and (5, 7). Furthermore, a single square (number 7) can be part of two different terms (AB and AC). Simplify by grouping the four 1s into pairs: (2, 6) and (5, 7). Read the map:

Both squares of the pair (2, 6) are in column B and row $\overline{C}$; thus (2, 6) represents $B\overline{C}$.

Both squares of the pair (5, 7) are in column A and row C; thus (5, 7) represents AC.

Therefore, the simplified expression is $AC + B\overline{C}$.

$$AB + AC + \overline{A}B\overline{C} = B\overline{C} + AC$$

$$\therefore \quad f_o = B\overline{C} + AC$$

Once the K-map is completed for an expression, the next step is to check for recognizable patterns. An isolated 1 represents a term containing all the variables. Two adjacent 1s represent a term containing all but one variable. The two 1s can be side by side or one on top of the other. However, diagonally positioned 1s are not significant and do not result in a reduction of variables. Four adjacent 1s represent a term containing all but two variables. The four 1s can be in a row or a column or a square. For our purpose of defining adjacent squares in K-maps, the top row is considered to be adjacent to the bottom row; and the left column is considered to be adjacent to the right column.

EXAMPLE 32–12   Determine a minimal expression for the K-map in Figure 32–27.

FIGURE 32–27 K-map for Example 32–12.

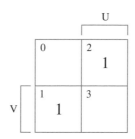

Solution   The K-map is for the two variables U and V. The two 1s are not adjacent; they are isolated. Therefore, there are two terms, each containing both variables.

$$\therefore \quad f_o = U\overline{V} + \overline{U}V$$

EXAMPLE 32–13   Determine a minimal expression for the K-map in Figure 32–28.

FIGURE 32–28   K-map for Example 32–13.

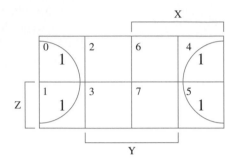

Solution   The K-map is for the three variables X, Y, and Z. The four 1s are adjacent; therefore, they represent one term in one variable.

$$\therefore \quad f_o = \overline{Y}$$

EXAMPLE 32–14   Determine a minimal expression for the K-map in Figure 32–29.

FIGURE 32–29   K-map for Example 32–14.

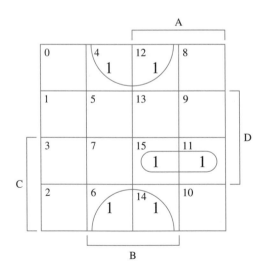

Solution   The K-map is for the four variables A, B, C, and D. Square 15 can be combined with 11 or 14. Since square 11 can be combined only with 15, we combine 11 and 15 into one term in three variables. The remaining four squares combine into one term in two variables.

$$\therefore \qquad f_o = ACD + B\overline{D}$$

Summary

The Karnaugh map was introduced in this section as a tool for simplifying logic circuits. Its relation to the truth table has been explored. Its usefulness resides in the ability of the user to recognize patterns. Table 32–1 and Guidelines 32–1 are designed to help in the pattern recognition process.

Table 32–1 gives the relationship between the number of adjacent entries in a K-map and the number of variables in the corresponding term for the three practical sized K-maps. The relationship in the table holds both for mapping an expression and for reading a K-map.

TABLE 32–1   Occupancy Rates for Karnaugh Maps

Number of Variables in Term	Number of Occupied Squares		
	Two-Variable Map	Three-Variable Map	Four-Variable Map
1	2	4	8
2	1	2	4
3		1	2
4			1

Guidelines 32–1 are useful in interpreting the patterns found within K-maps. Use the guidelines while working the exercise set at the end of this section.

GUIDELINES 32–1.   Interpreting Karnaugh Maps

1. The top row is adjacent to the bottom row.
2. The left column is adjacent to the right column.
3. Begin grouping the adjacent 1s with those that can be grouped only one way. A 1 can be used in more than one grouping.
4. Group as many adjacent 1s together as possible.
5. An isolated 1 represents a term with the same number of variables as are in the map. Thus an isolated 1 in a three-variable map represents a three-variable term.

(continued)

6. Two adjacent 1s represent a term with one less variable than is in the map. Thus two adjacent 1s in a three-variable map represent a two-variable term.

7. Three adjacent 1s are grouped as two overlapping sets of two adjacent 1s that are each treated as in Guideline 6.

8. Four adjacent 1s in a row, column, or square represent a term with two less variables than are in the map. Thus, four adjacent 1s in a square in a three-variable map represent a one-variable term.

9. Four adjacent 1s not in a row, column, or square are grouped as two sets of two adjacent 1s that are each treated as in Guideline 6. See Figure 32–26(b).

10. The term resulting from grouping a set of squares within a K-map contains the variable(s) common to each of the squares in the pattern. For example:

(a) In Figure 32–25(b): 2 adjacent 1s = 1 variable, B
(b) In Figure 32–26(b): 2 adjacent 1s = 2 variables, $B\overline{C}$
   2 adjacent 1s = 2 variables, $A\overline{C}$
(c) In Figure 32–27(b): an isolated 1 = 2 variables, $U\overline{V}$
   an isolated 1 = 2 variables, $\overline{U}V$
(d) In Figure 32–28: 4 adjacent 1s = 1 variable, $\overline{Y}$
(e) In Figure 32–29: 2 adjacent 1s = 3 variables, ACD
   4 adjacent 1s = 2 variables, $B\overline{D}$

EXERCISE 32–6

Create a Karnaugh map for each of the following expressions:

1. AB
2. $\overline{A}B$
3. $J\overline{K} + K$
4. $M\overline{N} + \overline{M}$
5. $\overline{P}QR + P\overline{R}$
6. UV + UW + VW

Use Karnaugh maps to simplify the following expressions:

7. $S\overline{T} + \overline{S}\,\overline{T}$
8. $\overline{X}Y + XY$
9. $H + \overline{G}H$
10. JK + $JK\overline{L}$
11. UV + UW + $U\overline{V}\,\overline{W}$
12. $\overline{P}\,\overline{Q} + PR + QR$
13. Create a K-map for the logic diagram in Figure 32–30.

FIGURE 32–30 Logic diagram for Exercise 32–6, problem 13.

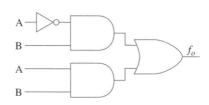

14. Create a K-map for the logic diagram in Figure 32–31.

15. Simplify the logic diagram in Figure 32–32.

16. Simplify the logic diagram in Figure 32–33.

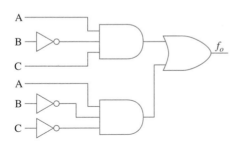

FIGURE 32–31 Logic diagram for Exercise 32–6, problem 14.

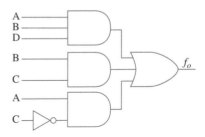

FIGURE 32–32 Logic diagram for Exercise 32–6, problem 15.

FIGURE 32–33 Logic diagram for Exercise 32–6, problem 16.

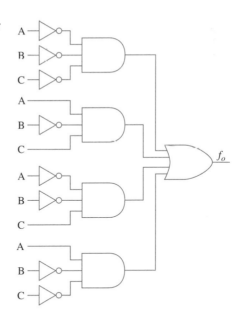

32–7 DEMORGAN'S THEOREM

In Section 32–2 the idea of the NOT operator was introduced along with its effect on a variable. In this section, DeMorgan's theorem is introduced to define the effect of the NOT operator on a logic expression. We will begin with the examination of the following two special cases.

EXAMPLE 32–15   Use a K-map to find another expression for $\overline{U \cdot V}$.

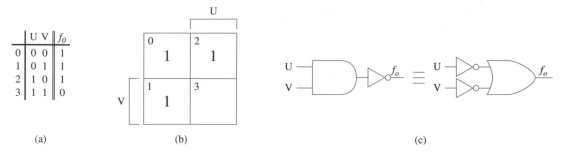

(a)   (b)   (c)

FIGURE 32–34   Illustration of DeMorgan's theorem: $\overline{U\,V} = \overline{U} + \overline{V}$.

Solution   The expression $\overline{U \cdot V}$ is read NOT the product U AND V.
Form a truth table as in Figure 32–34(a).
Create a K-map as in Figure 32–34(b).
Read the function from the K-map:

$$f_o = \overline{U} + \overline{V}$$

∴ $\overline{U \cdot V} = \overline{U} + \overline{V}$, as shown in Figure 32–34(c).

EXAMPLE 32–16   Use a K-map to find another expression for $\overline{Y + Z}$.

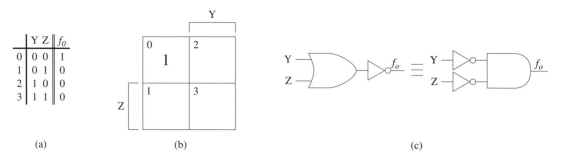

(a)   (b)   (c)

FIGURE 32–35   Illustration of DeMorgan's theorem: $\overline{Y + Z} = \overline{Y}\,\overline{Z}$.

Solution   The expression $\overline{Y + Z}$ is read NOT the sum Y OR Z.
Form a truth table as in Figure 32–35(a).
Create a K-map as in Figure 32–35(b).
Read the function from the K-map:

$$f_o = \overline{Y} \cdot \overline{Z}$$

∴ $\overline{Y + Z} = \overline{Y} \cdot \overline{Z}$, as shown in Figure 32–35(c).

The results of the preceding two examples are summarized as DeMorgan's theorem in Rule 32–1.

RULE 32–1. DeMorgan's Theorem

To invert a logic expression, invert each variable and exchange all AND and OR operators. Use signs of grouping to preserve the original order of operation. Remember, for any variable X, $\overline{\overline{X}} = X$. Thus, the inverse of

$$f_o = (A \cdot \overline{B}) + C$$

is given by

$$\overline{f_o} = (\overline{A} + B) \cdot \overline{C}$$

EXAMPLE 32–17   Invert $J \cdot \overline{K} + \overline{J} \cdot K \cdot L + \overline{L}$.

Solution   Use DeMorgan's Theorem:

$$J \cdot \overline{K} + \overline{J} \cdot K \cdot L + \overline{L}$$
$$(\overline{J} + K) \cdot (J + \overline{K} + \overline{L}) \cdot L$$

Observation   The parentheses are to retain the original order of operation.

EXAMPLE 32–18   Redraw the logic diagram in Figure 32–36(a) with the inverters on the input side of the gates.

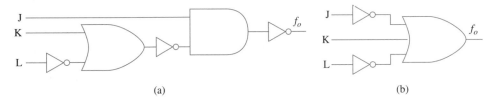

(a)   (b)

FIGURE 32–36   Logic diagrams for Example 32–18.

Solution   Determine the output function for the logic diagram:

$$f_o = \overline{\overline{J(K + \overline{L})}}$$

Let $M = (K + \overline{L})$ and apply DeMorgan's theorem:

$$f_o = \overline{\overline{JM}}$$
$$f_o = \overline{J} + M$$
$$f_o = \overline{J} + K + \overline{L}$$

∴   See Figure 32–36(b) for the new logic diagram.

As we saw in Section 32–2, inverting a variable twice results in the original variable. The same holds true for expressions; that is, $f_o = $ NOT NOT f_o. This provides the key to the technique for simplifying circuits using OR logic, as in the following example.

EXAMPLE 32–19   Simplify the logic diagram in Figure 32–37.

FIGURE 32–37   Logic diagram for Example 32–19.

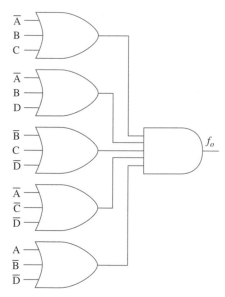

Solution   Determine the logic expression for the output function:

$$f_o = (\overline{A} + B + C)(\overline{A} + B + D)(\overline{B} + C + \overline{D})$$
$$\cdot\, (\overline{A} + \overline{C} + \overline{D})(A + \overline{B} + \overline{D})$$

Invert the function:

$$\overline{f}_o = \overline{(\overline{A} + B + C)(\overline{A} + B + D)(\overline{B} + C + \overline{D})}$$
$$\overline{\cdot\, (\overline{A} + \overline{C} + \overline{D})(A + \overline{B} + \overline{D})}$$
$$\overline{f}_o = A \cdot \overline{B} \cdot \overline{C} + A \cdot \overline{B} \cdot \overline{D} + B \cdot \overline{C} \cdot D$$
$$+\, A \cdot C \cdot D + \overline{A} \cdot B \cdot D$$

Form the truth table for $\overline{f}_o$ as in Figure 32–38.
Create the K-map for $\overline{f}_o$ as in Figure 32–39(a).

FIGURE 32–38 Truth table for Example 32–19.

LINE	A B C D	$A\overline{B}\,\overline{C}$	$A\overline{B}\,\overline{D}$	$B\overline{C}D$	ACD	$\overline{A}BD$	$\overline{f}_o$
0	0 0 0 0	0	0	0	0	0	0
1	0 0 0 1	0	0	0	0	0	0
2	0 0 1 0	0	0	0	0	0	0
3	0 0 1 1	0	0	0	0	0	0
4	0 1 0 0	0	0	0	0	0	0
5	0 1 0 1	0	0	1	0	1	1
6	0 1 1 0	0	0	0	0	0	0
7	0 1 1 1	0	0	0	0	1	1
8	1 0 0 0	1	1	0	0	0	1
9	1 0 0 1	1	0	0	0	0	1
10	1 0 1 0	0	1	0	0	0	1
11	1 0 1 1	0	0	0	1	0	1
12	1 1 0 0	0	0	0	0	0	0
13	1 1 0 1	0	0	1	0	0	1
14	1 1 1 0	0	0	0	0	0	0
15	1 1 1 1	0	0	0	1	0	1

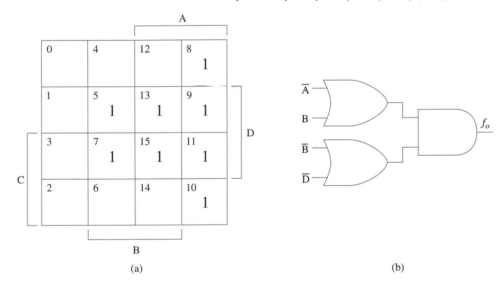

FIGURE 32–39 (a) K-map for inverted function; (b) simplified logic diagram of Figure 32–37.

Read a minimal form from the K-map:

$$\overline{f}_o = A \cdot \overline{B} + B \cdot D$$

Invert $\overline{f}_o$ for a minimal form for f_o:

$$\overline{\overline{f}} = \overline{A \cdot \overline{B} + B \cdot D}$$
$$f_o = (\overline{A} + B)(\overline{B} + \overline{D})$$

See Figure 32–39(b) for a simplified logic diagram.

The methods for simplifying logic diagrams of this and the previous section are summarized in the following procedure.

EXERCISE 32–7

Invert the following expressions.

1. $A \cdot B$ 2. $C \cdot \overline{D}$
3. $\overline{A} + B$ 4. $A + \overline{B} \cdot C$
5. $(U + \overline{V} + W)\overline{X}Y$ 6. $A(\overline{B} + C)\overline{D}$
7. $(A + \overline{B})(\overline{C} + B)(\overline{A} + D)$ 8. $XY + \overline{X}\overline{Z} + YZ$

Perform the indicated inversions for the following.

9. $\overline{X + \overline{Y \cdot Z}}$ 10. $\overline{\overline{U(V + W)}}$

11. $\overline{\overline{J} + \overline{KL}}$ 12. $\overline{\overline{A(B + \overline{CD})}}$

13. Simplify the logic diagram in Figure 32–40.
14. Simplify the logic diagram in Figure 32–41.

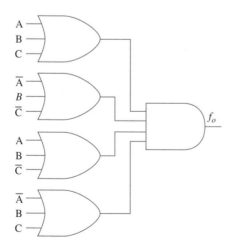

FIGURE 32–40   Logic diagram for Exercise 32–7, problem 13.

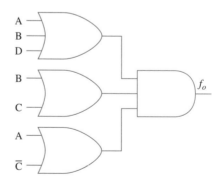

FIGURE 32–41   Logic diagram for Exercise 32–7, problem 14.

32–8 BOOLEAN THEOREMS

This section is a formal approach to the **Boolean algebra** for logic (two-state) variables. It provides an alternative approach for logic circuit simplification to the methods in the preceding sections.

Within the two rules given in this section, there is marked symmetry between the role of the AND operator and the OR operator. This symmetry is the basis of the concept of *duality*. The operator AND is the *dual* of the operator OR. The value 0 is the *dual* of the value 1. A variable, A, is the dual of its inverse, $\overline{A}$. Many of the properties listed in Rule 32–2 occur as pairs, that is, as a statement and its *dual*.

RULE 32–2. *Properties of Boolean Algebra for Logic Variables*

There are only two possible values: 0 and 1.

1. If $A \neq 0$, then $A = 1$. **2.** If $A \neq 1$, then $A = 0$.

The system is closed under conjunction, disjunction, and inversion.

3. $0 \cdot 0 = 0$ **4.** $1 + 1 = 1$

5. $0 + 0 = 0$ **6.** $1 \cdot 1 = 1$

7. $1 \cdot 0 = 0$ **8.** $1 + 0 = 1$

9. $\overline{0} = 1$ **10.** $\overline{1} = 0$

Simple results are independent of the order of operation.

11. $AB = BA$ **12.** $A + B = B + A$

13. $A(BC) = (AB)C$ **14.** $A + (B + C) = (A + B) + C$

However, the results of mixed operations are order dependent.

15. $A(B + C) = AB + AC$ **16.** $A + (BC) = (A + B)(A + C)$

In Boolean algebra, a second NOT counteracts the first NOT.

17. $\overline{\overline{A}} = A$

The preceding properties can be used directly in the process of logic expression simplification. However, it is easier to use them to develop a set of theorems for logic expression simplification.

EXAMPLE 32–20   Use Rule 32–2 to derive $A \cdot 0 = 0$.

Solution   Since A is a logic variable, then $A = 0$ or $A = 1$.
If $A = 0$, then $A \cdot 0 = 0 \cdot 0 = 0$, by Rule 32–2.3.
If $A = 1$, then $A \cdot 0 = 1 \cdot 0 = 0$, by Rule 32–2.7.

∴ $A \cdot 0 = 0$

EXAMPLE 32–21   Use DeMorgan's theorem to derive $A + 1 = 1$.

Solution   We just derived $A \cdot 0 = 0$ for any logic variable including $\overline{A}$. Thus

$$\overline{A} \cdot 0 = 0$$

Apply DeMorgan's theorem to both sides:

$$\overline{\overline{A} \cdot 0} = \overline{0}$$

$$\therefore \quad A + 1 = 1$$

The various theorems for simplifying logic expressions contained in the following rule can be derived from the properties listed in Rule 32–2 in a manner similar to that used in the preceding two examples.

RULE 32–3. *Some Theorems for Simplifying Logic Expressions*

1. $A \cdot 0 = 0$	**2.** $A + 1 = 1$
3. $A \cdot 1 = A$	**4.** $A + 0 = A$
5. $AA = A$	**6.** $A + A = A$
7. $A\overline{A} = 0$	**8.** $A + \overline{A} = 1$
9. $(A + B)A = A$	**10.** $AB + A = A$

In Rule 32–3, notice that theorem 1 is the dual of theorem 2; that is, by applying DeMorgan's theorem to theorem 1 we can derive theorem 2. In the same way, theorem 3 is the dual of theorem 4.

EXAMPLE 32–22   Show that $A(B + \overline{B}) = A$.

Solution   Use Rule 32–3.8:

$$B + \overline{B} = 1$$
$$A(B + \overline{B}) = A \cdot 1$$

Apply Rule 32–3.3:

$$A \cdot 1 = A$$

$$\therefore \quad A(B + \overline{B}) = A$$

EXERCISE 32–8

Use Rules 32–1 and 32–2 to derive the following relations.

1. $A \cdot 1 = A$ **2.** $A + 0 = A$ **3.** $AA = A$

4. $A + A = A$ **5.** $A\overline{A} = 0$ **6.** $A + \overline{A} = 1$

7. $(A + B)A = A$ **8.** $AB + A = A$ **9.** $A + (B\overline{B}) = A$

Use the previous results to simplify the following expressions.

10. $A(1 + C) + B$ **11.** $\overline{X}\,\overline{Y} + \overline{X}Y + XY$

12. $(UV + VW)(UW + VW)$ **13.** $D(DE + E)$

14. $JKL + JK + L$

32–9   APPLICATIONS

The goal of this section is to show how the concepts of Boolean algebra can be used in practical situations. To achieve this goal, we need the additional concept of *do-not-care* for logic circuit specification. In some applications, the output function for certain input combinations will not be specified because the output will not be used or because the input will not occur. The concept of *do-not-care* will be explored in the following example.

EXAMPLE 32–23   You have acquired a tuner-amplifier, a tape deck, and a CD player. You want to integrate them into a system. Unfortunately, the amplifier has only one pair of auxiliary input jacks. The problem is to design a switching circuit to connect the CD player or the tape deck automatically to the amplifier.

Solution   Begin by specifying the desired system operation as in Table 32–2.

TABLE 32–2   Operation Specification for Stereo System

Mode	Tuner-Amplifier Selector Switch	CD Player	Tape Deck	Auxiliary Input
Radio	AM or FM	Off	Off	Do-not-care
Tape from radio	AM or FM	Off	On	Do-not-care
Play a CD	Auxiliary	On	Off	CD player
Tape a CD	Auxiliary	On	On	CD player
Play a tape	Auxiliary	Off	On	Tape deck

Let A represent the selection of auxiliary input.
Let C represent that the CD player is on.
Let D represent that the tape deck is on.
Let f_o represent the connecting switch in Figure 32–42(a); $f_o = 1$ when the CD player is connected.

Create a truth table for the operation of the connecting switch in terms of A, C, and D. Enter a d for the output function for each

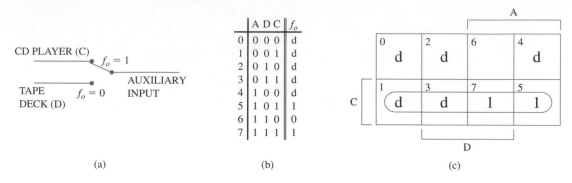

FIGURE 32–42 (a) Selection switch, (b) truth table, and (c) K-map for Example 32–23.

combination for which you *do-not-care* how the auxiliary input is configured. See Figure 32–42(b).

To map this system specification, create a three-variable K-map. Place a 1 in each square for $f_o = 1$. Place a d in each square for $f_o = $ *do-not-care*. See Figure 32–42(c). Read the map by grouping all the 1s and as many d's as convenient to make a maximal group (a set of two, four, or eight squares). This will give a minimal expression for f_o. Thus,

$$f_o = C$$

Therefore, use a 117-V ac relay with a contact configuration as in Figure 32–43 to perform the switching function. When the CD player is off, the relay is off, and the tape deck is connected to the auxiliary input.

FIGURE 32–43 Relay contact configuration for Example 32–23.

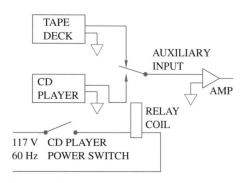

1. A certain brand of microwave oven requires that the main power switch be on, the door be latched, and the timer set to a time other than zero before the oven will operate. Draw a logic diagram for the operation of the oven.

2. There are two common mistakes made when a driver gets out of a car: (1) leaving the keys in the ignition and (2) leaving the lights on. Draw the logic diagram for a warning circuit to sound a buzzer when either mistake is made.

3. A novel lighter is being sold that uses a light beam to light the lighter. The lighter has a cutout on its side that has been fitted with a light source and a light sensor. Opening the lid of the lighter activates the light source; interrupting the light source causes the lighter to light. Draw a logic diagram for the operation of the lighter.

4. An automatic parking lot gate is shown in Figure 32–44. The ticket vendor is to extend a ticket when a car drives over sensor A and the driver presses button B. Draw a logic diagram to control the extending of a ticket.

5. Once the extended ticket (C = 1) of problem 4 is taken (C = 0) from dispenser C, then the arm is raised (E = 1) to let the car through. Draw a logic diagram to control the raising of the arm in Figure 32–44.

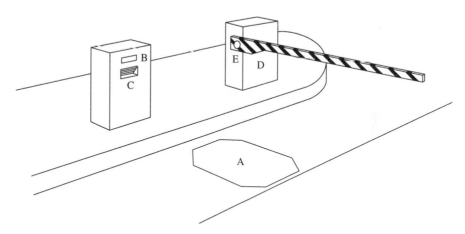

FIGURE 32–44 Entrance to a controlled parking lot. Sensor A detects the presence of a car. Pushing button B extends the ticket. Sensor C detects that the ticket is taken. Sensor D detects the presence of the car in the gate area. Sensor E detects gate position.

6. The raised arm of problem 5 is lowered (E = 0) without hitting the car. Draw a logic diagram to control the lowering of the arm. The lowering of the arm causes the system to be reset for the next car.

Boolean algebra   The formal concepts and theorems used to calculate with two-valued variables and functions.

Karnaugh map   A graphical presentation of a truth table used to simplify logic circuits.

truth table   An exhaustive technique used to list the input and output states for a Boolean or logic expression.

SECTION CHALLENGE

WEB CHALLENGE FOR CHAPTERS 31 AND 32

To evaluate your comprehension of Chapters 31 and 32, log on to **www.prenhall. com/harter** and take the online True/False and Multiple Choice assessments for each of the chapters.

SECTION CHALLENGE FOR CHAPTERS 31 AND 32*

The Internet Protocol (IP) defines the addressing system that connects all devices to the internet. The IP addressing scheme provides logical addressing of data packets so they can be routed both between local networks and throughout the internet.

An IP address is made up of two parts, the network portion and the host (device) portion. The network portion routes data packets between networks while the host portion determines the location of the particular device within a network.

An IP address is a 32-bit number that is divided into four groups called *octets*. Each octet contains eight binary numbers (a byte) consisting of ones and zeros as pictured in Figure C–19. Because of the difficulty in remembering the bit pattern of a 32-bit binary number, IP addressing is simplified through the use of a numbering system called *dotted decimal notation*.

The four decimal numbers in the dotted decimal notation are derived from the positional value of the 8 binary digits in each of the four octets; e.g., to change from 11000000 (the left most octet in Figure C–19) to dotted decimal notation, first assign powers of two positional values to each of the binary ones and then add their decimal values (128 + 64) to arrive at 192, the first decimal number in the dotted decimal 192.5.36.11. Your challenge is to convert the following IP addresses from one form to the other.

A. 00000001.01100110.00101101.10110001   B. 133.156.55.102

C. 01111011.00001100.00101101.01001101   D. 13.28.254.202

E. 00010101.00110100.10110001.10111100   F. 191.15.155.2

G. 11001011.00001010.11101001.00001111   H. 23.166.101.92

I. 01111001.00000010.11000111.01011000   J. 198.215.67.233

FIGURE C–19 An IP address contains four 1-byte (8-bit) octets.

8-bits octet	8-bits octet	8-bits octet	8-bits octet
11000000	00000101	00100100	00001011
192	5	36	11

*The solution to this Section Challenge is found in Appendix C.

GLOSSARY OF SELECTED TERMS

abscissa   The line drawn horizontally to fix a point on a graph; the x-coordinate.

absolute value   The value of a number without regard to its sign.

accuracy   The degree that a measurement or a calculation conforms to a recognized standard or a specified value.

admittance   The reciprocal of impedance; the vector quantity used to describe the ease of the passage of alternating current.

amplifier   A device used to increase the magnitude of an electric signal.

analog instrument   An instrument with a pointer to indicate values along a graduated scale.

angular velocity   The angular rate of travel of a rotating vector expressed in radians per second.

antilogarithm   The inverse of the logarithm function.

associative law of addition   Three terms may be grouped in either way to indicate which addition is performed first; thus, $(a + b) + c = a + (b + c)$.

attenuator   A resistive network that reduces the magnitude of an electric signal.

average power   The power (in watts) dissipated by the resistive component of the impedance in an ac circuit; also called true power.

AWG   American wire gauge; a table of standard wire diameters indicated by gauge numbers.

axis   A number line used as a basis for a coordinate system.

base   A number raised to a power.

base unit   The building blocks of the measurement system; a precise standard that gives the exact value of the unit.

binary   The number system based on the number 2.

Boolean algebra   The formal concepts and theorems used to calculate with two-valued variables and functions.

capacitive reactance   The opposition to alternating current by a capacitor; measured in ohms.

chain calculation The technique of solving arithmetic problems using a calculator without writing out all the intermediate results.

characteristic The integer part of a common logarithm.

coefficient of a product Any factor of a product; each factor is the coefficient of the other factor.

common denominator Equal denominators of several fractions.

common logarithm Base-ten logarithm.

commutative law of addition Two terms may be added in either order; thus, $a + b = b + a$.

complementary angles Two angles that sum to a right angle.

complements A method used to store negative numbers and to perform subtraction in computers.

complex conjugate Created by changing the sign of the imaginary component of a complex number in rectangular form; in polar form, a change in the sign of the argument results in a complex conjugate.

complex fraction Fraction with a numerator or denominator (or both) that contains a fraction.

complex number The sum of a real number and an imaginary number.

conditional equation True only for particular values of the variable in the equation.

conductance Measure of the ease with which a conductor carries an electric current; the reciprocal of resistance.

conductivity Indication of the ability of a material to pass an electric current.

conservation of energy Energy may be changed in form but energy can neither be created nor destroyed.

constant A numeric quantity that has a fixed value for a given problem.

conventional current A way of thinking of the direction of current flow in an electric circuit. Current is thought of as flowing from $+$ to $-$.

cosecant The reciprocal of the sine function.

cotangent The reciprocal of the tangent function.

current Flow of electric charge along a conductor in a closed circuit.

current divider Two or more resistances (loads) connected in parallel across a source will cause the source current to divide.

decibel A logarithmic expression that compares two power levels. The unit used to express gain, loss, and relative power levels.

dependent variable A variable that is the output of a function.

derived unit Formed from base units. Usually given names other than the derived unit names, e.g., watt, volt, or ohm.

determinant An operator that combines a square array of numbers into a single number; used in solving systems of linear equations.

deviation   Variation from a specified dimension or design requirement, typically defining upper and lower limits.

difference of two squares   A special product resulting from the multiplication of the sum and difference of two numbers.

direct proportion   Two ratios are in direct proportion when one ratio in the proportion increases and the other ratio also increases.

discontinuity   A break in the graph of a function.

discriminant   The radicand in the quadratic equation. The number of real roots depends on the sign of the radicand.

distributive law   When multiplying a monomial and a polynomial, each term in the polynomial is multiplied by the monomial.

domain   The list of all values of the independent variable for which the function is defined.

dynamic parameter   A nonlinear device parameter defined as the local slope of a second device parameter plotted against a third device parameter.

efficiency   A number that indicates how well an energy converter performs.

empirical data   Numerical results from measurements, as opposed to results from calculations.

engineering notation   Numbers written in powers of ten notation, having exponents that are multiples of 3.

equivalent angles   Angles that differ by an integral number of revolutions.

equivalent fractions   Fractions that have different forms but have the same value.

equivalent resistance   The combining of several resistances in a circuit into a single resistance called the equivalent resistance, R_T.

evaluating an expression   Substituting numbers for variables in an algebraic expression.

exact number   A number having no error.

exponent   In a power, the number of times the base is used as a factor in a product; the power to which a number or base is raised.

extraneous root   A solution of a mathematical equation that is not a solution to the physical problem.

factor   Each of the numbers forming a product.

factoring   The process of finding factors of a sum; finding factors of a polynomial or other expression.

formula   An equation expressing the relationship between physical quantities.

fractional equation   Equation that has a variable in the denominator of one or more terms.

fractional exponent   An exponent that is a fraction; an exponent of $1/n$ indicates the nth root.

fractional expression   Algebraic expression having a variable in the denominator of one or more terms.

frequency   The number of complete cycles of alternating current in 1 s; measured in hertz.

function   A mathematical relationship between a set of input values and a resulting output value.

graph   A visual representation of the relationship between two variables.

hexadecimal   The number system based on the number 16.

hypotenuse   The side opposite the right angle in a right triangle; also the longest side in a right triangle.

identical equation   An equation that is true for all values of the variable in the equation.

imaginary number   The square root of a negative number.

impedance   The opposition to the passage of alternating current through an electrical circuit; expressed in units of ohms.

implied exponent   Any number or variable without an exponent has an implied exponent of 1.

independent variable   A variable that is the input to a function.

indirect proportion   Two ratios in an indirect proportion where one ratio in the proportion increases and the other ratio decreases.

inductive reactance   The opposition to alternating current by an inductor; measured in ohms.

inequality   A statement that two numerical expressions do not have the same value.

instantaneous value   The value of current or voltage at a particular instant of time.

intercept   The point at which a graph crosses an axis.

intersect   Two lines or curves intersect where they have a point in common.

irrational expression   Algebraic expression containing a fractional exponent or a radical sign.

Karnaugh map   A graphical presentation of a truth table used to simplify logic circuits.

kilowatthour   Unit of energy used by power companies in the sale of energy to customers; not an SI unit.

like terms   Having the same literal factors but different numerical coefficients.

literal equation   Equation with one or more numbers represented by letters.

load line   A line drawn on the characteristic curve of a nonlinear device to locate the operating point on the circuit.

logarithm   The exponent to which the base must be raised to equal the original number.

mantissa   The fractional part of a common logarithm.

members of the equation   The expressions on either side of the equal sign.

mesh   A simple closed loop having no branches.

mesh currents   Currents with an assumed direction used to solve a network using mesh analysis.

mixed expression   Sum (difference) of a fraction and a polynomial.

mutual resistance The resistance shared by two mesh currents.

natural logarithm Base e (2.71828 . . .) logarithm.

natural numbers The set of numbers (1, 2, 3, etc.) used in counting; the counting numbers.

numerical equation A statement that declares two numerical expressions have the same value.

numerical expression A number or a list of numbers joined by arithmetic operators; e.g., 5 or $6 + 4$.

octal The number system based on the number 8.

Ohm's law Relationship between voltage and current formulated by Georg Simon Ohm; $E = IR$.

ordinate The line drawn vertically to fix a point on a graph; the y-coordinate.

parts per million Per one million; divided by 1 000 000.

percent Per one hundred; divided by 100.

perfect trinomial square A special product resulting from the multiplication of a binomial by itself.

period The reciprocal of frequency.

periodic behavior A behavior or pattern that repeats over and over again, especially in a graph.

phase angle The angle between the current and voltage phasor in a phasor diagram.

phasor A two-dimensional vector used to portray the phase relationship of voltage and current in an electrical circuit.

polar coordinates A coordinate system based on a set of concentric circles.

polynomial Rational expression having no variable as a divisor.

power Rate of converting (producing or using) energy; rate of doing work; J/s, watt.

power factor The ratio of resistance to impedance; the cosine of the circuit phase angle.

precision In measurement, the degree that individual measurements agree with each other, i.e., repeatability; in common use, it implies exactness.

principal square root The positive square root.

proportion Equation made up of two ratios.

quadratic equation An equation containing the square of the independent variable but no higher order terms.

radian A unit for measuring the size of angles.

radical An expression of the form $\sqrt[n]{a}$.

radix The base of a number system; two is the radix of the binary system.

range The list of all values that a function assumes over its domain.

rational expression Algebraic expression having whole numbers as exponents and containing no radical signs.

resistance A substance has a resistance of one ohm when the application of one volt results in a current of one ampere.

resistivity   Resistance between opposite parallel faces of a 1-m cube of a material.

revolution   A unit for measuring the size of angles; a complete circle equals 1 revolution.

right triangle   A triangle containing a right angle.

root of the equation   A value of the variable that makes the equation true; a solution.

roots   The abscissas where a graph intersects the x-axis.

scalar   A quantity having only magnitude.

scientific notation   Numbers written in powers of ten notation having the decimal point placed after the leftmost nonzero digit.

secant   The reciprocal of the cosine function.

SI   International System of Units; the metric system.

significant figure   Any figure (digit) needed to define a value in a calculation or a quantity in a measurement.

slope   The amount of change produced in the ordinate by a unit change of the abscissa.

static parameter   A parameter that does not change with time.

supplementary angles   Two angles that sum to a straight angle.

susceptance   The reciprocal of inductive or capacitive reactance.

symmetric property of equality   The property that allows the members of an equation to be interchanged.

system of linear equations   Set of linear equations, each involving the same variables.

term   Each of the numbers in a sum.

tolerance   The total amount of deviation permitted in a quantity.

transient   In an electric circuit, the time between the initial application of power and the settling to a steady-state condition.

trinomial product   The product of two binomials.

truth table   An exhaustive technique used to list the input and output states of a Boolean or logic expression.

undefined fraction   A fraction in which the denominator evaluates to zero.

unit analysis   Technique used in reduction within (or between) a system of measurements.

variable   A numeric quantity represented by a letter that can take on more than one value in an algebraic expression or equation.

vector   A quantity having both magnitude and direction.

vertex   The turning point of the graph of a quadratic equation.

vinculum   Sign of grouping; a bar drawn over terms to show they are treated as a unit.

voltage   Name given to energy added per unit of electric charge as the charge passes through a source.

voltage divider   Two or more resistances connected in series with a voltage source form a voltage divider.

APPENDIX A

Reference Tables

Symbols

Constants

Greek Alphabet

Selected Abbreviations

American Wire Gauge (AWG) for Solid, Annealed Copper Conductors at 20°C

Preferred Number Series for Standard Component Values for One Decade

Color Code for Specifying Preferred Resistor Values and Tolerances

Selected Identities and Conversion Factors

Prefixes and Symbols for Multiples and Submultiples of the SI Units

Development of the International System of Units

Symbols

$\times$	Multiply by, AND	{ }	Sign of grouping
$\div$	Divide by	$\therefore$	Therefore
$+$	Positive, add, OR	||	Parallel to
$-$	Negative, subtract	$\angle$	Angle
/	Divide by	$\|a\|$	Absolute value of a
$\pm$	Plus or minus	Δ	Interval
$\overline{}$	Overbar, NOT	Σ	Summation
$=$	Equals	%	Percent
$\equiv$	Identically equal to	$\sqrt{}$	Radical sign
$\Leftrightarrow$	Equivalent	$\llcorner$	Right angle
$\approx$	Approximately equal to	∞	Undefined, infinity
$\neq$	Not equal	$\Rightarrow$	Yields
$>$	Greater than	$f( )$	Function of
$\gg$	Much greater than	f_o	Logic equation output
$<$	Less than	$\|z$	Evaluated at z
$\ll$	Much less than	$x \rightarrow b$	As x goes to b
$\geq$	Greater than or equal to	$\rho\underline{/\theta}$	Polar coordinates
$\leq$	Less than or equal to	(x, y)	Rectangular coordinates
$-$	Bar, sign of grouping	$\begin{vmatrix} a & b \\ c & d \end{vmatrix}$	Determinant
( )	Sign of grouping		
[ ]	Sign of grouping		

Constants

e	2.718 281 828	$\pi/2$	1.570 796 327
π	3.141 592 654	$\sqrt{2}/2$	0.707 106 781
$\ln(10)$	2.302 585 093	j	$\sqrt{-1}$

Greek Alphabet

Name	Capital	Lowercase	Name	Capital	Lowercase
Alpha	A	α	Nu	N	ν
Beta	B	β	Xi	Ξ	ξ
Gamma	Γ	γ	Omicron	O	o
Delta	Δ	δ	Pi	Π	π
Epsilon	E	ϵ	Rho	P	ρ
Zeta	Z	ζ	Sigma	Σ	σ
Eta	H	η	Tau	T	τ
Theta	Θ	θ	Upsilon	Υ	υ
Iota	I	ι	Phi	Φ	ϕ
Kappa	K	κ	Chi	X	χ
Lambda	Λ	λ	Psi	Ψ	ψ
Mu	M	μ	Omega	Ω	ω

Selected Abbreviations

alternating current	ac	Kirchhoff's voltage law	KVL
American Wire Gauge	AWG	least significant digit	LSD
ampere	A	limit	lim
angular velocity	ω	logarithm (common)	log
antilogarithm (common)	antilog	logarithm (natural)	ln
antilogarithm (natural)	antiln	maximum	mx
audio frequency	AF	mega (1×10^6)	M
centimeter	cm	megabyte	MB
clockwise	cw	meter	m
cosecant	cps	micro (1×10^{-6})	μ
cosine	cos	mil (1×10^{-3} inch)	mil
cotangent	cot	milli (1×10^{-3})	m
coulomb	C	minute	min
counterclockwise	ccw	most significant digit	MSD
cycles per second (hertz is preferred unit)	cps	nano (1×10^{-9})	n
decibel	dB	ohm	Ω
degrees Celsius	°C	parts per million	ppm
degrees Fahrenheit	°F	peak-to-peak	pp
direct current	dc	pico (1×10^{-12})	p
double-in-line memory module	DIMM	radian	rad
		radio frequency	RF
electromotive force	emf	random access memory	RAM
equivalent	eq	resistive-capacitive	*RC*
farad	F	resistive-inductive	*RL*
flux density	B	revolution	rev
foot, feet	ft	revolutions per minute	RPM
giga (1×10^9)	G	root mean square	rms
gram	g	secant	sec
henry	H	second	s
hertz	Hz	siemens	S
horsepower	hp	sine	sin
hour	h	single-pole, double-throw	SPDT
hyperbolic cosine	cosh		
hyperbolic sine	sinh	single-pole, single-throw	SPST
hyperbolic tangent	tanh		
inch	in	tangent	tan
inductive-capacitive-resistive	LCR	tesla	T
		volt	V
International System of Units	SI	voltampere	VA
		watt	W
kelvin	K	watthour	Wh
kilo (1×10^3)	k	weber	Wb
kilowatthour	kWh		
Kirchhoff's current law	KCL		

American Wire Gauge (AWG) for Solid, Annealed Copper Conductors at 20°C				
	SI Metric Units		English Units	
Gauge No.	Dia. (mm)	Ω/km	Dia. (mils)*	Ω/1000 ft
0000	11.68	0.1608	460.0	0.0490
000	10.40	0.2028	409.6	0.0618
00	9.266	0.2557	364.8	0.0779
0	8.252	0.3224	324.9	0.0983
1	7.348	0.4066	289.3	0.1239
2	6.543	0.5127	257.6	0.1563
3	5.827	0.6465	229.4	0.1970
4	5.189	0.8152	204.3	0.2485
5	4.620	1.028	181.9	0.3133
6	4.115	1.296	162.0	0.3951
7	3.665	1.634	144.3	0.4982
8	3.264	2.061	128.5	0.6282
9	2.906	2.599	114.4	0.7921
10	2.588	3.277	101.9	0.9989
11	2.305	4.132	90.74	1.260
12	2.053	5.211	80.81	1.588
13	1.828	6.571	71.96	2.003
14	1.628	8.285	64.08	2.525
15	1.450	10.45	57.07	3.184
16	1.291	13.17	50.82	4.016
17	1.150	16.61	45.26	5.064
18	1.024	20.95	40.30	6.385
19	0.912	26.42	35.89	8.051
20	0.812	33.31	31.96	10.15
21	0.723	42.00	28.45	12.80
22	0.644	52.96	25.35	16.14
23	0.573	66.79	22.57	20.36
24	0.511	84.21	20.10	25.67
25	0.455	106.2	17.90	32.37
26	0.405	133.9	15.94	40.81
27	0.361	168.9	14.20	51.47
28	0.321	212.9	12.64	64.90
29	0.286	268.5	11.26	81.83
30	0.255	338.6	10.03	103.2
31	0.227	426.9	8.928	130.1
32	0.202	538.3	7.950	164.1
33	0.180	678.8	7.080	206.9
34	0.160	856.0	6.305	260.9
35	0.143	1079	5.615	329.0
36	0.127	1361	5.000	414.8
37	0.113	1716	4.453	523.1
38	0.101	2164	3.965	659.6
39	0.090	2729	3.531	831.8
40	0.080	3441	3.145	1049

*1 mil = 0.001 in

Preferred Number Series for Standard Component Values for One Decade*

E6 ± 20%	E12 ± 10% (bold) E24 ± 5% (all)	E96 ± 1% (bold) and E192 ± 0.1% (all numbers)							
1.0	**1.0**	**1.00**	1.01	**1.02**	1.04	**1.05**	1.06	**1.07**	1.09
	1.1	**1.10**	1.11	**1.13**	1.14	**1.15**	1.17	**1.18**	1.20
	1.2	**1.21**	1.23	**1.24**	1.26	**1.27**	1.29	**1.30**	1.32
	1.3	**1.33**	1.35	**1.37**	1.38	**1.40**	1.42	**1.43**	1.45
1.5	**1.5**	**1.47**	1.49	**1.50**	1.52	**1.54**	1.56	**1.58**	1.60
	1.6	**1.62**	1.64	**1.65**	1.67	**1.69**	1.72	**1.74**	1.76
	1.8	**1.78**	1.80	**1.82**	1.84	**1.87**	1.89	**1.91**	1.93
	2.0	**1.96**	1.98	**2.00**	2.03	**2.05**	2.08	**2.10**	2.13
2.2	**2.2**	**2.15**	2.18	**2.21**	2.23	**2.26**	2.29	**2.32**	2.34
	2.4	**2.37**	2.40	**2.43**	2.46	**2.49**	2.52	**2.55**	2.58
	2.7	**2.61**	2.64	**2.67**	2.71	**2.74**	2.77	**2.80**	2.84
	3.0	**2.87**	2.91	**2.94**	2.98	**3.01**	3.05	**3.09**	3.12
3.3	**3.3**	**3.16**	3.20	**3.24**	3.28	**3.32**	3.36	**3.40**	3.44
	3.6	**3.48**	3.52	**3.57**	3.61	**3.65**	3.70	**3.74**	3.79
	3.9	**3.83**	3.88	**3.92**	3.97	**4.02**	4.07	**4.12**	4.17
	4.3	**4.22**	4.27	**4.32**	4.37	**4.42**	4.48	**4.53**	4.59
4.7	**4.7**	**4.64**	4.70	**4.75**	4.81	**4.87**	4.93	**4.99**	5.05
	5.1	**5.11**	5.17	**5.23**	5.30	**5.36**	5.42	**5.49**	5.56
	5.6	**5.62**	5.69	**5.76**	5.83	**5.90**	5.97	**6.04**	6.12
	6.2	**6.19**	6.26	**6.34**	6.42	**6.49**	6.57	**6.65**	6.73
6.8	**6.8**	**6.81**	6.90	**6.98**	7.06	**7.15**	7.23	**7.32**	7.41
	7.5	**7.50**	7.59	**7.68**	7.77	**7.87**	7.96	**8.06**	8.16
	8.2	**8.25**	8.35	**8.45**	8.56	**8.66**	8.76	**8.87**	8.98
	9.1	**9.09**	9.20	**9.31**	9.42	**9.53**	9.65	**9.76**	9.88

*Note: Once a particular number series is selected, then the standard values within a specified range are determined from the 1 to 10 decade by multiplying by 0.01, 0.1, 10, 100, 1000, etc.

Color Code for Specifying Preferred Resistor Values and Tolerances

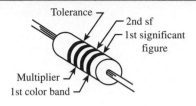

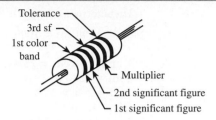

Four-band Color Code for E6, E12, and E24 Series of Preferred Values

Five-band Color Code for E96 and E192 Series of Preferred Values

Color	1st Band	2nd Band	3rd Band	4th Band	1st Band	2nd Band	3rd Band	4th Band	5th Band
Black	0	0	×1	—	0	0	0	×1	—
Brown	1	1	×10	—	1	1	1	×10	±1%(F)*
Red	2	2	×100	±2%(G)*	2	2	2	×100	—
Orange	3	3	×1000	—	3	3	3	×1000	—
Yellow	4	4	×10 000	—	4	4	4	×10 000	†
Green	5	5	×100 000	—	5	5	5	×100 000	±0.5%(D)
Blue	6	6	×1 000 000	—	6	6	6	×1 000 000	±0.25%(C)
Violet	7	7	—	—	7	7	7	—	±0.1%(B)
Gray	8	8	—	—	8	8	8	—	—
White	9	9	—	—	9	9	9	—	—
Gold	—	—	×0.1	±5%(J)	—	—	—	×0.1	—
Silver	—	—	×0.01	±10%(K)	—	—	—	×0.01	—
Plain	—	—	—	±20%(M)	—	—	—	—	—

* Tolerance code letters for specifying MIL-spec resistor tolerance.

† A yellow 5th band is sometimes used to indicate the reliability level of MIL-spec carbon-composition resistors, now obsolete.

Displacement (length):

1 m = 100 cm = 1000 mm = 3.2808 ft = 39.370 in
1 km = 0.621 37 mi = 3280.8 ft
1 in = 2.54 cm = 25.4 mm
1 mi = 5280 ft
1 yd = 3 ft = 36 in
1 ft = 12 in
1 revolution = 360° = 2π rad = 6.2832 rad

Area:

$1 \text{ m}^2 = 10.7639 \text{ ft}^2 = 1550.0 \text{ in}^2$ **$1 \text{ yd}^2 = 9 \text{ ft}^2$**
$1 \text{ cm}^2 = 0.155 \, 00 \text{ in}^2$ **$1 \text{ ft}^2 = 144 \text{ in}^2$**
$1 \text{ m}^2 = 10 \, 000 \text{ cm}^2$
$1 \text{ cm}^2 = 100 \text{ mm}^2$

Time:

1 h = 60 min = 3600 s
1 min = 60 s

Force:

1 N = 0.224 81 lbf   **1 lbf = 16 ozf**
1 lbf = 4.4482 N   **1 ton = 2000 lbf**

Mass:

1 kg = 0.068 522 slug   **1 kg = 1000 g**
1 slug = 14.594 kg   **1 g = 1000 mg**
1 kg = 2.2046 lbm

Velocity (speed):

1 m/s = 3.6 km/h = 2.2369 mi/h = 3.2808 ft/s
60 mi/h = 88 ft/s
1 ft/s = 0.3048 m/s
1 rev/min = 0.104 72 rad/s   **60 rev/min = 1 cps = 1 Hz**

Work (energy, torque):

1 J = 0.737 56 ft·lbf   1 ft·lbf = 1.3558 J
1 kWh = 3.6 MJ $= 2.655 \times 10^6$ ft·lbf   1 N·m = 0.737 56 lbf·ft
1 BTU = 1055.1 J = 778.17 ft·lbf

Power (mechanical):   **Power (electrical):**

1 hp = 550 ft·lbf/s = 33,000 ft·lbf/min = 745.70 W   **1 hp = 746 W**
1 W = 0.737 56 ft·lbf/s   1 kW = 1.3405 hp
1 kW = 1.3410 hp

*__Boldface__ quantities are exact.

Prefixes and Symbols for Multiples and Submultiples of the SI Units

Prefix	Symbol	Phonic	Multiple or Submultiple	Power of Ten
yotta	Y	yott′ə	1 000 000 000 000 000 000 000 000	10^{24}
zetta	Z	zett′ə	1 000 000 000 000 000 000 000	10^{21}
exa	E	ex′ə	1 000 000 000 000 000 000	10^{18}
peta	P	pet′ə	1 000 000 000 000 000	10^{15}
tera	T	ter′ə	1 000 000 000 000	10^{12}
giga	G	gig′ə	1 000 000 000	10^{9}
mega	M	meg′ə	1 000 000	10^{6}
kilo	k	ki′lō	1 000	10^{3}
hecto	h	hec′tō	100	10^{2}
deka	da	dek′ə	10	10^{1}
deci	d	des′ə	0.1	10^{-1}
centi	c	sen′tə	0.01	10^{-2}
milli	m	mil′ē	0.001	10^{-3}
micro	μ	mī′krō	0.000 001	10^{-6}
nano	n	nān′ō	0.000 000 001	10^{-9}
pico	p	pē′kō	0.000 000 000 001	10^{-12}
femto	f	fĕm′tō	0.000 000 000 000 001	10^{-15}
atto	a	ăt′tō	0.000 000 000 000 000 001	10^{-18}
zepto	z	zĕp′tō	0.000 000 000 000 000 000 001	10^{-21}
yocto	y	yŏc′tō	0.000 000 000 000 000 000 000 001	10^{-24}

Development of the International System of Units (Le Système International d'Unités)

Year	Event	Outcome
1790	French Revolution leads to reform of the system of weights and measures	Committee of the French Academy of Sciences is formed and agrees to a decimal system over a duodecimal system
1799	The Committee's work is completed and the metric system is established	The metric system becomes a reality with the meter as the unit of length and the gram as the unit of mass
1866	U.S. Congress legalizes the use of the metric system	By 1893, the international meter is the fundamental standard of length and the kilogram is the standard of mass in the U.S.
1870	International standardization begins with a meeting of 15 nations in Paris, France	This meeting leads to the establishment of a permanent International Bureau of Weights and Measures near Paris (BIPM, Bureau International des Poids et Mesures) and the 1875 International Metric Convention
1875	The General Conference on Weights and Measures (CGPM, Conférence Générale des Poids et Mesures) is constituted	The CGPM meets at least every six years in Paris, handles all international matters concerning the metric system, and controls the BIPM
1881	A unit of time is added to the meter and gram to produce the centimeter-gram-second (CGS) system	The International Electrical Congress adopts the CGS system
Circa 1900	Practical measurement using metric units begins to be based on the meter-kilogram-second (MKS) system	This leads the International Electrotechnical commission to recommend (in 1935) that the MKS system of mechanics be linked with electromagnetics by selecting the ampere as a base unit thereby defining the MKSA system
1954	The 10th CGPM adopts a rational and coherent system of units based on the four MKSA units plus the Kelvin and the candela	This sets the stage for the creation of the modern metric system
1960	The 11th CGPM gives the title of International System of Units (SI) to the metric system	The modern metric system is created, thereby deprecating the MKS, MKSA, and the CGS systems of measurement

APPENDIX B

Answers to Selected Problems

CHAPTER 1

EXERCISE 1–1

1. −7 **3.** +20 **5.** −8

7.

Test Score	55	65	90	40	80	50	75	95	30	65	100	45
Difference	−10	0	25	−25	15	−15	10	30	−35	0	35	−20

EXERCISE 1–2

1. **(a)** $10 + 5$ $10 - 5$ 10×5 $10 \div 5$

     (b) $16 + 8$ $16 - 8$ 16×8 $16 \div 8$

     (c) $20 + 4$ $20 - 4$ 20×4 $20 \div 4$

     (d) $60 + 12$ $60 - 12$ 60×12 $60 \div 12$

3. **(a)** $10 + 5 = 15$ $10 - 5 = 5$ $10 \times 5 = 50$ $10 \div 5 = 2$

     (b) $16 + 8 = 24$ $16 - 8 = 8$ $16 \times 8 = 128$ $16 \div 8 = 2$

     (c) $20 + 4 = 24$ $20 - 4 = 16$ $20 \times 4 = 80$ $20 \div 4 = 5$

     (d) $60 + 12 = 72$ $60 - 12 = 48$ $60 \times 12 = 720$ $60 \div 12 = 5$

5. $5 - 3$

7. $6 \div 3 = 2$

EXERCISE 1–3

1. true **3.** false, 13 **5.** true

7. true **9.** false, −6

Calculator Drill

11. 30 **12.** −12 **13.** 11

14. −3 **15.** 9 **16.** 6

17. −24 **18.** −12 **19.** 14

20. 29 **21.** −11 **22.** 5

23. −11 **24.** 23 **25.** 1

26. −10 **27.** 29 **28.** 5

29. 10 **30.** 0 **31.** 21

EXERCISE 1–4

1. $(6 + 3)$ or (2×2)

Calculator Drill

3. 1 **4.** 10 **5.** 2

6. 5 **7.** 4 **8.** 14

9. 13   10. 5   11. 77
12. 22   13. 6   14. 17
15. 194   16. 7   17. 2
18. 9   19. 3   20. 7
21. 6   22. 1

EXERCISE 1–5

1. $3 - 2$ 3. $-8 + 2$ 5. $2 - 3 - 7$ 7. $1 + 2 - 6$
9. $7 + 1 - 4$ 11. $-4 + 6 - 7$ 13. $-9 + 3 + 6$
15. $-7.3 + 6.1 + 3$ 17. $-1.4 + 5.3 - 6.7$ 19. $4.3 - 8.5 - 27$

EXERCISE 1–6

1. 3   3. 9   5. 8
7. 7   9. 1   10. 5

EXERCISE 1–7

1. 10   3. 5   5. 12
7. -23 9. -52 11. -4
13. -16 15. -13 17. -19
19. $21 - 16 = 5$ 21. $18 - 26 = -8$ 23. $-12 + 16 = 4$

EXERCISE 1–8

Calculator Drill

1. -20 2. 30   3. 16
4. -4 5. -19 6. 4
7. -36 8. 5   9. -4
10. -13 11. 8   12. 100
13. 48   14. 11   15. -46

EXERCISE 1–9

1. $7 < 10$ 3. $23 > 13$ 5. $3 = -10 + 13$
7. $2 + 7 > 3$ 9. $>$ 11. $>$
13. $<$ 15. $>$ 17. $>$

EXERCISE 1–10

1. true   3. false   5. true
7. false   9. false   10. true

1. −6	**3.** 10	**5.** −35
7. 42	**9.** −4	**11.** −90

Calculator Drill

13. −1240	**14.** −420	**15.** 192
16. −918	**17.** −9384	**18.** 936
19. 234	**20.** −2444	**21.** 1961
22. 1225	**23.** −1222	**24.** −518

EXERCISE 1–12

1. 5	**3.** −3	**5.** −2
7. 3	**9.** −4	**11.** 5

Calculator Drill

13. −3	**14.** −9	**15.** 2
16. −4	**17.** −3	**18.** −5
19. 3	**20.** 1	**21.** −5
22. 5	**23.** 1	**24.** 2
25. −2	**26.** −1	**27.** 2
28. 2	**29.** 0	**30.** 9

CHAPTER 2

EXERCISE 2–1

9. 3^3	**11.** 8^4	**13.** 5^3
15. 9^4	**17.** 7^7	**19.** 14^2

EXERCISE 2–2

Calculator Drill

1. 32	**2.** 64	**3.** 729
4. 125	**5.** 4096	**6.** 343
7. 7776	**8.** 243	**9.** 1000
10. 65 536	**11.** 128	**12.** 59 049
13. 0.25	**14.** 2	**15.** 3

EXERCISE 2–3

1. 75.4×10^7	**3.** 3.72×10^{-3}	**5.** 4.01×10^2
7. 5531×10^{-4}	**9.** 2.49×10^{-3}	**11.** 82.92×10^2
13. 28.6514×10^4	**15.** 1788.2×10^{-3}	**17.** 735×10^{-1}

EXERCISE 2–4

1. 3×10^2 **3.** 2.8×10^2 **5.** 5×10^{-1}
7. 3.6×10^{-1} **9.** 7.052×10^3 **11.** 6.3×10^{-3}
13. 1×10^{-3} **15.** 5.78×10^5 **17.** 5.55×10^3
19. 6.8×10^{-1} **21.** 1.313×10^{-2} **23.** -2.823×10^2
25. -8.25×10^{-3} **27.** -4.78×10^1

EXERCISE 2–5

1. 0.0635 **3.** 0.000 382 8 **5.** 0.009 935
7. 4.62 **9.** 88.31 **11.** $-183\,000$
13. 0.2882 **15.** $-0.000\,715$ **17.** 2.52

EXERCISE 2–6

Calculator Drill

1. 1.32×10^3 **2.** 564×10^3 or 0.564×10^6 **3.** 85×10^{-6}
4. 8.65×10^3 **5.** 4.7×10^{-3} **6.** 4.28×10^6
7. 8×10^{-3} **8.** 47.3×10^3 **9.** 0.572×10^{-3} or 572×10^{-6}

10. 0.027 **11.** 284 **12.** 7 320 000
13. 420 000 **14.** 0.000 000 008 20 **15.** 570 000
16. 0.000 001 25 **17.** 94 500 000 **18.** 0.000 39

EXERCISE 2–7

1. approximate **3.** exact **5.** approximate **7.** exact

EXERCISE 2–8

1. three **3.** four **5.** two
7. five **9.** four **11.** three

EXERCISE 2–9

Calculator Drill

1. 7.57×10^3 **2.** 991×10^{-3} **3.** 2.00×10^3
4. 39.2×10^{-3} **5.** 1.78×10^3 **6.** 727×10^3
7. 59.3×10^{-3} **8.** 100×10^3 **9.** 3.48×10^6
10. 14.1×10^3 **11.** 40.0×10^{-3} **12.** 149×10^3
13. 5.21×10^{-3} **14.** 56.6×10^0 **15.** 1.99×10^3
16. 70.0×10^{-3} **17.** 878×10^{-6} **18.** 48.0×10^3
19. 455×10^0 **20.** 24.9×10^{-3}

EXERCISE 2–10

1. units   **3.** thousandths   **5.** hundreds

7. ten-thousandths

9. 70.68   **11.** 0.9   **13.** 176.27

15. −14   **17.** 34×10^6 **19.** 15.400×10^6

EXERCISE 2–11

Calculator Drill

1. 4×10^2 **2.** 3.6314   **3.** 628×10^3

4. 39×10^3 **5.** 3.499   **6.** 6.608×10^{-2}

7. 0.24   **8.** 1889   **9.** 5×10^2

10. 1.9×10^2 **11.** 3.7×10^{-2} **12.** 804×10^6

13. 47.81   **14.** 0.20   **15.** 5.6×10^2

16. 18.7   **17.** 1.35×10^6 **18.** 0.911

19. 3.4   **20.** 180×10^3

EXERCISE 2–12

Calculator Drill

1. 19.29   **2.** 0.093 27   **3.** ± 0.8860

4. 9.996   **5.** 31.94   **6.** −8.119

7. 0.093 94   **8.** ± 0.8535 **9.** −1387

10. 1.448   **11.** −7.662   **12.** ± 0.6399

EXERCISE 2–13

Calculator Drill

1. 8.00×10^0 **2.** 5.00×10^0 **3.** 1.60×10^1

4. 4.00×10^0 **5.** 4.44×10^0 **6.** 4.38×10^1

7. 1.58×10^1 **8.** 7.58×10^0 **9.** 1.71×10^2

10. -1.36×10^0 **11.** 1.81×10^2 **12.** 4.00×10^0

EXERCISE 2–14

Calculator Drill

1. 1.56×10^1 **2.** 1.08×10^{-1} **3.** 7.24×10^0

4. 3.71×10^1 **5.** 3.15×10^0 **6.** 8.20×10^3

7. 1.20×10^{-1} **8.** 2.70×10^2 **9.** 4.51×10^3

10. 2.78×10^1 **11.** 4.57×10^1 **12.** 6.62×10^1

13. 3.86×10^1 **14.** 2.12×10^3 **15.** 2.32×10^0

EXERCISE 2–15

Calculator Drill

1. 200×10^{-3} 2. 250×10^{-3} 3. 167×10^{-3}
4. 111×10^{-3} 5. 333×10^{-3} 6. 10.8×10^{-3}
7. 12.0×10^{0} 8. 1.29×10^{-3} 9. 1.99×10^{-3}
10. 725×10^{0} 11. 159×10^{-6} 12. -1.09×10^{-3}
13. 24.4×10^{0} 14. 538×10^{0} 15. 141×10^{-6}
16. -21.1×10^{-3} 17. 43.1×10^{-3} 18. -108×10^{-3}

EXERCISE 2–16

Calculator Drill

1. 3.00 2. 14.0 3. 0.200 4. 1.67
5. 16.0 6. 4.00 7. 14.2 8. 61.6
9. 212 10. $0.001\ 19$ 11. 1050 12. 0.182

EXERCISE 2–17

Calculator Drill

1. 384×10^{-3} 2. 10.7×10^{0} 3. -1.38×10^{0} 4. -10.9×10^{0}
5. 3.70×10^{3} 6. 7.10×10^{3} 7. 11.2×10^{3} 8. 133×10^{-3}
9. 22.0×10^{3} 10. -393×10^{-3} 11. 1.00×10^{0} 12. 39.9×10^{0}
13. 12.7×10^{0} 14. 84.3×10^{0} 15. 32.9×10^{0} 16. 272×10^{-12}

EXERCISE 2–18

1. 10^{5} 2. 10^{-1} 3. 10^{-7} 4. 10^{4} 5. 10^{0}
6. 10^{0} 7. 10^{-18} 8. 10^{11} 9. 10^{6} 10. 10^{-2}
11. 30×10^{1} 12. 32×10^{2} 13. 9.0×10^{-2}
14. 10×10^{2} 15. 32×10^{6} 16. 18×10^{9}
17. 42 18. 36×10^{-10} 19. 25×10^{-6}
20. 49

EXERCISE 2–19

1. 10^{1} 2. 10^{-2} 3. 10^{1} 4. 10^{11} 5. 10^{5}
6. 5×10^{4} 7. 5×10^{1} 8. 3×10^{4} 9. 3×10^{9} 10. 7×10^{-3}
11. 2×10^{-7} 12. 2

EXERCISE 2–20

1. 10^{6} 2. 10^{15} 3. 10^{8} 4. 10^{4} 5. 10^{18} 6. 10^{-30}
7. 10^{-14} 8. 10^{9} 9. 10^{-12} 10. 10^{-24} 11. 10^{-10} 12. 10^{-27}

EXERCISE 2–21

1. 4×10^6 **2.** 9×10^4 **3.** 25×10^{-8} **4.** 36×10^{-10}
5. 8×10^{12} **6.** 27×10^{18} **7.** 64×10^{-12} **8.** 16×10^{-14}
9. 81×10^2 **10.** 32×10^{15} **11.** 49×10^{-10} **12.** 1000×10^{-12}

EXERCISE 2–22

1. 10^{-4} **2.** 10^6 **3.** 10^{-15} **4.** 10^{-32} **5.** 10^8
6. 10^{-5} **7.** 1 **8.** 10^{-24} **9.** 10^{-18} **10.** 10^6
11. 10^3 **12.** 10^{22} **13.** 4.00×10^{18} **14.** 81.0×10^{-12}

EXERCISE 2–23

1. 3×10^{-1} **3.** 4×10^{-1} **5.** 20×10^3
7. 70×10^{-3} **9.** 6 **11.** 60
13. 4×10^{-3} **15.** 30 **17.** 200
19. 0.8 **21.** 0.2

CHAPTER 3

EXERCISE 3–1

1. 0.003 125 N **3.** 12 000 s **5.** 234 000 J
7. 0.083 265 Ω/m **9.** 34 500 V

EXERCISE 3–2

1. f and l **2.** i **3.** k **4.** h
5. a and j **6.** l **7.** h **8.** b
9. a and j **10.** c **11.** d **12.** e

EXERCISE 3–3

1. 0.282 kΩ **3.** 176 μH **5.** 1.90 MW
7. 1.520 kHz **9.** 180.025 kΩ **11.** 6.28 m
13. 0.0170 J **15.** 0.000 902 S **17.** 854 W
19. 0.000 0470 F **21.** 208 700 J

Calculator Drill

23. 7.83 kV **24.** 83.3 μs **25.** 65.4 mH
26. 17.3 kW **27.** 152 mS **28.** 51.6 kHz
29. 681 kΩ **30.** 0.910 μF **31.** 20.4 mA

32. 0.362 kN	**33.** 50.3 kJ	**34.** 1.20 kV
35. 185 μA; 0.185 mA	**36.** 0.85 S	**37.** 1.53 MΩ
38. 70 μs	**39.** 1.11 V	**40.** 0.24 g
41. 612 mm; 0.612 m	**42.** 2.00 kW	**43.** 160 km; 0.160 Mm
44. 45.4 MHz	**45.** 0.30 mA	**46.** 8.84 V
47. 14.4 mm	**48.** 189.7 V	**49.** 24 A
50. 8.49 mW	**51.** 22 kA	**52.** 49 V
53. 68.0 mΩ	**54.** 0.17 kΩ	**55.** 5.2 H
56. 2.4 kJ	**57.** 46 W	**58.** 7.2 kΩ
59. 3.72 mV	**60.** 0.2 MHz	**61.** 24 N
62. 158 m	**63.** 0.28 kV	**64.** 6.3 Hz
65. 0.67 A	**66.** 14.1 J	**67.** 0.24 S

EXERCISE 3–4

1. 51.3 ft/s	**3.** 103 m/s	**5.** 8.57 mJ
7. 5.00 rev	**9.** 0.273 hp	**11.** 19.0 mm
13. 365.7 mi/h	**15.** 3.19×10^4 in	**17.** 4.447 kW
19. 137.7 kW	**21.** 1 m = 39.370 in	
23. 1 N = 3.5970 oz	**25.** 1 lbm/min = 0.007 5600 kg/s	

EXERCISE 3–5

1. 253 kWh	**3.** 5.97 kWh	**5.** 12.7 MJ
7. \$48.60	**9. (a)** \$11.25 and \$6.00	
	(b) \$5.25	

EXERCISE 3–6

1. 3.0×10^6 m^2	**3.** 718×10^{-6} m^3
5. 398.0×10^6 m^2	**7.** 63.4×10^{-9} m^3/s
9. 692×10^{-18} m^3	**11.** 0.820 m^2
13. 0.825 m^2	**15.** 0.0150 m^2
17. 733.3 km^2	**19.** 1129 km^2
21. 810 mm^3	**23.** 1382×10^{-6} m^2
25. 2285×10^6 m^2	**27.** 8.10×10^{-12} m^3
29. 112 in^2	**31.** 1.607×10^5 ft^3
33. 745 cm^2	**35.** 1 m^3/s = 2118.8 ft^3/min
37. 427 kPa	**39.** 0.614 lbm/in^2

CHAPTER 4

EXERCISE 4–3

1. rational polynomial   **3.** rational fractional expression
5. irrational   **7.** rational polynomial
9. irrational   **11.** irrational
13. rational polynomial   **15.** rational fractional expression

EXERCISE 4–4

1. 3   **3.** 1   **5.** 1
7. 2   **9.** $\frac{1}{3}$ **11.** $\frac{7}{4}$
13. 2   **15.** r^2 **17.** $5b$
19. $16d^2$ **21.** 7.4ω

EXERCISE 4–5

1. $5a$ **3.** $12x$ **5.** $-7c$
7. $-2b$ **9.** $5h$ **11.** $10A + 2B$
13. $-3c^2b$ **15.** $3x^2 + 3x + 2$ **17.** $-2x^3 + x^2 - 2x$
19. $-5x^2y$

EXERCISE 4–6

1. a   **3.** b   **5.** d
7. c   **9.** b   **11.** d

EXERCISE 4–7

11. $6y + 9$ **13.** $2x^2$
15. $3x + 2y + 4z$ **17.** $V^3 + 3V^2 + 5$
19. $-x - y$ **21.** $3n + 8$
23. $4x^3 - x^2 - 4x$ **25.** $7m + n$
27. $2x^2 - 5x$ **29.** $8\omega + 2.6\rho$
31. $3x^3 - 4x^2 - 12x + 8$ **33.** $-5a^3 + 2a + 5b^2$
35. 0   **37.** $-2\alpha + 15\beta - 9$
39. $-11x + 4y + 3z$

Calculator Drill

41. $5.45x^2 + 0.76x + 18.13$ **42.** $35.2x^2 - 1.5x - 27.7$
43. $29.8x^2 - 15.17x + 12.32$ **44.** $40.1x^2 - 19.7x - 28.2$

CHAPTER 5

EXERCISE 5–1

1. x^9 **3.** 1   **5.** b^2

7. $14y$ **9.** $-35ab$ **11.** 10^{-2}

13. $-x^4y^3$ **15.** $16\eta\theta\phi$ **17.** 16

19. $-8y^{-2}x^2$ **21.** $6x^2$ **23.** $-8y^3$

EXERCISE 5–2

1. 27   **3.** $2y - 6$ **5.** $3\mu - 15$

7. $-3Z + Z^2$ **9.** $\beta c - 4\beta$ **11.** $-3x + 3y$

13. $-2a^2 + 4ab$ **15.** $-14\theta\omega - 35\theta\phi$ **17.** $-y^7 - 4y^6$

19. $3cd^3 + 3c^3d$ **21.** $3s^3 + 12s^2$

EXERCISE 5–3

1. $12x + 8y + 20$ **3.** $21u - 14v + 7w$

5. $-40x + 5y - 15$ **7.** $6\mu - 12\theta + 18$

9. $-2x^2 - 4x + xy$ **11.** $12hj - 15h\theta + 3h$

13. $-6a^2b + 9ab^2 - 12abc$ **15.** $6x^4 - 21x^3 - 18x^2$

17. $-2a^3b + 2ab^3 - 2abc + 6ab$ **19.** $-a^2 - 2b + 3c - 5$

Calculator Drill

21. $7.55y^3 - 24.8xy + 15.7y$ **22.** $-68.1b^2 + 45.4bc - 105bd$

23. $-125\mu^3 + 225\mu^2 - 306\mu$ **24.** $-16.5a^3 - 62.9a^2b^2 + 32.3a^2$

EXERCISE 5–4

1. $a - 8$ **3.** $-a + 9c$

5. $-12x^2 - 6x + 4$ **7.** 2

9. $3\theta - \beta$ **11.** $-2y$

13. $2x + y$ **15.** $\mu + 2\pi$

17. $2V^2 - U$ **19.** $x^4 - 4x^2 + x - 2$

21. $-5y^2 + 4y + 2$

EXERCISE 5–5

1. $x^3 - 2x^2 - 4$ **3.** $-11y - 4y^2$ **5.** $\theta - 3\eta + 2$

7. $-14x^2 + 14x + 3$ **9.** $-15c^2 - 9c - 6ac$ **11.** $6\mu^2 - 7\mu + 7$

13. $-2\theta - 2$ **15.** $3x^3 + 6x^2 - 18x - 12$ **17.** $4a^2 - 8a + 6a^2b$

EXERCISE 5–6

1. a^4 **3.** c^2 **5.** μ^7

7. 3   **9.** $-3y^2z^5$ **11.** $2c^{-2}$

13. $-8a^2b^4$ **15.** $4c^2$

EXERCISE 5–7

1. $3x + 4$ **3.** $7 + 4y$ **5.** $3U + V$

7. $2b^3 + 4a^3$ **9.** $6x + 3y$ **11.** $a^3 + a$

13. $13c + 9 + 11c^2$ **15.** $x + 2xy + 3y^2$ **17.** $6b + 3$

19. $7\mu^2 + 4 + 5\mu^{-2}$

EXERCISE 5–8

1. $6(a + 2)$ **3.** $3(2 + 3b)$

5. $3(m + 1)$ **7.** $\omega(u + \mathbf{v})$

9. $7(\phi + \theta)$ **11.** $\rho(\rho + 5)$

13. $7y(x + 3)$ **15.** $8a^2(a + 2)$

17. $a(x + 3)$ **19.** $a^2b^3c(1 + a^2b^2c^2)$

21. $5(\alpha^2 + 3\alpha + 4)$ **23.** $ab(1 + ab + a^2b^2)$

25. $8\beta^2(2\pi^2 + \pi + 4)$ **27.** $5\eta\alpha(6\eta + 3 - 5\alpha)$

29. $5y^2z(yz + 3 + 2y^2)$ **31.** $3y^3x(7y + 6x + 9)$

EXERCISE 5–9

Calculator Drill

1. 12.0   **2.** 2.00   **3.** 5.00

4. 7.00   **5.** 80.0   **6.** 54.0

7. -2.00 **8.** 47.0   **9.** 6.00

10. 13.0   **11.** -1.00 **12.** -3.00

13. 1.00   **14.** 10.0   **15.** 20.0

16. 20.0   **17.** 2.00   **18.** 13.0

19. -0.284 **20.** 5.79   **21.** 7.75

22. -17.5 **23.** 584   **24.** 11.8

25. 34.4

CHAPTER 6

EXERCISE 6–1

1. true   **3.** false   **5.** true

7. true   **9.** true   **11.** true

13. true   **15.** false   **17.** true
19. false

EXERCISE 6–2

1. $x = 1$ **3.** $m = -8$ **5.** $x = -1$
7. $\alpha = 9$ **9.** $\phi = 4$

EXERCISE 6–3

1. $x = 2$ **3.** $x = 10$ **5.** $y = 8$
7. $\theta = 5$ **9.** $y = 3$ **11.** $n = -1$
13. $C = 4$ **15.** $\tau = -17$ **17.** $m = -13$
19. $x = 0$ **21.** $t = 20$

EXERCISE 6–4

1. $y = 4$ **3.** $\mu = 6$ **5.** $x = 15$
7. $y = 60$ **9.** $k = -6$ **11.** $P = -7$
13. $y = 96$ **15.** $\beta = -4$ **17.** $x = -63$
19. $\phi = 14$ **21.** $r = -2$ **23.** $y = 52$
25. $x = 4$ **27.** $k = -4$ **29.** $y = -3$

EXERCISE 6–5

1. $x = 2$ **3.** $x = 3$ **5.** $R = \frac{1}{2}$
7. $\phi = 3$ **9.** $m = 1$

EXERCISE 6–6

1. $y = 4$ **3.** $\phi = 1$ **5.** $R = -3$
7. $E = -12$ **9.** $\mu = -5$ **11.** $t = 4$
13. $u = 3$ **15.** $\eta = 0$

EXERCISE 6–7

1. $\Delta = 2$ **3.** $x = 2$ **5.** $y = -15$
7. $\pi = -1$ **9.** $\theta = 2$ **11.** $\eta = 5$
13. $y = 2$ **15.** $z = -\frac{10}{9}$

EXERCISE 6–8

1. $i = 11$ **3.** $\beta = -1$ **5.** $x = 3$
7. $R_1 = 5$ **9.** $W = 12$ **11.** $L = 18$

13. $R_1 = -8$ **15.** $\theta = 4$ **17.** $N = 0$

19. $V_t = 2$ **21.** $\alpha = -3$ **23.** $X_c = \frac{1}{3}$

EXERCISE 6–9

1. $\lambda = v/f$ **3.** $R = E/I$

5. $g_m = \mu/r_p$ **7.** $f_0 = BW/Q_0$

9. $E_1 = E_{BB} - I_B E_P$ **11.** $E_L = E_N - R_O I_L$

13. $R = \dfrac{t}{-xC}$ **15.** $I_2 = \dfrac{I_1 G}{R_1 + R_2}$

17. $\theta_{JA} = \dfrac{T_J - T_A}{P_D}$ **19.** $E_1 = E_T - I_1 R$

21. $Y_2 = \dfrac{V_1 Y^2}{V_1 Y_1 - i}$ **23.** $I_T = E_{th}/(R_{th} + R_L)$

25. $R_L = \dfrac{-A_v r_p}{A_v - \mu} = \dfrac{A_v r_p}{\mu - A_v}$ **27.** $Y = (BUG)/(2K)$

EXERCISE 6–10

1. 111 mA   **3.** 159 Ω **5.** 599°C

7. 6.8 kΩ **9.** 377 rad/s   **11.** 806 m

13. 7.0×10^{-2}

EXERCISE 6–11

1. $8x = 40, x = 5$ **3.** $6x - 5 = 25, x = 5$ **5.** $16 - x = 9, x = 7$

7. $2(2 + x) = 18, x = 7$ **9.** $6x - 2x = 52, x = 13$

EXERCISE 6–12

1. $x = 5$ **3.** $x = -1$ **5.** $x = -3$

7. 4336   **9.** $C = 35¢$ **11.** $V = 13$ min

13. resistor = 3¢,   **15.** diode = 170°C,
capacitor = 15¢   zener = 150°C

CHAPTER 7

EXERCISE 7–1

1. 27.8 mA   **3.** 150 C   **5.** 12.0 V

7. 9.00 V   **9.** 29.3 Ω

EXERCISE 7–2

1. **(a)** 1.0 kV   **3.** 0.60 kΩ   **5.** 8.0 A
   (b) 20.0 V
   (c) 29 V
   (d) 234 V
7. 2.5 Ω   **9.** 390 Ω

EXERCISE 7–3

1. 23.5 kΩ   **3.** 2.75 kΩ   **5. (a)** 273 kΩ
                                (b) 806 μA
                                (c) *See* Figure A1.

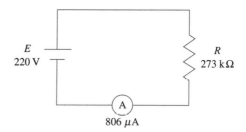

FIGURE A1 Solution to Exercise 7–3, problem 5c.

E
220 V

R
273 kΩ

A
806 μA

EXERCISE 7–4

1. 42 V   **3.** 37 V   **5.** 120 = 16 + 40 + 64

EXERCISE 7–5

1. **(a)** 14.5 kΩ   **3. (a)** 2.80 kΩ   **5. (a)** 20 mA
   (b) 10.0 mA   **(b)** 40.0 mA   **(b)** 68 V
   (c) 10.0 mA   **(c)** 28.0 V   **(c)** 1.1 kΩ
   (d) 30 V   **(d)** 4.5 kΩ
7. **(a)** 0.45 mA   **9.** $R_1 = 25$ kΩ, $R_2 = 25$ kΩ
   (b) 22 V $R_4 = 10$ kΩ

EXERCISE 7–6

1. 1.38 kW   **2.** 9.1 s   **3.** 9.37 MJ
4. 158 W   **5.** 1.53 MJ   **6. (a)** 14.0 Ω
                                     (b) 1.54 MJ
7. **(a)** 0.133 A   **8. (a)** 1.39 W   **9.** 100 V
   (b) 165 kJ   **(b)** 2.0 W

10. 30 mA   **11. (a)** 24.0 kW   **12. (a)** 58.8 mA
                            (b) 18.2 kW   **(b)** 215 mW, 315 mW, 176 mW
                                                       (c) 706 mW
                                                       (d) 706 mW = 706 mW

CHAPTER 8

EXERCISE 8–1

1. 0   **3.** -1 **5.** $\frac{1}{2}$
7. 5   **9.** 5

EXERCISE 8–2

1. $2/10 = 12/60$ **3.** $-3/-39 = 4/52$
5. $-33/77 = 9/-21$ **7.** $5m/5n = -8m/-8n$
9. $st/t^2 = -2st^2/-2t^3$ **11.** $-12xy/30x^2 = -6y^3/15xy^2$
13. $-2b^2/(2b^2 - 2ab)$

EXERCISE 8–3

1. $\frac{1}{2}$ **3.** $\frac{2}{5}$ **5.** $3b/2$
7. a **9.** $2dc/11$ **11.** x^5
13. $3a^2b^3/2$ **15.** $(a - 1)$ **17.** $t/4$
19. $1/(x + 1)$ **21.** $\theta\beta^2$

EXERCISE 8–4

1. $\frac{1}{2}$ **3.** $-\frac{2}{5}$ **5.** $-\frac{3}{20}$
7. -1 **9.** -6 **11.** $2(2x + y)/y$
13. $(b - c)/(c + b)$ **15.** $20b/3a^3$ **17.** $\pi r/4$
19. $\frac{1}{3}$

EXERCISE 8–5

1. $\frac{3}{2}$ **3.** $\frac{3}{4}$ **5.** $21/20$
7. 4   **9.** $1/y$ **11.** x/y
13. $-(a - 2) = 2 - a$ **15.** $(a - b)$

1. $\dfrac{\dfrac{8}{3}}{\dfrac{4}{5}}$ **3.** $\dfrac{\dfrac{-9}{2}}{\dfrac{3}{16}}$ **5.** $\dfrac{\dfrac{-6x}{7y}}{\dfrac{3x}{14y}}$

7. $\dfrac{\dfrac{R^2}{-t}}{\dfrac{2tR}{t^2}}$ **9.** $\dfrac{\dfrac{b}{b-3}}{\dfrac{b^2}{2}}$ **11.** 10

13. $\frac{3}{4}$ **15.** $\frac{1}{2}$ **17.** $\dfrac{x-y}{x+y}$

19. $a/(a-5)$

EXERCISE 8–7

1. $\dfrac{-1}{4}$ **3.** $-\frac{4}{5}$ **5.** $\dfrac{y-x}{3a}$

7. $\dfrac{-3y-4}{2x+1}$ **9.** $\dfrac{-2\pi R}{E+1}$ **11.** $\dfrac{5\omega+\beta}{6+\beta}$

EXERCISE 8–8

1. $\dfrac{2}{12},\dfrac{8}{12},\dfrac{5}{12}$ **3.** $\dfrac{-27}{36},\dfrac{12}{36},\dfrac{-20}{36}$ **5.** $\dfrac{14x}{42},\dfrac{12b}{42},\dfrac{35}{42}$

7. $\dfrac{ay}{xy},\dfrac{3}{xy},\dfrac{ax^2y}{xy}$ **9.** $\dfrac{x-b}{a-b},\dfrac{x}{a-b}$

EXERCISE 8–9

1. 2   **3.** -2 **5.** $10/a$

7. $2/(x+1)$ **9.** 2   **11.** $3/y$

EXERCISE 8–10

1. 6   **3.** 75   **5.** $2A$

7. $2\pi R$ **9.** $12\alpha\beta$ **11.** $8\alpha^2(\beta+1)$

13. $14(7K-1)$ **15.** $2s(t+1)$

EXERCISE 8–11

1. $\dfrac{2d+c}{d}$ **3.** $\dfrac{3x+5}{6x}$

5. $\dfrac{y + x}{xy}$

7. $\dfrac{mn + 1}{m}$

9. $\dfrac{17ab}{24}$

11. $\dfrac{my + 2nx}{xy}$

13. $2t/3$

15. $\dfrac{20a + 9b}{6}$

17. $\dfrac{-3x^2 + 9x - 1}{3x}$

19. $\dfrac{6a^2 - 5ab + 6b^2}{12ab}$

21. $-3/(x - 2)$

23. $\dfrac{-11b - 8}{18b(b - 1)}$

25. $\dfrac{-6m^2 - 12m + 60}{m(m + 5)(m - 5)}$

EXERCISE 8–12

1. $\frac{23}{4}$

3. $\frac{7}{3}$

5. $\dfrac{3m - 2}{m}$

7. $\dfrac{9a + 2}{3}$

9. $\dfrac{8x^2 - 7}{2x}$

11. $\dfrac{\omega CL + 3L - z}{L}$

13. $\dfrac{x^2 + 3x + 2}{x + 3}$

15. $\dfrac{x^2 + 10x + 9}{x + 2}$

EXERCISE 8–13

1. $\frac{3}{2}$

3. $\dfrac{ab - 1}{ab + 1}$

5. $\dfrac{1}{2a}$

7. $\dfrac{3a - b}{-2b}$

9. $\dfrac{2x - 5}{x - 2}$

11. $-\dfrac{b}{a}$

13. $\dfrac{3b - 1}{b + 8}$

15. $\dfrac{2(x - 1)(x + 3)}{2x - 7}$

CHAPTER 9

EXERCISE 9–1

1. $x = 2$

3. $a = \frac{3}{2}$

5. $R = -\frac{45}{7}$

7. $m = -1$

9. $V = \frac{16}{21}$

11. $I = -2.5$

13. $x = 30$

15. $a = -10$

17. $x = 10$

19. $I = 3$

21. $x = \frac{39}{7}$

EXERCISE 9–2

1. $x = 4$ **3.** $y = 1$ **5.** $x = 1$

7. $y = 2$ **9.** $y = -3$ **11.** $x = 7$

13. $x = 2$ **15.** $x = 6$ **17.** $y = -2$

19. $y = 2$

EXERCISE 9–3

1. $x = a/b$ **3.** $Z = 3a$ **5.** $y = \dfrac{2mn}{n + 2m}$

7. $W = \dfrac{b}{c\pi^2 k + 1}$ **9.** $s = 3/(D - C)$ **11.** $y = \dfrac{m - n}{ab + a}$

13. $x = \dfrac{3k(g - h)}{5k - 3g + 3h}$ **15.** $u = \dfrac{2mn}{m + n}$ **17.** $x = \dfrac{2b - ab}{b - a}$

19. $V = 3g$

EXERCISE 9–4

1. $F = \dfrac{9}{5}C + 32$ **3.** $f = \dfrac{1}{2\pi C X_C}$ **5.** $w = \dfrac{2kg}{v^2}$

7. $T_A = T_J - \theta_{JA} P_D$ **9.** $R_2 = \dfrac{E - C R_1}{C}$ **11.** $n = \dfrac{216}{108 - \theta}$

13. $S_2 = \dfrac{\eta F_1 S_1}{A_2}$ **15.** $d^2 = \dfrac{q_1 q_2}{4\pi\epsilon_0 F}$ **17.** $R_2 = \dfrac{R_T R_1}{R_1 - R_T}$

19. $\alpha = \dfrac{SR_E + SR_\beta - R_1 - R_2}{SR_\beta}$ **21.** $48.7\ \Omega$ **23.** $892\ \Omega$

25. $7.7\ \text{kHz}$ **27.** 65.7 **29.** $403\ \text{k}\Omega$

CHAPTER 10

EXERCISE 10–1

1. $V_1 = 0.90\ \text{V}$ **3.** $V_2 = 8.3\ \text{V}$ **5.** $E = 66.4\ \text{V}$

7. $\Sigma R = 1.4\ \text{M}\Omega$ **9.** $R_1 = 870\ \Omega$

EXERCISE 10–2

1. $0.30\ \text{kV}$ **3.** $30\ \text{V}$ **5.** $30\ \text{V}$

7. $26\ \text{V}$ **9.** $185\ \text{V}$

1. 17.9 mS   **3.** 1.10 mS   **5.** 161 μS
7. 0.370 mS   **9.** 0.122 μS   **11.** 1.79 S

1. 18 mS   **3.** 0.11 mS   **5.** 19 μS
7. 20.0 mS   **9.** 5.18 mS   **11.** 32 μS

1. 75.0 Ω **3.** 269 Ω **5.** 10 Ω
7. 1.3 kΩ **9.** 44.8 kΩ

1. 3.0 A   **3.** 6.40 mA   **5.** 4.72 mA
7. 0.82 A   **9.** 20 mA

1. 1.7 mA   **3.** 1.2 mA   **5.** 0.11 A
7. $I_1 = 3.3$ A, $I_2 = 1.7$ A   **9.** $I_3 = 1.67$ A, $I_4 = 3.33$ A

1. $I_N = 167$ mA, $R_N = 150\ \Omega$
3. $E_{Th} = 14$ V, $R_{Th} = 47\ \Omega$
5. $E_{Th} = 20$ V, $R_{Th} = 10\ \Omega$
7. $I_N = 0.22$ A, $R_N = 31\ \Omega$
9. $E_{Th} = 64$ V, $R_{Th} = 1.4$ kΩ

CHAPTER 11

1. 6   **3.** $15y^2$, 1
5. $x^2 + 2x + 1$ **7.** $x^2 + 5x + 6$
9. $x^2 - 8x + 15$ **11.** $x^2 + 10x + 24$
13. $x^2 - 12x + 20$ **15.** $x^2 - 3x - 10$

17. $6y^2 - 5y - 21$ **19.** $90y^2 - 29y + 2$

21. $-6x^2 + 17x - 12$ **23.** $-18x^2 + 36x - 16$

25. $4x^2 - 16$ **27.** $x^2 + \frac{3}{4}x + \frac{1}{8}$

29. $y^2 + \frac{2}{5}y - \frac{8}{25}$

EXERCISE 11–2

1. $x^2 - 1$ **3.** $x^2 - 25$ **5.** $4x^2 - 1$

7. $25x^2 - 25$ **9.** $16x^2 - 49$ **11.** $y^2 - a^2$

13. $4y^2 - c^2$ **15.** $9y^2 - 4t^2$ **17.** $4x^2 - \frac{1}{4}$

19. $\frac{4}{25}y^2 - \frac{1}{64}$

EXERCISE 11–3

1. $x^2 + 6x + 9$ **3.** $x^2 - 4x + 4$ **5.** $x^2 + 10x + 25$

7. $4x^2 + 8x + 4$ **9.** $16x^2 - 16x + 4$ **11.** $4x^2 + 4x + 1$

13. $16x^2 - 24x + 9$ **15.** $x^2 + \frac{2}{5}x + \frac{1}{25}$ **17.** $x^2 + x + \frac{1}{4}$

19. $\frac{1}{16}x^2 + \frac{1}{8}x + \frac{1}{16}$ **21.** $\frac{16}{25}x^2 + \frac{1}{5}x + \frac{1}{64}$

EXERCISE 11–4

1. $(x - 2)(x + 2)$ **3.** $(2x - 8)(2x + 8)$

5. $(y - c)(y + c)$ **7.** $(1 - 4c)(1 + 4c)$

9. $(2c - 3y)(2c + 3y)$ **11.** $(ab - x)(ab + x)$

13. $(\frac{1}{4} - x)(\frac{1}{4} + x)$ **15.** $(\frac{1}{3} - 5z)(\frac{1}{3} + 5z)$

EXERCISE 11–5

1. $(x - 3)^2$ **3.** $(x - 2)^2$ **5.** $(y + 1)^2$

7. $(3y - 5)^2$ **9.** $(2x - 5c)^2$ **11.** $(5x - 3y)^2$

13. $(4y + 3)^2$ **15.** $(8x + y)^2$ **17.** $(11z + 4)^2$

19. $(0.9u - v)^2$

EXERCISE 11–6

1. $(3 - x)(a + 2)$ **3.** $(m - 7)(x - 5)$

5. $(c - a)(x + y)$ **7.** $(z^2 + 1)(z + 1)$

9. $(b - 3)(b + 3)(4b + 5)$ **11.** $(x - 1)(x + 1)(b - 1)$

13. $(2x - 3)(0.5y + a)$ or **15.** $(w + 1)(w - 1)(w + 3)$
$(x - 1.5)(y + 2a)$

EXERCISE 11–7

1. $3(a + 1)$ **3.** $(y^2 + c^2)(y + c)(y - c)$

5. $3(a - 1)(a + 1)$ **7.** $8y(y - 2)(y + 2)$

9. $5(y + 2)^2$ **11.** $2(2a + b)^2$

13. $5(y - m - n)(y + m + n)$ **15.** $(x - 4)(x + 3)$

17. $(x^2 + 1)(x + 1)$ **19.** $(2y - a - c)(2y - a + c)$

EXERCISE 11–8

1. $x = 4/(1 - a)$ **3.** $x = b/(5 - c)$

5. $x = 1$ **7.** $x = 2$

9. $x = b - 3$ **11.** $x = b + 4$

13. $E = R/(2I + 3)$ **15.** $\alpha = \beta/(\beta + 1)$

17. $R = i_1 Z/(\beta i - i_1)$ **19.** $n = IZ/(V - IR)$

21. $R_1 = \dfrac{R_E(1 - S)}{S(1 - \alpha) - 1}$ **23.** $R_E = \dfrac{R_1 R_2 - SR_1 R_2 (1 - \alpha)}{S(R_1 + R_2) - (R_1 + R_2)}$

25. $I_B = \dfrac{V_{CC}}{R_1 + R_E(\beta + 1)}$ **27.** 23.9 mS

29. 738 Ω

CHAPTER 12

EXERCISE 12–1

1. $\frac{1}{4}$ **3.** $\frac{2}{3}$ **5.** $\frac{2}{3}$

7. 4.21/1   **9.** $\frac{3}{400}$ **11.** $\frac{23}{19}$

EXERCISE 12–2

1. -7.00% **3.** no   **5.** 800.000 kHz

7. -5.36% **9.** 9.9992 V   **11.** 216/224 Ω, 209/231 Ω, 198/242 Ω

EXERCISE 12–3

1. 15 ppm   **3.** 136 ppm   **5.** 220 ppm

7. 0.1%   **9.** 0.01%   **11.** 0.405%

13. $\pm 0.005\%$ **15.** 99.8 kΩ

EXERCISE 12–4

    1. 50.0 ± 4.0 A   **3.** $560 \, \Omega \pm 56.0 \, \Omega$ **5.** 236.0 W ± 35.4 W

    7. $\pm 0.010\%$ **9.** $\pm 0.360\%$ **11.** $\pm 2.8\%$

    13. 426/473 V   **15.** 114 W $\pm 13\%$ **17.** 49.0/50.8 mA

    19. $2.14/2.28 \, k\Omega$

EXERCISE 12–5

    1. (a) 4327 W   **3.** 0.88   **5.** 71 W

       (b) 4.33 kW

    7. (a) 392 J   **9. (a)** 7.9 kW   **11.** 1.19 kW

       (b) 74%   **(b)** $27.00

EXERCISE 12–6

    1. $x = 4$ **3.** $x = 3$ **5.** $x = 1.86$

    7. true   **9.** false   **11.** true

EXERCISE 12–7

    1. $1.51 \, \Omega$ **3.** $1.68 \, \Omega$ **5.** 468 m

EXERCISE 12–8

    1. $0.967 \, \Omega$ **3.** $130 \, m\Omega$ **5. (a)** 1.01 mm

                                            (b) #18

EXERCISE 12–9

    1. $80.7 \, m\Omega$ **3.** 3.16 m   **5.** $0.0523 \times 10^{-6} \, \Omega \cdot m$

EXERCISE 12–10

    1. $1.79 \, \Omega$ **3.** 59°C

CHAPTER 13

EXERCISE 13–1

	Constant	Dependent Variable	Independent Variables
1.	k	F	V, m, r
3.	G	F	d, m_1, m_2
5.	R	V	T, M

    1. 7   **3.** 17   **5.** -13

    7. $-5b - 3$ **9.** $x = g(y)$ **11.** $z = f(v)$

    13. $f(2) = 16, f(-3) = 41$ **15.** $Q(1) = -1.5, Q(0.5) = -1.07$

           $f(0.5) = 2.5$ $Q(5) = 0.833$

    1. (a) doubled   **3. (a)** one-fourth   **5. (a)** half

        (b) one-third   **(b)** doubled   **(b)** doubled

        (c) increase 5 times   **(c)** not changed   **(c)** one-fourth

        (d) inversely

    7. (a) doubled

        (b) multiplied by 0.866

    1. $r_p \gg R_0$ **3.** $G_1 \gg G_2$ **5.** $K \gg 1$

    7. $A'R_f \gg (r_0 + R_f + R_L)$ **9.** $4k^2 \gg 1$ **11.** $I \gg I'$

CHAPTER 14

    5. zero   **7.** x-coordinate   **9.** II

    11. *See* Figure A2.

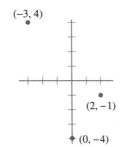

FIGURE A2 Solution to Exercise 14–1, problem 11.

    1. *See* Figure A3.

      FIGURE A3   Solution to Exercise
14–2, problem 1.

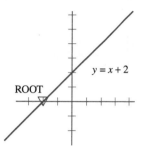

    3. *See* Figure A4.   **5.** *See* Figure A5.

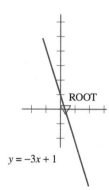

FIGURE A4   Solution to Exercise 14–2, problem 3.   **FIGURE A5**   Solution to Exercise 14–2, problem 5.

7. *See* Figure A6.

9. *See* Figure A7.

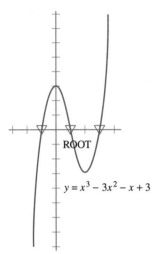

$$y = x^3 - 3x^2 - x + 3$$

ROOT

FIGURE A6   Solution to Exercise 14–2, problem 7.

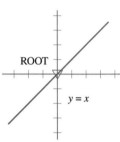

ROOT

$$y = x$$

FIGURE A7   Solution to Exercise 14–2, problem 9.

11. *See* Figure A8.

13. *See* Figure A9.

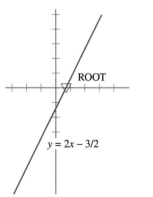

ROOT

$$y = 2x - 3/2$$

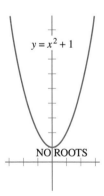

$$y = x^2 + 1$$

NO ROOTS

FIGURE A8   Solution to Exercise 14–2, problem 11.

FIGURE A9   Solution to Exercise 14–2, problem 13.

15. *See* Figure A10.　　　　　　**17.** *See* Figure A11.

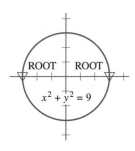

FIGURE A10　Solution to Exercise 14–2, problem 15.

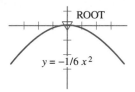

FIGURE A11　Solution to Exercise 14–2, problem 17.

EXERCISE 14–3

1. (a) negative
(b) −1
(c) (0, 5)

3. (a) negative
(b) −2
(c) (0, −4)

5. (a) positive
(b) 4
(c) (0, 0)

7. (a) negative
(b) $-\frac{2}{3}$
(c) (0, 12)

9. (a) positive
(b) 1
(c) (0, 0)

11. positive slope, y-intercept at (0, 3), root at (−3/10, 0)

13. positive slope, y-intercept at (0, 0), root at (0, 0)

15. positive slope, y-intercept at (0, 0), root at (0, 0)

17. negative slope, y-intercept at (0, 6), root at (12, 0)

19. positive slope, y-intercept at (0, −2), root at $(\frac{2}{5}, 0)$

EXERCISE 14–4

1. $y = -x + 9$

3. $y = -\frac{1}{4}x + 4.5$

5. $y = \frac{2}{5}x - 2$

7. $y = -x + 7$

9. $y = -\frac{6}{11}x + \frac{3}{11}$

11. $y = 5$

1. *See* Figure A12.   **3.** *See* Figure A13.

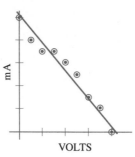

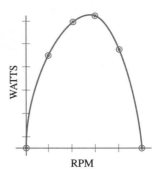

FIGURE A12   Solution to Exercise 14–5, problem 1.

FIGURE A13   Solution to Exercise 14–5, problem 3.

5. *See* Figure A14.

FIGURE A14   Solution to Exercise 14–5, problem 5.

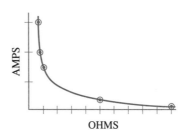

CHAPTER 15

EXERCISE 15–1

1. 0.3 W   **3.** 0.4 A   **5.** −50°C   **7.** 75
9. 3.2 mm   **11.** 30 mA   **13.** 3.7 h/d

EXERCISE 15–2

1. 1 A   **3.** 125°C/W   **5.** ≈430 μS

EXERCISE 15–3

1. intersection at $(1, 1)$ **3.** intersection at $(-3, -6)$
5. intersection at $(3, 1)$ **7.** intersection at $(2, 3)$
9. intersection at $(2, -2)$

EXERCISE 15–4

1. 18 mA, 1.4 V   **3.** 75 mA, 3.7 V   **5.** $I = 0.43$ A, $V_L = 25$ V, $V_R = 35$ V

CHAPTER 16

EXERCISE 16–1

1. $x = 5, y = -1$ **3.** $x = 2, y = 7$ **5.** $x = 3, y = 11$
7. $x = 4, y = 8$ **9.** $x = 0, y = 1$

EXERCISE 16–2

1. $x = 1, y = 1$ **3.** $x = 4, y = -1$ **5.** $x = -4, y = \frac{5}{2}$
7. $x = 8, y = 1$ **9.** $x = 0, y = 1$ **11.** $x = 1, y = 6$
13. $x = -1, y = 5$

EXERCISE 16–3

1. $E = 1.11 E_{\text{av}}$ **3.** $L = X_L/\omega$ **5.** $w = I^2 L$

7. $Q_0 = X_L/R$ **9.** $R_L = \dfrac{X_L Q_0' Q_0}{Q_0 - Q_0'}$

EXERCISE 16–4

1. $\Delta = 1 + 2 = 3$ **3.** $\Delta = 4 - 4 = 0$ **5.** $x = 3, y = 1$
7. $x = 5, y = 3$ **9.** $x = 1, y = 8$ **11.** $x = 3, y = -\frac{2}{3}$
13. $x = 2, y = -2$ **15.** $x = 10, y = -2$ **17.** $x = 20, y = 21$
19. $x = 6, y = -10$

EXERCISE 16–5

1. 4   **3.** 0
5. $x = 1, y = 3, z = 5$ **7.** $x = 4, y = -1, z = 3$
9. $x = -2, y = 4, z = 1$ **11.** $I_A = 0.40, I_B = 0.80, I_C = 0.40$

CHAPTER 17

EXERCISE 17–1

1. $I = 2.22$ mA in a   **3.** $I = 8.89$ mA in a
ccw direction   cw direction

5. $I = 0.667$ mA in a ccw direction   **7.** $I = 1.33$ mA in a cw direction   **9.** $V_3 = 8$ V

EXERCISE 17–2

1. 1 A down   **3.** 2 A up   **5.** 9 A up

7. 6 A down   **9.** 2 A down

EXERCISE 17–3

1. $I_1 = -5$ A, $I_2 = -5$ A

3. $I_1 = -2.28$ A, $I_2 = -2.76$ A

5. $I_1 = -46.3$ mA, $I_2 = -36.8$ mA

7. $I_1 = 0.793$ A, $I_2 = -0.975$ A, $I_3 = -0.994$ A

9. $I_1 = 2.68$ A, $I_2 = 1.36$ A, $I_3 = 0.989$ A

11. $I_{R2} = 0.366$ A to the right

13. $I_{RL} = 25.5$ A; battery is being charged

15. $I_{R3} = 6.14$ A to the right

CHAPTER 18

EXERCISE 18–1

1. ± 5 **3.** ± 2.83 **5.** ± 7 **7.** ± 4.24

9. ± 10 **11.** ± 2 **13.** ± 0.167 **15.** ± 3

17. ± 2.28 **19.** ± 4.24

EXERCISE 18–2

1. $(x + 5)(x + 2)$ **3.** $(a + 7)(a + 1)$

5. $(x + 1)(x + 1)$ **7.** $(a - 4)(a - 1)$

9. $(y - 4)(y - 3)$ **11.** $(x - 6)(x - 4)$

13. $(a - 9)(a + 2)$ **15.** $(x + 8)(x - 5)$

17. $(2x + 3)(x + 1)$ **19.** $(3x - 2)(x + 3)$

21. $(2b + 5)(2b - 3)$ **23.** $(5a - 7)(a - 2)$

25. $(3x + 2)(2x + 7)$ **27.** $(9a + 4)(2a - 3)$

29. $(4b - 3)(b - 2)$

EXERCISE 18–3

1. $y = -2$ and $y = -4$ **3.** $x = -5$ and $x = -1$

5. $y = 1$ and $y = 5$ **7.** $x = -1$

9. $y = 3$ and $y = -1$ **11.** $x = 7$ and $x = -2$

13. $y = \frac{2}{5}$ and $y = 1$ **15.** $x = -\frac{1}{3}$ and $x = 5$

17. $x = -\frac{5}{3}$ and $x = \frac{3}{4}$ **19.** $x = \frac{9}{8}$ and $x = -\frac{4}{3}$

EXERCISE 18–4

1. $x = 0.5$ and $x = -3$ **3.** $x = 0.667$ and $x = -3$

5. $y = 1.5$ and $y = -0.444$ **7.** $x = -1.67$ and $x = 0.75$

9. $y = 0.966$ and $y = 0.438$

EXERCISE 18–5

1. (a) up **3.** (a) up
   (b) one real   (b) none
   (c) $(0, 0)$ (c) $(0, 2)$
   (d) $(0, 0)$ (d) none

5. (a) down **7.** (a) up
   (b) two real   (b) none
   (c) $(2, 4)$ (c) $(-0.833, 0.917)$
   (d) $x = 0$ and $x = 4$ (d) none

9. (a) down **11.** (a) up
   (b) none   (b) two real
   (c) $(-1, -1)$ (c) $(6, -13)$
   (d) none   (d) $x = 11.1$ and $x = 0.9$

EXERCISE 18–6

1. $R_1 = 20\ \Omega, R_2 = 22\ \Omega$ **3.** $V = 3.0\ \text{V}$ **5.** $E = 2.4\ \text{V}$

7. $R_1 = 300\ \Omega, R_2 = 600\ \Omega$ **9.** $R_1 = 38.9\ \Omega, R_2 = 28.9\ \Omega$ **11.** $200\ \Omega$
                                         $R_3 = 77.8\ \Omega$

CHAPTER 19

EXERCISE 19–1

1. a^8 **3.** $8b^3$ **5.** c

7. $-49c^2$ **9.** y^{4x} **11.** $8x^3/(27a^3)$

13. a^{bc} **15.** $-2x^4y^3$ **17.** y^{a+2}

19. $8x^9y^3$ **21.** $25S^4/T^2$

EXERCISE 19–2

1. 1   **3.** 4   **5.** 1   **7.** 1   **9.** 0.25

11. 9   **13.** 2   **15.** 0.5   **17.** x/y^2 **19.** $2/(ba^3)$

21. $3ab^2$ **23.** y^3/x^2 **25.** $\dfrac{y^2}{x^2}$ **27.** $\dfrac{1}{27b^6c^3}$

EXERCISE 19–3

Calculator Drill

1. 5   **2.** 6   **3.** 12

4. 3   **5.** 4   **6.** 2

7. 8   **8.** 25   **9.** 4

10. 25.2   **11.** 5.54   **12.** 239

13. 0.25   **14.** 0.296   **15.** 0.396

16. 0.0442   **17.** 2.45   **18.** 125

19. 5.76   **20.** 9   **21.** 3.17

23. $\sqrt[3]{a}$ **25.** $x^2\sqrt{x}$ **27.** $7a\sqrt[3]{a}$

29. $-b\sqrt[3]{7b^2}$ **31.** $2^{1/2}$ **33.** $-2a^{1/2}$

35. $-5bc^{2/5}d^{1/5}$ **37.** $(T+s)^{1/3}$ **39.** $-(R-t)^{1/5}$

41. $a^{1/3}$ **43.** x **45.** $a^{7/12}$

EXERCISE 19–4

1. 7   **3.** -6^3 **5.** x^{14}

7. $\sqrt[3]{a}\sqrt[3]{b}$ **9.** $-\sqrt[4]{c}\sqrt[4]{d}$ **11.** $\sqrt[7]{4}\sqrt[7]{m}\sqrt[7]{n}$

13. $\sqrt{2}/\sqrt{3}$ **15.** $\sqrt{a}/\sqrt{b}$ **17.** $-\sqrt{3x}/\sqrt{5y}$

19. $\sqrt[6]{7}$ **21.** $\sqrt[6]{a}$ **23.** $-\sqrt[10]{aby}$

25. $3^{2/3}$ **27.** $6^{5/2}$ **29.** $-a^{7/9}$

EXERCISE 19–5

1. $8xy^2\sqrt{xy}$ **3.** $3x^3y^2\sqrt[3]{2x^2y^2}$ **5.** $-3a^3b^4\sqrt{2}$ **7.** $2a^2\sqrt{3a}/(3b^5)$

9. $\sqrt[3]{4}$ **11.** $\sqrt{2}$ **13.** $\sqrt[4]{9a}$ **15.** $\sqrt{15}/5$

17. $\sqrt{15}/(3x)$ **19.** $\sqrt[6]{108a^5}/(3a)$ **21.** $\frac{1}{2}$ **23.** $\frac{1}{2}$

25. $2\sqrt{y^2+9}$ **27.** $8a^6b^2c\sqrt{c}$ **29.** $\dfrac{2x+3}{2x}\sqrt{2x}$ **31.** $\dfrac{\sqrt[3]{ay^3}-24}{2y}$

EXERCISE 19–6

1. $x=49$ **3.** $x=16$ **5.** $x=4$

7. $a=62$ **9.** $x=6$ **11.** $y=3$

13. $a = 2.25$ **15.** $a = 8.17$ **17.** $c = 2.20$

19. $a = 2$ **21.** $y = 0$ **23.** $x = 0.75$

25. $y = 48$ **27.** $g = \dfrac{2S}{t^2}$ **29.** $V = \dfrac{4\pi r^3}{3}$

CHAPTER 20

EXERCISE 20–1

1. 3 **3.** 4 **5.** 2.5

Calculator Drill

7. 0.3010 **8.** 0.6021 **9.** 0.9031

10. 1.2041 **11.** 1.5051 **12.** 0.3222

13. 0.4914 **14.** 0.6128 **15.** 0.7076

16. 0.7853 **17.** 0.8513 **18.** 0.7924

19. 0.5328 **20.** 0.5775 **21.** 0.7723

22. 0.0212 **23.** 0.4698 **24.** 0.9036

25. 0.8820 **26.** 0.9614 **27.** 0.6637

EXERCISE 20–2

1. **(a)** 0, 0.8451, 0.8451   **3.** **(a)** -2, 0.8751, -1.1249

     (b) 1, 0.8451, 1.8451   **(b)** -1, 0.8751, -0.1249

     (c) 2, 0.8451, 2.8451   **(c)** 0, 0.8751, 0.8751

     (d) 3, 0.8451, 3.8451   **(d)** 1, 0.8751, 1.8751

5. **(a)** 0, 0.2095, 0.2095   **7.** **(a)** 0, 0.7443, 0.7443

     (b) 1, 0.2095, 1.2095   **(b)** 1, 0.7443, 1.7443

     (c) 2, 0.2095, 2.2095   **(c)** 2, 0.7443, 2.7443

     (d) 3, 0.2095, 3.2095   **(d)** 3, 0.7443, 3.7443

9. **(a)** -3, 0.0212, -2.9788

     (b) -2, 0.0212, -1.9788

     (c) -1, 0.0212, -0.9788

     (d) 0, 0.0212, 0.0212

EXERCISE 20–3

Calculator Drill

1. 2.25   **2.** 9.35

3. 1.44   **4.** 5.06

5. 3.33   **6.** 4.10×10^{-1}

7. 8.86×10^{-2}

8. 2.48×10^1

9. 2.93×10^1

10. 6.84×10^2

11. $2, 2 \times 10^1, 2 \times 10^2$

12. $3, 3 \times 10^1, 3 \times 10^2$

13. $4, 4 \times 10^2, 4 \times 10^4$

14. $5, 5 \times 10^2, 5 \times 10^4$

15. $7 \times 10^{-1}, 7 \times 10^{-2}, 7 \times 10^{-3}$

EXERCISE 20–4

1. $1.1761 = 0.4771 + 0.6990$

3. $1.7745 = 1.2304 + 0.5441$

5. $0.9542 \approx 1.2553 - 0.3010$

7. false

9. false

11. 28

13. 4

15. 6.01

EXERCISE 20–5

1. $0.9542 = 2(0.4771)$

3. $2.9248 = 2(1.4624)$

5. $0.9542 \approx (1.9085)/2$

7. $0.4515 = 3(0.3010)/2$

9. 3.28×10^4

11. 1.87

13. 1.59

15. 4.06

EXERCISE 20–6

1. 1.2

3. 0.75

5. -4.1

7. a

9. $-z$

Calculator Drill

11. 0.2231

12. 1.1506

13. 1.3863

14. 1.6094

15. 1.7918

16. 1.9459

17. -1.8579

18. -3.5405

19. -4.1997

20. 2.9124

21. 6.8320

22. 8.8594

23. 4.9642

24. 3.7121

25. 11.0588

26. 6.9518

27. -0.1851

28. -3.2391

29. 4.3732

30. 7.6014

31. 2.718

32. 1.00

33. 10.0

34. 19.0

35. 9.01

36. 5.10×10^{-3}

37. 2.72×10^{-2}

38. 0.250

39. 1.69

40. 9.22×10^{-44}

41. 0.943

42. 1.12×10^2

43. 8.55×10^{-4}

44. 2.40

45. 9.37×10^{-3}

46. 4.97×10^{-2}

47. 1.67×10^6

48. 1.19×10^{-2}

49. 8.90×10^{-8}

50. 1.75

EXERCISE 20–7

1. 1.3863 **3.** −0.7134 **5.** −0.2357

7. 1.7781 **9.** 0.7559 **11.** −0.5376

EXERCISE 20–8

1. $3.9318 = 2.8332 + 1.0986$ **3.** $2.7881 = 4.1744 - 1.3863$

5. $5.4161 \approx 2(2.7081)$ **7.** $2.0794 \approx (4.1589)/2$

9. 2.33 **11.** 292

13. 2.65 **15.** 8.44

EXERCISE 20–9

1. $x = 3.35$ **3.** $x = 10.0$

5. $x = 0.465$ **7.** $x = 8.48$

9. $5 \log (\sqrt{x}) = \ln (x) + 1.59$ **11.** $\ln (V) = \ln (17) - 4.5$

$\quad 5 \log (x)/2 = 2.3026 \log (x) + 1.59$ $\quad\quad V = 0.189$

$\quad\quad \log (x) \quad = 1.59/(2.5 - 2.3026)$

$\quad\quad\quad\quad x = 1.13 \times 10^8$

13. $2 \ln (V) - \ln (V) = 3.29$ **15.** $\log (Z) = \dfrac{1}{2.3026}$

$\quad\quad\quad \ln (V) = 3.29$ $\quad\quad Z = 10^{1/2.3026} = 2.72$

$\quad\quad\quad\quad V = \text{antiln } 3.29$

$\quad\quad\quad\quad V = e^{3.29} = 26.8$

17. $P_S/P_N = 10^{N_{dB}/10}$

$\quad\quad P_N = P_S 10^{-(N_{dB}/10)}$

19. $N \log (2) = \log (f_2) - \log (f_1)$

$\quad\quad \log (f_2) = \log f_1 + \log (2^N)$

$\quad\quad \log (f_2) = \log (f_1 2^N)$

$\quad\quad\quad\quad f_2 = f_1 2^N$

EXERCISE 20–10

1. $x = 1.1461$ **3.** $y = 1.9542$ **5.** $s = 0.3424$

7. $x = 1.3863$ **9.** $x = -4.8416$ **11.** $t = 17.93$

13. $t = 3.249$ **15.** $t = 9.226$ **17.** $y = 5.000$

19. $a = -1.951$

EXERCISE 20–11

1. *See* Figure A15.

3. *See* Figure A16.

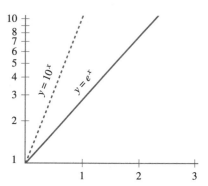

FIGURE A15 Solution to Exercise 20–11, problems 1 and 2.

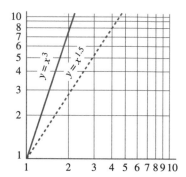

FIGURE A16 Solution to Exercise 20–11, problems 3 and 4.

5. *See* Figure A17.

FIGURE A17 Solution to Exercise 20–11, problem 5.

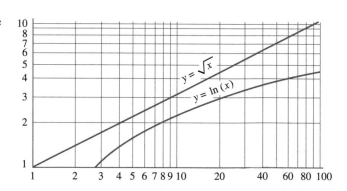

7. *See* Figure A18.

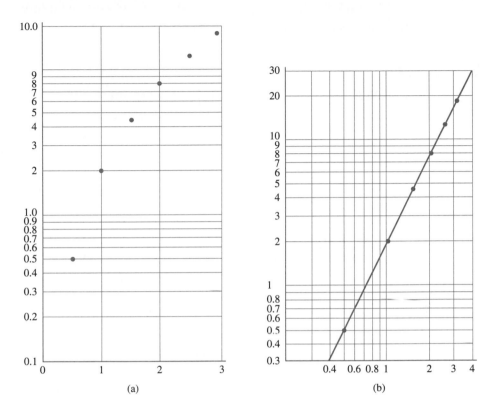

(a)

(b)

FIGURE A18   Solution to Exercise 20–11, problems 7a and b.

EXERCISE 20–12

1. 3.95   **3.** −17 dB   **5.** 1.3 kΩ

7. 950 μH   **9.** 1.6 MHz

CHAPTER 21

EXERCISE 21–1

1. 24.26 dB   **3.** 0.34 W   **5.** 46.02 dB

7. 3.6 V   **9.** 70.46 dB

EXERCISE 21–2

1. (a) 11.76 dBm   **(b)** 23.01 dBm   **(c)** −17.45 dBm

(d) 33.42 dBm   **(e)** −2.15 dBm   **(f)** 19.40 dBm

3. (a) 50.10 dBmV   **(b)** -16.48 dBmV   **(c)** 71.60 dBmV

    (d) 37.15 dBmV   **(e)** 58.49 dBmV   **(f)** -31.37 dBmV

5. 13.40 dBmV   **7.** 18.80 dBmV   **9.** 49 V

EXERCISE 21–3

1. 8.26 A

3. (a) $i = 216$ mA

    (b) $v_L = 6.77$ V

    (c) $v_R = 43.2$ V

5. $i = 0.276$ A, $v_R = 15.5$ V   **7.** $L = 16.6$ H, $i = 875$ mA

    $v_L = 2.5$ V $v_R = 7.0$ V

EXERCISE 21–4

1. 10.9 μA   **3.** 1.48 V

5. $t = 1.54$ s, $i = 1.8$ mA   **7.** $i = 0.17$ mA, $v_R = 17$ V

    $v_C = 7.0$ V $v_C = 84$ V

EXERCISE 21–5

1. E24   **3.** E12   **5.** E6

7. E192   **9.** 2.87

11. 1.000, 1.012, 1.024, 1.037, 1.049, 1.062, 1.075, 1.088

13. 1.82, 2.18, 2.67, 3.27

15. 8.870, 9.085, 9.311, 9.536

CHAPTER 22

EXERCISE 22–1

1. 60 rev   **3.** 0.80 rev, 288°   **5.** 90°

7. 317°   **9.** 446°   **11.** 25.8°

13. 0.833 rev   **15.** 2.40 rev   **17.** 4.71 rev

19. 2.64 rev   **21.** 2.00 rad   **23.** 4.71 rad

25. 3.05 rad   **27.** 2.36 rad

EXERCISE 22–2

1. complementary   **3.** supplementary   **5.** complementary

7. neither   **9.** supplementary   **11.** 36°

13. 99.3°   **15.** 90°   **17.** 0.132^r

19. 0.532^r **21.** $32.7°$ **23.** $45°$

25. $40.1°$ **27.** 0.501^r

EXERCISE 22–3

1. $\angle\gamma = 101°, c$ **3.** $\angle\beta = 49°, a$

5. $\angle\gamma = 100°, c$ **7.** $\angle\gamma = 1.14$ rad, a

9. $\angle\alpha = 1.14$ rad, a **11.** $\angle\beta = \pi/4, a$

13. $\angle\gamma = \pi/3$, all same size   **15.** $\angle\gamma = 65.4°, c$

17. $\angle\beta = 20.9°, c$ **19.** $\angle\alpha = 90°, a$

EXERCISE 22–4

1. $c = 2.5$ m, $\angle\beta = 53.1°$ **3.** $b = 35.6$ m, $\angle\alpha = 24.2°$

5. $c = 90.1$ m, $\angle\alpha = 60.6°$ **7.** $a = 3.05$ m, $\angle\beta = 40.4°$

9. $b = 11$ m, $\angle\beta = 10.4°$

EXERCISE 22–5

Calculator Drill

	s	c	t		s	c	t
1.	0.0175,	0.9998,	0.0175	**2.**	0.4226,	0.9063,	0.4663
3.	0.9816,	0.1908,	5.1446	**4.**	0.2588,	0.9659,	0.2679
5.	0.9998,	0.0175,	57.290	**6.**	0.0,	1.0,	0.0
7.	0.7826,	0.6225,	1.2572	**8.**	0.9449,	0.3272,	2.8878
9.	0.1977,	0.9803,	0.2016	**10.**	0.2974,	0.9548,	0.3115
11.	0.0009,	1.0000,	0.0009	**12.**	0.0936,	0.9956,	0.0940
13.	0.9893,	0.1461,	6.7720	**14.**	0.9770,	0.2130,	4.5864
15.	0.7570,	0.6534,	1.1585	**16.**	0.8980,	0.4399,	2.0413
17.	0.7683,	0.6401,	1.2002	**18.**	0.9478,	0.3190,	2.9714
19.	0.3633,	0.9317,	0.3899	**20.**	0.6211,	0.7837,	0.7926
21.	0.9975,	0.0707,	14.101	**22.**	0.2474,	0.9689,	0.2553
23.	0.7104,	0.7038,	1.0092	**24.**	0.0170,	0.9999,	0.0170
25.	0.0,	1.0,	0.0	**26.**	0.4794,	0.8776,	0.5463
27.	0.0998,	0.9950,	0.1003	**28.**	0.9320,	0.3624,	2.5722
29.	0.9927,	0.1205,	8.2381	**30.**	0.6210,	0.7838,	0.7923
31.	0.1583,	0.9874,	0.1604	**32.**	0.3127,	0.9499,	0.3292
33.	0.9208,	0.3902,	2.3600	**34.**	0.1987,	0.9801,	0.2027
35.	1.000,	0.0008,	1255.8	**36.**	0.9425,	0.3342,	2.8198

	s	c	t			s	c	t
37.	0.8260,	0.5636,	1.4655		**38.**	0.7229,	0.6909,	1.0463
39.	0.3476,	0.9376,	0.3707		**40.**	0.5818,	0.8133,	0.7154

EXERCISE 22–6

1. $\angle\beta = 70°$, $a = 3.42$ m, $b = 9.40$ m

3. $\angle\beta = 72.5°$, $a = 7.88$ m, $b = 25$ m

5. $\angle\beta = 61.6°$, $b = 36.1$ m, $c = 41.0$ m

7. $\angle\beta = 57.7°$, $b = 4.41$ m, $c = 5.22$ m

9. $\angle\beta = 75.9°$, $a = 3.54$ m, $c = 14.5$ m

11. $\angle\beta = 45°$, $a = 0.70$ km, $c = 0.99$ km

13. $\angle\alpha = 74.7°$, $a = 289$ m, $b = 79.2$ m

15. $\angle\alpha = 39.2°$, $a = 553$ mm, $b = 678$ mm

17. $\angle\alpha = 53.8°$, $b = 43.5$ m, $c = 73.6$ m

19. $\angle\alpha = 14.4°$, $a = 0.275$ km, $c = 1.10$ km

EXERCISE 22–7

Calculator Drill

1. **(a)** 6.14°   **(b)** 52.3°   **(c)** 14.8°
   (d) 34.6°   **(e)** 30.0°   **(f)** 23.3°
   (g) 39.9°   **(h)** 53.7°   **(i)** 24.3°
   (j) 90.0°   **(k)** 2.46°   **(l)** 67.07°

2. **(a)** 0.107 rad   **(b)** 0.912 rad   **(c)** 0.258 rad
   (d) 0.604 rad   **(e)** 0.524 rad   **(f)** 0.406 rad
   (g) 0.697 rad   **(h)** 0.937 rad   **(i)** 0.424 rad
   (j) 1.57 rad   **(k)** 0.043 rad   **(l)** 1.17 rad

3. **(a)** 83.9°   **(b)** 37.7°   **(c)** 75.2°
   (d) 55.4°   **(e)** 60.0°   **(f)** 66.7°
   (g) 50.1°   **(h)** 36.3°   **(i)** 65.7°
   (j) 0.000°   **(k)** 87.5°   **(l)** 22.9°

4. **(a)** 1.46 rad   **(b)** 0.658 rad   **(c)** 1.31 rad
   (d) 0.967 rad   **(e)** 1.05 rad   **(f)** 1.16 rad
   (g) 0.874 rad   **(h)** 0.633 rad   **(i)** 1.15 rad
   (j) 0.000 rad   **(k)** 1.53 rad   **(l)** 0.40 rad

5. **(a)** 84.3°   **(b)** 35.3°   **(c)** 72.4°
   (d) 23.9°   **(e)** 57.5°   **(f)** 45.0°
   (g) 67.0°   **(h)** 16.2°   **(i)** 80.6°
   (j) 0.688°   **(k)** 86.0°   **(l)** 82.0°

6. (a) 1.47 rad   (b) 0.615 rad   (c) 1.26 rad
    (d) 0.418 rad   (e) 1.00 rad   (f) 0.785 rad
    (g) 1.17 rad   (h) 0.282 rad   (i) 1.41 rad
    (j) 0.012 rad   (k) 1.5 rad   (l) 1.43 rad

EXERCISE 22–8

1. $b = 20$ m, $\angle \alpha = 36.9°$, $\angle \beta = 53.1°$
3. $b = 36.2$ mm, $\angle \alpha = 25.8°$, $\angle \beta = 64.2°$
5. $a = 7.0$ m, $\angle \alpha = 16.3°$, $\angle \beta = 73.7°$
7. $a = 31.3$ km, $\angle \alpha = 54.8°$, $\angle \beta = 35.2°$
9. $c = 21$ mm, $\angle \alpha = 48.8°$, $\angle \beta = 41.2°$
11. $c = 1000$ m, $\angle \alpha = 30.0°$, $\angle \beta = 60.0°$

CHAPTER 23

EXERCISE 23–1

1. $5°$ 3. $-175°$ 5. $90°$
7. $61°$ 9. $22°$ 11. $-155°$

EXERCISE 23–2

1. I   3. III   5. III
7. I   9. I   11. I

Calculator Drill

13. 0.966   14. -0.829 15. -0.866
16. 0.017   17. -0.643 18. 0.829
19. 0.500   20. 0.970   21. 0.259
22. -0.985 23. 0.999   24. 0.707
25. 0.259   26. 0.559   27. -0.500
28. -1.00 29. -0.766 30. 0.559
31. 0.866   32. -0.242 33. 0.966
34. -0.174 35. 0.052   36. -0.707
37. 3.73   38. -1.48 39. 1.73
40. -0.017 41. 0.839   42. 1.48
43. 0.577   44. -4.01 45. 0.268
46. 5.67   47. 19.1   48. -1.00

EXERCISE 23–3

1. *See* Figure A19.

FIGURE A19 Solution to Exercise 23–3, problem 1.

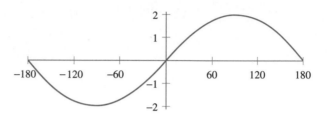

3. *See* Figure A20.

FIGURE A20 Solution to Exercise 23–3, problem 3.

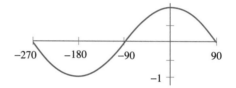

5. *See* Figure A21.

FIGURE A21 Solution to Exercise 23–3, problem 5.

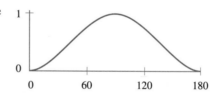

7. *See* Figure A22.

FIGURE A22 Solution to Exercise 23–3, problem 7.

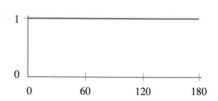

EXERCISE 23–4

Calculator Drill

1. 45.0°   **2.** 60.0°   **3.** 60.0°

4. 30.0°   **5.** 45.0°   **6.** 35.0°

7. −30.0°   **8.** −45.0°   **9.** −60.0°

10. $-20.0°$ 11. $120°$ 12. $150°$
13. $135°$ 14. $110°$ 15. $-45.0°$
16. $-57.0°$ 17. $-35°$ 18. $-11.3°$
19. $-80°$ 20. $-89°$ 21. $89.9°$

EXERCISE 23–5

1. $\angle\alpha = 36.9°, \angle\beta = 53.1°, \angle\gamma = 90°$
3. $\angle\alpha = 41.3°, \angle\beta = 94.1°, \angle\gamma = 44.6°$
5. $\angle\alpha = 36.9°, \angle\beta = 63.1°, c = 57.4$ mm
7. $\angle\beta = 51.3°, \angle\gamma = 105.4°, a = 40.3$ m
9. $\angle\gamma = 110°, b = 31$ m, $c = 58.5$ m
11. $\angle\gamma = 115.7°, a = 2.62$ m, $b = 11.1$ m

EXERCISE 23–6

1. $1.5\underline{/30°}$ 3. $2\underline{/-30°}$ 5. $3.7\underline{/-47°}$
7. $3.9\underline{/143°}$ 9. $2\underline{/150°}$ 11. $3\underline{/-175°}$
13. $4.5\underline{/-83°}$ 15. $4.6\underline{/73°}$ 17. $4.2\underline{/55°}$
19. $4.2\underline{/88°}$ 21. $4.5\underline{/125°}$ 23. $3.5\underline{/-100°}$

EXERCISE 23–7

Calculator Drill

1. $15.0\underline{/0°}$ 2. $3.57\underline{/-90°}$ 3. $8.70\underline{/180°}$
4. $1.41\underline{/45°}$ 5. $10.0\underline{/60.1°}$ 6. $7.61\underline{/-33.6°}$
7. $6.57\underline{/-50.8°}$ 8. $205\underline{/-31.8°}$ 9. $154\underline{/97.5°}$
10. $29.8\underline{/154.6°}$ 11. $1.00\underline{/-150°}$ 12. $32.7\underline{/61.5°}$
13. $44.1\underline{/-27.1°}$ 14. $34.2\underline{/-117.3°}$ 15. $5.76\underline{/135°}$
16. $15.2\underline{/-7.54°}$ 17. $14.5\underline{/157.3°}$ 18. $262\underline{/59.2°}$
19. $3.67\underline{/110°}$ 20. $1036\underline{/-43°}$ 21. $2.33\underline{/-37.2°}$
22. $(0.00, 17)$ 23. $(-16, 0.00)$ 24. $(0.00, -12.7)$
25. $(64.3, 76.6)$ 26. $(21.7, 12.5)$ 27. $(6.99, -26.1)$
28. $(53.0, 18.6)$ 29. $(4.64, 4.22)$ 30. $(6.77, -79.0)$
31. $(0.365, -0.449)$ 32. $(-314, -314)$ 33. $(-68.2, 30.3)$
34. $(1.12, -12.8)$ 35. $(-11.8, 13.1)$ 36. $(-9.18, -4.88)$
37. $(10.1, 19.9)$ 38. $(-6.55, 2.33)$ 39. $(29.8, -7.39)$
40. $(47.7, -70.0)$ 41. $(18.4, -76.1)$ 42. $(-379, -352)$

CHAPTER 24

EXERCISE 24–1

    1. $j3$ **3.** $j6$ **5.** $-j5$

    7. $j13.5$ **9.** $-j12$ **11.** $-j10.4$

  13. $j0.935$ **15.** $-j7.27$

EXERCISE 24–2

    1. $15 + j0$ **3.** $0 + j1.7$ **5.** $-5 - j4$

    7. $-1 + j0$ **9.** $17 - j1$ **11.** complex

  13. real   **15.** real   **17.** real

EXERCISE 24–3

    1. *See* Figure A23.   **3.** *See* Figure A23.   **5.** *See* Figure A23.

    7. *See* Figure A23.   **9.** *See* Figure A23.   **11.** $0 + j0$

  13. $-4 - j2$ **15.** $-1 - j2$ **17.** $-1 + j0$

  19. $2 + j0$ **21.** $-2 + j4$

FIGURE A23 Solution to Exercise 24–3, problems 1, 3, 5, 7, and 9.

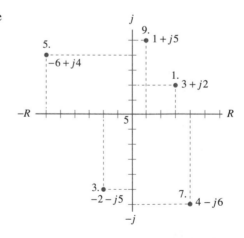

EXERCISE 24–4

    1. $3\underline{/120°}$ **3.** $1\underline{/150°}$ **5.** $3\underline{/-90°}$

    7. $2\underline{/30°}$ **9.** $3\underline{/-150°}$ **11.** *See* Figure A24.

  13. *See* Figure A24.   **15.** *See* Figure A24.   **17.** *See* Figure A24.

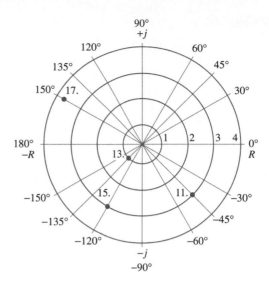

FIGURE A24 Solution to Exercise 24–4, problems 11, 13, 15, and 17.

EXERCISE 24–5

1. $10.0\underline{/53.1°}$

3. $4.79\underline{/116.6°}$

5. $13.1\underline{/-\,103.8°}$

7. $6.61\underline{/54.8°}$

9. $6.36\underline{/-60.3°}$

11. $24.1\underline{/-126.1°}$

13. $17.1\underline{/-4.2°}$

15. $4.21\underline{/-97.3°}$

17. $0.931\underline{/76.0°}$

19. $1.33 + j4.82$

21. $2.34 + j6.67$

23. $6.11 - j1.30$

25. $-4.26 + j11.5$

27. $1.18 - j1.84$

29. $-3.47 - j6.95$

31. $583 + j581$

33. $-0.133 + j0.136$

35. $-8.40 - j296$

37. $-125 + j28.1$

39. $0.0421 + j0.356$

EXERCISE 24–6

1. $8\underline{/0°}$

3. $10\underline{/0°}$

5. $10\underline{/-90°}$

7. $16\underline{/45°}$

9. $30\underline{/90°}$

11. $9.9\underline{/92°}$

13. $3.2\underline{/90°}$

15. $13\underline{/1.38^r}$

17. $10.4\underline{/-0.181^r}$

CHAPTER 25

EXERCISE 25–1

1. $7.0 + j2.0,\ 7\underline{/16°}$

3. $-12 + j1.0,\ 12\underline{/175°}$

5. $16 + j8.0,\ 17\underline{/27°}$

7. $16 + j0.33,\ 16\underline{/1.2°}$

9. $35 + j22,\ 41\underline{/32°}$

11. $2.66 + j0.94,\ 2.8\underline{/19°}$

13. $-1.2 + j0.98,\ 1.6\underline{/141°}$

15. $-24 - j7.5,\ 25\underline{/-163°}$

17. $58 + j40,\ 70\underline{/34°}$

19. $-84.08 - j101.4,\ 131.8\underline{/-2.26^r}$

    1. $12\underline{/90°} = 0 + j12$

    3. $47\underline{/67°} = 18 + j43$

    5. $27\underline{/-31°} = 23 - j14$

    7. $15\underline{/27°} = 13 + j7$

    9. $12\underline{/-65°} = 5 - j11$

  11. $646\underline{/33°} = 542 + j352$

  13. $10.9 \times 10^3\underline{/98°} = -1.52 \times 10^3 + j10.8 \times 10^3$

  15. $2.21 \times 10^3\underline{/37°} = 1.76 \times 10^3 + j1.33 \times 10^3$

  17. $364\underline{/-3.5°} = 363 - j22.2$

  19. $2.82\underline{/-56.9°} = 1.54 - j2.36$

  21. $197.1\underline{/3.7^r} = -167.2 - j104.4$

  23. $11\ 855\underline{/30.7°}$, $10\ 194 + j6052$

    1. $5 + j3$ **3.** $12.5 + j4$ **5.** $-9.4 - j10.6$

    7. $0.72\underline{/3°} = 0.72 + j0.04$ **9.** $0.35\underline{/82°} = 0.05 + j0.35$

  11. $4.0\underline{/162°} = -3.8 + j1.2$ **13.** $3.0\underline{/60°} = 1.5 + j2.6$

  15. $3.0\underline{/-40°} = 2.3 - j1.9$ **17.** $2.0\underline{/149°} = -1.7 + j1.0$

  19. $0.413\underline{/-90.5°} = -0.004 - j0.413$

  21. $3.1\underline{/0.36^r} = 2.9 + j1.1$

  23. $0.589\underline{/-4.64^r} = -0.043 + j0.587$

    1. $9\underline{/20°}$ **3.** $16\underline{/40°}$ **5.** $144\underline{/-34°}$

    7. $204.5\underline{/112.4°}$ **9.** $300\underline{/-28°}$ **11.** $7\underline{/20°}$

  13. $5\underline{/-32°}$ **15.** $1.65\underline{/8.8°}$ **17.** $4.05\underline{/56°}$

  19. $9.05\underline{/-85.6°}$ **21.** $4.52\underline{/-27.5°}$ **23.** $0.918\underline{/0.286^r}$

CHAPTER 26

    1. 17 Hz   **3.** 5 Hz   **5.** 35.7 Hz

    7. 60.0 Hz   **9.** 16.7 ms   **11.** 83.3 μs

  13. 2.4 s   **15.** 12.3 ms   **17.** 200 rad/s

  19. 1.9 krad/s   **21.** 37.7 krad/s   **23.** 314 rad/s

EXERCISE 26–2

1. $10\,\Omega$ **3.** $107\,\Omega$ **5.** $35.3\,\text{k}\Omega$

7. $164\,\text{m}\Omega$ **9.** $62.5\,\text{V}$ **11.** $39.6\,\text{V}$

13. $368\,\mu\text{A}$ **15.** $1.74\,\text{A}$

EXERCISE 26–3

1. (a) $1.51\,\text{k}\Omega$ **3.** (a) $94.2\,\Omega$ **5.** (a) $4.52\,\text{k}\Omega$

(b) $126\,\text{k}\Omega$ (b) $7.85\,\text{k}\Omega$ (b) $377\,\text{k}\Omega$

(c) $50.3\,\text{M}\Omega$ (c) $3.14\,\text{M}\Omega$ (c) $151\,\text{M}\Omega$

7. (a) $10\,\Omega$ **9.** (a) $2.5\,\text{k}\Omega$ **11.** $200\,\text{V}$

(b) $1.6\,\text{mH}$ (b) $400\,\text{mH}$

13. $22.3\,\text{mA}$ **15.** $8.34\,\text{mA}$

EXERCISE 26–4

1. (a) $177\,\Omega$ **3.** (a) $5.31\,\text{k}\Omega$ **5.** (a) $2.65\,\Omega$

(b) $2.12\,\Omega$ (b) $63.6\,\Omega$ (b) $31.8\,\text{m}\Omega$

(c) $5.31\,\text{m}\Omega$ (c) $159\,\text{m}\Omega$ (c) $79.6\,\mu\Omega$

7. (a) $32.3\,\Omega$ **9.** (a) $3.8\,\Omega$ **11.** $5.53\,\text{V}$

(b) $9.87\,\mu\text{F}$ (b) $85\,\mu\text{F}$

13. $490\,\text{mA}$ **15.** $1.79\,\text{A}$

EXERCISE 26–5

1. $70.7\underline{/45°}\ \text{V} = 50 + j50\ \text{V}$ **3.** $5.39\underline{/21.8°}\ \text{V} = 5.0 + j2.0\ \text{V}$

5. $30\underline{/90°}\ \text{V} = 0 + j30\ \text{V}$ **7.** $20\underline{/90°}\ \text{V} = 0 + j20\ \text{V}$

9. $76.7\underline{/12°}\ \text{V} = 75 + j16\ \text{V}$ **11.** $3.0 + j16.7\ \text{V}$

13. $6.9 - j4.0\ \text{V}$ **15.** $5.6 + j63.76\ \text{V}$

17. $9.0 - j28\ \text{V}$ **19.** $27.5 - j47.6\ \text{V}$

21. $5.8\underline{/-31°}\ \text{V} = 5.0 - j3.0\ \text{V}$ **23.** $13\underline{/39°}\ \text{V} = 10 + j8.0\ \text{V}$

25. $21\underline{/90°}\ \text{V} = 0 + j21\ \text{V}$ **27.** (a) $V_R = 1.0\underline{/0°}\ \text{kV}$,

$V_L = 754\underline{/90°}\ \text{V}$

(b) $E = 1.25\underline{/37°}\ \text{kV}$

29. (a) $V_L = 339\underline{/90°}\ \text{V}$,

$V_C = 239\underline{/-90°}\ \text{V}$

(b) $E = 100\underline{/90°}\ \text{V}$

EXERCISE 26–6

1. $2.0 - j5.0\ \text{A} = 5.4\underline{/-68°}\ \text{A}$ **3.** $0.30 - j1.2\ \text{A} = 1.2\underline{/-76°}\ \text{A}$

5. $0 - j1.6\ \text{A} = 1.6\underline{/-90°}\ \text{A}$ **7.** $0 - j6.0\ \text{A} = 6.0\underline{/-90°}\ \text{A}$

9. $0.510 + j0.462$ A $= 0.688\underline{/42.2°}$ A

11. $3.7 + j11$ mA

13. $390 - j134$ mA

15. $38 + j55\ \mu\text{A}$

17. $1.6 + j0.13$ A

19. $0.33 + j2.7$ A

21. $3.0 - j2.0$ A $= 3.6\underline{/-34°}$ A

23. $8 + j7$ A $= 11\underline{/41°}$ A

25. $3.7 + j1.2$ A $= 3.9\underline{/18°}$ A

27. (a) $I_R = 21.3\underline{/0°}$ mA

           $I_L = 32.7\underline{/-90°}$ mA

    (b) $I = 39\underline{/-57°}$ mA

29. (a) $I_L = 53\underline{/-90°}$ mA

       $I_C = 120\underline{/90°}$ mA

    (b) $I = 67\underline{/90°}$ mA

CHAPTER 27

EXERCISE 27–1

1. $40\underline{/20°}\ \Omega$

3. $56\underline{/0°}$

5. $17\underline{/-8°}\ \Omega$

7. resistive

9. resistive and inductive

11. resistive and capacitive

13. $10 + j0 \Rightarrow 10\underline{/0°}\ \Omega$

15. $5 - j2 \Rightarrow 5.4\underline{/-22°}\ \Omega$

17. $14\underline{/90°} \Rightarrow 0 + j14\ \Omega$

19. $R = 490\ \Omega,\ X_L = 140\ \Omega$

21. $R = 3.1\ \text{k}\Omega,\ X_C = 1.0\ \text{k}\Omega$

23. $R = 32.4\ \Omega,\ X_L = 76.4\ \Omega$

25. $139\underline{/-78.8°}\ \Omega$ and
$R = 27\ \Omega,\ C = 20\ \mu\text{F}$

27. $852\underline{/-54.1°}\ \Omega$ and
$R = 500\ \Omega,\ C = 0.58\ \mu\text{F}$

29. $3.26\underline{/-47.5°}\ \text{k}\Omega$ and
$R = 2.2\ \text{k}\Omega,\ C = 0.066\ \mu\text{F}$

EXERCISE 27–2

1. (a) $1880 - j1522 \Rightarrow 2.42\underline{/-39°}\ \text{k}\Omega$

    (b) $-39°$

    (c) 41.3 mA

    (d) $R_T = 1.88\ \text{k}\Omega$ and $C_T = 1.74\ \mu\text{F}$

3. (a) $51 - j291 \Rightarrow 295\underline{/-80°}\ \Omega$

    (b) $-80°$

    (c) 407 mA

    (d) $R_T = 51\ \Omega$ and $C_T = 9.12\ \mu\text{F}$

5. (a) $62 - j79 \Rightarrow 101\underline{/-52°}$

    (b) $-52°$

    (c) 356 mA

    (d) $R_T = 62\ \Omega$ and $C_T = 5.0\ \mu\text{F}$

7. (a) 21.99 kΩ

    (b) $34.8\underline{/39.2°}\ \text{k}\Omega$

    (c) $39.2°$

    (d) $287\ \mu\text{A}$

(e) 7.75 V$\underline{/0°}$

(f) 6.31 V$\underline{/90°}$

(g) 7.75 + j6.31 = 10$\underline{/39.2°}$ V

EXERCISE 27–3

1. 100$\underline{/-15°}$ mS   **3.** 1.0$\underline{/-82°}$ mS   **5.** 18$\underline{/19°}$ mS

7. 8.3$\underline{/90°}$ mS   **9.** 1.1$\underline{/5°}$ mS   **11.** 667$\underline{/90°}$ μS

13. 2.4$\underline{/-90°}$ mS   **15.** 14.7$\underline{/90°}$ μS   **17.** 1.96$\underline{/90°}$ μS

19. 1.2$\underline{/17°}$ mS   **21.** 8.08$\underline{/-61°}$ μS   **23.** 79.4$\underline{/15°}$ mS

EXERCISE 27–4

1. 335 pF

5. 108 Ω

9. (1) 969$\underline{/35.5°}$ μS

    (2) 2.38$\underline{/0°}$ mS

13. Y_T = 340$\underline{/-73.5°}$ μS

    Z_T = 2.94$\underline{/73.5°}$ kΩ

17. R_p = 214 Ω, L_p = 35.1 mH

21. **(a)** Y_1 = 4.76$\underline{/-81°}$ mS

      Y_2 = 1.33$\underline{/-51°}$ mS

  (b) 5.95$\underline{/-74.6°}$ mS

  (c) 168$\underline{/74.6°}$ Ω

  (d) 595$\underline{/-74.6°}$ mA

3. 39 μH

7. (1) 6.66$\underline{/-60°}$ mS

    (2) 6.25$\underline{/63.2°}$ mS

11. Y_T = 6.47$\underline{/60.5°}$ mS

    Z_T = 155$\underline{/-60.5°}$ Ω

15. Y_T = 2.55$\underline{/-17.9°}$ mS

    Z_T = 392$\underline{/17.9°}$ Ω

19. R_P = 342 Ω, C_P = 0.954 μF

CHAPTER 28

EXERCISE 28–1

1. 17.9 V   **3.** −0.576 mA   **5.** −10.3 A

7. 3.4 V   **9.** −259 V   **11.** 258°

13. 92.3°   **15.** 123°

EXERCISE 28–2

1. 10.1 V   **3.** 12.2 V   **5.** 127.3 V

7. 11.1 A   **9.** 5.73 mA   **11.** 1.87 A

13. 25.5 mA   **15.** 21.2 V   **17.** 62.2 V

19. 38 W   **21.** 29 W   **23.** 0.832 lead

25. 0.371 unknown   **27.** 557 mW   **29.** $R_T = 3.45$ kΩ and
$$L_T = 1.19 \text{ H at}$$
$$\omega = 5027 \text{ rad/s}$$

EXERCISE 28–3

1. $i = I_{mx} \sin (900t)$ A
$e = E_{mx} \sin (900t + 10°)$ V

3. $e = E_{mx} \sin (377t)$ V
$i = I_{mx} \sin (377t + 65°)$A

5. i lags e by 38°

7. e lags i by 40°

9. **(a)** $i = 42.4 \times 10^{-3} \sin (2513t)$ A
$e = 3.54 \sin (2513t^r + 48°)$ V

(b) $i = 5.31$ mA

(c) $P = 50.2$ mW

CHAPTER 29

EXERCISE 29–1

Calculator Drill

1. 1.14	**2.** 1.56	**3.** 1.15
4. −1.15	**5.** −1.50	**6.** 2.01
7. 1.18	**8.** 1.00	**9.** 1.01
10. 1.01	**11.** −2.37	**12.** −1.41
13. 1.01	**14.** −3.58	**15.** −1.10
16. −1.32	**17.** 4.51	**18.** 1.30
19. 0.268	**20.** −0.466	**21.** −3.02
22. −0.642	**23.** 0.458	**24.** 0.574

EXERCISE 29–3

1. Begin with the right member:

$$\cos (\theta)/\sin (\theta) = (x/z)/(y/z)$$
$$= x/y$$
$$= \cot (\theta)$$

which is the left member.

3. Begin with the right member:

$$\cot (\theta) \sec (\theta) = (x/y)(z/x)$$
$$= z/y$$
$$= \csc (\theta)$$

which is the left member.

5. Begin with the left member:

$$\tan(\theta) = \sin(\theta)\sec(\theta) \text{ (See problem 2)}$$
$$= \sec(\theta)/\csc(\theta)$$

which is the right member.

7. Begin with the left member:

$$\tan(\theta)\cot(\theta) = [\sin(\theta)/\cos(\theta)][\cos(\theta)/\sin(\theta)]$$
$$= 1$$

which is the right member.

9. Begin with the left member:

$$\tan(\theta) + \cot(\theta) = \frac{\sin(\theta)}{\cos(\theta)} + \frac{\cos(\theta)}{\sin(\theta)}$$
$$= \frac{\sin^2(\theta) + \cos^2(\theta)}{\cos(\theta)\sin(\theta)}$$
$$= \frac{1}{\cos(\theta)\sin(\theta)}$$
$$= \sec(\theta)\csc(\theta)$$

which is the right member.

11. Begin with the left member:

$$\sin(\theta)\cos(\theta)\sec(\theta)\csc(\theta) = \frac{\sin(\theta)\cos(\theta)}{\sin(\theta)\cos(\theta)}$$
$$= 1$$

which is the right member.

13. Begin with the right member:

$$\sec(\theta) = 1/\cos(\theta)$$
$$= \frac{\sin^2(\theta) + \cos^2(\theta)}{\cos(\theta)}$$
$$= \frac{\sin(\theta)}{\cos(\theta)}\sin(\theta) + \cos(\theta)$$
$$= \tan(\theta)\sin(\theta) + \cos(\theta)$$

which is the left member.

15. Begin with the right member:

$$\cot (\theta) \sec^2 (\theta) = \cot (\theta)(1 + \tan^2 (\theta)) \text{ (by problem 4)}$$
$$= \cot (\theta) + \cot (\theta) \tan^2 (\theta)$$
$$= \cot (\theta) + \tan (\theta) \text{ (by problem 7)}$$

which is the left member.

17. Begin with the left member:

$$\frac{1}{1 - \sin (\theta)} + \frac{1}{1 + \sin (\theta)} = \frac{1 + \sin (\theta)}{1 - \sin^2 (\theta)} + \frac{1 - \sin (\theta)}{1 - \sin^2 (\theta)}$$
$$= \frac{2}{1 - \sin^2 (\theta)}$$
$$= 2/\cos^2 (\theta)$$
$$= 2 \sec^2 (\theta)$$

which is the right member.

19. Begin with the right member:

$$\frac{\sin (\theta)}{1 + \cos (\theta)} = \frac{\sin (\theta)}{1 + \cos (\theta)} \times \frac{1 - \cos (\theta)}{1 - \cos (\theta)}$$
$$= \frac{\sin (\theta)(1 - \cos (\theta))}{1 - \cos^2 (\theta)}$$
$$= \frac{\sin (\theta)(1 - \cos (\theta))}{\sin^2 (\theta)}$$
$$= \frac{1 - \cos (\theta)}{\sin (\theta)}$$

which is the left member.

21. Begin with the left member:

$$\frac{1}{\sin \theta} - \sin \theta = \frac{1 - \sin^2 \theta}{\sin \theta}$$
$$= \frac{\cos \theta}{\sin \theta} \cos \theta$$
$$= \cot \theta \cos \theta$$

which is the right member.

23. Begin with the left member:

$$\frac{\sin \theta}{1 + \cos \theta} + \frac{1 + \cos \theta}{\sin \theta} = \frac{\sin^2 \theta}{(1 + \cos \theta) \sin \theta} + \frac{(1 + \cos \theta)^2}{(1 + \cos \theta) \sin \theta}$$

$$= \frac{\sin^2 \theta + 1 + 2 \cos \theta + \cos^2 \theta}{(1 + \cos \theta) \sin \theta}$$

$$= 2 \frac{(1 + \cos \theta)}{(1 + \cos \theta) \sin \theta}$$

$$= 2 \csc \theta$$

which is the right member.

EXERCISE 29–4

Calculator Drill

1. (a) -1.175 **2. (a)** 1.543 **3. (a)** -0.762
    (b) -0.521 **(b)** 1.128 **(b)** -0.462
    (c) 0.000 **(c)** 1.000 **(c)** 0.000
    (d) 0.637 **(d)** 1.185 **(d)** 0.537
    (e) 1.509 **(e)** 1.811 **(e)** 0.834
    (f) 3.627 **(f)** 3.762 **(f)** 0.964

EXERCISE 29–5

1. *See* Figure A25.   **3.** *See* Figure A26.

FIGURE A25 Solution to Exercise 29–5, problem 1.

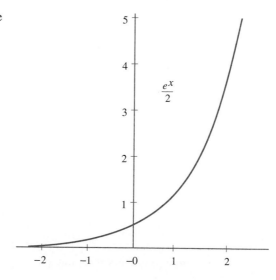

FIGURE A26 Solution to Exercise 29–5, problem 3.

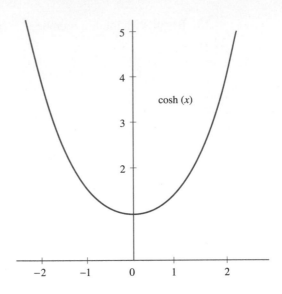

cosh (x)

EXERCISE 29–6

1. Left member:

$$\sinh(x) + \cosh(x) = \frac{e^x - e^x}{2} + \frac{e^x + e^{-x}}{2}$$
$$= e^x$$

which is the right member.

3. Begin with the right member:

$$\sinh(x)/\cosh(x) = \left(\frac{e^x - e^{-x}}{2}\right)\bigg/\left(\frac{e^x + e^{-x}}{2}\right)$$
$$= (e^x - e^{-x})/(e^x + e^{-x})$$
$$= \tanh(x)$$

which is the left member.

5. Begin with the right member:

$$-\sinh(-x) = -(e^{-x} - e^x)/2$$
$$= (e^x - e^{-x})/2$$
$$= \sinh(x)$$

which is the left member.

7. Begin with the left member:

$$\cosh (2x) = (e^{2x} + e^{-2x})/2$$

Right member

$$\cosh^2 (x) + \sinh^2 (x) = \left(\frac{e^x + e^{-x}}{2}\right)^2 + \left(\frac{e^x - e^{-x}}{2}\right)^2$$

$$= \frac{e^{2x} + 2 + e^{-2x}}{4} + \frac{e^{2x} - 2 + e^{-2x}}{4}$$

$$= (e^{2x} + e^{-2x})/2$$

9. Left member:

$$\tanh^2 (x) + 1/\cosh^2 (x) = \sinh^2 (x)/\cosh^2 (x) + 1/\cosh^2 (x)$$

$$= (\sinh^2 (x) + 1)/\cosh^2 (x))$$

$$= \left[\left(\frac{e^x - e^{-x}}{2}\right)^2 + 1\right] / \left(\frac{e^x + e^{-x}}{2}\right)^2$$

$$= \left(\frac{e^{2x} - 2 + e^{-2x}}{4} + \frac{4}{4}\right) / \left(\frac{e^{2x} + 2 + e^{-2x}}{4}\right)$$

$$= \left(\frac{e^{2x} + 2 + e^{-2x}}{4}\right) / \left(\frac{e^{2x} + 2 + e^{-2x}}{4}\right)$$

$$= 1$$

which is the right member.

EXERCISE 29–7

1. 0.75   **3.** −2.50   **5.** 1.00

7. 1.40   **9.** 0.55   **11.** −2.00

CHAPTER 30

EXERCISE 30–1

1. $-\infty < \theta < \infty$ **3.** $-\infty < X < \infty$ **5.** $-\infty < X < \infty$

$-1 \leqslant Y \leqslant 1$ $-\infty < Y < \infty$ $-1 < Y < 1$

7. $-\infty < U < \infty$ **9.** $-\infty < s < \infty$

$0 < Z < \infty$ $-\infty < t < \infty$

1. continuous   **3.** essential discontinuities

5. continuous   **7.** essential discontinuities

9. Sawtooth curve with a step discontinuity at x equal an integer

EXERCISE 30–3

1. does not exist   **3.** 0   **5.** ∞

7. ∞ **9.** 2

EXERCISE 30–4

1. $y = 0$ for $1 \ll x$ **3.** $s = 0$ for $1 \ll |t|$

5. $u = |v|$ for $1 \ll V$ **7.** $R = 2$ for $2 \ll |r|$

CHAPTER 31

EXERCISE 31–1

1. (a) 100   **3. (a)** 5000   **5. (a)** 7   **7. (a)** 6

    (b) 1000   **(b)** 0.5   **(b)** 700   **(b)** 5

    (c) 0.01   **(c)** 50   **(c)** 0.7   **(c)** 6

    (d) 0.1   **(d)** 5   **(d)** 70   **(d)** 2

    (e) 0.01   **(e)** 0.0005   **(e)** 70   **(e)** 5

    (f) 100   **(f)** 50   **(f)** 7   **(f)** 2

EXERCISE 31–2

1. 4, 1, 0.25   **3.** 1536, 40   **5.** 256, 0.0625

7. 0.0625, 3584   **9.** 3072, 144

EXERCISE 31–3

1. 6   **3.** 48   **5.** 0.75

7. 2.75   **9.** 3.3125   **11.** 72

13. 184   **15.** 0.140 625   **17.** 62.25

19. 5.4375   **21.** 272   **23.** 624

25. 0.066 406 25   **27.** 118.125   **29.** 240.796 875

EXERCISE 31–4

1. (a) 10011_2, 23_8, 13_{16} **3. (a)** 0.1_2, 0.4_8, 0.8_{16}

    (b) 10110_2, 26_8, 16_{16} **(b)** 0.01_2, 0.2_8, 0.4_{16}

(c) 1111_2, 17_8, F_{16} (c) 0.11_2, 0.6_8, $0.C_{16}$

(d) $100\ 100_2$, 44_8, 24_{16} (d) 0.001_2, 0.1_8, 0.2_{16}

(e) $110\ 000_2$, 60_8, 30_{16} (e) 0.0001_2, 0.04_8, 0.1_{16}

(f) $100\ 111_2$, 47_8, 27_{16} (f) 0.00101_2, 0.12_8, 0.28_{16}

(g) $1\ 000\ 000_2$, 100_8, 40_{16} (g) $0.001\ 010\ 0_2$, 0.11770_8, $0.27EFA_{16}$

(h) $1\ 100\ 100_2$, 144_8, 64_{16} (h) 1.00000_2, 0.77534_8, $0.FEB85_{16}$

5. 30.3_8

EXERCISE 31–5

1. $1\ 010\ 100_2$, 54_{16} **3.** $10\ 000\ 000\ 001_2$, 401_{16}

5. $1\ 0010\ 0100_2$, 444_8 **7.** $1010\ 1011_2$, 253_8

9. 13.4_8, B.8   **11.** 525_8, 155_{16}

EXERCISE 31–6

All numbers are binary.

1. 1010   **3.** 11 010   **5.** 11 110

7. 1110   **9.** 10   **11.** 1000

13. 111   **15.** 101

EXERCISE 31–7

All numbers are octal.

1. 10   **3.** 7   **5.** 5

7. 11   **9.** 14   **11.** 51

13. 7775   **15.** 60 416   **17.** 3

19. 2   **21.** 25   **23.** 47

25. 6341   **27.** 62 457   **29.** 7777

EXERCISE 31–8

All numbers are hexadecimal.

1. E   **3.** D   **5.** 8

7. 69   **9.** 15   **11.** CE

13. 1326   **15.** 3000   **17.** 9

19. 3   **21.** 3   **23.** 3

25. 7   **27.** 56   **29.** 76D

EXERCISE 31–9

1. $1000\ 0000_2$ **3.** $1001\ 1001_2$ **5.** $1000\ 1110_2$

7. $1357\ 6420_8$ **9.** $7666\ 6666_8$ **11.** $5326\ 0777_8$

13. $8253\ 7998_{10}$	**15.** $0746\ 1320_{10}$	**17.** $0000\ 0000_{10}$
19. FEEE EEEE_{16}	**21.** E054 67DB_{16}	**23.** CFA9 8321_{16}

EXERCISE 31–10

1. $1000\ 0001_2$	**3.** $1010\ 1001_2$	**5.** $1001\ 0100_2$
7. $5126\ 7001_8$	**9.** $6034\ 6667_8$	**11.** $7666\ 6667_8$
13. $7438\ 9223_{10}$	**15.** $9000\ 0000_{10}$	**17.** $6509\ 8373_{10}$
19. 8E51 9BD1_{16}	**21.** C274 36FB_{16}	**23.** DA9E F889_{16}

EXERCISE 31–11

All numbers are binary.

1. $0111\ 1001$	**3.** $0001\ 0101$	**5.** $1111\ 0110$
7. $0010\ 0010$	**9.** $0110\ 1010$	**11.** $0111\ 1001$
13. $0001\ 0100$	**15.** $1111\ 0111$	**17.** $0010\ 0010$
19. $0110\ 1011$		

EXERCISE 31–12

All numbers are decimal.

1. 100	**3.** -23	**5.** 41
7. 30	**9.** 12	**11.** -53
13. -5	**15.** 60	**17.** 163
19. 39	**21.** 49	**23.** 6998

CHAPTER 32

EXERCISE 32–1

1. NOT B   **3.** A AND B   **5.** C OR A AND B

7. A AND NOT B AND C   **9.** NOT A OR NOT B

11. A OR B AND C OR A AND NOT B

EXERCISE 32–2

1. 0   **3.** 0   **5.** $\overline{A}$

7. 0   **9.** B   **11.** 1

EXERCISE 32–3

1. 1   **3.** 0   **5.** 0   **7.** 0

9. *See* Figure A27.   **11.** *See* Figure A28.

13. *See* Figure A29.

15.

	A	B	AB
0	0	0	0
1	0	1	0
2	1	0	0
3	1	1	1

FIGURE A27   Solution to Exercise 32–3, problem 9.

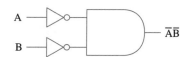

FIGURE A28   Solution to Exercise 32–3, problem 11.

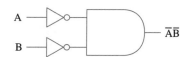

FIGURE A29   Solution to Exercise 32–3, problem 13.

17. (a) 8   **(b)**

	A	B	C	$\overline{A}$	$\overline{A}BC$
0	0	0	0	1	0
1	0	0	1	1	0
2	0	1	0	1	0
3	0	1	1	1	1
4	1	0	0	0	0
5	1	0	1	0	0
6	1	1	0	0	0
7	1	1	1	0	0

(c) *See* Figure A30.

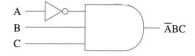

FIGURE A30   Solution to Exercise 32–3, problem 17c.

EXERCISE 32–4

1. 1   **3.** 1   **5.** 1   **7.** 1

9. *See* Figure A31.   **11.** *See* Figure A32.   **13.** *See* Figure A33.

15.

	A	B	A+B
0	0	0	0
1	0	1	1
2	1	0	1
3	1	1	1

FIGURE A31 Solution to Exercise 32–4, problem 9.

FIGURE A32 Solution to Exercise 32–4, problem 11.

FIGURE A33 Solution to Exercise 32–4, problem 13.

17. (a) 8

(b)

	A	B	C	$\overline{A}$	$\overline{A} + B + C$
0	0	0	0	1	1
1	0	0	1	1	1
2	0	1	0	1	1
3	0	1	1	1	1
4	1	0	0	0	0
5	1	0	1	0	1
6	1	1	0	0	1
7	1	1	1	0	1

(c) *See* Figure A34.

FIGURE A34 Solution to Exercise 32–4, problem 17c.

1. *See* Figure A35.   3. *See* Figure A36.   5. *See* Figure A37.

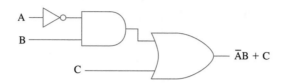

FIGURE A35   Solution to Exercise 32–5, problem 1.

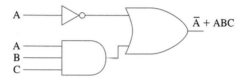

FIGURE A36   Solution to Exercise 32–5, problem 3.

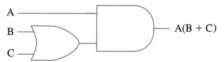

FIGURE A37   Solution to Exercise 32–5, problem 5.

7.

	A	B	C	$\overline{C}$	$A\overline{C}$	$A\overline{C} + B$
0	0	0	0	1	0	0
1	0	0	1	0	0	0
2	0	1	0	1	0	1
3	0	1	1	0	0	1
4	1	0	0	1	1	1
5	1	0	1	0	0	0
6	1	1	0	1	1	1
7	1	1	1	0	0	1

9.

	A	B	C	$\overline{C}$	$B\overline{C}$	$C + B\overline{C}$	$A(C + B\overline{C})$
0	0	0	0	1	0	0	0
1	0	0	1	0	0	1	0
2	0	1	0	1	1	1	0
3	0	1	1	0	0	1	0
4	1	0	0	1	0	0	0
5	1	0	1	0	0	1	1
6	1	1	0	1	1	1	1
7	1	1	1	0	0	1	1

11. True **13.** True **15.** False

17.

	A	B	C	$A + B$	$\overline{C}$	$(A + B)\overline{C}$
0	0	0	0	0	1	0
1	0	0	1	0	0	0
2	0	1	0	1	1	1
3	0	1	1	1	0	0
4	1	0	0	1	1	1
5	1	0	1	1	0	0
6	1	1	0	1	1	1
7	1	1	1	1	0	0

EXERCISE 32–6

1. *See* Figure A38. **3.** *See* Figure A39. **5.** *See* Figure A40.

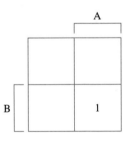

FIGURE A38 Solution to Exercise 32–6, problem 1.

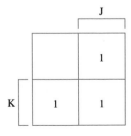

FIGURE A39 Solution to Exercise 32–6, problem 3.

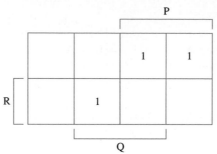

FIGURE A40 Solution to Exercise 32–6, problem 5.

7. $\overline{T}$ **9.** H **11.** U

13. *See* Figure A41. **15.** *See* Figure A42.

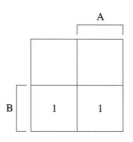

FIGURE A41 Solution to Exercise 32–6, problem 13.

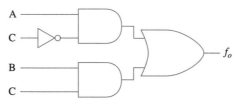

FIGURE A42 Solution to Exercise 32–6, problem 15.

EXERCISE 32–7

1. $\overline{A} + \overline{B}$ **3.** $A\overline{B}$ **5.** $\overline{U}V\overline{W} + X + \overline{Y}$

7. $\overline{A}B + C\overline{B} + A\overline{D}$ **9.** $\overline{X}YZ$ **11.** $\overline{J}KL$

13. *See* Figure A43.

B ——— f_o **FIGURE A43** Solution to Exercise 32–7, problem 13.

EXERCISE 32–8

1. $A = 0$ or $A = 1$
   If $A = 0$, then $A \cdot 1 = 0 \cdot 1 = 0 = A$
   If $A = 1$, then $A \cdot 1 = 1 \cdot 1 = 1 = A$
   $\therefore \quad A \cdot 1 = A$

3. $A = 0$ or $A = 1$
   If $A = 0$, then $AA = 0 \cdot 0 = 0 = A$
   If $A = 1$, then $AA = 1 \cdot 1 = 1 = A$
   $\therefore \quad AA = A$

5. $A = 0$ or $A = 1$
   If $A = 0$, then $A\overline{A} = 0 \cdot 1 = 0$
   If $A = 1$, then $A\overline{A} = 1 \cdot 0 = 0$
   $\therefore \quad A\overline{A} = 0$

7. $B = 0$ or $B = 1$
   For $B = 0$:
   $(A + B)A = (A + 0)A = AA = A$
   For $B = 1$:
   $(A + B)A = (A + 1)A = 1 \cdot A = A$
   $\therefore \quad (A + B)A = A$

9. By problems 5, $B\overline{B} = 0$; thus:
   $A + (B\overline{B}) = A + 0 = A$

11. $\overline{X}\,\overline{Y} + (\overline{X}Y + \overline{X}Y) = XY$
    $\overline{X}(Y + \overline{Y}) + Y(X + \overline{X})$
    $\overline{X} + Y$

13. DE

EXERCISE 32–9

1. *See* Figure A44, where:
$P = 1$	Main power ON
$L = 1$	Door latched
$T = 1$	Timer $\neq 0$
$f_o = 1$	Oven ON

FIGURE A44 Solution to Exercise 32–9, problem 1.

3. *See* Figure A45, where:
$L = 1$	Lid UP
$B = 0$	Light beam OFF
$f_o = 1$	Lighter lit

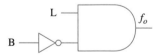

FIGURE A45 Solution to Exercise 32–9, problem 3.

5. *See* Figure A46, where:

C = 0 Ticket taken
D = 0 No car near gate
f_o = 1 Gate UP

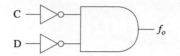

FIGURE A46 Solution to Exercise 32–9, problem 5.

Solutions to Section Challenges

SOLUTION TO SECTION CHALLENGE FOR CHAPTERS 1, 2, AND 3

A. 21 N   B. 0.12 kA

C. 16.4 lbf · ft or 22.2 N · m   D. 35.0 μA

SOLUTION TO SECTION CHALLENGE FOR CHAPTERS 4, 5, 6, AND 7

A. $E = V_{60RE} + V_{BE} + V_{RE}$

Substitute $I_B (60R_E)$ for V_{60RE} and $I_E R_E$ for V_{RE}:

$E = I_B (60R_E) + V_{BE} + I_E R_E$

Subtract V_{BE} from both members:

$E - V_{BE} = I_B (60R_E) + I_E R_E$

Factor R_E from the right member:

$E - V_{BE} = R_E (60I_B + I_E)$

Interchange members and divide each member by $(60I_B + I_E)$:

$$R_E = \frac{E - V_{BE}}{60I_B + I_E}$$

B. $R_E = \dfrac{E - V_{BE}}{60I_B + I_E}$

$$R_E = \frac{3.00 - 0.640}{60 \times 80.0 \times 10^{-6} + 7.00 \times 10^{-3}}$$

$R_E = 200 \ \Omega$

$60R_E = 60 \times 200 = 12.0 \text{ k}\Omega$

SOLUTION TO SECTION CHALLENGE FOR CHAPTERS 8, 9, AND 10

$$R_T = \frac{1}{\dfrac{1}{R_1} + \dfrac{1}{R_2}}$$

Add the fractions in the denominator of the right member:

$$R_T = \frac{1}{\dfrac{R_2 + R_1}{R_1 R_2}}$$

Invert and multiply:

$$R_T = 1 \times \frac{R_1 R_2}{R_1 + R_2}$$

$$R_T = \frac{R_1 R_2}{R_1 + R_2}$$

A. 333 Ω   B. 2.58 kΩ   C. 96.0 kΩ   D. 19.9 Ω

E. 67.8 kΩ   F. 13.0 kΩ   G. 79.1 Ω   H. 29.7 kΩ

I. 427 kΩ   J. 4.76 kΩ

SOLUTION TO SECTION CHALLENGE FOR CHAPTERS 11 AND 12

$$R_1 = \frac{R_A (R_A R_2/R_3)}{R_A + (R_A R_2/R_1) + (R_A R_2/R_3)}$$

Factor out R_A and simplify:

$$R_1 = \frac{R_A (R_A R_2/R_3)}{R_A (1 + R_2/R_1 + R_2/R_3)} = \frac{R_A R_2/R_3}{1 + R_2/R_1 + R_2/R_3}$$

Form a common denominator and then invert and multiply:

$$R_1 = \frac{R_A R_2/R_3}{\dfrac{R_1 R_3 + R_2 R_3 + R_2 R_1}{R_1 R_3}} = R_A \frac{R_2}{R_3} \times \frac{R_1 R_3}{R_1 R_3 + R_2 R_3 + R_2 R_1}$$

Simplify and solve for R_A:

$$R_A = \frac{R_1 R_2 + R_1 R_3 + R_2 R_3}{R_2}$$

$$R_B = \frac{R_1 R_2 + R_1 R_3 + R_2 R_3}{R_1}$$

$$R_C = \frac{R_1 R_2 + R_1 R_3 + R_2 R_3}{R_3}$$

A. $R_A = 5.50$ Ω $R_B = 11.0$ Ω $R_C = 3.67$ Ω

B. $R_A = 43.3$ Ω $R_B = 65.0$ Ω $R_C = 32.5$ Ω

C. $R_A = 20.0$ Ω $R_B = 20.0$ Ω $R_C = 20.0$ Ω

With $R_A' = R_B' = R_C' = 12.0$ Ω:

D. $R_{ac} = (R_A \| R_A') \| [(R_B \| R_B') + (R_C \| R_C')] = 3.771 \| 8.562 = 2.62$ Ω

E. $R_{ac} = 6.28$ Ω   F. $R_{ac} = 5.00$ Ω

SOLUTION TO SECTION CHALLENGE FOR CHAPTERS 13, 14, AND 15

Gain transfer function:

$$K = \frac{\Delta c}{\Delta r}$$

Select two points (250 gal/min, 15.0 lbf/in^2) and (0 gal/min, 3.0 lbf/in^2):

$\Delta c = 250 - 0 = 250$ gal/min
$\Delta r = 15.0 - 3.0 = 12.0$ lbf/in^2

Substitute and solve for the gain:

$$K = 250/12.0 = 20.8 \ \frac{\text{gal/min}}{\text{lbf/in}^2}$$

The equation for output flow rate, c, in terms of the input pneumatic pressure, r, is:

$$c = Kr + b$$

For the pneumatically operated control valve where $K = 20.8 \ \dfrac{\text{gal/min}}{\text{lbf/in}^2}$, and $b = -62.45$ gal/min:

$$c = 20.8r - 62.45 \ (\text{gal/min})$$

A. $c = 94$ gal/min when $r = 7.5$ lbf/in^2
B. $c = 42$ gal/min when $r = 5.0$ lbf/in^2
C. $c = 208$ gal/min when $r = 13.0$ lbf/in^2

SOLUTION TO SECTION CHALLENGE FOR CHAPTERS 16 AND 17

A. $C_1 = 8.11$ pF $C_2 = 0.94$ pF $C_3 = 2.89$ pF
B. Start your solution by assuming a clockwise direction for the mesh currents, I_1 and I_2, as shown in Figure C–8:
   1. $39.0 \times 10^3 I_1 + 5.60 \times 10^3 I_1 - 5.60 \times 10^3 I_2 = 12.0$
   2. $5.60 \times 10^3 I_2 - 5.60 \times 10^3 I_1 + 910 I_2 = -0.650$

Combine like terms:
   1. $44.6 \times 10^3 I_1 - 5.60 \times 10^3 I_2 = 12.0$
   2. $-5.60 \times 10^3 I_1 + 6.51 \times 10^3 I_2 = -0.650$

Solve for the mesh currents:

$$I_1 = 287.58 \ \mu A \qquad I_2 = 147.54 \ \mu A$$

Determine the device currents:

$$I_{R1} = I_1 = 288 \ \mu A \qquad I_{R2} = I_1 - I_2 = 140 \ \mu A \qquad I_{RE} = I_2 = 148 \ \mu A$$

C. $L = 231.0 \ \mu H \qquad C = 200.0 \ pF$

D. State the conditions of the circuit as a set of equations:

$$R_{sm} + R_{md} + R_{lg} = 68.0$$
$$2R_{sm} = (R_{md} + R_{lg}) - 29.0$$
$$3R_{md} = 5/4 \ (R_{sm} + R_{lg})$$

Write the equations in the general form:

$$
\begin{array}{rrrrrrr}
R_{sm} & + & R_{md} & + & R_{lg} & = & 68.0 \\
2R_{sm} & - & R_{md} & - & R_{lg} & = & -29.0 \\
-5R_{sm} & + & 12R_{md} & - & 5R_{lg} & = & 0
\end{array}
$$

Solve using determinants:

$$R_{sm} = 663/51 = 13.0 \ \Omega$$
$$R_{md} = 1020/51 = 20.0 \ \Omega$$
$$R_{lg} = 1785/51 = 35.0 \ \Omega$$

SOLUTION TO SECTION CHALLENGE FOR CHAPTERS 18 AND 19

The current in the load, R_L, is:

$$I = \frac{E_{Th}}{R_{Th} + R_L}$$

The power in the load, P_L, is:

$$P_L = I^2 R_L \text{ and } I^2 = \left(\frac{E_{Th}}{R_{Th} + R_L}\right)^2$$

$$P_L = \left(\frac{E_{Th}}{R_{Th} + R_L}\right)^2 R_L$$

Express R_L in the form of a complete quadratic, $aR_L^2 + bR_L + c = 0$:

$$\frac{P_L}{R_L} = \frac{E_{Th}^2}{(R_{Th} + R_L)^2} = \frac{E_{Th}^2}{R_{Th}^2 + 2R_{Th}R_L + R_L^2}$$

Invert each member:

$$\frac{R_L}{P_L} = \frac{R_{Th}^2 + 2R_{Th}R_L + R_L^2}{E_{Th}^2}$$

$$\frac{R_{Th}^2 + 2R_{Th}R_L + R_L^2}{E_{Th}^2} - \frac{R_L}{P_L} = 0$$

Place each term over the common denominator, $P_L E_{Th}^2$:

$$\frac{P_L R_{Th}^2}{P_L E_{Th}^2} + \frac{2P_L R_{Th}R_L}{P_L E_{Th}^2} + \frac{P_L R_L^2}{P_L E_{Th}^2} - \frac{R_L E_{Th}^2}{P_L E_{Th}^2} = 0$$

Collect like terms and simplify:

$$\frac{R_L^2}{E_{Th}^2} + \left(\frac{2P_L R_{Th}R_L}{P_L E_{Th}^2} - \frac{R_L E_{Th}^2}{P_L E_{Th}^2}\right) + \frac{R_{Th}^2}{E_{Th}^2} = 0$$

$$\frac{R_L^2}{E_{Th}^2} + R_L \frac{(2P_L R_{Th} - E_{Th}^2)}{P_L E_{Th}^2} + \frac{R_{Th}^2}{E_{Th}^2} = 0$$

Apply the quadratic formula and solve for R_L when $E_{Th} = 10.00$ V, $R_{Th} = 5.00$ Ω, and $P_L = 4.734$ W:

$$\frac{R_L^2}{100.0} + \left(\frac{2(4.734)(5.00) - 100.0}{4.734(100.0)}\right)R_L + \frac{25.0}{100.0} = 0$$

$$0.0100R_L^2 - 0.1112R_L + 0.250 = 0$$

$$R_L = \frac{0.1112 \pm \sqrt{0.1112^2 - 4(0.0100)(0.250)}}{2(0.0100)}$$

A. $R_L = 7.99$ Ω and $R_L = 3.13$ Ω
B. $R_L = 5.00$ Ω

Each answer checks and each is a point on the power curve of Figure S–1.

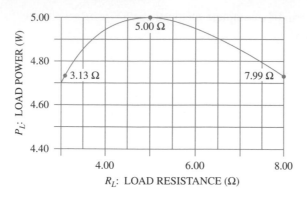

FIGURE S–1 Power curve for R_L of Figure C–9, Section Challenge for Chapters 18 and 19.

SOLUTION TO SECTION CHALLENGE FOR CHAPTERS 20 AND 21

A. $e^{-t/\tau} = 1 - \dfrac{\Delta T}{P/\delta}$

Take the natural log of both members and solve for τ:

$$\tau = \frac{-t}{\ln\left(1 - \dfrac{\Delta T}{P/\delta}\right)} = \frac{-15.0}{\ln\left(1 - \dfrac{12.0}{182/8.00}\right)} = 20.0 \text{ s}$$

B. $v_r = v_{or}/10^{Ndb/20} = (2.50 \times 10^{-3})/10^{-50.0/20} = 790.6 \text{ mV}$

$r = v_r/V_{dc} = (790.6 \times 10^{-3})/14.5 = 0.0545$

$C = 2400/(rR) = 2400/(0.0545 \times 19.33) = 2280 \ \mu\text{F}$

SOLUTION TO SECTION CHALLENGE FOR CHAPTERS 22, 23, 24, AND 25

A. $L = 38.0 + \text{side } AC + 38.0$

$L = 38.0 + 12.0/\tan(22.5°) + 38.0$

$L = 38.0 + 29.0 + 38.0$

$L = 105.0 \text{ cm}$

B. $OB = r/\sin(B/2) = 1.50/\sin(60.0/2) = 3.00 \text{ cm}$

$DB = r/\tan(B/2) = 1.50/\tan(60.0/2) = 2.60 \text{ cm}$

$AB = 8.56 + DB = 8.56 + 2.60 = 11.16 \text{ cm}$

$BC^2 = a^2 = b^2 + c^2 - 2bc \cos(A)$

$BC^2 = 10.00^2 + 11.16^2 - 2(10.00)(11.16)\cos(45.0)$

$BC = \sqrt{66.72} = 8.17 \text{ cm}$

C.

Vector		x-component	y-component
OA	$12.3\underline{/28.5°}$	10.81	5.87
OB	$6.84\underline{/90°}$	0	6.84
OC	$5.85\underline{/147.4°}$	−4.93	3.15
OD	$14.7\underline{/-135.3°}$	−10.45	−10.34
OE	$8.34\underline{/-64.2°}$	3.63	−7.51
		$R_x = -0.94$	$R_y = -1.99$

$\rho\underline{/\theta} = R_x, R_y = -0.94, -1.99 = 2.20\underline{/-115.3°}$

SOLUTIONS TO SECTION CHALLENGE FOR CHAPTERS 26, 27, AND 28

A1. $I_R = E/R = 48.0\underline{/0°}/(1.2 \times 10^3) = 40.0\underline{/0°}$ mA
$I_C = E/X_C = E\omega C = (48.0\underline{/0°})(2\pi400)(0.82 \times 10^{-6}\underline{/90°}) = 98.9\underline{/90°}$ mA
$I_L = E/X_L = E/(\omega L) = 48.0\underline{/0°}/(2\pi400 \times 0.750\underline{/90°}) = 25.5\underline{/-90°}$ mA

A2. $I = I_R + I_C + I_L$
$\begin{aligned} I_R &= 40.0\underline{/0°} \text{ mA} &&= 40.0 + j0 \text{ mA} \\ I_C &= 98.9\underline{/90°} \text{ mA} &&= 0 + j98.9 \text{ mA} \\ I_L &= 25.0\underline{/-90°} \text{ mA} &&= 0 - j25.5 \text{ mA} \\ &&I = 40.0 + 73.4 \text{ mA} &= 83.6\underline{/61.4°} \text{ mA} \end{aligned}$

A3.

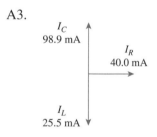

FIGURE C–15(b)   **FIGURE C–15(c)**

A4. $e = \sqrt{2}E\sin(\omega t) = 67.9 \sin(2513t)$ V
$i = \sqrt{2}I\sin(\omega t + \theta) = 118 \sin(2513t + 61.4°)$ mA

B1. $Z_1 = X_C + X_L = -j500 + j180 = -j320 = 320\underline{/-90°}$
$Z_2 = R + X_C = 300 - j400 = 500\underline{/-53.1°}$
$Z_3 = R + X_L = 200 + j320 = 377.4\underline{/58.0°}$

B2. $Y_1 = 1/Z_1 = 1/320\underline{/-90°} = 3.13\underline{/90°}$ mS

$Y_2 = 1/Z_2 = 1/500\underline{/-53.1°} = 2.00\underline{/53.1°}$ mS

$Y_3 = 1/Z_3 = 1/377.4\underline{/58.0°} = 2.65\underline{/-58.0°}$ mS

B3. $I_1 = E/Z_1 = EY_1 = (15.0\underline{/0°})(3.13 \times 10^{-3}\underline{/90°}) = 47.0\underline{/90°}$ mA

$I_2 = E/Z_2 = EY_2 = (15.0\underline{/0°})(2.00 \times 10^{-3}\underline{/53.1°}) = 30.0\underline{/53.1°}$ mA

$I_3 = E/Z_3 = EY_3 = (15.0\underline{/0°})(2.65 \times 10^{-3}\underline{/-58.0°}) = 39.8\underline{/-58.0°}$ mA

B4. $Z_T = 1/Y_T$ and $Y_T = Y_1 + Y_2 + Y_3$

$Y_1 = 3.13\underline{/90°}$ mS $\quad = \quad 0 + j3.13$ mS

$Y_2 = 2.00\underline{/53.1°}$ mS $\quad = 1.20 + j1.60$ mS

$Y_3 = 2.65\underline{/-58.0°}$ mS $\quad = 1.40 - j2.25$ mS

$\overline{\qquad\qquad\qquad Y_T = 2.60 + j2.48}$ mS $= 3.59\underline{/43.6°}$ mS

$Z_T = 1/Y_T = 1/3.59\underline{/43.6°} = 279\underline{/-43.6°}$ Ω

B5. $I = E/Z_T = EY_T = (15.0\underline{/0°})(3.59 \times 10^{-3}\underline{/43.6°}) = 53.9\underline{/43.6°}$ mA

B6. $P = EI\cos(\theta) = (15.0)(53.9 \times 10^{-3})\cos(43.6°) = 586$ mW

SOLUTION TO SECTION CHALLENGE FOR CHAPTERS 29 AND 30

A. Replace $\sin(\theta) \csc(\theta)$ with one, 1, $(\sin(\theta) \csc(\theta) = 1)$:

$$\eta = \frac{\sec(\theta)}{\mu\csc(\theta) + \sec(\theta)}$$

In the denominator, replace $\csc(\theta)$ with its identity $1/\sin(\theta)$ and replace $\sec(\theta)$ with its identity $1/\cos(\theta)$:

$$\eta = \frac{\sec(\theta)}{\dfrac{\mu}{\sin(\theta)} + \dfrac{1}{\cos(\theta)}}$$

Place the terms in the denominator over a common denominator, replace $\sec(\theta)$ with its identity $1/\cos(\theta)$, and simplify:

$$\eta = \frac{\sec(\theta)}{\dfrac{\mu\cos(\theta) + \sin(\theta)}{\sin(\theta)\cos(\theta)}} = \frac{1}{\cos(\theta)} \times \frac{\sin(\theta)\cos(\theta)}{\mu\cos(\theta) + \sin(\theta)} = \frac{\sin(\theta)}{\mu\cos(\theta) + \sin(\theta)}$$

$$\eta = \frac{\sin(\theta)}{\mu\cos(\theta) + \sin(\theta)} = \frac{\sin(11°)}{0.58\cos(11.0°) + \sin(11.0°)} = 0.25$$

B. Replace $2 \sin(\theta) \cos(\theta)$ with $\sin(2\theta)$ and solve for θ:

$$R = \frac{v_0^2 \sin(2\theta)}{g} \quad \text{and} \quad \sin(2\theta) = \frac{Rg}{v_0^2}$$

$$\theta = \tfrac{1}{2} \sin^{-1}\left(\frac{Rg}{v_0^2}\right) = \tfrac{1}{2} \sin^{-1}\left(\frac{1650(9.81)}{320^2}\right) = 4.548°$$

$$\theta = 4.55° \quad \text{or} \quad \theta = 85.5°$$

SOLUTION TO SECTION CHALLENGE FOR CHAPTERS 31 AND 32

A. 1.102.45.177   B. 10000101.10011100.00110111.01100110

C. 123.12.45.77   D. 00001101.00011100.11111110.11001010

E. 21.52.177.188   F. 10111111.00001111.10011011.00000010

G. 203.10.233.15   H. 00010111.10100110.01100101.01011100

I. 121.2.199.88   J. 11000110.11010111.01000011.11101001

INDEX

Abbreviations, 748

Abscissa, 308, 310, 315, 336, 346, 540

Absolute error, 265, 268–70

Absolute value, 12, 13, 24, 44

Accuracy, 36, 60, 259, 263, 266

Acute angles, 498, 502

Addition
associative law of, 98, 101
commutative law of, 98, 101
equation transformation, 121–22
of like terms, 96–97
of phasors (vectors), 553–54
of polynomials, 98–99, 110–11
of signed numbers, 13–16
order of operations, 7–8
symbol (operator) for, 5–6

Admittance, 612–21
computing, 614
computing branch, 617–18
computing circuit, 618–19
concepts of, 612–15
definition of, 612, 623
diagrams of, 612–15
equations for, 612, 614, 618
polar form of, 613, 614
rectangular form of, 613, 614, 620

Algebraic equations (See Equations)

Algebraic expressions, 93–94, 96
evaluation of, 116–17, 118

Algebraic symbols, 5–6, 10–11, 21, 90, 92

Alternating-current circuits
parallel (See Parallel circuits, ac)
series (See Series circuits, ac)

Alternating currents
current wave, 573, 625
determining resistive load of, 574
effective values of, 630–31
instantaneous values of, 625–28, 636–39
maximum value of, 625–28, 631
notation of, 631
phasor diagrams of, 573–74, 576–77, 580–81, 590–93
sinusoidal properties of, 570–72, 625–26
terminology of, 570–72
voltage wave, 573, 625, 626

Alternating currents, effects of
on capacitance, 583
on inductance, 583
on resistance, 583

Alternating currents, voltage of
applied to capacitive load, 579–81
applied to inductive load, 577
applied to resistive load, 573

American wire gauge, 167, 283–84, 289, 749

Ammeter, 145

Amperes, 63, 65, 144

Amplifier, 468, 473, 489

Analog instruments, 265, 289

AND, 708–10
gate, 709–11
operator, 706, 708

Angles, 493–95, 497–98
acute, 498, 502
complementary, 497, 502, 516
definition of, 493
direction, 518
equivalent, 519, 520, 535
notation, 494
obtuse, 498
of any magnitude, 518–19
right, 497, 501, 502
straight, 497
supplementary, 497, 516
units of, 494–96

Angular displacement, 625–26

Angular velocity, 571–72, 595, 625
symbol for, 571

Antilogarithms, 457, 466
common, 444–45
natural (exponential), 452–53

Approximations, 58–59, 303

Arccosine function, 511, 524

Arcsine function, 510–11, 524

Arctangent function, 511, 524

Arithmetic symbols, 5–6, 10–11, 21, 92

Associative law, 98, 103
Asymptotes, 664–66
Attenuator, 469–70, 473, 489
Averaging, 326–27
Axis, 308, 329

Base
    of logarithms, 439, 451, 454,
        457
    of number systems, 3, 670,
        672–73
    of powers, 26, 60, 426, 439
Base unit, 63, 85
Binary numbers, 3, 671–73, 703,
        705, 710, 720
    addition of, 685–86
    bits, 3, 671
    calculating decimal value of,
        673–75
    converting to hexadecimal, 684
    converting to octal, 683
    digits, 3, 671
    one's complement of, 692
    subtraction of, 686–87
    true (two's) complement of,
        694–95
Binomial factors, sign patterns
        of, 403
Binomials, 97
    in multiplication, 105–7,
        241–43, 557
    squaring, 245–46
Bits, 3, 671
Boolean Algebra, 705–6, 738
    properties of, 733
    theorems, 734
Borrow, definition of, 685
Branch currents, 381, 590–93
    determining, 229–30,
        385–86, 391

Capacitance, 578–81, 582–83,
        590, 597
    equation for, 578
Capacitive reactance, 578–81, 612
    definition of, 579, 595
    equation for, 579, 583
    frequency dependence of,
        579–80, 583
    plotting, 600

Capacitive susceptance, 612–13
    equation for, 613
    forms of, 614
    plotting, 614
Capacitors, 578–81, 616
    reactive properties of, 579–81
Carry, 685
    end-around, 698
    end-off, 699
Cartesian coordinate system (See
        Rectangular coordinate
        system)
Chain calculations, 8, 24, 50–2,
        221
Characteristic of the common
        logarithm, 441–42
    definition of, 431, 466
Circuits (See Electrical circuits;
        Logic circuits)
Circular functions, 519–24, 646
        (See also Trigonometric
        functions)
    application to electronics, 537
    cosine, 520, 522
    definitions, 520
    graphs of, 521–24
    inverse (See Inverse circular
        functions)
    periodic behavior of, 521
    signs of, 520
    sine, 520, 521
    tangent, 520, 523–24
Coefficients, 95
    implied, 95
    numerical, 95, 96
    of products, 95, 101
Coincident, 343
Combined operations, 50–52
Common denominators,
        189–96, 200
    adding fractions, 190–91
    finding, 192–93
    subtracting fractions, 190
Common factors, 115–16, 178–79
Common logarithms, 439–43,
        444, 466 (See also
        Logarithms, properties of)
    base of, 439, 451, 454, 457
    definition of, 439

finding, 441
    graph of, 443, 452
    parts of, 441
Commutative law, 98, 103
Complements, 691–701
    addition of, 697–99
    base minus one, 692
    definition, 691, 703
    eight's, 694–95
    F's, 692, 693
    largest-digit, 692–93
    nine's, 692, 693
    of logic levels, 708
    one's 692, 696–99
    seven's, 692–93
    sixteen's, 694–95
    subtraction of, 698–701
    ten's, 694–95
    true, 694–95
    two's, 694–95, 696, 698–701
Complex conjugate, 560, 565
Complex fractions, 184–86,
        198–99, 200
Complex numbers, 538, 539–44,
        551, 553, 591
    concepts associated with, 539
    definition of, 539
    imaginary part of, 539–40, 553
    multiplication of, 557
    plotting, 540–42
    real part of, 539–40, 553
Complex numbers, polar form
        of, 543–44, 601, 613
    converting to rectangular form
        of, 546
Complex numbers, rectangular
        form of, 539, 540, 543,
        601, 613
    converting to polar form of,
        545–46
Complex plane, 537–38, 539,
        541, 543, 600
    constructing, 540–41
Compound inequalities, 18
Conditional equations, 120, 142
Conductance, 66, 148, 219, 238,
        612, 613, 614
    in parallel circuits, 220–22,
        228–30

Conductivity, 148, 171
Conductors (*See* Electrical conductors)
Constants, 294, 306, 407
  in literal equations, 207–8
  logic, 705
  table of, 747
Convention
  definition of, 7
  order of logic operators, 716
  order of operators, 7, 50–51
Conversion factors, 75, 752
Coordinate systems
  Cartesian, 309
  log-log, 461–63
  polar (*See* Polar coordinate system)
  rectangular (*See* Rectangular system)
  semilog, 461–62
Coordinates, 310, 312, 313, 319, 336, 343, 346
  definition of, 308
  of vertex, 413–14
  polar, 530, 532, 535, 543
  rectangular, 308–10, 532, 540
Cosecant function, 643, 644, 655
Cosine function, 504–6, 520, 644
  graph of, 522, 646
Cotangent function, 644, 655
Coulomb, definition of, 65, 144
Counting numbers (*See* Natural numbers)
Current dividers, 225–29, 230, 238
Current phasor, 590
Currents (*See* Electrical currents)

Decibel millivolt, 472, 477
  computing actual voltage, 474–75
  definition of, 474
Decibel milliwatt, 472, 477
  computing actual power, 473–74
  definition of, 473
Decibels
  definition of, 468, 489
  from voltage ratio, 471–72

expressing power gain or loss, 469–72
Decimal multiples and submultiples, 69–72
Decimal notation, 26, 29, 31, 32–33, 39
Decimal numbers, 2, 3, 670–71, 672, 673
  converting fractions, 680–82
  converting whole numbers, 680
  in engineering notation, 34
  in power of ten notation, 29–31
  in scientific notation, 31–32
  nine's complements of, 693
  ten's complements of, 694–95
  value of digits in, 671
Decimal point, 670, 671, 672
  placement of, 28–31, 33, 670–71
Decimal value, 673–75, 676–77
Degrees
  as unit of angular measure, 494
  conversion factors for, 496
DeMorgan's theorem, 727–30, 734
Denominators, 21, 22, 175–81, 184–86
  common, 189–96
  sign of, 187–88
  unequal, 194–96
Derived units, 63–64, 85
Determinants, 361–71
  definition of, 361, 372
  in mesh analysis, 383–84, 387–91
  second order, 361–65
  third order, 366–71
Deviation, 289
  percent, 259–60
Difference of two squares, 247, 255 (*See also* Special products)
Digits 3, 672, 673
  definition of 2, 670
Discontinuities, 523, 658–61
  definition of, 658, 666
  essential, 660–61
  removable, 658–59
  step, 659–60

Discriminant, 414–15, 422
Distributive law (*See* Multiplication, distributive law of)
Division
  by complex numbers, 560–62
  by fractions, 49–50, 182–83
  by multiplication, 49–50, 182–83
  equation transformation, 125
  of fractions, 182–83
  of monomials, 111–13
  of phasors, 560–62
  of polynomials, 113–14
  of signed numbers, 21–23
  order of operations, 6–7
  symbol (operator) for, 5–6
  with powers of ten, 54
Domain, 524, 657, 666
Do-not-care, 735–36
Double signs, 11, 15
Duality, concept of, 733–34
Dynamic parameters, 350
  graphic estimation of, 336–41

Efficiency, 271–77, 289
  formulas for, 272
  of a system, 275–77
  per unit, 272
  percent, 273–77
Electric charge, 65, 144
Electrical circuits, 144–61
  parallel (*See* Parallel circuits)
  parameters of, 144, 150
  power in, 166–69
  series (*See* Series circuits)
Electrical conductors, 148, 150–51, 279–86
  cross-sectional area of, 278, 281–82
  length of, 279–80
  material of, 279, 285–87
  resistance of, 279–88
  temperature of, 279, 288
  voltage generation in, 570
Electrical currents, 65, 144–45, 148–50, 171, 219. (*See also* Alternating currents)
  formula for, 151

Electrical currents, *Continued*
  in parallel circuits, 225–31
  in series circuits, 155, 162,
    216, 345
Electrical currents, direction of,
    374–75
  conventional flow, 374–77, 395
  electron flow, 374–75
  in branches, 385–86
  in loops, 375–79
  in meshes, 382–86
Electrical energy, 66–67, 76,
    78–81, 168, 271, 276
  cost of, 78–81
  formula for, 78
  usage, 79–80
Electrical formulas (*See*
  Formulas)
Electrical system, 67, 275
Electromotive force, 66, 144,
    145, 150–51
Electronic systems
  gains and losses, 468–77
  signal levels, 474–75
Electronic technology
  physical quantities used in,
    65–76, 135
Empirical data
  definition of, 325, 329
  graphing, 325–28
Energy, 66–67, 165
  conversion, 168, 271–72,
    275–76
  electrical (*See* Electrical
    energy)
  law of conservation of, 271,
    289
  wasted, 217–72
Engineering notation, 26, 34–35,
    37, 39, 60, 70
Equal sign, 17, 120
Equations, 5–6, 17, 120–33
  conditional, 120
  definition of, 120
  exponential (*See* Exponential
    equations)
  forming, 138, 139
  fractional (*See* Fractional
    equations)
  graphing, 311–21

identical, 74, 120, 142,
    646–47, 651–52
instantaneous, 625, 627, 628,
    629–30, 636–39
KVL (*See* Kirchoff's voltage
    law, equations)
linear (*See* Linear equations)
literal (*See* Literal equations)
logarithmic (*See* Logarithmic
    equations)
members of, 120, 142
numerical, 5, 13, 24
quadratic (*See* Quadratic
    equations)
radical (*See* Radical
    equations)
root of, 121, 142, 399, 401,
    404, 409, 414–16
simultaneous (*See*
    Simultaneous equations)
solving, 120–33, 202–6,
    399–409, 458, 459–61
systems of linear (*See* Systems
    of linear equations)
types of, 120
Equivalent equations (*See*
  Equations, solving)
Equivalent fractions, 176,
    177–78, 189, 200
Equivalent resistance, 156,
    158–60, 171, 216, 222–24
  computing, 222–24
  formulas, 222–23
  in parallel circuits, 222–24
Errors
  absolute, 265–70
  in products and quotients,
    267–68
  in sums and differences,
    268–70
  relative, 265–70
  percent, 259
Exponential equations, 468
  definition of, 459
  for charging of RC series
    circuits, 482–84
  for charging of RL series
    circuits, 478–81
  general forms, 484
  solving, 459–61

Exponential functions, 452–53,
    478–79, 482. (*See also*
    Hyperbolic functions)
Exponents, 26–8, 436
  definition of, 26, 60
  fractional, (*See* Fractional
    exponents)
  implied, 112, 118
  in division, 111
  in multiplication, 103
  laws of, 424
  negative, 425–26
  zero, 112, 425
Expressions, evaluating, 116, 118
Extraneous roots, 205, 214

Factoring, 115–16, 118, 178–79,
    247, 255
  by grouping, 249–50, 251
  combining types of, 250–51
  difference of two squares,
    247–48, 251
  in solving literal equations,
    252–53
  in solving quadratic equations,
    399, 401, 403–5
  perfect trinomial square,
    247–48, 251
  trinomials, 401–5
Factors, 20, 26, 55, 101
  common, 178–79
  definition of, 19, 24
  literal, 95
  numerical, 95
Force, 64, 67
Formulas
  definition of, 131, 142
  deriving, 360
  evaluation, 135–36, 209–10
  simplifying, 302–5
  solving, 131–33
Fractional equations,
  definition of, 204, 214
  solving, 204–6
Fractional exponents, 46–47,
    426–28, 434, 436
Fractional expressions, 93, 101
Fractions, 21–23, 28
  addition of, 187, 189,
    190–91, 194–96

algebraic, 175
approximations with, 58
arithmetic, 175
changing signs of, 187–88
complex, 184–86, 198–99
dividing, 49–50, 176, 182–83
equivalent (*See* Equivalent
  fractions)
introductory concepts of,
  175–76
multiplying, 176, 180–81
powers of, 57
signs of, 175, 187
simplifying, 178–79, 198–99
subtraction of, 187, 190–91
undefined, 175, 200
unit, 73–74
Frequency, 64, 572, 575, 579, 583
  definition of, 570, 595
Functional notation, 295–97
Functional variation, 297–99
Functions
  definition of, 293, 306
  evaluating, 296–99
  graphing, 314–16
  notation of (*See* Functional
    notation)
  variation of (*See* Functional
    variation)

General numbers, 93–4
Graphs
  construction of, 311–21
  definition of, 308, 312, 329
  of circular functions, 521–23
  of linear equations, 316–24
  scales of, 461–62
Graphs, use of
  determining dynamic
    parameters, 336–41
  determining static
    parameters, 331–35
  solving equations, 315, 399,
    409–12
Greek letters, 90, 747

Heat energy, 168, 271–72
Hexadecimal numbers, 3,
  671–73, 703
  addition of, 689–90

calculating decimal value of,
  673–77
converting to binary, 683–84
converting to octal, 684
digits, 672
F's complements of, 693
sixteen's complements of, 694
subtraction of, 689–91
Horsepower, 274
Hyperbolic functions, 648–51
  cosine, 649
  graphing, 651
  inverse (*See* Inverse
    hyperbolic functions)
  sine, 648
  tangent, 649–50
Hyperbolic identities, 651–52
Hypotenuse, 504, 510
  definition of, 501, 516

Identical equations, 74, 120, 142
Identities, 74, 76, 655 (*See also*
  Identical equations)
  hyperbolic, 651–52
  trigonometric, 646–47
Imaginary numbers, 48, 538–40,
  551, 556–59
  in division, 560
  in multiplication, 557
Impedance, 597–604, 612, 632
  computing, 597–99, 618–19
  definition of, 597, 623
  diagram, 600–1
  equation for, 597
  forms of 601–2, 604
Index of radical, 45–46
  implied, 45
Inductance, 575–78, 582, 583,
  590, 597
  equation for, 575
Inductive reactance, 575–78,
  595, 612
  equation for, 575
  frequency dependence of,
    575, 583
  plotting, 600
Inductive susceptance, 612–13
  equation for, 613, 614
  forms of, 614
  plotting, 614

Inductors, 575–78, 616
Inequality, 17–18, 24
Infinity, 662
Instantaneous equations, 480, 483,
  625–28, 629–30, 636–39
Instantaneous values, 478, 484,
  489, 625
Intercept, 318, 329
International System of Units
  (*See* SI Metric system)
Intersect, 315, 318, 329, 343,
  346–49
Inverse circular functions (*See
  also* Inverse trigonometric
  functions)
  principal values of, 524
Inverse hyperbolic functions
  formulas for 652–54
Inverse trigonometric functions,
  510–15
  definitions of, 510–11
  in solving right triangles,
    512–15
Inverter, 707
Irrational expressions, 93, 94,
  96, 101

Joule, 66–67, 78, 165

Karnaugh maps, 720–26, 738
  do-not-care, 735–36
  interpreting, 723, 725–26
  relationship to truth tables,
    720–22
  simplifying logic diagrams,
    720, 721–22, 725
  with OR logic, 730–32
Kelvin, 67–68
Kilowatthour, 78–81, 85
Kirchoff's current law, 226,
  591–93
Kirchoff's voltage law, 157–58,
  162, 374
  applied to ac series circuits,
    584–88
Kirchoff's voltage law,
  equations, 374–79
  for a simple loop, 377–79
  in mesh analysis, 381, 382, 387
K-maps (*See* Karnaugh maps)

KVL equations (*See* Kirchoff's
voltage law, equations)

Lag, 580, 583, 590, 598, 633
Law of Cosines, 525–29
Law of Sines, 525–29
Lead, 576–77, 583, 590, 598, 633
Least significant digit, 670, 673,
676–77, 678–81
Like terms, 96–97, 101
Limits, 662–68
Linear equations
definition of, 316
deriving from graphs, 322–27
graphing, 316–21
systems of (*See* Systems of
linear equations)
Lines, concept of, 493
Literal equations
application of factoring,
252–53
constants in, 207–8
definition of, 207, 214
variables in, 207–8
Literal numbers (*See* General
numbers)
Load line, 346–48, 350
Logarithmic equations
computing power gain or loss
with, 468–77
computing voltage with,
474–77
definition of, 458
solving, 458–59
Logarithms, 439, 458, 466
base of, 439, 451, 454, 457
changing base, 454–55, 457
common (*See* Common
logarithms)
graph of, 443, 452
natural (*See* Natural
logarithms)
of powers, 448, 457
of products, 445–46, 457
of quotients, 444–45, 457
of radicals, 448–50, 457
properties of, 443, 445–50, 457
Log-log plots, 461–63
Logic circuits, 706, 707, 710

minimum, 720
simplification, 721–22, 725,
733
Logic diagrams, 715–16, 718, 721
simplifying, 731–32
symbols, 707–8, 709, 713–14
Logic expressions, 705–6,
715–16, 718
equivalent, 716–17
simplifying, 733–34
Logic operators, 706
conjunction (AND), 709–12
disjunction (OR), 712–13
inversion (NOT), 707–8
order of precedence, 716
Loop currents, 375–79
LSD (*See* Least significant
digit)

Mantissas of common
logarithms, 441–43
definition of, 441, 466
Measurement, systems of (*See* SI
Metric system)
Mechanical energy, 272, 276
Meshes
analysis of, 381–92
currents, 343–44, 381, 395
currents, solving for, 381–86,
387–92
definition of, 381, 395
Metal conductors (*See* Electrical
conductors)
Metric system (*See* SI Metric
system)
Mixed expressions
changing to fractions, 197
definition of, 197, 200
Monomials, 97
in division, 113–14
in multiplication, 103–7
Most significant digit, 670,
676–77, 678–81
MSD (*See* Most significant
digit)
Multiplication
approximations in, 58–59
associative law of, 103
commutative law of, 103

distributive law of, 105–7,
115, 118, 129–31, 241–43,
557–58
equation transformation, 124
in algebra, 92, 241
of binomials, 105–6, 241–43
of complex numbers, 556–59
of fractions, 176, 177–78,
180–81
of monomials, 103–7
of phasors, 556–59
of polynomials, 107, 110–11
of signed numbers, 19–21
order of operations, 7–8
symbol (operator) for, 5–6, 92
with powers of ten, 53
Mutual resistance, 381–86,
387, 395

Natural logarithms, 451–52 (*See
also* Logarithms, properties
of)
definition of, 451, 466
graph of, 452
properties of, 455–57
Natural numbers, definition of,
2, 24
Negative numbers, 3, 11, 15, 20
Networks, solving, 374, 381
by mesh analysis, 387–92
theorems, 232–37
Newton, 67
Newton's second law, 67
Nomographs, 464–65
Nonlinear series circuits
definition of, 345
graphic analysis of, 345–48
mathematical analysis of, 345
Norton's theorem, 232–37
equivalent circuit, 233
equivalent current source, 233
equivalent resistance, 233
NOT, 707–8
Number systems, 2–3, 670,
671–73
base, 3, 670–72
binary (*See* Binary numbers)
comparison of, 672
decimal (*See* Decimal numbers)

digits, 3, 670–72
hexadecimal (*See*
    Hexadecimal numbers)
octal (*See* Octal numbers)
radix, 670
radix point, 672
Numbers approximate, 26,
    36–38, 40–43
adding, 40–41
dividing, 42–43
multiplying, 42–43
subtracting, 40–41
Numbers, exact, 26, 35, 60
Numerator, 21, 22, 175, 178,
    184–85, 190
sign of, 187–88
Numerical equation, 5, 13, 24
Numerical expressions, 5–6,
    9–10, 24
Numerical factor, 95

Obtuse angles, 498
Octal numbers, 3, 672, 703
addition of, 687–88
calculating decimal value of,
    673–75
converting to binary, 682
converting to decimal, 678
converting to hexadecimal, 684
eight's (true) complements
    of, 694
seven's complements of,
    692–93
subtraction of, 687–89
Ohm, definition of, 66
Ohm's law, 150–53, 158, 171,
    216, 228, 573
Operations
calculator, 2
combined, 50–52
Operators, 5, 10 (*See also* Logic
    operators; Relational
    operators)
order of, 6–8, 9–10, 50–51
order of logic, 716
Ordered pairs, 308, 310, 311,
    326, 538, 540
Ordinate, 308, 310, 336, 540
Origin, 308–10

Parallel circuits, 219–32
application of Ohm's law,
    228, 230
branch currents in, 225–26,
    228–30
computing current of, 228–31
computing equivalent
    resistance of, 222–24, 230
conductance of, 219, 220,
    229, 230
current division in, 225–29
equivalent resistance of,
    222–24, 230, 303
formulas for, 230
Parallel circuits, ac, 590–93
admittance of, 612, 614,
    616–21
current in, 590
current phasor in, 590–93
forming equivalent
    components of, 616–17
voltage in, 590
Parallel equivalent circuits,
    620–21
computing, 620–21
forming, 620
Parallel lines, 342
Parts per million, 262–64, 287
converting from percent to, 262
converting to percent, 263
guidelines for applying, 262–63
Percent, 259–60, 289
converting to parts per
    million, 262
deviation, 259–60, 263
efficiency, 273
error, 259
guidelines for applying, 262–63
tolerance, 259
Perfect trinomial square, 246,
    247–48, 251, 255
Period, definition of, 521, 571,
    595
Periodic behavior, 521, 535
Phase angles, 577, 580, 582,
    598–99, 632
definition of, 573, 595
Phase shift, 577
Phasors, 542–44, 553, 593

current, 590
definition of, 543, 551
diagrams of, 573–74, 576–77,
    580–81, 582–84, 590–93,
    599, 626
graphing, 543, 547
polar form of, 543–44, 563,
    564
rectangular form of, 544,
    553–54
resolving systems of, 547–50
voltage, 582–86
Phasors, mathematics of, 553–65
addition, 553–54, 584–87,
    591–93
division, 560–62
multiplication, 556–59
powers of, 563
roots of, 564–65
subtraction, 554–55
Physical quantities, 62–64, 82
Plus or minus, 26, 45, 406–9
Points
complex plane, 541, 544
concepts of, 493
empirical data, 325–28
graphical analysis, 342–43, 346
graphical estimation, 336–41
graphs, 311–15, 317–18
phasors, 543
polar coordinates, 530–31
rectangular coordinates,
    309–10
slope, 317, 323
Polar coordinate system,
    530–34, 535, 543
argument, 530–32, 533
basic components of, 530
converting to rectangular,
    532–34, 546
magnitude, 530–32, 533
pole, 530, 535
Polynomials, 93–94, 97–99
addition of, 98–99, 110–11
definition of, 97, 101
factoring of, 115–16, 247–52
in division, 113–14
in multiplication, 107,
    110–11, 113

Polynomials, *Continued*
  subtraction of, 108–9
  types of, 97
Positive numbers, 3, 11, 20, 44
Power, 67, 85, 142, 165–68
  average, 630, 632, 640
  computing, 631, 633–35
  dissipation, 168, 169, 633
  electrical, 67, 274
  formulas for, 166–67
  in a reactive load, 632
  in a resistive load, 629–31
  mechanical, 67, 274–75
Power factor, 632–35
  definition of, 632, 640
Powers, 26, 46
  of a fraction, 57
  of a power, 55
  of a product, 56, 82
  of units, 82–83, 84
Powers of eight, 674
Powers of sixteen, 674–75
Powers of ten notation, 28–30, 32
  division of, 54–55
  multiplication of, 53
Powers of two, 674
Precision, 36–38, 40–43, 60
Preferred number series,
    485–88, 750
  base numbers, 487
Primes, 91
Products, 20, 26, 95, 96
  definition of, 19
  power of, 56
Proportion, 503
  definition of, 278, 289
  direct, 279, 289
  indirect, 281–82, 289
Pythagorean theorem, 501–2, 525

Quadrants, 308–10, 520
  signs in, 309
Quadratic equations, 399, 422
  determining characteristics of
    roots, 414–16
  determining shape of curve,
    409–12
  determining vertex of curve,
    412–13

discriminant of, 414–15
general form of, 401, 406
pure (*See* Quadratic equations,
    incomplete)
roots of, 414–16
Quadratic equations, complete
  definition of, 399, 401
  solving, 399, 401, 403–5
Quadratic equations, incomplete
  definition of, 399
  solving, 399–98
Quadratic equations solving
  by factoring (*See* Quadratic
    equations, complete)
  by graphing, 399, 409–16
  by quadratic formula, 399,
    406–9
  by square root (*See* Quadratic
    equations, incomplete)
  in electronic problems, 417–20
Quadratic formula, 399, 406,
    407
  in solving quadratic
    equations, 406–9
Quotient, 22, 96, 257, 258

Radians, 495–96, 516
  conversion factors for, 496
  definition of, 495
Radical equations
  definition of, 434
  solving, 434–35
Radical sign, 45–46
Radicals, 45–46, 60
  definition of, 45
  index of, 45–6, 564
  laws of, 429
Radicals, simplifying, 426–28,
    430–32
  by lowering the index,
    430–31
  by rationalizing the
    denominator, 431–32
  by removing factors, 430
Radicand, definition of, 45–46
Radix, 670, 672, 673, 694, 703
Radix point, 672–75
  implied, 673
Range, 524, 657, 666

Rational expressions, 93–94,
    96, 101
  fractional, 93–94
  polynomial, 93–94
Ratios (*See also* Percent), 495,
    503–5, 510–11
  definition of, 257
  equivalent, 258
  expressing, 257, 304
Real numbers, 538
  in division, 560
  in multiplication, 557–58
Reciprocals, 48
  use in division, 49–50
Rectangular coordinate system,
    308–10, 519
  converting to polar, 532–34,
    545–46
  terms of, 310
Relational operators, 17–18
Resistance, 66, 144, 148–51,
    171, 219, 573–74, 597
  equivalent, 156, 158, 162,
    222–23, 230, 302–3, 616
  formula for, 151
  in a series circuit, 154–56,
    158, 162
  mutual, 381–86
  of metal conductors (*See*
    Electrical conductors,
    resistance of)
  plotting, 600
  specific (*See* Resistivity)
  total, 156
Resistivity, 285–86, 289
Resistors, 149–50
  color bands, 150, 160, 259, 751
  power considerations for, 168
  preferred number series, 159
  tolerance, 160, 259
Revolutions, 494–96
  conversion factors for, 496
  definition of, 494, 516
  fractional parts of, 494, 495,
    498
Right angles, 497–98, 501–2
Right triangles, 501–2, 516
  inverse trigonometric
    functions in, 510–11

relation to Pythagorean
theorem, 501–2
solving, 506–9, 512–15
Roots, 121, 142, 315, 318, 329,
399, 409, 416
characteristics of, 414–15
Rounding, 38–39

Scalars, definition of, 537, 551
Scientific calculators (*See*
Calculators)
Scientific notation, 26, 31–32,
37, 39, 54, 60, 441–43
Secant function, 643–46, 655
Semiconductor parameters,
approximation of, 336–41
Semilog plots, 462
Series circuits, 154–64, 216–19
application of Ohm's law,
157–63, 216
characteristics of, 155, 157,
162, 216
definition of, 154
equivalent, 156, 162–63,
222–24
nonlinear (*See* Nonlinear
series circuits)
resistance in, 154, 155–56,
158, 162, 217
voltage division in, 217–19
Series circuits, ac, 582–88, 597,
602–4
current in, 582
impedance of (*See*
Impedance)
solving, 606–11
voltage phasor in, 582–84
Series equivalent circuits, ac, 598
determining components of,
602–4, 606–7
Siemens, 66, 219, 226, 612
Sign bit, 696–701
Sign change, 15
Signal power, 474
Signal voltage, 475
Signed numbers, 2, 3–4
absolute value of, 12
addition of, 13–14, 16
division of, 22–23

implied sign, 3
multiplication of, 19–21
negative, 3
opposite, 13
positive, 3
subtraction of, 13, 14–15
use of calculator with,
16–17
Significant figures, 36–38, 40,
42, 60
Signs (*See also* Symbols)
double, 11, 15
implied, 3, 175, 187
Signs of grouping (*See* Symbols
of grouping)
SI Metric system, 62–71, 73, 85
base units, 62–63
conversion, 73–77
derived units, 63–64
physical quantities, 63–64
physical symbols, 63–64
prefixes, 69
unit multiples and
submultiples, 62, 69–72
Similar triangles, 503
ratio of two sides of, 504
Simultaneous equations, 354
deriving formulas from, 360
techniques (*See* Systems of
linear equations, solving)
Sine function, 504–7, 520, 521,
524, 644
Sinusoidal alternating currents
(*See* Alternating currents,
sinusoidal properties of)
Slope, 317–18, 329, 338, 342, 347
local, 339
Special products, forming, 244
Square root, 44–48
approximations in, 58
definition of, 44, 60
in solving quadratic equations,
399–400
negative, 45
of negative numbers, 47–48,
538–39
positive (principal), 45, 46, 60
Squaring, 26
approximations in, 58

Static parameters, 331, 350
graphic estimation of, 331–35
Subscripts, 91
Subtraction
equation transformation, 123
of like terms, 96–97
of polynomials, 108–9
of signed numbers, 13,
14–15
order of operations, 7
symbol (operator) for, 5–6
Sum, 13
Susceptance, 612–14
definition of, 612, 623
Symbols, 2–3, 5, 26, 90, 747
(*See also* Operators)
double meaning of, 10–11
logic diagram, 707, 709, 713
of physical quantities, 63–64
of unit prefixes, 69
of units, 63–64
Symbols of grouping, 2, 9–10,
109, 110, 729 (*See also*
Vinculum)
bars, 21–22, 45, 51
braces, 9
brackets, 9
parentheses, 9, 51, 129–31
Symmetric property of equality,
127, 142
Systems of linear equations,
354, 372
Systems of linear equations,
solving, 354–61
by addition method, 354–57
by graphic method, 342–44
by second-order determinants,
361, 363–65
by substitution method, 357–59
by subtraction method, 354–57
by third-order determinants,
368–71
Systems of linear equations, use
of, 374–92 (*See also*
Networks, solving)
solving networks, 381

Tangent function, 505–6, 520,
523–24, 644
Temperature

coefficient, 264
    effect on resistance, 288
    unit of measurement, 67–68
Terms, 13, 24
    combining like, 96–97, 126
Thévenin's Theorem, 232–35
    equivalent circuit, 233
    equivalent resistance, 233
    equivalent voltage source, 233
Time, 68, 625–32
Tolerance, 36, 265, 289, 487
    color codes, 259, 751
    percent, 259
Total resistance (*See also*
    Equivalent resistance)
    in series circuits, 155–56
Transient, 489
    behavior, 478
    general equations, 484
    in RC circuits, 482–84
    in RL circuits, 478–81
Triangles
    definition of, 499
    finding third angle of, 500–1
    properties of, 499
    right (*See* Right triangles)
    similar (*See* Similar triangles)
    solving, 525–29
Trigonometric functions, 503–9,
    519, 643–46
    cosecant, 643, 644
    cosine, 504–6, 644
    cotangent, 644
    definitions of, 505, 644
    in solving right triangles, 506–9
    in solving triangles, 525–29
    inverse (*See* Inverse
        trigonometric functions)
    secant, 643–46

sine, 504–7, 644
    table of, 506
    tangent, 505–6, 644
Trigonometric identities, 646–47
Trinomial products, 241, 245–46
    definition of, 255
    terms of, 241–43
Trinomials, 97
    factoring, 401–5
True complements, 694
Truth tables, 705–6, 711, 739
    relation to Karnaugh maps,
        720–22

Undefined fractions, 175, 200
Unit analysis, 73–77, 78–81, 85
Unit prefixes, 62, 69

Variables, 90–91, 95, 293–94
    definition of, 90, 101, 306
    dependent, 293–94, 298, 306,
        309
    dividing powers of, 111–12
    in algebraic expressions, 116
    in literal equations, 207–8
    independent, 293, 298, 302,
        306, 309
    logic, 705
    multiplying powers of, 103–5
Variation
    direct, 298, 300
    inverse, 299
Vectors, 547, 553, 597, 612 (*See
    also* Phasors)
    definition of, 537, 551
    rotating, 542
    three-dimensional, 537
    two-dimensional, 537
Vertex, 412–13, 415–16, 422

Vinculum, 175, 184, 200 (*See
    also* Symbols of grouping,
    bars)
    changing the sign of, 187–88
    sign of, 187
Volt, 66
Voltage, 145–48, 151, 158, 162,
    171
    dividers, 217–19, 238
    effective values of, 630–32
    Kirchoff's law, 157
    phasor diagram of, 573–74,
        576–77, 580–81
    wiring loss, 266
Voltage drop, 66, 145, 157, 162,
    374
    formulas for, 150, 160,
        216–19, 220, 230
Voltage rise, 66, 145, 157, 374
    formula for, 150
Voltmeter, 145, 147

Wattage ratings, 80, 168
Watts, 64, 67, 166, 168, 274
Whole numbers, 2, 673
Wire (*See* Electrical conductors)
Wiring loss voltages, 266
Word problems, solving, 138–41
    procedures for, 139
Work, 66, 165

X-axis, 308, 310, 315, 329
X-coordinate, 308–10, 315
X-intercept, 318, 329
Y-axis, 308, 310, 329
Y-coordinate, 308–10
Y-intercept, 318, 322–24, 329

Zero, 2, 3, 37, 112, 175, 683–84
Zero power, 53, 112, 425

$$y = \frac{\begin{vmatrix} a_1 & k_1 & c_1 \\ a_2 & k_2 & c_2 \\ a_3 & k_3 & c_3 \end{vmatrix}}{\Delta} \qquad (16\text{–}10)$$

$$z = \frac{\begin{vmatrix} a_1 & b_1 & k_1 \\ a_2 & b_2 & k_2 \\ a_3 & b_3 & k_3 \end{vmatrix}}{\Delta} \qquad (16\text{–}11)$$

$$\Sigma V = \Sigma E \qquad (17\text{–}1)$$

$$V_1 + V_2 + V_3 = E_1 + E_2 + E_3 \qquad (17\text{–}2)$$

$$IR_1 + IR_2 + IR_3 = E_1 + E_2 + E_3 \qquad (17\text{–}3)$$

$$ax^2 + bx + c = 0 \qquad (18\text{–}1)$$

$$x = -\frac{b}{2a} \pm \sqrt{\left(\frac{b}{2a}\right)^2 - \frac{c}{a}}$$

$$\text{or } x = \frac{-b}{2a} \pm \sqrt{\left(\frac{-b}{2a}\right)^2 - \frac{c}{a}} \qquad (18\text{–}2)$$

$$x = -\frac{b}{2a} \pm \sqrt{\frac{b^2}{4a^2} - \frac{4ac}{4a^2}}$$

$$= -\frac{b}{2a} \pm \sqrt{\frac{b^2 - 4ac}{4a^2}}$$

$$= -\frac{b}{2a} \pm \frac{\sqrt{b^2 - 4ac}}{2a}$$

$$x = \frac{-b \pm \sqrt{b^2 - 4ac}}{2a} \qquad (18\text{–}3)$$

$$y = ax^2 + bx + c \qquad (18\text{–}4)$$

$$x_0 = -\frac{b}{2a} \qquad (18\text{–}5)$$

$$y_0 = ax_0^2 + bx_0 + c \qquad (18\text{–}6)$$

$$N_{dB} = 10 \log\left(\frac{P_{out}}{P_{in}}\right) \quad \text{dB} \qquad (21\text{–}1)$$

$$N_{dB} = 10 \log\left(\frac{E_{out}^2/R_{out}}{E_{in}^2/R_{out}}\right)$$

$$= 10 \log\left[\left(\frac{E_{out}}{E_{in}}\right)^2\right]$$

$$= 2(10) \log\left(\frac{E_{out}}{E_{in}}\right)$$

$$N_{dB} = 20 \log\left(\frac{E_{out}}{E_{in}}\right) \quad \text{dB} \qquad (21\text{–}2)$$

$$N_{dBm} = 10 \log\left(\frac{P}{1 \text{ mW}}\right) \quad \text{dBm} \qquad (21\text{–}3)$$

$$N_{dBmV} = 20 \log\left(\frac{V}{1 \text{ mV}}\right) \quad \text{dBm V} \qquad (21\text{–}4)$$

$$i = \frac{E}{R}[1 - e^{-t/(L/R)}] \qquad (21\text{–}5)$$

$$v_L = Ee^{-t/(L/R)} \qquad (21\text{–}6)$$

$$v_R = E[1 - e^{-t/(L/R)}] \qquad (21\text{–}7)$$

$$i = \frac{E}{R}e^{-t/RC} \qquad (21\text{–}8)$$

$$v_C = E(1 - e^{-t/RC}) \qquad (21\text{–}9)$$

$$v_R = Ee^{-t/RC} \qquad (21\text{–}10)$$

$$y = Y_{mx}(1 - e^{-t/r}) \qquad (21\text{–}11)$$

$$y = Y_{mx}e^{-t/\tau} \qquad (21\text{–}12)$$

$$N_{x+1} = 10^{x/n} \qquad (21\text{–}13)$$

$$T = 1/f \quad \text{(seconds)} \qquad (26\text{–}1)$$